Base 2 Exponential Function: $f(x) = 2^x$

x	-2	-1	0	1	2
2^x	$\frac{1}{4}$	$\frac{1}{2}$	1	2	4

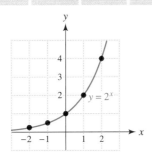

Common Logarithmic Function: $f(x) = \log x$

x	0.1	1	4	7	10
$\log x$	-1	0	$\log 4$	$\log 7$	1

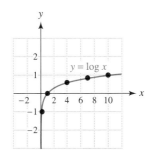

Families of Functions

Some examples of families of functions are linear and quadratic functions. The formulas and graphs of some families of functions are shown.

Constant Functions: $f(x) = b$

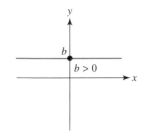

Linear Functions: $f(x) = mx + b$

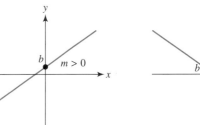

Quadratic Functions: $f(x) = ax^2 + bx + c$

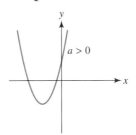
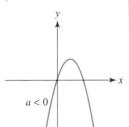

Exponential Functions: $f(x) = Ca^x$

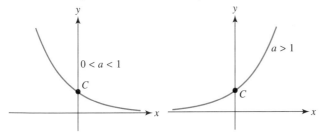

Logarithmic Functions: $f(x) = \log_a x$

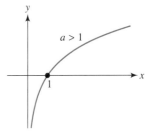

INTERMEDIATE ALGEBRA
through Modeling and Visualization

Gary Rockswold
Minnesota State University, Mankato

Boston • San Francisco • New York • London • Toronto
Sydney • Tokyo • Singapore • Madrid • Mexico City
Munich • Paris • Cape Town • Hong Kong • Montreal

To my daughter, Jessica

Publisher	Jason A. Jordan
Sponsoring Editor	Jennifer Crum
Project Manager	Ruth Berry
Editorial Assistant	Sharon Smith
Managing Editor	Ron Hampton
Production Supervisor	Kathleen A. Manley
Production Services	Kathy Diamond
Art Editors	Jennifer Bagdigian and Naomi Kornhauser
Marketing Manager	Dona Kenly
Illustrators	Techsetters, Inc., Bob Giuliani, and Darwen and Vally Hennings
Text and Cover Design	Dennis Schaefer
Prepress Supervisor	Caroline Fell
Compositor	Nesbitt Graphics, Inc.
Print Buyer	Evelyn Beaton

Photo Credits
Cover: Stone/Tim Bradley, 2000 (table), Imagefinder/Index Stock Imagery (background); p.1: Kaluzny/Thatcher/Stone; p. 3: NASA; p. 72: Jonathan Kim/Liaison; p. 140: PhotoDisc; p. 198: Daniel J. Cox/Liaison; p. 274: ©Duomo/CORBIS; p. 277: NASA; p. 339: Richard R. Hanson/Photo Researchers; p. 399: Glen Allison/Stone; p. 467: Billy Hustace/Stone; p. 527: Grant V. Faint/The Image Bank; p. 573: Eric Sander/Liaison; p. 628: Gary Gladstone/The Image Bank

Library of Congress Cataloging-in-Publication Data
Rockswold, Gary K.
    Intermediate algebra through modeling and visualization / Gary Rockswold.
       p. cm.
Includes bibliographical references and index.
    ISBN 0-201-61073-6 (alk. paper)
 1. Algebra. I. Title.
 QA152.2 .R632 2001
 512.9--dc21
                                        00-054306

Copyright © 2002 Pearson Education, Inc.
All rights reserved. No part of this publication may be reproduced, stored in a retrieval system, or transmitted, in any form or by any means, electronic, mechanical, photocopying, recording or otherwise, without the prior written permission of the publisher. Printed in the United States of America.

Reprinted with corrections.

For information on obtaining permission for the use of material from this work, please submit a written request to Pearson Education, Inc., Rights and Contracts Department, 75 Arlington Street, Suite 300, Boston, MA 02116 or fax your request to (617) 848-7047.

2 3 4 5 6 7 8 9 10—QWT—05 04 03 02

Contents

List of Applications		vii
Preface		xi
Walkthrough		xix

1 Real Numbers and Algebra — 1

1.1	Describing Data with Sets of Numbers	2
	Group Activity: Working with Real Data	12
1.2	Operations on Real Numbers	13
	Checking Basic Concepts for Sections 1.1 and 1.2	24
1.3	Integer Exponents	24
	Group Activity: Working with Real Data	40
1.4	Modeling Data with Formulas	41
	Checking Basic Concepts for Sections 1.3 and 1.4	50
1.5	Visualization of Data	50
	Checking Basic Concepts for Section 1.5	64
	CHAPTER 1 SUMMARY	65
	CHAPTER 1 REVIEW EXERCISES	67
	CHAPTER 1 TEST	70
	CHAPTER 1 EXTENDED AND DISCOVERY EXERCISES	71

2 Linear Functions and Models — 72

2.1	Functions and Their Representations	73
2.2	Linear Functions	88
	Checking Basic Concepts for Sections 2.1 and 2.2	101
2.3	The Slope of a Line	102
	Group Activity: Working with Real Data	115
2.4	Equations of Lines and Linear Models	116
	Checking Basic Concepts for Sections 2.3 and 2.4	130
	CHAPTER 2 SUMMARY	131
	CHAPTER 2 REVIEW EXERCISES	133
	CHAPTER 2 TEST	136
	CHAPTER 2 EXTENDED AND DISCOVERY EXERCISES	138

3 Linear Equations and Inequalities — 140

3.1	Linear Equations	141
	Group Activity: Working with Real Data	155

3.2	Linear Inequalities	155
	Checking Basic Concepts for Sections 3.1 and 3.2	167
3.3	Compound Inequalities	167
	Group Activity: Working with Real Data	178
3.4	Absolute Value Equations and Inequalities	178
	Checking Basic Concepts for Sections 3.3 and 3.4	188
	CHAPTER 3 SUMMARY	188
	CHAPTER 3 REVIEW EXERCISES	190
	CHAPTER 3 TEST	195
	CHAPTER 3 EXTENDED AND DISCOVERY EXERCISES	196

4

Systems of Linear Equations

198

4.1	Systems of Linear Equations in Two Variables	199
4.2	The Substitution and Elimination Methods	211
	Checking Basic Concepts for Sections 4.1 and 4.2	222
4.3	Systems of Linear Inequalities	223
	Group Activity: Working with Real Data	232
4.4	Systems of Equations in Three Variables	232
	Checking Basic Concepts for Sections 4.3 and 4.4	244
4.5	Matrix Solutions of Linear Systems	244
	Group Activity: Working with Real Data	256
4.6	Determinants	257
	Checking Basic Concepts for Sections 4.5 and 4.6	265
	CHAPTER 4 SUMMARY	265
	CHAPTER 4 REVIEW EXERCISES	267
	CHAPTER 4 TEST	271
	CHAPTER 4 EXTENDED AND DISCOVERY EXERCISES	272

5

Polynomial Expressions and Functions

274

5.1	Polynomial Expressions	275
	Group Activity: Working with Real Data	286
5.2	Polynomial Functions and Models	286
	Checking Basic Concepts for Sections 5.1 and 5.2	296
5.3	Multiplication of Polynomials	296
5.4	Factoring Polynomials	305
	Checking Basic Concepts for Sections 5.3 and 5.4	313
5.5	Factoring Trinomials	313
	Group Activity: Working with Real Data	322
5.6	Special Types of Factoring	323
	Checking Basic Concepts for Sections 5.5 and 5.6	330
	CHAPTER 5 SUMMARY	330
	CHAPTER 5 REVIEW EXERCISES	332
	CHAPTER 5 TEST	335
	CHAPTER 5 EXTENDED AND DISCOVERY EXERCISES	336

6

Quadratic Functions and Equations

339

6.1	Quadratic Functions and Their Graphs	340
	Group Activity: Working with Real Data	355
6.2	Quadratic Equations	356
	Checking Basic Concepts for Sections 6.1 and 6.2	368

6.3	The Quadratic Formula	368
6.4	Quadratic Inequalities	378
	Checking Basic Concepts for Sections 6.3 and 6.4	388
	CHAPTER 6 SUMMARY	389
	CHAPTER 6 REVIEW EXERCISES	391
	CHAPTER 6 TEST	395
	CHAPTER 6 EXTENDED AND DISCOVERY EXERCISES	397

7

Rational Expressions and Functions

399

7.1	Rational Functions and Equations	400
	Group Activity: Working with Real Data	411
7.2	Multiplication and Division of Rational Expressions	412
	Checking Basic Concepts for Sections 7.1 and 7.2	424
7.3	Addition and Subtraction of Rational Expressions	424
7.4	Modeling with Proportions and Variation	434
	Checking Basic Concepts for Sections 7.3 and 7.4	449
7.5	Division of Polynomials	449
	Checking Basic Concepts for Section 7.5	457
	CHAPTER 7 SUMMARY	457
	CHAPTER 7 REVIEW EXERCISES	459
	CHAPTER 7 TEST	463
	CHAPTER 7 EXTENDED AND DISCOVERY EXERCISES	464

8

Radical Expressions and Functions

467

8.1	Radical Expressions and Rational Exponents	468
8.2	Radical Functions	479
	Checking Basic Concepts for Sections 8.1 and 8.2	489
8.3	Operations on Radical Expressions	489
	Group Activity: Working with Real Data	496
8.4	Equations Involving Radical Expressions	497
	Checking Basic Concepts for Sections 8.3 and 8.4	510
8.5	Complex Numbers	510
	Checking Basic Concepts for Section 8.5	519
	CHAPTER 8 SUMMARY	519
	CHAPTER 8 REVIEW EXERCISES	522
	CHAPTER 8 TEST	524
	CHAPTER 8 EXTENDED AND DISCOVERY EXERCISES	525

9

Conic Sections

527

9.1	Parabolas and Circles	528
	Group Activity: Working with Real Data	540
9.2	Ellipses and Hyperbolas	540
	Checking Basic Concepts for Sections 9.1 and 9.2	551
9.3	Nonlinear Systems of Equations and Inequalities	552
	Checking Basic Concepts for Section 9.3	563
	CHAPTER 9 SUMMARY	564
	CHAPTER 9 REVIEW EXERCISES	566
	CHAPTER 9 TEST	569
	CHAPTER 9 EXTENDED AND DISCOVERY EXERCISES	571

10 — Exponential and Logarithmic Functions — 573

10.1	Exponential Functions	574
	Group Activity: Working with Real Data	587
10.2	Logarithmic Functions	588
	Checking Basic Concepts for Sections 10.1 and 10.2	599
10.3	Properties of Logarithms	599
10.4	Exponential and Logarithmic Models	607
	Checking Basic Concepts for Sections 10.3 and 10.4	620

CHAPTER 10 SUMMARY — 620
CHAPTER 10 REVIEW EXERCISES — 622
CHAPTER 10 TEST — 625
CHAPTER 10 EXTENDED AND DISCOVERY EXERCISES — 626

11 — Sequences and Series — 628

11.1	Sequences	629
11.2	Arithmetic and Geometric Sequences	638
	Checking Basic Concepts for Sections 11.1 and 11.2	649
11.3	Series	650
	Group Activity: Working with Real Data	660
11.4	The Binomial Theorem	661
	Checking Basic Concepts for Sections 11.3 and 11.4	667

CHAPTER 11 SUMMARY — 667
CHAPTER 11 REVIEW EXERCISES — 669
CHAPTER 11 TEST — 671
CHAPTER 11 EXTENDED AND DISCOVERY EXERCISES — 672

Answers to Selected Exercises — A-1
Glossary — G-1
Index — I-1
Bibliography — B-1
Graphing Calculator Laboratory Manual Topics — GC-1

List of Applications

Astronomy
Bode's law, 672
Distant galaxies, 166
Escape velocity, 49
Halley's comet, 550
Magnitude of a star, 598
Modeling brightness, 466
Moons of Jupiter, 3, 525
Orbit
   of Explorer VII, 550
   of Mars, 569
   of Mercury, 544
   of Pluto, 550
   of Uranus, 570
Planetary orbits, 337
Radio telescopes, 538, 540, 572
Size in the universe, 24
Sizes of objects, 588
Speed
   of a comet, 539
   of Earth, 69
   of light, 39
Trajectories of comets, 539, 546
Walk to the moon, 40

Biology
Allometric growth, 465
Animals and trotting speeds, 49
Antelope populations, 234, 239
Area of skin, 526
Birds and wing size, 49
Blow-fly population, 336
Body mass index, 71
Breathing and carbon dioxide, 173
Decline of the bluefin tuna, 618
E. coli bacteria, 586
Estimating a bear's size, 198
Estimating the chest size of a bear, 270
Fertilizer use, 619
Fish population, 462
Heart beat in animals, 69
Insect population, 411, 574, 618, 629, 633, 638, 672
Life span of robins, 610, 618
Lynx and hares, 336
Manatee deaths, 100, 294
Modeling
   bacteria growth, 624, 625, 634
   bird populations, 614
   blood flow in animals, 626
   tree density, 581, 587
   weight of bears, 250–251
   wing span, 488
   wood in a tree, 526
   yeast growth, 627
Photosynthesis and temperature, 397
Predicting fawns, 243
Pulse rate in animals, 49, 488
Seedling growth, 354, 366
Size and weight of a fish, 166
Stepping frequency, 43
Surface area of wings, 485
Survival rate of birds, 397
Tent worms, 672
Walking speed of animals, 43
Weight and wing area, 499
Weight of
   a bass, 100
   a bear, 255, 273
   a bird, 507
   a fish, 194
   a small fish, 140
Wing span and weight, 48
Worms and moisture, 329

Business and Finance
Airline complaints, 394
Amway sales, 23
Annuities, 654
Appreciation of lake property, 649
Buying a computer, 23
Calculating interest, 39, 69
Checking account, 15
Cigar sales, 334
Commercial banks, 114
Compact disc sales, 145
Compound interest, 577, 585
Depreciation, 638, 660
Determining costs, 270
Fast-food sales, 212
Foreign currency, 39
Growth in salary, 597
Home mortgage payments, 671
Interest, 165, 193, 305, 585, 609
   and investments, 256
Investment
   account, 626
   amounts, 252
   mixture, 243
Leasing a car, 23
LP record sales, 145
Maximizing revenue, 343, 355, 394
Median home price, 672
Microsoft stocks, 295
Minimizing cost, 526
Minimum wage, 356
Mutual fund growth, 617, 624
Salaries, 648, 659, 670
Sales tax, 75
Straight-line depreciation, 660
Sum-of-the-years'-digits, 660
SUV sales, 115, 295
Ticket prices, 322

Education
Bachelor's degrees awarded, 2
Classroom ventilation, 644
College tuition, 438, 447
Compact disc sales, 232
Degrees awarded, 56
Educational attainment, 176, 394
Grades, 154
Higher education enrollment, 11
Income and education, 77
Modeling tuition, 216

Private college tuition, 95, 171
Public college tuition, 52, 108, 129
School bus deaths, 193
School enrollment, 176
Student aid, 295
Student loans, 210, 221
Students and exams, 173
Tuition, 93, 271
Women in college, 322

Entertainment
Broadway revenue, 154
Compact disc sales, 64, 145, 165
Concert tickets, 222
Cost of CDs, 242
LP record sales, 145, 165
Movie box office, 40

Environment
Acid rain, 626
Antarctic ozone layer, 129
Carbon dioxide
  emitters, 637
  levels, 573
Endangered species, 19
Farm pollution, 130
Garbage and household size, 255
Global warming, 71, 573
Greenhouse gases, 587, 619
Interpreting carbon dioxide levels, 139
Manatee deaths, 100
Municipal waste, 129
Polar ice caps, 71
Rain forests, 100, 670
Solid waste, 129, 637
Unhealthy air quality, 135

General Interest
Auditorium seating, 672
Bicycle speed records, 48
Calculating π, 673
Calories and land ownership, 597, 618
Candy and peanuts, 223
Charitable giving, 256
Cigarette butts, 295
Classroom ventilation, 644
Cost of driving, 129
Designing a paper cup, 497
Diving and water pressure, 48
Finding costs, 242
Flying a kite, 395
Forms of communication, 140
Funeral costs, 216–218
Gasoline prices, 55
Liver transplants, 618
Mayors' salaries, 69
Mathematics in newspapers, 93
Meeting a future mate, 154
Missing adults, 5–6
Number of cars waiting, 461
Old Faithful geyser, 255
Passports, 210
People waiting in line, 411, 423
Phone bills, 221
Population growth, 551, 582
Predicting home prices, 243
Probability, 305, 334, 411
Salty snacks, 154
School bus deaths, 193
Skydiving, 508
Thanksgiving travel, 334
Theater seating, 649
Ticket cost, 315–316
Tickets and soft drinks, 202–204
Time spent in line, 410–411, 461, 464
Working together, 462, 464

Geometry
Area, 395
Area of
  a circle, 49
  a cone, 563
  a cylinder, 552, 569
  a frame, 335
  a quadrangle, 264
  a room, 313
  an ellipse, 569
  triangles, 49, 259, 264, 270
Baseball diamond dimensions, 523
Circumference of a circle, 49
Dimensions of
  a box, 570
  a cone, 395
  a cube, 523
  a cylinder, 388
  a pen, 388
  an ellipse, 550
Equilateral triangles, 49
Fencing, 153
Graphing a rectangle, 138
Height of a tree, 154, 436–437, 446
Maximizing area, 354
Perimeter of an ellipse, 569
Pool dimensions, 570
Rectangles, 165, 177, 194
Room dimensions, 148, 153
Screen dimensions, 395
Spheres, 285
Squares and circles, 285
Stacking logs, 659
Surface area of a cone, 524
Triangles, 49, 242–243, 256, 271
Volume of
  a cone, 563
  a cylinder, 552, 569
  a room, 285

Government
Budget for National Parks, 355
Cost of a stamp, 12
Federal
  budget, 12
  debt, 36, 40, 164, 585, 586
  regulations, 221
Government costs, 154
Head Start, 64
Income tax, 85
Income tax rates, 99
Medicaid recipients, 64
Military personnel spending, 61
Minimum wage, 115
Per capita income, 138
State and federal inmates, 129, 154
Trade deficit, 367
U.S. unemployment, 61
Welfare beneficiaries, 64, 137

Highway Design/Construction
Airplane curves, 356
Arch bridge, 551
Braking distance, 48, 367–368, 394, 396
Building construction, 383–384
Construction, 366
Corrosion in airplanes, 518
Cost of driving, 85
Curve design, 504
Design of open channels, 488
Downhill highway grade, 461
Fence construction, 396

List of Applications   ix

Flood control, 508
Guy wire, 509
Passing distance, 317, 322
Highway
   curves, 410, 509, 523
   design, 338
   grade, 405
   hill, 326
   valley, 329
Modeling
   stopping distance, 377
   traffic flow, 580, 587
Rain gutter, 393
Reaction distance, 368
Road maps, 272–273
Rolling resistance of cars, 446
Roof trusses, 210
Runway length, 594, 596, 613, 618
Safe curve speed, 366
Sag curve, 329, 381
Skid marks, 509
Slippery roads, 411–412
Stopping distance, 368, 371
Stopping on a hill, 399, 405–406
Strength of a beam, 443, 448
Taxiway, 340, 344–345, 362
Traffic flow, 272
Train track curves, 403–404, 410, 538
Transportation costs, 447
Uphill highway grade, 410

Medicine/Health
AIDS
   cases, 290, 294, 312, 349–350, 650–651
   deaths, 154, 275, 294
   research, 166
Centenarians, 154
Classroom ventilation, 153
Crutch length, 82
Deaths from lung cancer, 35
Doctors in private sector, 93, 101
Fat grams, 115, 135
Hepatitis C research, 166
HIV infection rates, 130
HIV infections, 137, 147–148
Hospitals, 378
Lead poisoning, 154
Medicaid recipients, 64
Medicare costs, 176

Modeling AIDS deaths, 377
Older mothers, 114
Smoking in the U.S., 149
Tetanus cases, 62, 109, 164
Weight and height, 231–232

Science
Aerial photographs, 423
Air filtration, 656
Airplane speed, 221, 271
Boat speed, 270
Celsius temperature, 4, 70
Comparing ages, 197
Converting units, 85
Cruise control, 99
Dating artifacts, 586
Decibels, 591
Distance, 41–42, 48, 88, 108, 128, 137
   and time, 153, 158, 163, 165, 175, 177, 187, 193, 194
   between bicyclists, 193
   to the horizon, 508
Earthquakes, 597, 619, 624
Earth's gravity, 465
Electrical resistance, 429, 433, 448
Electricity, 429
Emptying a pool, 418
Error in measurement, 187, 194
Fahrenheit temperature, 4, 70
Falling object, 359–360, 367, 638, 649
Falling time, 523
Fat grams, 115, 135
Filling a water tank, 423
Flow rates, 135
Football hang time, 479–480, 487, 523
Headlight, 571
Height
   of a building, 462, 464
   of a stone, 394, 397
   of a tree, 154, 436–437, 446
Hooke's law, 447
Leverage and wrenches, 440–441
Lightbulbs, 462
Mixing
   acid, 221
   antifreeze, 221, 270
Mixture problem, 256, 270

Modeling, 113, 128
   a pendulum, 465
   motion, 69, 367
   sound levels, 591, 596, 619
   water flow, 377
Mowing the lawn, 423
Musical notes, 478–479
Musical scales, 478–479
Pendulum, 523–524
Photography, 433, 434
Pumping water, 423
Radioactive carbon dating, 524, 626
Recording music, 197, 462
Resistance and current, 448
River current, 221
Room ventilation, 648
Sketching a model, 114
Solar heat, 571
Speed of
   a boat, 423, 462
   an airplane, 423
Strength of a beam, 443, 448
Television dimensions, 502, 508
Temperature scales, 4, 137, 177, 194
Thermostat, 99
Tightening lug nuts, 447
Two-cycle engines, 115
Visualization, 140
Voltage, 429
Water flow, 128
Wind speed, 423

Sports
Athlete's running speed, 205
Average baseball salaries, 295
Baseball on the moon, 355
Basketball court, 222
Bouncing ball, 649, 660
Burning calories, 270, 271
Burning fat calories, 221
Drinking fluids and exercise, 196
Flight of a baseball, 309, 312
Football
   hang time, 479–480, 487, 523
   tickets, 243
Height of a baseball, 355
Home runs, 210
Inline skating, 50
Jogging speeds, 256

Men's Olympic times, 114
Modeling heart rate, 291, 294
Pulse rate and exercise, 274
Running, 419, 423
Sport drinks, 196
Swimming pool maintenance, 587, 645, 648
Target heart rate, 226, 231
Tickets, 270
Weight lifting, 100
Weight machines, 551
Women on the run, 284
Women's Olympic times, 100
Youth soccer, 155

Technology
Aerial photographs, 423
Boolean algebra and computers, 467
Cable modems, 23
Cars and horsepower, 284
Cellular phone
  path loss, 598
  subscribers, 20
  technology, 607
  transmission, 470, 478
  use, 281, 586
Computer
  memory, 26, 40
  sales, 221
Cordless phone sales, 59
DDT, 339
DVD and picture dimensions, 508
Digital
  cameras, 285
  images, 11
  pictures, 276–277
  subscriber lines, 23
Early cellular phone growth, 197
Equations and supercomputers, 261–262, 264
Global Positioning System, 39
GPS clocks, 39
Image resolution, 2
Internet in Europe, 362–363
Internet use, 64, 116, 118, 366, 586
Invention of computer, 198
Laser print, 34
Music storage, 434
PC computers, 285
Pagers, 270
Path loss for cellular phones, 598
Pixels, 2–3, 276
Recording music, 197, 446, 462
Screen dimensions, 377
Space Shuttle, 312
Square view rectangle, 124
Video games, 398
Wind power, 448, 482–483, 487

U.S. and World Demographics
Accidental deaths, 86, 388
Age
  at first marriage, 134
  in the U.S., 100, 194
Aging in America, 338
Alcohol consumption, 70
Asian-American population, 64, 136
Birth rate, 135
Born outside the U.S., 178
Captured prison escapees, 659
Federal debt, 36, 40, 164, 585, 586
Handguns manufactured, 210
Highest elevations, 187
Marriages, 135
Median
  home prices, 177
  income, 153
Memorial Day travelers, 137
Modeling population, 586, 624
Motor vehicle registration, 86
Number of radio stations, 114
Pedestrian fatalities, 269
Per capita income, 212
Population
  growth, 524
  of urban regions, 598, 619
Prison escapees, 659
Remaining life expectancy, 138
Union membership, 154
U.S. energy consumption, 394
U.S. median family income, 115
U.S. population, 36, 61, 130, 135, 165, 366
Weapons convictions, 23
Western population, 129
Women
  in politics, 130
  in the workforce, 337
  officers, 194
World population, 34, 40, 285

Weather
Air temperature, 334
  and altitude, 447, 462
Altitude
  and dew point, 165, 464
  and temperature, 160–161, 165, 170, 177, 194
Average temperature, 313
Cloud formation, 160–161
Clouds and temperature, 196
Deserts, grasslands, and forests, 231
Dew point, 336,
  and altitude, 165, 464
First frost, 51
Global warming, 71
Hurricanes, 597, 619
Modeling
  ocean temperature, 291, 294
  wind speed, 624
Monthly average precipitation
  Houston, TX, 194
  Las Vegas, NV, 87
Monthly average temperature
  Boston, MA, 178, 183
  Buenos Aires, Argentina, 187
  Chesterfield, Canada, 187
  Marquette, MI, 187
  Memphis, TN, 187
  Washington, D.C., 78
Ozone and UV radiation, 448
Predicting wind speed, 597
Skin cancer and UV radiation, 463
Ultraviolet radiation, 72
Water content in snow, 435, 446
Wind speeds
  Hilo, HI, 94
  Louisville, KY, 87
  Myrtle Beach, SC, 99
  San Francisco, CA, 51

PREFACE

Intermediate Algebra through Modeling and Visualization offers an innovative approach to the intermediate algebra curriculum that allows students to gain both skills and understanding. This text not only demonstrates the relevance of mathematics but it also prepares students for future courses. The early introduction of functions and graphs allows the instructor to use applications and visualization to present mathematical topics. Real data, graphs, and tables play an important role in the course, giving meaning to the numbers and equations that students encounter. This approach increases student interest, motivation, and the likelihood for success.

Approach

A comprehensive curriculum is presented with a *balanced* and *flexible* approach that is essential for today's intermediate algebra courses. Instructors have the flexibility to strike their own balance with regard to emphasis on skills, rule of four, applications, modeling, and graphing calculator technology. With this approach to the rule of four (verbal, graphical, numerical, and symbolic methods), instructors can easily emphasize one rule more than another to meet their students' needs. This flexibility also extends to modeling, applications, and graphing calculator use. The text contains numerous practical applications, including modeling of real-world data with functions and word problems. Instructors have the freedom to fully integrate graphing calculators throughout the course. However, regular graphing calculator use is not a requirement of the text.

In *Intermediate Algebra through Modeling and Visualization* mathematical concepts are introduced by moving from the concrete to the abstract. Relevant applications underscore mathematical concepts. This text includes a diverse collection of unique, up-to-date applications that answer the commonly asked question: "When will I ever use this?" Modeling, visualization, and the rule of four allow greater access to mathematics for students with different learning styles. However, the primary purpose of this text is to teach mathematical concepts and skills. Standard mathematical definitions, theorems, symbolism, and rigor are maintained.

Organization

This text consists of 11 chapters and 50 sections. Most sections represent one class day.

Chapter 1: Real Numbers and Algebra
This chapter reviews several topics from beginning algebra, including numbers, operations on numbers, and integer exponents. Formulas are introduced as a way to describe data, and the xy-plane is used to plot and visualize data.

Chapter 2: Linear Functions and Models
Basic concepts about functions, along with their verbal, symbolic, graphical, and numerical representations, are discussed in this chapter. It features linear functions, linear models, slope, and lines. The applications presented make the concepts more concrete and relevant to students.

Chapter 3: Linear Equations and Inequalities
This chapter covers solving linear equations and inequalities. Symbolic, graphical, and numerical techniques for solving equations are presented. Many of the concepts presented in this chapter are used to solve equations and inequalities in subsequent chapters. Real-world applications and models promote student learning and understanding. Compound inequalities and absolute value inequalities are included.

Chapter 4: Systems of Linear Equations
The concepts presented in Chapter 3 are extended to systems of linear equations and inequalities. Emphasis is given to systems involving two equations in two variables. Matrix methods and determinants are included at the end of the chapter.

Chapter 5: Polynomial Expressions and Functions
Polynomials and polynomial functions are introduced, and a discussion about factoring is included. This chapter extends many of the techniques introduced in Chapters 2 and 3 to polynomials. Functions, applications, and real-world data continue to motivate many of the discussions.

Chapter 6: Quadratic Functions and Equations
Building on material from Chapter 5, this chapter discusses quadratic equations and functions. Quadratic functions and their graphs are used to model nonlinear data. Quadratic equations are solved with standard symbolic techniques, along with graphical and numerical methods.

Chapter 7: Rational Expressions and Functions
Rational expressions and functions are introduced in this chapter. Arithmetic operations on rational expressions are included. Applications occur throughout, along with a special section on proportions and variation.

Chapter 8: Radical Expressions and Functions
Radical notation and some basic functions, such as the square root and cube root functions, are discussed. Presentation of properties of rational exponents was delayed until this chapter so that students can immediately apply them. The chapter concludes with a section on complex numbers.

Chapter 9: Conic Sections
This chapter introduces parabolas, circles, ellipses, and hyperbolas and includes applications. The chapter concludes with a section on solving nonlinear equations and inequalities.

Chapter 10: Exponential and Logarithmic Functions
Logarithms, properties of logarithms, exponential and logarithmic functions, and exponential and logarithmic equations are covered in this chapter. Linear growth and exponential growth are compared so that students grasp the fundamental difference be-

tween the two types of growth. The chapter contains many real-world applications and models.

Chapter 11: Sequences and Series
This chapter introduces the basic concepts of both sequences and series, concentrating on arithmetic and geometric sequences and series. Both graphical and symbolic representations of sequences are discussed, along with applications and models. The chapter concludes with the binomial theorem.

Features

Applications and Models
Interesting, straightforward applications are a strength of this textbook, helping students become more effective problem solvers. Applications are intuitive and not overly technical so that they can be introduced in a minimum of class time. Current data are utilized to create meaningful mathematical models, exposing students to a wealth of real-world uses of mathematics. A unique feature of this text is that the applications and models are woven into both the discussions and the exercises. Students can more easily learn how to solve applications if they are discussed within the text. (See pages 94, 108, and 290.)

Graphing Calculator Technology
The use of a graphing calculator with this text is optional, but for instructors who want to fully integrate graphing technology into their courses, this text integrates the graphing calculator thoroughly and seamlessly, without sacrificing development of traditional algebraic skills. Students are encouraged to solve problems with multiple methods, learning to evaluate graphing calculator techniques against other problem-solving methods. An GCLM icon next to selected examples refers students to the *Graphing Calculator Lab Manual* that accompanies the text, in which the examples from the book are used to further instruct students in how to use the graphing calculator. (See pages 82, 144, and 350.)

Putting It All Together
This helpful feature occurs at the end of each section and summarizes techniques and reinforces the mathematical concepts presented in the section. It is given in an easy-to-follow grid format. (See pages 37, 218, and 311.)

Section Exercise Sets
The exercise sets are at the heart of any mathematics text, and this text includes a wide variety of exercises that are instructive for student learning. Each exercise set contains exercises involving basic concepts, skill-building, and applications. In addition, many exercises ask students to read and interpret graphs. Writing About Mathematics exercises are also included at the end of every exercise set. The exercise sets are carefully graded and categorized by topic, making it easier for an instructor to select an appropriate assignment. Each exercise set concludes with an assortment of appealing applications. (See pages 97, 126, and 151.)

Checking Basic Concepts
Provided after every two sections, this feature presents a brief set of exercises that students can use for review purposes or group activities. These exercises require 10–20

minutes to complete and can be used during class if time permits. (See pages 130, 296, and 449.)

Group Activities: Working with Real Data
This feature occurs after selected sections (1 or 2 per chapter) and provides an opportunity for students to work collaboratively on a problem that involves real-world data. Most activities can be completed with limited use of class time. (See pages 115, 322, and 496.)

Chapter and Section Introductions
Many intermediate algebra students have little or no understanding of what mathematics is about. Chapter and section introductions present and explain some of the reasons for studying mathematics. They provide insights into the relevance of mathematics to many aspects of real life. (See pages 50, 72, and 274.)

Chapter Summaries
Chapter summaries are presented in an easy-to-read grid format for students to use in reviewing the important topics in the chapter. (See pages 132, 330, and 389.)

Chapter Review Exercises
Chapter Review Exercises contain both skill-building exercises, which are keyed to the appropriate sections within the chapter, and application exercises, which stress the relevance of mathematical concepts to the real world. The Chapter Review Exercises stress different techniques for solving problems and provide students with the review necessary to successfully pass a chapter test. (See pages 134, 190, and 391.)

Chapter Tests
A test is provided in the end-of-chapter review material so that students can practice their knowledge and skills. (See pages 70, 138, and 195.)

Extended and Discovery Exercises
These exercises occur at the end of each chapter and are usually more complex than the Review Exercises, requiring extension or discovery of a topic presented in the chapter. They can be utilized for either collaborative learning or extra homework assignments. (See pages 71, 140, and 336.)

Making Connections
This feature occurs throughout the text and helps students see how previously learned concepts are related to new concepts. (See pages 93, 171, and 172.)

Critical Thinking
At least one Critical Thinking exercise is included in most sections and poses a question that can be used for either classroom discussion or a homework assignment. These exercises typically ask students to extend a mathematical concept beyond what has already been discussed. (See pages 27, 173, and 346.)

Technology Notes
Occurring throughout the text, Technology Notes offer students guidance, suggestions, and cautions on the use of the graphing calculator. (See pages 124, 147, and 474.)

Preface

Sources

For the numerous real-world applications that appear throughout the text, genuine sources are cited to help establish the practical applications of mathematics in real life. (See pages 100, 349, and 403.) In addition, a comprehensive bibliography appears at the end of the text.

Supplements for the Student

Graphing Calculator Lab Manual (0-201-72619-X)

Written specifically by mathematics instructor Joe May to accompany *Intermediate Algebra through Modeling and Visualization*, the *Graphing Calculator Lab Manual* is organized in a just-in-time format, with new keystrokes and skills introduced as needed in the text. The manual is divided into two parts, with the first half covering the TI–83 Plus and the second half covering the TI–86. Extra emphasis is placed on illustrating the uses and skills associated with setting a window. An icon GCLM is placed in the text next to every example that appears in the manual, so students immediately know when they can reference the manual for extra help. There is a list at the end of this text of all the topics covered in the manual.

Student's Solutions Manual (0-201-72616-5)

This manual, written by mathematics instructor Terry Krieger, contains solutions to the odd-numbered exercises for each section (excluding Writing About Mathematics and Group Activity exercises) and solutions to all Checking Basic Concepts exercises, Chapter Review Exercises, and Chapter Test questions.

InterAct Math® Tutorial Software (0-201-72620-3)

Available on a dual-platform, Windows/Macintosh CD-ROM, this interactive tutorial software provides algorithmically generated practice exercises that are correlated at the objective level to the content of the text. Every exercise in the program is accompanied by an example and a guided solution designed to involve students in the solution process. For Windows users, selected problems also include a video clip to provide additional instruction and help students visualize concepts. The software recognizes common student errors and provides appropriate feedback, and it also tracks student activity and scores and can generate printed summaries of students' progress. Instructors can use the InterAct Math® Plus course-management software to create, administer, and track tests and monitor student performance during practice sessions (see the description on page xvii).

InterAct MathXL: www.mathxl.com

InterAct MathXL is a Web-based tutorial system that helps students prepare for tests by allowing them to take practice tests and receive personalized study plans based on their results. Practice tests are correlated directly to the section objectives in the text, and once a student has taken an on-line practice test, the software scores the test and generates a study plan that identifies strengths, pinpoints topics where more review is needed, and links directly to the appropriate section(s) of the InterAct Math® tutorial software for additional practice and review. A course-management feature allows instructors to create and administer tests, view students' test results, study plans, and practice work. Students gain access to the InterAct MathXL Web site through a password-protected subscription; subscriptions can either be bundled free with new copies of the text or purchased separately with a used book.

Videotapes (0-201-70487-0)

Created specifically to accompany *Intermediate Algebra through Modeling and Visualization*, these videotapes cover every section of every chapter, and the lecturers present examples that are taken directly from the text. Each video segment includes a "stop the tape" feature that encourages students to pause the video, work through the example presented on their own, and then resume play to watch the video instructor go over the solution.

Digital Video Tutor (0-201-72618–1)

This supplement provides the entire set of videotapes for this text in digital format on CD-ROM, making it easy and convenient for students to watch video segments displayed on a computer, either at home or on campus. Available for student purchase with the text at minimal cost, the Digital Video Tutor is ideal for distance learning and supplemental instruction.

Addison-Wesley Math Tutor Center

The Addison-Wesley Math Tutor Center is staffed by qualified mathematics instructors who provide students with tutoring on examples and odd-numbered exercises from their textbooks. Tutoring is provided via toll-free telephone, fax, or e-mail, and White Board technology allows tutors and students to actually see problems being worked while they "talk" in real time over the Internet during their tutoring sessions. The Math Tutor Center is accessed through a registration number that may be bundled free with a new textbook or purchased separately with a used book.

Web Site: www.MyMathLab.com

Ideal for lecture-based, lab-based, and on-line courses, this state-of-the-art Web site provides students with a centralized point of access to the wide variety of on-line resources available with this text. The pages of the actual book are loaded into MyMathLab.com, and as students work through a section of the on-line text, they can link directly from the pages to supplementary resources (e.g., tutorial software, interactive animations, and audio and video clips) that provide instruction, exploration, and practice beyond that offered in the printed book. MyMathLab.com generates personalized study plans for students and allows instructors to track all student work on tutorials, quizzes, and tests. Complete course-management capabilities, including a host of communication tools for course participants, are provided to create a user-friendly and interactive on-line learning environment.

Supplements for the Instructor

Annotated Instructor's Edition (0-201-71967-3)

The Annotated Instructor's Edition contains Teaching Tips and provides answers to every exercise in the text except the Writing About Mathematics exercises. Answers that do not fit on the same page as the exercises themselves are supplied in the Instructor's Answers section at the back of the text.

Instructor's Solutions Manual (0-201-72615-7)

This manual provides solutions to all section-level exercises (excluding Writing About Mathematics exercises), Critical Thinking exercises, Checking Basic Concepts exercises, Chapter Review Exercises, Extended and Discovery Exercises, Group Activity exercises, and Chapter Test questions.

Printed Test Bank and Instructor's Resource Guide (0-201-72622-X)
The Printed Test Bank portion of this manual contains

- three free-response test forms per chapter, one of which (Form C) places stronger emphasis on applications and graphing calculator technology than the other test forms;
- one multiple-choice test form per chapter; and
- one free-response and one multiple-choice final exam.

The Instructor's Resource Guide portion of the manual contains

- four sets of cumulative review exercises that cover Chapters 1–3, 1–6, 1–9, and 1–11;
- transparency masters consisting of tables, figures, and examples from the text;
- teaching notes designed to assist instructors with presenting graphing calculator topics in class; and
- supplemental graphing calculator activities.

TestGen-EQ with QuizMaster-EQ (0-201-72623-8)
Available on a dual-platform, Windows/Macintosh CD-ROM, this fully networkable software enables instructors to create, edit, and administer tests by using a computerized test bank of questions organized according to the chapter content of the text. Six question formats are available, and a built-in question editor allows the instructor to create graphs, import graphics, and insert mathematical symbols and templates, variable numbers, or text. An Export to HTML feature allows practice tests to be posted to the Internet, and instructors can use QuizMaster-EQ to post quizzes to a local computer network so that students can take them on-line. QuizMaster-EQ automatically grades the quizzes, stores results, and lets the instructor view or print a variety of reports for individual students or for an entire class or section.

InterAct Math® Plus (0-201-72140-6)
This networkable software provides course-management capabilities and network-based test administration for Addison-Wesley's InterAct Math® Tutorial Software (see the description on page xv). InterAct Math® Plus enables instructors to create and administer on-line tests, summarize students' results, and monitor students' progress with the tutorial software, providing an invaluable teaching and tracking resource.

Web Site: www.MyMathLab.com
In addition to providing a wealth of resources for lecture-based courses, this state-of-the-art Web site gives instructors a quick and easy way to create a complete on-line course based on *Intermediate Algebra through Modeling and Visualization*. Hosted nationally at no cost to instructors, students, or schools, MyMathLab.com provides access to an interactive learning environment wherein all content is keyed directly to the text. A customized version of Blackboard, Inc.™ provides the course-management platform. MyMathLab.com lets instructors administer pre-existing tests and quizzes or create their own, provides detailed tracking of all student work, and offers a wide array of communication tools for course participants. Within MyMathLab.com, students link directly from on-line pages of their text to supplementary resources such as tutorial software, interactive animations, and audio and video clips. Instructors can access on-line versions of the *Printed Test Bank and Instructor's Resource Guide*, which includes PowerPoint slides. MyMathLab.com is accessed through a registration number that may be bundled free with a new textbook or purchased separately with a used book.

Acknowledgments

Many individuals contributed to the development of this textbook. I would like to thank the following reviewers, whose comments and suggestions were invaluable in preparing *Intermediate Algebra through Modeling and Visualization*:

Gisela Acosta, *Valencia Community College, East Campus*
Daniel D. Anderson, *University of Iowa*
Marie Aratari, *Oakland Community College*
Rick Armstrong, *St. Louis Community College, Florissant Valley*
Michael Butler, *College of the Redwoods*
Daniel P. Fahringer, *Harrisburg Area Community College*
Donna S. Fatheree, *University of Southwestern Louisiana*
Eugenia E. Fitzgerald, *Phoenix College*
Debbie Garrison, *Valencia Community College, East Campus*
James W. Harris, *John A. Logan College*
Margaret Hathaway, *Kansas City Community College*
Jeannie Hollar, *Lenoir-Rhyne College*
Joel K. Haack, *University of Northern Iowa*
Sally Jackman, *Richland College*
S. Marie Kunkle, *Indiana University—Kokomo*
Greg Langkamp, *Seattle Central Community College*
Lew Ludwig, *Ohio University*
Jeanette O'Rourke, *Middlesex County College*
Don Reichman, *Mercer County Community College*
Elaine Robinson, *Motlow State Community College*
Jolene Rhodes, *Valencia Community College, East Campus*
David Shellabarger, *Lane Community College*
Gregory Sliwa, *Broome Community College*
Sanford O. Smith, *J. Sargeant Reynolds Community College*
Sam Tinsley, *Richland College*
David Wasilewski, *Luverne County Community College*

I would also like to recognize the efforts of Terry Krieger from Winona State University for assembling the answers and solutions manuals. Donna Foster of Piedmont Technical College, Paul Lorczak, Beth Marsh, and John Morin deserve special credit for their help with accuracy checking. Without the excellent cooperation from the professional staff at Addison-Wesley, this project would have been impossible. Thanks go to Jason Jordan for his support of the project. Particular recognition is due Jennifer Crum and Ruth Berry who gave advice, support, assistance, and encouragement. The outstanding contributions of Adam Hamel, Dona Kenly, Kathy Manley, Tricia Mescall, Sharon Smith, Joe Vetere, Kathy Diamond, Jennifer Bagdigian, and Naomi Kornhauser are also appreciated. Thanks go to Wendy Rockswold, who proofread the manuscript several times and was instrumental in the success of this project. A special thank you goes to the many students and instructors who class-tested this edition. Their suggestions were insightful. Please feel free to write to Gary Rockswold, Department of Mathematics, Minnesota State University, Mankato, MN 56001 with your comments. Your opinion is important.

Gary Rockswold

WALKTHROUGH

Chapter Openers

Each chapter opener describes an applied, real-life example to motivate students by giving them insight into the relevance of that chapter's central mathematical concepts.

Rule of Four

Throughout the text, concepts are consistently presented through verbal, graphical, numerical, and symbolic representations to support multiple learning styles and methods of problem solving. Many sections are divided into sub-sections that present, in turn, graphical, numerical, and symbolic solutions to the same types of examples. This structure provides a flexible approach to the rule of four that lets instructors easily emphasize whichever method(s) they prefer.

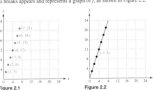

WALKTHROUGH

Applications and Models

Both the exposition and exercises contain unique applications that truly model real-world data. Examples often begin with concrete applications that are used to derive the abstract mathematical concepts—an approach that motivates students by illustrating the relevance of the math from the very start. Application headings in the exercise sets call out the real-world topics presented, and sources for data are cited throughout.

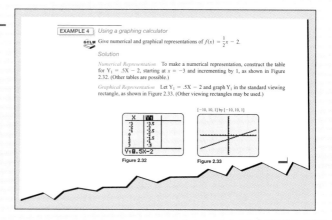

Graphing Calculator Technology

Graphing calculator use is optional, but for instructors who wish to fully integrate graphing calculators, this text thoroughly and seamlessly integrates graphing technology without sacrificing traditional algebraic skills. Students are encouraged to solve problems with multiple methods, learning to evaluate graphing calculator techniques against other problem-solving methods. An icon next to selected examples refers students to the *Graphing Calculator Lab Manual* that accompanies the text, where they find that same example used to further instruct them in the use of the graphing calculator.

Making Connections

This feature occurs throughout the text and helps students see how topics are interrelated. It also helps motivate students by calling out connections between mathematics and the real world.

Critical Thinking

The *Critical Thinking* feature appears in most sections and poses a question that can be used for either classroom discussion or homework. Critical Thinking questions typically ask students to extend a mathematical concept beyond what has already been discussed in the text.

Technology Notes

Occurring throughout the text, *Technology Notes* offer students guidance and suggestions for using the graphing calculator to solve problems and explore concepts.

WALKTHROUGH

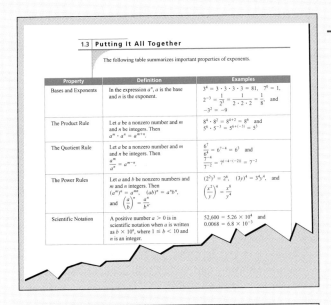

Putting It All Together

This helpful feature occurs at the end of each section to summarize techniques and reinforce concepts. These unique summaries give students a consistent, visual study aid for every section of the text.

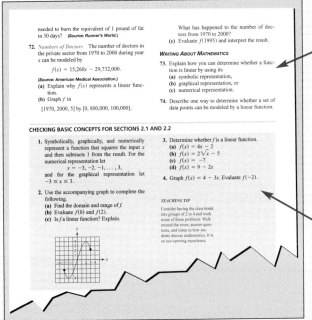

Section Exercise Sets

Each exercise set contains a wide variety of problems involving basic concepts, skill-building, and applications, and many exercises ask students to read and interpret graphs. Exercise sets are carefully graded and categorized according to the topics within sections, making it easier for instructors to select appropriate assignments. Each set concludes with "Writing About Mathematics" exercises to help students verbalize and synthesize concepts.

Checking Basic Concepts

Provided after every two sections, *Checking Basic Concepts* can be used for review or for group activities. These brief exercise sets require 10–20 minutes to complete and can be used during class if time permits.

Group Activities: Working with Real Data

Group Activities (1–2 per chapter) ask students to work collaboratively to solve a problem involving real data. Most activities can be completed with limited use of class time.

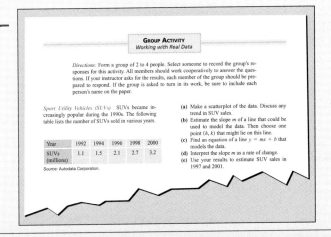

WALKTHROUGH

For Extra Help

Found at the beginning of each exercise set, these boxes direct students to the various supplementary resources that are available to them for extra help and practice.

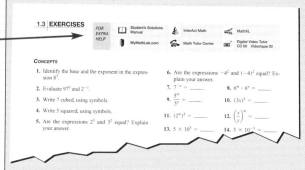

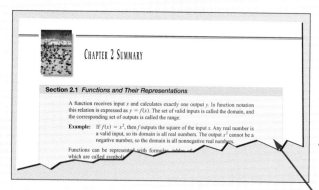

End-of-Chapter Material

The end-of-chapter material includes the following features to ensure that students have ample opportunities to practice skills, synthesize concepts, and explore material in greater depth:

- A *Chapter Summary* to present the key concepts from each section

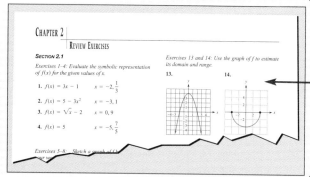

- Extensive *Chapter Review Exercises* that include both skill-building exercises and applications, and stress different problem-solving techniques

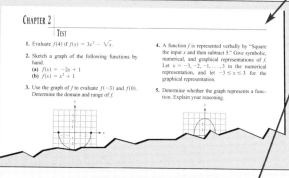

- The *Chapter Test* to give students a chance to practice their knowledge and skills

- *Extended and Discovery Exercises*, which are ideal for collaborative learning or extra-credit homework assignments, allowing students to extend their knowledge of a particular topic.

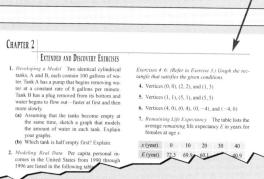

CHAPTER 1 | REAL NUMBERS AND ALGEBRA

"Why do I need to learn math?" This question is commonly asked by students across the country. A look at history provides some reasons for studying mathematics.

All of us have ancestors who were not only illiterate but who also were unable to perform even basic arithmetic. They did not lack intelligence—rather, they lacked education. In about A.D. 400, Saint Augustine declared that people who could add and subtract were in conspiracy with the devil. One hundred years ago many Americans were educated in rural schools, where it was common for school boards to have difficulty finding teachers who could multiply and divide. In the 1940s and 1950s, many mathematics teachers had no training beyond algebra. Today, knowledge of algebra, trigonometry, and calculus is essential for many occupations.

Mathematics opens doors in our society. Without mathematical skills, opportunities are lost. Most vocations and professions require a higher level of mathematical understanding than in the past. Seldom are the mathematical expectations of employees lowered, as the work place becomes more technical—not less. Mathematics is the *language of technology.* The Department of Education cites mathematics as a key to achievement in our society. Students with a solid mathematics background earned, on average, 38 percent more per hour than their peers without such a background.

Decisions in high school or college to avoid mathematics classes have lifelong ramifications that affect vocations, incomes, and lifestyles. Switching majors to escape taking mathematics may prevent us from pursuing our dreams. Like reading and writing, mathematics is a necessary component for us to reach our full potentials.

What skills will be needed in society during the next 50 years? Although no one can answer this question with certainty, history demonstrates that the importance of mathematics will not diminish, but will likely increase.

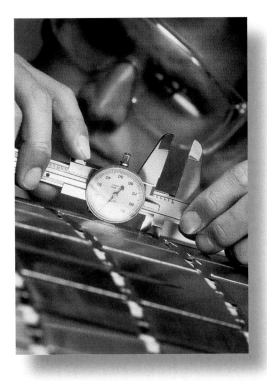

It is not enough to have a good mind; the main thing is to use it well.
—René Descartes

Sources: A. Toffler and H. Toffler, *Creating a New Civilization; USA Today.*

1.1 Describing Data with Sets of Numbers

Natural and Whole Numbers ~ Integers and Rational Numbers ~ Real Numbers ~ Properties of Real Numbers

Introduction

The need for numbers has existed in nearly every society. Numbers first occurred in the measurement of time, currency, goods, and land. As the complexity of a society increased, so did its numbers. One tribe that lived near New Guinea counted only to 6. Any number higher than 6 was referred to as "many." This number system met the needs of that society. However, a highly technical society could not function with so few numbers.

In this section we discuss numbers that are vital to our technological society. We also show how different types of data can be described with sets of numbers. (*Source: Historical Topics for the Mathematics Classroom, Thirty-first Yearbook*, NCTM.)

Natural and Whole Numbers

One important set of numbers found in most societies is the set of **natural numbers.** These numbers comprise the *counting numbers* and may be expressed as

$$N = \{1, 2, 3, 4, 5, 6, \ldots\}.$$

Set braces, { }, are used to enclose the elements of a set. Because there are infinitely many natural numbers, three dots show that the list continues in the same pattern without end. A second set of numbers, called the **whole numbers,** is given by

$$W = \{0, 1, 2, 3, 4, 5, \ldots\}.$$

Natural numbers and whole numbers can be used when data are not broken into fractional parts. For example, Table 1.1 lists the number of bachelor's degrees awarded during selected academic years. Note that either natural numbers or whole numbers are appropriate to describe the data because a fraction of a degree cannot be awarded.

TABLE 1.1 Bachelor's Degrees Awarded

Year	1969–1970	1979–1980	1989–1990	1997–1998
Degrees	792,317	929,417	1,051,344	1,169,121

Source: Department of Education.

EXAMPLE 1 *Describing image resolution*

The screens for computer terminals or graphing calculators are made up of tiny squares called *pixels*. A rectangular screen on a graphing calculator might be 95 pixels across and 63 pixels high, whereas a computer terminal could be 2048 by 2048 pixels. Figure 1.1 shows a graphing calculator screen. If we look closely, we can see that each symbol and number is made up of several pixels. Fractional parts of a pixel do not occur. As a result, natural numbers are appropriate to describe them.

(a) Find the number of pixels in a graphing calculator screen that is 95 by 63 pixels.

(b) The photograph in Figure 1.2 shows an image of Jupiter and two of its moons, Io and Europa, taken by *Voyager 1*. This photograph is 820 by 540 pixels. How many pixels are there?

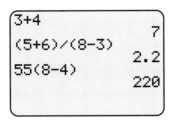

Figure 1.1

Figure 1.2 Jupiter and Two Moons
Source: NASA.

Solution

(a) There are $95 \times 63 = 5985$ pixels.
(b) There are $820 \times 540 = 442{,}800$ pixels.

INTEGERS AND RATIONAL NUMBERS

The set of **integers** is given by

$$I = \{\ldots, -3, -2, -1, 0, 1, 2, 3, \ldots\}.$$

The integers include both the natural numbers and the whole numbers. During the eighteenth century, negative numbers were not readily accepted by mathematicians. Such numbers did not seem to have real meaning. However, today when a person opens a personal checking account for the first time, negative numbers quickly take on meaning. There is a difference between a positive and a negative balance.

A **rational number** is any number that can be expressed as the ratio of two integers $\frac{p}{q}$, where $q \neq 0$. Rational numbers can be written as fractions and include all integers. Some examples of rational numbers are

$$\frac{8}{1}, \quad \frac{2}{3}, \quad -\frac{3}{5}, \quad -\frac{7}{2}, \quad \frac{22}{7}, \quad 1.2, \quad \text{and} \quad 0.$$

Note that 1.2 and 0 are both rational numbers because they can be written as $\frac{12}{10}$ and $\frac{0}{1}$.

Rational numbers may be expressed in decimal form that either *repeats* or *terminates*. The fraction $\frac{1}{3}$ may be expressed as $0.\overline{3}$, a repeating decimal, and the fraction $\frac{1}{4}$ may be expressed as 0.25, a terminating decimal. The overbar indicates that $0.\overline{3} = 0.3333333\ldots$.

Integers and rational numbers are used to describe things such as temperature. Table 1.2 on the next page lists equivalent temperatures in both degrees Fahrenheit and degrees Celsius. Note that both positive and negative numbers are used to describe temperature.

TABLE 1.2 Fahrenheit and Celsius Temperature

°F	°C	Observation
−89	−67.$\overline{2}$	Alcohol freezes
−40	−40	Mercury freezes
0	−17.$\overline{7}$	Snow and salt mixture
32	0	Water freezes
100	37.$\overline{7}$	A very warm day
212	100	Water boils

EXAMPLE 2 *Classifying numbers*

Classify each real number as a natural number, whole number, integer, or rational number.

(a) $\frac{6}{3}$ **(b)** -1 **(c)** 0 **(d)** $-\frac{11}{3}$

Solution

(a) Because $\frac{6}{3} = 2$, the number $\frac{6}{3}$ is a natural number, a whole number, an integer, and a rational number.
(b) The number -1 is an integer and a rational number but not a natural or a whole number.
(c) The number 0 is a whole number, an integer, and a rational number but not a natural number.
(d) The fraction $-\frac{11}{3}$ is a rational number as it is the ratio of two integers. However, it is not a natural number, a whole number, or an integer.

REAL NUMBERS

Real numbers can be represented by decimal numbers. Every fraction has a decimal form, so real numbers include rational numbers. However, some real numbers cannot be expressed by fractions. They are called **irrational numbers.** The numbers $\sqrt{2}$, $\sqrt{15}$, and π are examples of irrational numbers. They can be expressed by decimals but not by decimals that either repeat or terminate. Examples of real numbers include

$$2, \quad -10, \quad 151\frac{1}{4}, \quad -131.37, \quad \frac{1}{3}, \quad -\sqrt{5}, \quad \text{and} \quad \sqrt{11}.$$

Any real number may be approximated by a terminating decimal. We use the symbol ≈, which means **approximately equal,** to denote an approximation. Each of the following real numbers has been approximated to three *decimal places*.

$$\pi \approx 3.142, \quad \frac{2}{3} \approx 0.667, \quad \sqrt{200} \approx 14.142$$

Figure 1.3 shows the relationship between the different sets of numbers. Note that each real number is either a rational number or an irrational number but not both. The natural numbers, whole numbers, and integers are contained in the set of rational numbers.

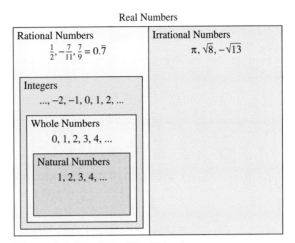

Figure 1.3 The Set of Real Numbers

EXAMPLE 3 *Classifying numbers*

Classify each real number as a natural number, an integer, a rational number, or an irrational number.

$$5, \quad -1.2, \quad \frac{13}{7}, \quad -\sqrt{7}, \quad -12, \quad \sqrt{16}$$

Solution

Natural numbers: 5 and $\sqrt{16} = 4$

Integers: 5, -12, and $\sqrt{16} = 4$

Rational numbers: 5, -1.2, $\frac{13}{7}$, -12, and $\sqrt{16} = 4$

Irrational number: $-\sqrt{7}$

Even though a data set may contain only integers, decimals are often needed to describe it. One common way to do so is to find the **average**.

EXAMPLE 4 *Analyzing numbers of missing adults*

Table 1.3 (on the next page) lists numbers of missing adults nationwide that are being actively investigated by police departments. Find the average number of missing adults during this 5-year period. Is the result a natural, a rational, an irrational, or a real number?

TABLE 1.3 **Missing Adults Nationwide**

Year	1995	1996	1997	1998	1999
Missing Adults	31,364	34,478	36,937	35,946	38,209

Source: The FBI's National Crime Information Center.

Solution

The average number of missing adults was

$$\frac{31{,}364 + 34{,}478 + 36{,}937 + 35{,}946 + 38{,}209}{5} = \frac{176{,}934}{5} = 35{,}386.8.$$

The average of these five natural numbers is both a rational number *and* a real number but neither a natural number nor an irrational number.

Critical Thinking

Is the sum of two irrational numbers ever a rational number? Explain.

PROPERTIES OF REAL NUMBERS

Several properties of real numbers are used in algebra. They are the identity properties, the commutative properties, the associative properties, and the distributive properties.

IDENTITY PROPERTIES The **identity property of 0** states that, if 0 is added to any real number a, the result is a. The number 0 is called the **additive identity**. For example,

$$-3 + 0 = -3 \quad \text{and} \quad 0 + 18 = 18.$$

The **identity property of 1** states that, if any number a is multiplied by 1, the result is a. The number 1 is called the **multiplicative identity**. Examples include

$$-7 \cdot 1 = -7 \quad \text{and} \quad 1 \cdot 9 = 9.$$

We can summarize these results as follows.

Identity Properties

For any real number a,

$$a + 0 = 0 + a = a$$

and

$$a \cdot 1 = 1 \cdot a = a.$$

COMMUTATIVE PROPERTIES The **commutative property for addition** states that two numbers, a and b, can be added in any order and the result is the same. That is, $a + b = b + a$. For example, if a person is paid \$5 and then \$7 or paid \$7 and then \$5, the result is the same. Either way the person was paid a total of

$$5 + 7 = 7 + 5 = 12 \text{ dollars}.$$

1.1 Describing Data with Sets of Numbers

There is also a **commutative property for multiplication.** It states that two numbers, a and b, can be multiplied in any order and the result is the same. That is, $a \cdot b = b \cdot a$. For example, 3 groups of 5 people or 5 groups of 3 people both contain

$$3 \cdot 5 = 5 \cdot 3 = 15 \text{ people.}$$

We can summarize these results as follows.

Commutative Properties

For any real numbers a and b,

$$a + b = b + a$$

and

$$a \cdot b = b \cdot a.$$

ASSOCIATIVE PROPERTIES The commutative properties allow us to reverse the order of two numbers when we add or multiply. The associative properties allow us to change how numbers are grouped. For example, we may add the numbers 3, 4, and 5 as follows.

$$(3 + 4) + 5 = 7 + 5 = 12$$
$$3 + (4 + 5) = 3 + 9 = 12$$

In either case we obtain the same answer, which is the result of the **associative property for addition.** Note that we did not change the order of the numbers; rather we only changed how the numbers were grouped. There is also an **associative property for multiplication,** which is illustrated as follows.

$$(3 \cdot 4) \cdot 5 = 12 \cdot 5 = 60$$
$$3 \cdot (4 \cdot 5) = 3 \cdot 20 = 60$$

We can summarize these results as follows.

Associative Properties

For any real numbers a, b, and c,

$$(a + b) + c = a + (b + c)$$

and

$$(a \cdot b) \cdot c = a \cdot (b \cdot c).$$

Note: Sometimes we omit the multiplication dot. Thus $a \cdot b = ab$ and $5 \cdot x = 5x$.

EXAMPLE 5 *Identifying properties of real numbers*

State the property of real numbers that justifies each statement.

(a) $4 \cdot (3x) = (4 \cdot 3)x$ **(b)** $(1 \cdot 5) \cdot 4 = 5 \cdot 4$ **(c)** $5 + ab = ab + 5$

Solution

(a) This equation illustrates the associative property for multiplication, with the grouping of the numbers changed. That is, $4 \cdot (3x) = (4 \cdot 3)x = 12x$.
(b) This equation illustrates the identity property for 1 because $1 \cdot 5 = 5$.
(c) This equation illustrates the commutative property for addition, the order of the terms 5 and ab changed.

DISTRIBUTIVE PROPERTIES The **distributive properties** are used frequently in algebra to simplify expressions. An example of the distributive property is

$$3(6 + 5) = 3 \cdot 6 + 3 \cdot 5.$$

It is important to multiply the 3 by *both* the 6 and 5—not just the 6. This distributive property is valid when addition is replaced by subtraction. For example,

$$3(6 - 5) = 3 \cdot 6 - 3 \cdot 5.$$

We can summarize these results as follows.

Distributive Properties

For any real numbers a, b, and c,

$$a(b + c) = ab + ac$$

and

$$a(b - c) = ab - ac.$$

Note: Because multiplication is commutative, the distributive properties may be written as

$$(b + c)a = ba + ca \quad \text{and} \quad (b - c)a = ba - ca.$$

EXAMPLE 6 *Applying the distributive properties*

Apply a distributive property to each expression.

(a) $5(4 + x)$ (b) $10 - (1 + a)$ (c) $9x - 5x$

Solution

(a) $5(4 + x) = 5 \cdot 4 + 5 \cdot x = 20 + 5x$
(b) $10 - (1 + a) = 10 - 1(1 + a) = 10 - (1 \cdot 1) - (1 \cdot a) = 9 - a$
(c) $9x - 5x = (9 - 5)x = 4x$

1.1 Putting It All Together

Data and numbers play a central role in a diverse, technological society. Because of the variety of data, it has been necessary to develop different sets of numbers. Without numbers, data could be described qualitatively but not quantitatively. For example, we might say that the day seems hot, but we would not be able to give an actual number for the temperature. The following table summarizes some of the sets of numbers.

Concept	Examples	Comments
Natural Numbers	$1, 2, 3, 4, 5, \ldots$	Sometimes referred to as the *counting numbers*
Whole Numbers	$0, 1, 2, 3, 4, \ldots$	Includes the natural numbers
Integers	$\ldots, -2, -1, 0, 1, 2, \ldots$	Includes the natural numbers and the whole numbers
Rational Numbers	$\frac{1}{2}, -3, \frac{128}{6}, -0.335, 0,$ $0.25 = \frac{1}{4}, 0.\overline{3} = \frac{1}{3},$ and all fractions	Includes integers and all fractions $\frac{p}{q}$, where p and $q \neq 0$ are integers, and all repeating and terminating decimals
Real Numbers	$\pi, \sqrt{3}, -\frac{4}{7}, 0, -10,$ $0.\overline{6} = \frac{2}{3}, 1000, \sqrt{15}$	Any number that can be expressed in decimal form, including the rational numbers

The real numbers have several important properties, which are summarized in the following table.

Property	Definition	Examples
Identity (0 and 1)	The identity element for addition is 0 and the identity element for multiplication is 1. For any real number a, $a + 0 = a$ and $a \cdot 1 = a$.	$5 + 0 = 5$ and $5 \cdot 1 = 5$
Commutative	For any real numbers a and b, $a + b = b + a$ and $a \cdot b = b \cdot a$.	$4 + 6 = 6 + 4$ and $4 \cdot 6 = 6 \cdot 4$
Associative	For any real numbers a, b, and c, $(a + b) + c = a + (b + c)$ and $(a \cdot b) \cdot c = a \cdot (b \cdot c)$.	$(3 + 4) + 5 = 3 + (4 + 5)$ and $(3 \cdot 4) \cdot 5 = 3 \cdot (4 \cdot 5)$
Distributive	For any real numbers a, b, and c, $a(b + c) = ab + ac$ and $a(b - c) = ab - ac$.	$5(x + 2) = 5x + 10$ and $5(x - 2) = 5x - 10$

1.1 EXERCISES

CONCEPTS

1. Which numbers are in the set of natural numbers and which are in the set of whole numbers?

2. Which numbers are in the set of integers? Give an example of an integer that is not a natural number.

3. Which numbers are in the set of rational numbers? Give an example.

4. Which numbers are in the set of real numbers? If a number is a real number but not a rational number, what type of number must it be? Give an example of a real number that is not a rational number.

5. Give an example of the commutative property for addition.

6. Give an example of the commutative property for multiplication.

7. Why is the number 1 called the multiplicative identity?

8. Why is the number 0 called the additive identity?

9. Give an example of the associative property for multiplication.

10. Give an example of a distributive property.

CLASSIFYING NUMBERS

Exercises 11–18: Classify the number as one or more of the following: natural number, integer, rational number, or real number.

11. 12,914 (Number of Subway franchises in 1997)

12. -14.5 (Percent change in the number of unhealthy days in Los Angeles from 1995 to 1996)

13. $\frac{89}{3687}$ (Fraction of 18- to 19-year-old males who are married in the U.S.)

14. 4 (Pounds of garbage the average person produces each day)

15. 7.5 (Average number of gallons of water used each minute while taking a shower)

16. 2.64 (Average number of people per household in 1997)

17. $90\sqrt{2}$ (Distance in feet from home plate to second base in baseball)

18. -100 (Wind chill when the temperature is $-30°F$ and the wind speed is 40 miles per hour)

Exercises 19–22: Classify each real number as one or more of the following: natural number, whole number, integer, rational number, or irrational number.

19. $-5, 6, \frac{1}{7}, \sqrt{7}, 0.2$ 20. $-3, \frac{2}{9}, \sqrt{9}, -1.37$

21. $\frac{3}{1}, -\frac{5}{8}, \sqrt{5}, 0.\overline{45}, \pi$

22. $0, \frac{50}{10}, -\frac{23}{27}, 0.\overline{6}, -\sqrt{3}$

Exercises 23–28: For the measured quantity, state the set of numbers that is most appropriate to describe it. Choose from the natural numbers, integers, or rational numbers. Explain your answer.

23. Shoe sizes

24. Populations of states

25. Gallons of gasoline

26. Speed limits

27. Temperatures given in a winter weather forecast in Montana

28. Number of compact discs sold

PROPERTIES OF REAL NUMBERS

Exercises 29–38: State whether the equation is the result of an identity, a commutative, an associative, or a distributive property.

29. $b + 0 = b$
30. $1 \cdot 5 = 5$
31. $4 + a = a + 4$
32. $(5 + 1) + 8 = 5 + (1 + 8)$
33. $8(9x - 3) = 8 \cdot 9x - 8 \cdot 3$
34. $4(3 + 5a) = 4 \cdot 3 + 4 \cdot 5a$
35. $4 \cdot (10 \cdot 6) = (4 \cdot 10) \cdot 6$
36. $x \cdot 5 = 5x$
37. $3 \cdot (6 \cdot 2) = (6 \cdot 2) \cdot 3$
38. $bac = abc$

Exercises 39–42: Use a commutative property to write an equivalent expression.

39. $4 + a$
40. ba
41. $a \cdot \dfrac{1}{3}$
42. $100 + x$

Exercises 43–46: Use an associative property to write an equivalent expression.

43. $4 + (5 + b)$
44. $(x + 2) + 3$
45. $5(10x)$
46. $(a \cdot 5) \cdot 4$

Exercises 47–56: Use a distributive property to write an equivalent expression.

47. $4(x + y)$
48. $-3(a + 5)$
49. $(x - 7)5$
50. $(11 + b)a$
51. $-(x + 1)$
52. $-(a - 2)$
53. $ax - ay$
54. $4a + 4b$
55. $12 + 3x$
56. $2 - 4x$

Exercises 57–60: Use a distributive property to evaluate the expression two different ways.

57. $13(16 + 23)$
58. $4(8 - 12)$
59. $-5(19 - 7)$
60. $(8 + 12)(-9)$

Exercises 61–66: Calculate the average of the list of numbers. Classify the result as a natural number, an integer, or a rational number.

61. 3, 4, 5, 8
62. 5, 8, 10, 23, 9
63. 45, 33, 52
64. 3.2, 7.5, 8.1, 12.8, 13.4
65. 121.5, 45.7, 99.3, 45.9
66. 99.88, 39.11, 85.67, 23.86, 19.11

APPLICATIONS

67. *Digital Images* If an image downloaded from the Internet is 240 pixels across by 360 pixels high and a different image is 360 pixels across by 240 pixels high, how do the total numbers of pixels in the images compare? What property of real numbers does this result illustrate?

68. *Digital Images* The dimensions of digital images on the Internet can vary greatly in size. Calculate the number of pixels in each image.
    (a) 760 by 480 pixels
    (b) 64 by 128 pixels

69. *Higher Education* The following table lists the total higher education enrollment.

Year	1994	1995	1996	1997
Students (millions)	14.2	14.3	13.9	14.1

Source: Department of Education.

(a) What was the enrollment in 1996?
(b) Mentally estimate the average enrollment for this 4-year period.
(c) Calculate the average enrollment. Is your estimate from part (b) in reasonable agreement with your calculated result?

70. *Federal Budget* The following table lists the spending by the federal government for selected years.

Year	1994	1995	1996	1997	1998
Budget ($ trillions)	1.46	1.52	1.56	1.60	1.67

Source: Office of Management and Budget.

(a) What was the budget in 1996?
(b) Mentally estimate the average budget for this five-year period.
(c) Calculate the average budget. Is your estimate from part (b) in reasonable agreement with your calculated result?

WRITING ABOUT MATHEMATICS

71. Is subtraction either commutative or associative? Explain, using examples.

72. Is division either commutative or associative? Explain, using examples.

GROUP ACTIVITY
Working with Real Data

Directions: Form a group of 2 to 4 people. Select someone to record the group's responses for this activity. All members of the group should work cooperatively to answer the questions. If your instructor asks for your results, each member of the group should be prepared to respond. If the group is asked to turn in its work, be sure to include each person's name on the paper.

1. *Cost of a Stamp* The following table shows the cost of a first-class stamp for selected years.

Year	1958	1968	1974	1981
Cost (¢)	4	6	10	18

Year	1988	1991	1995	1999
Cost (¢)	25	29	32	33

Source: U.S. Postal Service.

(a) Discuss how the cost has changed over the years.
(b) Would it be logical to use the table to estimate the year when a stamp cost 32.5¢? Explain.

2. *Cost of a Stamp in Today's Money* The following table shows the cost of a first-class stamp over the years, adjusted to 1999 dollars.

Year	1958	1968	1974	1981
Cost (¢)	23	29	34	33

Year	1988	1991	1995	1999
Cost (¢)	36	36	35	33

Source: Department of Commerce.

(a) Discuss how the cost has changed over the years in terms of 1999 dollars.
(b) The cost increased to 34¢ in 2001. Was this rate increase reasonable? Explain your answer.

1.2 OPERATIONS ON REAL NUMBERS

The Real Number Line ~ Arithmetic Operations ~ Data and Number Sense

INTRODUCTION

Real numbers are used to describe data. To obtain information from data we frequently perform operations on real numbers. For example, exams are often assigned a score between 0 and 100. This step reduces each exam to a number or data point. We might obtain more information about the exams by calculating the average score. To do so we perform the arithmetic operations of addition and division.

In this section we discuss operations on real numbers and provide examples of where these computations occur in real life.

THE REAL NUMBER LINE

You can visualize the real number system by using a number line, as shown in Figure 1.4. Each real number corresponds to a point on the number line. The point associated with the real number 0 is called the **origin.**

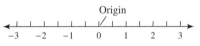
Figure 1.4 The Number Line

Figure 1.5

If a real number a is located to the left of a real number b on the number line, we say that a is **less than** b and write $a < b$. Similarly, if a real number a is located to the right of a real number b, we say that a is **greater than** b and write $a > b$. Thus $-3 < 2$ because -3 is located to the left of 2, and $2 > -3$ because 2 is located to the right of -3.

The **absolute value** of a real number a, written $|a|$, is equal to its distance from the origin on the number line. Distance may be either a positive number or zero, but it cannot be a negative number. As the points corresponding to 2 and -2 are both 2 units from the origin, $|2| = 2$ and $|-2| = 2$, as shown in Figure 1.5. The absolute value of a real number is *never* negative.

EXAMPLE 1 *Finding the absolute value of a real number*

Find the absolute value of each real number.

(a) 9.12
(b) $-\pi$
(c) $-a$, if a is a positive number
(d) a, if a is a negative number

Solution

(a) $|9.12| = 9.12$
(b) $|-\pi| = \pi$ because $\pi \approx 3.14$ is a positive number.
(c) If a is positive, $-a$ is negative, and $|-a| = a$.
(d) If a is negative, $-a$ is positive. Thus $a < 0$ implies that $|a| = -a$. For example, if we let $a = -5$, then $|-5| = -(-5) = 5$.

From Example 1 note that

$$|a| = a \quad \text{if } a \text{ is positive, and}$$
$$|a| = -a, \quad \text{if } a \text{ is negative.}$$

ARITHMETIC OPERATIONS

The four arithmetic operations are addition, subtraction, multiplication, and division.

ADDITION AND SUBTRACTION OF REAL NUMBERS. In an addition problem, the two numbers added are called **addends,** and the answer is called the **sum.** In the addition problem $3 + 5 = 8$, the numbers 3 and 5 are the addends and 8 is the sum.

The **additive inverse** or **opposite** of a real number a is $-a$. For example, the additive inverse of 5 is -5 and the additive inverse of -1.6 is 1.6. When we add opposites, the result is 0. That is, $a + (-a) = 0$ for every real number a.

EXAMPLE 2 Finding additive inverses

Find the additive inverse or opposite of each number. Then find the sum of the number and its opposite.

(a) 10,961 (b) π (c) $-\dfrac{3}{4}$

Solution

(a) The opposite of 10,961 is $-10,961$. Their sum is $10,961 + (-10,961) = 0$.
(b) The opposite of π is $-\pi$. Their sum is $\pi + (-\pi) = 0$.
(c) The opposite of $-\dfrac{3}{4}$ is $\dfrac{3}{4}$. Their sum is $-\dfrac{3}{4} + \dfrac{3}{4} = 0$.

When you add real numbers, it may be helpful to think of money. A positive number represents being paid an amount of money, whereas a negative number indicates a debt owed. The sum

$$8 + (-6) = 2$$

would represent being paid \$8 and owing \$6, resulting in \$2 being left over. Similarly,

$$-7 + (-5) = -12$$

would represent owing \$7 and \$5, resulting in a debt of \$12.

To add two real numbers we may use the following rules.

> **Addition of Real Numbers**
>
> To add two numbers that are either *both positive* or *both negative*, add their absolute values. Their sum has the same sign as the two numbers.
>
> To add two numbers with *opposite signs*, find the absolute value of each number. Subtract the smaller absolute value from the larger. The sum has the same sign as the sign of the number with the largest absolute value. If the two numbers are opposites, their sum is 0.

The next example illustrates addition of real numbers.

EXAMPLE 3 *Adding real numbers*

Evaluate each expression.

(a) $-3 + (-5)$ **(b)** $-4 + 7$ **(c)** $8.4 + (-9.5)$

Solution

(a) The addends are both negative, so we add the absolute values $|-3|$ and $|-5|$ to obtain 8. As the signs of the addends are both negative, the answer is -8. That is, $-3 + (-5) = -8$. If we owe \$3 and then owe an additional \$5, the total amount owed is \$8.

(b) The addends have opposite signs, so we subtract their absolute values to obtain 3. The answer is positive because $|7|$ is greater than $|-4|$. That is, $-4 + 7 = 3$. If we owe \$4 and are paid \$7, the net result is that we have \$3 to keep.

(c) $8.4 + (-9.5) = -1.1$ because $|-9.5|$ is 1.1 more than $|8.4|$. If we are paid \$8.40 and owe \$9.50, we still owe \$1.10.

Addition of positive and negative numbers occurs at banks if we let deposits be represented by positive numbers and withdrawals be represented by negative numbers.

EXAMPLE 4 *Balancing a checking account*

The initial balance in a checking account is \$157. Find the final balance if the following represents a list of withdrawals and deposits: $-55, -19, 123, -98$.

Solution

We find the sum.

$$157 + (-55) + (-19) + 123 + (-98) = 102 + (-19) + 123 + (-98)$$
$$= 83 + 123 + (-98)$$
$$= 206 + (-98)$$
$$= 108$$

The final balance is \$108. This result may be supported by evaluating the expression with a graphing calculator. See Figure 1.6 on the following page.

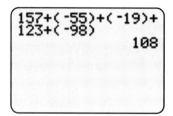

Figure 1.6

Technology Note *Subtraction and Negation*

On graphing calculators, two *different* keys typically represent subtraction and negation. Be sure to press the negation key instead of the subtraction key when working Example 4.

The answer to a subtraction problem is called the **difference.** Addition and subtraction of real numbers occurs in everyday life at grocery stores, where the costs of various items are added to the total and discounts from coupons are subtracted. When you're subtracting two real numbers, it sometimes helps to change the subtraction problem to an addition problem.

> **Subtraction of Real Numbers**
>
> For any real numbers a and b,
> $$a - b = a + (-b).$$
> To subtract b from a, add a and the opposite of b.

EXAMPLE 5 *Subtracting real numbers*

Evaluate each expression by hand.

(a) $-12 - 7$ (b) $-5.1 - (-10.6)$ (c) $\dfrac{1}{2} - \left(-\dfrac{2}{3}\right)$

Solution

(a) $-12 - 7 = -12 + (-7) = -19$
(b) $-5.1 - (-10.6) = -5.1 + 10.6 = 5.5$
(c) $\dfrac{1}{2} - \left(-\dfrac{2}{3}\right) = \dfrac{1}{2} + \dfrac{2}{3} = \dfrac{3}{6} + \dfrac{4}{6} = \dfrac{7}{6}$

MULTIPLICATION AND DIVISION OF REAL NUMBERS. In a multiplication problem, the two numbers multiplied are called the **factors,** and the answer is called the **product.** In the problem $3 \cdot 5 = 15$, the numbers 3 and 5 are factors and 15 is the product. The **multiplicative inverse** or **reciprocal** of a nonzero number a is $\dfrac{1}{a}$. The product of a nonzero number a and its reciprocal is $a \cdot \dfrac{1}{a} = 1$. For example, the reciprocal of -5 is

$-\frac{1}{5}$ because $-5 \cdot -\frac{1}{5} = 1$, and the reciprocal of $\frac{2}{3}$ is $\frac{3}{2}$ because $\frac{2}{3} \cdot \frac{3}{2} = 1$.

To multiply positive or negative numbers we may use the following rules.

> **Multiplication of Real Numbers**
>
> The product of two numbers with *like* signs is positive. The product of two numbers with *unlike* signs is negative.

EXAMPLE 6 *Multiplying real numbers*

Evaluate each expression.

(a) $-11 \cdot 8$ **(b)** $\frac{3}{5} \cdot \frac{4}{7}$ **(c)** $-1.2(-10)$ **(d)** $1.2 \cdot 5 \cdot (-7)$

Solution

(a) The product is negative because the factors -11 and 8 have unlike signs. Thus $-11 \cdot 8 = -88$.
(b) The product is positive because both factors are positive. Thus
$$\frac{3}{5} \cdot \frac{4}{7} = \frac{3 \cdot 4}{5 \cdot 7} = \frac{12}{35}.$$
(c) As both factors are negative, the product is positive. Thus $-1.2(-10) = 12$.
(d) $1.2 \cdot 5 \cdot (-7) = 6 \cdot (-7) = -42$

In the division problem $20 \div 4 = 5$, the number 20 is the **dividend,** 4 is the **divisor,** and 5 is the **quotient.** This division problem can be written as $\frac{20}{4} = 5$. Division of real numbers can be defined in terms of multiplication and reciprocals.

> **Division of Real Numbers**
>
> For real numbers a and b, with $b \neq 0$,
> $$\frac{a}{b} = a \cdot \frac{1}{b}.$$
> That is, to divide a by b, multiply a by the reciprocal of b.

EXAMPLE 7 *Dividing real numbers*

Evaluate each expression.

(a) $-12 \div \frac{1}{2}$ **(b)** $\frac{\frac{2}{3}}{-7}$ **(c)** $\frac{-4}{-24}$ **(d)** $6 \div 0$

Solution

(a) $-12 \div \dfrac{1}{2} = -12 \cdot \dfrac{2}{1} = -24$

(b) $\dfrac{\frac{2}{3}}{-7} = \dfrac{2}{3} \div (-7) = \dfrac{2}{3} \cdot \left(-\dfrac{1}{7}\right) = -\dfrac{2}{21}$

(c) $\dfrac{-4}{-24} = -4 \cdot \left(-\dfrac{1}{24}\right) = \dfrac{4}{24} = \dfrac{1}{6}$

(d) $6 \div 0 = \dfrac{6}{0}$ is undefined because division by 0 is not possible. Try using your calculator to divide 6 by 0 and observe the result. (See Figures 1.7 and 1.8.)

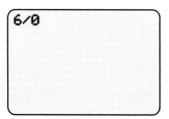
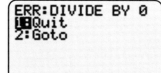

Figure 1.7 **Figure 1.8**

Many graphing calculators have the capability to perform arithmetic on fractions and express the answer as either a decimal or a fraction. The next example illustrates this capability.

EXAMPLE 8 *Performing arithmetic operations with technology*

GCLM Use a graphing calculator to evaluate each expression. Express your answer as a decimal and as a fraction.

(a) $\dfrac{1}{3} + \dfrac{2}{5} - \dfrac{4}{9}$ (b) $\left(\dfrac{4}{9} \cdot \dfrac{3}{8}\right) \div \dfrac{2}{3}$

Solution

(a) From Figure 1.9,

$$\dfrac{1}{3} + \dfrac{2}{5} - \dfrac{4}{9} = 0.2\overline{8}, \text{ or } \dfrac{13}{45}.$$

Note: Generally it is a good idea to put parentheses around fractions when you are using a graphing calculator.

(b) From Figure 1.10, $\left(\dfrac{4}{9} \cdot \dfrac{3}{8}\right) \div \dfrac{2}{3} = 0.25$, or $\dfrac{1}{4}$.

Technology Note *Fractions and Graphing Calculators*

Many graphing calculators have the capability to express decimals as fractions. This capability is illustrated in Figures 1.9 and 1.10.

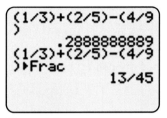
Figure 1.9

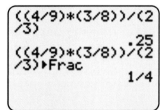
Figure 1.10

Data and Number Sense

In everyday life we commonly make approximations involving a variety of data. To make estimations we often use arithmetic operations on real numbers. Sometimes we have to ask ourselves if a result looks reasonable.

EXAMPLE 9 *Determining a reasonable answer*

Table 1.4 lists the number of endangered species reported for selected years.

TABLE 1.4 **Endangered Species**

Year	1995	1996	1997	1998
Species	1599	1523	1640	1694

Source: Fish and Wildlife Service.

Determine mentally which of the following values represents the average number of endangered species from 1995 to 1998: 1225, 1614, or 1694. Explain your reasoning.

Solution

The average value would lie between the maximum and minimum numbers of endangered species. The value of 1225 is less than the smallest value in Table 1.4 and 1694 is equal to the largest value in the table. The only reasonable choice is 1614 because it is located in the middle of the four data items. This conclusion is supported by Figure 1.11, which shows the computed average. Note the parentheses around the numerator.

Figure 1.11

EXAMPLE 10 Determining a reasonable answer

It is 2823 miles from New York to Los Angeles. Determine mentally which of the following would best estimate the actual number of hours required to drive this distance in a car: 50, 100, or 120 hours.

Solution

Speed equals distance divided by time. Dividing by 100 is easy, so start by dividing 100 hours into 2800 miles. The average speed would be $\frac{2800}{100} = 28$ miles per hour, which is too slow for most drivers. A more reasonable choice is 50 hours because then the average speed would be double, or about 56 miles per hour.

EXAMPLE 11 Estimating a numeric value

Table 1.5 lists the number of subscribers of cellular telephones in selected years. If trends continued, estimate the number of subscribers in 1998.

TABLE 1.5 **Cellular Phone Subscribers**

Year	1994	1995	1996	1997	1998
Subscribers (millions)	24.1	33.8	44.0	55.3	?

Source: Cellular Telecommunications Industry Association.

Solution

The data show that the number of subscribers has increased each year by about 10 or 11 million. A reasonable estimate might be $55 + 11 = 66$ million. Estimates may vary slightly.

1.2 Putting It All Together

The number line may be used to visualize the set of real numbers. Each real number corresponds to a point on the number line. To obtain information from data we often perform arithmetic operations on those data. There are four basic arithmetic operations: addition, subtraction, multiplication, and division. The absolute value of a real number a cannot be a negative number and is equal to the distance between a and the origin on the number line. The following table summarizes some of the information presented in this section.

Operation	Definition	Examples
Absolute Value of a Real Number	For any real number a, $\|a\| = a$ if a is positive, and $\|a\| = -a$ if a is negative.	$\|-5\| = 5$, $\|3.7\| = 3.7$, and $\|-4 + 4\| = 0$

1.2 Operations on Real Numbers

Operation	Definition	Examples
Addition of Real Numbers	See the highlighted box: Addition of Real Numbers on page 15.	$3 + (-6) = -3$, $-2 + (-10) = -12$, $-11 + 3 = -8$, and $18 + 11 = 29$
Subtraction of Real Numbers	We can transform a subtraction problem into an addition problem: $a - b = a + (-b)$.	$4 - 6 = 4 + (-6) = -2$, $-7 - (-8) = -7 + 8 = 1$ $-8 - 5 = -8 + (-5) = -13$, and $9 - (-1) = 9 + 1 = 10$
Multiplication of Real Numbers	The product of two numbers with *like* signs is positive. The product of two numbers with *unlike* signs is negative.	$3 \cdot 5 = 15$, and $-4 \cdot (-7) = 28$, $-8 \cdot 7 = -56$, and $5 \cdot (-11) = -55$
Division of Real Numbers	For real numbers a and b with $b \neq 0, \dfrac{a}{b} = a \cdot \dfrac{1}{b}$.	$-3 \div \dfrac{3}{4} = -3 \cdot \dfrac{4}{3} = -4$ and $\dfrac{5}{2} \div \left(-\dfrac{7}{4}\right) = \dfrac{5}{2} \cdot \left(-\dfrac{4}{7}\right) = -\dfrac{10}{7}$

1.2 EXERCISES

FOR EXTRA HELP: Student's Solutions Manual, MyMathLab.com, InterAct Math, Math Tutor Center, MathXL, Digital Video Tutor CD 1 Videotape 1

CONCEPTS

1. If $a > b$, then a is located to the _____ of b on the number line.

2. If $a < 0$, then a is located to the _____ of the origin on the number line.

3. The product of two negative numbers is a _____ number.

4. The sum of two negative numbers is a _____ number.

5. The quotient of a positive number and a negative number is a _____ number.

6. If $a < 0$ and $b > 0$, then $a - b$ is a _____ number.

7. The additive inverse of a is _____.

8. The multiplicative inverse, or reciprocal, of $\dfrac{a}{b}$ is _____.

THE REAL NUMBER LINE AND ABSOLUTE VALUE

Exercises 9–12: Plot each number on a number line. Be sure to mark an appropriate scale on the number line.

9. $0, 4, -3, 1.5$

10. $10, 20, -5, -15$

11. $100, 300, -200, 50$

12. $-0.4, 0.2, 0.5, -0.1, 0$

Exercises 13–20: Evaluate the absolute value.

13. $|-6.1|$

14. $|17|$

15. $|8 - 11|$

16. $|2 \cdot 8 - 23|$

17. $|x|$, where $x > 0$

18. $|x|$, where $x < 0$

19. $|x - y|$, where $x > 0$ and $y < 0$

20. $|x + y|$, where $x < 0$ and $y < 0$

Exercises 21–26: State whether only positive numbers are typically used to measure the given quantity or whether both positive and negative numbers are used. Explain your reasoning.

21. Area
22. Distance
23. Temperature
24. A person's net worth
25. Gas mileage of a car
26. Elevation relative to sea level

ARITHMETIC OPERATIONS

Exercises 27–34: Find the additive inverse, or opposite, for each number or expression.

27. 56
28. $-\dfrac{5}{7}$
29. -6.9
30. 12.8
31. $-\pi + 2$
32. $a + b$
33. $a - b$
34. $-x + 3$

Exercises 35–42: Find the multiplicative inverse for each number or expression.

35. 3
36. $\dfrac{3}{8}$
37. $-\dfrac{2}{3}$
38. 1.5
39. π
40. $-\dfrac{x}{y}$
41. $a + 3$
42. $x - 1$

Exercises 43–70: Perform the following arithmetic operations without a calculator. If you are uncertain about an answer, check it by using a calculator.

43. $4 + (-6)$
44. $-10 + 14$
45. $-7.4 + (-9.2)$
46. $-8.4 - 10.3$
47. $-\dfrac{3}{4} - \left(-\dfrac{1}{4}\right)$
48. $\dfrac{1}{5} - \left(-\dfrac{3}{10}\right)$
49. $-18 + (-3) + 1$
50. $-9 + 1 + (-2) + 5$
51. $-(-3 + 6)$
52. $-5 + 7 - (-2) + 3$
53. $-9 \cdot 7$
54. $-6 \cdot -12$
55. $-(8 \cdot -4)$
56. $\dfrac{4}{5} \cdot \dfrac{3}{8}$
57. $-\dfrac{1}{2} \cdot \dfrac{5}{7} \cdot \dfrac{1}{3}$
58. $-\dfrac{6}{7} \cdot -\dfrac{5}{3}$
59. $-\dfrac{1}{2} \div -\dfrac{3}{4}$
60. $-5 \div \dfrac{4}{5}$
61. $\dfrac{3}{4} \div (-2)$
62. $-\dfrac{4}{5} \div \dfrac{7}{10}$
63. $-\dfrac{8}{2}$
64. $-\dfrac{45}{9}$
65. $-25 \div -5$
66. $8 \div 0$
67. $-4 \cdot 7 \cdot (-5)$
68. $-2 \cdot -6 \cdot 6 \cdot -1$
69. $6 \cdot \dfrac{2}{3} \cdot 3 \cdot \left(-\dfrac{1}{6}\right)$
70. $\left(\dfrac{4}{5} \cdot \dfrac{5}{8}\right) \div \dfrac{1}{2}$

Exercises 71–78: Use a calculator to evaluate the expression.

71. $-23.1 + 45.7 - (-34.6)$
72. $102 - (-341) + (-112)$
73. $\dfrac{1}{2} + \dfrac{2}{3} - \dfrac{5}{7}$
74. $-\dfrac{8}{13} + \dfrac{1}{2} - \dfrac{2}{5}$
75. $-\dfrac{3}{4} \cdot \dfrac{4}{5} \div \dfrac{5}{3}$
76. $\left(\dfrac{3}{4} \div (-11)\right) - \dfrac{2}{5}$
77. $\dfrac{1}{2}\left(\dfrac{4}{11} + \dfrac{2}{5}\right)$
78. $-\dfrac{5}{13} + \left(\dfrac{3}{5} \div \dfrac{2}{17}\right)$

DATA AND NUMBER SENSE

Exercises 79–82: Mentally estimate the average of the group of numbers. Check your estimate by finding the average with a calculator.

79. $9, 5, 15, -9$
80. $12, 8, 27, -7$
81. $101, 98, -42, 82$
82. $-4, 3, 5, -8, 5, -24$

Exercises 83–86: Mentally evaluate the expression.

83. $\dfrac{1}{5} \cdot \dfrac{2}{3} \cdot \dfrac{1}{7} \cdot \dfrac{1}{9} \cdot 5 \cdot \dfrac{3}{2} \cdot 7 \cdot 9$

84. $\dfrac{1}{2} \cdot \dfrac{1}{3} \cdot \dfrac{1}{4} \cdot (-4) \cdot (-3) \cdot (-2)$

85. $\left(\dfrac{1}{2} - \dfrac{1}{3}\right) + \left(\dfrac{1}{3} - \dfrac{1}{4}\right) + \left(\dfrac{1}{4} - \dfrac{1}{5}\right) + \left(\dfrac{1}{5} - \dfrac{1}{6}\right)$

86. $\dfrac{1}{2} \div \dfrac{1}{3} \cdot \dfrac{1}{3} \div \dfrac{1}{4} \cdot \dfrac{1}{4} \div \dfrac{1}{5} \cdot \dfrac{1}{5}$

APPLICATIONS

87. *Buying a Computer* An advertisement for a computer states that it costs $202 down and $98.99 per month for 24 months. Mentally estimate which of the following represents the cost of the computer if it is purchased under these terms: $2600, $3200, or $3800. Explain your reasoning. Use a calculator to find the actual cost.

88. *Leasing a Car* To lease a car costs $1497 down and $249 per month for 36 months. Mentally estimate the cost of the lease. Use a calculator to find the actual cost.

89. *Amway Sales* The following table lists worldwide retail sales for Amway products.

Year	1988	1990	1992
Sales ($ billions)	1.8	2.1	3.9

Year	1994	1996	1998
Sales ($ billions)	5.3	6.8	5.7

Source: Amway.

(a) What were Amway sales in 1994?
(b) Estimate sales in 1993. Explain how you arrived at your estimate and compare it to the actual value of $4.5 billion.
(c) Estimate sales in 1997. Actual sales were $7.0 billion in 1997. Discuss the difficulty of obtaining an accurate estimate from only the data given in the table.

90. *Weapons Convictions* The following table lists state and federal weapons convictions from 1992 through 1996.

Year	1992	1994	1996
Convictions	30,243	34,084	36,370

Source: Bureau of Justice *Statistics Bulletin*.

(a) Discuss any trends in weapons convictions from 1992 through 1996.
(b) Mentally estimate which of the following best represents the average number of weapons convictions for these three years: 32,000, 34,000, or 36,000.
(c) Check your estimate in part (b) by calculating the actual average.

91. *Cable Modems* Cable modems provided by the cable TV industry give high-speed access to the Internet. The following table lists the expected numbers of cable modem subscribers for selected years.

Year	1999	2000	2001	2002
Subscribers (millions)	1.0	2.0	3.0	4.3

Source: Yankee Group.

(a) Discuss any trends in cable modem subscribers from 1999 through 2002.
(b) Estimate the number of subscribers in 2003. Explain your reasoning.

92. *Digital Subscriber Lines* Local phone companies provide digital subscriber lines (DSL) for high-speed Internet access. This technology competes directly with cable modem service provided by cable TV companies. The following table lists the expected numbers of digital subscriber lines for selected years.

Year	1999	2000	2001	2002
DSL users (millions)	0.3	0.7	1.5	2.7

Source: Yankee Group.

(a) Discuss any trends in DSL from 1999 through 2002.
(b) Estimate the number of subscribers in 2003. Explain your reasoning.

WRITING ABOUT MATHEMATICS

93. Suppose that $a \neq b$. Explain how a number line can be used to determine whether a is greater than b or whether a is less than b. Give an example of each situation.

94. Explain why a positive number times a negative number is a negative number.

CHECKING BASIC CONCEPTS FOR SECTIONS 1.1 AND 1.2

1. Identify each number as a natural number, an integer, a rational number, or a real number.
   (a) -9
   (b) $\frac{8}{4}$
   (c) $\sqrt{5}$
   (d) 0.5

2. Identify the property of real numbers that each equation illustrates. Write the property, using variables.
   (a) $3 + 4 = 4 + 3$
   (b) $-5 \cdot (4 \cdot 8) = (-5 \cdot 4) \cdot 8$
   (c) $4(5 + 2) = 4 \cdot 5 + 4 \cdot 2$

3. Evaluate each expression.
   (a) $-3 + 4 + (-5)$
   (b) $5.1 \cdot (-4) \cdot 2$
   (c) $-\frac{2}{3} \cdot \left(\frac{1}{4} \div \frac{2}{5}\right)$

4. A small lake covers 200 acres, has an average depth of 15 feet, and contains about 2 billion gallons of water. Estimate the number of gallons in a lake that covers 600 acres and has an average depth of 30 feet.

1.3 INTEGER EXPONENTS

Bases and Positive Exponents ~ Zero and Negative Exponents ~ Product, Quotient, and Power Rules ~ Order of Operations ~ Scientific Notation

INTRODUCTION

Technology has brought with it the need for both small and large numbers. The size of an average virus is 5 millionths of a centimeter, whereas the distance to the nearest star, Alpha Centauri, is 25 trillion miles. To represent such numbers we often use exponents. In this section we discuss properties of integer exponents and some of their applications. (Source: C. Ronan, *The Natural History of the Universe*.)

BASES AND POSITIVE EXPONENTS

The area of a square that is 8 inches on a side is given by the expression

$$8 \cdot 8 = 8^2 = 64 \text{ square inches.}$$

The expression 8^2 is an **exponential expression** with **base** 8 and **exponent** 2. Exponential expressions occur frequently in a variety of applications. For example, suppose that an investment doubles its initial value 3 times. Then its final value is

$$2 \cdot 2 \cdot 2 = 2^3 = 8$$

times larger than its original value. Table 1.6 contains examples of exponential expressions.

TABLE 1.6

Expression	Base	Exponent
$2 \cdot 2 \cdot 2 = 2^3$	2	3
$6 \cdot 6 \cdot 6 \cdot 6 = 6^4$	6	4
$7 = 7^1$	7	1
$0.5 \cdot 0.5 = 0.5^2$	0.5	2
$x \cdot x \cdot x = x^3$	x	3

Read 0.5^2 as "0.5 squared," 2^3 as "2 cubed," and 6^4 as "6 to the fourth power." The terms *squared* and *cubed* come from geometry. If the length of a side of a square is 4, then its area is

$$4 \cdot 4 = 4^2 = 16$$

square units, as illustrated in Figure 1.12. Similarly, if the length of an edge of a cube is 4, then its volume is

$$4 \cdot 4 \cdot 4 = 4^3 = 64$$

cubic units, as shown in Figure 1.13.

Figure 1.12   **Figure 1.13**
4 Squared   4 Cubed

EXAMPLE 1   *Writing numbers in exponential notation*

GCLM Using the given base, write each number as an exponential expression. Check your results with a calculator.

(a) 10,000   (base 10)
(b) 27   (base 3)
(c) 32   (base 2)

Solution

(a) $10{,}000 = 10 \cdot 10 \cdot 10 \cdot 10 = 10^4$
(b) $27 = 3 \cdot 3 \cdot 3 = 3^3$
(c) $32 = 2 \cdot 2 \cdot 2 \cdot 2 \cdot 2 = 2^5$

These values are supported in Figure 1.14, where exponential expressions are evaluated with a graphing calculator, using the (^) key.

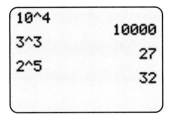

Figure 1.14

Technology Note *Powers of a Number*

On some calculators you may find the square of a number by using either a special squaring key or by using (^). However, for other powers you usually have to use (^). These alternatives are illustrated in Figure 1.15, where $8^2 = 64$ is entered two different ways.

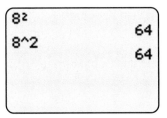

Figure 1.15

Computer memory is often measured in bytes. A *byte* is capable of storing one letter of the alphabet. For example, the word "math" requires four bytes to store in a computer. Bytes of computer memory are often manufactured in amounts equal to powers of 2, as illustrated in the next example.

EXAMPLE 2 *Analyzing computer memory using exponents*

In computer technology, 1 K (kilobyte) of memory is equal to 2^{10} bytes, and 1 MB (megabyte) of memory is equal to 2^{20} bytes. Determine whether 1 K of memory is equal to 1000 bytes and whether 1 MB is equal to 1,000,000 bytes. **(Source: D. Horn, *Basic Electronics Theory*.)**

Solution

Figure 1.16 shows that $2^{10} = 1024$ and that $2^{20} = 1,048,576$. Thus 1 K represents slightly more than 1000 bytes and 1 MB is more than 1,000,000 bytes.

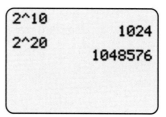

Figure 1.16

ZERO AND NEGATIVE EXPONENTS

Exponents can be defined for any integer. If a is any nonzero real number, we define

$$a^0 = 1.$$

For example, $3^0 = 1$ and $\left(\dfrac{1}{7}\right)^0 = 1$. We can also define negative integer exponents as follows.

$$5^{-4} = \frac{1}{5^4}, \quad y^{-2} = \frac{1}{y^2}, \quad \left(\frac{1}{2}\right)^{-5} = \left(\frac{1}{1/2}\right)^5 = 2^5$$

Powers of 10 are important because they are used frequently in mathematics to express numbers that are either small or large in absolute value. Table 1.7 may be used to simplify powers of 10. Note that, if the power decreases by 1, the result is decreased by $\dfrac{1}{10}$.

TABLE 1.7 Powers of Ten

Power of 10	Value
10^3	1000
10^2	100
10^1	10
10^0	1
10^{-1}	$\dfrac{1}{10} = 0.1$
10^{-2}	$\dfrac{1}{100} = 0.01$
10^{-3}	$\dfrac{1}{1000} = 0.001$

Critical Thinking

Use Table 1.7 to explain why it is reasonable for 10^0 to equal 1.

The following summarizes integer exponents with a nonzero base.

Integer Exponents

Let a be a nonzero real number and n be a positive integer. Then

$$a^n = a \cdot a \cdot a \cdot \ldots \cdot a \quad (n \text{ factors of } a),$$
$$a^0 = 1, \quad \text{and}$$
$$a^{-n} = \frac{1}{a^n}.$$

EXAMPLE 3 *Evaluating numbers in exponential notation*

Evaluate each expression by hand and then check your result with a calculator when possible.

(a) 3^{-4} (b) $\left(\frac{5}{7}\right)^{-2}$ (c) $\frac{1}{2^{-3}}$ (d) $(xy)^{-1}$

Solution

(a) $3^{-4} = \frac{1}{3^4} = \frac{1}{3 \cdot 3 \cdot 3 \cdot 3} = \frac{1}{81}$ (b) $\left(\frac{5}{7}\right)^{-2} = \frac{1}{(5/7)^2} = \frac{1}{25/49} = \frac{49}{25}$

(c) $\frac{1}{2^{-3}} = \frac{1}{1/2^3} = 2^3 = 8$ (d) $(xy)^{-1} = \frac{1}{(xy)^1} = \frac{1}{xy}$

The results for parts (a)–(c) are supported by Figure 1.17.

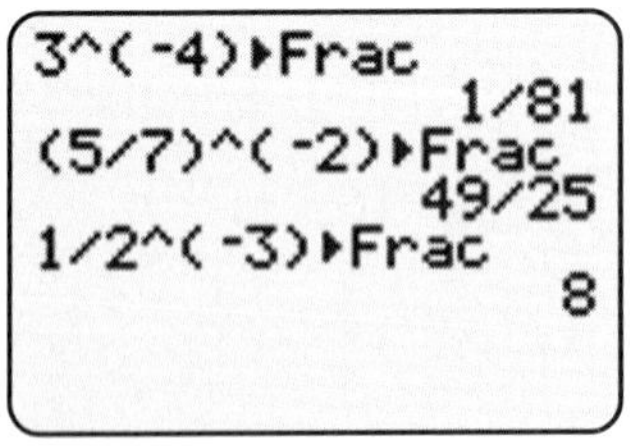

```
3^(-4)▶Frac
              1/81
(5/7)^(-2)▶Frac
              49/25
1/2^(-3)▶Frac
                 8
```

Figure 1.17

PRODUCT, QUOTIENT, AND POWER RULES

We can calculate products and quotients of exponential expressions *provided their bases are the same*. For example,

$$3^2 \cdot 3^3 = (3 \cdot 3) \cdot (3 \cdot 3 \cdot 3) = 3^5.$$

This expression has a total of $2 + 3 = 5$ factors of 3, so the result is 3^5. To multiply exponential expressions with like bases, add exponents.

The Product Rule

For any nonzero number a and integers m and n,
$$a^m \cdot a^n = a^{m+n}.$$

The product rule holds for negative exponents. For example,
$$10^5 \cdot 10^{-2} = 10^{5+(-2)} = 10^3.$$

EXAMPLE 4 *Using the product rule*

Multiply and simplify.

(a) $10^2 \cdot 10^4$ **(b)** $7^3 \cdot 7^{-4}$ **(c)** $x^3 \cdot x^{-2} \cdot x^4$ **(d)** $3y^2 \cdot 2y^{-4}$

Solution

(a) $10^2 \cdot 10^4 = 10^{2+4} = 10^6 = 1{,}000{,}000$

(b) $7^3 \cdot 7^{-4} = 7^{3+(-4)} = 7^{-1} = \dfrac{1}{7}$

(c) $x^3 \cdot x^{-2} \cdot x^4 = x^{3+(-2)+4} = x^5$

(d) $3y^2 \cdot 2y^{-4} = 3 \cdot 2 \cdot y^2 \cdot y^{-4} = 6y^{2+(-4)} = 6y^{-2} = \dfrac{6}{y^2}$

Note that 6 is not raised to the power of -2 in the expression $6y^{-2}$.

Consider division of exponential expressions using the following example.
$$\frac{6^5}{6^3} = \frac{6 \cdot 6 \cdot \cancel{6}^1 \cdot \cancel{6}^1 \cdot \cancel{6}^1}{\cancel{6} \cdot \cancel{6} \cdot \cancel{6}} = 6 \cdot 6 = 6^2$$

After canceling three 6s, two 6s are left in the numerator, and the result is $6^{5-3} = 6^2 = 36$. To divide exponential expressions with like bases, subtract exponents.

The Quotient Rule

For any nonzero number a and integers m and n,
$$\frac{a^m}{a^n} = a^{m-n}.$$

The quotient rule holds true for negative exponents. For example,
$$\frac{2^{-6}}{2^{-4}} = 2^{-6-(-4)} = 2^{-2} = \frac{1}{2^2} = \frac{1}{4}.$$

This result is supported by Figure 1.18, shown on the following page.

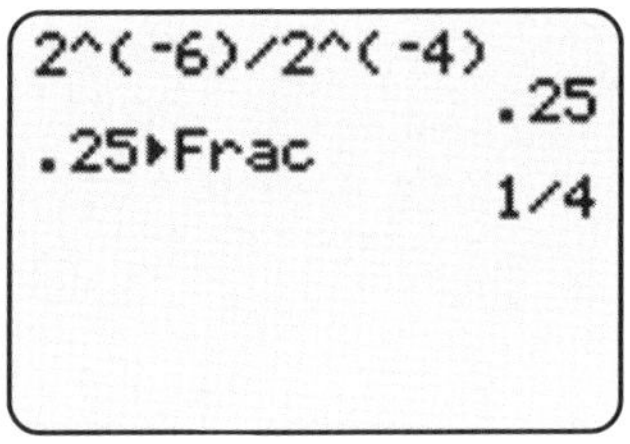

```
2^(-6)/2^(-4)
                .25
.25▶Frac
                1/4
```

Figure 1.18

EXAMPLE 5 *Using the quotient rule*

Simplify each expression. Use positive exponents.

(a) $\dfrac{10^4}{10^6}$ (b) $\dfrac{x^5}{x^2}$ (c) $\dfrac{15x^2y^3}{5x^4y}$

Solution

(a) $\dfrac{10^4}{10^6} = 10^{4-6} = 10^{-2} = \dfrac{1}{10^2}$

(b) $\dfrac{x^5}{x^2} = x^{5-2} = x^3$

(c) $\dfrac{15x^2y^3}{5x^4y} = \dfrac{15}{5} \cdot \dfrac{x^2}{x^4} \cdot \dfrac{y^3}{y^1} = 3 \cdot x^{(2-4)}y^{(3-1)} = 3x^{-2}y^2 = \dfrac{3y^2}{x^2}$

How should we evaluate $(4^3)^2$? To answer this question consider

$$(4^3)^2 = 4^3 \cdot 4^3 = 4^{3+3} = 4^6.$$

Similarly,

$$(x^4)^3 = x^4 \cdot x^4 \cdot x^4 = x^{4+4+4} = x^{12}.$$

These results suggest that, to raise a power to a power, multiply the exponents.

> **Raising Powers to Powers**
>
> For any real number a and integers m and n,
>
> $$(a^m)^n = a^{mn}.$$

EXAMPLE 6 *Raising powers to powers*

Simplify each expression. Use positive exponents.

(a) $(5^2)^3$ (b) $(2^4)^{-2}$ (c) $(b^{-7})^5$

Solution

(a) $(5^2)^3 = 5^{2 \cdot 3} = 5^6$

(b) $(2^4)^{-2} = 2^{4 \cdot (-2)} = 2^{-8} = \dfrac{1}{2^8}$

(c) $(b^{-7})^5 = b^{-7 \cdot 5} = b^{-35} = \dfrac{1}{b^{35}}$

How can we simplify the expression $(2x)^3$? Consider the following.
$$(2x)^3 = 2x \cdot 2x \cdot 2x = (2 \cdot 2 \cdot 2) \cdot (x \cdot x \cdot x) = 2^3 x^3$$
This result suggests that, to cube a product, cube each factor.

Raising Products to Powers

For any real numbers a and b and integer n,
$$(ab)^n = a^n b^n.$$

EXAMPLE 7 *Raising products to powers*

Simplify each expression.

(a) $(6y)^2$ **(b)** $(x^2 y)^{-2}$ **(c)** $(2xy^3)^4$

Solution

(a) $(6y)^2 = 6^2 y^2 = 36 y^2$

(b) $(x^2 y)^{-2} = (x^2)^{-2} y^{-2} = x^{-4} y^{-2} = \dfrac{1}{x^4 y^2}$

(c) $(2xy^3)^4 = 2^4 x^4 (y^3)^4 = 16 x^4 y^{12}$

To simplify a power of a quotient use the following rule.

Raising Quotients to Powers

For nonzero numbers a and b and any integer n,
$$\left(\dfrac{a}{b}\right)^n = \dfrac{a^n}{b^n}.$$

EXAMPLE 8 *Raising quotients to powers*

Simplify each expression.

(a) $\left(\dfrac{3}{x}\right)^3$ **(b)** $\left(\dfrac{1}{2^3}\right)^{-2}$ **(c)** $\left(\dfrac{3x^{-3}}{y^2}\right)^4$

Solution

(a) $\left(\dfrac{3}{x}\right)^3 = \dfrac{3^3}{x^3} = \dfrac{27}{x^3}$

(b) $\left(\dfrac{1}{2^3}\right)^{-2} = (2^3)^2 = 2^6 = 64$

(c) $\left(\dfrac{3x^{-3}}{y^2}\right)^4 = \dfrac{3^4 (x^{-3})^4}{(y^2)^4} = \dfrac{81 x^{-12}}{y^8} = \dfrac{81}{x^{12} y^8}$

ORDER OF OPERATIONS

When evaluating the expression 3 + 4 · 5, is the result 35 or 23? Figure 1.19 shows that a calculator gives a result of 23. This is because multiplication is performed before addition.

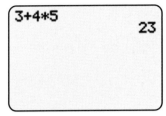

Figure 1.19

Since it is important that we evaluate arithmetic expressions consistently, the following rules are used.

Order of Operations

1. Perform all calculations within parentheses (using the following order of operations).
2. Evaluate all exponential expressions. Do any negations *after* evaluating exponents.
3. Do all multiplication and division from *left to right*.
4. Do all addition and subtraction from *left to right*.

Be sure to evaluate exponents before negation is performed. For example,
$$-2^4 = -(2 \cdot 2 \cdot 2 \cdot 2) = -16, \quad \text{but}$$
$$(-2)^4 = (-2) \cdot (-2) \cdot (-2) \cdot (-2) = 16.$$
These results are supported by Figure 1.20.

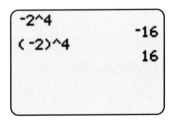

Figure 1.20

EXAMPLE 9 *Evaluating arithmetic expressions*

Evaluate each expression by hand. Use a calculator to support your results.

(a) $5 - 3 \cdot 2 - (4 + 5)$ (b) $-3^2 + \dfrac{5 + 7}{2 + 1}$ (c) $4^3 - 5(2 - 6 \cdot 2)$

Solution

(a) $5 - 3 \cdot 2 - (4 + 5) = 5 - 3 \cdot 2 - 9$
$$= 5 - 6 - 9$$
$$= -1 - 9$$
$$= -10$$

(b) $-3^2 + \dfrac{5 + 7}{2 + 1} = -3^2 + \dfrac{(5 + 7)}{(2 + 1)}$
$$= -3^2 + \dfrac{12}{3}$$
$$= -9 + \dfrac{12}{3}$$
$$= -9 + 4$$
$$= -5$$

Assume that both the numerator and the denominator of a fraction have parentheses around them.

(c) $4^3 - 5(2 - 6 \cdot 2) = 4^3 - 5(2 - 12)$
$$= 4^3 - 5(-10)$$
$$= 64 - 5(-10)$$
$$= 64 + 50$$
$$= 114$$

These results are supported by Figure 1.21.

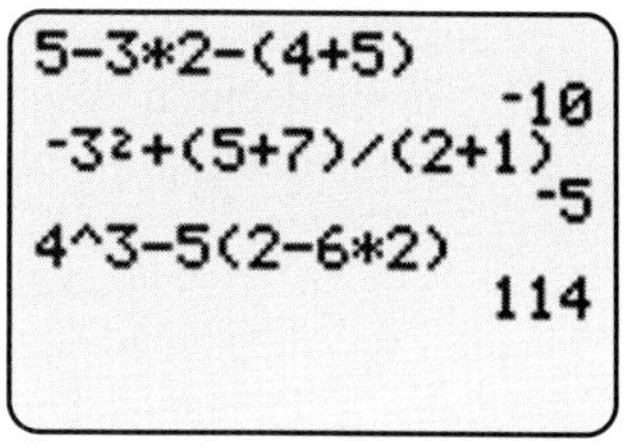

```
5-3*2-(4+5)
              -10
-3²+(5+7)/(2+1)
              -5
4^3-5(2-6*2)
             114
```

Figure 1.21

SCIENTIFIC NOTATION

Numbers that are large or small in absolute value occur frequently in applications. For simplicity these numbers are often expressed in scientific notation. As mentioned in the introduction to the section, the distance to the nearest star is 25 trillion miles. This number can be expressed in scientific notation as

$$25{,}000{,}000{,}000{,}000 = 2.5 \times 10^{13}.$$

In contrast, a typical virus is about 5 millionths of a centimeter in diameter. In scientific notation this number can be written as

$$0.000005 = 5 \times 10^{-6}.$$

A calculator set in *scientific mode* expresses these numbers in scientific notation, as illustrated in Figure 1.22 on the following page. The letter E denotes a power of 10. That is, $2.5\text{E}13 = 2.5 \times 10^{13}$ and $5\text{E}-6 = 5 \times 10^{-6}$.

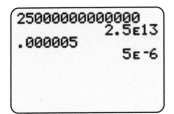

Figure 1.22

Scientific Notation

A real number a is in **scientific notation** when a is written as $b \times 10^n$, where $1 \leq |b| < 10$ and n is an integer.

Use the following steps to express a positive number a in scientific notation.

Writing a Positive Number in Scientific Notation

1. Move the decimal point in a number a until it represents a number b such that $1 \leq b < 10$.
2. Count the number of decimal places that the decimal point was moved. Let this positive integer be n.
3. If the decimal point was moved to the left, then $a = b \times 10^n$.
   If the decimal point was moved to the right, then $a = b \times 10^{-n}$.

Note: The scientific notation for a negative number a is the additive inverse of the scientific notation of $|a|$. For example, $5200 = 5.2 \times 10^3$, so $-5200 = -5.2 \times 10^3$.

Table 1.8 shows the values of some important powers of 10.

TABLE 1.8 Important Powers of 10

Number	10^3	10^6	10^9	10^{12}
Value	Thousand	Million	Billion	Trillion

EXAMPLE 10 *Writing a number in scientific notation*

Express each number in scientific notation.

(a) 360,000 (Dots in 1 square inch of some types of laser print)
(b) 0.00000538 (Time in seconds for light to travel 1 mile)
(c) 10,000,000,000 (Estimated world population in 2050)

Solution

(a) Move the assumed decimal point in 360,000 five places to the *left* to obtain 3.6.

$$3.\underbrace{6\ 0\ 0\ 0\ 0}.$$

The scientific notation for 360,000 is 3.6×10^5.

(b) Move the decimal point in 0.00000538 six places to the *right* to obtain 5.38.

$$0.\underbrace{0\ 0\ 0\ 0\ 0\ 5}{,}38$$

The scientific notation for 0.00000538 is 5.38×10^{-6}.

(c) Move the decimal point in 10,000,000,000 ten places to the *left* to obtain 1.

$$1.\underbrace{0\ 0\ 0\ 0\ 0\ 0\ 0\ 0\ 0\ 0}.$$

The scientific notation for 10,000,000,000 is 1×10^{10}. These results are supported by Figure 1.23, where the calculator is in *scientific mode*. Note that positive powers of 10 indicate a large number, whereas negative powers of 10 indicate a small number.

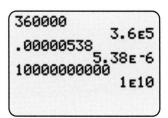

Figure 1.23

In the next example we convert numbers from scientific form to **standard form.**

EXAMPLE 11 *Writing a number in standard form*

Write the number in standard form. Support your answer with a calculator.

(a) 2×10^8 (Number of years for the sun to complete one orbit in the Milky Way)

(b) 2.1×10^{-2} (Fraction of deaths worldwide caused by lung cancer in 1997)

Solution

(a) Move the assumed decimal point in 2 to the *right* 8 places to obtain 200,000,000.

(b) Move the decimal point in 2.1 to the *left* 2 places to obtain 0.021.

These results are supported by Figure 1.24.

Figure 1.24

Arithmetic can be performed on expressions in scientific notation. For example, multiplication of the expressions 8×10^4 and 4×10^2 may be done by hand.

$$(8 \times 10^4) \cdot (4 \times 10^2) = (8 \cdot 4) \times (10^4 \cdot 10^2) \quad \text{Properties of real numbers}$$
$$= 32 \times 10^6 \quad \text{Add exponents and simplify.}$$
$$= 3.2 \times 10^7 \quad \text{Write in scientific notation.}$$

Division may be performed as follows.

$$\frac{8 \times 10^4}{4 \times 10^2} = \frac{8}{4} \times \frac{10^4}{10^2} \quad \text{Property of fractions}$$
$$= 2 \times 10^2 \quad \text{Subtract exponents.}$$

These results are supported by Figure 1.25, where the calculator is set in scientific mode.

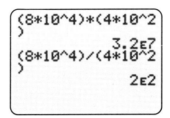

Figure 1.25

The next example illustrates how scientific notation is used in applications.

EXAMPLE 12   *Analyzing the federal debt*

In 1997, the federal debt held by the public was 3.77 trillion dollars, and the population of the United States was 268 million. Approximate the national debt per person.

Solution

In scientific notation 3.77 trillion equals 3.77×10^{12} and 268 million equals 268×10^6, or 2.68×10^8. The per person debt held by the public is given by

$$\frac{3.77 \times 10^{12}}{2.68 \times 10^8} \approx \$14{,}067.$$

Figure 1.26 supports this result.

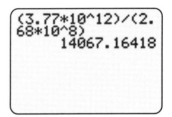

Figure 1.26

Critical Thinking

Estimate the number of seconds that you have been alive.

1.3 Putting It All Together

The following table summarizes important properties of exponents.

Property	Definition	Examples
Bases and Exponents	In the expression a^n, a is the base and n is the exponent.	$3^4 = 3 \cdot 3 \cdot 3 \cdot 3 = 81$, $7^0 = 1$, $2^{-3} = \dfrac{1}{2^3} = \dfrac{1}{2 \cdot 2 \cdot 2} = \dfrac{1}{8}$, and $-3^2 = -9$
The Product Rule	Let a be a nonzero number and m and n be integers. Then $a^m \cdot a^n = a^{m+n}$.	$8^4 \cdot 8^2 = 8^{4+2} = 8^6$ and $5^6 \cdot 5^{-3} = 5^{6+(-3)} = 5^3$
The Quotient Rule	Let a be a nonzero number and m and n be integers. Then $\dfrac{a^m}{a^n} = a^{m-n}$.	$\dfrac{6^7}{6^4} = 6^{7-4} = 6^3$ and $\dfrac{7^{-4}}{7^{-2}} = 7^{(-4-(-2))} = 7^{-2}$
The Power Rules	Let a and b be nonzero numbers and m and n integers. Then $(a^m)^n = a^{mn}$, $(ab)^n = a^n b^n$, and $\left(\dfrac{a}{b}\right)^n = \dfrac{a^n}{b^n}$.	$(2^2)^3 = 2^6$, $(3y)^4 = 3^4 y^4$, and $\left(\dfrac{x^2}{y}\right)^4 = \dfrac{x^8}{y^4}$
Scientific Notation	A positive number a is in scientific notation when a is written as $b \times 10^n$, where $1 \le b < 10$ and n is an integer.	$52{,}600 = 5.26 \times 10^4$ and $0.0068 = 6.8 \times 10^{-3}$

1.3 EXERCISES

FOR EXTRA HELP: Student's Solutions Manual, MyMathLab.com, InterAct Math, Math Tutor Center, MathXL, Digital Video Tutor CD 1 Videotape 1

CONCEPTS

1. Identify the base and the exponent in the expression 8^3.

2. Evaluate 97^0 and 2^{-1}.

3. Write 7 cubed, using symbols.

4. Write 5 squared, using symbols.

5. Are the expressions 2^3 and 3^2 equal? Explain your answer.

6. Are the expressions -4^2 and $(-4)^2$ equal? Explain your answer.

7. $7^{-n} = $ _____.

8. $6^m \cdot 6^n = $ _____.

9. $\dfrac{5^m}{5^n} = $ _____.

10. $(3x)^k = $ _____.

11. $(2^m)^k = $ _____.

12. $\left(\dfrac{x}{y}\right)^m = $ _____.

13. $5 \times 10^3 = $ _____.

14. $5 \times 10^{-3} = $ _____.

PROPERTIES OF EXPONENTS

Exercises 15–20: (Refer to Example 1.) Write the number as an exponential expression, using the base shown. Check your result with a calculator.

15. 8 (base 2)
16. 1000 (base 10)
17. 256 (base 4)
18. $\dfrac{1}{16}$ (base 2)
19. 1 (base 6)
20. $\dfrac{1}{125}$ (base 5)

Exercises 21–30: Evaluate the expression by hand. Check your result with a calculator.

21. 4^2
22. 2^{-3}
23. -3^4
24. $(-3)^4$
25. 5^0
26. $\left(-\dfrac{2}{3}\right)^{-3}$
27. $\left(\dfrac{2}{3}\right)^3$
28. $\dfrac{1}{4^{-2}}$
29. $\left(-\dfrac{1}{2}\right)^4$
30. $\left(-\dfrac{3}{4}\right)^3$

Exercises 31–42: Use the product rule to simplify the expression.

31. $3^5 \cdot 3^{-3}$
32. $10^2 \cdot 10^5$
33. $10^{-5} \cdot 10^2$
34. $x^2 \cdot x^3$
35. $y^2 \cdot y^{-3}$
36. $2x^4 \cdot 4x^{-4}$
37. $10^0 \cdot 10^6 \cdot 10^2$
38. $y^3 \cdot y^{-5} \cdot y^4$
39. $5^{-2} \cdot 5^3 \cdot 2^{-4} \cdot 2^3$
40. $2^{-3} \cdot 3^4 \cdot 3^{-2} \cdot 2^5$
41. $2a^3 \cdot b^2 \cdot a^{-4} \cdot 4b^{-5}$
42. $3x^{-4} \cdot 2x^2 \cdot 5y^4 \cdot y^{-3}$

Exercises 43–54: Use the quotient rule to simplify the expression. Use positive exponents to write your answer.

43. $\dfrac{4^3}{4^2}$
44. $\dfrac{5^4}{5^{-7}}$
45. $\dfrac{10^{-3}}{10^{-5}}$
46. $\dfrac{6^{-5}}{6}$
47. $\dfrac{b^{-3}}{b^2}$
48. $\dfrac{x^0}{x^{-5}}$
49. $\dfrac{24x^3}{6x}$
50. $\dfrac{10x^5}{5x^{-3}}$
51. $\dfrac{12a^2b^3}{18a^4b^2}$
52. $\dfrac{-6x^7y^3}{3x^2y^{-5}}$
53. $\dfrac{21x^{-3}y^4}{7x^4y^{-2}}$
54. $\dfrac{32x^3y}{-24x^5y^{-3}}$

Exercises 55–64: Use the power rules to simplify the expression. Use positive exponents to write your answer.

55. $(3^2)^4$
56. $(-2^2)^3$
57. $(x^3)^{-2}$
58. $(xy)^3$
59. $(4y^2)^3$
60. $(-2xy^3)^{-4}$
61. $\left(\dfrac{4}{x}\right)^3$
62. $\left(\dfrac{-3}{x^3}\right)^2$
63. $\left(\dfrac{2x}{z^4}\right)^{-5}$
64. $\left(\dfrac{2xy}{3z^5}\right)^{-1}$

ORDER OF OPERATIONS

Exercises 65–76: (Refer to Example 9.) Evaluate each expression by hand. Check your answer with a calculator.

65. $4 + 5 \cdot 6$
66. $4 - 5 - 9$
67. $2(4 + (-8))$
68. $500 - 10^3$
69. $5 \cdot 2^3$
70. $\dfrac{4+8}{2} - \dfrac{6+1}{3}$
71. $\dfrac{-2^4 - 3^2}{4} + \dfrac{1+2}{4}$
72. $\dfrac{(-4^2 + 1)}{\tfrac{2}{3}}$
73. $\dfrac{1 - 2 \cdot 4^2}{5^{-1}}$
74. $6 \div 4 \div 2$
75. $4 + 6 - 3 \cdot 5 \div 3$
76. $-3(25 - 2 \cdot 5^2) \div 5$

SCIENTIFIC NOTATION

Exercises 77–84: Write the number in scientific notation.

77. 2,322,000   (U.S. deaths in 1996)

78. 102,000   (New York City AIDS cases in 1997)

79. 26.9 billion   (Dollars spent on health care in 1960)

80. 1.035 trillion   (Dollars spent on health care in 1996)

81. 0.051   (Fraction of the population expected to spend time in prison)

82. 0.156   (Fraction of people without health insurance)

83. 0.000001   (Approximate wavelength of light in meters)

84. 0.00138   (Fraction of the people who died from a heart attack in 1995)

Exercises 85–92: Write the number in standard form.

85. 5×10^5

86. -7.85×10^3

87. 9.3×10^6

88. 2.961×10^2

89. -6×10^{-3}

90. 4.1×10^{-2}

91. 5.876×10^{-5}

92. 9.9×10^{-1}

Exercises 93–98: Evaluate the expression. Write your answer in both scientific notation and standard form.

93. $(2 \times 10^4)(3 \times 10^2)$

94. $(5 \times 10^{-4})(4 \times 10^6)$

95. $(4 \times 10^{-4})(2 \times 10^{-2})$

96. $\dfrac{6 \times 10^4}{2 \times 10^2}$

97. $\dfrac{6.2 \times 10^3}{3.1 \times 10^{-2}}$

98. $\dfrac{2 \times 10^{-2}}{8 \times 10^{-5}}$

APPLICATIONS

99. *GPS Clocks*   The Global Positioning System (GPS) is made up of 24 satellites that allow individuals with a GPS receiver to pinpoint their positions on Earth. Every 1024 weeks the clocks in the GPS satellites reset to zero. The first time this resetting occurred was on August 21, 1999.   (*Source:* Associated Press.)
    (a) Find an exponent k so that $2^k = 1024$.
    (b) Estimate the number of years in 1024 weeks.

100. *Foreign Currency*   During May 1999, one dollar was equivalent to 1818 Italian lira. (*Source:* Lipper.)
    (a) Write 1818 in scientific notation.
    (b) At that time 1,000,000 lira was equivalent to about how many dollars?

101. *Calculating Interest*   If P dollars is deposited in a savings account paying 5% annual interest, the amount A in the account after x years is given by the formula $A = P(1.05)^x$. Find A for the given values of P and x.
    (a) $P = \$500$; $x = 2$ years
    (b) $P = \$1000$, $x = 4$ years

102. *Calculating Interest*   If P dollars are deposited in a savings account paying r percent annual interest, then the amount A in the account after x years is given by the formula $A = P\left(1 + \dfrac{r}{100}\right)^x$. Find A for the given values of P, r, and x.
    (a) $P = \$200$, $r = 10\%$, $x = 7$ years
    (b) $P = \$1500$, $r = 8\%$, $x = 15$ years

103. *Astronomy*   Light travels at 186,000 miles per second. The distance that light travels in 1 year is called a *light-year*.
    (a) Calculate the number of miles in 1 light-year. Write your answer in scientific notation.
    (b) Express your answer from part (a) in standard notation.
    (c) Except for the sun, the nearest star is Alpha Centauri. Its distance from Earth is about 4.27 light-years. How many miles is this?
    (d) If a rocket flew at 50,000 miles per hour, how many years would it take to reach Alpha Centauri?

104. *Movie Box Office* The *Star Wars* prequel *Episode I: The Phantom Menace* had a record-breaking first day, grossing $28,500,000.
    (a) Write this number in scientific notation.
    (b) If the average cost of a ticket was $6, estimate the number of people that attended the movie on its first day.

105. *Federal Debt* (Refer to Example 12.) In 1990, the federal debt held by the public was $2.19 trillion, and the population of the United States was 249 million. Approximate the national debt per person.

106. *Federal Debt* In 1980, the federal debt held by the public was $710 billion, and the debt per person was $3127. Approximate the population of the United States in 1980.

107. *Computer Memory* (Refer to Example 2.) If a computer has 256 MB of memory, how many bytes is this? Express your answer in standard form.

108. *Computer Memory* One gigabyte of computer memory equals 2^{30} bytes. Write the number of bytes in 1 gigabyte, using standard notation.

109. *World Population* The following table lists populations of selected countries in 1996 and their projected populations in 2025. Rewrite the table, expressing each population in scientific notation.

Country	1996	2025
China	1,255,100,000	1,480,000,000
Germany	82,400,000	80,900,000
India	975,800,000	1,330,200,000
Mexico	95,800,000	130,200,000
United States	265,000,000	332,500,000

Source: United Nations Population Fund.

110. *World Population* If current trends continue, world population P in billions may be modeled by the equation $P = 6(1.014)^x$, where x is in years and $x = 0$ corresponds to the year 2000. Estimate the world population in 2010 and 2025. (*Source:* United Nations Population Fund.)

WRITING ABOUT MATHEMATICS

111. A student evaluates three expressions:

    -4^2 as 16; $6 + 4 \cdot 2$ as 20; and $20 \div 4 \div 2$ as 10.

    Correct the errors and explain the student's mistakes.

112. Give the product and quotient rules for exponents and an example of each.

GROUP ACTIVITY
Working with Real Data

Directions: Form a group of 2 to 4 people. Select someone to record the group's responses for this activity. All members of the group should work cooperatively to answer the questions. If your instructor asks for your results, each member of the group should be prepared to respond. If the group is asked to turn in its work, be sure to include each person's name on the paper.

1. *Walk to the Moon* The distance to the moon is about 2.37×10^5 miles. Walking at 4 miles per hour, estimate the number of hours it would take to travel this distance. How many years is this?

2. *Salary* Suppose that for full-time work a person earns 1¢ for the first week, 2¢ for the second week, 4¢ for the third week, 8¢ for the fourth week, and so on for 1 year.
    (a) Discuss whether you think this pay scale would be a good deal.
    (b) Estimate how much this person would make the last week of a 52-week year.

1.4 MODELING DATA WITH FORMULAS

Data, Variables, and Formulas ~ Modeling Data ~
Tables and Graphing Calculators

INTRODUCTION

Mathematics can be abstract or applied. Abstract mathematics focuses on the properties of numbers, theorems, and proofs. It is not concerned with applications nor is it based on scientific experiments. Theorems that were derived centuries ago are still valid today. One example is the Pythagorean theorem. In this sense, abstract mathematics transcends time. Yet, even though mathematics can be developed in an abstract setting—separate from science and all measured data—it has countless applications. Mathematics is a fuel that powers society's ability to create new products.

In this section we discuss how mathematics can be used to model the real world. We return to this topic throughout the text.

DATA, VARIABLES, AND FORMULAS

Suppose that we want to calculate the distance traveled by a car moving at a constant speed of 30 miles per hour. One method would be to make a table of values, as shown in Table 1.9.

TABLE 1.9

Elapsed time (hours)	1	2	3	4	5	6
Distance (miles)	30	60	90	120	150	180

Note that for each 1-hour increase the distance increases by 30 miles. Many times it is not possible to list all relevant values in a table. Instead, we use *variables* to describe data. For example, we might let elapsed time be represented by the variable t and let distance be represented by the variable d. If $t = 2$, then $d = 60$; if $t = 5$, then $d = 150$. In this example, the value of d is always equal to the value of t multiplied by 30. We can model this situation by using the *equation* or *formula* $d = 30t$. The values for distance in Table 1.9 can be calculated by letting $t = 1, 2, 3, 4, 5, 6$.

A **variable** is a symbol, such as x, y, or z, used to represent any unknown number or quantity. An **equation** is a statement that says two mathematical expressions are equal. Examples of equations include

$$3 + 6 = 9, \quad x + 1 = 4, \quad d = 30t, \quad \text{and} \quad x + y = 20.$$

The first equation contains only constants, the second equation contains one variable, and both the third and fourth equations contain two variables. A **formula** is an equation that can be used to calculate one quantity by using a known value of another quantity. (Formulas can also contain known values of more than one quantity.) The formula $y = \dfrac{x}{3}$ computes the number of yards in x feet. If $x = 15$, then $y = \dfrac{15}{3} = 5$. That is, in 15 feet there are 5 yards.

EXAMPLE 1   Writing and using a formula

If a car travels at a constant speed of 70 miles per hour, write a formula that calculates the distance d that the car travels in t hours. Evaluate your formula when $t = 1.5$ and interpret the result.

Solution

Traveling at 70 miles per hour, the car will travel a distance of $d = 70t$ miles in t hours. Evaluating this formula at $t = 1.5$ results in

$$d = 70(1.5) = 105.$$

After 1.5 hours the car has traveled 105 miles.

MODELING DATA

Faster moving automobiles require more distance to stop. For example, at 60 miles per hour it takes more than twice the distance to stop than it does at 30 miles per hour. Highway engineers have developed formulas to estimate the braking distance of a car.

EXAMPLE 2   Calculating braking distance

The braking distances in feet for a typical car traveling on wet, level pavement are shown in Table 1.10. Distances have been rounded to the nearest foot.

TABLE 1.10

Speed (miles per hour)	10	20	30	40	50	60	70
Distance (feet)	11	44	100	178	278	400	544

Source: L. Haefner, *Introduction to Transportation Systems.*

(a) If a car doubles its speed, what happens to the braking distance?
(b) If the speed is represented by the variable x and the braking distance by the variable d, then the braking distance may be calculated by the formula $d = \frac{x^2}{9}$. Verify the distance values in Table 1.10 for $x = 10, 30, 60$.
(c) Calculate the braking distance for a car traveling at 90 miles per hour. If a football field is 300 feet long, how many football field lengths does this braking distance represent?

Solution

(a) When the speed increases from 10 to 20 miles per hour, the stopping distance increases by a factor of $\frac{44}{11} = 4$. Similarly, if the speed doubles from 30 to 60 miles per hour, the distance increases by a factor of $\frac{400}{100} = 4$. Thus it appears that, if the speed of a car doubles, the braking distance quadruples.

(b) Let $x = 10, 30, 60$ in the formula $d = \frac{x^2}{9}$. Then

$$d = \frac{10^2}{9} = \frac{100}{9} \approx 11 \text{ feet},$$

$$d = \frac{30^2}{9} = \frac{900}{9} = 100 \text{ feet, and}$$

$$d = \frac{60^2}{9} = \frac{3600}{9} = 400 \text{ feet}.$$

These values agree with the values in Table 1.10.

(c) If $x = 90$, then $d = \frac{90^2}{9} = 900$ feet. At 90 miles per hour the braking distance equals three football fields stretched end to end.

The number b is a **square root** of a number a if $b^2 = a$. For example, one square root of 9 is 3 because $3^2 = 9$. The other square root of 9 is -3 because $(-3)^2 = 9$. We use the symbol $\sqrt{9}$ to denote the *positive* or **principal square root** of 9. That is, $\sqrt{9} = 3$. The following are examples of how to evaluate the square root symbol. A calculator is sometimes needed to approximate square roots.

$$\sqrt{16} = 4, \quad -\sqrt{100} = -10, \quad \sqrt{3} \approx 1.732, \quad \pm\sqrt{4} = \pm 2$$

The symbol "$\pm$" is read "plus or minus." Note that ± 2 represents the numbers 2 or -2.

The number b is a **cube root** of a number a if $b^3 = a$. The cube root of 8 is 2 because $2^3 = 8$, which may be written as $\sqrt[3]{8} = 2$. Similarly, $\sqrt[3]{-27} = -3$ because $(-3)^3 = -27$. Each real number has exactly one cube root.

■ **MAKING CONNECTIONS**

Square Roots and Cube Roots

The square root of a negative number is not a real number. However, the cube root of a negative number is a real number. For example, $\sqrt{-8}$ is not a real number, whereas $\sqrt[3]{-8} = -2$ is a real number.

Roots of numbers often occur in biology, as illustrated in the next example.

EXAMPLE 3 *Analyzing the walking speed of animals*

When smaller animals walk, they tend to take faster, shorter steps, whereas larger animals tend to take slower, longer steps. For example, a hyena is about 0.8 meters high at the shoulder and takes roughly 1 step per second when walking, whereas an elephant 3 meters high at the shoulder takes 1 step every 2 seconds. If an animal is h meters high at the shoulder, then the frequency F in steps per second while it is walking can be estimated with the formula $F = \frac{0.87}{\sqrt{h}}$. The value of F is referred to as the animal's *stepping frequency*. (**Source:** C. Pennycuick, *Newton Rules Biology*.)

(a) A Thomson's gazelle is about 0.6 meter high at the shoulder. Estimate its stepping frequency.
(b) A giraffe is about 2.7 meters high at the shoulder. Estimate its stepping frequency.
(c) What happens to the stepping frequency as h increases?

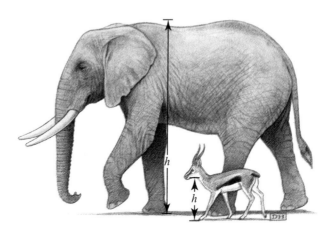

Solution

(a) $F = \dfrac{0.87}{\sqrt{0.6}} \approx 1.12$. A Thomson's gazelle takes about 1.12 steps per second when walking.

(b) $F = \dfrac{0.87}{\sqrt{2.7}} \approx 0.53$. A giraffe takes roughly half a step per second when walking, or 1 step every 2 seconds.

(c) As h increases, the denominator of $\dfrac{0.87}{\sqrt{h}}$ also increases, so the ratio becomes smaller. Thus, as h increases, the stepping frequency decreases.

In the next example we find a formula that models a data table.

EXAMPLE 4 *Modeling data with a formula*

The data in Table 1.11 can be modeled by the formula $y = ax$. Find a.

TABLE 1.11

x	1	2	3	4	5
y	3	6	9	12	15

Solution

Each value of y is 3 times the corresponding value of x, so $a = 3$. We can also find a symbolically. If $x = 1$, then $y = 3$. We can substitute these values into the equation.

$y = ax$ Given equation
$3 = a \cdot 1$ Let $x = 1$ and $y = 3$.
$3 = a$ $a \cdot 1 = a$

Thus $a = 3$. The formula $y = 3x$ models the data in Table 1.11.

Critical Thinking

Write a formula that calculates the time T for a bike rider to travel 100 miles moving at x miles per hour. Test your formula for different values of x.

TABLES AND GRAPHING CALCULATORS

Many graphing calculators are able to generate tables. We can check our results in Example 4 with this feature. First, we enter the formula $Y_1 = 3X$, as shown in Figure 1.27. Then to generate a table, we specify the starting x-value (TblStart) and the increment (ΔTbl) between x-values. The graphing calculator calculates the required table automatically. See Figures 1.28 and 1.29.

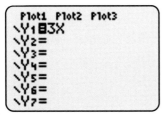
Figure 1.27

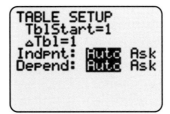
Figure 1.28

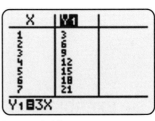
Figure 1.29

To verbally describe the table shown in Figure 1.29, we use the phrase "Table $Y_1 = 3X$ starting at $x = 1$, incrementing by 1."

EXAMPLE 5 Using the table feature

Table $Y_1 = X^2/9$ starting at $x = 10$, incrementing by 10. Compare this table to Table 1.10 in Example 2.

Solution

In Figures 1.30–1.32 the desired table is generated. Note that, if values are rounded to the nearest foot, the values in Figure 1.32 agree with those in Table 1.10.

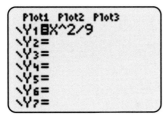
Figure 1.30

Figure 1.31

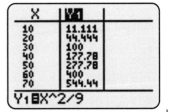
Figure 1.32

1.4 Putting It All Together

A *variable* is a symbol used to represent any unknown number or quantity. An *equation* is a statement that two mathematical expressions are equal. A *formula* is an equation that provides a way to calculate one quantity from the value of another quantity (or values of other quantities). The table feature on a graphing calculator can be used to create a table of values efficiently. Usually, we must specify the equation, the starting value of *x*, and the increment between *x*-values in the table.

The following table summarizes the square root and cube root.

Roots of Numbers	Definition	Examples
Square Root	The positive or principal square root of a is written $\sqrt{a}$.	$\sqrt{25} = 5$ $\pm\sqrt{100} = \pm 10$ $-\sqrt{16} = -4$
Cube Root	The cube root of a is written $\sqrt[3]{a}$.	$\sqrt[3]{-8} = -2$ because $(-2)^3 = -8$. $\sqrt[3]{64} = 4$ because $4^3 = 64$.

1.4 EXERCISES

FOR EXTRA HELP: Student's Solutions Manual, MyMathLab.com, InterAct Math, Math Tutor Center, MathXL, Digital Video Tutor CD 1 Videotape 2

CONCEPTS

1. Identify the variables in the equation $x^2 + y^2 = 9$.

2. Give an example of an equation containing no variables.

3. Give an example of an equation containing one variable.

4. Give an example of an equation containing two variables.

WRITING FORMULAS

Exercises 5–10: Write a formula that does the following.

5. Converts *x* miles to *y* feet

6. Converts *x* quarts to *y* gallons

7. Finds the area *A* of a square with a side of length *s*

8. Finds the surface area *A* of a cube with a side of length *s*

9. Determines the number of seconds *y* in *x* hours

10. Determines the gas mileage *G* of a car that travels *x* miles on *y* gallons of gas.

USING DATA, VARIABLES, AND FORMULAS

Exercises 11–24: Evaluate the formula for the given value of the variable.

11. $y = 5x$ $x = 6$

12. $y = \dfrac{x}{10}$ $x = 30$

13. $y = x + 5$ $x = -3.1$

14. $d = 5 - 4t$ $t = -1.5$

15. $d = t^2 + 1$ $t = -3$

16. $z = 3k^2 - \dfrac{3}{4}$ $k = \dfrac{1}{4}$

17. $z = \sqrt{2k}$ $k = 18$

18. $y = \sqrt{5-x}$ $x = 1$

19. $y = -\frac{1}{2}\sqrt[3]{x}$ $x = \frac{1}{8}$

20. $M = \sqrt[3]{1-p}$ $p = 65$

21. $N = 3h^3 - 1$ $h = \frac{1}{3}$

22. $S = 1 - \frac{1}{2}w^3$ $w = -2$

23. $P = |5 - w|$ $w = 4.7$

24. $D = |2t - 5|$ $t = 2.5$

Exercises 25–30: Select the formula that best models the data in the table.

25.

x	1	2	3	4	5
y	2	4	6	8	10

(i) $y = x + 2$, (ii) $y = 2x$, (iii) $y = 4x - 2$

26.

x	−2	−1	0	1	2
y	4	1	0	1	4

(i) $y = x^2$, (ii) $y = x + 6$, (iii) $y = 2x$

27.

x	0	1	4	9	16
y	0	1	2	3	4

(i) $y = \sqrt[3]{x}$, (ii) $y = -\frac{1}{6}x^2 + \frac{7}{6}x$, (iii) $y = \sqrt{x}$

28.

x	−27	−1	0	8	64
y	−3	−1	0	2	4

(i) $y = \sqrt[3]{x}$, (ii) $y = \frac{x}{4}$, (iii) $y = \sqrt{x}$

29.

x	−4	−2	0	2	4
y	4	2	0	2	4

(i) $y = x^2$, (ii) $y = x$, (iii) $y = |x|$

30.

x	0	1	2	3	4
y	−1	0	3	8	15

(i) $y = x^3 - 1$, (ii) $y = x^2 - 1$, (iii) $y = x - 1$

Exercises 31–36: (Refer to Example 4.) Find a value of the variable a so that the equation models the data.

31. $y = ax$

x	−2	−1	0	1	2
y	6	3	0	−3	−6

32. $y = ax$

x	−10	−5	5	10	15
y	−15	−7.5	7.5	15	22.5

33. $d = t - a$

t	0	1	2	3	4
d	−2	−1	0	1	2

34. $d = t - a$

t	3	6	9	12	15
d	8	11	14	17	20

35. $N = aw^2$

w	−2	−1	0	1	2
N	8	2	0	2	8

36. $N = aw^2$

w	0	2	4	6	8
N	0	−4	−16	−36	−64

Exercises 37–42: Complete the table, using the formula.

37. $y = \frac{5}{2}x - \frac{1}{2}$

x	0	2	4	6	8
y					

38. $y = \frac{1}{2}x^2$

x	0	2	4	6	8
y					

39. $y = |5x|$

x	−3	−1	1	3
y				

48 CHAPTER 1 Real Numbers and Algebra

40. $y = \sqrt{x + 2}$

x	−2	−1	2	7
y				

41. $y = \sqrt[3]{x} - 2$

x	−1	0	1	8
y				

42. $y = x^3 - 4x$

x	−2	0	1	2
y				

APPLICATIONS

43. *Modeling Motion* The following table lists the distance y traveled by a car in t hours. Find an equation that models these data.

Elapsed time (hours)	1	2	3	4
Distance (miles)	60	120	180	240

44. *Braking Distance* (Refer to Example 2.) The braking distance d for a car on dry, level pavement traveling at x miles per hour is given by $d = \dfrac{x^2}{12}$. (*Source:* L. Haefner.)
  (a) Make a table of braking distances for speeds of 10 to 70 miles per hour in increments of 10 miles per hour.
  (b) What is the braking distance for a car traveling at 40 miles per hour?
  (c) What happens to the braking distance when the speed doubles?
  (d) At 60 miles per hour how much farther does it take the car to stop on wet pavement than on dry pavement? $\left(\text{Hint: For wet pavement } d = \dfrac{x^2}{9}.\right)$

45. *Diving and Water Pressure* The world's record for descending below the surface of the ocean on a single breath is in excess of 400 feet by Francisco Ferreras. During his descent his heart rate slows from 60 beats per minute on the surface to 4 beats per minute at 400 feet. He is able to hold his breath for 7 minutes. These dives are dangerous because of the extreme water pressure. The water pressure P in pounds per square inch at a depth of x feet can be calculated with the formula $P = 0.445x$.
  (a) Calculate the water pressure at a depth of 400 feet.
  (b) The world's deepest diving mammals are sperm whales, which can dive to a depth of 7000 feet. Calculate the water pressure at this depth. (*Source:* G. Carr, *Mechanics of Sport.*)

46. *Bicycle Speed Records* The fastest speed attained on a bicycle is 223.3 feet per second. This speed was attained by the bicyclist following a pace vehicle. The bicyclist thus experienced less wind resistance and was carried along by the draft or "suction" created by the pace vehicle. The formula $M = \dfrac{15}{22}x$ converts feet per second to miles per hour. Find the speed of this record in miles per hour. (*Source:* G. Carr.)

47. *Wing Span and Weight* A bird's weight W is frequently related to the length L of its wing span. For one species of bird the formula $W = 1.1L^3$ could be used to predict a bird's weight W in kilograms for a wing span of L meters. (*Source:* C. Pennycuick.)
  (a) If a bird has a wing span of 0.75 meter, estimate its weight.
  (b) If a bird has a wing span of 1.5 meters, estimate its weight.
  (c) If the wing span of a bird doubles, what happens to its weight?

48. *Birds* The surface area of a bird's wings S is frequently related to its weight W. For one species of bird, the formula $S = 0.11 \sqrt[3]{W^2}$ could be used to predict the surface area S in square meters of a bird's wings for a weight W in kilograms. (*Source:* **C. Pennycuick.**)
(a) If a bird weighs 0.5 kilogram, estimate the area of its wings.
(b) If a bird's weight doubles, does the area of its wings also double?

49. *Escape Velocity* To escape the gravity of a planet or a moon, a spacecraft must reach the *escape velocity*, which we denote E. The larger the planet or moon, the greater the escape velocity is. But for a spacecraft simply to attain a circular orbit, a slower velocity C is all that is necessary. The relationship between E and C is modeled by $E = \sqrt{2}C$. Use this formula to approximate the missing values in the following table. (*Source:* **H. Karttunen,** *Fundamental Astronomy.*)

Planet	Venus	Earth	Moon	Mars
C (miles per hour)	16,260			8050
E (miles per hour)		25,040	5360	

Source: M. Zeilik, *Introductory Astronomy and Astrophysics.*

50. *Escape Velocity* (Refer to the previous exercise.) The escape velocity for the largest planet in our solar system, Jupiter, is 136,000 miles per hour. Calculate the velocity necessary for a circular orbit around Jupiter. (*Source:* **M. Zeilik.**)

51. *Escape Velocity* (Refer to Exercises 49 and 50.) The speed necessary for a circular orbit around Saturn is 57,000 miles per hour. Find the escape velocity for Saturn. (*Source:* **M. Zeilik.**)

52. *Animals and Trotting Speeds* (Refer to Example 3.) The relationship between the shoulder height h and an animal's stepping frequency F in steps per second while *trotting* is given by the formula $F = \dfrac{1.84}{\sqrt{h}}$. (*Source:* **C. Pennycuick.**)
(a) Estimate the stepping frequency for a trotting buffalo that is 1.5 meters high at the shoulders.
(b) Discuss what happens to an animal's stepping frequency while trotting as its shoulder height increases.

53. *Pulse Rate in Animals* According to one model, the rate at which an animal's heart beats varies with its weight. Smaller animals tend to have faster pulses, whereas larger animals tend to have slower pulses. The pulse rate of an animal can be modeled by the equation $N = \dfrac{885}{\sqrt{W}}$, where N is the number of beats per minute and W is the animal's weight in pounds. (*Source:* **C. Pennycuick.**)
(a) Estimate the pulse for a 30-pound dog.
(b) Estimate the pulse for a 2000-pound elephant.

54. *Area of an Equilateral Triangle* An equilateral triangle has three sides equal in length. Its area A is given by $A = \dfrac{\sqrt{3}}{4}s^2$, where s is the length of a side, as shown in the accompanying figure. Calculate the areas for the given values of s. Approximate your answers to the nearest hundredth.

(a) $s = 1.25$ feet
(b) $s = 5.58$ meters

55. *Circumference of a Circle* The circumference C of a circle is given by $C = 2\pi r$, where r is the radius. Calculate the circumference of each circle with the given radius. Approximate your answer to the nearest tenth.
(a) $r = 14.1$ inches
(b) $r = 1.37$ miles

56. *Area of a Circle* The area A of a circle is given by $A = \pi r^2$, where r is the radius. Calculate the area of each circle with the given radius. Approximate your answer to the nearest tenth.
(a) $r = 12$ inches
(b) $r = 6.1$ feet

57. *Inline Skating* During a strenuous skating workout an athlete can burn 336 calories in 40-minutes. (*Source: Runner's World.*)
  (a) Write a formula that calculates the calories C burned from skating 40 minutes a day for x days.
  (b) How many calories could be burned in 30 days?

58. *Inline Skating* (Refer to the previous exercise.) If a person loses 1 pound for every 3500 calories burned, write a formula that gives the number of pounds P lost in x days from skating 40 minutes per day. How many pounds could be lost in 100 days?

WRITING ABOUT MATHEMATICS

59. Explain the difference between a mathematical expression and an equation.

60. Give an example of a formula that models an application and identify each variable. Explain how to use the formula.

CHECKING BASIC CONCEPTS FOR SECTIONS 1.3 AND 1.4

1. Evaluate each expression.
   (a) 2^4
   (b) $3^{-2} \cdot 2^0$
   (c) $\dfrac{2^4}{2^2 \cdot 2^{-3}}$
   (d) $x^3 \cdot x^{-4} \cdot x^2$
   (e) $\left(\dfrac{2x^3}{y^{-4}}\right)^2$

2. Evaluate each expression by hand. Check your result with a calculator.
   (a) $4 + 5 \cdot (-2)$
   (b) $\dfrac{1 + 3}{-4 + 3}$
   (c) $2^3 - 5(2 - 3 \cdot 4)$

3. Express each number in scientific notation.
   (a) 103,000
   (b) 0.000523
   (c) 6.7

4. Express each number in standard form.
   (a) 5.43×10^6
   (b) 9.8×10^{-3}

5. *Indoor Air Pollution* Ventilation is an effective method for removing indoor air pollution. The formula $y = 900x$ calculates the cubic feet per hour of air that should be circulated in a classroom containing x people. Make a table showing the ventilation necessary for classes containing 10, 20, 30, and 40 people. How much ventilation is necessary per person? (*Source: American Society of Heating, Refrigerating, and Air-Conditioning Engineers, ASHRAE.*)

1.5 VISUALIZATION OF DATA

Relations ~ The Rectangular Coordinate System ~ Scatterplots and Line Graphs ~ The Viewing Rectangle ~ Visualizing Data with a Graphing Calculator

INTRODUCTION

Computers, the Internet, and other types of electronic communication are creating large amounts of data. The challenge for society is to use these data to solve important problems and create new knowledge. Before conclusions can be drawn, data must be analyzed. A powerful tool in this step is visualization. Pictures are capable of communicating large quantities of information in short periods of time. A full page of computer graphics typically contains a hundred times more information than a page of text.

The map in Figure 1.33 shows the average date of the first 32°F temperature in autumn. Imagine trying to describe this map by using *only* words. In this section we discuss how graphs are used to visualize data. **(Source: J. Williams, *The Weather Almanac 1995*.)**

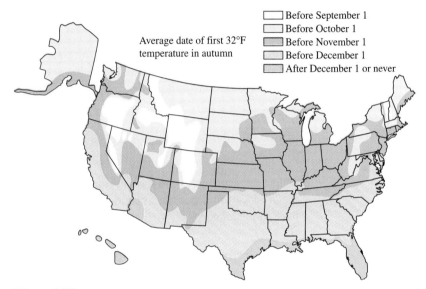

Figure 1.33

RELATIONS

Table 1.12 lists the monthly average wind speeds in miles per hour for San Francisco. In this table January corresponds to 1, February to 2, and so on, until December is represented by 12. For example, in April the average wind speed is 12 miles per hour.

TABLE 1.12 **Average Wind Speeds in San Francisco**

Month	1	2	3	4	5	6	7	8	9	10	11	12
Wind Speed (miles per hour)	7	9	11	12	13	14	14	13	11	9	8	7

Source: J. Williams, *The Weather Almanac 1995.*

If we let x be the month and y be the wind speed, then the **ordered pair** (x, y) represents the average wind speed y during the month x. For example, the ordered pair $(2, 9)$ indicates that in February the average wind speed is 9 miles per hour, whereas the ordered pair $(9, 11)$ indicates that the average wind speed in September is 11 miles per hour. *Order is important* in an ordered pair.

The data in Table 1.12 establish a relation; that is, each month is associated with a wind speed in an ordered pair (month, wind speed). This relation can be represented by a set S, which contains 12 ordered pairs:

$$S = \{(1, 7), (2, 9), (3, 11), (4, 12), (5, 13), (6, 14),$$
$$(7, 14), (8, 13), (9, 11), (10, 9), (11, 8), (12, 7)\}.$$

Relation

A **relation** is a set of ordered pairs.

If we denote the ordered pairs in a relation (x, y), then the set of all x-values is called the **domain** of the relation and the set of all y-values is called the **range**. In Table 1.12 the domain is

$$D = \{1, 2, 3, 4, 5, 6, 7, 8, 9, 10, 11, 12\},$$

which corresponds to the 12 months. The range is

$$R = \{7, 8, 9, 11, 12, 13, 14\},$$

which corresponds to the monthly average wind speeds. Note that an average wind speed of 14 miles per hour occurs more than once in Table 1.12, but it is listed only once in the range set R.

EXAMPLE 1 *Finding the domain and range of a relation*

Find the domain and range for the relation given by

$$S = \{(-1, 5), (0, 1), (2, 4), (4, 2), (5, 1)\}.$$

Solution

The domain D is determined by the first element in each ordered pair, or

$$D = \{-1, 0, 2, 4, 5\}.$$

The range R is determined by the second element in each ordered pair, or

$$R = \{1, 2, 4, 5\}.$$

EXAMPLE 2 *Finding the domain and range of a relation*

Table 1.13 lists the average cost of tuition and fees at public colleges from 1994 through 1997. Express this table as a relation S. Identify the domain and range of S.

TABLE 1.13 **Tuition and Fees at Public Colleges**

Year	1994	1995	1996	1997
Cost	$2705	$2811	$2966	$3111

Source: The College Board.

1.5 Visualization of Data

Solution

Let the year be the first element in the ordered pair and the cost of tuition be the second element. Then relation S is given by the following set of ordered pairs.

$$S = \{(1994, 2705), (1995, 2811), (1996, 2966), (1997, 3111)\}$$

The domain of S is

$$D = \{1994, 1995, 1996, 1997\}$$

and the range of S is

$$R = \{2705, 2811, 2966, 3111\}.$$

It is possible for a relation to contain infinitely many ordered pairs. For example, let the equation $y = 2x$ define a relation S, where x is any real number. Then S contains infinitely many ordered pairs of the form $(x, 2x)$, such as $(-2, -4)$, $(3, 6)$, and $(0.1, 0.2)$.

THE RECTANGULAR COORDINATE SYSTEM

We can use the **Cartesian coordinate plane,** or *xy*-plane, to visualize a relation. The horizontal axis is the ***x*-axis** and the vertical axis is the ***y*-axis.** The axes intersect at the **origin** and determine four regions called **quadrants.** They are numbered I, II, III, and IV counterclockwise, as illustrated in Figure 1.34. We can plot the ordered pair (x, y) by using the *x*-axis and the *y*-axis. For example, the point $(1, 2)$ is located in quadrant I, 1 unit to the right of the origin and 2 units above the *x*-axis, as shown in Figure 1.35.

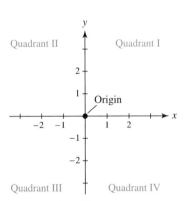

Figure 1.34 The *xy*-plane

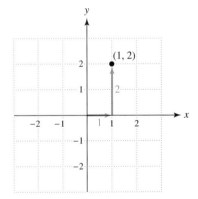

Figure 1.35 Plotting a Point

Similarly, the ordered pair $(-2, 3)$ is located in quadrant II, $(-3, -3)$ is in quadrant III, and $(3, -2)$ is in quadrant IV. A point lying on a coordinate axis does not belong to any quadrant. The point $(-2, 0)$ is located on the x-axis, whereas the point $(0, -2)$ lies on the y-axis. See Figure 1.36.

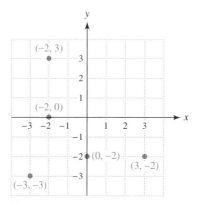

Figure 1.36 Plotting Points

EXAMPLE 3 *Plotting points*

Plot the data listed in Table 1.14. If possible, state the quadrant containing each point.

TABLE 1.14

x	-3	0	1	4
y	1	4	-2	3

Solution

We plot the points $(-3, 1)$, $(0, 4)$, $(1, -2)$, and $(4, 3)$ in the xy-plane, as shown in Figure 1.37. The point $(-3, 1)$ is in quadrant II, $(1, -2)$ is in quadrant IV, and $(4, 3)$ is in quadrant I. The point $(0, 4)$ lies on the y-axis and does not belong to any quadrant.

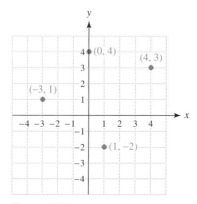

Figure 1.37

1.5 Visualization of Data

> ■ **MAKING CONNECTIONS**
>
> *The Number Line and the xy-Plane*
>
> Real numbers may be plotted on *one* number line that is equivalent to the *x*-axis. Ordered pairs may be plotted by using *two* number lines, one for the *x*-axis and one for the *y*-axis. Later you may learn that points located on objects such as cubes and spheres may be plotted by using *three* number lines: for the *x*-, *y*-, and *z*-axes. This situation frequently occurs in three-dimensional (3D) computer graphics.

SCATTERPLOTS AND LINE GRAPHS

If distinct points are plotted in the *xy*-plane, the resulting graph is called a **scatterplot.** Figure 1.37 is an example. A different scatterplot is shown in Figure 1.38, where the points (1, 2), (2, 4), (3, 5), (4, 6), (5, 4), and (6, 3) have been plotted.

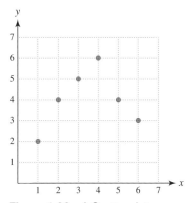

Figure 1.38 A Scatterplot

The next example illustrates how to make a scatterplot from real data.

EXAMPLE 4   *Making a scatterplot of gasoline prices*

Table 1.15 lists the average price of a gallon of gasoline for selected years. Make a scatterplot of these data.

TABLE 1.15   **Average Prices of Gasoline**

Year	1955	1965	1975	1985	1995
Cost (per gallon)	29¢	31¢	57¢	120¢	121¢

Source: Department of Energy.

Solution

Plot the points (1955, 29), (1965, 31), (1975, 57), (1985, 120), and (1995, 121). The *x*-values vary from 1955 to 1995, so we label the *x*-axis from 1950 to 2000. The *y*-values vary from 29 to 121, so we label the y-axis from 0 to 150. (Note that labels on the x- and y-axes may vary.) Figure 1.39 shows these points as plotted and labeled.

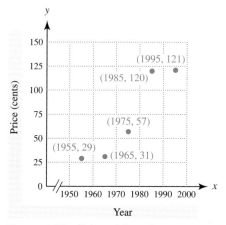

Figure 1.39 Price of Gasoline

Sometimes it is helpful to connect the data points in a scatterplot with straight line segments. This type of graph emphasizes changes in the data and is called a **line graph.**

EXAMPLE 5 *Interpreting a line graph*

The line graph shown in Figure 1.40 depicts the total number of all types of college degrees awarded in millions for selected years. **(Source: Department of Education.)**

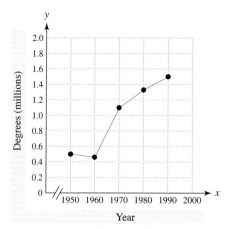

Figure 1.40 College Graduates

(a) Did the number of graduates ever decrease during this time period? Explain.
(b) Approximate the number of college graduates in the year 1970.
(c) Determine the 10-year period when the increase in the number of college graduates was greatest. What was this increase?

Solution

(a) Yes, the number decreased slightly between 1950 and 1960. For this time period, the line segment slopes slightly downward from left to right.

(b) In 1970, about 1.1 million degrees were awarded.

(c) The greatest increase corresponds to the line segment that slopes upward most from left to right. This increase occurred between 1960 and 1970 and was about $1.1 - 0.5 = 0.6$ million graduates.

Critical Thinking

Discuss ways that the following list of test scores could be visualized.

Test Scores: 10, 10, 8, 7, 6, 6, 10, 5, 8, 6

THE VIEWING RECTANGLE

Graphing calculators provide several features beyond those found on scientific calculators. Graphing calculators have additional keys that can be used to create tables, scatterplots, and graphs.

The **viewing rectangle,** or **window,** on a graphing calculator is similar to the viewfinder in a camera. A camera cannot take a picture of an entire scene. The camera must be centered on some object and can photograph only a portion of the available scenery. A camera can capture different views of the same scene by zooming in and out, as can graphing calculators. The xy-plane is infinite, but the calculator screen can show only a finite, rectangular region of the xy-plane. The viewing rectangle must be specified by setting minimum and maximum values for both the x- and y-axes before a graph can be drawn.

We use the following terminology regarding the size of a viewing rectangle. **Xmin** is the minimum x-value along the x-axis, and **Xmax** is the maximum x-value. Similarly, **Ymin** is the minimum y-value along the y-axis, and **Ymax** is the maximum y-value. Most graphs show an x-scale and a y-scale with tick marks on the respective axes. Sometimes the distance between consecutive tick marks is 1 unit, but at other times it might be 5 or 10 units. The distance represented by consecutive tick marks on the x-axis is called **Xscl,** and the distance represented by consecutive tick marks on the

y-axis is called **Yscl** (see Figure 1.41). This information about the viewing rectangle can be written concisely as [Xmin, Xmax, Xscl] by [Ymin, Ymax, Yscl]. For example, $[-10, 10, 1]$ by $[-10, 10, 1]$ means that Xmin $= -10$, Xmax $= 10$, Xscl $= 1$, Ymin $= -10$, Ymax $= 10$, and Yscl $= 1$. This setting is referred to as the **standard viewing rectangle.** The viewing rectangle in Figure 1.41 is $[-3, 3, 1]$ by $[-3, 3, 1]$.

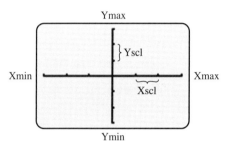

Figure 1.41

EXAMPLE 6 *Setting the viewing rectangle*

 Show the standard viewing rectangle and the viewing rectangle $[-2, 3, 0.5]$ by $[-100, 200, 50]$ on your calculator.

Solution

The window settings and viewing rectangles are displayed in Figures 1.42–1.45. Note that in Figure 1.43 there are 10 tick marks on the positive x-axis because its length is 10 units and the distance between consecutive tick marks is 1 unit. In contrast, in Figure 1.45 there are 6 tick marks on the positive x-axis because its length is 3 units and the distance between consecutive tick marks is 0.5 unit.

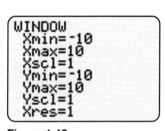

Figure 1.42

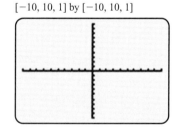

$[-10, 10, 1]$ by $[-10, 10, 1]$

Figure 1.43

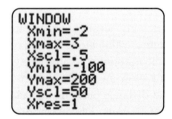

Figure 1.44

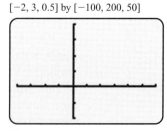

$[-2, 3, 0.5]$ by $[-100, 200, 50]$

Figure 1.45

Visualizing Data with a Graphing Calculator

Many graphing calculators have the capability to create scatterplots and line graphs. The next example illustrates how to make a scatterplot with a graphing calculator.

EXAMPLE 7 *Making a scatterplot with a graphing calculator*

 Plot the points $(-2, -2)$, $(-1, 3)$, $(1, 2)$, and $(2, -3)$ in $[-4, 4, 1]$ by $[-4, 4, 1]$.

Solution

The points $(-2, -2)$, $(-1, 3)$, $(1, 2)$, and $(2, -3)$ have been entered in Figure 1.46 using the STAT EDIT feature. The variable L1 represents the x-values, and the variable L2 represents the y-values. In Figure 1.47 the graphing calculator has been set to make a scatterplot using the STATPLOT feature, and in Figure 1.48 the points have been plotted. If you have a different model of calculator you may need to consult your owner's manual.

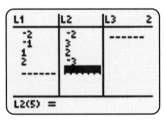
Figure 1.46

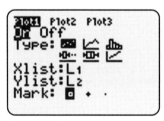
Figure 1.47

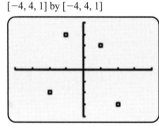
Figure 1.48

In the next example, a graphing calculator is used to create a line graph of sales of cordless telephones.

EXAMPLE 8 *Making a line graph with a graphing calculator*

 Table 1.16 lists numbers of cordless telephones sold for selected years from 1987 through 2000. Make a line graph of these sales in an appropriate viewing rectangle. Then interpret the line graph.

TABLE 1.16 **Cordless Phone Sales**

Year	1987	1990	1993	1996	2000
Phones (millions)	6.2	9.9	18.7	22.8	33.3

Source: Cellular Telecommunications Industry Association.

Solution

Plot the points (1987, 6.2), (1990, 9.9), (1993, 18.7), (1996, 22.8), and (2000, 33.3). The x-values vary from 1987 to 2000, and the y-values vary between 6.2 and 33.3. We select the viewing rectangle [1985, 2002, 5] by [0, 40, 10], although other viewing rectangles are possible. The viewing rectangle should be large enough to show all five data points without being too large. A line graph can be created by selecting this option on the graphing calculator. Figures 1.49 and 1.50 show the data entries and plotting scheme. Figure 1.51 shows the resulting graph. It reveals that sales have increased dramatically during this time period.

[1985, 2002, 5] by [0, 40, 10]

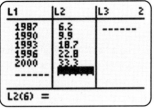

Figure 1.49

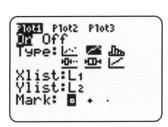

Figure 1.50

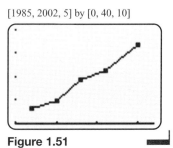

Figure 1.51

1.5 Putting It All Together

Visualization is frequently used in mathematics, science, and business as a way to summarize and understand data better. The xy-plane is commonly used to visualize relations. A relation is a set of ordered pairs.

Concept	Explanation	Example
Relation	A set of ordered pairs	$S = \{(1, 2), (-2, 3), (4, 2)\}$
Domain and Range	If a relation consists of a set of ordered pairs (x, y), then the set of x-values is the domain and the set of y-values is the range.	If $S = \{(1, 2), (-2, 3), (4, 2)\}$, then $D = \{-2, 1, 4\}$ and $R = \{2, 3\}$.
Scatterplot	A scatterplot results when individual points are plotted in the xy-plane.	
Line Graph	A line graph is similar to a scatterplot except that line segments are drawn between consecutive points.	

1.5 EXERCISES

CONCEPTS

1. What is a relation?

2. What are the domain and range of a relation?

3. Sketch the *xy*-plane and identify each of the following: the *x*-axis, the *y*-axis, the origin, and the four quadrants.

4. Sketch an example of a scatterplot and a line graph.

RELATIONS AND THE RECTANGULAR COORDINATE SYSTEM

Exercises 5–10: Identify the domain and range of the relation S.

5. $S = \{(1, 2), (3, -4), (5, 6)\}$

6. $S = \{(0, 4), (0, 6), (3, -1), (4, 0)\}$

7. $S = \{(-2, 3), (-1, 2), (0, 1), (1, 0), (2, 1)\}$

8. $S = \left\{ \left(\frac{1}{2}, -\frac{3}{4}\right), \left(-\frac{5}{8}, \frac{4}{7}\right), \left(\frac{1}{2}, \frac{3}{4}\right), \left(\frac{8}{7}, \frac{3}{4}\right) \right\}$

9. $S = \{(41, 67), (87, 53), (41, 88), (96, 24)\}$

10. $S = \{(-1.2, -1.1), (0.8, 2.5), (1.5, -0.6)\}$

Exercises 11–14: Express the relation S in the table as a set of ordered pairs. Then identify the domain and range of S.

11.

x	1	3	5	7	9
y	3	7	11	15	19

12.

x	−2.1	−1.5	0.7	1.3	2.9
y	9.6	7.4	3.3	−2.0	−8.8

13. U.S. unemployment rate in percent

x	1994	1995	1996	1997	1998	1999
y	6.1	5.6	5.5	4.9	4.5	4.1

Source: Department of Labor.

14. U.S. population in millions

x	1800	1840	1880	1920	1960	2000
y	5	17	50	106	179	281

Source: Bureau of the Census.

Exercises 15–20: Express the relation shown in the graph as a set of ordered pairs.

15.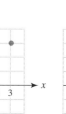

16.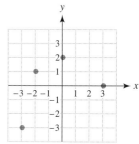

17.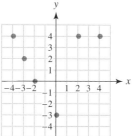

18.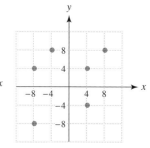

19. Billions of dollars spent on military personnel in the United States. Answers may vary slightly. (*Source:* Department of Defense.)

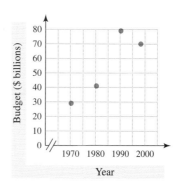

20. Cases of Tetanus in the United States. Answers may vary slightly. (*Source:* Department of Health and Human Services.)

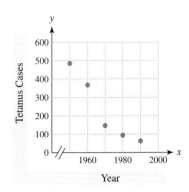

Exercises 21–24: Plot the points in the table in the xy-plane. If possible, state the quadrant containing each point.

21.

x	1	−3	0	−1
y	2	0	−2	3

22.

x	2	−4	−2	0
y	6	−4	0	−5

23.

x	10	−30	50	−20
y	50	20	−25	−25

24.

x	0.2	0.4	0.6	0.8
y	3	1	−1	−3

SCATTERPLOTS AND LINE GRAPHS

Exercises 25–28: Complete the following.
(a) Find the domain and range of the relation.
(b) Determine the minimum and maximum of the x-values; of the y-values.
(c) Label appropriate scales on the x- and y-axes.
(d) Make a scatterplot of the data by hand.

25. $\{(0, 2), (-3, 4), (-2, -4), (1, -3), (0, 0)\}$

26. $\{(1, 1), (3, 0), (-4, -4), (5, -2), (0, 3)\}$

27. $\{(10, 50), (-30, 40), (20, -50), (30, 20)\}$

28. $\{(5, 15), (25, 20), (10, 10), (-10, 30), (-20, -10)\}$

Exercises 29–32: Make a line graph from the ordered pairs.

29. $(0, 2), (1, 4), (2, 5), (4, 4), (5, 2)$

30. $(-2, 4), (-1, 1), (0, 0), (1, 1), (2, 4)$

31. $(4, 4), (8, -4), (12, 8), (16, 0), (20, -8)$

32. $(10, 20), (20, 30), (30, 40), (40, 60), (50, 30)$

Exercises 33–36: Make a line graph from the table of data.

33.

x	0	1	2	3
y	−2	−1	0	3

34.

x	−2	−1	0	1	2
y	0	3	4	3	0

35.

x	0	1	2	3
y	−3	0	3	0

36.

x	0	1	4	9
y	0	1	2	3

GRAPHING CALCULATORS

Exercises 37–42: Show the given viewing rectangle on your graphing calculator. Predict the number of tick marks on the positive x-axis and the positive y-axis.

37. Standard viewing rectangle

38. $[-12, 12, 2]$ by $[-8, 8, 2]$

39. $[0, 100, 10]$ by $[-50, 50, 10]$

40. $[-30, 30, 5]$ by $[-20, 20, 5]$

41. $[1980, 1995, 1]$ by $[12{,}000, 16{,}000, 1000]$

42. $[1900, 1990, 10]$ by $[1700, 2800, 100]$

Exercises 43–46: Match the viewing rectangle with the correct figure.

43. $[-5, 5, 1]$ by $[-5, 5, 1]$

44. $[-5, 5, 1]$ by $[-2, 2, 1]$

45. $[-100, 100, 50]$ by $[-100, 100, 10]$

46. $[-2, 8, 1]$ by $[-3, 7, 1]$

a.

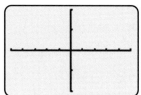

b.

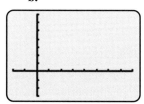

c.

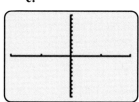

d.

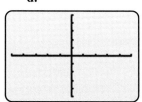

Exercises 47–52: Express the relation shown in the graph as a set of ordered pairs.

47.
$[-3, 3, 1]$ by $[-2, 2, 1]$

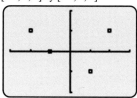

48.
$[-6, 6, 1]$ by $[-4, 4, 1]$

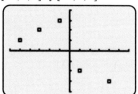

49.
$[-5, 5, 1]$ by $[-3, 3, 1]$

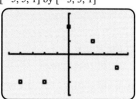

50.
$[-9, 9, 2]$ by $[-6, 6, 2]$

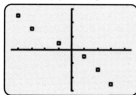

51.
$[1980, 2000, 5]$ by $[0, 60, 10]$

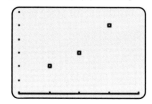

52.
$[1900, 2000, 20]$ by $[0, 500, 100]$

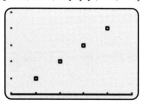

Exercises 53–58: Use your calculator to make a scatterplot of the relation after determining an appropriate viewing rectangle.

53. $\{(4, 3), (-2, 1), (-3, -3), (5, -2)\}$

54. $\{(5, 5), (2, 0), (-2, 7), (2, -8), (-1, -5)\}$

55. $\{(20, 40), (-25, -15), (-20, 25), (15, -25)\}$

56. $\{(-13, 12), (3, 10), (-15, -4), (12, -9)\}$

57. $\{(3.1, 6.2), (-5.1, 10.1), (-0.7, -1.4)\}$

58. $\{(-1.2, 0.6), (1.0, -0.5), (-0.4, 0.2),$
    $(-2.8, 1.4), (2.8, -1.4)\}$

Exercises 59–64: Graphing Real Data Each table contains real data.
   (a) Determine an appropriate viewing rectangle.
   (b) Make a line graph of the data.
   (c) Comment on any trends in the data.

59. Head Start participation y in thousands during year x

x	1970	1980	1990	1997
y	480	380	540	790

    Source: Department of Health and Human Services.

60. Sales y of CDs in millions during year x

x	1988	1990	1992	1994
y	150	290	410	660

    Source: Recording Industry Association of America.

61. Welfare beneficiaries y in millions during year x

x	1991	1993	1995	1997
y	12.6	14.1	13.7	10.9

    Source: Administration for Children and Families.

62. Medicaid recipients y in millions during year x

x	1975	1981	1990	1996
y	3.6	3.4	3.2	4.3

    Source: Health Care Financing Administration.

63. Projected Asian-American population y in millions during year x

x	1998	2000	2002	2004
y	10.5	11.2	12.0	12.8

    Source: Bureau of the Census.

64. Internet usage in millions of users y during year x

x	1989	1991	1993	1995
y	1.6	7.5	20.1	49.6

    Source: The Internet Society.

WRITING ABOUT MATHEMATICS

65. Explain how the domain and range of a relation can be used to determine an appropriate viewing rectangle for a scatterplot.

66. Explain the difference between a scatterplot and a line graph. Give an example of each.

CHECKING BASIC CONCEPTS FOR SECTION 1.5

1. State the domain and range of the relation $S = \{(-5, 3), (1, 4), (2, 3), (1, -1)\}$.

2. Plot the following points in the xy-plane. If possible, state the quadrant containing each point.
   (a) $(1, 4)$
   (b) $(0, -3)$
   (c) $(2, -2)$
   (d) $(-2, 3)$

3. The following table lists the number of people in millions living below the poverty level for selected years. Make a line graph of these data. Comment on any trends in the data.

Year	1960	1970	1980	1990
Number	40	25	29	34

   Source: Bureau of the Census.

Chapter 1 Summary

Section 1.1 *Describing Data with Sets of Numbers*

Some basic sets of numbers include natural numbers N, whole numbers W, and integers I. In set notation they can be written as

$$N = \{1, 2, 3, 4, 5, 6, \ldots\}, \quad W = \{0, 1, 2, 3, 4, 5, \ldots\}, \text{ and}$$
$$I = \{\ldots, -2, -1, 0, 1, 2, \ldots\}.$$

Rational numbers, irrational numbers, and real numbers are also sets of numbers. A rational number can be written as a fraction $\frac{p}{q}$, where p and q are integers and $q \neq 0$.

Rational numbers include natural numbers, whole numbers, and integers. Real numbers include all numbers that can be expressed in decimal form and comprise the rational numbers and irrational numbers. If a real number is not a rational number, it is an irrational number. Examples of irrational numbers are $\sqrt{2}$, $\sqrt{7}$, and π.

Important properties of real numbers include the identity, commutative, associative, and distributive properties. Let a, b, and c be real numbers.

Identity Properties
$$a + 0 = a \qquad a \cdot 1 = a$$

Commutative Properties
$$a + b = b + a \qquad a \cdot b = b \cdot a$$

Associative Properties
$$(a + b) + c = a + (b + c) \qquad (a \cdot b) \cdot c = a \cdot (b \cdot c)$$

Distributive Properties
$$a(b + c) = ab + ac \qquad a(b - c) = ab - ac$$

Section 1.2 *Operations on Real Numbers*

The number line is used to graph real numbers. The number 0 is located at the origin. The absolute value of a number equals its distance from the origin. For example $|-3| = 3$ because -3 is located 3 units from the origin.

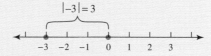

The four arithmetic operations on real numbers are addition, subtraction, multiplication, and division.

Section 1.3 *Integer Exponents*

Exponents occur frequently in formulas. In the expression 2^3, 2 is the base and 3 is the exponent. The expression 2^3 evaluates to 8 because $2^3 = 2 \cdot 2 \cdot 2$. Exponents may be positive, negative, or zero. Any nonzero base to the zero power equals 1, such as $5^0 = 1$. The following important properties may be used to simplify expressions, where a and b are nonzero real numbers and m and n are integers.

$$a^m \cdot a^n = a^{m+n}. \quad \text{Product rule}$$

$$\frac{a^m}{a^n} = a^{m-n}. \quad \text{Quotient rule}$$

$$(a^m)^n = a^{mn} \quad \text{Power rules}$$
$$(ab)^n = a^n b^n$$
$$\left(\frac{a}{b}\right)^n = \frac{a^n}{b^n}$$

Scientific notation is used to express numbers that are either large or small in absolute value. For example, 50,000 can be written as 5×10^4 and 0.005 can be written as 5×10^{-3}.

Section 1.4 *Modeling Data with Formulas*

Data may be modeled using variables and formulas. A formula can sometimes be used to summarize data efficiently. For example, the formula $A = 10t$ calculates the amount earned by a person working for t hours at $10 per hour. If $t = 8$, then the person earns $A = 10 \cdot 8 = \$80$ in 8 hours.

Section 1.5 *Visualization of Data*

Data can sometimes be represented by a relation. A relation S is a set of ordered pairs such as

$$S = \{(-2, 4), (-1, 6), (1.5, 6), (5, -9.1)\}.$$

If the ordered pairs in a relation are in the form (x, y), the set of all x-values is called the domain D and the set of all y-values is called the range R. In the preceding relation

$$D = \{-2, -1, 1.5, 5\} \quad \text{and} \quad R = \{-9.1, 4, 6\}.$$

Relations can be visualized by plotting points in the xy-plane, which results in a scatterplot. If the points are connected by line segments, it becomes a line graph. Graphing calculators are capable of creating both scatterplots and line graphs.

Chapter 1
Review Exercises

Section 1.1

Exercises 1 and 2: Classify each real number as one or more of the following: natural number, whole number, integer, rational number, or irrational number.

1. $-2, 9, \frac{2}{5}, \sqrt{11}, \pi, 2.68$

2. $\frac{6}{2}, -\frac{2}{7}, \sqrt{6}, 0.\overline{3}, \frac{0}{4}$

Exercises 3–6: State whether the equation illustrates an identity, commutative, associative, or distributive property.

3. $a \cdot 1 = a$
4. $4 \cdot x = x \cdot 4$
5. $(a + 1) + 4 = a + (1 + 4)$
6. $a(b + 2) = a \cdot b + a \cdot 2$

7. Use identity properties to simplify the expression $1 \cdot (a + 0)$.

8. Use a commutative property to write $x \cdot \frac{1}{4}$ as an equivalent expression.

9. Use an associative property to write $8(10x)$ as an equivalent expression.

10. Use a distributive property to write $a(10 + b)$ as an equivalent expression.

Exercises 11 and 12: Evaluate the expression two different ways by applying a distributive property.

11. $5(8 + 11)$
12. $3(9 - 5)$

Exercises 13 and 14: Calculate the average of the list of numbers.

13. $6, 9, 3, 11, 5, 20$
14. $3.2, 6.8, 6.1, 10.8, 1.7$

Section 1.2

15. Plot the numbers $-3, 0, 2,$ and $\frac{7}{2}$ on a number line.

16. Evaluate the expression $|-7.2 + 4|$

17. Find the additive inverse of $-\frac{2}{3}$.

18. Find the multiplicative inverse of $\frac{4}{5}$.

Exercises 19–22: Evaluate the expression without a calculator. Check any results about which you are uncertain with a calculator.

19. $-5 + (-7) + 8$
20. $-9 + 11$
21. $-12 - (-8)$
22. $\frac{1}{2} + (-2) + \frac{3}{4}$

Exercises 23–26: Evaluate the expression with a calculator.

23. $98 - (-45) + (-107)$

24. $\frac{1}{3} + \frac{2}{5} - \frac{5}{2}$
25. $\frac{2}{3} \div (-4) - \frac{1}{3}$

26. $-\frac{7}{11} + \frac{\frac{1}{5}}{\frac{2}{9}}$

Section 1.3

27. Identify the base and the exponent in the expression 4^{-2}.

28. Use a calculator to determine whether 3^π and π^3 are equal.

Exercises 29–34: Evaluate the expression by hand. Check your result with a calculator.

29. 5^2
30. 3^{-2}
31. -2^4
32. $(-2)^4$
33. 9^0
34. $\left(\frac{2}{3}\right)^{-3}$

Exercises 35–46: Simplify the expression. Write the result using positive exponents.

35. $4^3 \cdot 4^{-5}$

36. $10^4 \cdot 10^{-2}$

37. $x^7 \cdot x^{-2}$

38. $\dfrac{3^4}{3^{-7}}$

39. $\dfrac{5a^{-4}}{10a^2}$

40. $\dfrac{15a^4 b^3}{3a^2 b^6}$

41. $(2^2)^4$

42. $(x^{-3})^5$

43. $(4x^{-2}y^3)^2$

44. $(4a)^5$

45. $\left(\dfrac{5x^3}{3z^4}\right)^3$

46. $\left(\dfrac{-3x^4 y^3}{z}\right)^{-2}$

Exercises 47–52: Evaluate each expression by hand. Check your answer with a calculator.

47. $2 + 3 \cdot 9$

48. $4 - 1 - 6$

49. $5 \cdot 2^3$

50. $\dfrac{2+4}{2} + \dfrac{3-1}{3}$

51. $20 \div 4 \div 2$

52. $\dfrac{3^3 - 2^4}{4 - 3}$

Exercises 53 and 54: Write the number in scientific notation.

53. 186,000

54. 0.00034

Exercises 55 and 56: Write the number in standard form.

55. 4.5×10^4

56. 9.23×10^{-3}

Section 1.4

Exercises 57–60: Evaluate the formula for the given value of the variable.

57. $y = 12x$, $\quad x = 3$

58. $d = \sqrt{t - 3}$, $\quad t = 67$

59. $N = h^2 - \dfrac{3}{4}$, $\quad h = \dfrac{3}{2}$

60. $P = w^3 - 2$, $\quad w = -2$

61. Select the formula that best models the data in the table.

x	1	2	3	4	5
y	-1	1	3	5	7

(i) $y = x - 2$, (ii) $y = 3x - 4$, (iii) $y = 2x - 3$

62. Find a value for a so that $y = ax$ models the data.

x	-2	0	2	4	6
y	-3	0	3	6	9

Section 1.5

63. Identify the domain and range of the relation $S = \{(-1, 1), (2, 3), (3, -6), (3, 7)\}$.

64. Express the relation shown in the graph as a set of ordered pairs.

$[-12, 12, 4]$ by $[-8, 8, 4]$

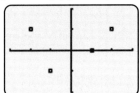

Exercises 65 and 66: Plot the points in the table. If possible, state the quadrant containing each point.

65.

x	-2	-1	0	2
y	2	-3	1	-1

66.

x	-15	-5	10	20
y	-5	0	20	-10

Exercises 67 and 68: Show the viewing rectangle on your graphing calculator. Predict the number of tick marks on the positive x-axis and the positive y-axis.

67. $[-9, 9, 1]$ by $[-6, 6, 3]$

68. $[-20, 20, 5]$ by $[-12, 12, 4]$

Exercises 69 and 70: Use your calculator to make a scatterplot of the relation after determining an appropriate viewing rectangle.

69. $\{(0, 4), (-1, 2), (3, 3), (4, -1), (2, 0)\}$

70. $\{(-10, 10), (50, 20), (-20, -30), (45, -25)\}$

APPLICATIONS

71. *Mayors' Salaries* The following table lists the four highest mayors' salaries in 1997.

City	Chicago	New York
Salary	$170,000	$165,000

City	Newark	Detroit
Salary	$147,000	$143,000

Source: U.S. Conference of Mayors.

(a) Mentally estimate which of the following represents the average salary: $141,500, $156,250, or $168,750.
(b) Check your estimate by calculating the actual average with a calculator.

72. *Calculating Interest* If P dollars is deposited in a savings account paying 8% annual interest, then the amount A in the account after t years is given by the formula $A = P(1.08)^t$. Find A for the given values of P and t.
(a) $P = \$2500$, $t = 10$ years
(b) $P = \$800$, $t = 7$ years

73. *Speed of Earth* Earth orbits the sun in a nearly circular orbit with a radius of 93,000,000 miles.
(a) Calculate the distance in miles traveled by Earth in 1 year. Write your answer in scientific notation. (*Hint:* The circumference of a circle is given by $C = 2\pi r$.)
(b) Determine the speed of Earth around the sun in miles per hour.

74. *Modeling Motion* The following table lists the distance in miles traveled by a car for various elapsed times. Find an equation that models these data.

Elapsed Time (hr)	2	4	6	8
Distance (mi)	80	160	240	320

75. *Heart Beat in Animals* The rate N at which an animal's heart beats varies with its weight. This relation can be modeled by the equation $N = \frac{885}{\sqrt{W}}$, where N is in beats per minute and W is the animal's weight in pounds. Estimate the pulse for a 10-pound cat and a 160-pound person. **(Source: C. Pennycuick, *Newton Rules Biology*.)**

76. *Graphing Real Data* The following data show the poverty threshold y for a single person from 1960 through 1997. Make a line graph and comment on any trends in the data.

x	1960	1970	1980	1990	1997
y	$1490	$1954	$4190	$6652	$8178

Source: Bureau of the Census.

CHAPTER 1 TEST

1. Classify each real number as one or more of the following: natural number, whole number, integer, rational number, or irrational number.

    $-5, \dfrac{2}{3}, -\dfrac{1}{\sqrt{5}}, \sqrt{9}, \pi, -1.83$

2. State whether each equation illustrates an identity, commutative, associative, or distributive property.
    (a) $a + 0 = a$
    (b) $4(12x) = 48x$
    (c) $5(2 + 3x) = 5(3x + 2)$
    (d) $a(x - y) = ax - ay$

3. Calculate the average of the list of numbers: 34, 15, 96, 11, 0

4. Plot the numbers $-1.5, 0, 3$, and $\dfrac{3}{2}$ on a number line.

5. Evaluate the expression $\left| \dfrac{1}{2} + \dfrac{2}{3} - \dfrac{8}{3} \right|$

6. Find the multiplicative inverse of $-\dfrac{5}{4}$.

7. Evaluate each expression without a calculator.
    (a) $-\dfrac{1}{2} + \dfrac{2}{3} \div 3$
    (b) $-4 + \dfrac{\frac{2}{3}}{-\frac{1}{4}}$
    (c) $5 - 2 \cdot 5^2 \div 5$

8. Evaluate each expression without a calculator.
    (a) 5^{-2}
    (b) π^0
    (c) $\left(-\dfrac{2}{5} \right)^4$

9. Simplify each expression. Use positive exponents to write the result.
    (a) $x^6 \cdot x^{-4} \cdot y^3$
    (b) $\dfrac{16x^{-2}y^8}{6xy^{-7}}$
    (c) $(2yz^{-2})^3$
    (d) $\left(\dfrac{15x^4}{10xy^{-2}} \right)^{-2}$

10. Write 5.2×10^{-4} in standard form.

11. Express the relation shown in the graph as a set of ordered pairs. Identify the domain and range.
    $[-500, 500, 100]$ by $[-300, 300, 100]$

    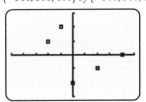

12. Make a scatterplot of the data. Write an equation that models the data.

x	1	2	3	4
y	1.25	2.50	3.75	5.00

13. Use your calculator to make a scatterplot of the relation S after determining an appropriate viewing rectangle.

    $S = \{(-15, 5), (10, 20), (-10, 15), (0, 0), (5, 35)\}$

14. *Temperature* The formula $C = \dfrac{5}{9}(F - 32)$ can be used to convert degrees Fahrenheit, F, to degrees Celsius, C. If the outside temperature is 5°F, find the equivalent temperature in Celsius.

15. *Alcohol Consumption* In 1994, about 211 million people in the United States were age 14 or over. They consumed, on average, 2.21 gallons of alcohol per person. Use scientific notation to estimate the total gallons of alcohol consumed by this age group. (*Source: Department of Health and Human Services.*)

16. *Radius of a Circle* If a circle has an area of A square units, its radius r is given by $r = \sqrt{\dfrac{A}{\pi}}$. Find the radius of a circle with an area of 25 square feet. Approximate this radius to the nearest hundredth of a foot.

Chapter 1
Extended and Discovery Exercises

1. *Global Warming* If the global climate were to warm significantly as a result of the greenhouse effect or other climatic change, the Arctic ice cap would start to melt. This ice cap contains an estimated 680,000 cubic miles of water. More than 200 million people currently live on land that is less than 3 feet above sea level. In the United States several large cities have low average elevations. Three examples are Boston (14 feet), New Orleans (4 feet), and San Diego (13 feet). In this exercise you are to estimate the rise in sea level if this cap were to melt and determine whether this event would have a significant impact on people.
   (a) The surface area of a sphere is given by the formula $4\pi r^2$, where r is its radius. Although the shape of the earth is not exactly spherical, it has an average radius of 3960 miles. Estimate the surface area of the earth.
   (b) Oceans cover approximately 71% of the total surface area of the earth. How many square miles of the earth's surface are covered by oceans?
   (c) Approximate the potential rise in sea level by dividing the total volume of the water from the ice cap by the surface area of the oceans. Convert your answer from miles to feet.
   (d) Discuss the implications of your calculation. How would cities such as Boston, New Orleans, and San Diego be affected?
   (e) The Antarctic ice cap contains 6,300,000 cubic miles of water. Estimate how much sea level would rise if this ice cap melted. (*Source:* **Department of the Interior, Geological Survey.**)

2. *China's Gross Domestic Product* The following table shows the gross domestic product (GDP) for China in selected years. (For comparison purposes the GDP for the United States in 1998 was $8.5 trillion.)

Year	1977	1987	1997	1998
GDP ($ billions)	172	268	902	961

Sources: World Bank; Department of Commerce.

   (a) Make a scatterplot of the data. Discuss how the GDP has changed over this time period.
   (b) Estimate China's GDP in 1982. Explain your reasoning.
   (c) Assuming that trends continue, estimate China's GDP in 2007. Explain your reasoning.

3. *Body Mass Index* Many studies have tried to find a recommended relationship between a person's height and weight. The following steps may be used to compute the body mass index (BMI). Federal guidelines suggest that $19 \leq BMI \leq 25$ is desirable. (*Source:* **Associated Press.**)
   Step 1: Multiply a person's weight W in pounds by 0.455.
   Step 2: Multiply a person's height H in inches by 0.0254.
   Step 3: Square the result in Step 2.
   Step 4: Divide the answer in Step 1 by the answer in Step 3.
   The result is the person's BMI.
   Write a formula to calculate the BMI given the height H and weight W of a person. Let y represent the BMI.

Exercises 4–6: Body Mass Index (*Refer to the previous exercise.*) Compute the BMI for each individual.

4. 119 pounds, 5 feet 9 inches (Steffi Graf, tennis player) (*Source:* **Monroe, J.,** *Steffi Graf*).

5. 153 pounds, 5 feet 10 inches (Jackie Joyner-Kersee, track and field athlete) (*Source:* **Goldstein, M. and J. Larson,** *Jackie Joyner-Kersee Superwoman.*)

6. 300 pounds, 7 feet 1 inch (Shaquille O'Neal, professional basketball player) (*Source:* **The Topps Company, Inc.**)

CHAPTER 2
LINEAR FUNCTIONS AND MODELS

Mathematics is a unique subject. Although mathematics is not dependent on people making observations about the real world, it is frequently used to describe the real world. Mathematics has become an invaluable tool for modeling phenomena occurring in everyday life. For example, sunbathing is a popular pastime. Ultraviolet light from the sun is responsible for both tanning and burning exposed skin. With mathematics, we can use numbers to describe the intensity of ultraviolet light. The following table shows the maximum ultraviolet intensity measured in milliwatts per square meter for various latitudes and dates.

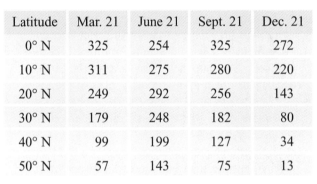

Latitude	Mar. 21	June 21	Sept. 21	Dec. 21
0° N	325	254	325	272
10° N	311	275	280	220
20° N	249	292	256	143
30° N	179	248	182	80
40° N	99	199	127	34
50° N	57	143	75	13

If a student from Chicago, Illinois, located at a latitude of 42° N, spends spring break in Hawaii at a latitude of 20° N, the sun's ultraviolet rays in Hawaii will be approximately $\frac{249}{99}$, or 2.5, times more intense than in Chicago. Suppose that you travel to the equator for spring break. How much more intense will the sun be than where you presently live?

Education is not the filling of a pail, but the lighting of a fire.
—William Butler Yeats

Source: J. Williams, *The USA Today Weather Almanac 1995.*

2.1 FUNCTIONS AND THEIR REPRESENTATIONS

Basic Concepts ~ Representations of a Function ~
Definition of a Function ~ Identifying a Function ~
Tables, Graphs, and Graphing Calculators

INTRODUCTION

In Chapter 1 we learned how to use numbers to describe data. For example, instead of simply saying that it is *hot* outside, we might use the number 102°F to describe the temperature. We also learned that data can be modeled with formulas and graphs. Formulas and graphs are sometimes used to represent functions, which are important in mathematics. In this section we introduce functions and their representations.

BASIC CONCEPTS

Functions are used to calculate many important quantities. For example, suppose that a person works for $7 per hour. Then we could use a function f to calculate the amount of money someone earned after working x hours simply by multiplying the **input** x by 7. The result y is called the **output**. This concept is shown visually in the following diagram.

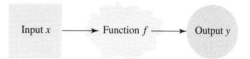

For each valid input x, a function computes *exactly one* output y, which may be represented by the ordered pair (x, y). If the input is 5 hours, f outputs $7 \cdot 5 = \$35$; if the input is 8 hours, f outputs $7 \cdot 8 = \$56$. These results can be represented by the ordered pairs (5, 35) and (8, 56). Sometimes an input may not be valid. For example, if $x = -3$, there is no reasonable output because a person cannot work -3 hours.

We say that *y is a function of x* because the output y is determined by and depends on the input x. As a result, y is called the **dependent variable** and x is the **independent variable.** To emphasize that y is a function of x, we use the notation $y = f(x)$. The symbol $f(x)$ does not represent multiplication of a variable f and a variable x. The notation $y = f(x)$ is called **function notation,** is read "y equals f of x," and means that function f with input x produces output y. For example, if $x = 3$ hours, $y = f(3) = \$21$.

REPRESENTATIONS OF A FUNCTION

A function f forms a relation between inputs x and outputs y that can be represented verbally, numerically, symbolically, and graphically. Functions can also be represented with diagrams. We begin by considering a function f that converts yards to feet.

VERBAL REPRESENTATION (WORDS). To convert x yards to y feet we must multiply x by 3. Therefore, if function f computes the number of feet in x yards, a **verbal representation** of f is given by "Multiply the input x in yards by 3 to obtain the output y in feet."

NUMERICAL REPRESENTATION (TABLE OF VALUES). A function f that converts yards to feet is shown in Table 2.1, where $y = f(x)$.

TABLE 2.1

x (yards)	1	2	3	4	5	6	7
y (feet)	3	6	9	12	15	18	21

A *table of values* is called a **numerical representation** of a function. Many times it is impossible to list all possible inputs x in a table. On the one hand, if a table does not contain every x-input, it is a *partial* numerical representation. On the other hand, a *complete* numerical representation includes all possible inputs. Table 2.1 is a partial numerical representation of f because many valid inputs, such as $x = 10$ or $x = 5.3$, are not shown in it.

SYMBOLIC REPRESENTATION (FORMULA). A *formula* provides a **symbolic representation** of a function. The computation performed by f to convert x yards to y feet is expressed by $y = 3x$. A formula for f is $f(x) = 3x$, where $y = f(x)$. We say that function f is *represented by or given by* $f(x) = 3x$.

GRAPHICAL REPRESENTATION (GRAPH). A **graphical representation**, or **graph**, visually associates an x-input with a y-output. The ordered pairs

$$(1, 3), (2, 6), (3, 9), (4, 12), (5, 15), (6, 18), \text{ and } (7, 21)$$

from Table 2.1 are plotted in Figure 2.1. This scatterplot suggests a line for the graph f. If we restrict inputs to $x \geq 0$ and plot all ordered pairs (x, y) satisfying $y = 3x$, a line with no breaks appears and represents a graph of f, as shown in Figure 2.2.

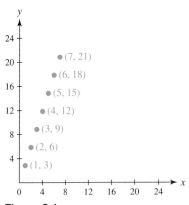
Figure 2.1

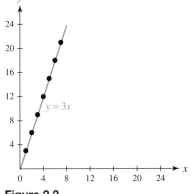
Figure 2.2

DIAGRAMMATIC REPRESENTATION (DIAGRAM). Functions may be represented by **diagrams.** Figure 2.3 is a diagram of a function, where an arrow is used to identify the output y associated with input x. For example, input 2 results in output 6, which is written in function notation as $f(2) = 6$. That is, 2 yards are equivalent to 6 feet. In contrast, Figure 2.4 shows a relation that is not a function because input 2 results in two different outputs, 5 and 6. Remember, a function produces exactly one output for each valid input.

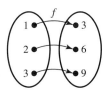
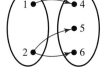

Figure 2.3 Function   **Figure 2.4** Not a Function

EXAMPLE 1 *Calculating sales tax*

Let a function f compute a sales tax of 7% on a purchase of x dollars. Use the given representation to evaluate $f(2)$.

(a) *Verbal Representation*   Multiply a purchase of x dollars by 0.07 to obtain a sales tax of y dollars.

(b) *Numerical Representation*   Shown in Table 2.2

TABLE 2.2

x	$1.00	$2.00	$3.00	$4.00
$f(x)$	$0.07	$0.14	$0.21	$0.28

(c) *Symbolic Representation* $f(x) = 0.07x$

(d) *Graphical Representation*   Shown in Figure 2.5

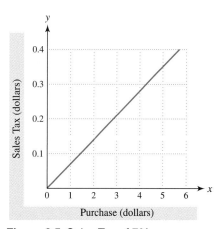

Figure 2.5 Sales Tax of 7%.

(e) *Diagrammatic Representation*   Shown in Figure 2.6

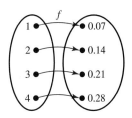

Figure 2.6

Solution

(a) Multiply the input 2 by 0.07 to obtain 0.14. The sales tax on $2.00 is $0.14.
(b) From Table 2.2, $f(2) = \$0.14$.
(c) Because $f(x) = 0.07x, f(2) = 0.07(2) = 0.14$, or $0.14.
(d) To evaluate $f(2)$ with a graph, first find 2 on the x-axis. Then move vertically upward until you reach the graph of f. The point on the graph may be estimated as $(2, 0.14)$, meaning that $f(2) = 0.14$ (see Figure 2.7).

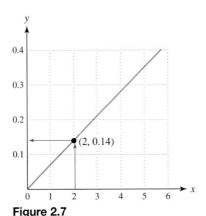

Figure 2.7

(e) In Figure 2.6, follow the arrow from $x = 2$ to $y = 0.14$. Thus $f(2) = 0.14$.

EXAMPLE 2 Evaluating symbolic representations

Evaluate each function f at the given value of x.

(a) $f(x) = 3x - 7$ $x = -2$

(b) $f(x) = \dfrac{x}{x + 2}$ $x = 0.5$

(c) $f(x) = \sqrt{x - 1}$ $x = 10$

Solution

(a) $f(-2) = 3(-2) - 7 = -6 - 7 = -13$

(b) $f(0.5) = \dfrac{0.5}{0.5 + 2} = 0.2$

(c) $f(10) = \sqrt{10 - 1} = \sqrt{9} = 3$

DEFINITION OF A FUNCTION

A function is a fundamental concept in mathematics. Its definition should allow for all representations of a function. A function receives an input x and produces exactly one output y, which can be expressed as an ordered pair,

$$(x, y).$$
     Input   Output

A relation is a set of ordered pairs, and a function is a special type of relation.

2.1 Functions and Their Representations

Solution

The input x is the month and the output y is the monthly average temperature. The set S is a function because each month x is paired with exactly one monthly average temperature y. Note that, even though an average temperature of 37°F occurs in both February and December, S is nonetheless a function.

EXAMPLE 6 *Determining whether a table of values represents a function*

Determine whether Table 2.3 represents a function.

TABLE 2.3

x	1	2	3	1	4
y	−4	8	2	5	−6

Solution

The table does not represent a function because input $x = 1$ produces two outputs: −4 and 5.

VERTICAL LINE TEST. To decide whether a graph represents a function, we must be convinced that it is impossible for two distinct points with the same x-coordinate to lie on the graph. For example, the ordered pairs (4, 2) and (4, −2) are distinct points with the same x-coordinate. These two points could not lie on the graph of the same function because input 4 would result in *two* outputs, ±2. When the points (4, 2) and (4, −2) are plotted, they lie on the same vertical line, as shown in Figure 2.11. A graph passing through these points intersects the vertical line twice, as illustrated in Figure 2.12.

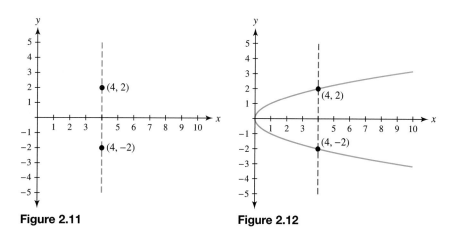

Figure 2.11   **Figure 2.12**

To determine whether a graph represents a function, simply visualize vertical lines in the xy-plane. If each vertical line intersects a graph at most once, it is a graph of a function. This test is called the **vertical line test** for a function. The graph shown in Figure 2.12 fails the vertical line test.

EXAMPLE 7 Determining whether a graph represents a function

Determine whether the graphs shown in Figures 2.13 and 2.14 represent functions.

(a)
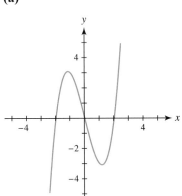

Figure 2.13

(b)
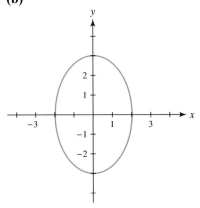

Figure 2.14

Solution

(a) Any vertical line will cross the graph at most once, as depicted in Figure 2.15. Therefore the graph does represent a function.

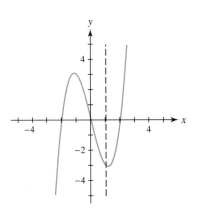

Figure 2.15 Passes Vertical Line Test

Figure 2.16 Fails Vertical Line Test

(b) The graph does not represent a function because some vertical lines can intersect the graph twice, as shown in Figure 2.16.

TABLES, GRAPHS, AND GRAPHING CALCULATORS

We can use graphing calculators to create graphs and tables, usually more efficiently and reliably than pencil-and-paper techniques. However, a graphing calculator uses the same techniques that we might use to sketch a graph. For example, one way to sketch a graph of $y = 2x - 1$ is first to make a table of values, as shown in Table 2.4.

TABLE 2.4

x	0	1	2	3
y	−1	1	3	5

We can plot these points in the *xy*-plane, as shown in Figure 2.17. Next we might connect the points, as shown in Figure 2.18.

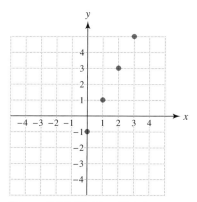

Figure 2.17 Plotting Points

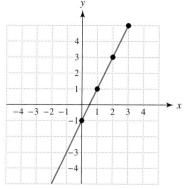

Figure 2.18 Graphing a Line

In a similar manner a graphing calculator plots numerous points and connects them to make a graph. To create a similar graph with a graphing calculator, we enter the formula $Y_1 = 2X - 1$, set an appropriate viewing rectangle, and graph (see Figures 2.19–2.21).

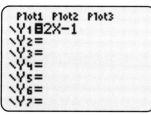

Figure 2.19

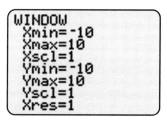

Figure 2.20

$[-10, 10, 1]$ by $[-10, 10, 1]$

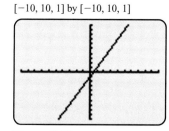

Figure 2.21

We can also use a graphing calculator to create a table of values, as illustrated in Figures 2.22 and 2.23.

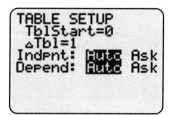

Figure 2.22

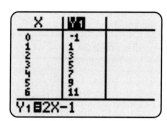

Figure 2.23

EXAMPLE 8　Using a graphing calculator

People who sustain leg injuries often require crutches. A proper crutch length can be estimated without using trial and error. The function f given by $f(x) = 0.72x + 2$ can compute an appropriate crutch length in inches for a person x inches tall. (*Source: Journal of the American Physical Therapy Association.*)

(a) Graph f in the viewing rectangle [60, 72, 2] by [40, 60, 5]. Interpret the graph.
(b) Construct the table for f starting at $x = 60$ and incrementing by 2. What is the proper crutch length for a person 6 feet tall?

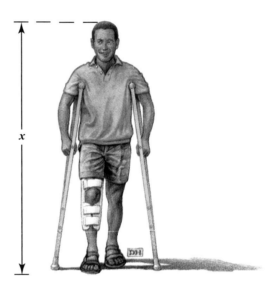

Solution

(a) To graph f let $Y_1 = .72X + 2$ and set the viewing rectangle (see Figures 2.24 and 2.25). As the x-values increase so do the y-values, indicating that taller people need longer crutches.

[60, 72, 2] by [40, 60, 5]

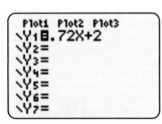

Figure 2.24

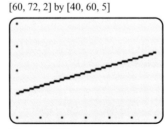

Figure 2.25

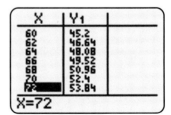

Figure 2.26

(b) The required table is shown in Figure 2.26. A 6-foot, or 72-inch, tall person needs a crutch that is about $53.84 \approx 54$ inches long.

Making Connections

Relations and Functions

A relation can be thought of as a set of input–output pairs. A function is a special type of relation whereby each input results in exactly one output.

2.1 Putting It All Together

One of the most important concepts in mathematics is that of a function. A function produces exactly one output for each valid input. The domain of a function f consists of the set of valid inputs, and the range is the set of corresponding outputs. The following table summarizes the different representations of a function.

Type of Representation	Explanation	Comments
Verbal	Precise word description of what is computed	May be oral or written Must be stated *precisely*
Symbolic	Mathematical formula	Efficient and concise way of representing a function (e.g., $f(x) = 2x - 3$)
Numerical	List of specific inputs and their outputs	May be in the form of a table or an explicit set of ordered pairs
Graphical, diagrammatic	Shows inputs and outputs visually	No words, formulas, or tables Many types of graphs and diagrams are possible.

2.1 EXERCISES

CONCEPTS

1. The set of valid inputs for a function is called the _____.

2. The set of outputs for a function is called the _____.

3. A function computes _____ output for each valid input.

4. What is the vertical line test used for?

5. Name four types of representations for a function.

6. Explain what the notation $f(x)$ means.

Exercises 7–10: Determine whether the phrase describes a function.

7. Calculating the square of a number

8. Determining your age to the nearest whole number

84   CHAPTER 2   Linear Functions and Models

9. Outputting the students who passed a given math exam

10. Outputting the children of parent x

REPRESENTING AND EVALUATING FUNCTIONS

Exercises 11–20: Evaluate the symbolic representation $f(x)$ at the given values of x.

11. $f(x) = 4x - 2$ $x = -1, 0$

12. $f(x) = 5 - 3x$ $x = -4, 2$

13. $f(x) = \sqrt{x}$ $x = 0, \dfrac{9}{4}$

14. $f(x) = \sqrt[3]{x}$ $x = -1, 27$

15. $f(x) = x^2$ $x = -5, \dfrac{3}{2}$

16. $f(x) = x^3$ $x = -2, 0.1$

17. $f(x) = 3$ $x = -8, \dfrac{7}{3}$

18. $f(x) = x^2 + 5$ $x = -\dfrac{1}{2}, 6$

19. $f(x) = \dfrac{2}{x+1}$ $x = -5, 4$

20. $f(x) = \dfrac{x}{x-4}$ $x = -3, 1$

Exercises 21–28: Sketch a graph of f by hand. Support your result with a graphing calculator.

21. $f(x) = -x + 3$ 22. $f(x) = -2x + 1$

23. $f(x) = 2x$ 24. $f(x) = \dfrac{1}{2}x - 2$

25. $f(x) = 4 - x$ 26. $f(x) = 6 - 3x$

27. $f(x) = x^2$ 28. $f(x) = \sqrt{x}$

Exercises 29–34: Use the graph of f to evaluate the given expressions.

29. $f(0)$ and $f(2)$.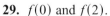

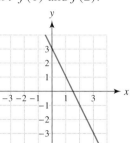

30. $f(-2)$ and $f(2)$.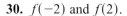

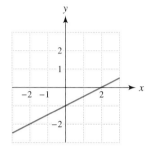

31. $f(-1)$ and $f(0)$.   32. $f(-2)$ and $f(1)$.

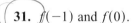

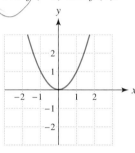 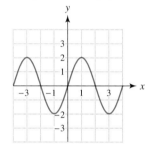

33. $f(1)$ and $f(2)$.   34. $f(-1)$ and $f(4)$.

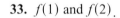

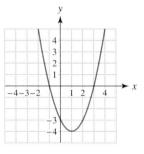 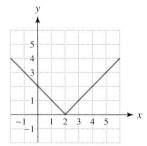

Exercises 35 and 36: Use the table to evaluate the given expression.

35. $f(0)$ and $f(2)$.

x	0	1	2	3	4
$f(x)$	5.5	4.3	3.7	2.5	1.9

36. $f(-10)$ and $f(5)$.

x	-10	-5	0	5	10
$f(x)$	23	96	-45	-33	23

Exercises 37 and 38: Use the diagram to evaluate f(1990). Interpret your answer.

37. The function f computes average fuel efficiency of new U.S. passenger cars in miles per gallon during year x. **(Source: Department of Transportation.)**

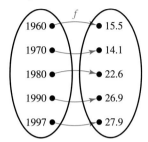

38. The function f computes average cost of tuition at public colleges and universities during academic year x. **(Source: The College Board.)**

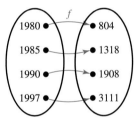

Exercises 39–42: Express the verbal representation for the function f numerically, symbolically, and graphically. For the numerical representation let $x = -3, -2, -1, \ldots, 3$, and for the graphical representation let $-3 \leq x \leq 3$.

39. Add 5 to the input x to obtain the output y.

40. Square the input x to obtain the output y.

41. Multiply the input x by 5 and then subtract 2 to obtain the output y.

42. Divide the input x by 2 and then add 3 to obtain the output y.

Exercises 43–46: Converting Units Express the verbal representation for the function f symbolically and graphically. Assume that $y = f(x)$ and let $0 \leq x \leq 10$ when graphing.

43. To convert x gallons to y liters, multiply x by 3.785.

44. To convert x pounds to y kilograms, divide x by 2.205.

45. To convert x kilometers to y miles, divide x by 1.609.

46. To convert x acres to y square feet, multiply x by 43,560.

Exercises 47–50: Give a verbal representation for $f(x)$.

47. $f(x) = x - \dfrac{1}{2}$

48. $f(x) = \dfrac{3}{4}x$

49. $f(x) = \dfrac{x}{3}$

50. $f(x) = x^2 + 1$

51. *Cost of Driving* In 1997, the average cost of driving a new car in the United States was 41 cents per mile. Symbolically, graphically, and numerically represent a function f that computes the cost in dollars of driving x miles. For the numerical representation let $x = 10, 20, 30, \ldots, 70$. **(Source: Associated Press.)**

52. *Federal Income Taxes* In 1999, the lowest U.S. income tax rate was 15 percent. Symbolically, graphically, and numerically represent a function f that computes the tax on a taxable income of x dollars. For the numerical representation let $x = 1000, 2000, 3000, \ldots, 7000$, and for the graphical representation let $0 \leq x \leq 30,000$. **(Source: Internal Revenue Service.)**

Exercises 53–56: Graph f in $[-4.7, 4.7, 1]$ by $[-3.1, 3.1, 1]$.
(a) Use the trace feature on your calculator to estimate $f(-1)$.
(b) Evaluate $f(-1)$ symbolically.

53. $f(x) = 3x + 1$

54. $f(x) = 1 - x$

55. $f(x) = 0.5x^2$

56. $f(x) = \dfrac{5}{2}$

IDENTIFYING DOMAINS AND RANGES

Exercises 57–64: Use the graph of f to estimate its domain and range.

57.
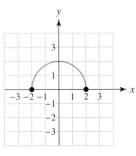

58.
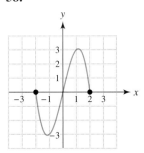

59.
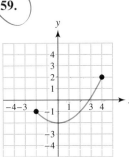

60.
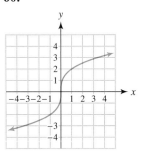

61.
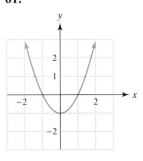

62.
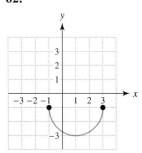

63.
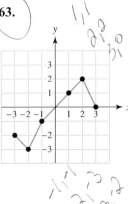

64.
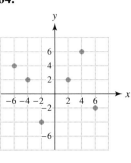

Exercises 65 and 66: Use the diagram to find the domain and range of f.

65.
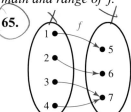

66.

12 → 26
13 → 32
15 → 9

67. *Accidental Deaths* The function f computes the number of accidental deaths y per 100,000 people during year x. **(Source: Department of Health and Human Services.)**

$f = \{(1910, 84.4), (1930, 80.5), (1950, 60.3),$
$(1970, 56.2), (1990, 36.9)\}$

(a) Evaluate $f(1950)$ and interpret the result.
(b) Identify the domain and range of f.
(c) Describe the trend in accidental deaths from 1910 through 1990.

68. *Motor Vehicle Registrations* The following table lists motor vehicle registrations y in millions during year x. Let $y = f(x)$.

x	1910	1930	1950	1970	1990
$f(x)$	0.5	26.7	49.1	108.4	188.7

Source: American Automobile Manufacturers Association.

(a) Evaluate $f(1930)$ and interpret the result.
(b) Identify the domain and range of f.
(c) Represent f with a diagram.

IDENTIFYING A FUNCTION

69. *Average Precipitation* The following table lists the monthly average precipitation P in Las Vegas, Nevada, where $x = 1$ corresponds to January and $x = 12$ corresponds to December.

x (month)	1	2	3	4	5	6
P (inches)	0.5	0.4	0.4	0.2	0.2	0.1

x (month)	7	8	9	10	11	12
P (inches)	0.4	0.5	0.3	0.2	0.4	0.3

Source: J. Williams.

(a) Determine the value of P during May.
(b) Is P a function of x? Explain.
(c) If $P = 0.4$, find x.

70. *Wind Speeds* The following table lists the monthly average wind speed W in Louisville, Kentucky, where $x = 1$ corresponds to January and $x = 12$ corresponds to December.

x (month)	1	2	3	4	5	6
W (miles per hour)	10.4	12.7	10.4	10.4	8.1	8.1

x (month)	7	8	9	10	11	12
W (miles per hour)	6.9	6.9	6.9	8.1	9.2	9.2

Source: J. Williams.

(a) Determine the month with the highest average wind speed.
(b) Is W a function of x? Explain.
(c) If $W = 6.9$, find x.

Exercises 71–74: Determine whether the diagram could represent a function.

71. **72.**

73. **74.**

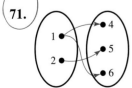
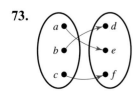

Exercises 75–84: Determine whether the graph represents a function.

75. **76.**

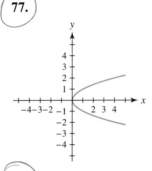
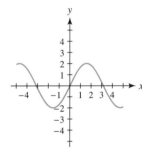

77. **78.**

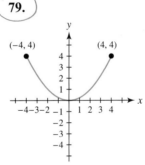
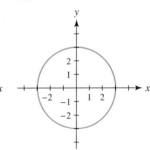

79. **80.**

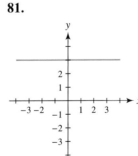
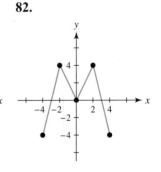

81. **82.**

83.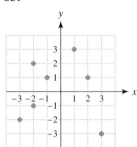

84.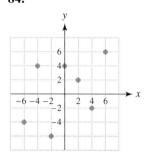

87. S is given by the following table.

x	5	10	5
y	2	1	0

88. S is given by the following table.

x	−3	−2	−1
y	10	10	10

Exercises 85–88: Determine whether S defines a function.

85. $S = \{(1, 2), (4, 5), (7, 8), (5, 4), (2, 2)\}$

86. $S = \{(4, 7), (-2, 1), (3, 8), (4, 9)\}$

WRITING ABOUT MATHEMATICS

89. Give an example of a function. Identify the domain and range of your function. Symbolically, graphically, and numerically represent your function.

90. Explain in your own words what a function is. How is a function different from a relation?

2.2 LINEAR FUNCTIONS

Basic Concepts ~ Representations of Linear Functions ~ Modeling Data with Linear Functions

INTRODUCTION

In applied mathematics people in various fields frequently use functions to model real-world phenomena, such as electricity, weather, and the economy. Because there are so many different applications of mathematics, people have created a wide assortment of functions. In fact, new functions are invented every day for use in business, education, and government. In this section we discuss an important type of function called a *linear function*.

BASIC CONCEPTS

Suppose that a car is initially located 100 miles north of the Texas border, traveling north on Interstate 35 at 50 miles per hour. Distances between the automobile and the border are listed in Table 2.5 for various times.

TABLE 2.5

Elapsed Time (hours)	0	1	2	3	4	5
Distance (miles)	100	150	200	250	300	350

The car is moving at a constant speed, so the distance increases by 50 miles every hour. The scatterplot shown in Figure 2.27 suggests that a line might model these data.

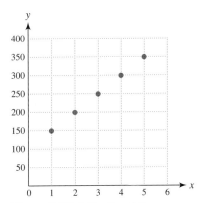

Figure 2.27 A Scatterplot

If the car travels for x hours, we find the distance from the border by multiplying x by 50 and then adding the initial distance of 100 miles. This calculation is modeled by $f(x) = 50x + 100$; for example, if $x = 1.5$,

$$f(1.5) = 50(1.5) + 100 = 175$$

means that the car is 175 miles from the Texas border after 1.5 hours. A graph of $f(x) = 50x + 100$ is shown in Figure 2.28. We call f a linear function because its graph is a line.

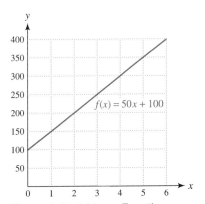

Figure 2.28 A Linear Function

Linear Function

A function f represented by $f(x) = ax + b$, where a and b are constants, is a **linear function**.

For $f(x) = 50x + 100$, we have $a = 50$ and $b = 100$. The value of a represents the speed of the car, and b corresponds to the initial distance of the car from the border.

EXAMPLE 1 Identifying linear functions

Determine whether f is a linear function. If it is, find values for a and b so that $f(x) = ax + b$.

(a) $f(x) = 4 - 3x$ **(b)** $f(x) = 8$ **(c)** $f(x) = 2x^2 + 8$

Solution

(a) Let $a = -3$ and $b = 4$. Then $f(x) = -3x + 4$, and f is a linear function.
(b) Let $a = 0$ and $b = 8$. Then $f(x) = 0x + 8$, and f is a linear function.
(c) Function f is not linear because its formula contains x^2.

REPRESENTATIONS OF LINEAR FUNCTIONS

The graph of a linear function is a line. To graph a linear function f we usually start by making a table of values and then plot two or more points. We can then sketch the graph of f by drawing a line through these points, as demonstrated in the next example.

EXAMPLE 2 Graphing a linear function by hand

Sketch a graph of $f(x) = x - 1$. Support your result by using a graphing calculator.

Solution

Begin by making a table of values. Pick convenient values of x, such as $x = -1, 0, 1$.

$$f(-1) = -1 - 1 = -2$$
$$f(0) = 0 - 1 = -1$$
$$f(1) = 1 - 1 = 0$$

Display the results, as shown in Table 2.6.

TABLE 2.6

x	-1	0	1
y	-2	-1	0

Plot the points $(-1, -2)$, $(0, -1)$, and $(1, 0)$. Then sketch a line through the points to obtain the graph of f. A graph of a line results when infinitely many points are plotted, as in Figure 2.29. This result is supported by Figure 2.30, where $Y_1 = X - 1$ has been graphed.

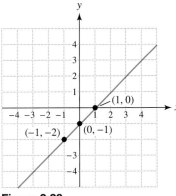

Figure 2.29

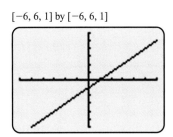

Figure 2.30

Critical Thinking

Two points determine a line. Why is it a good idea to plot at least three points when graphing a linear function by hand?

EXAMPLE 3 *Representing a linear function*

A linear function is given by $f(x) = -3x + 2$.

(a) Give a verbal representation of f.
(b) Make a numerical representation of f by letting $x = -1, 0, 1$.
(c) Plot the points in the table from part (b). Then sketch a graph of f.

Solution

(a) *Verbal Representation* Multiply the input x by -3 and then add 2 to obtain the output.
(b) *Numerical Representation* Evaluate $f(x) = -3x + 2$ at $x = -1, 0, 1$, which results in Table 2.7.

TABLE 2.7

x	−1	0	1
$f(x)$	5	2	−1

(c) *Graphical Representation* To make a graphical representation of f by hand, plot the points $(-1, 5)$, $(0, 2)$, and $(1, -1)$ from Table 2.7. Then draw a line passing through these points, as shown in Figure 2.31 on the following page.

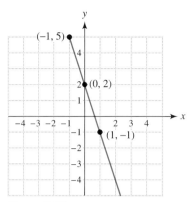

Figure 2.31

In the next example a graphing calculator is used to create graphical and numerical representations of a linear function.

EXAMPLE 4 *Using a graphing calculator*

Give numerical and graphical representations of $f(x) = \frac{1}{2}x - 2$.

Solution

Numerical Representation To make a numerical representation, construct the table for $Y_1 = .5X - 2$, starting at $x = -3$ and incrementing by 1, as shown in Figure 2.32. (Other tables are possible.)

Graphical Representation Let $Y_1 = .5X - 2$ and graph Y_1 in the standard viewing rectangle, as shown in Figure 2.33. (Other viewing rectangles may be used.)

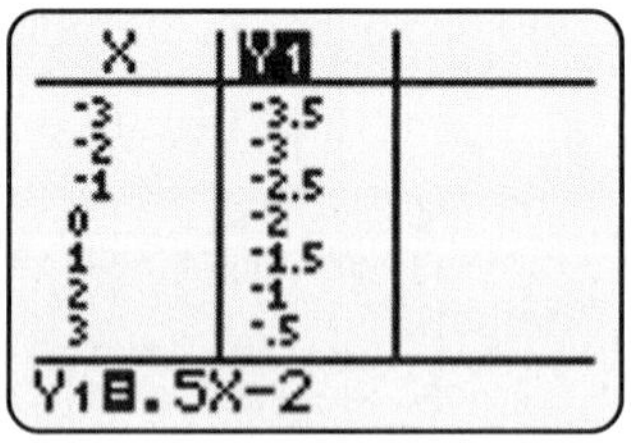

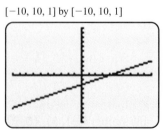

Figure 2.32

$[-10, 10, 1]$ by $[-10, 10, 1]$

Figure 2.33

■ MAKING CONNECTIONS

Mathematics in Newspapers

Think of the mathematics that you see in newspapers. Often percentages are described *verbally*, numbers are displayed in *tables*, and data are shown in *graphs*. Seldom are *formulas* given, which is an important reason not to study only symbolic representations.

MODELING DATA WITH LINEAR FUNCTIONS

A distinguishing feature of a linear function is that each time the input x increases by 1 unit, the output $f(x) = ax + b$ always changes by an amount equal to a. For example, the number of doctors in the private sector from 1970 to 2000 can be modeled by

$$f(x) = 15,260x - 2,973,000,$$

where x is the year. The value $a = 15,260$ indicates that the number of doctors has increased by about 15,260 each year. **(Source:** American Medical Association.**)**

The following are other examples of quantities that are modeled by linear functions. Try to determine the value of the constant a.

- The wages earned by an individual working x hours at $8 per hour
- The distance traveled by light in x seconds if the speed of light is 186,000 miles per second
- The cost of tuition and fees when registering for x credits if each credit costs $80 and the fees are fixed at $50

EXAMPLE 5 *Modeling the cost of tuition*

Suppose that tuition costs $80 per credit and that student fees are fixed at $50. Give symbolic, graphical, and numerical representations for a linear function f that models the cost of registering for x credits.

Solution

Symbolic Representation The function multiplies the input x by 80 and then adds 50. Let $f(x) = 80x + 50$.

Graphical Representation Let $Y_1 = 80X + 50$ and graph Y_1, as shown in Figure 2.34. Note that the graph is a line.

Numerical Representation To create a numerical representation of f, construct the table for Y_1, as shown in Figure 2.35. (Other tables are possible.) Note that each time x increases by 1 credit, Y_1 increases by 80 dollars.

[0, 20, 5] by [0, 2000, 500]

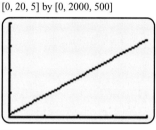
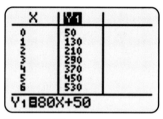

Figure 2.34   Figure 2.35

In the next example we consider a simple linear function that models wind speed.

EXAMPLE 6 *Modeling wind speed with a constant linear function*

The monthly average wind speeds in miles per hour at Hilo, Hawaii, from May through December are listed in Table 2.8. The months have been assigned the usual numbers.

TABLE 2.8 Average Wind Speeds at Hilo, Hawaii

Month	5	6	7	8	9	10	11	12
Wind Speed (miles per hour)	7	7	7	7	7	7	7	7

Source: Williams, J., *The Weather Almanac 1995.*

(a) What was the average wind speed in October? Discuss the average wind speed from May through December at Hilo.
(b) Symbolically represent a function f that models these data.
(c) Sketch a graph of f together with the data.

Solution

(a) October corresponds to the 10th month. The average wind speed during October is 7 miles per hour. The monthly average wind speed is 7 miles per hour from May to December.
(b) Regardless of the input, the output is always 7. Thus let $f(x) = 7$.
(c) Graph $y = 7$ with the data points

$$\{(5, 7), (6, 7), (7, 7), (8, 7), (9, 7), (10, 7), (11, 7), (12, 7)\}$$

to obtain Figure 2.36.

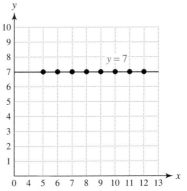

Figure 2.36 Modeling Wind Speed

The function represented by $f(x) = 7$ is an example of a **constant linear function,** or **constant function.** The following are applications of constant functions.

- A thermostat computes a constant function regardless of the weather outside by maintaining a set temperature.
- A cruise control in a car computes a constant function by maintaining a fixed speed regardless of the type of road or terrain.

Tuition at colleges and universities has risen dramatically during the past two decades. The next example illustrates how tuition can be modeled with a linear function.

EXAMPLE 7 *Modeling tuition with a linear function*

Table 2.9 lists the average tuition at private colleges and universities for selected years.

TABLE 2.9

Year	1980	1985	1990	1995
Tuition	$3617	$6121	$9340	$12,432

Source: The College Board.

These data may be modeled by $f(x) = 593.3x - 1{,}171{,}268$, where x is the year. A scatterplot of the data and a graph of f are shown in Figure 2.37.

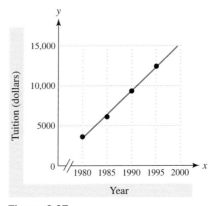

Figure 2.37

(a) Evaluate $f(1991)$ and interpret the result.
(b) Estimate private tuition in the year 2000.
(c) The value of a in the formula for f is 593.3. Interpret this value.

Solution

(a) $f(1991) = 593.3(1991) - 1{,}171{,}268 \approx 9992$. This model estimates that tuition in 1991 at private colleges and universities was \$9992. (The actual value was \$10,017.)

(b) To estimate tuition in the year 2000 evaluate $f(2000)$:

$$f(2000) = 593.3(2000) - 1{,}171{,}268 \approx \$15{,}332.$$

(c) Tuition has risen by about \$593.30 each year.

2.2 Putting It All Together

A linear function is a relatively simple type of function frequently used to model real data. Essential to understanding functions in general is to first understand linear functions. The following table summarizes symbolic, verbal, graphical, and numerical representations of a linear function.

Type of Representation	Comments	Example
Symbolic	Mathematical formula in the form $f(x) = ax + b$	$f(x) = 2x + 1$, where $a = 2$ and $b = 1$
Verbal	Multiply the input x by a and add b.	Multiply the input x by 2 and add 1 to obtain the output.
Graphical	The graph of a linear function is a line.	
Numerical (table)	A table of values. For each unit increase in x, the output of $f(x) = ax + b$ changes by an amount equal to a.	x: 0, 1, 2, 3 $f(x)$: 1, 3, 5, 7

2.2 EXERCISES

CONCEPTS

1. The formula for a linear function is given by $f(x) = $ _____.

2. The formula for a constant linear function is given by $f(x) = $ _____.

3. The graph of a linear function is a _____.

4. The graph of a constant linear function is a _____.

5. If $f(x) = 7x + 5$, each time x increases by 1 unit, $f(x)$ increases by _____ units.

6. If $f(x) = 5$, each time x increases by 1 unit, $f(x)$ increases by _____ units.

IDENTIFYING LINEAR FUNCTIONS

Exercises 7–14: Determine whether f is a linear function. If f is linear, give values for a and b so that f may be expressed as $f(x) = ax + b$.

7. $f(x) = \frac{1}{2}x - 6$
8. $f(x) = x$
9. $f(x) = \frac{5}{2} - x^2$
10. $f(x) = \sqrt{x} + 3$
11. $f(x) = -9$
12. $f(x) = 1.5 - 7.3x$
13. $f(x) = -9x$
14. $f(x) = \frac{1}{x}$

Exercises 15–18: Determine whether the graph represents a linear function.

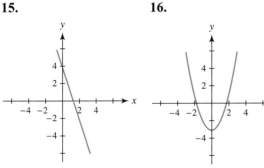

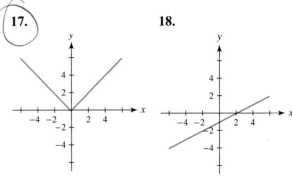

Exercises 19–22: Determine whether f could be a linear function.

19.
x	1	2	3	4
$f(x)$	0	2	4	6

20.
x	1	2	3	4
$f(x)$	0	1	3	7

21.
x	-2	-1	0	1	2
$f(x)$	-5	0	20	40	100

22.
x	-2	-1	0	1	2
$f(x)$	6	3	0	-3	-6

EVALUATING LINEAR FUNCTIONS

Exercises 23–28: Evaluate $f(x)$ at the given values of x.

23. $f(x) = 4x$ $x = -4, 5$
24. $f(x) = -2x + 1$ $x = -2, 3$
25. $f(x) = 5 - x$ $x = -\frac{2}{3}, 3$
26. $f(x) = \frac{1}{2}x - \frac{1}{4}$ $x = 0, \frac{1}{2}$
27. $f(x) = -22$ $x = -\frac{3}{4}, 13$
28. $f(x) = 9x - 7$ $x = -1.2, 2.8$

Exercises 29–32: Use the graph to evaluate the given expressions.

29. $f(-1)$ and $f(0)$ **30.** $f(-2)$ and $f(2)$

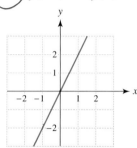 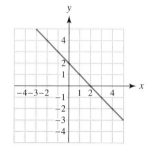

31. $f(-2)$ and $f(4)$ **32.** $f(0)$ and $f(3)$

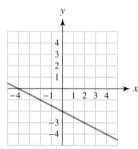 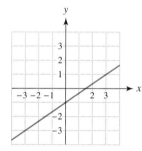

Exercises 33 and 34: Use the table to evaluate the given expressions.

33. $f(1)$ and $f(3)$

x	1	2	3	4
f(x)	−2	0	2	4

34. $f(-10)$ and $f(5)$

x	−10	−5	0	5	10
f(x)	100	90	80	70	60

Exercises 35–38: Use the verbal description of f to evaluate $f(3)$.

35. Multiply the input by 6.

36. Multiply the input by −3 and add 7.

37. Output 8.7 for every input.

38. Divide the input by 6 and subtract $\frac{1}{2}$.

REPRESENTING LINEAR FUNCTIONS

Exercises 39–42: Match $f(x)$ with its graph (a.–d.) without using a graphing calculator.

39. $f(x) = 3x$ **40.** $f(x) = -2x$

41. $f(x) = x - 2$ **42.** $f(x) = 2x + 1$

a. 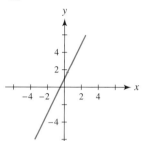   **b.**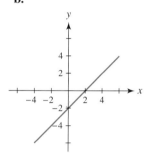

c. 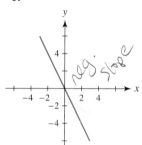   **d.**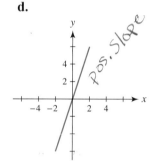

Exercises 43–50: (Refer to Example 2.)
(a) Complete the table for the given $f(x)$.

x	−2	0	2
f(x)	?	?	?

(b) Plot the points in the table.
(c) Sketch a graph of f.
(d) Use a graphing calculator to support your result in part (c).

43. $f(x) = 2x$ **44.** $f(x) = -\frac{1}{2}x$

45. $f(x) = x + 1$ **46.** $f(x) = x - 2$

47. $f(x) = 3x - 3$ **48.** $f(x) = -2x + 1$

49. $f(x) = 3 - x$ **50.** $f(x) = \frac{1}{4}x + 2$

Exercises 51–56: Write a symbolic representation for a linear function f that computes the following.

51. The number of pounds in x ounces

52. The number of dimes in x dollars

53. The distance traveled by a car moving at 65 miles per hour for t hours

54. The long-distance phone bill *in dollars* for calling t minutes at 10 cents per minute and a fixed fee of $4.95

55. The total number of hours in a day during day x

56. The cost of downhill skiing x times with a $500 season pass

APPLICATIONS

Exercises 57–60: Match the situation with the graph (a.–d.) that models it best, where x-values represent time.

57. The cost of college tuition from 1990 to 2000.

58. The cost of 1 megabyte of computer memory during the past 5 years.

59. The distance between New York City and Denver, Colorado, during the past 10 years.

60. The total distance traveled by a car moving at 60 miles per hour after x hours

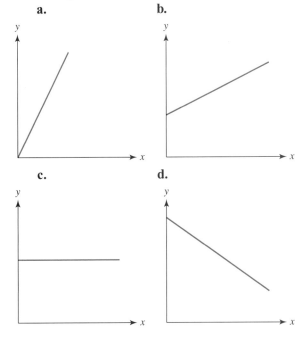

61. *Thermostat* Let $y = f(x)$ describe the temperature y of a room that is kept at 70°F for x hours.
    (a) Represent f symbolically and graphically over a 24-hour period for $0 \leq x \leq 24$.
    (b) Construct a table of f for $x = 0, 4, 8, 12, \ldots, 24$.
    (c) What type of function is f?

62. *Cruise Control* Let $y = f(x)$ describe the speed y of an automobile after x minutes if the cruise control is set at 60 miles per hour.
    (a) Represent f symbolically and graphically over a 15-minute period for $0 \leq x \leq 15$.
    (b) Construct a table of f for $x = 0, 1, 2, \ldots, 6$.
    (c) What type of function is f?

63. *Income Tax Rates* The following table lists the lowest personal income tax rate between 1994 and 1999.

Year	1994	1995	1996
Rate (%)	15	15	15
Year	1997	1998	1999
Rate (%)	15	15	15

Source: Internal Revenue Service.

(a) Find a symbolic representation for a function f that models these data.
(b) Assume that the domain of f is $D = \{1994, 1995, \ldots, 1999\}$. Graph f.

64. *Average Wind Speed* The following table lists the average wind speed at Myrtle Beach, South Carolina. The months are numbered 1–12.

Month	1	2	3	4	5	6
Wind (miles per hour)	7	8	8	8	7	7
Month	7	8	9	10	11	12
Wind (miles per hour)	7	7	7	6	6	6

Source: J. Williams.

(a) Could these data be modeled exactly by a constant function f?
(b) Find a symbolic representation for a constant function that models this data approximately.
(c) Graph f and the data.

65. *Manatee Deaths* Collisions with boats kill more manatees than any other known cause. In 1998 run-ins with boats killed a record number of manatees. The following table lists the deaths blamed on boat collisions from 1993 through 1998.

Year	1993	1994	1995
Deaths	35	49	42

Year	1996	1997	1998
Deaths	60	54	66

Source: Florida Department of Environmental Protection.

(a) Make a line graph of the data.
(b) Could the data be modeled accurately with a linear function? Explain your answer.

66. *Estimating the Weight of a Bass* Sometimes the weight of a fish can be estimated by measuring its length. The following table lists typical weights of bass having various lengths.

Length (inches)	12	14	16	18	20	22
Weight (pounds)	1.0	1.7	2.5	3.6	5.0	6.6

Source: Minnesota Department of Natural Resources.

(a) Let x be the length and y be the weight. Make a line graph of the data.
(b) Could the data be modeled accurately with a linear function? Explain your answer.

67. *Solid Waste* In 1960, the average American disposed of 2.7 pounds of garbage per day, whereas today this amount is about 4.3 pounds per day. **(Source: Environmental Protection Agency.)**
(a) Find $f(x) = ax$ so that f models the amount of garbage disposed of by a person today after x days.
(b) Construct the table for f for
$$x = 10, 20, 30, \ldots, 70.$$
Evaluate $f(60)$ and interpret the result.

68. *Rain Forests* Rain forests are defined as forests that grow in regions receiving more than 70 inches of rain per year. The world is losing an estimated 49 million acres of rain forests each year. **(Source: New York Times Almanac, 1999.)**
(a) Find $f(x) = ax$ so that f models the acres of rain forest lost in x years.
(b) Construct the table for f for
$$x = 1, 2, 3, \ldots, 7.$$
Evaluate $f(7)$ and interpret your result.

69. *Age in the United States* The median age of the population for each year x between 1820 and 1995 can be approximated by
$$f(x) = 0.09x - 147.1$$
(Source: Bureau of the Census.)
(a) Graph f in the viewing rectangle
$$[1820, 1995, 20] \text{ by } [0, 40, 10].$$
Discuss any trends shown in the graph.
(b) Construct the table for f starting at $x = 1820$, incrementing by 20. Use the table to evaluate $f(1900)$ and interpret the result.
(c) The value of a in the formula for $f(x)$ is 0.09. Interpret this value.

70. *Women's Olympic Times* The winning times in seconds for the women's 200-meter dash at the Olympic Games may be modeled by $f(x) = -0.0635x + 147.9$, where x is the year with $1948 \leq x \leq 1996$. **(Source: Olympic Committee.)**
(a) Graph f in the viewing rectangle
$$[1948, 1996, 4] \text{ by } [21, 25, 1].$$
Discuss what happened to the winning times during this time period.
(b) Evaluate $f(1996)$ and interpret the result.
(c) The value of a in the formula for $f(x)$ is -0.0635. Interpret this value.

71. *Weight Lifting* Lifting weights can increase a person's muscle mass. Each additional pound of muscle burns an extra 40 calories per day. Write a linear function that models the number of calories burned each day by x pounds of muscle. By burning an extra 3500 calories a person can lose 1 pound of fat. How many pounds of muscle are needed to burn the equivalent of 1 pound of fat in 30 days? *(Source: Runner's World.)*

72. *Numbers of Doctors* The number of doctors in the private sector from 1970 to 2000 during year x can be modeled by

$$f(x) = 15{,}260x - 29{,}732{,}000.$$

(Source: American Medical Association.)

(a) Explain why $f(x)$ represents a linear function.
(b) Graph f in
[1970, 2000, 5] by [0, 800,000, 100,000].

What has happened to the number of doctors from 1970 to 2000?
(c) Evaluate $f(1995)$ and interpret the result.

WRITING ABOUT MATHEMATICS

73. Explain how you can determine whether a function is linear by using its
    (a) symbolic representation,
    (b) graphical representation, or
    (c) numerical representation.

74. Describe one way to determine whether a set of data points can be modeled by a linear function.

CHECKING BASIC CONCEPTS FOR SECTIONS 2.1 AND 2.2

1. Symbolically, graphically, and numerically represent a function that squares the input x and then subtracts 1 from the result. For the numerical representation let
$$x = -3, -2, -1, \ldots, 3,$$
and for the graphical representation let $-3 \leq x \leq 3$.

2. Use the accompanying graph to complete the following.
   (a) Find the domain and range of f.
   (b) Evaluate $f(0)$ and $f(2)$.
   (c) Is f a linear function? Explain.

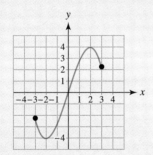

3. Determine whether f is a linear function.
   (a) $f(x) = 4x - 2$
   (b) $f(x) = 2\sqrt{x} - 5$
   (c) $f(x) = -7$
   (d) $f(x) = 9 - 2x$

4. Graph $f(x) = 4 - 3x$. Evaluate $f(-2)$.

2.3 The Slope of a Line

Slope ~ Slope–Intercept Form of a Line ~ Interpreting Slope in Applications

Introduction

Figures 2.38–2.40 show some graphs of lines, where the *x*-axis represents time.

Which graph might represent the amount of water coming out of a faucet?

Which graph might represent the height of a young adult?

Which graph might represent the amount of gas in your car's tank while you are driving?

To answer these questions, you probably used the concept of slope. In mathematics, slope is a real number that measures the "tilt" of a line in the *xy*-plane. We assume throughout the text that lines are always straight. In this section we discuss how slope relates to the graph of a linear function and how slope occurs in applications.

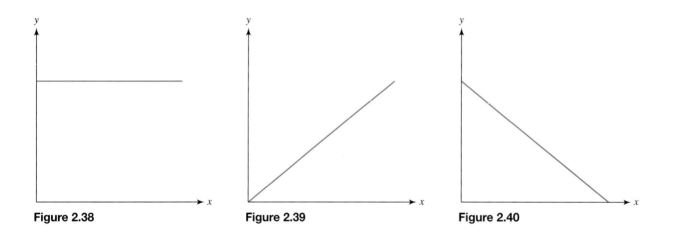

Figure 2.38 **Figure 2.39** **Figure 2.40**

Slope

The graph shown in Figure 2.41 illustrates the cost of buying *x* gallons of gasoline. The graph tilts upward from left to right, which indicates that the cost increases as the number of gallons purchased increases. Note that for every 2 gallons purchased the cost increases by $3. We say that the graph *rises* 3 units for every 2 units of *run*. The ratio $\frac{\text{rise}}{\text{run}}$ equals the *slope* of the line. The slope of this line is $\frac{3}{2}$, or 1.5. That is, for every unit of run along the *x*-axis the graph rises 1.5 units. A slope of 1.5 indicates that gasoline costs $1.50 per gallon.

2.3 The Slope of a Line 103

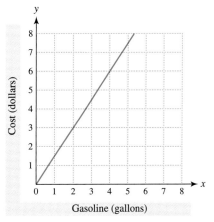

Figure 2.41 Cost of Gasoline

A more general case is shown in Figure 2.42 where a line passes through the points (x_1, y_1) and (x_2, y_2). The **rise** or *change in y* is $y_2 - y_1$, and the **run** or *change in x* is $x_2 - x_1$. The slope m is given by $m = \dfrac{y_2 - y_1}{x_2 - x_1}$.

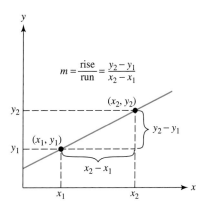

Figure 2.42

Slope

The **slope** m of the line passing through the points (x_1, y_1) and (x_2, y_2) is

$$m = \frac{y_2 - y_1}{x_2 - x_1},$$

where $x_1 \neq x_2$. That is, slope equals rise over run.

EXAMPLE 1 *Calculating the slope of a line*

Find the slope of the line passing through the points $(-4, 1)$ and $(2, 4)$. Plot these points and the line. Interpret the slope.

Solution

Begin by letting $(x_1, y_1) = (-4, 1)$ and $(x_2, y_2) = (2, 4)$. The slope is

$$m = \frac{y_2 - y_1}{x_2 - x_1} = \frac{4 - 1}{2 - (-4)} = \frac{3}{6} = \frac{1}{2}.$$

A graph of the line passing through these two points is shown in Figure 2.43. A slope of $\frac{1}{2}$ indicates that the line rises 1 unit for every 2 units of run.

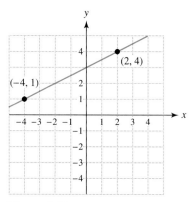

Figure 2.43

We would get the same slope as in the preceding example if we let $(x_1, y_1) = (2, 4)$ and $(x_2, y_2) = (-4, 1)$. In this case the calculation would be

$$m = \frac{y_2 - y_1}{x_2 - x_1} = \frac{1 - 4}{-4 - 2} = \frac{-3}{-6} = \frac{1}{2}.$$

If the slope of a line has **positive slope,** the line *rises* from left to right. In Figure 2.44 the rise is 2 units for each unit of run, so the slope is 2. If the slope has **negative slope,** the line *falls* from left to right. In Figure 2.45 the line *falls* 1 unit for every 2 units of run, so the slope is $-\frac{1}{2}$. Slope 0 indicates that a line is horizontal, as shown in Figure 2.46. If (x_1, y_1) and (x_2, y_2) are two points on a vertical line, $x_1 = x_2$. In Figure 2.47 the run is $x_2 - x_1 = 0$, so the slope of a vertical line is always undefined.

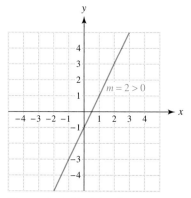

Figure 2.44 Positive Slope

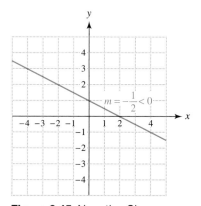

Figure 2.45 Negative Slope

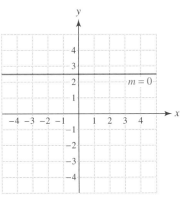

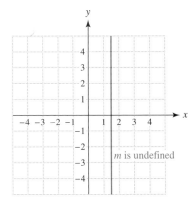

Figure 2.46 Zero Slope

Figure 2.47 Undefined Slope

EXAMPLE 2 *Sketching a line with a given slope*

Sketch a line passing through the point (0, 4) and having slope $-\frac{2}{3}$.

Solution

Start by plotting the point (0, 4). A slope of $-\frac{2}{3}$ indicates that the *y*-values *decrease* 2 units each time the *x*-values increase by 3 units. That is, the line *falls* 2 units for every 3-unit increase in the run. The line passes through (0, 4), so a 2-unit decrease in *y* and a 3-unit increase in *x* results in the line passing through the point $(0 + 3, 4 - 2)$ or (3, 2), as shown in Figure 2.48.

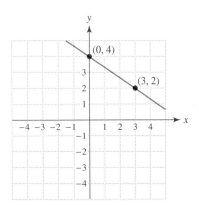

Figure 2.48

SLOPE–INTERCEPT FORM OF A LINE

The graph of $f(x) = 2x + 3$ is a line rising to the right that passes through (0, 3) and (1, 5), as shown in Figure 2.49 on the next page. Therefore the slope of this line is

$$m = \frac{5 - 3}{1 - 0} = 2.$$

Note that slope 2 equals the coefficient of x in the formula $f(x) = 2x + 3$. In general, if $f(x) = ax + b$, the slope of the graph of f is $m = a$. For example, the graph of $f(x) = 6x - 5$ has slope $m = 6$, and the graph of $f(x) = -\frac{4}{5}x + 1$ has slope $m = -\frac{4}{5}$.

The point $(0, 3)$ lies on the graph of $f(x) = 2x + 3$ and is located on the y-axis. The y-value of 3 is called the *y-intercept*. A **y-intercept** is the y-coordinate of a point where a graph intersects the y-axis. To find a y-intercept let $x = 0$ in $f(x)$. If $f(x) = ax + b$, then

$$f(0) = a(0) + b = b.$$

Thus if $f(x) = -4x + 3$, the y-intercept is 3, and if $f(x) = \frac{1}{2}x - 8$, the y-intercept is -8.

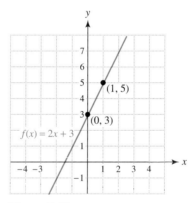

Figure 2.49

Because $y = f(x)$, any linear function may be represented by $y = mx + b$, where m is the slope and b is the y-intercept. The form $y = mx + b$ is called the *slope–intercept form*.

Slope–Intercept Form

The line with slope m and y-intercept b is given by

$$y = mx + b,$$

the **slope–intercept form** of a line.

EXAMPLE 3 *Sketching a line*

Sketch a line with slope -2 and y-intercept 3. Write its slope–intercept form.

2.3 The Slope of a Line

Solution

For the y-intercept of 3, plot the point $(0, 3)$. Because the slope is -2, each 1-unit increase in x results in a decrease in y of 2 units. Thus the line passes through the point $(0 + 1, 3 - 2)$ or $(1, 1)$. The slope–intercept form of this line is $y = -2x + 3$. See Figure 2.50.

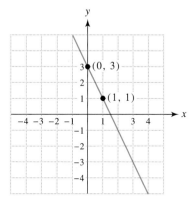

Figure 2.50

EXAMPLE 4 *Using a graph to write the slope–intercept form*

For each graph shown in Figures 2.51 and 2.52, write the slope–intercept form of the line.

(a)   (b)

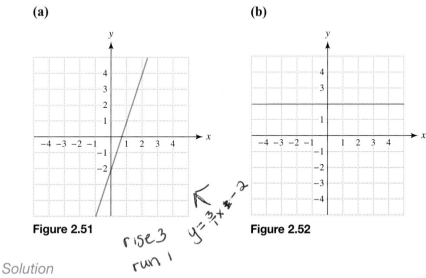

Figure 2.51   **Figure 2.52**

Solution

(a) The graph passes through $(0, -2)$, so the y-intercept is -2. Because the graph rises 3 units for each unit increase in x, the slope is 3. The slope–intercept form is $y = 3x - 2$.

(b) The graph passes through $(0, 2)$, so the y-intercept is 2. Because the graph is a horizontal line, its slope is 0. The slope–intercept form is $y = 0x + 2$, or more simply, $y = 2$.

Interpreting Slope in Applications

When a linear function is used to model physical quantities in the real world, the slope of its graph provides certain information. Slope can be interpreted as a **rate of change** of a quantity, which we illustrate in the next three examples.

EXAMPLE 5 *Interpreting slope*

The distance y in miles that an athlete training for a bicycle race is from home after x hours is shown in Figure 2.53.

(a) Find the y-intercept. What does the y-intercept represent?
(b) The graph passes through the point (2, 6). Discuss the meaning of this point.
(c) Find the slope–intercept form of this line. Interpret the slope as a rate of change.

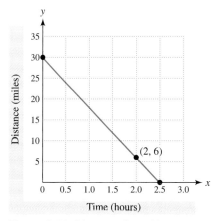

Figure 2.53 Distance from Home

Solution

(a) From the graph, the y-intercept is 30, which indicates that the athlete is initially 30 miles from home.
(b) The point (2, 6) means that after 2 hours the athlete is 6 miles from home.
(c) The line passes through the points (0, 30) and (2, 6). Thus its slope is

$$m = \frac{6 - 30}{2 - 0} = -12,$$

and the slope–intercept form is $y = -12x + 30$. A slope of -12 indicates that the athlete is traveling at 12 miles per hour *toward* home. The negative sign indicates that the distance between the athlete and home is decreasing.

EXAMPLE 6 *Interpreting slope*

The formula $f(x) = 123x + 786$ models tuition at four-year public colleges and universities from 1981 through 1995, where $x = 1$ represents 1981, $x = 2$ represents 1982, and so on.

(a) Find the slope of the graph of f.
(b) Interpret the slope as a rate of change.

2.3 The Slope of a Line

Solution

(a) The slope of the graph of $f(x) = 123x + 786$ is $m = 123$.

(b) A slope of $m = 123$ means that the *rate of change* in the tuition has been, on average, $123 per year.

EXAMPLE 7 *Analyzing tetanus cases*

Table 2.10 lists the number of reported cases of tetanus for selected years.

TABLE 2.10

Year	1950	1960	1970	1980	1990
Cases of Tetanus	486	368	148	95	64

Source: Department of Health and Human Services.

(a) Make a line graph of the data.
(b) Find the slope of each line segment in the graph.
(c) Interpret these slopes as rates of change.

Solution

(a) A line graph connecting the points (1950, 486), (1960, 368), (1970, 148), (1980, 95), and (1990, 64) is shown in Figure 2.54.

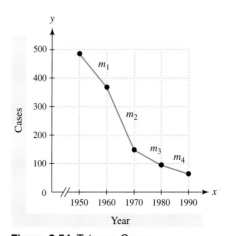

Figure 2.54 Tetanus Cases

(b) The slope of each line segment is

$$m_1 = \frac{368 - 486}{1960 - 1950} = -11.8;$$

$$m_2 = \frac{148 - 368}{1970 - 1960} = -22.0;$$

$$m_3 = \frac{95 - 148}{1980 - 1970} = -5.3;$$

$$m_4 = \frac{64 - 95}{1990 - 1980} = -3.1.$$

(c) The slope $m_1 = -11.8$ means that, on average, the number of tetanus cases between 1950 and 1960 *decreased* by 11.8 cases per year. The other three slopes can similarly be interpreted as rates of change.

Critical Thinking

An athlete runs away from home at 10 miles per hour for 30 minutes and then jogs back home at 5 miles per hour. Sketch a graph that shows the distance between the athlete and home.

2.3 Putting It All Together

The graph of a linear function is a line. The "tilt" of a line is called the slope and equals rise over run. A positive slope indicates that the line *rises* from left to right, whereas a negative slope indicates that the line *falls* from left to right. A horizontal line has slope 0 and a vertical line has undefined slope. The slope of the graph of $f(x) = ax + b$ is $m = a$ and its y-intercept is b. When a linear function is being used to model physical quantities, slope indicates a rate of change in the quantity. The following table summarizes some basic concepts of slope and the slope–intercept form.

Type of Representation	Comments	Example
Slope	For two points, (x_1, y_1) and (x_2, y_2), slope is given by $$m = \frac{y_2 - y_1}{x_2 - x_1}.$$	The slope of the line passing through $(-2, 3)$ and $(1, 5)$ is $$m = \frac{5 - 3}{1 - (-2)} = \frac{2}{3}.$$ If the x-values increase by 3 units, the y-values increase by 2 units.
Slope–Intercept Form for a Line $y = mx + b$	The slope equals m, and the y-intercept equals b.	If $y = \frac{1}{2}x + 1$, then the slope is $\frac{1}{2}$ and the y-intercept is 1.

2.3 EXERCISES

SLOPE

Exercises 1–6: Use the concept of rise over run to find the slope of the line.

1.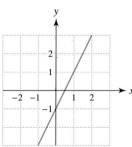

2.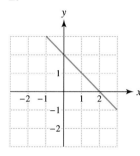

3.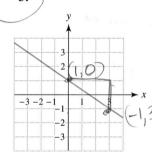

4.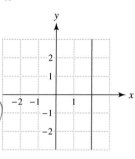

5.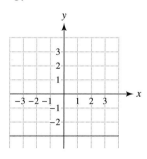

6.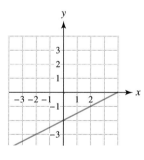

Exercises 7–18: Find the slope of the line passing through the two points.

7. $(1, 2), (2, 4)$

8. $(-3, 2), (2, -3)$

9. $(2, 1), (-1, 3)$

10. $\left(-\dfrac{3}{5}, -\dfrac{4}{5}\right), \left(\dfrac{4}{5}, \dfrac{3}{5}\right)$

11. $(-3, 6), (4, 6)$

12. $(-3, 5), (-5, 5)$

13. $\left(\dfrac{1}{2}, -\dfrac{2}{7}\right), \left(\dfrac{1}{2}, \dfrac{13}{17}\right)$

14. $(-1, 6), (-1, -4)$

15. $(1989, 10), (1999, 16)$

16. $(1950, 6.1), (2000, 10.6)$

17. $(2.1, 3.6), (-1.2, 4.3)$

18. $(12, -34), (14, 64)$

Exercises 19–26: Complete the following.
(a) Find the slope and y-intercept of the graph of f.
(b) Sketch a graph of f by hand. Use a graphing calculator to check your work.

19. $f(x) = -3x + 2$ **20.** $f(x) = \dfrac{1}{2}x - 1$

21. $f(x) = \dfrac{1}{3}x$ **22.** $f(x) = -2x$

23. $f(x) = 2$ **24.** $f(x) = -3$

25. $f(x) = -x + 3$ **26.** $f(x) = \dfrac{2}{3}x - 2$

Exercises 27–30: Let f be a linear function. Use the table to find the slope and y-intercept of the graph of f.

27.

x	0	1	2	3
$f(x)$	-2	0	2	4

28.

x	-1	0	1	2
$f(x)$	5	10	15	20

29.

x	-2	-1	0	1	2
$f(x)$	18	11	4	-3	-10

30.

x	-4	-2	0	2	4
$f(x)$	6	3	0	-3	-6

SLOPE–INTERCEPT FORM

Exercises 31–36: Use the graph to express the line in slope–intercept form.

31.
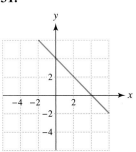

32.
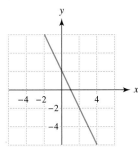

33.
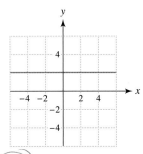

34.
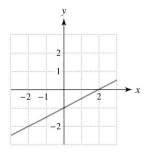

35.
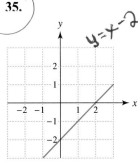

36.
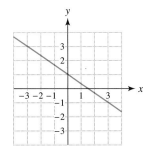

Exercises 37–40: Write the slope–intercept form for a line satisfying the following conditions.

37. Slope 3, y-intercept -5

38. Slope $-\frac{2}{3}$, y-intercept 7

39. Passing through $(0, 4)$ and $(-2, 0)$

40. Passing through $(1, 0)$ and $\left(0, -\frac{3}{2}\right)$

Exercises 41–44: Let $f(x)$ represent a linear function.
  (a) Find the missing value in the table.
  (b) Write the slope–intercept form for f.

41.

x	0	1	2
f(x)	−1	1	?

42.

x	1	2	3
f(x)	12	8	?

43.

x	−2	0	4
f(x)	2	5	?

44.

x	0	5	20
f(x)	0	10	?

INTERPRETING SLOPE

Exercises 45–48: **Modeling** *Choose the graph (a.–d.) that models the situation best.*

45. Money that a person earns after x hours, working for $10 per hour

46. Total acres of rain forests in the world during the past 20 years

47. World population from 1980 to 2000

48. Square miles of land in Nevada from 1950 to 2000

a.
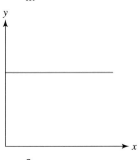

b.
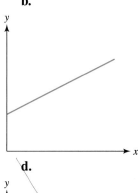

c.
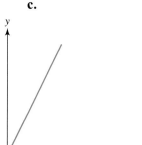

d.
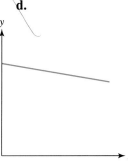

2.3 The Slope of a Line 113

Exercises 49–52: Modeling The line graph represents the gallons of water in a small swimming pool after x hours. Assume that a pump at the pool can either add water to or remove water from the pool.
  (a) Estimate the slope of each line segment.
  (b) Interpret each slope as a rate of change.
  (c) Describe what happened to the amount of water in the pool.

49.

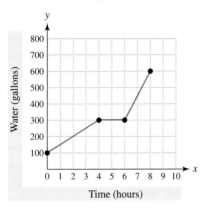

50.

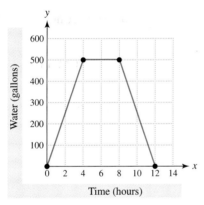

51.

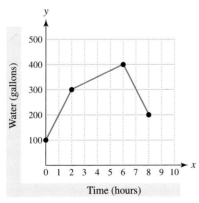

52.

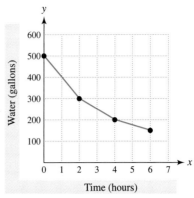

Exercises 53–56: Modeling Distance An individual is driving a car along a straight road. The graph shows the distance that the driver is from home after x hours.
  (a) Find the slope of each line segment in the graph.
  (b) Interpret each slope as a rate of change.
  (c) Describe both the motion and location of the car.

53.

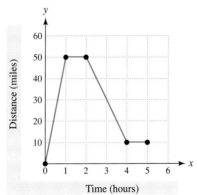

54.

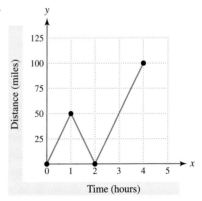

55.

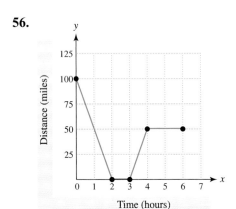

56.

Exercises 57–60: Sketching a Model Sketch a graph that models the given situation.

57. The distance that a bicycle rider is from home if the rider is initially 20 miles away from home and arrives home after riding at a constant speed for 2 hours

58. The distance that an athlete is from home if the athlete runs away from home at 8 miles per hour for 30 minutes and then walks back home at 4 miles per hour

59. The distance that a person is from home if this individual drives to a mall, stays 2 hours, and then drives home, assuming that the distance to the mall is 20 miles and that the trip takes 30 minutes

60. The amount of water in a 10,000-gallon swimming pool that is filled at the rate of 1000 gallons per hour, left full for 10 hours, and then drained at the rate of 2000 gallons per hour

APPLICATIONS

61. *Older Mothers* The number of children in thousands born to mothers 40 years old or older is modeled by $f(x) = 4x - 7910$, where x is the year with $1990 \leq x \leq 2000$. **(Source: National Center for Health Statistics.)**
  (a) Estimate the number of children born to older mothers in 1993.
  (b) Graph f in [1990, 2000, 1] by [0, 100, 10]. Describe the graph.
  (c) What is the slope of the graph of f?
  (d) Interpret the slope as a rate of change.

62. *Commercial Banks* From 1987 to 1997 the number of federally insured commercial banks could be modeled by
$$f(x) = -458x + 923{,}769,$$
where x is the year. **(Source: Federal Deposit Insurance Corporation.)**
  (a) Graph $f(x)$ in the viewing rectangle [1987, 1997, 2] by [8000, 15,000, 1000]. What happened to the number of banks during this time period?
  (b) What is the slope of the graph of f?
  (c) Interpret the slope as a rate of change.

63. *Number of Radio Stations* The number of radio stations on the air from 1950 to 1995 may be modeled by
$$f(x) = 214.2x - 415{,}368,$$
where x is the year. **(Source: National Association of Broadcasters.)**
  (a) How many radio stations were on the air in 1980?
  (b) What is the slope of the graph of f?
  (c) Interpret the slope as a rate of change.

64. *Men's Olympic Times* The winning times in seconds for the men's 400-meter hurdles at the Olympic Games can be approximated by $f(x) = -0.0975x + 240.8$, where x is the year for $1900 \leq x \leq 1996$. **(Source: Olympic Committee.)**
  (a) Graph $f(x)$ in the viewing rectangle [1900, 1996, 4] by [45, 60, 1]. What happened to the winning times during this time period?
  (b) What is the slope of the graph of f?
  (c) Interpret the slope as a rate of change.

65. *Two-Cycle Engines* Two-cycle engines used in snowmobiles, jet skis, chain saws, and outboard motors require a mixture of gas and oil to run properly. For certain engines the amount of oil in pints that should be added to x gallons of gasoline is computed by $f(x) = \frac{4}{25}x$. (*Source:* Johnson Outboard Motor Company.)
    (a) How many pints of oil should be added to 6 gallons of gasoline?
    (b) Graph f for $0 \leq x \leq 25$.
    (c) What is the slope of the graph of f?
    (d) Interpret the slope as a rate of change.

66. *Fat Grams* Some slices of pizza contain 10 grams of fat.
    (a) Find a formula $f(x)$ that calculates the number of fat grams in x slices of pizza.
    (b) Graph f for $0 \leq x \leq 6$.
    (c) What is the slope of the graph of f?
    (d) Interpret the slope as a rate of change.

67. *U.S. Median Family Income* In 1990, the median family income was about $41,000, and in 1999 it was about $55,000. (*Source:* Department of the Treasury.)
    (a) Let $x = 0$ represent 1990, $x = 1$ represent 1991, and so on. Find values for a and b so that $f(x) = ax + b$ models the data. (*Hint:* Substitute $x = 0$ into $f(x)$ to determine b.)
    (b) Estimate the median family income in 1995.

68. *Minimum Wage* In 1980, the minimum wage was about $3.10 per hour, and in 1999 it was $5.15. (*Source:* Department of Labor.)
    (a) Let $x = 0$ represent 1980, $x = 1$ represent 1981, until $x = 19$ represents 1999. Find values for a and b so that $f(x) = ax + b$ models the data.
    (b) If this trend continues, what will be the minimum wage in 2003?

WRITING ABOUT MATHEMATICS

69. Describe the information that the slope m of a line gives. Be as complete as possible.

70. Could one line have two different slope–intercept forms? Explain your answer.

GROUP ACTIVITY
Working with Real Data

Directions: Form a group of 2 to 4 people. Select someone to record the group's responses for this activity. All members should work cooperatively to answer the questions. If your instructor asks for the results, each member of the group should be prepared to respond. If the group is asked to turn in its work, be sure to include each person's name on the paper.

Sport Utility Vehicles (SUVs) SUVs became increasingly popular during the 1990s. The following table lists the number of SUVs sold in various years.

Year	1992	1994	1996	1998	2000
SUVs (millions)	1.1	1.5	2.1	2.7	3.2

Source: Autodata Corporation.

(a) Make a scatterplot of the data. Discuss any trend in SUV sales.
(b) Estimate the slope m of a line that could be used to model the data. Then choose one point (h, k) that might lie on this line.
(c) Find an equation of a line $y = mx + b$ that models the data.
(d) Interpret the slope m as a rate of change.
(e) Use your results to estimate SUV sales in 1997 and 2001.

2.4 Equations of Lines and Linear Models

Point–Slope Form ~ Horizontal and Vertical Lines ~
Parallel and Perpendicular Lines

Introduction

In 1999, there were approximately 100 million Internet users in the United States, and this number was expected to grow to about 144 million in 2002. This growth is illustrated in Figure 2.55, where the line passes through the points (1999, 100) and (2002, 144). In this section we discuss how to find the equation of the line that models these data. To do so we need to discuss the point–slope form of a line.

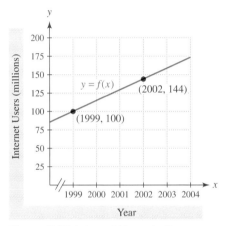

Figure 2.55 Internet Users in the United States

Point–Slope Form

If we know the slope and y-intercept of a line, we can write its slope–intercept form, $y = mx + b$. The slope–intercept form is an example of an **equation of a line.** The point–slope form is a different form of the equation of a line.

Suppose that a (nonvertical) line with slope m passes through the point (h, k). If (x, y) is a different point on this line, then $m = \dfrac{y - k}{x - h}$ (see Figure 2.56).

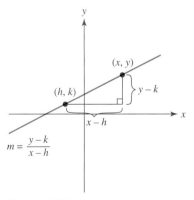

Figure 2.56

We can use this slope formula to find the point–slope form.

$$m = \frac{y - k}{x - h} \qquad \text{Slope formula}$$

$$m(x - h) = y - k \qquad \text{Multiply both sides by } (x - h).$$

$$y - k = m(x - h) \qquad \text{Rewrite the equation.}$$

$$y = m(x - h) + k \qquad \text{Add } k \text{ to both sides.}$$

The equation $y - k = m(x - h)$ is traditionally called the *point–slope form*. We think of y as being a function of x, written $y = f(x)$, so the equivalent form $y = m(x - h) + k$ is also referred to as the point–slope form.

> **Point–Slope Form**
>
> The line with slope m passing through the point (h, k) is given by
>
> $$y = m(x - h) + k,$$
>
> or equivalently,
>
> $$y - k = m(x - h),$$
>
> the **point–slope form** of a line.

EXAMPLE 1 *Using the point–slope form*

Find the point–slope form of a line passing through the point $(1, 2)$ with slope -3. Does the point $(2, -1)$ lie on this line?

Solution

Let $m = -3$ and $(h, k) = (1, 2)$ in the point–slope form.

$$y = m(x - h) + k \qquad \text{Point–slope form}$$

$$y = -3(x - 1) + 2 \qquad \text{Substitute.}$$

To determine whether the point $(2, -1)$ lies on the line, substitute $x = 2$ and $y = -1$ into the equation.

$$-1 \stackrel{?}{=} -3(2 - 1) + 2 \qquad \text{Let } x = 2 \text{ and } y = -1.$$

$$-1 \stackrel{?}{=} -3 + 2 \qquad \text{Simplify.}$$

$$-1 = -1 \qquad \text{The point satisfies the equation.}$$

The point $(2, -1)$ lies on the line because it satisfies the point–slope form.

We can use the point–slope form to find the equation of a line passing through two points.

EXAMPLE 2 *Finding the equation of a line*

Use the point–slope form to find an equation of the line passing through $(-2, 3)$ and $(6, -1)$.

Solution

Before we can apply the point–slope form, we must find the slope.

$$m = \frac{y_2 - y_1}{x_2 - x_1} \quad \text{Slope formula}$$

$$= \frac{-1 - 3}{6 - (-2)} \quad \text{Substitute.}$$

$$= -\frac{1}{2} \quad \text{Simplify.}$$

We can use either $(-2, 3)$ or $(6, -1)$ for (h, k) in the point–slope form. If we let $(h, k) = (-2, 3)$, the point–slope form becomes the following.

$$y = m(x - h) + k \quad \text{Point–slope form}$$

$$y = -\frac{1}{2}(x - (-2)) + 3 \quad \text{Substitute.}$$

$$y = -\frac{1}{2}(x + 2) + 3 \quad \text{Simplify.}$$

If we let $(h, k) = (6, -1)$, the point–slope form becomes

$$y = -\frac{1}{2}(x - 6) - 1.$$

Note that, although the two point–slope forms are different, they are equivalent because their graphs are identical.

Example 2 illustrates the fact that the point–slope form *is not* unique for a given line. However, the slope–intercept form *is* unique because each line has a unique slope and a unique y-intercept. If we simplify both point–slope forms in Example 2, they reduce to the same slope–intercept form.

$$y = -\frac{1}{2}(x + 2) + 3 \qquad y = -\frac{1}{2}(x - 6) - 1 \quad \text{Point–slope forms}$$

$$y = -\frac{1}{2}x - 1 + 3 \qquad y = -\frac{1}{2}x + 3 - 1 \quad \text{Distributive property}$$

$$y = -\frac{1}{2}x + 2 \qquad y = -\frac{1}{2}x + 2 \quad \text{Identical slope–intercept forms}$$

In the next example we model the data presented in the introduction to this section.

EXAMPLE 3 *Modeling growth in Internet usage*

In 1999, there were approximately 100 million Internet users in the United States, and this number was expected to grow to about 144 million in 2002 (see Figure 2.55).

(a) Find values for m, h, and k so that $f(x) = m(x - h) + k$ models these data.
(b) Interpret m as a rate of change.
(c) Use f to estimate Internet usage in 2003.

Solution

(a) For the data points (1999, 100) and (2002, 144) the slope of the line passing through them is

$$m = \frac{144 - 100}{2002 - 1999} = \frac{44}{3}.$$

Thus we can write

$$f(x) = \frac{44}{3}(x - 1999) + 100.$$

(b) Slope $m = \frac{44}{3} \approx 14.7$ indicates that the number of Internet users is increasing by about 14.7 million users per year.

(c) $f(2003) = \frac{44}{3}(2003 - 1999) + 100 \approx 159$ million.

If a line intersects the y-axis at the point (0, 6), the y-intercept is 6. Similarly, if the line intersects the x-axis at the point (4, 0), the *x-intercept* is 4. This line and its intercepts are illustrated in Figure 2.57. The x-coordinate of a point where a graph intersects the x-axis is called the **x-intercept.** The next example interprets intercepts in a physical situation.

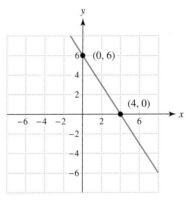

Figure 2.57 x-intercept 4; y-intercept 6

EXAMPLE 4 Modeling water in a pool

A small swimming pool containing 6000 gallons of water is emptied by a pump removing 1200 gallons per hour.

(a) How long did it take to empty the pool?
(b) Sketch a linear function f that models the amount of water in the pool after x hours.
(c) Identify the x-intercept and the y-intercept. Interpret each intercept.
(d) Find the slope–intercept form of the line in the graph. Interpret the slope as a rate of change.

Solution

(a) The time to empty the pool was $\frac{6000}{1200} = 5$ hours.

(b) Initially the pool contained 6000 gallons, and after 5 hours the pool was empty. Therefore the graph of f is a line passing through the points $(0, 6000)$ and $(5, 0)$, as shown in Figure 2.58.

(c) The x-intercept is 5, which means that after 5 hours the pool is empty. The y-intercept of 6000 means that initially (when $x = 0$) the pool contained 6000 gallons of water.

(d) To find the equation of the line shown in Figure 2.58, we first find the slope of the line passing through the points $(0, 6000)$ and $(5, 0)$.

$$m = \frac{y_2 - y_1}{x_2 - x_1} \quad \text{Slope formula}$$

$$= \frac{0 - 6000}{5 - 0} \quad \text{Substitute.}$$

$$= -1200 \quad \text{Simplify.}$$

The slope is -1200 and the y-intercept is 6000, so the slope–intercept form is

$$y = -1200x + 6000.$$

A slope of -1200 indicates that the pump *removed* water at the rate of 1200 gallons per hour.

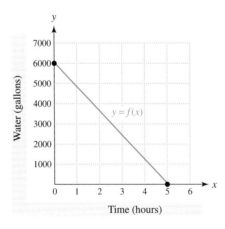

Figure 2.58 Water in a Pool

Critical Thinking

Can the graph of a function have more than one y-intercept? Explain.

Can the graph of a function have more than one x-intercept? Explain.

HORIZONTAL AND VERTICAL LINES

The graph of a constant function is a horizontal line. For example, the graph of $f(x) = 3$ is a horizontal line with y-intercept 3, as shown in Figure 2.59. Its equation may be expressed as $y = 3$, so every point on the line has a y-coordinate of 3. In general, the equation $y = b$ represents a horizontal line with y-intercept b, as shown in Figure 2.60.

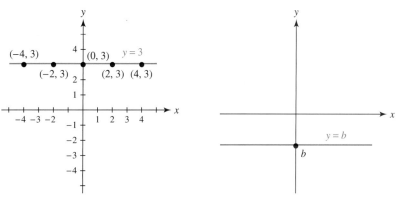

Figure 2.59   **Figure 2.60**

A vertical line cannot be represented by a function because different points on a vertical line have the same x-coordinate. The equation of the vertical line depicted in Figure 2.61 is $x = 3$. In general, an equation of a vertical line with x-intercept h is $x = h$, as shown in Figure 2.62.

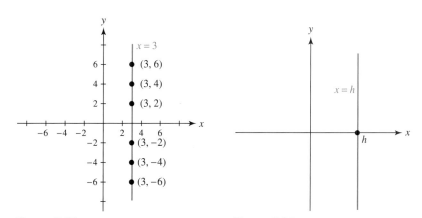

Figure 2.61   **Figure 2.62**

Equations of Horizontal and Vertical Lines

The equation of a horizontal line with y-intercept b is $y = b$.

The equation of a vertical line with x-intercept h is $x = h$.

EXAMPLE 5 *Finding equations of horizontal and vertical lines*

GCLM Find equations of vertical and horizontal lines passing through the point $(-3, 4)$. Graph these lines by hand and with a graphing calculator.

Solution

The *x*-coordinate of the point $(-3, 4)$ is -3. The vertical line $x = -3$ passes through every point in the *xy*-plane with an *x*-coordinate of -3, including the point $(-3, 4)$.

Similarly, the horizontal line $y = 4$ passes through every point with a *y*-coordinate of 4, including the point $(-3, 4)$. The lines $x = -3$ and $y = 4$ are graphed in Figures 2.63 and 2.64.

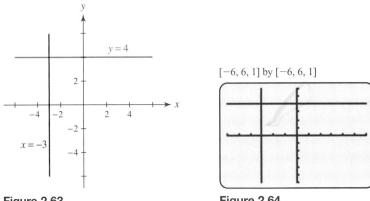

Figure 2.63   Figure 2.64

Technology Note: *Graphing Vertical Lines*

The equation of a vertical line is $x = h$ and cannot be expressed on a graphing calculator in the form "$Y_1 =$". Some graphing calculators graph a vertical line by accessing the DRAW menu.

PARALLEL AND PERPENDICULAR LINES

Slope is an important concept in determining whether two lines are parallel. For example, the lines $y = 2x$ and $y = 2x + 1$ are parallel because they both have slope 2.

Parallel Lines

Two lines with the same slope are parallel.

Two nonvertical parallel lines have the same slope.

EXAMPLE 6 *Finding parallel lines*

Find the slope–intercept form of a line parallel to $y = -2x + 5$, passing through the point $(-2, 3)$. Support your result graphically.

2.4 Equations of Lines and Linear Models

Solution

Because the line $y = -2x + 5$ has slope -2, any parallel line also has slope -2. The line passing through $(-2, 3)$ with slope -2 is determined as follows.

$$y = -2(x + 2) + 3 \quad \text{Point–slope form}$$
$$y = -2x - 4 + 3 \quad \text{Distributive property}$$
$$y = -2x - 1 \quad \text{Slope–intercept form}$$

Graphical support is shown in Figure 2.65, where the lines $Y_1 = -2X + 5$ and $Y_2 = -2X - 1$ are parallel. Y_2 passes through $(-2, 3)$.

[−6, 6, 1] by [−6, 6, 1]

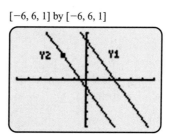

Figure 2.65

Two lines with nonzero slopes are perpendicular if the product of their slopes equals -1.

Perpendicular Lines

Two lines with nonzero slopes m_1 and m_2 are perpendicular if $m_1 m_2 = -1$.

If two lines have slopes m_1 and m_2 such that $m_1 \cdot m_2 = -1$, then they are perpendicular.

Table 2.11 shows examples of slopes m_1 and m_2 that result in perpendicular lines. Note that m_1 and m_2 are negative reciprocals of each other; that is, $m_2 = -\dfrac{1}{m_1}$.

TABLE 2.11 Slopes of Perpendicular Lines

m_1	1	$-\dfrac{1}{2}$	-4	$\dfrac{2}{3}$	$\dfrac{3}{4}$
m_2	-1	2	$\dfrac{1}{4}$	$-\dfrac{3}{2}$	$-\dfrac{4}{3}$

EXAMPLE 7 *Finding perpendicular lines*

 Find the slope–intercept form of the line perpendicular to $y = -\frac{1}{2}x + 1$, passing through the point $(4, -1)$. Graph the lines in the viewing rectangle $[-15, 15, 1]$ by $[-10, 10, 1]$.

Solution

The line $y = -\frac{1}{2}x + 1$ has slope $m_1 = -\frac{1}{2}$. The slope of a perpendicular line is $m_2 = 2$, because

$$m_1 m_2 = -\frac{1}{2} \cdot 2 = -1.$$

A slope–intercept form of the line having slope 2 and passing through $(4, -1)$ may be found as follows.

$$y = 2(x - 4) - 1 \quad \text{Point–slope form}$$
$$y = 2x - 8 - 1 \quad \text{Distributive property}$$
$$y = 2x - 9 \quad \text{Slope–intercept form}$$

Graph $Y_1 = -.5X + 1$ and $Y_2 = 2X - 9$, as shown in Figure 2.66. The perpendicular lines intersect at $(4, -1)$.

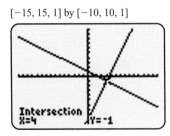

Figure 2.66

Technology Note *Square Viewing Rectangles*

Figure 2.66 shows a square viewing rectangle, in which perpendicular lines intersect at 90°. Try graphing the two perpendicular lines in Example 7 by using the viewing rectangle $[-6, 6, 1]$ by $[-10, 10, 1]$. Do the lines appear perpendicular? For many graphing calculators a square viewing rectangle results when the distance along the y-axis is about $\frac{2}{3}$ the distance along the x-axis. On many graphing calculators you can create a square viewing rectangle automatically by using the ZOOM menu.

2.4 Putting It All Together

The following table shows important forms of an equation of a line.

Concept	Comments	Example
Point–Slope Form $y = m(x - h) + k$ or $y - k = m(x - h)$	Used to find the equation of a line, given two points or one point and the slope	Given two points $(1, 2)$ and $(3, 5)$, first compute $$m = \frac{5 - 2}{3 - 1} = \frac{3}{2}.$$ An equation of this line is $$y = \frac{3}{2}(x - 1) + 2.$$
Slope–Intercept Form $y = mx + b$	A unique equation for a line, determined by the slope m and the y-intercept b	An equation of the line with slope $m = 3$ and y-intercept $b = -5$ is $$y = 3x - 5.$$

The graph of a linear function f is a line. Therefore linear functions can be represented by

$$f(x) = mx + b \quad \text{or} \quad f(x) = m(x - h) + k.$$

The following table summarizes the important concepts involved with special types of lines.

Concept	Equation(s)	Example
Horizontal Line	$y = b$, where b is a constant	A horizontal line with y-intercept 5 has the equation $y = 5$.
Vertical Line	$x = h$, where h is a constant	A vertical line with x-intercept -3 has the equation $x = -3$.
Parallel Lines	$y = m_1 x + b_1$ and $y = m_2 x + b_2$, where $m_1 = m_2$	The lines $y = 2x - 1$ and $y = 2x + 5$ are parallel because both have slope 2.
Perpendicular Lines	$y = m_1 x + b_1$ and $y = m_2 x + b_2$, where $m_1 m_2 = -1$	The lines $y = 3x - 5$ and $y = -\frac{1}{3}x + 2$ are perpendicular because $m_1 m_2 = 3\left(-\frac{1}{3}\right) = -1$.

2.4 EXERCISES

CONCEPTS

1. How many lines are determined by two distinct points?

2. How many lines are determined by a point and a slope?

3. Give the slope–intercept form of a line.

4. Give the point–slope form of a line.

5. Give an equation of a horizontal line with y-intercept b.

6. Give an equation of a vertical line with x-intercept h.

7. If two parallel lines have slopes m_1 and m_2, what can be said about m_1 and m_2?

8. If two perpendicular lines have slopes m_1 and m_2, then $m_1 \cdot m_2 =$ _____.

Exercises 9–12: Match the equation with its graph (a.–d.), where m and b are constants.

9. $y = mx + b,\ m > 0$

10. $y = mx + b,\ m < 0$ and $b \neq 0$

11. $y = mx,\ m < 0$

12. $y = b$

a.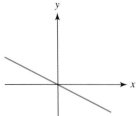

b.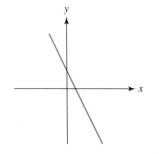

c.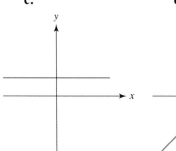

d.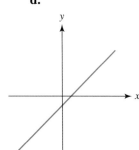

Exercises 13–18: Determine whether the given point lies on the line.

13. $(-3, 4)$ $\quad y = -\dfrac{2}{3}x + 2$

14. $(4, 2)$ $\quad y = \dfrac{1}{4}x - 1$

15. $(0, 4)$ $\quad y = -(x - 1) + 3$

16. $(2, -11)$ $\quad y = -2(x + 3) - 1$

17. $(-4, 3)$ $\quad y = \dfrac{1}{2}(x + 4) + 2$

18. $(1, -10)$ $\quad y = 3(x - 5) - 1$

EQUATIONS OF LINES

Exercises 19–22: Use the labeled point to write a point–slope form of the line.

19.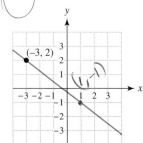

20.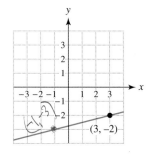

21.

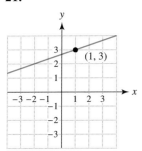

22.

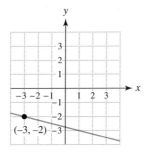

Exercises 23–30: Find a point–slope form of the line satisfying the given conditions.

23. Slope -2, passing through $(2, -3)$

24. Slope $\frac{1}{2}$, passing through $(-4, 1)$

25. Slope 1.3, passing through $(1990, 25)$

26. Slope 45, passing through $(1999, 103)$

27. Passing through $(1, 3)$ and $(-5, -1)$

28. Passing through $(-2, 4)$ and $(3, -1)$

29. Passing through $(6, 0)$ and $(0, 4)$

30. Passing through $(1980, 16)$ and $(2000, 66)$

Exercises 31–36: Write the equation in slope–intercept form.

31. $y = 2(x - 1) - 2$

32. $y = -3(x + 2) + 5$

33. $y = \frac{1}{2}(x + 4) + 1$

34. $y = -\frac{2}{3}(x - 3) - 6$

35. $y = 22(x - 1.5) - 10$

36. $y = -30(x + 3) + 106$

Exercises 37–46: Find the slope–intercept form of the line satisfying the given conditions.

37. Slope $-\frac{1}{3}$, passing through $(0, -5)$

38. Slope 5, passing through $(-1, 4)$

39. Passing through $(3, -2)$ and $(2, -1)$

40. Passing through $(8, 3)$ and $(-7, 3)$

41. x-intercept 2, y-intercept $-\frac{2}{3}$

42. x-intercept -3, y-intercept 4

43. Parallel to $y = 4x - 2$, passing through $(1, 3)$

44. Parallel to $y = -\frac{2}{3}x$, passing through $(0, -10)$

45. Perpendicular to $y = -\frac{1}{3}x + 4$, passing through $(-3, 5)$

46. Perpendicular to $y = \frac{3}{4}(x - 2) + 1$, passing through $(-2, -3)$

Exercises 47–54: Find an equation of a line satisfying the given conditions.

47. Vertical, passing through $(-1, 6)$

48. Vertical, passing through $(2, -7)$

49. Horizontal, passing through $\left(\frac{3}{4}, -\frac{5}{6}\right)$

50. Horizontal, passing through $(5.1, 6.2)$

51. Perpendicular to $y = \frac{1}{2}$, passing through $(4, -9)$

52. Perpendicular to $x = 2$, passing through $(3, 4)$

53. Parallel to $x = 4$, passing through $\left(-\frac{2}{3}, \frac{1}{2}\right)$

54. Parallel to $y = -2.1$, passing through $(7.6, 3.5)$

GRAPHICAL INTERPRETATION

55. *Distance and Speed* A person is driving a car along a straight road. The following graph shows the distance y in miles that the driver is from home after x hours.
  (a) Is the person traveling toward or away from home?
  (b) The graph passes through $(1, 35)$ and $(3, 95)$. Discuss the meaning of these points.
  (c) Find the slope–intercept form of the line. Interpret the slope as a rate of change.
  (d) Use the graph to estimate the y-coordinate of the point $(4, y)$ lying on the graph. Then find y, using your formula from part (c).

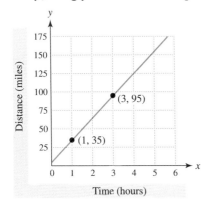

56. *Water Flow* The following graph shows the amount of water y in an 80-gallon tank after x minutes have elapsed.
  (a) Is water entering or leaving the tank? How much water is in the tank after 3 minutes?
  (b) Find both the x- and y-intercepts. Explain their meanings.
  (c) Find a slope–intercept form of the line. Interpret the slope as a rate of change.
  (d) Use the graph to estimate the y-coordinate of the point $(6, y)$ that lies on the line. Then find y, using your formula from part (c).

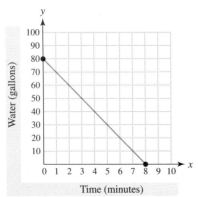

Exercises 57–62: Modeling Match the situation to the graph (a.–f.) that models it best.

57. The total federal debt y from 1985 to 1990.

58. The approximate distance y from New York City to Seattle, Washington, during year x.

59. The amount of money y earned working for x hours at a fixed hourly rate.

60. The sales of 8-millimeter movie projectors from 1970 to 1990.

61. The yearly average temperature y in degrees Celsius at the South Pole for year x.

62. The height above sea level of a rocket launched from a submarine during the first minute of the rocket's flight.

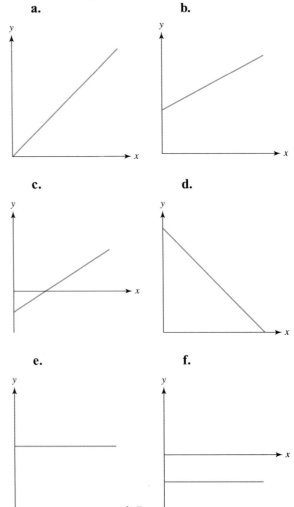

Exercises 63–64: Decide whether the points in the table lie on a line. If they do, find the slope–intercept form of the line.

63.
x	1	2	3	4
y	−4	0	4	8

64.
x	−1	0	1	2
y	8	5	4	1

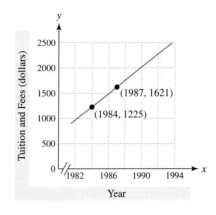

APPLICATIONS

65. *State and Federal Inmates* From 1988 to 1995, state and federal prison inmates in thousands can be modeled by

$$f(x) = 70(x - 1988) + 628,$$

where x is the year. (**Source:** Department of Justice.)
   (a) Find the number of inmates in 1988.
   (b) Graph f in the viewing rectangle [1988, 1995, 1] by [600, 1200, 100].
   (c) What is the slope of the graph of f? Interpret the slope as a rate of change.

66. *Municipal Waste* From 1960 to 1995, municipal solid waste in millions of tons can be modeled by $f(x) = 3.4(x - 1960) + 87.8$, where x is the year. (**Source:** Environmental Protection Agency.)
   (a) Find the tons of waste in 1960.
   (b) Graph f in the viewing rectangle [1960, 1995, 5] by [60, 220, 20].
   (c) What is the slope of the graph of f? Interpret the slope as a rate of change.

67. *Tuition and Fees* The graph models average tuition and fees in dollars at public four-year colleges and universities from 1981 through 1994. See the accompanying figure. (**Source:** The College Board.)
   (a) The graph passes through the points (1984, 1225) and (1987, 1621). Explain the meaning of each point.
   (b) Find a point–slope form of this line. Interpret the slope as a rate of change.

68. *Western Population* In 1950, the western region of the United States had a population of 20.2 million, and in 1960 it was 28.4 million.
   (a) Find values for m, h, and k so that $f(x) = m(x - h) + k$ models this population during year x.
   (b) Estimate the western population in 1990.

69. *Cost of Driving* The cost of driving a car includes both fixed costs and mileage costs. Assume that it costs $189.20 per month for insurance and car payments and it costs $0.30 per mile for gasoline, oil, and routine maintenance.
   (a) Find values for a and b so that $f(x) = ax + b$ models the monthly cost of driving the car x miles.
   (b) What does the y-intercept on the graph of f represent?

70. *Antarctic Ozone Layer* The ozone layer occurs in the atmosphere between altitudes of 12 and 18 miles and it is an important filter of ultraviolet light from the sun. The thickness of the ozone layer is frequently measured in Dobson units. An average value is 300 Dobson units. In 1991 the reported minimum in the antarctic *ozone hole* was about 110 Dobson units. (**Source:** Huffman, R., *Atmospheric Ultraviolet Remote Sensing*.)
   (a) The equation $y = 0.01x$ describes the thickness y in millimeters of an ozone layer that is x Dobson units. How many millimeters thick was the ozone layer over the antarctic in 1991?
   (b) Was the ozone hole in the antarctic actually a *hole* in the ozone layer? Explain.
   (c) What is the average thickness of the ozone layer?

71. *HIV Infection Rates* In 1996, there were an estimated 22 million HIV infections worldwide, with an annual infection rate of 3.1 million. (*Source:* Centers for Disease Control and Prevention.)
   (a) Find values for m, h, and k so that $f(x) = m(x - h) + k$ models the number of HIV infections in year x.
   (b) Estimate the number of HIV infections in 2003.

72. *Farm Pollution* In 1979, the number of farm pollution incidents reported in England and Wales was 1480, and in 1988 it was 4000. (*Source:* Mason, C., *Biology of Freshwater Pollution*.)
   (a) Determine a line in the form
   $$y = m(x - h) + k$$
   that models these data, where y represents the number of pollution incidents during year x.
   (b) Estimate the number of incidents in 1986.

73. *U.S. Population* The population for selected years is shown in the following table.

Year	1960	1970	1980	1990
Population (millions)	179	203	227	249

Source: Bureau of the Census.

   (a) Make a scatterplot of the data.
   (b) Find values for m, h, and k so that
   $$f(x) = m(x - h) + k$$
   models the data. (Answers may vary.)
   (c) Use f to estimate the population in 2000.

74. *Women in Politics* The following table lists percentages of women in state legislatures for selected years.

Year	1981	1983	1985	1987
Percent	12.1	13.3	14.8	15.7

Year	1989	1991	1993	1995
Percent	17.0	18.3	20.5	20.7

Source: National Women's Political Caucus.

   (a) Make a scatterplot of the data.
   (b) Find values for m, h, and k so that $f(x) = m(x - h) + k$ models these data. (Answers may vary.)
   (c) Use f to estimate the percentage of women in state legislatures in 2000.

WRITING ABOUT MATHEMATICS

75. Explain how you can recognize equations of parallel lines. How can you recognize equations of perpendicular lines?

76. Suppose that some real data can be modeled by a linear function f. Explain what the slope of the graph of f indicates about the data.

CHECKING BASIC CONCEPTS FOR SECTIONS 2.3 AND 2.4

1. Find the slope and a point on each line.
   (a) $y = -3(x - 5) + 7$
   (b) $y = 10$
   (c) $x = -5$
   (d) $y = 5x + 3$

2. (a) Calculate the slope of the line passing through the points $(2, -4)$ and $(5, 2)$.
   (b) Find a point–slope form and the slope–intercept form of the line.
   (c) Find the x-intercept and the y-intercept of the line.

3. The following graph shows the distance that a car is from home after x hours.
   (a) Is the car moving toward or away from home?
   (b) Find the slope of the line. Interpret the slope as a rate of change.
   (c) Find the x-intercept and the y-intercept. Explain the meaning of each.
   (d) Determine a and b so that
   $$f(x) = ax + b$$
   models this situation.

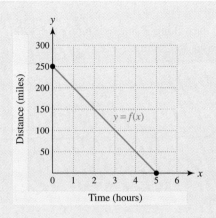

4. Find the equation of a vertical line and the equation of a horizontal line passing through $(-2, 5)$.

5. Find equations of lines that are perpendicular and parallel to $y = -\frac{1}{2}x + 3$, passing through $(2, -4)$.

Chapter 2 Summary

Section 2.1 Functions and Their Representations

A function receives input x and calculates exactly one output y. In function notation this relation is expressed as $y = f(x)$. The set of valid inputs is called the domain, and the corresponding set of outputs is called the range.

Example: If $f(x) = x^2$, then f outputs the square of the input x. Any real number is a valid input, so its domain is all real numbers. The output x^2 cannot be a negative number, so the domain is all nonnegative real numbers.

Functions can be represented with formulas, tables of values, graphs, and words, which are called symbolic, numerical, graphical, and verbal representations, respectively. The vertical line test is used to determine whether a graph represents a function.

Section 2.2 Linear Functions

There are many different types of functions. A linear function may be expressed symbolically by $f(x) = ax + b$. Its graph is a line with slope $m = a$ and y-intercept b. Slope a means that, for each unit increase in the input x, the output $f(x)$ changes by a fixed amount equal to a.

Example: If $f(x) = 2x - 3$, then the graph of f has slope 2 and y-intercept -3. Each time x increases by 1 unit, y increases by 2 units.

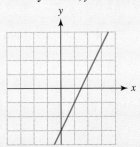

Section 2.3 *The Slope of a Line*

Slope is the ratio of rise over run and indicates a rate of change. For any two points (x_1, y_1) and (x_2, y_2) on a line, the slope of the line is $m = \dfrac{y_2 - y_1}{x_2 - x_1}$. A positive slope indicates a line rising to the right, a negative slope indicates a line falling to the right, and a zero slope indicates a horizontal line. A vertical line has an undefined slope.

Example: A line passing through $(3, 4)$ and $(-2, 6)$ has slope $m = \dfrac{6 - 4}{-2 - 3} = -\dfrac{2}{5}$.

This ratio indicates that, for every 5-unit increase in x, y decreases by 2 units.

The slope–intercept form of a line is unique and given by $y = mx + b$. For example, $y = -4x + 3$ represents a line with slope -4 and y-intercept 3.

Section 2.4 *Equations of Lines and Linear Models*

Two basic forms of the equation of a line are the slope–intercept form and the point–slope form. If a line has slope m and passes through the point (h, k), then the point–slope form is given by $y = m(x - h) + k$. Any point on the line may be used for (h, k), so the point–slope form is not unique.

Example: The equation of a line passing through $(3, 4)$ with slope $-\dfrac{1}{4}$ is

$$y = -\dfrac{1}{4}(x - 3) + 4.$$

A vertical line is given by $x = h$, and a horizontal line is given by $y = k$. Examples include $x = -1$ and $y = 5$. If parallel lines have slopes m_1 and m_2, then $m_1 = m_2$. If two (nonvertical) lines are perpendicular then their slopes satisfy $m_1 \cdot m_2 = -1$, or equivalently, $m_1 = -\dfrac{1}{m_2}$.

Parallel Lines: $\quad y = 2x - 5 \quad$ and $\quad y = 2x + 7 \quad\quad$ Both lines have slope 2.

Perpendicular Lines: $\quad y = \dfrac{2}{3}x + 1 \quad$ and $\quad y = -\dfrac{3}{2}x - 3 \quad\quad$ The product of their slopes is $\dfrac{2}{3}\left(-\dfrac{3}{2}\right) = -1.$

Chapter 2
Review Exercises

Section 2.1

Exercises 1–4: Evaluate the symbolic representation of $f(x)$ for the given values of x.

1. $f(x) = 3x - 1$ $\quad x = -2, \frac{1}{3}$

2. $f(x) = 5 - 3x^2$ $\quad x = -3, 1$

3. $f(x) = \sqrt{x - 2}$ $\quad x = 0, 9$

4. $f(x) = 5$ $\quad x = -5, \frac{7}{5}$

Exercises 5–8: Sketch a graph of f by hand. Check your work with a graphing calculator.

5. $f(x) = -2x$

6. $f(x) = \frac{1}{2}x - \frac{3}{2}$

7. $f(x) = x^2 - 1$

8. $f(x) = \sqrt{x + 1}$

Exercises 9 and 10: Use the graph of f to evaluate the given expressions.

9. $f(0)$ and $f(-3)$

10. $f(-2)$ and $f(1)$

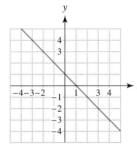
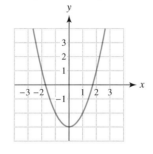

11. Evaluate $f(-1)$ and $f(3)$.

x	−1	1	3	5
f(x)	7	3	−1	−5

12. A function f is represented verbally by "Multiply the input x by 3 and then subtract 2." Give numerical, symbolic, and graphical representations for f. Let $x = -3, -2, -1, \ldots, 3$ in the numerical representation, and let $-3 \leq x \leq 3$ for the graphical representation.

Exercises 13 and 14: Use the graph of f to estimate its domain and range.

13.
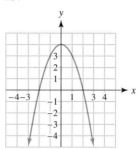

14.
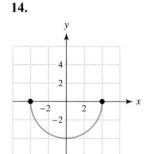

Exercises 15 and 16: Determine whether the graph represents a function.

15.
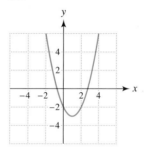

16.
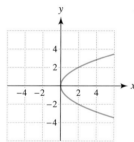

Exercises 17 and 18: Find the domain and range of S. Then state whether S defines a function.

17. $S = \{(-3, 4), (-1, 4), (2, 3), (4, -1)\}$

18. $S = \{(-1, 5), (0, 3), (1, -2), (-1, 2), (2, 4)\}$

Section 2.2

Exercises 19 and 20: Determine whether the graph represents a linear function.

19.
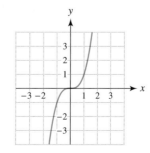

20.
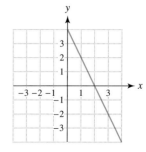

Exercises 21–24: Determine the slope of the line in the graph.

21.
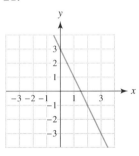

22.
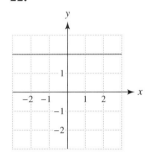

23.
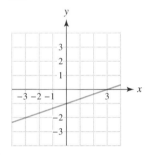

24.
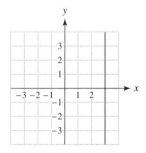

Section 2.3

Exercises 25–28: Calculate the slope of the line passing through the given points.

25. $(-1, 2), (3, 8)$

26. $\left(-3, \frac{5}{2}\right), \left(1, -\frac{1}{2}\right)$

27. $(3, -4), (5, -4)$

28. $(-2, 6), (-2, 8)$

29. Let $f(x) = \frac{1}{2}x - 2$.
   (a) Find the slope and y-intercept of the graph of f.
   (b) Sketch a graph of f by hand. Use a graphing calculator to check your work.

Section 2.4

30. Let f be a linear function. Find the slope, x-intercept, and y-intercept of the graph of f.

x	−1	0	1	2
$f(x)$	6	4	2	0

Exercises 31–34: Write a point–slope form and the slope–intercept form of a line satisfying the given conditions.

31. x-intercept 2, y-intercept -3

32. Passing through $(-1, 4)$ and $(2, -2)$

33. Parallel to $y = 4x - 3$, passing through $\left(-\frac{3}{5}, \frac{1}{5}\right)$

34. Perpendicular to $y = \frac{1}{2}x$, passing through $(-1, 1)$

Exercises 35 and 36: Find the slope–intercept form of the line.

35.
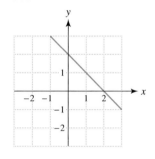

36.
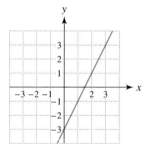

Exercises 37–38: Determine whether the given point lies on the line.

37. $(2, -1)$ $y = -2(x - 1) + 3$

38. $\left(4, -\frac{5}{2}\right)$ $y = \frac{1}{4}(x + 2) - 4$

Exercises 39–40: Find an equation of a line satisfying the given conditions.

39. Vertical, passing through $(-4, 14)$

40. Horizontal, passing through $\left(\frac{11}{13}, -\frac{7}{13}\right)$

Exercises 41 and 42: Decide whether the points in the table lie on a line. If they do, find the slope–intercept form of the line.

41.
x	1	2	3	4
y	6	7	8	7

42.

x	−2	0	2	4
y	1	5	9	13

APPLICATIONS

43. *Modeling* The following line graph shows the population of the United States in millions.
    (a) Find the slope of each line segment.
    (b) Interpret each slope as a rate of change.

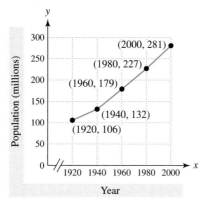

44. *Flow Rates* A water tank has an inlet pipe with a flow rate of 10 gallons per minute and an outlet pipe with a flow rate of 6 gallons per minute. A pipe can be either completely closed or open. The following graph shows the number of gallons of water in the tank after x minutes. Interpret the graph.

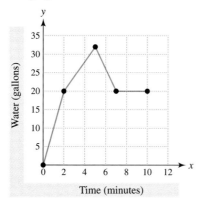

45. *Age at First Marriage* The median age at the first marriage for men from 1890 to 1960 can be modeled by $f(x) = -0.0492x + 119.1$, where x is the year. (*Source:* National Center of Health Statistics.)
    (a) Find the median age in 1910.
    (b) Graph f in [1885, 1965, 10] by [22, 26, 1]. What happened to the median age during this time period?

(c) What is the slope of the graph of f? Interpret the slope as a rate of change.

46. *Marriages* From 1980 to 1997 the number of U.S. marriages in millions could be modeled by $f(x) = 2.4$, where x is the year.
    (a) Estimate the number of marriages in 1991.
    (b) What information does f give about the number of marriages during this time period?

47. *Fat Grams* One cup of whole milk contains 8 grams of fat.
    (a) Give a formula for $f(x)$ that calculates the number of fat grams in x cups of milk.
    (b) What is the slope of the graph of f?
    (c) Interpret the slope as a rate of change.

48. *Birth Rate* The U.S. birth rate per 1000 people from 1990 through 1997 is shown in the following table.

Year	1990	1991	1992	1993
Birth Rate	16.7	16.3	16.0	15.7

Year	1994	1995	1996	1997
Birth Rate	15.3	14.8	14.7	14.5

Source: Bureau of the Census.

(a) Make a scatterplot of the data.
(b) Find values for m, h, and k so that $f(x) = m(x - h) + k$ models these data. (Answers may vary.)
(c) Use f to estimate the birth rate in 2000.

49. *Unhealthy Air Quality* The Environmental Protection Agency (EPA) monitors air quality in U.S. cities. The function f, represented by the following table, gives the annual number of days with unhealthy air quality in Los Angeles, California, from 1992 through 1996.

x	1992	1993	1994	1995	1996
f(x)	185	146	136	103	88

Source: Environmental Protection Agency.

(a) Find $f(1995)$ and interpret your result.
(b) Identify the domain and range of f.
(c) Discuss the trend of air pollution in Los Angeles.

50. *Temperature Scales* The following table shows equivalent temperatures in degrees Celsius and degrees Fahrenheit.

°C	−40	0	15	35	100
°F	−40	32	59	95	212

(a) Plot the data. Let the x-axis correspond to the Celsius temperature and the y-axis correspond to the Fahrenheit temperature. What type of relation exists between the data?
(b) Find $f(x) = ax + b$ so that f receives the Celsius temperature x as input and outputs the corresponding Fahrenheit temperature.
(c) If the temperature is 20°C, what is this temperature in degrees Fahrenheit?

51. *Graphical Model* A 500-gallon water tank is initially full and then emptied at a constant rate of 50 gallons per minute. Ten minutes after the tank is empty, it is filled by a pump that outputs 25 gallons per minute. Sketch a graph that depicts the amount of water in the tank after x minutes.

52. *HIV Infections* In 1996, there were about 775,000 HIV infections in the United States, with an annual infection rate of 40,000. (*Source:* Centers for Disease Control and Prevention.)
(a) Assuming that this trend continues, find $f(x) = m(x - h) + k$ so that f models the number of HIV infections during year x.
(b) Find $f(2000)$ and interpret the result.

53. *Asian-American Population* In 1998, the Asian-American segment of the U.S. population was 10.5 million and was predicted to be 12.5 million in 2003.
(a) Find values for m, h, and k so that $f(x) = m(x - h) + k$ models this population in millions.
(b) Evaluate $f(2002)$ and interpret the result.

54. Determine the slope–intercept form of the line that passes through these data points.

x	−2	−1	0	1	2
y	5.4	4.2	3	1.8	0.6

Chapter 2

Test

1. Evaluate $f(4)$ if $f(x) = 3x^2 - \sqrt{x}$.

2. Sketch a graph of the following functions by hand.
(a) $f(x) = -2x + 1$
(b) $f(x) = x^2 + 1$

3. Use the graph of f to evaluate $f(-3)$ and $f(0)$. Determine the domain and range of f.

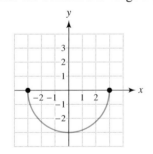

4. A function f is represented verbally by "Square the input x and then subtract 5." Give symbolic, numerical, and graphical representations of f. Let $x = -3, -2, -1, \ldots, 3$ in the numerical representation, and let $-3 \le x \le 3$ for the graphical representation.

5. Determine whether the graph represents a function. Explain your reasoning.

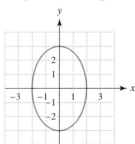

6. Determine the slope of the line shown in the graph.

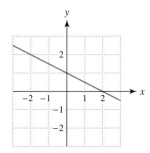

7. Find the slope–intercept form of the line passing through the points $(1, -2)$ and $\left(-5, \frac{3}{2}\right)$. What are its x- and y-intercepts?

8. Let f be a linear function. Find the slope, x-intercept, and y-intercept of the graph of f.

x	−2	−1	0	2
f(x)	9	7	5	1

9. Give a point–slope form and the slope–intercept form of a line parallel to $y = 1 - 3x$, passing through $\left(\frac{1}{3}, 2\right)$.

10. Find the slope–intercept form for the line shown in the graph.

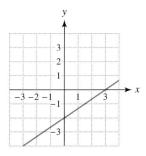

11. Find an equation of a vertical line passing through the point $\left(\frac{2}{3}, -\frac{1}{7}\right)$.

12. *Modeling* The following line graph shows the number of welfare beneficiaries in millions for selected years. (*Source:* Administration for Children and Families.)
    (a) Find the slope of each line segment.
    (b) Interpret each slope as a rate of change.

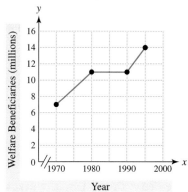

13. *Distance from Home* Starting at home, a driver travels away from home on a straight highway for 2 hours at 60 miles per hour, stops for 1 hour, and then drives home at 40 miles per hour. Sketch a graph that shows the distance between the driver and home.

14. *Memorial Day Travelers* The following table shows the number of travelers on Memorial Day for selected years.

Year	1990	1992	1994
Travelers (millions)	24.8	27.0	29.2

Year	1996	1998	2000
Travelers (millions)	30.3	32.0	34.4

Source: American Automobile Association.

(a) Make a scatterplot of the data.
(b) Find values for m, h, and k so that $f(x) = m(x - h) + k$ models these data.
(c) Use f to estimate the number of travelers in 2002.

Chapter 2

Extended and Discovery Exercises

1. *Developing a Model* Two identical cylindrical tanks, A and B, each contain 100 gallons of water. Tank A has a pump that begins removing water at a constant rate of 8 gallons per minute. Tank B has a plug removed from its bottom and water begins to flow out—faster at first and then more slowly.
   (a) Assuming that the tanks become empty at the same time, sketch a graph that models the amount of water in each tank. Explain your graphs.
   (b) Which tank is half empty first? Explain.

2. *Modeling Real Data* Per capita personal incomes in the United States from 1990 through 1996 are listed in the following table.

Year	1990	1991	1992	1993
Income	$18,666	$19,091	$20,105	$20,800

Year	1994	1995	1996
Income	$21,809	$23,359	$24,436

   Source: Department of Commerce.

   (a) Make a scatterplot of the data.
   (b) Find a function f that models the data. Explain your reasoning.
   (c) Use f to estimate per capita personal income in 1998.

3. *Graphing a Rectangle* One side of a rectangle has vertices (0, 0) and (5, 3). Use your graphing calculator to graph four lines that outline the boundary of the rectangle if the point (0, 5) lies on one side of this rectangle. Your rectangle should look like the one in the figure.

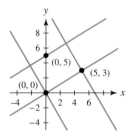

Exercises 4–6: (Refer to Exercise 3.) Graph the rectangle that satisfies the given conditions.

4. Vertices (0, 0), (2, 2), and (1, 3)

5. Vertices (1, 1), (5, 1), and (5, 5)

6. Vertices (4, 0), (0, 4), (0, −4), and (−4, 0)

7. *Remaining Life Expectancy* The table lists the average *remaining* life expectancy E in years for females at age x.

x (year)	0	10	20	30	40
E (year)	72.3	69.9	60.1	50.4	40.9

x (year)	50	60	70	80
E (year)	31.6	23.1	15.5	9.2

   Source: Department of Health and Human Services.

   (a) Make a line graph of the data.
   (b) Assume that the graph represents a function f. Calculate the slopes of each line segment and interpret each slope as a rate of change.
   (c) Determine the life expectancy (not the remaining life expectancy) of a 20-year-old woman. What is the life expectancy of a 70-year-old woman? Discuss reasons why these two expectancies are not equal.

8. *Weight of a Small Fish* The figure shows a graph of a function f that models the weight in milligrams of a small fish, *Lebistes reticulatus*, during the first 14 weeks of its life. (**Source:** D. Brown and P. Rothery, *Models in Biology.*)

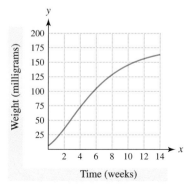

(a) Estimate the weight of the fish when it hatches, at 6 weeks, and at 12 weeks.

(b) If (x_1, y_1) and (x_2, y_2) are points on the graph of a function, the *average rate of change of f from x_1 to x_2* is given by $\dfrac{y_2 - y_1}{x_2 - x_1}$. Approximate the average rates of change of f from hatching to 6 weeks and from 6 weeks to 12 weeks.

(c) Interpret these rates of change.

(d) During which time period does the fish gain weight the fastest?

9. *Interpreting Carbon Dioxide Levels* Carbon dioxide gas is a greenhouse gas that may cause Earth's climate to warm. Plants absorb carbon dioxide during daylight and release carbon dioxide at night. The burning of fossil fuels, such as gasoline, produces carbon dioxide. At Mauna Loa, Hawaii, atmospheric carbon dioxide levels in parts per million have been measured regularly since 1958. The accompanying figure shows a graph of the carbon dioxide levels between 1960 and 1995. (**Source:** Nilsson, A., *Greenhouse Earth*.)

(a) What is the overall trend in the carbon dioxide levels?

(b) Discuss what happens to the carbon dioxide levels each year.

(c) Give an explanation for the shape of this graph.

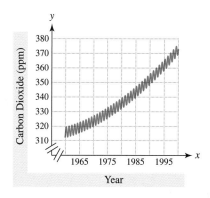

10. (Refer to Exercise 9.) The atmospheric carbon dioxide levels at Barrow, Alaska, in parts per million from 1970 to 1995 are shown in the accompanying figure. (**Source:** Zeilik, M., S. Gregory, and D. Smith, *Introductory Astronomy and Astrophysics*.)

(a) Compare this graph with the graph in Exercise 9.

(b) Discuss possible reasons for their similarities and differences.

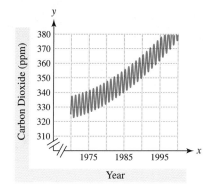

CHAPTER 3: Linear Equations and Inequalities

In this chapter we use technology to help us solve equations visually. A graph produced on a graphing calculator contains approximately the same amount of information as a page of text. However, many people find the information in a graph easier to absorb. The reason is that much of the human brain is devoted to processing visual information. The accompanying table lists the approximate age in years of several forms of communication. Note that eyesight has had a considerably longer time to evolve than other forms.

Form of Communication	Age (years)
Eyesight	500,000,000
Oral communication	50,000
Written words	5000
Printed books	500
Television	60
Scientific visualization	20

*Say it, I'll forget.
Demonstrate it, I may recall.
But if I'm involved, I'll understand.
—Old Chinese Proverb—*

Sources: M. Friedhoff and W. Benzon, *The SECOND Computer Revolution VISUALIZATION*; G. Nielson and B. Shriver, Editors, *Visualization in Scientific Computing*.

3.1 LINEAR EQUATIONS

Equations ~ Numerical Solutions ~ Graphical Solutions ~
Symbolic Solutions ~ Using More Than One Method ~ Percentages

INTRODUCTION

A primary objective of mathematics is solving equations. Billions of dollars are spent each year to solve equations that hold the answers for creating better products. The ability to solve equations has resulted in televisions, CD players, satellites, fiber optics, CAT scans, computers, and accurate weather forecasts. In this section we discuss linear equations and their applications. Linear equations are one of the simplest types of equations. They can be solved numerically, graphically, and symbolically.

EQUATIONS

In Chapter 2 we discussed modeling data with linear functions. For example, we used $f(x) = 50x + 100$ to model the distance in miles that a car was from the Texas border after x hours. How could we use $f(x)$ to determine when the car was 300 miles from the border? We could solve the equation

$$50x + 100 = 300.$$

This is an example of a *linear equation* in one variable.

Linear Equation in One Variable

A **linear equation** in one variable is an equation that can be written in the form

$$ax + b = 0,$$

where $a \neq 0$.

Examples of linear equations include

$$2x - 1 = 0, \quad -5x = 10 + x, \quad \text{and} \quad 3x + 8 = 2.$$

Although the second and third equations do not appear to be in the form $ax + b = 0$, they can be transformed by using properties of algebra, which we discuss later in this section.

To *solve* an equation means we find all values for a variable that make the equation a true statement. Such values are called **solutions,** and the set of all solutions is called the **solution set.** For example, substituting $x = 2$ into the equation $3x - 1 = 5$ results in $3(2) - 1 = 5$, which is a true statement. The value $x = 2$ *satisfies* the equation $3x - 1 = 5$ and is the only solution. The solution set is $\{2\}$. Two equations are *equivalent* if they have the same solution set. *Linear equations have exactly one solution.*

■ MAKING CONNECTIONS

Linear Functions and Equations

A linear function can be written as $f(x) = ax + b$.

A linear equation can be written as $ax + b = 0$ with $a \neq 0$.

NUMERICAL SOLUTIONS

Suppose that we want to solve the linear equation $50x + 100 = 300$. We could use trial and error, substituting different values for x into the equation. For example, if we let $x = 1$, then

$$50(1) + 100 = 150 \neq 300.$$

Thus $x = 1$ is *not* a solution. We could repeat this for other values of x, which results in Table 3.1.

TABLE 3.1

x	1	2	3	4	5	6
$50x + 100$	150	200	250	300	350	400

Table 3.1 indicates that the solution to the equation $50x + 100 = 300$ is $x = 4$. This table gives a **numerical solution** to this equation. Graphing calculators have the capability to create such tables, as illustrated in the next example.

EXAMPLE 1 *Solving a linear equation numerically*

 Solve $2 - \frac{1}{2}x = 1$ numerically.

Solution

Begin by letting $Y_1 = 2 - .5X$ and $Y_2 = 1$. Figure 3.1 reveals that, when $x = 2$, $Y_1 = Y_2 = 1$. Therefore the solution is $x = 2$.

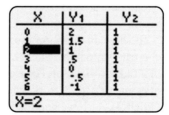

Figure 3.1 Numerical Solution

GRAPHICAL SOLUTIONS

One disadvantage of solving equations numerically is that constructing a table relies on trial and error. Finding a solution, especially if it is a fraction, might take a long time. We can also solve the equation

$$2 - \frac{1}{2}x = 1$$

graphically, rather than numerically. The graphs of $y_1 = 2 - \frac{1}{2}x$ and $y_2 = 1$ intersect at (2, 1), as shown in Figure 3.2. We are seeking an x-value that satisfies $2 - \frac{1}{2}x = 1$, so $x = 2$ is the solution.

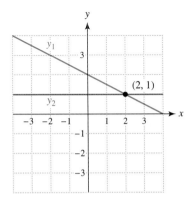

Figure 3.2 Graphical Solution

EXAMPLE 2 *Solving a linear equation graphically*

Figure 3.3 shows graphs of $y_1 = 2x + 1$ and $y_2 = -x + 4$. Use the graph to solve the equation

$$2x + 1 = -x + 4.$$

Check your answer.

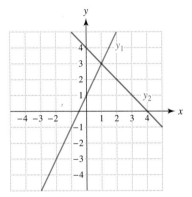

Figure 3.3

Solution

The graphs of y_1 and y_2 intersect at the point (1, 3). Therefore $x = 1$ is the solution. We can check this solution by substituting $x = 1$ into the equation.

$$2x + 1 = -x + 4 \quad \text{Given equation}$$
$$2(1) + 1 \stackrel{?}{=} -1 + 4 \quad \text{Let } x = 1.$$
$$3 = 3 \quad \text{The answer checks.}$$

In the next example we show how a graphing calculator can be used to solve a linear equation graphically.

EXAMPLE 3 *Solving a linear equation graphically*

 Solve $3(1 - x) = 2$ graphically.

Solution

We begin by graphing $Y_1 = 3(1 - X)$ and $Y_2 = 2$, as shown in Figure 3.4. Their graphs intersect near the point (0.3333, 2). Because $\frac{1}{3} = 0.\overline{3}$, $x = \frac{1}{3}$ appears to be the solution. We can verify this result as follows.

$$3(1 - x) = 2 \quad \text{Given equation}$$
$$3\left(1 - \frac{1}{3}\right) \stackrel{?}{=} 2 \quad \text{Substitute } x = \frac{1}{3}.$$
$$2 = 2 \quad \text{It checks.}$$

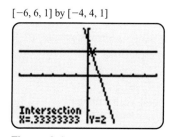

Figure 3.4

Technology Note: *Points of Intersection*

You can find points of intersection in different ways with a graphing calculator. One way is to find points of intersection by zooming in sufficiently. Many types of graphing calculators have the capability to find points of intersection. Other types have equation solvers.

In the next example, we solve a real-life application graphically.

EXAMPLE 4 *Solving a linear equation graphically*

From 1985 to 1990, sales of compact discs in millions in the United States can be modeled by $y_1 = 51.6(x - 1985) + 9.1$, and sales of vinyl LP records in millions can be modeled by $y_2 = -31.9(x - 1985) + 167.7$. Estimate graphically the year x when sales of LP records and compact discs were equal. Interpret the slopes of the graphs of y_1 and y_2 as rates of change. (*Source:* Recording Industry Association of America.)

Solution

To solve this problem graphically we let $Y_1 = 51.6(X - 1985) + 9.1$ and $Y_2 = -31.9(X - 1985) + 167.7$, as shown in Figures 3.5 and 3.6. Their graphs intersect near (1986.9, 107.1). An x-value of 1986.9 corresponds to the year that CD and LP record sales were equal. A y-value of 107.1 represents the number (in millions) of each that were sold. Rounded to the nearest million and year, CD and LP record sales were both 107 million in 1987.

The slope of the graph of y_1 is 51.6. This slope indicates that sales of CDs from 1985 to 1990 were *increasing,* on average, by 51.6 million per year. In contrast, sales of vinyl LP records were *decreasing,* on average, by 31.9 million per year.

Figure 3.5

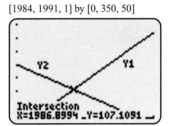

[1984, 1991, 1] by [0, 350, 50]

Figure 3.6

Symbolic Solutions

Linear equations can also be solved symbolically. One advantage of a symbolic method is that the solution is always exact. To solve an equation symbolically, we write a sequence of equivalent equations, using algebraic properties. For example, to solve $3x - 5 = 0$, we might add 5 to both sides of the equation and then divide both sides by 3 to obtain $x = \frac{5}{3}$.

$$3x - 5 = 0 \quad \text{Given equation}$$
$$3x = 5 \quad \text{Add 5 to both sides.}$$
$$\frac{3x}{3} = \frac{5}{3} \quad \text{Divide both sides by 3.}$$
$$x = \frac{5}{3} \quad \text{Simplify.}$$

Adding 5 to both sides is an example of the *addition property of equality* and dividing both sides by 3 is an example of the *multiplication property of equality*. Note that dividing both sides by 3 is equivalent to multiplying both sides by $\frac{1}{3}$.

Properties of Equality

Addition Property of Equality

If a, b, and c are real numbers, then

$$a = b \quad \text{is equivalent to} \quad a + c = b + c.$$

Multiplication Property of Equality

If a, b, and c are real numbers with $c \neq 0$, then

$$a = b \quad \text{is equivalent to} \quad ac = bc.$$

The addition property states that an equivalent equation results if the same number is added to (or subtracted from) both sides of an equation. Similarly, the multiplication property states that an equivalent equation results if both sides of an equation are multiplied (or divided) by the same nonzero number.

EXAMPLE 5 *Solving a linear equation symbolically*

Solve $2 - \frac{1}{2}x = 1$ symbolically.

Solution

To solve an equation symbolically, we write a sequence of equivalent equations by applying the properties of equality.

$$2 - \frac{1}{2}x = 1 \qquad \text{Given equation}$$

$$-2 + 2 - \frac{1}{2}x = 1 + -2 \qquad \text{Add } -2 \text{ to both sides.}$$

$$-\frac{1}{2}x = -1 \qquad \text{Simplify.}$$

$$-2 \cdot -\frac{1}{2}x = -1 \cdot -2 \qquad \text{Multiply both sides by } -2, \text{ the reciprocal of } -\frac{1}{2}.$$

$$x = 2 \qquad \text{Simplify.}$$

In the next example we use the distributive property to solve a linear equation.

EXAMPLE 6 *Solving a linear equation symbolically*

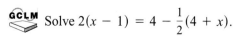 Solve $2(x - 1) = 4 - \frac{1}{2}(4 + x)$.

Solution

We begin by applying the distributive property.

$$2(x - 1) = 4 - \frac{1}{2}(4 + x) \quad \text{Given equation}$$

$$2x - 2 = 4 - 2 - \frac{1}{2}x \quad \text{Distributive property}$$

$$2x - 2 = 2 - \frac{1}{2}x \quad \text{Simplify.}$$

Next, we transpose (or move) the constant terms to the right and the x-terms to the left.

$$2x - 2 + 2 = 2 - \frac{1}{2}x + 2 \quad \text{Add 2 to both sides.}$$

$$2x = 4 - \frac{1}{2}x \quad \text{Simplify.}$$

$$2x + \frac{1}{2}x = 4 - \frac{1}{2}x + \frac{1}{2}x \quad \text{Add } \frac{1}{2}x \text{ to both sides.}$$

$$\frac{5}{2}x = 4 \quad \text{Simplify.}$$

Finally, we multiply by $\frac{2}{5}$, which is the reciprocal of $\frac{5}{2}$.

$$\frac{2}{5} \cdot \frac{5}{2}x = 4 \cdot \frac{2}{5} \quad \text{Multiply by } \frac{2}{5}.$$

$$x = \frac{8}{5} = 1.6 \quad \text{Simplify.}$$

Technology Note *Finding a Numerical Solution*

We can solve Example 6 numerically by letting

$$Y_1 = 2(X - 1) \quad \text{and} \quad Y_2 = 4 - .5(4 + X),$$

as illustrated in Figure 3.7, where $Y_1 = Y_2$ when $x = 1.6$. Note that to find the solution we must increment x by 0.1 rather than by 1.

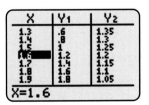

Figure 3.7

In the next example we solve a linear equation to estimate the year when a cumulative total of 900 thousand HIV infections were reached in the United States.

EXAMPLE 7 Modeling HIV infections

The cumulative number of HIV infections in thousands for the United States in year x may be modeled by $f(x) = 40x - 79{,}065$. Estimate when this total reached 900 thousand. (*Source:* Centers for Disease Control and Prevention.)

Solution

We need to determine when $f(x)$ equals 900 thousand.

$$40x - 79{,}065 = 900 \quad \text{$f(x) = 900$}$$
$$40x - 79{,}065 + 79{,}065 = 900 + 79{,}065 \quad \text{Add 79,065 to both sides.}$$
$$40x = 79{,}965 \quad \text{Simplify.}$$
$$\frac{40x}{40} = \frac{79{,}965}{40} \quad \text{Divide both sides by 40.}$$
$$x = 1999.125 \quad \text{Simplify.}$$

According to this model, the total of 900 thousand HIV infections was reached during 1999.

USING MORE THAN ONE METHOD

We can solve an equation by using more than one method. In the next example we use symbolic, graphical, and numerical methods to solve a linear equation.

EXAMPLE 8 Solving a linear equation symbolically, graphically, and numerically

The length of a rectangular room is 2 feet more than its width. If the perimeter of the room is 80 feet, write an equation whose solution gives the width of the room. Solve the equation symbolically, graphically, and numerically.

Solution

Let x represent the width of the room. The length is given by $x + 2$ (see Figure 3.8), and the perimeter P is represented by

$$P = x + (x + 2) + x + (x + 2).$$

This equation simplifies to $P = 4x + 4$, so substitute 80 for P and solve $4x + 4 = 80$.

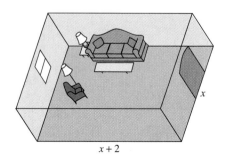

Figure 3.8

Symbolic Solution Start by subtracting 4 from each side of the equation.

$$4x + 4 = 80 \qquad \text{Equation to be solved}$$
$$4x + 4 - 4 = 80 - 4 \qquad \text{Subtract 4.}$$
$$4x = 76 \qquad \text{Simplify.}$$
$$\frac{4x}{4} = \frac{76}{4} \qquad \text{Divide both sides by 4.}$$
$$x = 19 \qquad \text{Simplify.}$$

The width of the room is 19 feet, and the length is 21 feet.

Graphical Solution Graph $Y_1 = 4x + 4$ and $Y_2 = 80$, as shown in Figure 3.9. Their graphs intersect at the point (19, 80). Thus $x = 19$ is the width.

Numerical Solution Construct a table for $Y_1 = 4X + 4$ and $Y_2 = 80$, as shown in Figure 3.10. Note that $Y_1 = Y_2 = 80$ when $x = 19$.

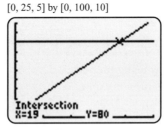

Figure 3.9

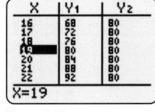

Figure 3.10

Critical Thinking

If the same linear equation is solved with symbolic, numerical, and graphical methods, how should the answers compare? Can you think of a situation wherein the answers may not agree exactly?

PERCENTAGES

Applications involving percentages often make use of linear equations. Taking P percent of x is given by Px, where P is written in decimal form. For example, to calculate 35% of x, we compute $0.35x$. As a result, 35% of $150 is $0.35(150) = 52.5$, or $52.50.

EXAMPLE 9 *Analyzing smoking data*

In 1998, an estimated 27.7% of Americans aged 12 and older, or 60 million people, were cigarette smokers. Use these data to estimate the number of Americans that are aged 12 and older. *(Source: Department of Health and Human Services.)*

Solution

Let x be the number of Americans aged 12 and older. Then 27.7% of x equals 60 million, so we must solve the equation

$$0.277x = 60.$$

To solve this equation, we divide both sides by 0.277.

$$\frac{0.277x}{0.277} = \frac{60}{0.277}$$

$$x \approx 216.6 \qquad \text{Approximate.}$$

In 1998, approximately 216.6 million Americans were aged 12 and older.

3.1 Putting It All Together

A linear equation in one variable can be written in the form $ax + b = 0$, where $a \neq 0$, and has exactly one solution. Linear equations can be solved with symbolic, graphical, and numerical methods. The following table summarizes each method.

Method	Comments	Example
Symbolic	Solve the equation by using properties of equations. The solution is exact.	$2x - 1 = 5$ $2x = 6$ Add 1 to both sides. $\frac{2x}{2} = \frac{6}{2}$ Divide both sides by 2. $x = 3$ Simplify.
Graphical	Let Y_1 be the left side of the equation and Y_2 be the right side. Graph Y_1 and Y_2. The x-value at their point of intersection is the solution.	To solve $2x - 1 = 5$ let $Y_1 = 2X - 1$ and $Y_2 = 5$. The solution is $x = 3$.
Numerical (table)	Let Y_1 be the left side of the equation and Y_2 be the right side. Construct tables of Y_1 and Y_2. Find the x-value where $Y_1 = Y_2$.	To solve $2x - 1 = 5$ let $Y_1 = 2X - 1$ and $Y_2 = 5$. The solution is $x = 3$.

3.1 EXERCISES

CONCEPTS

1. Give the general form of a linear equation.
2. How many solutions does a linear equation have?
3. Is $x = 1$ the solution for $4x - 1 = 3x$?
4. Are $3x = 6$ and $x = 2$ equivalent equations?
5. What symbol must occur in every equation?
6. Name three methods for solving a linear equation.

Exercises 7–10: Decide whether the given value is a solution to the equation.

7. $x - 6 = -2$ $x = 5$
8. $-\frac{1}{2}x + 1 = \frac{1}{3}x - \frac{2}{3}$ $x = 2$
9. $3(2t + 3) = \frac{13}{3} - t$ $t = -\frac{2}{3}$
10. $t - 1 = 2 - (t + 1)$ $t = 2$

NUMERICAL SOLUTIONS

Exercises 11–14: Use the table to solve the equation. Check the solution.

11. $2x + 6 = 0$
12. $5 - 2x = -1$
13. $4 - 2x = 7$
14. $2(x - 2) = 5.2$

Exercises 15–16: Use the table to solve the given equation, where Y_1 equals the left side of the equation and Y_2 equals the right side. Check the solution.

15. $5 - 2x = x + 2$
16. $2x + 1 = -x + 7$

Exercises 17–20: Complete the table. Then use the table to solve the equation.

17. $-4x + 8 = 0$

x	1	2	3	4	5
$-4x + 8$	4				

18. $3x + 2 = 5$

x	-2	-1	0	1	2
$3x + 2$	-4				

19. $4 - 2x = x + 7$

x	-2	-1	0	1	2
$4 - 2x$	8				0
$x + 7$	5				9

20. $3(x - 1) = -2(1 - x)$

x	-2	-1	0	1	2
$3(x - 1)$	-9				
$-2(1 - x)$	-6				

Exercises 21–26: Solve the linear equation numerically.

21. $x - 3 = 7$
22. $2x + 1 = 6$
23. $2x - \frac{1}{2} = 3$
24. $7 - 4x = 3.8$

25. $-3x - 6 = x + 2$ 26. $3(x - 1) + \dfrac{3}{2} = 2x$

GRAPHICAL SOLUTIONS

Exercises 27–30: A linear equation is solved graphically by letting y_1 equal the left side of the equation and y_2 equal the right side of the equation. Find the solution.

27.
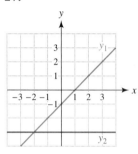

28.
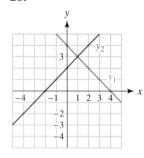

29.
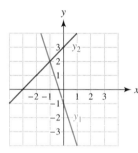

30.
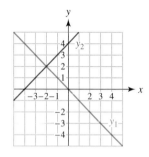

Exercises 31–38: Solve the equation graphically.

31. $5 - 2x = 7$ 32. $-2x + 3 = -7$

33. $\dfrac{7}{2}x - (x + 8) = 0$ 34. $3x + 1 = \dfrac{3}{2}x - 4$

35. $5(x + 2) + 1 = -3x$

36. $\dfrac{x + 5}{3} = \dfrac{2x + 2}{4}$

37. $5(x - 1990) + 15 = 100$

38. $10(x - 1985) - 20 = 55$

SYMBOLIC SOLUTIONS

Exercises 39–56: Solve the equation symbolically.

39. $3x - 7 = 8$ 40. $5 - 2x = -2$

41. $2x = 8 - \dfrac{1}{2}x$ 42. $-7 = 3.5x$

43. $3x - 1 = 11(1 - x)$

44. $4 - 3x = -5(1 + 2x)$

45. $x + 4 = 2 - \dfrac{1}{3}x$ 46. $2x - 5 = 6 - \dfrac{5}{2}x$

47. $2(x - 1) = 5 - 2x$

48. $-(x - 4) = 4(x + 1) + 3(x - 2)$

49. $\dfrac{2x + 1}{3} = \dfrac{2x - 1}{2}$ (*Hint:* Multiply both sides by 6.)

50. $\dfrac{3 - 4x}{5} = \dfrac{3x - 1}{2}$

51. $4.2x - 6.2 = 1 - 1.1x$

52. $8.4 - 2.1x = 1.4x$ 53. $\dfrac{1}{2}x - \dfrac{3}{2} = 4$

54. $5 - \dfrac{1}{3}x = x - 3$

55. $4(x - 1980) + 6 = 18$

56. $-5(x - 1900) - 55 = 145$

SOLVING LINEAR EQUATIONS BY MORE THAN ONE METHOD

Exercises 57–60: Solve the equation
  (a) numerically,
  (b) graphically, and
  (c) symbolically.
Do your answers agree?

57. $2x - 1 = 13$ 58. $9 - x = 3x + 1$

59. $3x - 5 - (x + 1) = 0$

60. $\dfrac{1}{2}x - 1 = \dfrac{5 - x}{2}$

WRITING AND SOLVING EQUATIONS

Exercises 61–70: Complete the following.
  (a) Translate the sentence into an equation, using the variable x.
  (b) Solve the resulting equation.

61. The sum of a number and 2 is 12.

62. Twice a number plus 7 equals 9.

63. A number divided by 5 equals the number increased by 1.

64. 25 times a number is 125.

65. If a number is increased by 5 and then divided by 2, the result equals 7.

66. A number subtracted from 8 is 5.

67. The quotient of a number and 2 is 17.

68. The product of 5 and a number equals 95.

69. The sum of three consecutive integers is 30.

70. A rectangle that is 5 inches longer than it is wide has a perimeter of 60 inches.

APPLICATIONS

71. *Geometry* Find the value of x if the perimeter of the room is 48 feet.

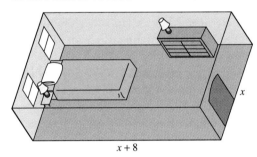

72. *Fencing* Three identical pens for dogs are to be fenced with 120 feet of fence, as illustrated in the accompanying figure. If the length of each pen is twice its width plus 2 feet, find the dimensions of each pen.

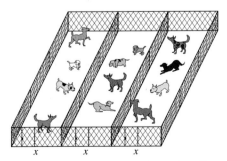

73. *Distance and Time* A train 100 miles west of St. Louis, Missouri, is traveling east at 60 miles per hour. How long will it take for the train to be 410 miles east of St. Louis?

74. *Distance and Time* At first an athlete jogs at 6 miles per hour and then jogs at 5 miles per hour, traveling 7 miles in 1.3 hours. How long did the athlete jog at each speed? (*Hint:* Let x represent the time spent running at 6 miles per hour and $1.3 - x$ represent the time spent running at 5 miles per hour.)

75. *U.S. Median Income* In 1980, median family income was \$21,153, and it increased by \$1322 per year until 1993. How many years did it take for median family income to reach \$27,763? (*Source:* **Department of Commerce.**)

76. *Room Dimensions* The length of a room is 4 feet more than its width. If the perimeter of the room is 120 feet, find the dimensions of the room.

77. *Classroom Ventilation* Ventilation is an effective method for removing indoor air pollutants. A classroom should have a ventilation rate of 900 cubic feet of air per hour for each person in the classroom. (*Source:* **ASHRAE.**)

(a) What ventilation rate should a classroom containing 40 people have?
(b) If a classroom ventilation system moves 60,000 cubic feet of air per hour, determine the maximum number of people that should be in the classroom.

78. *Lead Poisoning* According to the EPA, the maximum amount of lead that can be ingested by a person without becoming ill is 36.5 milligrams per year. (*Source: Nemerow, N., and Dasgupta, A., Industrial and Hazardous Waste Treatment.*)
   (a) Write an expression that gives the maximum amount of lead in milligrams that can be "safely" ingested over x years.
   (b) Under these guidelines what is the minimum number of years that the consumption of 500 milligrams should be spread over?

79. *Height of a Tree* In the accompanying figure, a person 6 feet tall casts a shadow 5 feet long. A nearby tree casts a shadow 43 feet long. Find the height of the tree.

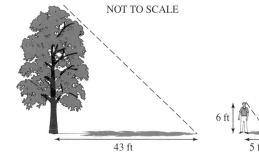

NOT TO SCALE

80. *Grades* To receive an A in a college course a student must average 90 percent correct on four exams of 100 points each and on a final exam of 200 points. If a student scores 83, 87, 94, and 91 on the 100-point exams, what minimum score on the final exam is necessary for the student to receive an A?

81. *AIDS Deaths in the United States* There were 17,047 AIDS deaths in 1998, which represented 34.5% of the 1995 AIDS deaths. Determine the number of AIDS deaths in 1995. (*Source: Centers for Disease Control and Prevention.*)

82. *Salty Snacks* The average American ate 22 pounds of salty snacks in 1994, up 25.7% from 1988. What was the average consumption in 1988? (*Source: Snack Foods Association.*)

83. *Meeting a Future Mate* According to a survey, about 32% of adult Americans believe that people will meet their mates on the Internet during this century. In this survey 480 respondents held this belief. Determine the number of people participating in the survey. (*Source: Men's Health.*)

84. *Union Membership* In 1998, 6.9 million, or 37.5%, of all government workers were unionized. How many government workers were there in 1998? (*Source: Department of Labor.*)

85. *State and Federal Inmates* From 1988 to 1995 the number of state and federal prison inmates in thousands during year x can be modeled by $f(x) = 70x - 138{,}532$. (*Source: Department of Justice.*)
   (a) Determine graphically the year when there were 908 thousand inmates.
   (b) Solve part (a) numerically.
   (c) Solve part (a) symbolically.

86. *U.S. Government Costs* From 1960 to 1990 the cost (in 1960 dollars) to administer social and economic programs rose from $1.9 billion to $16 billion. This cost in billions of dollars can be modeled by $f(x) = 0.47x - 919.3$, where x is the year. (*Source: Center for the Study of American Business.*)
   (a) Estimate graphically when the cost was $6.6 billion.
   (b) Solve part (a) numerically.
   (c) Solve part (a) symbolically.

87. *Broadway Revenue* Record Broadway ticket revenues were recorded for the 1999 season. These ticket revenues in millions of dollars have grown at a constant rate during the past decade and may be modeled by

$$f(x) = 32.8(x - 1999) + 588,$$

where x is the year. (*Source: The League of American Theatres and Producers, Inc.*)
   (a) Estimate ticket revenues in 1995 and 1999.
   (b) Estimate numerically the year when they were $424 million.
   (c) Solve part (b) symbolically.

88. *Centenarians* In 1990 there were 37,306 centenarians, people aged 100 or older, in the

United States. The number of centenarians is estimated to increase by 8260 per year. (*Source: Bureau of the Census.*)

(a) Find values for m, h, and k so that $f(x) = m(x - h) + k$ models the number of centenarians during year x.
(b) Graphically estimate the year when there will be 450,000 centenarians.
(c) Solve part (b) symbolically.

WRITING ABOUT MATHEMATICS

89. Explain each of the following terms: linear equation, solution, solution set, and equivalent equations. Give an example of each.

90. Explain how to solve the equation $ax + b = 0$ symbolically and graphically. Use both methods to solve the equation $5x + 10 = 0$.

GROUP ACTIVITY
Working with Real Data

Directions: Form a group of 2 to 4 people. Select someone to record the group's responses for this activity. All members of the group should work cooperatively to answer the questions. If your instructor asks for your results, each member of the group should be prepared to respond. If the group is asked to turn in its work, be sure to include each person's name on the paper.

U.S. Youth Soccer Soccer has become an increasingly popular sport for young people in the United States. The table lists organized soccer participation for youths under the age of 19.

Year	1980	1986	1992	1998
Number (millions)	0.9	1.5	2.3	3.6

Source: Soccer Industry Council of America.

(a) Make a scatterplot of the data. Discuss any trend in soccer participation.
(b) Find a linear function f given by
$$f(x) = m(x - h) + k$$
that models the data.
(c) Interpret m as a rate of change.
(d) Estimate the year when participation was 2 million.
(e) If trends continue, predict the year when participation may reach 5 million.

3.2 LINEAR INEQUALITIES

Properties of Inequalities ~ Numerical Solutions ~ Graphical Solutions ~ Symbolic Solutions

INTRODUCTION

On a freeway, the speed limit might be 75 miles per hour. A driver traveling x miles per hour is obeying the speed limit if $x \leq 75$ and breaking the speed limit if $x > 75$. A speed of $x = 75$ represents the boundary between obeying the speed limit and breaking it. A posted speed limit, or *boundary,* allows drivers to easily determine whether they are speeding.

Solving linear inequalities is closely related to solving linear equations because equality is the boundary between *greater than* and *less than*. In this section we discuss techniques needed to solve linear inequalities.

PROPERTIES OF INEQUALITIES

An **inequality** results whenever the equals sign in an equation is replaced with any one of the symbols $<$, $\leq$, $>$, or $\geq$. Examples of linear equations include

$$2x + 1 = 0, \quad 1 - x = 6, \quad \text{and} \quad 5x + 1 = 3 - 2x,$$

and, therefore, examples of linear inequalities include

$$2x + 1 < 0, \quad 1 - x \geq 6, \quad \text{and} \quad 5x + 1 \leq 3 - 2x.$$

A **solution** to an inequality is a value of the variable that makes the statement true. The set of all solutions is called the **solution set.** Two inequalities are *equivalent* if they have the same solution set. Inequalities frequently have infinitely many solutions. For example, the solution set to the inequality $x - 5 > 0$ includes all real numbers greater than 5, which can be written as $x > 5$. Using **set builder notation,** we can write the solution set as $\{x \mid x > 5\}$. This expression is read as "the set of all real numbers x such that x is greater than 5."

Linear Inequality in One Variable

A **linear inequality** in one variable is an inequality that can be written in the form

$$ax + b > 0,$$

where $a \neq 0$. (The symbol $>$ may be replaced with $\geq$, $<$, or $\leq$.)

A connection among linear functions, linear equations, and linear inequalities is as follows.

■ **MAKING CONNECTIONS**

Linear functions, equations, and inequalities

There are similarities among linear functions, equations, and inequalities. A *linear function* is given by $f(x) = ax + b$, a *linear equation* by $ax + b = 0$, and a *linear inequality* by $ax + b > 0$.

■

NUMERICAL SOLUTIONS

In Section 3.1 we solved linear equations with numerical, graphical, and symbolic methods. We can also use these methods to solve linear inequalities.

EXAMPLE 1 *Solving an inequality numerically*

 Solve $2x - 6 > 0$, using a table.

Solution

Begin by evaluating $2x - 6$ for various values of x, as shown in Table 3.2.

TABLE 3.2

x	0	1	2	3	4	5	6
$2x - 6$	-6	-4	-2	0	2	4	6

The solution to the equation $2x - 6 = 0$ is $x = 3$. Again, for linear inequalities, equality is the boundary between *greater than* and *less than*. Here $2x - 6 > 0$ for x-values satisfying $x > 3$, so the solution set is $\{x \mid x > 3\}$. Figure 3.11 shows Table 3.2 as generated with a graphing calculator.

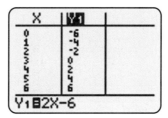

Figure 3.11

EXAMPLE 2 | *Solving an inequality numerically*

Solve $5 - 3x \leq x - 3$.

Solution

Begin by evaluating $5 - 3x$ and $x - 3$ for various values of x, as shown in Table 3.3.

TABLE 3.3

x	-1	0	1	2	3	4	5
$5 - 3x$	8	5	2	-1	-4	-7	-10
$x - 3$	-4	-3	-2	-1	0	1	2

From the table, $5 - 3x = x - 3$ when $x = 2$ and $5 - 3x < x - 3$ when $x > 2$. The solution set is $\{x \mid x \geq 2\}$. Figure 3.12 shows Table 3.3 as generated with a graphing calculator.

Figure 3.12

Note: Numerical solutions can sometimes be difficult to find if the *x*-value that determines equality is not an integer. Graphical and symbolic methods work well in such situations.

GRAPHICAL SOLUTIONS

Figure 3.13 shows the distance that two cars are from Chicago, Illinois, after x hours while traveling in the same direction on a freeway. The distance for Car 1 is denoted y_1, and the distance for Car 2 is denoted y_2.

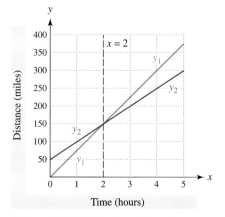

Figure 3.13 Distances of Two Cars

After $x = 2$ hours, $y_1 = y_2$ and both cars are 150 miles from Chicago. To the left of the dashed vertical line $x = 2$, the graph of y_1 is below the graph of y_2, so Car 1 is closer to Chicago than Car 2. Thus

$$y_1 < y_2 \quad \text{when} \quad x < 2.$$

To the right of the dashed vertical line $x = 2$, the graph of y_1 is above the graph of y_2, so Car 1 is farther from Chicago than Car 2. Thus

$$y_1 > y_2 \quad \text{when} \quad x > 2.$$

These concepts can be used to solve the linear inequality from Example 2 graphically.

EXAMPLE 3 *Solving an inequality graphically*

Solve $5 - 3x \leq x - 3$.

Solution

The graphs of $y_1 = 5 - 3x$ and $y_2 = x - 3$ intersect at the point $(2, -1)$, as shown in Figure 3.14. Equality, or $y_1 = y_2$, occurs when $x = 2$ and the graph of y_1 is below the graph of y_2 when $x > 2$. Thus $5 - 3x \leq x - 3$ is satisfied when $x \geq 2$. The solution set is $\{x \mid x \geq 2\}$. Figure 3.15 shows the same graph generated with a graphing calculator.

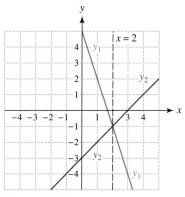

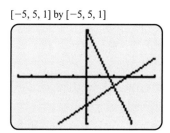

$[-5, 5, 1]$ by $[-5, 5, 1]$

Figure 3.14

Figure 3.15

Symbolic Solutions

If we have $3 < 5$, then $3 + 1 < 5 + 1$. That is, we can add the same number to both sides of an inequality. This is an example of one property of inequalities. The following properties are used to solve inequalities symbolically.

Properties of Inequalities

Let a, b, and c be real numbers.

1. $a < b$ and $a + c < b + c$ are equivalent.
   (The same number may be added to or subtracted from both sides of an inequality.)
2. If $c > 0$, then $a < b$ and $ac < bc$ are equivalent.
   (Both sides of an inequality may be multiplied or divided by the same positive number.)
3. If $c < 0$, then $a < b$ and $ac > bc$ are equivalent.
   (Both sides of an inequality may be multiplied or divided by the same negative number provided the inequality symbol is reversed.)

Similar properties exist for the $\leq$ and $\geq$ symbols.

To solve an inequality we apply properties of inequalities to find a simpler, equivalent inequality.

EXAMPLE 4 *Solving an inequality symbolically*

Solve $2x - 1 > 4$.

Solution

Begin by adding 1 to both sides of the inequality.

$$2x - 1 > 4 \quad \text{Given inequality}$$
$$2x - 1 + 1 > 4 + 1 \quad \text{Add 1 to both sides.}$$
$$2x > 5 \quad \text{Simplify.}$$
$$\frac{2x}{2} > \frac{5}{2} \quad \text{Divide both sides by 2.}$$
$$x > \frac{5}{2} \quad \text{Simplify.}$$

The solution set is $\left\{x \mid x > \frac{5}{2}\right\}$.

In the next example we use Property 3 of inequalities to solve the inequality that was solved numerically and graphically in Examples 2 and 3.

EXAMPLE 5 *Solving an inequality symbolically*

Solve $5 - 3x \leq x - 3$.

Solution

Begin by subtracting 5 from both sides of the inequality.

$$5 - 3x \leq x - 3 \quad \text{Given inequality}$$
$$5 - 3x - 5 \leq x - 3 - 5 \quad \text{Subtract 5 from both sides.}$$
$$-3x \leq x - 8 \quad \text{Simplify.}$$
$$-3x - x \leq x - 8 - x \quad \text{Subtract } x \text{ from both sides.}$$
$$-4x \leq -8 \quad \text{Simplify.}$$

Next divide both sides by -4. Since we are dividing by a *negative* number, Property 3 requires reversing the inequality by changing $\leq$ to $\geq$.

$$\frac{-4x}{-4} \geq \frac{-8}{-4} \quad \text{Divide by } -4; \text{ reverse the inequality.}$$
$$x \geq 2 \quad \text{Simplify.}$$

The solution set is $\{x \mid x \geq 2\}$.

In the lower atmosphere, the air generally becomes colder as the altitude increases. One mile above Earth's surface the temperature is about 29°F colder than the ground-level temperature. As the air temperature cools, the chance of clouds forming increases. In the next example we estimate the altitudes at which clouds will not form.

EXAMPLE 6 *Finding the altitude of clouds*

If ground temperature is 90°F, the temperature T above Earth's surface is modeled by $T(x) = 90 - 29x$, where x is the altitude in miles. Suppose that clouds will form only if the temperature is 53°F or colder.

(a) Determine symbolically the altitudes at which there are no clouds.
(b) Give graphical support for your answer.

(*Source:* Miller, A., and R. Anthes, *Meteorology.*)

Solution

(a) *Symbolic Solution* Clouds will not form at altitudes at which the temperature is greater than 53°F. Thus we must solve the inequality $T(x) > 53$.

$$90 - 29x > 53 \qquad \text{Inequality to be solved}$$
$$90 - 29x - 90 > 53 - 90 \qquad \text{Subtract 90 from both sides.}$$
$$-29x > -37 \qquad \text{Simplify.}$$
$$\frac{-29x}{-29} < \frac{-37}{-29} \qquad \text{Divide by } -29; \text{ reverse inequality.}$$
$$x < \frac{37}{29} \qquad \text{Simplify.}$$

The result, $\frac{37}{29} \approx 1.28$, indicates that clouds will not form below about 1.28 miles. Note that models are usually not exact, so rounding values is appropriate.

(b) *Graphical Solution* Graph $Y_1 = 90 - 29X$ and $Y_2 = 53$. In Figure 3.16 their graphs intersect near the point (1.28, 53). The graph of Y_1 is above the graph of Y_2 when $x < 1.28$.

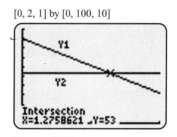

Figure 3.16

Critical Thinking

A linear equation has one solution that can be checked by substituting it into the given equation. A linear inequality has infinitely many solutions. Discuss ways that the solution set for a linear inequality could be checked.

3.2 Putting It All Together

The solution set for a linear inequality can be found numerically, graphically, and symbolically. One strategy for solving a linear inequality is first to locate the *x*-value that results in equality. This *x*-value represents the *boundary* between *greater than* and *less than*. Using this value, we can find the solution set for the linear inequality. The table on the following page summarizes methods for solving linear inequalities in the form $y_1 > y_2$. Inequalities involving $\geq$, $<$, or $\leq$ may be solved similarly.

Method	Description
Numerical Method	Set Y_1 equal to the left side of the equation and Y_2 equal to the right side of the equation. The solution set for $Y_1 > Y_2$ includes all x-values where the values for Y_1 are greater than the values for Y_2.
Graphical Method	Set Y_1 equal to the left side of the inequality and Y_2 equal to the right side of the inequality. Graph Y_1 and Y_2 and find their point of intersection. The solution set for $Y_1 > Y_2$ includes all x-values where the graph of Y_1 is above the graph of Y_2.
Symbolic Method	Solve linear inequalities similarly to the solution of linear equations. Move x-terms to one side of the inequality and constant terms to the other side, using properties of inequalities. Be sure to *reverse the inequality symbol* when multiplying or dividing both sides by a negative number.

3.2 EXERCISES

CONCEPTS

1. Give an example of a linear inequality.

2. Can a linear inequality have infinitely many solutions? Explain.

3. Are $2x > 6$ and $x > 3$ equivalent inequalities? Explain.

4. Are $-4x < 8$ and $x < -2$ equivalent inequalities? Explain.

5. Give one difference between an equation and an inequality.

6. Name three methods for solving a linear inequality.

Exercises 7–10: Decide whether the given value for the variable is a solution to the inequality.

7. $x - 5 \le 3$ $x = 2$

8. $\dfrac{3}{2}x - \dfrac{1}{2} \ge 1 - x$ $x = -2$

9. $2t - 3 > 5t - (2t + 1)$ $t = 5$

10. $2(z - 1) < 3(z + 1)$ $z = \pi$

NUMERICAL SOLUTIONS

Exercises 11–14: Use the table to solve the inequality.

11. $3x + 6 > 0$

12. $6 - 3x \le -3$

13. $-2x + 7 > 5$

14. $5(x - 3) \le 4$

3.2 Linear Inequalities

Exercises 15–16: Use the table to solve the inequality; Y_1 is the left side of the inequality, and Y_2 is the right side.

15. $5 - 2x \geq x + 2$ **16.** $2x + 1 > -x + 7$

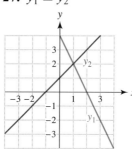 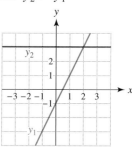

Exercises 17–20: Complete the table. Then use the table to solve the inequality.

17. $-2x + 6 \leq 0$

x	1	2	3	4	5
$-2x + 6$	4				-4

18. $3x - 1 < 8$

x	0	1	2	3	4
$3x - 1$					

19. $5 - x > x + 7$

x	-3	-2	-1	0	1
$5 - x$	8				4
$x + 7$	4				8

20. $2(3 - x) \geq -3(x - 2)$

x	-2	-1	0	1	2
$2(3 - x)$					
$-3(x - 2)$					

Exercises 21–26: Solve the inequality numerically.

21. $x - 3 > 0$

22. $2x < 0$

23. $2x - 1 \geq 3$

24. $2 - 3x \leq -4$

25. $x + 1 < 3x - 1$

26. $3(x - 1) + 1 \geq 7x$

GRAPHICAL SOLUTIONS

Exercises 27–30: Use the graph to solve the inequality.

27. $y_1 \leq y_2$ **28.** $y_2 > y_1$

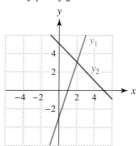 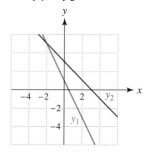

29. $y_1 > y_2$ **30.** $y_1 \leq y_2$

31. *Distance Between Cars* Car 1 and Car 2 are both traveling in the same direction. Their distances in miles north of St. Louis, Missouri, after x hours are shown in the following graph, where $0 \leq x \leq 8$.

(a) Which car is traveling faster? Explain.

(b) How many hours elapse before the two cars are the same distance from St. Louis? How far are they from St. Louis when this equality occurs?

(c) During what time interval is Car 2 farther from St. Louis than Car 1?

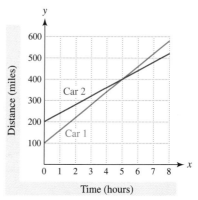

164 CHAPTER 3 Linear Equations and Inequalities

32. Use the following graph to solve each inequality.
    (a) $f(x) > g(x)$
    (b) $f(x) \leq g(x)$

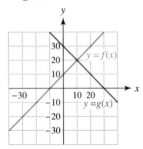

Exercises 33–40: Solve the inequality graphically.

33. $4 - 2x \leq 8$

34. $-2x + 3 < -3$

35. $x - (2x + 4) > 0$

36. $2x + 1 \geq 3x - 4$

37. $2(x + 2) + 5 < -x$

38. $\dfrac{x + 5}{3} \geq \dfrac{1 - x}{2}$

39. $25(x - 1995) + 100 \leq 0$ (*Hint:* Use the viewing rectangle $[1985, 2000, 5]$ by $[-50, 50, 10]$.)

40. $5(x - 1980) - 20 > 50$

SYMBOLIC SOLUTIONS

Exercises 41–54: Solve the inequality symbolically.

41. $x + 3 \leq 5$

42. $x - 5 \geq -3$

43. $\dfrac{1}{4}x > 9$

44. $14 < -3.5x$

45. $4x - 2 \geq \dfrac{5}{2}$

46. $4 - 3x \leq -\dfrac{2}{3}$

47. $x - \dfrac{3}{2} < 7 - \dfrac{1}{2}x$

48. $4x - 6 > 12 - 10x$

49. $\dfrac{5}{2}(2x - 3) < 6 - 2x$

50. $1 - \left(\dfrac{3}{2}x - 4\right) > \dfrac{1}{2}(x + 1)$

51. $\dfrac{3x - 2}{2} \leq \dfrac{x - 4}{5}$

52. $\dfrac{5 - 2x}{2} \geq \dfrac{2x + 1}{4}$

53. $3(x - 2000) + 15 < 45$

54. $-2(x - 1990) + 75 > 25$

SOLVING LINEAR INEQUALITIES BY MORE THAN ONE METHOD

Exercises 55–58: Solve the inequality
    (a) *numerically,*
    (b) *graphically, and*
    (c) *symbolically.*

55. $5x - 2 < 8$

56. $3x - 3 \geq 2x$

57. $2x - 4 - 3(x + 3) \leq 0$

58. $x - 1 > \dfrac{4 - x}{2}$

APPLICATIONS

59. *Federal Debt* The line graph shows the total federal debt in billions of dollars from 1940 through 2000. Estimate the years when the deficit was less than $1 trillion dollars.

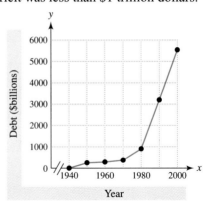

60. *Tetanus Cases* The line graph shows the numbers of reported tetanus cases in the United States from 1950 through 1990. Estimate when the reported cases were 100 or less.

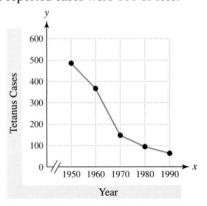

61. *Interest* The following graph shows the annual interest y on a loan of x dollars with an interest rate of 10%. Determine the loan amounts that result in the following.
(a) An annual interest equal to $100
(b) An annual interest of more than $100
(c) An annual interest of $100 or less

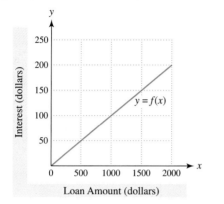

62. *U.S. Population* The following graph models the population in millions from 1960 through 2000. Estimate when each of the following occurred.
(a) A population equal to 225 million
(b) A population of 225 million or less
(c) A population of 225 million or more

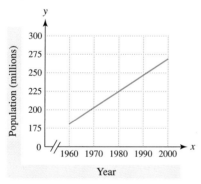

63. *Geometry* Find values for x so that the perimeter of the figure is less than 50 feet.

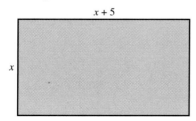

64. *Geometry* A rectangle is twice as long as it is wide. If the rectangle is to have a perimeter of 36 inches or less, what values for the width are possible?

65. *Distance and Time* Two cars are traveling in the same direction along a freeway. After x hours, the first car's distance from a rest stop is given by $y_1 = 70x$ and the second car's distance is given by $y_2 = 60x + 35$.
(a) What is the speed of each car?
(b) When are the cars the same distance from the rest stop?
(c) At what times is the first car farther from the rest stop than the second? Assume that $x \geq 0$.

66. *Sales of CDs and LP Records* (Refer to Example 4, Section 3.1.) From 1985 to 1990 sales of CDs in millions in the United States can be modeled by

$$y_1 = 51.6(x - 1985) + 9.1,$$

and sales of vinyl LP records can be modeled by

$$y_2 = -31.9(x - 1985) + 167.7.$$

(a) Graph y_1 and y_2 in the viewing rectangle [1984, 1991, 1] by [0, 350, 50].
(b) Estimate the years when CD sales were greater than or equal to LP record sales.
(c) What happened to sales of CDs and LP records after 1990?

67. *Altitude and Temperature* (Refer to Example 6.) If the temperature on the ground is 60°F, the air temperature x miles high is given by $T(x) = 60 - 29x$. Determine symbolically and graphically the altitudes at which the air temperature is greater than 0°F. (*Source:* A. Miller.)

68. *Altitude and Dew Point* If the dew point on the ground is 70°F, then the dew point x miles high is given by $D(x) = 70 - 5.8x$. (*Source:* A. Miller.)
(a) For each 1-mile increase in altitude, how much does the dew point change?
(b) Determine symbolically the altitudes at which the dew point is greater than 30°F.
(c) Solve part (b) graphically.

69. *AIDS Research* AIDS research funding in billions of dollars from 1994 to 2000 in the United States can be modeled by
$$f(x) = 0.083(x - 1994) + 1.3,$$
where x is the year. (*Source: National Institutes of Health.*)
  (a) Did funding increase or decrease during this time period? Explain.
  (b) Estimate when AIDS funding was about $1.55 billion.
  (c) When was funding greater than or equal to $1.55 billion?

70. *Hepatitis C Research* The hepatitis C virus (HCV) can live in a person for years without any symptoms after the individual was initially infected with a tainted blood transfusion. An estimated 4 million Americans have HCV, and some 10,000 people die from it each year. From 1994 to 2000 research funding for Hepatitis C in millions of dollars may be modeled by $f(x) = 4.43(x - 1994) + 7$, where x is the year. Determine when this funding was less than or equal to $20.3 million. (*Source: National Institutes of Health.*)

71. *Size and Weight of a Fish* If a bass has a length of x inches, where $20 \leq x \leq 25$, its weight W in pounds can be estimated from the formula $W(x) = 0.96x - 14.4$.
  (a) What length of bass is likely to weigh 6.7 pounds?
  (b) What lengths of bass are likely to weigh less than 6.7 pounds?

72. *Distant Galaxies* In the late 1920s the famous observational astronomer Edwin P. Hubble (1889–1953) determined both the distance to several galaxies and the velocity at which they were receding from Earth. The following graph shows four galaxies with their distances x in light-years from Earth and velocities y in miles per second that they are moving away from Earth. (*Sources: A. Acker and C. Jaschek, Astronomical Methods and Calculations; A. Sharov and I. Novikov, Edwin Hubble, The Discoverer of the Big Bang Universe.*)

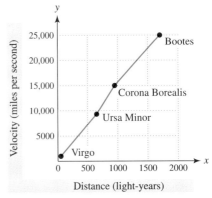

  (a) What relationship exists between distance and velocity of a galaxy?
  (b) A galaxy is determined to be receding at more than 12,500 miles per second. What can be said about its distance from Earth?

WRITING ABOUT MATHEMATICS

73. Explain the following terms and give an example of each.
  (a) Linear function
  (b) Linear equation
  (c) Linear inequality

74. Suppose that a student says that a linear equation and a linear inequality can be solved symbolically in exactly the same way. How would you respond?

CHECKING BASIC CONCEPTS FOR SECTIONS 3.1 AND 3.2

1. Solve $6 - 4x = 5$ symbolically.

2. Solve the linear equation
$$2(3x + 4) + 3 = -1$$
   (a) symbolically,
   (b) graphically, and
   (c) numerically.
   Do your answers agree?

3. Solve the linear inequality $4 - 3x < \frac{1}{2}x$ symbolically.

4. Solve the linear inequality $x + 4 \geq 1 - 2x$
   (a) symbolically,
   (b) graphically, and
   (c) numerically.
   Do your answers agree?

3.3 COMPOUND INEQUALITIES

Basic Concepts ~ Symbolic Solutions and Number Lines ~
Numerical and Graphical Solutions ~ Interval Notation

INTRODUCTION

A person weighting 143 pounds and needing to purchase a life vest for white-water rafting is not likely to find one designed exactly for this weight. Life vests are manufactured to support a range of body weights. A vest approved for weights between 100 and 160 pounds might be appropriate for this person. In other words, if a person's weight is x, this life vest is safe if $x \geq 100$ *and* $x \leq 160$. This example illustrates the concept of a *compound inequality*.

BASIC CONCEPTS

A **compound inequality** consists of two inequalities joined by the words *and* or *or*. The following are two examples of compound inequalities.

$$2x \geq -3 \quad \text{and} \quad 2x < 5$$
$$x + 2 \geq 3 \quad \text{or} \quad x - 1 < -5$$

If a compound inequality contains the word *and*, a solution must satisfy *both* inequalities. For example, $x = 1$ is a solution of the first compound inequality because

$$\underset{\text{True}}{2(1) \geq -3} \quad \text{and} \quad \underset{\text{True}}{2(1) < 5}$$

are both true statements.

If a compound inequality contains the word *or*, a solution must satisfy *at least one* of the two inequalities. Thus $x = 5$ is a solution to the second compound inequality.

$$\underset{\text{True}}{5 + 2 \geq 3} \quad \text{or} \quad \underset{\text{False}}{5 - 1 < -5}$$

The first statement is true. Note that $x = 5$ does not need to satisfy the second statement for this compound inequality to be true.

EXAMPLE 1   Determining solutions to compound inequalities

Determine whether the given x-values are solutions to the compound inequalities.

(a) $x + 1 < 9$ and $2x - 1 > 8$ $x = 5, -5$
(b) $5 - 2x \leq -4$ or $5 - 2x \geq 4$ $x = 2, -3$

Solution

(a) Substitute $x = 5$ into the given compound inequality.

$$5 + 1 = 6 < 9 \quad \text{and} \quad 2(5) - 1 = 9 > 8$$
$$\text{True} \qquad\qquad\qquad\qquad \text{True}$$

Both inequalities are true, so $x = 5$ is a solution.

Now substitute $x = -5$.

$$-5 + 1 = -4 < 9 \quad \text{and} \quad 2(-5) - 1 = -11 > 8$$
$$\text{True} \qquad\qquad\qquad\qquad \text{False}$$

Both inequalities are not true, so $x = -5$ is not a solution.

(b) Substitute $x = 2$ into the given compound inequality.

$$5 - 2(2) = 1 \leq -4 \quad \text{or} \quad 5 - 2(2) = 1 \geq 4$$
$$\text{False} \qquad\qquad\qquad\qquad \text{False}$$

Neither inequality is true, so $x = 2$ is not a solution.

Now substitute $x = -3$.

$$5 - 2(-3) = 11 \leq -4 \quad \text{or} \quad 5 - 2(-3) = 11 \geq 4$$
$$\text{False} \qquad\qquad\qquad\qquad \text{True}$$

At least one of the two inequalities is true, so $x = -3$ is a solution.

SYMBOLIC SOLUTIONS AND NUMBER LINES

We can use a number line to graph solutions to compound inequalities, such as

$$x \leq 6 \quad \text{and} \quad x > -4.$$

The solution set for $x \leq 6$ is shaded to the left of 6, with a bracket placed at $x = 6$, as shown in Figure 3.17. The solution set for $x > -4$ can be shown by shading a different number line to the right of -4 and placing a left parenthesis at -4. Because the inequalities are connected by *and*, the solution set consists of all numbers that are shaded on both number lines. The final number line represents the **intersection** of the two solution sets. That is, the solution set includes where the graphs "overlap."

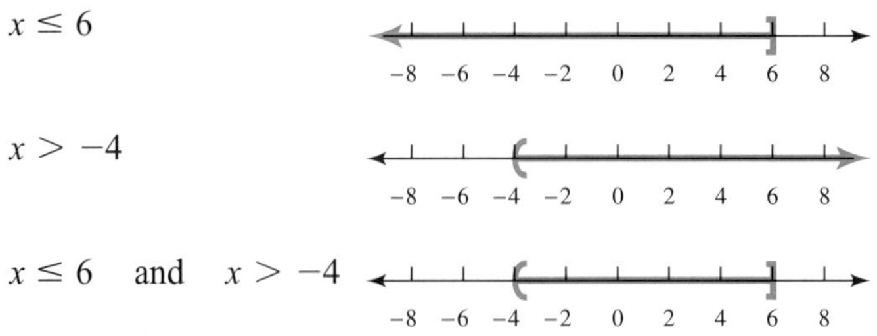

Figure 3.17

Critical Thinking

Graph the following inequalities and discuss your results.

1. $x < 2$ and $x > 5$
2. $x > 2$ or $x < 5$

In the next example we use a number line to help solve a compound inequality symbolically.

EXAMPLE 2 *Solving a compound inequality containing "and"*

Solve $2x + 4 > 8$ and $5 - x < 9$. Graph the solution.

Solution

First solve each linear inequality separately.

$$2x + 4 > 8 \quad \text{and} \quad 5 - x < 9$$
$$2x > 4 \quad \text{and} \quad -x < 4$$
$$x > 2 \quad \text{and} \quad x > -4$$

Graph the two inequalities on two different number lines. On a third number line, shade solutions that appear on both of the first two number lines. As shown in Figure 3.18, the solution set is $\{x \mid x > 2\}$.

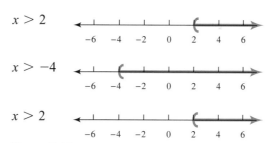

Figure 3.18

Sometimes a compound inequality containing the word *and* can be combined into a three-part inequality. For example, rather than writing

$$x > 5 \quad \text{and} \quad x \leq 10,$$

we could write the **three-part inequality**

$$5 < x \leq 10.$$

This three-part inequality is represented by the number line shown in Figure 3.19.

Figure 3.19

Three-part inequalities occur frequently in applications. In the next example we find altitudes at which the air temperature is within a certain range.

EXAMPLE 3 *Solving a three-part inequality*

Air temperature is colder at higher altitudes. If the ground-level temperature is 80°F, the air temperature x miles above Earth's surface can be modeled by $T(x) = 80 - 29x$. Find the altitudes at which the air temperature ranges from 22°F to -7°F. (*Source:* A. Miller and R. Anthes, *Meteorology*.)

Solution

We write and solve the three-part inequality $-7 \leq T(x) \leq 22$.

$$-7 \leq T(x) \leq 22 \quad \text{Original three-part inequality}$$
$$-7 \leq 80 - 29x \leq 22 \quad \text{Substitute for } T(x).$$
$$-87 \leq -29x \leq -58 \quad \text{Subtract 80 from each part.}$$
$$\frac{-87}{-29} \geq x \geq \frac{-58}{-29} \quad \text{Divide by } -29; \text{ reverse inequality symbols.}$$
$$3 \geq x \geq 2 \quad \text{Simplify.}$$
$$2 \leq x \leq 3 \quad \text{Rewrite inequality.}$$

The air temperature ranges from 22°F to -7°F for altitudes between 2 and 3 miles.

We can also solve compound inequalities containing the word *or*.

EXAMPLE 4 *Solving a compound inequality with "or"*

Solve $x + 2 < -1$ or $x + 2 > 1$.

Solution

We first solve each linear inequality.

$$x + 2 < -1 \quad \text{or} \quad x + 2 > 1 \quad \text{Given compound inequality}$$
$$x < -3 \quad \text{or} \quad x > -1 \quad \text{Subtract 2.}$$

We can graph the simplified inequalities on different number lines, as shown in Figure 3.20. A solution must satisfy at least one of the two inequalities. Thus the solution set for the compound inequality results from taking the **union** of the first two number lines. We can write the solution, using set-builder notation, as $\{x \mid x < -3\} \cup \{x \mid x > -1\}$ or $\{x \mid x < -3 \text{ or } x > -1\}$.

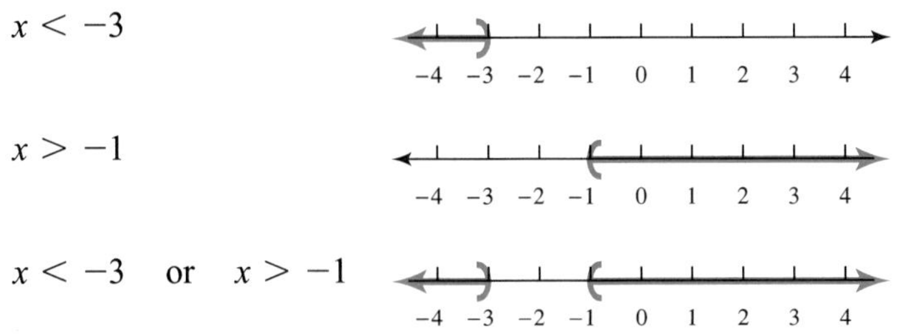

$x < -3$

$x > -1$

$x < -3 \quad \text{or} \quad x > -1$

Figure 3.20

Making Connections

Writing Three-Part Inequalities

The inequality $-2 < x < 1$ means that $x > -2$ *and* $x < 1$. A three-part inequality should *not* be used when *or* connects a compound inequality. Writing $x < -2$ or $x > 1$ as $1 < x < -2$ is incorrect because it states that x must be both greater than 1 *and* less than -2. This statement is impossible for any value of x to satisfy.

Numerical and Graphical Solutions

Compound inequalities can also be solved graphically and numerically, as illustrated in the next example.

EXAMPLE 5 *Solving a compound inequality numerically and graphically*

Tuition at private colleges and universities from 1980 to 1997 can be modeled by $f(x) = 575(x - 1980) + 3600$. Estimate when the average tuition was between $8200 and $10,500.

Solution

Numerical Solution Let $Y_1 = 575(X - 1980) + 3600$. Make a table of values, as shown in Figure 3.21. In 1988, the average tuition was $8200 and in 1992 it was $10,500. Therefore from 1988 to 1992 the average tuition ranged from $8200 to $10,500.

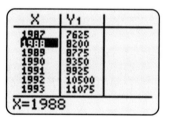

Figure 3.21

Graphical Solution Let $Y_1 = 575(X - 1980) + 3600$, $Y_2 = 8200$, and $Y_3 = 10,500$. We must find x-values for $Y_2 \leq Y_1 \leq Y_3$. Figures 3.22 and 3.23 show that Y_1 is between Y_2 and Y_3 when $1988 \leq x \leq 1992$.

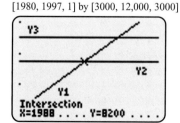

Figure 3.22

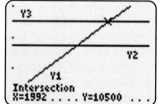

Figure 3.23

INTERVAL NOTATION

The solution set in Example 3 was $\{x \mid 2 \leq x \leq 3\}$. This solution set can be graphed on a number line, as shown in Figure 3.24.

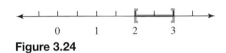

Figure 3.24

A convenient notation for number line graphs is called **interval notation.** Instead of drawing the entire number line as in Figure 3.24, the solution set can be expressed as [2, 3] in interval notation. Because the solution set includes the endpoints 2 and 3, brackets are used. A solution set that includes all real numbers satisfying $-2 < x < 3$ can be expressed as $(-2, 3)$. Parentheses indicate that the endpoints are not included. The interval $0 \leq x < 4$ is represented by [0, 4).

Table 3.4 provides some examples of interval notation. The symbol ∞ refers to infinity, and it does not represent a real number. The interval $(5, \infty)$ represents $x > 5$, which has no maximum x-value, so ∞ is used for the right endpoint. The symbol $-\infty$ may be used similarly.

TABLE 3.4 **Interval Notation**

Inequality	Interval Notation	Number Line Graph
$-1 < x < 3$	$(-1, 3)$	
$-3 < x \leq 2$	$(-3, 2]$	
$-2 \leq x \leq 2$	$[-2, 2]$	
$x < -1$ or $x > 2$	$(-\infty, -1) \cup (2, \infty)$ ($\cup$ is the union symbol.)	
$x > -1$	$(-1, \infty)$	
$x \leq 2$	$(-\infty, 2]$	

■ **MAKING CONNECTIONS**

Points and Intervals

The expression (1, 2) may represent a point in the xy-plane or the interval $1 < x < 2$. To alleviate confusion, phrases such as "the point (1, 2)" or "the interval (1, 2)" are used.

EXAMPLE 6 Solving an inequality symbolically, numerically, and graphically

Solve $2x + 1 \leq -1$ or $2x + 1 \geq 3$ symbolically. Write the solution in interval notation and give graphical and numerical support to your answer.

Solution

Symbolic Solution First solve each linear inequality.

$2x + 1 \leq -1$	or	$2x + 1 \geq 3$	Given compound inequality
$2x \leq -2$	or	$2x \geq 2$	Subtract 1.
$x \leq -1$	or	$x \geq 1$	Divide by 2.

The solution set may be written as $(-\infty, -1] \cup [1, \infty)$.

Numerical Solution Construct the table for $Y_1 = 2X + 1$ for $x = -3, -2, -1, \ldots, 3$, as shown in Figure 3.25. This table supports the symbolic solution because $Y_1 \leq -1$ when $x \leq -1$ or $Y_1 \geq 3$ when $x \geq 1$.

Graphical Solution Graph $Y_1 = 2X + 1$, $Y_2 = -1$, and $Y_3 = 3$. From Figures 3.26 and 3.27, the slanted graph of Y_1 either intersects or is below the graph of Y_2 when $x \leq -1$, and the graph of Y_1 either intersects or is above the graph of Y_3 when $x \geq 1$.

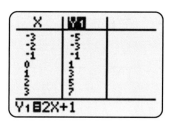

Figure 3.25

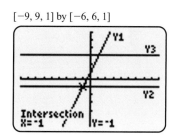

[−9, 9, 1] by [−6, 6, 1]

Figure 3.26

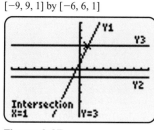

[−9, 9, 1] by [−6, 6, 1]

Figure 3.27

Critical Thinking

Carbon dioxide is emitted when human beings breathe. In one study of college students, the amount of carbon dioxide exhaled in grams per hour was measured during both lectures and exams. The average amount exhaled during lectures L satisfied $25.33 \leq L \leq 28.17$, whereas the average amount exhaled during exams E satisfied $36.58 \leq E \leq 40.92$. What do these results indicate? Explain.
(Source: Wang, T., ASHRAE Trans.)

3.3 Putting It All Together

Two inequalities joined by the words *and* or *or* are called a compound inequality. Compound inequalities may be solved symbolically, numerically, and graphically. A number line is a useful visual tool for solving compound inequalities. Some compound inequalities containing the word *and* may be written as a three-part inequality. For example, $x > -4$ and $x \leq 6$ may be written as $-4 < x \leq 6$. Interval notation is an efficient way to express the solution set to an inequality (see Table 3.4). The following table lists basic concepts related to solving compound inequalities.

Type of Inequality	Method to Solve Inequality
Solving a compound inequality with *and*	*Step 1:* First solve each inequality individually. *Step 2:* The solution set includes values that satisfy both inequalities from Step 1.
Solving a compound inequality with *or*	*Step 1:* First solve each inequality individually. *Step 2:* The solution set includes values that satisfy at least one of the inequalities from Step 1.
Solving a three-part inequality	Work on all three parts at the same time. Be sure to perform the same steps on all three parts. Continue until the inequality is in the form $a \leq x \leq b$, where a and b are real numbers.

3.3 EXERCISES

FOR EXTRA HELP: Student's Solutions Manual, MyMathLab.com, InterAct Math, Math Tutor Center, MathXL, Digital Video Tutor CD 2 Videotape 4

CONCEPTS

1. Give an example of a compound inequality containing the word *and*.

2. Give an example of a compound inequality containing the word *or*.

3. Is $x = 1$ a solution of the compound inequality $x > 3$ and $x \leq 5$?

4. Is $x = 1$ a solution of the compound inequality $x < 3$ or $x \geq 5$?

5. Is the compound inequality $x \geq -5$ and $x \leq 5$ equivalent to $-5 \leq x \leq 5$?

6. Name three methods for solving a compound inequality.

Exercises 7–12: Determine whether the given values of x are solutions to the compound inequality.

7. $x - 1 < 5$ and $2x > 3$ $x = 2, x = 6$

8. $2x + 1 \geq 4$ and $1 - x \leq 3$ $x = -2, x = 3$

9. $3x < -5$ or $2x \geq 3$ $x = 0, x = 3$

10. $x + 1 \leq -4$ or $x + 1 \geq 4$ $x = -5, x = 2$

11. $2 - x > -5$ and $2 - x \leq 4$ $x = -3, x = 0$

12. $x + 5 \geq 6$ or $3x \leq 3$ $x = -1, x = 1$

SYMBOLIC SOLUTIONS

Exercises 13–20: Solve the compound inequality. Graph the solution set, using a number line.

13. $x \leq 3$ and $x \geq -1$

14. $x \geq 5$ and $x > 6$

15. $2x < 5$ and $2x > -4$

16. $2x + 1 < 3$ and $x - 1 \geq -5$

17. $x \leq -1$ or $x \geq 2$

18. $2x \leq -6$ or $x \geq 6$

19. $5 - x > 1$ or $x + 3 \geq -1$

20. $1 - 2x > 3$ or $2x - 4 \geq 4$

3.3 Compound Inequalities

Exercises 21–30: Solve the compound inequality.

21. $x - 3 \leq 4$ and $x + 5 \geq -1$

22. $2x \geq -10$ and $x < 8$

23. $3x - 1 > -1$ and $2x - \frac{1}{2} > 6$

24. $2(x + 1) < 8$ and $-2(x - 4) > -2$

25. $x - 4 \geq -3$ or $x - 4 \leq 3$

26. $1 - 3x \geq 6$ or $1 - 3x \leq -4$

27. $-x < 1$ or $5x + 1 < -10$

28. $7x - 6 > 0$ or $-\frac{1}{2}x \leq 6$

29. $1 - 7x < -48$ and $3x + 1 \leq -9$

30. $3x - 4 \leq 8$ or $4x - 1 \leq 13$

Exercises 31–38: Solve the three-part inequality.

31. $-27 \leq 3x \leq 9$
32. $-16 \leq -4x \leq 8$
33. $-1 < 2x - 1 < 3$
34. $2 \leq 4x + 5 \leq 6$
35. $-2 \leq 5 - \frac{1}{3}x < 2$
36. $-\frac{3}{2} < 4 - 2x < \frac{7}{2}$
37. $100 \leq 10(5x - 2) \leq 200$
38. $-15 < 5(x - 1990) < 30$

NUMERICAL AND GRAPHICAL SOLUTIONS

Exercises 39–42: Use the table to solve the three-part inequality.

39. $-3 \leq 3x \leq 6$
40. $-5 \leq 2x - 1 \leq 1$

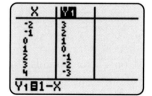
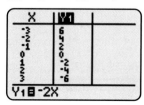

41. $-1 < 1 - x < 2$
42. $-2 \leq -2x < 4$

Exercises 43–46: Use the graph to solve the compound inequality.

43. $-2 \leq y_1 \leq 2$
44. $1 \leq y_1 < 3$

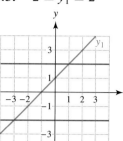

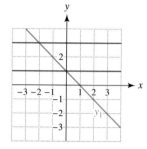
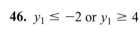

45. $y_1 < -2$ or $y_1 > 2$
46. $y_1 \leq -2$ or $y_1 \geq 4$

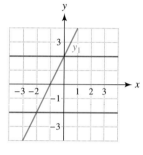
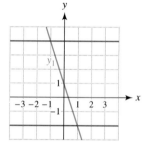

47. *Distance* The function f computes the distance y in miles between a car and the city of Omaha, Nebraska, after x hours, where $0 \leq x \leq 6$. The graphs of f and the horizontal lines $y = 100$ and $y = 200$ are shown in the following figure.
    (a) Is the car moving toward or away from Omaha? Explain.
    (b) Determine the times when the car is 100 miles or 200 miles from Omaha.
    (c) When is the car from 100 to 200 miles from Omaha?
    (d) When is the car's distance from Omaha greater than or equal to 200 miles?

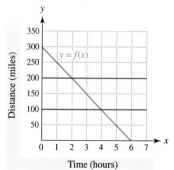

176 CHAPTER 3 Linear Equations and Inequalities

48. Use the following figure to solve each equation or inequality.
(a) $y_1 = y_2$
(b) $y_2 = y_3$
(c) $y_1 \leq y_2 \leq y_3$
(d) $y_2 < y_3$

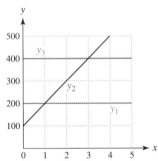

Exercises 49–52: (Refer to Example 5.) Solve the compound inequality numerically and graphically.

49. $-2 \leq 2x - 4 \leq 4$ 50. $-1 \leq 1 - x \leq 3$

51. $x + 1 < -1$ or $x + 1 > 1$

52. $2x - 1 < -3$ or $2x - 1 > 5$

INTERVAL NOTATION

Exercises 53–68: Write the inequality in interval notation.

53. $2 \leq x \leq 10$ 54. $-1 < x < 5$

55. $5 < x \leq 8$ 56. $-\dfrac{1}{2} \leq x \leq \dfrac{5}{6}$

57. $x < 4$ 58. $x \leq -3$

59. $x > -2$ 60. $x \geq 6$

61. $x \leq -2$ or $x \geq 4$ 62. $x \leq -1$ or $x > 6$

63. $x < 1$ or $x \geq 5$ 64. $x < -3$ or $x > 3$

65. ←—|—|—(—|—|—]—|—→
     -6 -4 -2 0 2 4 6

66. ←—|—|—|—[—|—|—→
     -6 -4 -2 0 2 4 6

67. ←—|—|—)—|—|—|—→
     -6 -4 -2 0 2 4 6

68. ←—|—[—|—|—]—|—→
     -6 -4 -2 0 2 4 6

USING MORE THAN ONE METHOD

Exercises 69–72: Solve the compound inequality symbolically. Support your results graphically and numerically. Write the solution set in interval notation.

69. $4 \leq 5x - 1 \leq 14$ 70. $-4 < 2x < 4$

71. $4 - x \geq 1$ or $4 - x < 3$

72. $x + 3 \geq -2$ or $x + 3 \leq 1$

APPLICATIONS

73. *Educational Attainment* The following line graph shows the percentage of the population that completed 4 years or more of college for selected years. Estimate the years when this percentage ranged from 6% to 17%. **(Source: Bureau of the Census.)**

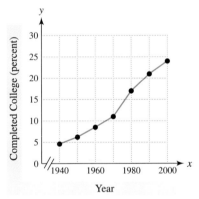

74. *School Enrollment* The line graph shows the school enrollment in millions from kindergarten through university level. Estimate the years when the enrollment was from 57 to 65 million students. **(Source: Department of Education.)**

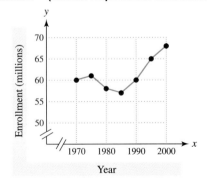

75. *Medicare Costs* Based on current trends, future Medicare costs in billions of dollars may be

modeled by $f(x) = 18x - 35{,}750$, where $1995 \leq x \leq 2007$. (*Source:* Office of Management and Budget.)
(a) Graphically estimate the years when Medicare costs will be from 250 to 340 billion dollars.
(b) Solve part (a) symbolically.

76. *Median Home Prices* The median price P of a single-family home from 1980 to 1991 may be modeled by $P(x) = 3400(x - 1980) + 61{,}000$, where x is the year. (*Source:* Department of Commerce.)
(a) Determine numerically the years when the median price ranged from \$78,000 to \$95,000.
(b) Solve part (a) symbolically.

77. *Geometry* For what values of x is the perimeter of the rectangle from 40 to 60 feet?

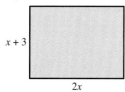

78. *Geometry* A rectangle is three times as long as it is wide. If the rectangle is to have a perimeter from 100 to 160 inches, what values for the width are possible?

79. *Distance and Time* A car's distance in miles from a rest stop is given by $f(x) = 70x + 50$, where x is time in hours.
(a) Construct the table for f for $x = 4, 5, 6, \ldots, 10$ and use the table to solve the inequality $470 \leq f(x) \leq 680$. Explain what your result means.
(b) Solve the inequality in part (a) symbolically.

80. *Altitude and Temperature* If the air temperature at ground level is 70°F, the air temperature x miles high is given by $T(x) = 70 - 29x$. Determine the altitudes at which the air temperature is from 26.5°F to −2.5°F. (*Source:* A. Miller and R. Anthes, *Meteorology*.)

81. *Temperature Scales* The formula
$$F = \frac{9}{5}C + 32$$
may be used to convert Celsius temperatures to Fahrenheit temperatures. The greatest temperature ranges on Earth are recorded in Siberia where the temperature has varied from −90°F to 98°F. Find this temperature range in degrees Celsius, using the formula.

82. *Temperature Scales* The formula
$$C = \frac{5}{9}(F - 32)$$
may be used to convert Fahrenheit temperatures to Celsius temperatures. If the Celsius temperature ranged from 5°C to 20°C, use this formula to find the corresponding temperature range in degrees Fahrenheit.

WRITING ABOUT MATHEMATICS

83. Suppose that the solution set for a compound inequality can be written as $x < -3$ or $x > 2$. A student writes it as $2 < x < -3$. Is the student's three-part inequality correct? Explain your answer.

84. How can you determine whether an x-value is a solution to a compound inequality connected by the word *and*? Give an example. Repeat the question for a compound inequality connected by the word *or*.

GROUP ACTIVITY
Working with Real Data

Directions: Form a group of 2 to 4 people. Select someone to record the group's responses for this activity. All members of the group should work cooperatively to answer the questions. If your instructor asks for your results, each member of the group should be prepared to respond. If the group is asked to turn in its work, be sure to include each person's name on the paper.

Born Outside the United States The foreign-born portion of the population increased from 4.7% in 1970 to 9.3% in 1998. This increase could be modeled by a linear function. Use these data to estimate the years when this percentage was from 6% to 7%. (*Source:* Bureau of the Census.)

3.4 ABSOLUTE VALUE EQUATIONS AND INEQUALITIES

Basic Concepts ~ Absolute Value Equations ~
Absolute Value Inequalities

INTRODUCTION

Monthly average temperatures can vary greatly from one month to another, whereas yearly average temperatures remain fairly constant from one year to the next. In Boston, Massachusetts, the yearly average temperature is 50°F, but monthly average temperatures can vary from 28°F to 72°F. Because 50°F − 28°F = 22°F and 72°F − 28°F = 22°F, the monthly average temperatures are always within 22°F of the yearly average temperature of 50°F. If T represents a monthly average temperature, we can model this situation by using the absolute value inequality

$$|T - 50| \le 22.$$

The absolute value is necessary because a monthly average temperature T may be either greater than or less than 50°F. In this section we discuss how to solve absolute value equations and inequalities symbolically, graphically, and numerically. (*Source:* A. Miller and J. Thompson, *Elements of Meteorology.*)

BASIC CONCEPTS

In Chapter 1 we discussed the absolute value of a number. We can define a function called the **absolute value function** given by $f(x) = |x|$. To graph $y = |x|$, we begin by making a table of values, as shown in Table 3.5.

TABLE 3.5

x	−2	−1	0	1	2
$|x|$	2	1	0	1	2

Next we plot these points and then sketch the graph shown in Figure 3.28. Note that the graph is V-shaped.

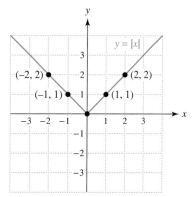

Figure 3.28

ABSOLUTE VALUE EQUATIONS

An equation that contains an absolute value is called an **absolute value equation.** Examples include

$$|x| = 2, \quad |2x - 1| = 5, \quad \text{and} \quad |5 - 3x| - 3 = 1.$$

Consider the absolute value equation $|x| = 2$. This equation has *two* solutions: $|2| = 2$ and $|-2| = 2$, so the solution set is $\{-2, 2\}$. We can also demonstrate this result with a numerical table or a graph. In Table 3.5 $|x| = 2$ when $x = -2$ or $x = 2$. In Figure 3.29 the graph of $y_1 = |x|$ intersects the graph of $y_2 = 2$ at the points $(-2, 2)$ and $(2, 2)$. The x-values at these points of intersection correspond to the solutions $x = -2$ and $x = 2$.

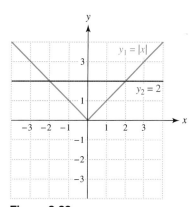

Figure 3.29

We generalize this discussion in the following manner.

Solving $|x| = k$

1. If $k > 0$, then $|x| = k$ is equivalent to $x = k$ or $x = -k$.
2. If $k = 0$, then $|x| = k$ is equivalent to $x = 0$.
3. If $k < 0$, then $|x| = k$ has no solutions.

EXAMPLE 1 *Solving an absolute value equation*

Solve each equation.

(a) $|x| = 20$ (b) $|x| = -5$

Solution

(a) The solutions are $x = -20$ and $x = 20$.
(b) There are no solutions because $|x|$ is never negative.

We can solve other absolute value equations similarly.

EXAMPLE 2 *Solving an absolute value equation*

 Solve $|2x - 5| = 3$.

Solution

If $|2x - 5| = 3$, then either $2x - 5 = 3$ or $2x - 5 = -3$. Solve each equation separately.

$$2x - 5 = 3 \quad \text{or} \quad 2x - 5 = -3$$
$$2x = 8 \quad \text{or} \quad 2x = 2$$
$$x = 4 \quad \text{or} \quad x = 1$$

The solution set is $\{1, 4\}$.

Numerical support can be given for the results in the preceding example. Construct the table for $Y_1 = \text{abs}(2X - 5)$ and $Y_2 = 3$, as shown in Figure 3.30. It indicates that $Y_1 = Y_2 = 3$ when $x = 1$ or $x = 4$. Graphical support can also be given. Note that the V-shaped graph of Y_1 intersects the horizontal graph of Y_2 at the points $(1, 3)$ and $(4, 3)$ in Figures 3.31 and 3.32.

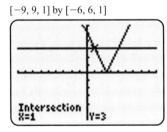

Figure 3.30

Figure 3.31

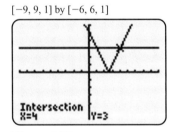

Figure 3.32

This discussion leads to the following result.

Absolute Value Equations

Let $k > 0$ be a positive number. Then

$$|ax + b| = k$$

is equivalent to

$$ax + b = k \quad \text{or} \quad ax + b = -k.$$

EXAMPLE 3 — Solving an absolute value equation

Solve $|5 - x| - 2 = 8$.

Solution

Start by adding 2 to both sides to obtain
$$|5 - x| = 10.$$
This equation is satisfied if any of the following expressions hold.

$$5 - x = 10 \quad \text{or} \quad 5 - x = -10$$
$$-x = 5 \quad \text{or} \quad -x = -15$$
$$x = -5 \quad \text{or} \quad x = 15$$

The solution set is $\{-5, 15\}$.

Critical Thinking

Discuss the solutions to the equation $|ax + b| = k$ if $k = 0$ or if $k < 0$.

ABSOLUTE VALUE INEQUALITIES

Like other inequalities, we can solve absolute value inequalities graphically, numerically, and symbolically. For example, to solve $|x| < 3$ graphically, let $y_1 = |x|$ and $y_2 = 3$ (see Figure 3.33). Their graphs intersect at the points $(-3, 3)$ and $(3, 3)$. The graph of y_1 is below the graph of y_2 for x-values between, but not including, $x = -3$ and $x = 3$. The solution set is $\{x \mid -3 < x < 3\}$ and is shaded on the x-axis.

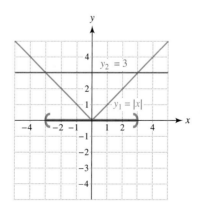

Figure 3.33

Other absolute value inequalities can be solved graphically in a similar way. In Figure 3.34 on the following page the solutions to $|2x - 1| = 3$ are $x = -1$ and

$x = 2$. The V-shaped graph of $y_1 = |2x - 1|$ is below the horizontal line $y_2 = 3$ when $-1 < x < 2$. Thus $|2x - 1| < 3$ whenever $-1 < x < 2$. The solution set is shaded on the x-axis.

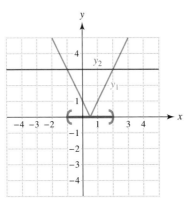

Figure 3.34   **Figure 3.35**

In Figure 3.35 the V-shaped graph of $y_1 = |2x - 1|$ is above the horizontal line $y_2 = 3$ both to the left of -1 and to the right of 2. That is, $|2x - 1| > 3$ whenever $x < -1$ or $x > 2$. The solution set is shaded on the x-axis.

This discussion is summarized as follows.

Absolute Value Inequalities

Let the solutions to $|ax + b| = k$ be c and d, where $c < d$ and $k > 0$.

1. $|ax + b| < k$ is equivalent to $c < x < d$.
2. $|ax + b| > k$ is equivalent to $x < c$ or $x > d$.

Similar statements can be made for inequalities involving $\leq$ or $\geq$.

■ **MAKING CONNECTIONS**

Graphs of $y = |ax + b|$

The graph of $y = |ax + b|$, $a \neq 0$, is V-shaped and intersects a horizontal line above the x-axis twice. This graph can be used to help visualize the solution to either $|ax + b| < k$ or $|ax + b| > k$.

EXAMPLE 4   Solving absolute value equations and inequalities

Solve each absolute value equation and inequality.

(a) $|2 - 3x| = 4$ (b) $|2 - 3x| < 4$ (c) $|2 - 3x| > 4$

Solution

(a) Solve each of the following equations.

$$2 - 3x = 4 \quad \text{or} \quad 2 - 3x = -4$$
$$-3x = 2 \quad \text{or} \quad -3x = -6$$
$$x = -\frac{2}{3} \quad \text{or} \quad x = 2$$

The solution set is $\left\{-\frac{2}{3}, 2\right\}$.

(b) Solutions to $|2 - 3x| < 4$ include x-values between, but not including, $-\frac{2}{3}$ and 2. Thus the solution set is $\left\{x \mid -\frac{2}{3} < x < 2\right\}$.

(c) Solutions to $|2 - 3x| > 4$ include x-values to the left of $x = -\frac{2}{3}$ or to the right of $x = 2$. Thus the solution set is $\left\{x \mid x < -\frac{2}{3} \text{ or } x > 2\right\}$.

EXAMPLE 5 *Modeling temperature in Boston*

GCLM In the introduction to this section we discussed how the inequality $|T - 50| \leq 22$ models the range for the monthly average temperatures T in Boston, Massachusetts.

(a) Solve this inequality and interpret the result.
(b) Give graphical support for part (a).

Solution

(a) *Symbolic Solution* Start by solving $|T - 50| = 22$.

$$T - 50 = 22 \quad \text{or} \quad T - 50 = -22$$
$$T = 72 \quad \text{or} \quad T = 28 \qquad \text{Add 50 to both sides}$$

Thus the solution set to $|T - 50| \leq 22$ is $\{T \mid 28 \leq T \leq 72\}$. Monthly average temperatures in Boston vary from 28°F to 72°F.

(b) *Graphical Solution* The graphs of $Y_1 = \text{abs}(X - 50)$ and $Y_2 = 22$ intersect at the points (28, 22) and (72, 22), as shown in Figures 3.36 and 3.37. The V-shaped graph of Y_1 intersects the horizontal graph of Y_2, or is below it, when $28 \leq x \leq 72$. Thus the solution set is $\{T \mid 28 \leq T \leq 72\}$. This result agrees with the symbolic result.

[0, 100, 10] by [0, 70, 10]

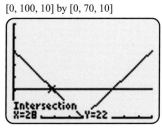

Figure 3.36

[0, 100, 10] by [0, 70, 10]

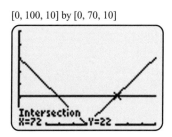

Figure 3.37

Critical Thinking

Find the solution set for the following inequalities. Discuss your results.

1. $|2x - 5| > -3$
2. $|2x - 5| < -3$

3.4 Putting It All Together

The following table summarizes methods for solving absolute value equations and inequalities involving $<$ and $>$ symbols. Inequalities containing $\leq$ and $\geq$ are solved similarly.

Problem	Symbolic Solution	Graphical Solution
$\|ax + b\| = k, k > 0$	Solve the equations $$ax + b = k$$ and $$ax + b = -k.$$	Graph $Y_1 = \|ax + b\|$ and $Y_2 = k$. Find the x-values of the two points of intersection.
$\|ax + b\| < k, k > 0$	If the solutions to $$\|ax + b\| = k$$ are c and d, $c < d$, then the solutions to $$\|ax + b\| < k$$ satisfy $$c < x < d.$$	Graph $Y_1 = \|ax + b\|$ and $Y_2 = k$. Find the x-values of the two points of intersection. The solutions are between these x-values, where the graph of Y_1 lies below the graph of Y_2.
$\|ax + b\| > k, k > 0$	If the solutions to $$\|ax + b\| = k$$ are c and d, $c < d$, then the solutions to $$\|ax + b\| > k$$ satisfy $$x < c \quad \text{or} \quad x > d.$$	Graph $Y_1 = \|ax + b\|$ and $Y_2 = k$. Find the x-values of the two points of intersection. The solutions are outside these x-values on the number line, where the graph of Y_1 is above the graph of Y_2.

3.4 EXERCISES

CONCEPTS

1. Give an example of an absolute value equation.
2. Give an example of an absolute value inequality.
3. Is $x = -3$ a solution to $|x| = 3$?
4. Is $x = -4$ a solution to $|x| > 3$?
5. Is $|x| = 5$ equivalent to $x = -5$ or $x = 5$?
6. Is $|x| < 3$ equivalent to $x < -3$ or $x > 3$? Explain.

Exercises 7–12: Determine whether the given values of x are solutions to the absolute value equation or inequality.

7. $|2x - 5| = 1$ $\quad x = -3, x = 3$
8. $|5 - 6x| = 1$ $\quad x = 1, x = 0$
9. $|7 - 4x| \leq 5$ $\quad x = -2, x = 2$
10. $|2 + x| < 2$ $\quad x = -4, x = -1$
11. $|7x + 4| > -1$ $\quad x = -\frac{4}{7}, x = 2$
12. $|12x + 3| \geq 3$ $\quad x = -\frac{1}{4}, x = 2$

Exercises 13–14: Use the graph of y_1 to solve the equation and inequalities.

13. (a) $y_1 = 2$ (b) $y_1 < 2$ (c) $y_1 > 2$

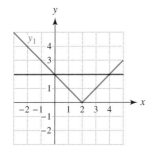

14. (a) $y_1 = 3$ (b) $y_1 \leq 3$ (c) $y_1 \geq 3$

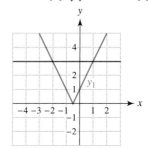

SYMBOLIC SOLUTIONS

Exercises 15–30: Solve the absolute value equation.

15. $|x| = 7$
16. $|x| = 4$
17. $|x| = 0$
18. $|x| = -6$
19. $|4x| = 9$
20. $|-3x| = 7$
21. $|-2x| - 6 = 2$
22. $|5x| + 1 = 5$
23. $|2x + 1| = 11$
24. $|1 - 3x| = 4$
25. $|-2x + 3| + 3 = 4$
26. $|6x + 2| - 2 = 6$
27. $\left|\frac{1}{2}x - 1\right| = 5$
28. $\left|6 - \frac{3}{4}x\right| = 3$
29. $|2x - 6| = -7$
30. $\left|1 - \frac{2}{3}x\right| + 2 = 0$

Exercises 31–34: Solve the absolute value equation and inequalities.

31. (a) $|2x| = 8$
    (b) $|2x| < 8$
    (c) $|2x| > 8$

32. (a) $|3x - 9| = 6$
    (b) $|3x - 9| \leq 6$
    (c) $|3x - 9| \geq 6$

33. (a) $|5 - 4x| = 3$
    (b) $|5 - 4x| \leq 3$
    (c) $|5 - 4x| \geq 3$

34. (a) $\left|\dfrac{x-5}{2}\right| = 2$

(b) $\left|\dfrac{x-5}{2}\right| < 2$

(c) $\left|\dfrac{x-5}{2}\right| > 2$

Exercises 35–48: Solve the absolute value inequality.

35. $|2x| > 7$
36. $|-12x| < 30$
37. $|-4x + 4| < 16$
38. $|-5x - 8| > 2$
39. $2|x + 5| \geq 8$
40. $-3|x - 1| \geq -9$
41. $|8 - 6x| - 1 \leq 2$
42. $4 - \left|\dfrac{2x}{3}\right| < -7$
43. $5 + \left|\dfrac{2-x}{3}\right| \leq 9$
44. $\left|\dfrac{x+3}{5}\right| \leq 12$
45. $|2x - 1| \leq -3$
46. $|x + 6| \geq -5$
47. $|x + 1| - 1 > -3$
48. $-2|1 - 7x| \geq 2$

NUMERICAL AND GRAPHICAL SOLUTIONS

Exercises 49–52: Use the table of $Y_1 = |ax + b|$ to solve the equation and inequality.

49. (a) $Y_1 = 2$ (b) $Y_1 < 2$

50. (a) $Y_1 = 1$ (b) $Y_1 > 1$

51. (a) $Y_1 = 3$ (b) $Y_1 \geq 3$

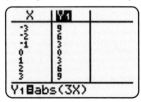

52. (a) $Y_1 = 6$ (b) $Y_1 \leq 6$

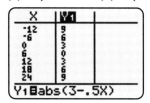

Exercises 53 and 54: Use the V-shaped graph of Y_1 to solve the equation or inequality.

53. (a) $Y_1 = 1$ (b) $Y_1 \leq 1$ (c) $Y_1 \geq 1$
$[-3, 3, 1]$ by $[-2, 2, 1]$ $[-3, 3, 1]$ by $[-2, 2, 1]$

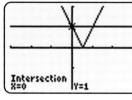 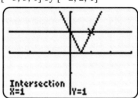

54. (a) $Y_1 = 3$ (b) $Y_1 < 3$ (c) $Y_1 > 3$
$[-6, 6, 1]$ by $[-4, 4, 1]$ $[-6, 6, 1]$ by $[-4, 4, 1]$

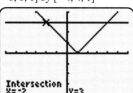 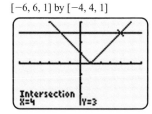

Exercises 55–64: Solve the inequality graphically.

55. $|x| \geq 1$
56. $|x| < 2$
57. $|x - 1| \leq 3$
58. $|x + 5| \geq 2$
59. $|4 - 2x| > 2$
60. $|10 - 3x| < 4$
61. $|1.5x - 3| \geq 6$
62. $|7 - 4x| \leq 2.5$
63. $|8.1 - x| > -2$
64. $\left|\dfrac{5x - 9}{2}\right| \leq -1$

USING MORE THAN ONE METHOD

Exercises 65–68: Solve the absolute value inequality
  (a) symbolically,
  (b) graphically, and
  (c) numerically.

65. $|3x| \leq 9$
66. $|5 - x| \geq 3$
67. $|2x - 5| > 1$
68. $|-8 - 4x| < 6$

APPLICATIONS

Exercises 69–72: Average Temperatures (Refer to Example 5.) The given inequality models the range for the monthly average temperatures T in degrees Fahrenheit at the location specified.
  (a) Solve the inequality.
  (b) Give a possible interpretation of the inequality.

69. $|T - 43| \leq 24$, Marquette, Michigan
70. $|T - 62| \leq 19$, Memphis, Tennessee
71. $|T - 10| \leq 36$, Chesterfield, Canada
72. $|T - 61.5| \leq 12.5$, Buenos Aires, Argentina

73. *Highest Elevations* The table lists the highest elevation in each continent.

Continent	Elevation (feet)
Asia	29,028
S. America	22,834
N. America	20,320
Africa	19,340
Europe	18,510
Antarctica	16,066
Australia	7,310

Source: National Geographic.

  (a) Calculate the average A of these elevations.
  (b) Which continents have their highest elevations within 1000 feet of A?
  (c) Which continents have their highest elevations within 5000 feet of A?

74. *Distance* Suppose that two cars, both traveling at a constant speed of 60 miles per hour, approach each other on a straight highway.
  (a) If they are initially 4 miles apart, sketch a graph of the distance between the cars after x minutes, where $0 \leq x \leq 4$.
  (b) Write an absolute value equation whose solution gives the times when the cars were 2 miles apart.
  (c) Solve your equation from part (b).

75. *Error in Measurements* Products are often manufactured to be a given size or shape to within a certain tolerance. For instance, if an aluminum can is supposed to have a diameter of 2.5 inches, either 2.501 inches or 2.499 inches might be acceptable. If the maximum error in the diameter of the can is restricted to 0.002 inch, an acceptable diameter d must satisfy the absolute value inequality

$$|d - 2.5| \leq 0.002.$$

Solve this inequality for d and interpret the result.

76. *Error in Measurements* (Refer to Exercise 75.) Suppose that a person can operate a stopwatch accurately to within 0.02 second. If a runner's time in the 400-meter dash is recorded as 51.67 seconds, what range of values could the true time assume?

WRITING ABOUT MATHEMATICS

77. If $a \neq 0$, how many solutions are there to the equation $|ax + b| = k$ when
  (a) $k > 0$,
  (b) $k = 0$, and
  (c) $k < 0$?
  Explain each answer by using a graph. (*Hint:* The graph of $y = |ax + b|$ is V-shaped.)

78. Suppose that you know two solutions to $|ax + b| = k$. How can you use these solutions to solve the inequalities $|ax + b| < k$ and $|ax + b| > k$? Give an example.

CHECKING BASIC CONCEPTS FOR SECTIONS 3.3 AND 3.4

1. (a) Is $x = 3$ a solution of the compound inequality $x + 2 < 4$ or $2x - 1 \geq 3$?
   (b) Is $x = 3$ a solution of the compound inequality $x + 2 < 4$ and $2x - 1 \geq 3$?

2. Solve the following compound inequalities.
   (a) $-5 \leq 2x + 1 \leq 3$
   (b) $1 - x \leq -2$ or $1 - x > 2$

3. Solve $|3 - x| = 5$
   (a) symbolically,
   (b) graphically, and
   (c) numerically.

4. Solve the absolute value equation and inequalities.
   (a) $|3x - 6| = 8$
   (b) $|3x - 6| < 8$
   (c) $|3x - 6| > 8$

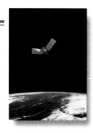

CHAPTER 3 SUMMARY

Section 3.1 Linear Equations

A linear equation can be written in the form $ax + b = 0$, where $a \neq 0$. A linear equation has exactly one solution, which can be found symbolically, graphically, or numerically.

Example: $3x - 1 = 5$ implies that $3x = 6$ or $x = 2$.

[−9, 9, 1] by [−6, 6, 1]

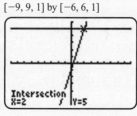

A Graphical Solution

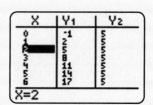

A Numerical Solution

Section 3.2 Linear Inequalities

A linear inequality results when the equals sign in a linear equation is replaced with $<, >, \leq$, or $\geq$. Many inequalities have infinitely many solutions. Like linear equations, linear inequalities may be solved symbolically, graphically, and numerically. Solving a linear inequality requires reversing the inequality symbol when both sides of an inequality are multiplied or divided by a negative number.

Example: $-3x - 8 < 1$ Given inequality
$-3x < 9$ Add 8.
$x > -3$ Divide both sides by -3; reverse inequality.

Section 3.3 Compound Inequalities

A compound inequality consists of two inequalities that are joined by the words *and* or *or*. Examples of compound inequalities are

$$2x > -4 \quad \text{and} \quad 5x - 2 < 9$$
$$3x - 1 < -4 \quad \text{or} \quad 3x - 1 > 4.$$

If a compound inequality is connected by the word *and*, a solution must satisfy both inequalities. If a compound inequality is connected by the word *or*, a solution must satisfy at least one of the inequalities. Number lines are valuable in determining solutions of compound inequalities. Interval notation may be used to express solutions to inequalities. Compound inequalities of the form $x > c$ and $x < d$ may be written as the three-part inequality $c < x < d$ provided $c < d$.

Section 3.4 Absolute Value Equations and Inequalities

The graph of $y = |ax + b|, a \neq 0$, is V-shaped and intersects the horizontal line $y = k$ twice if $k > 0$. In this case there are two solutions to the equation $|ax + b| = k$ determined by $ax + b = \pm k$.

Example: The equation $|2x - 1| = 5$ has two solutions.

$$2x - 1 = 5 \quad \text{or} \quad 2x - 1 = -5$$
$$2x = 6 \quad \text{or} \quad 2x = -4$$
$$x = 3 \quad \quad \quad x = -2$$

[−9, 9, 1] by [−6, 6, 1]

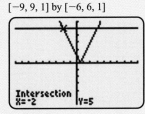

A Graphical Solution

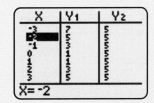

A Numerical Solution

Section 3.4 Absolute Value Equations and Inequalities (continued)

If the solutions to $|ax + b| = k$ are c and d with $c < d$, then the solution set for $|ax + b| < k$ is $\{x \mid c < x < d\}$, and the solution set for $|ax + b| > k$ is $\{x \mid x < c \text{ or } x > d\}$.

Example: The solutions to $|2x - 1| = 5$ are $x = -2$ and $x = 3$, so the solution set for $|2x - 1| < 5$ is $\{x \mid -2 < x < 3\}$, and the solution set for $|2x - 1| > 5$ is $\{x \mid x < -2 \text{ or } x > 3\}$.

CHAPTER 3
REVIEW EXERCISES

SECTION 3.1

Exercises 1 and 2: Complete the table and then use the table to solve the equation.

1. $3x - 6 = 0$

x	0	1	2	3	4
$3x - 6$	-6				6

2. $5 - 2x = 3$

x	-1	0	1	2	3
$5 - 2x$					

3. Solve $3 - 5x = 18$ numerically.

4. Use the graph to solve the equation $y_1 = y_2$.

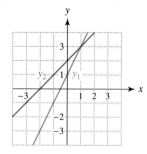

5. Solve $5 - 2x = \dfrac{1}{2} + x$ graphically.

6. Solve $4x - 3 = 5 + 2x$ graphically or numerically.

Exercises 7–12: Solve the equation symbolically.

7. $2x - 7 = 21$

8. $1 - 7x = -\dfrac{5}{2}$

9. $-2(4x - 1) = 1 - x$

10. $-\dfrac{3}{4}(x - 1) + 5 = 6$

11. $\dfrac{x - 4}{3} = 2$

12. $\dfrac{2x - 3}{2} = \dfrac{x + 3}{5}$

Exercises 13 and 14: Translate the sentence into an equation and then solve the equation.

13. The sum of twice x and 25 is 19.

14. If 5 is subtracted from twice x, it equals x plus 1.

Section 3.2

15. Use the table to solve $Y_1 \geq Y_2$, where Y_1 and Y_2 are linear functions.

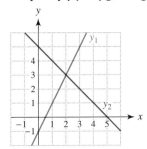

16. Solve $2(3 - x) + 4 < 0$ numerically.

17. Solve the inequality $y_1 \geq y_2$, using the graph.

18. Solve $5 - 4x \leq -2$ graphically.

Exercises 19–22: Solve the inequality symbolically.

19. $-2x + 1 \leq 3$ **20.** $x - 5 \geq 2x + 3$

21. $\dfrac{3x - 1}{4} > \dfrac{1}{2}$ **22.** $-3.2(x - 2) < 1.6x$

Section 3.3

Exercises 23–26: Solve the compound inequality. Graph the solution set on a number line.

23. $x + 1 \leq 3$ and $x + 1 \geq -1$

24. $2x + 7 < 5$ and $-2x \geq 6$

25. $5x - 1 \leq 3$ or $1 - x < -1$

26. $3x + 1 > -1$ or $3x + 1 < 10$

27. Use the table to solve $-2 \leq 2x + 2 \leq 4$

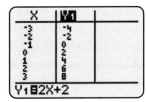

28. Use the following figure to solve each equation and inequality.
(a) $y_1 = y_2$
(b) $y_2 = y_3$
(c) $y_1 \leq y_2 \leq y_3$
(d) $y_2 < y_3$

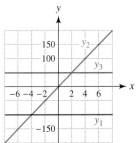

29. The graphs of y_1 and y_2 are shown in the following figure. Solve each equation and inequality.
(a) $y_1 = y_2$
(b) $y_1 < y_2$
(c) $y_1 > y_2$

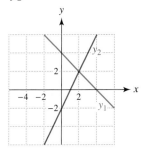

30. The graphs of three linear functions f, g, and h are shown in the following figure. Solve each equation and inequality.
(a) $f(x) = g(x)$
(b) $g(x) = h(x)$
(c) $f(x) < g(x) < h(x)$

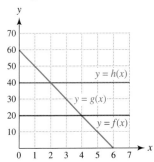

Exercises 31–36: Write the inequality in interval notation.

31. $-3 \leq x \leq \dfrac{2}{3}$ **32.** $-6 < x \leq 45$

33. $x < \dfrac{7}{2}$

34. $x \geq 1.8$

35. $x > -3$ and $x < 4$

36. $x < 4$ or $x > 10$

Exercises 37–40: Solve the compound inequality symbolically and support your result graphically or numerically. Write the solution set in interval notation.

37. $-4 < x + 1 < 6$

38. $20 \leq 2x + 4 \leq 60$

39. $-3 < 4 - \dfrac{1}{3}x < 7$

40. $2 \leq \dfrac{1}{2}x - 2 \leq 12$

Section 3.4

Exercises 41–44: Determine whether the given values of x are solutions to the absolute value equation or inequality.

41. $|12x - 24| = 24 \quad x = -3, x = 2$

42. $|5 - 3x| > 3 \quad x = \dfrac{4}{3}, x = 0$

43. $|3x - 6| \leq 6 \quad x = -3, x = 4$

44. $|2 + 3x| + 4 < 11 \quad x = -3, x = \dfrac{2}{3}$

45. Use the accompanying table to solve each equation and inequality.
    (a) $Y_1 = 2$ (b) $Y_1 < 2$ (c) $Y_1 > 2$

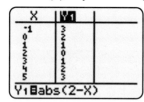

46. Use the graph of Y_1 and the horizontal line $y = 3$ to solve each equation and inequality.
    (a) $Y_1 = 3$ (b) $Y_1 \leq 3$ (c) $Y_1 \geq 3$

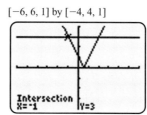 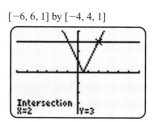

Exercises 47–50: Solve the absolute value equation.

47. $|x| = 22$

48. $|2x - 9| = 7$

49. $\left|4 - \dfrac{1}{2}x\right| = 17$

50. $\dfrac{1}{3}|3x - 1| + 1 = 9$

Exercises 51–52: Solve each absolute value equation and inequality.

51. (a) $|x + 1| = 7$
    (b) $|x + 1| \leq 7$
    (c) $|x + 1| \geq 7$

52. (a) $|1 - 2x| = 6$
    (b) $|1 - 2x| \leq 6$
    (c) $|1 - 2x| \geq 6$

Exercises 53–56: Solve the absolute value inequality.

53. $|x| > 3$

54. $|-5x| < 20$

55. $|4x - 2| \leq 14$

56. $\left|1 - \dfrac{4}{5}x\right| \geq 3$

Exercises 57–58: Solve the inequality graphically.

57. $|2x| \geq 3$

58. $\left|\dfrac{1}{2}x - 1\right| \leq 2$

APPLICATIONS

59. *Distance Between Bicyclists* The following graph shows the distance between two bicyclists traveling toward each other along a straight road after x hours.
  (a) After how long did the bicycle riders meet?
  (b) When were they 20 miles apart?
  (c) Find the times when they were less than 20 miles apart.
  (d) Estimate the sum of the speeds of the two bicyclists.

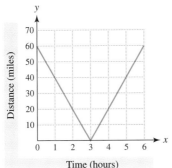

60. *Interest* As shown in the following graph, the function f computes the annual interest y on a loan of x dollars with an interest rate of 15%. Determine the loan amounts that result in the following.
  (a) An annual interest equal to $300
  (b) An annual interest of more than $300
  (c) An annual interest of less than $300

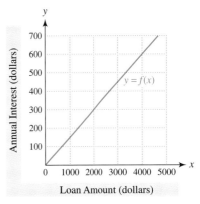

61. *Distance* Two cars, Car 1 and Car 2, are traveling in the same direction on a straight highway. Their distances in miles south of Minneapolis, Minnesota, after x hours are shown in the following graph.
  (a) Which car is traveling faster? Explain.
  (b) How many hours elapse before the two cars are the same distance from Minneapolis? How far are they from Minneapolis?
  (c) During what time interval is Car 1 closer to Minneapolis than Car 2?

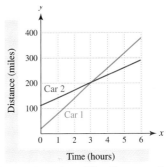

62. *School Bus Deaths* The line graph shows the number of deaths occuring from school bus crashes. For which years were there more than 10 deaths? (*Source:* National Center for Statistics and Analysis.)

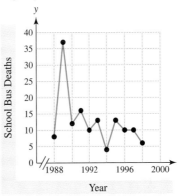

63. *Age in the United States* The median age of the population for each year x between 1820 and 1995 can be approximated by

$$f(x) = 0.09x - 147.1.$$

   (*Source:* **Bureau of the Census.**)
   (a) Graph f in [1820, 1995, 20] by [0, 40, 10]. Interpret the slope of the graph as a rate of change.
   (b) Estimate graphically when the median age was 23 years.
   (c) Solve part (b) symbolically.

64. *Women Officers* The number of women officers in the Marine Corps from 1960 to 1997 may be modeled by $f(x) = 17.7x - 34{,}636$, where x represents the year. (*Source:* **Department of Defense.**)
   (a) Estimate graphically the years when the number of women officers was from 490 to 670.
   (b) Solve part (a) symbolically.

65. *Weight of a Fish* If a walleye has a length of x inches, where $30 \le x \le 35$, its weight W in pounds can be estimated from the formula $W(x) = 1.11x - 23.3$. (*Source:* **Minnesota Department of Natural Resources.**)
   (a) What length of walleye is likely to weigh 12 pounds?
   (b) What lengths of walleye are likely to weigh less than 12 pounds?

66. *Air Temperature* If the temperature on the ground is 60°F, the air temperature x miles high is given by $T(x) = 60 - 29x$. Determine the altitudes at which the air temperature is from 40°F to 20°F symbolically and graphically. (*Source:* **A. Miller.**)

67. *Distance and Time* A car is 200 miles west of Rapid City, South Dakota, traveling east at 70 miles per hour. How long will it take for the car to be 395 miles east of Rapid City?

68. *Geometry* A rectangle is 5 feet longer than twice its width. If the rectangle has a perimeter of 88 feet, what are the dimensions of the rectangle?

69. *Temperature Scales* The formula

$$F = \frac{9}{5}C + 32$$

may be used to convert Fahrenheit temperature to Celsius temperature. The temperature range at Houghton Lake, Michigan, has varied between −48°F and 107°F. Find this temperature range in Celsius.

70. *Error in Measurements* A square garden is being fenced along its 160-foot perimeter. If the length L of the fence must be within 1 foot of the garden's perimeter, write an absolute value inequality that gives acceptable values for L. Solve your inequality.

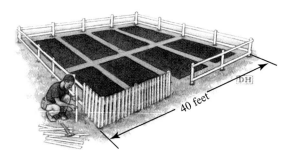

71. *Average Precipitation* The average rainfall in Houston, Texas, is 3.9 inches per month. Each month's average A is within 1.7 inches of 3.9 inches. (*Source:* **J. Williams, The Weather Almanac 1995.**)
   (a) Write an absolute value inequality that models this situation.
   (b) Solve the inequality.

Chapter 3 Test

1. Solve $3 - 5x = 18$ symbolically. Check your answer numerically.

2. Use the accompanying graph to solve each equation and inequality.
   (a) $y_1 = y_2$
   (b) $y_1 \geq y_2$
   (c) $y_1 \leq y_2$

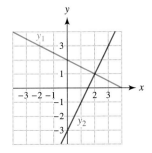

3. Solve $2 - \dfrac{1}{4}x = 1 + 2x$ graphically.

4. Solve $-\dfrac{2}{3}(3x - 2) + 1 = x$ symbolically.

5. Translate the following sentence into an equation and then solve the equation for x. "If 2 is added to 5 times x, it equals x minus 4."

6. Solve $5 - 6x \leq -2x$ graphically.

7. Solve each inequality symbolically.
   (a) $-\dfrac{5}{2}x + \dfrac{1}{2} \leq 2$
   (b) $3.1(3 - x) < 2x$

8. Graph the solution set to the compound inequality on a number line.
   $$2x + 6 < 2 \text{ and } -3x \geq 3$$

9. Use the table to solve the compound inequality $-3x < -3$ or $-3x > 6$.

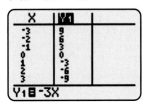

10. Use the following figure to solve each equation and inequality.
    (a) $y_1 = y_2$
    (b) $y_2 = y_3$
    (c) $y_1 \leq y_2 \leq y_3$
    (d) $y_2 < y_3$

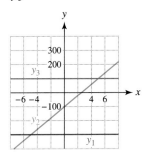

11. Solve the compound inequality
    $$-2 < 2 + \dfrac{1}{2}x < 2$$
    symbolically and write the solution set in interval notation.

12. Solve the equation $\left|2 - \dfrac{1}{3}x\right| = 6$ symbolically and graphically.

13. Solve each equation and inequality, using a method of your choice.
    (a) $|1 - 5x| = 3$
    (b) $|1 - 5x| \leq 3$
    (c) $|1 - 5x| \geq 3$

14. *Sport Drinks* The following graph shows a relationship between the calories in an 8-ounce serving of a sport drink and the corresponding grams of carbohydrates. Estimate the number of calories x in 8 ounces of a sport drink with the following grams of carbohydrates. (*Source: Runner's World.*)
 (a) 16 grams of carbohydrates
 (b) More than 16 grams of carbohydrates
 (c) Less than 16 grams of carbohydrates

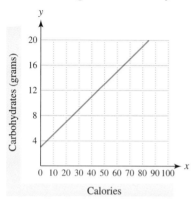

15. *Drinking Fluids and Exercise* To determine the number of ounces of fluid that a person should drink in a day, divide his or her weight by 2 and then add 0.4 ounce for every minute of exercise.
 (a) Write a function that gives the fluid requirements for a person weighing 150 pounds and exercising x minutes a day.
 (b) If a 150-pound runner needs 89 ounces of fluid each day, determine symbolically the runner's daily minutes of exercise.
 (c) Solve part (b) graphically.

CHAPTER 3

EXTENDED AND DISCOVERY EXERCISES

1. *Clouds and Temperature* If the air temperature is greater than the dew point, clouds do not form. If the air temperature cools to the dew point, fog or clouds may appear. Suppose that the temperature x miles high is given by $T(x) = 90 - 29x$ and that the dew point x miles high is given by $D(x) = 70 - 5.8x$. (*Source: A. Miller.*)
 (a) Find the temperature and dew point at ground level.
 (b) At what altitude are the air temperature and dew point equal?
 (c) Determine the altitudes at which clouds do not form.
 (d) Determine the altitudes at which clouds may form.

2. *Critical Thinking* (Refer to Exercise 1.) For each 1-mile increase in altitude, the air temperature decreases 29°F, whereas the dew point decreases 5.8°F. At ground level the dew point is usually less than the air temperature. However, if the air temperature reaches the dew point, fog or clouds may form.
 (a) Discuss why clouds are likely to form at some altitude in the sky.
 (b) Suppose that the dew point is slightly less than the air temperature near the ground. How does that affect the altitude at which clouds form?

3. *Recording Music* A compact disc (CD) can hold approximately 600 million bytes. One *byte* is capable of storing one letter of the alphabet. For example, the word "function" requires 8 bytes to store in computer memory. One million bytes is commonly referred to as a *megabyte* (MB). Recording music requires an enormous amount of memory. The accompanying table lists the megabytes x needed to record y seconds (sec) of music.

x (MB)	0.129	0.231	0.415	0.491
y (sec)	6.010	10.74	19.27	22.83

x (MB)	0.667	1.030	1.160	1.260
y (sec)	31.00	49.00	55.25	60.18

Source: Gateway 2000 System CD.

(a) Make a scatterplot of the data.
(b) What type of relationship seems to exist between x and y? Why does this relationship seem reasonable?
(c) Find the slope-intercept form of a line that models the data. Interpret the slope of this line as a rate of change.
(d) Check your answer in part (c) by graphing the line and data in the same graph.
(e) Write a linear equation whose solution gives the megabytes needed to record 120 seconds of music.
(f) Solve the equation in part (e) graphically and symbolically.

4. *Early Cellular Phone Growth* Cellular phone use has grown dramatically in the United States. When there were 25,000 customers in New York City, the investment cost per cellular site was $12 million. (A cellular site would include such things as a relay tower to transmit signals between cellular phones.) When the number of customers rose to 100,000, the investment cost per cellular site rose to $96 million. Although cost usually decreases with additional customers, such was not the case for early cellular technology. Instead, cost increased as a result of having to purchase expensive real estate and establish communications links among a large number of cellular sites. The relationship between cellular sites and investment cost per site was approximately linear. (*Source:* M. Paetsch, *Mobile Communications in the US and Europe.*)

(a) Find values for m, h, and k so that $f(x) = m(x - h) + k$ models the cost per cellular site in millions of dollars when there were x customers.
(b) Graph f in an appropriate viewing rectangle for the data points (25,000, 12) and (100,000, 96).
(c) Write an inequality whose solution set gives the numbers of customers when the investment cost per cellular site was between $28.8 million and $51.2 million.
(d) Solve the inequality in part (c) graphically and symbolically.

5. *Comparing Ages* The age of the Universe is about 20 billion years, whereas the age of Earth is about 4.5 billion years. The earliest evidence of dinosaurs dates back 200 million years, whereas the earliest evidence for *Homo sapiens* dates back 300,000 years. If the age of the Universe is condensed into 1 year, determine the approximate times when Earth, the dinosaurs, and *Homo sapiens* first appeared.

CHAPTER 4
SYSTEMS OF LINEAR EQUATIONS

In 1940, a physicist named John Atanasoff at Iowa State University needed to solve 29 equations with 29 variables simultaneously. This task was too difficult to do by hand, so he and a graduate student invented the first fully electronic digital computer. Thus the desire to solve a mathematical problem led to one of the most important inventions of the twentieth century. Today people can solve thousands of equations with thousands of variables. Solutions to such equations have resulted in better airplanes, cars, electronic devices, weather forecasts, and medical equipment.

Equations are also widely used in biology. The following table contains the weight W, neck size N, and chest size C for three black bears. Suppose that park rangers find a bear with a neck size of 22 inches and a chest size of 38 inches. Can they use the data in the table to estimate the bear's weight? Using systems of linear equations, they can answer this question.

W (pounds)	N (inches)	C (inches)
80	16	26
344	28	45
416	31	54
?	22	38

Education is what survives when what has been learned has been forgotten.
—B. F. Skinner

Sources: A. Tucher, *Fundamentals of Computing*; M. Triola, *Elementary Statistics*; Minitab, Inc.

4.1 SYSTEMS OF LINEAR EQUATIONS IN TWO VARIABLES

Basic Concepts ~ Graphical and Numerical Solutions ~
Types of Linear Systems

INTRODUCTION

Many formulas involve more than one variable. For example, to calculate the heat index we need to know both the air temperature and the humidity. To calculate monthly car payments we need the loan amount, interest rate, and duration of the loan. Some applications involve large numbers of variables. To design new aircraft it is necessary to solve equations containing thousands of variables. In this section we consider systems of equations containing only two linear equations in two variables. However, the concepts discussed in this section are used to solve larger systems of equations.

BASIC CONCEPTS

Families commonly take vacations. A total of 184 million American adults took a family vacation trip in either 1998 or 1999. Two million more adults took a trip in 1999 than in 1998. Is it possible to determine the number of adults that traveled each year?
(*Source:* USA Today.)

We can solve this problem by letting x represent the number of adults that traveled in 1999 and y represent the number of adults that traveled in 1998. Then we can model the given information with the following two equations.

$$x + y = 184 \quad \text{The total is 184 million.}$$
$$x - y = 2 \quad \text{The difference is 2 million.}$$

Each equation contains two variables, x and y. These two equations form a **system of two linear equations in two variables**. An ordered pair (x, y) is a **solution** to a system of equations if the values for x and y make *both* equations true. Any system of two linear equations in two variables may be written in **standard form** as

$$ax + by = c$$
$$dx + ey = k,$$

where a, b, c, d, e, and k are constants.

EXAMPLE 1   *Testing for solutions*

Determine which ordered pair is a solution to the system of equations: $(0, 3)$ or $(-1, 2)$.

$$-x + 4y = 9$$
$$3x - 3y = -9$$

Solution

For $(0, 3)$ to be a solution, the values of $x = 0$ and $y = 3$ must satisfy both equations.

$$-0 + 4(3) \stackrel{?}{=} 9 \quad \text{False}$$
$$3(0) - 3(3) \stackrel{?}{=} -9 \quad \text{True}$$

Both equations are not true, so (0, 3) is *not* a solution. To test (−1, 2) substitute $x = -1$ and $y = 2$ into each equation.

$$-(-1) + 4(2) \stackrel{?}{=} 9 \quad \text{True}$$
$$3(-1) - 3(2) \stackrel{?}{=} -9 \quad \text{True}$$

Both equations are true, so (−1, 2) is a solution.

GRAPHICAL AND NUMERICAL SOLUTIONS

Graphical, numerical, and symbolic techniques can be used to solve systems of equations. In this section we focus on graphical and numerical techniques and delay discussion of symbolic techniques until the next section. In the next example we determine the number of adults that took a vacation trip in 1998 and in 1999 by solving the system of equations presented earlier.

EXAMPLE 2 *Solving a system of equations graphically and numerically*

 Solve the system of equations

$$x + y = 184$$
$$x - y = 2$$

graphically and numerically. Interpret your results.

Solution

Graphical Solution Before we can graph the system, we need to solve each equation for *y*.

$$x + y = 184 \quad \text{First equation}$$
$$y = 184 - x \quad \text{Subtract } x \text{ from both sides.}$$

Solve the second equation for *y*.

$$x - y = 2 \quad \text{Second equation}$$
$$-y = 2 - x \quad \text{Subtract } x \text{ from both sides.}$$
$$y = -2 + x \quad \text{Multiply both sides by } -1.$$

Graph $Y_1 = 184 - X$ and $Y_2 = -2 + X$, as shown in Figure 4.1. Their graphs intersect at the point (93, 91), which is the solution. Thus 93 million adults took a vacation trip in 1999, and 91 million adults did so in 1998. Note that $93 + 91 = 184$ and $93 - 91 = 2$.

Numerical Solution Construct the table for Y_1 and Y_2, starting at $x = 90$ and incrementing by 1. Figure 4.2 reveals that $Y_1 = Y_2 = 91$ when $x = 93$. Thus (93, 91) is a solution.

[0, 200, 50] by [0, 200, 50]

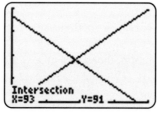

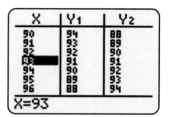

Figure 4.1   **Figure 4.2**

Note: Solving a system of equations numerically may take some trial and error to find an appropriate starting value and increment for *x*. In some situations numerically solving a system exactly may *not* be possible. However, the important mathematical concept to remember is that you are looking for an *x*-value where $Y_1 = Y_2$.

■ **MAKING CONNECTIONS**

Intersection of Graphs Method of Solving Systems of Equations

In Example 2 we solved a system of two linear equations in two variables, using the intersection of graphs method. We did so by first solving each equation for *y* and then graphing both equations. A point of intersection corresponds to a solution.

EXAMPLE 3 *Solving a system of equations graphically and numerically*

Solve the system of equations

$$2x - 3y = 6$$
$$4x + y = 5$$

graphically and numerically.

Solution

Graphical Solution Solve the first equation for *y*.

$$2x - 3y = 6 \qquad \text{First equation}$$
$$-3y = -2x + 6 \qquad \text{Subtract } 2x \text{ from both sides.}$$
$$y = \frac{-2x + 6}{-3} \qquad \text{Divide both sides by } -3.$$
$$= \frac{2}{3}x - 2 \qquad \text{Divide } -3 \text{ into each term in the numerator.}$$

Next, solve the second equation for *y*.

$$4x + y = 5 \qquad \text{Second equation}$$
$$y = -4x + 5 \qquad \text{Subtract } 4x \text{ from both sides.}$$

Let $Y_1 = (2/3)X - 2$ and $Y_2 = -4X + 5$. The graphs of Y_1 and Y_2 intersect at the point $(1.5, -1)$ in Figure 4.3, which is the solution.

Numerical Solution Construct the table for Y_1 and Y_2, starting at $x = 0$ and incrementing by 0.5. Figure 4.4 shows that $Y_1 = Y_2 = -1$ when $x = 1.5$. Thus $(1.5, -1)$ is a solution.

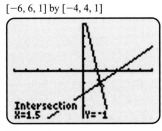

Figure 4.3

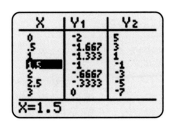

Figure 4.4

Types of Linear Systems

A system of linear equations that has at least one solution is a **consistent system**; otherwise, it is an **inconsistent system**. A system of linear equations in two variables can be represented graphically by two lines in the *xy*-plane. Three different situations involving the graphs of two lines are illustrated in Figures 4.5–4.7. In Figure 4.5 the lines intersect at a single point, which represents a *unique solution*. In Figure 4.6 the two lines are identical, which occurs when the two equations are equivalent. For example, the equations $x + y = 1$ and $2x + 2y = 2$ are equivalent. If we divide the second equation by 2 we obtain the first equation. As a result, their graphs are identical and every point on the line represents a solution. Thus there are infinitely many solutions, and the system of equations is a **dependent system**. Finally, in Figure 4.7 the lines are parallel and do not intersect. There are no solutions, so the system is inconsistent.

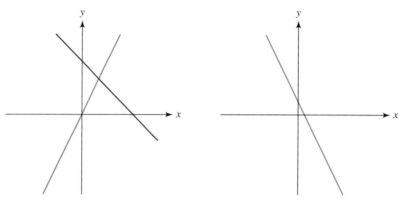

Figure 4.5 Unique Solution **Figure 4.6** Dependent

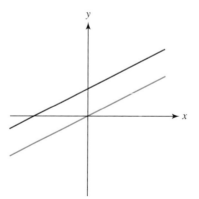

Figure 4.7 Inconsistent

In the next three examples we illustrate each of these situations.

EXAMPLE 4 *Solving a linear system with a unique solution*

Suppose that two groups of students go to a football game. The first group buys 3 tickets and 3 soft drinks for $36, and the second group buys 4 tickets and 2 soft drinks for $44. Find the price of a ticket and the price of a soft drink graphically.

4.1 Systems of Linear Equations in Two Variables

Solution

Let x represent the cost of a ticket and y represent the cost of a soft drink. Then $3x + 3y = 36$ represents 3 tickets and 3 soft drinks costing \$36 and $4x + 2y = 44$ represents 4 tickets and 2 soft drinks costing \$44. This information can be written as a system of equations.

$$3x + 3y = 36$$
$$4x + 2y = 44$$

To find a solution graphically, begin by solving each equation for y.

$$y = \frac{-3x + 36}{3}$$

$$y = \frac{-4x + 44}{2}$$

These equations can be written in *slope-intercept form* as

$$y = -x + 12$$
$$y = -2x + 22.$$

Graph $Y_1 = -X + 12$ and $Y_2 = -2X + 22$, as shown in Figure 4.8. The lines intersect at $(10, 2)$. Thus a ticket costs \$10, and a soft drink costs \$2.

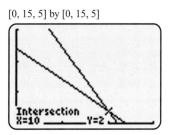

[0, 15, 5] by [0, 15, 5]

Figure 4.8

EXAMPLE 5 *Solving a dependent linear system*

Now suppose that two groups of students go to a different football game. The first group buys 4 tickets and 2 soft drinks for \$34, and the second group buys 2 tickets and 1 soft drink for \$17. If possible, find the price of a ticket and the price of a soft drink graphically.

Solution

Let x represent the cost of a ticket and y represent the cost of a soft drink. Then this situation can be modeled by the following system of equations.

$$4x + 2y = 34$$
$$2x + y = 17$$

Solve each equation for y.

$$y = \frac{-4x + 34}{2}$$

$$y = -2x + 17$$

The graphs of $Y_1 = (-4X + 34)/2$ and $Y_2 = -2X + 17$ are shown in Figure 4.9, and they are *identical*. The system is dependent because not enough information is available to determine a unique solution. Note that the second group bought exactly half what the first group bought and paid half as much. As a result, the two equations contain essentially the same information and are *equivalent*. We can demonstrate this result by dividing the first equation ($4x + 2y = 34$) by 2 to obtain the second equation ($2x + y = 17$). Thus there are infinitely many solutions. For example, a ticket could cost \$8 and a soft drink could cost \$1, or a ticket could cost \$7 and a soft drink could cost \$3. The solution set can be expressed as $\{(x, y) \mid 2x + y = 17\}$.

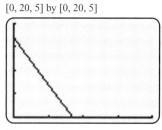

Figure 4.9

Critical Thinking

Suppose that a system of two linear equations with two variables is dependent. If you try to solve this system numerically by using a table, how could you recognize that the system is indeed dependent? Explain your answer.

EXAMPLE 6 *Recognizing an inconsistent linear system*

Now suppose that two groups of students go to a concert. The first group buys 4 tickets and 2 soft drinks for \$44, and the second group buys 2 tickets and 1 soft drink for \$20. If possible, find the price of a ticket and the price of a soft drink graphically.

Solution

Let x be the cost of a ticket and y be the cost of a soft drink. Then the following system models the data.

$$4x + 2y = 44$$
$$2x + y = 20$$

Solving for y, we can write each equation in slope-intercept form.

$$y = -2x + 22$$
$$y = -2x + 20$$

The graphs of $Y_1 = -2X + 22$ and $Y_2 = -2X + 20$ are parallel lines with slope -2 and thus do not intersect (see Figure 4.10). The linear system is *inconsistent*, and there are no solutions. Note that the second group purchased exactly half what the first group purchased. If pricing had been consistent, the second group would have paid half, or $22, instead of $20. *Inconsistent pricing* resulted in an *inconsistent linear system*.

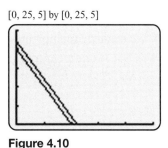

[0, 25, 5] by [0, 25, 5]

Figure 4.10

We can summarize these results as follows.

> ### A System of Two Linear Equations with Two Variables
>
> A system involving two linear equations and two variables can have no solutions, one solution, or infinitely many solutions. Its graph consists of two lines.
>
> 1. If the two lines are parallel, the system of equations is *inconsistent* and there are no solutions.
> 2. If the lines intersect at a single point, there is one solution.
> 3. If the lines are identical, the equations are *dependent* and there are infinitely many solutions.

EXAMPLE 7 *Finding an athlete's running speeds*

An athlete jogs at a faster pace for 30 minutes and then jogs at a slower pace for 90 minutes. The first pace is 4 miles per hour faster than the second pace, and the athlete covers a total distance of 18 miles.

(a) Write a linear system whose solution gives the athlete's running speeds.
(b) Solve the resulting system graphically.

Solution

(a) Let x be the faster speed of the runner and y be the slower speed. The athlete runs $\frac{1}{2}$ hour at x miles per hour and $\frac{3}{2}$ hours at y miles per hour. Rate times time equals distance and the total distance traveled is 18 miles, so

$$\frac{1}{2}x + \frac{3}{2}y = 18.$$

Because the first pace is 4 miles per hour faster than the second pace, we know that

$$x - y = 4.$$

Thus we need to solve the linear system of equations

$$\frac{1}{2}x + \frac{3}{2}y = 18 \quad \text{First equation}$$

$$x - y = 4 \quad \text{Second equation}$$

(b) We begin by multiplying by 2 to clear fractions and then solve for y.

$$x + 3y = 36 \quad \text{Multiply first equation by 2.}$$
$$3y = -x + 36 \quad \text{Subtract } x.$$
$$y = -\frac{x}{3} + 12 \quad \text{Divide by 3.}$$

Solving the second equation for y gives

$$y = x - 4.$$

We graph $Y_1 = -X/3 + 12$ and $Y_2 = X - 4$, as shown in Figure 4.11. The graphs intersect at the point (12, 8). Thus the athlete ran at 12 miles per hour and then slowed to 8 miles per hour.

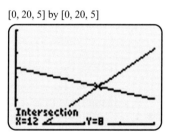

Figure 4.11

4.1 Putting It All Together

A system of two linear equations in two variables may be written in standard form as

$$ax + by = c$$
$$dx + ey = k,$$

where a, b, c, d, e, and k are all constants. Linear systems in two variables may be solved graphically or numerically by first solving each equation for y and then graphing or constructing a table of the equations to find any solutions. The following table summarizes the types of systems of equations and the number of solutions that may be encountered.

System	Solution	Graph
Unique Solution $x + y = 3$ $x - y = 1$	There is one solution: $x = 2$ and $y = 1$. This solution may be written as the ordered pair $(2, 1)$. **Check:** $2 + 1 = 3$ True $2 - 1 = 1$ True	Graph $Y_1 = 3 - X$ and $Y_2 = X - 1$. Their graphs intersect at $(2, 1)$.
Dependent $x + y = 1$ $2x + 2y = 2$	There are infinitely many solutions, such as $(2, -1)$ and $(0, 1)$. **Solution Set:** $\{(x, y) \mid x + y = 1\}$	Graph $Y_1 = 1 - X$ and $Y_2 = (-2X + 2)/2$. The graphs are identical.
Inconsistent $x + y = 1$ $x + y = 2$	There are no solutions.	Graph $Y_1 = 1 - X$ and $Y_2 = 2 - X$. The lines are parallel with slope -1 and do not intersect.

4.1 EXERCISES

CONCEPTS

1. Can a system of linear equations have exactly two solutions? Explain.

2. Give an example of a system of linear equations.

3. Name two ways to solve a system of linear equations.

4. If the graphical solution yields two different intersecting lines, what does this result indicate about the number of solutions?

5. If the graphical solution yields two parallel lines, what does this result indicate about the number of solutions?

6. How many solutions does a dependent linear system have? How can you recognize a dependent linear system graphically?

Exercises 7–10: (Refer to Example 1.) Decide which of the ordered pairs is a solution for the linear system of equations.

7. $(1, -2), (4, 4)$
$2x - y = 4$
$3x + y = 1$

8. $(-3, -1), (3, 1)$
$x - 3y = 0$
$3x + y = 10$

9. $(4, 6), \left(-1, \dfrac{13}{3}\right)$
$x - 3y = -14$
$4x + 3y = 9$

10. $(4, 0), (3, 5)$
$5x - 4y = 20$
$-x + 4y = -4$

GRAPHICAL SOLUTIONS

Exercises 11–14: A system of two linear equations has been solved graphically. Use the graph to find any possible solutions.

11.
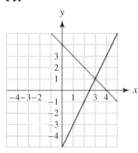

12.
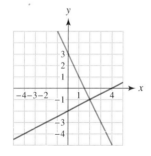

13.
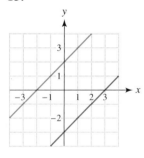

14.
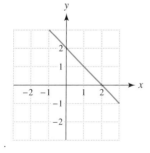

Exercises 15–20: A system of linear equations has been solved graphically.
(a) Use the graph to find any possible solutions.
(b) Write the system of equations in standard form.

15. $Y_1 = 3 - X, Y_2 = X + 1$
$[-6, 6, 1]$ by $[-4, 4, 1]$
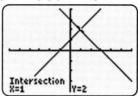

16. $Y_1 = 1 - 2X, Y_2 = 2 - X$
$[-6, 6, 1]$ by $[-4, 4, 1]$
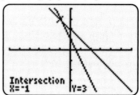

17. $Y_1 = (40 - 2X)/3, Y_2 = 20 + X$
$[-50, 50, 10]$ by $[-50, 50, 10]$
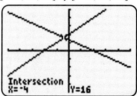

18. $Y_1 = 6X - 2, Y_2 = 5 - 2X$
$[-9, 9, 1]$ by $[-6, 6, 1]$
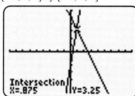

19. $Y_1 = X - 2, Y_2 = X + 2$
$[-9, 9, 1]$ by $[-6, 6, 1]$
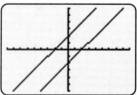

20. $Y_1 = 1 - 2X, Y_2 = -2X - 3$
$[-9, 9, 1]$ by $[-6, 6, 1]$
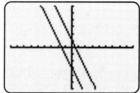

4.1 Systems of Linear Equations in Two Variables

Exercises 21–32: Determine graphically whether the system is dependent, inconsistent, or has a unique solution. Find any possible solutions.

21. $x + y = 6$
    $x - y = 2$

22. $x + y = 9$
    $x - y = 3$

23. $x - y = 4$
    $2x - 2y = 4$

24. $2x + y = 5$
    $4x + 2y = 10$

25. $6x - 4y = -2$
    $-3x + 2y = 1$

26. $4x - 3y = 3$
    $-8x + 6y = 1$

27. $4x + 3y = 2$
    $5x + 2y = 6$

28. $-3x + 2y = 4$
    $4x - y = 3$

29. $\frac{1}{4}x + \frac{1}{2}y = 2$
    $\frac{1}{8}x - y = -\frac{3}{2}$

30. $\frac{1}{2}x - \frac{3}{8}y = \frac{1}{8}$
    $\frac{1}{3}x - \frac{1}{2}y = -\frac{1}{12}$

31. $0.1x + 0.2y = 0.5$
    $0.7x - 0.3y = 0.1$

32. $2.3x + 4.3y = 5.63$
    $1.1x - 3.6y = 0.43$

Numerical Solutions

Exercises 33–36: A system of two linear equations has been solved numerically. Find any possible solutions.

33. $Y_1 = 5 - X,$
    $Y_2 = 2X - 1$

34. $Y_1 = 1 - 2X,$
    $Y_2 = -1 - 3X$

35. $Y_1 = 2 - X,$
    $Y_2 = X$

36. $Y_1 = 2X - 1,$
    $Y_2 = -3 + 2X$

Exercises 37–42: Solve the system of linear equations numerically.

37. $x + y = 3$
    $x - y = 7$

38. $2x + y = 3$
    $3x - y = 7$

39. $3x + 2y = 5$
    $-x - y = -5$

40. $2x + 3y = 3.5$
    $3x + 2y = 6.5$

41. $0.5x - 0.1y = 0.1$
    $0.1x - 0.3y = -0.4$

42. $\frac{x}{3} + \frac{y}{6} = -\frac{1}{2}$
    $\frac{x}{6} + \frac{y}{3} = \frac{5}{2}$

Using More Than One Method

Exercises 43–50: If possible, solve the system of linear equations graphically and numerically.

43. $2x + 2y = 4$
    $x - 3y = -2$

44. $6x - y = 3$
    $x - 2y = -5$

45. $-x - y = 3$
    $2x - y = 6$

46. $\frac{1}{2}x + \frac{1}{8}y = 1$
    $-\frac{1}{2}x + \frac{5}{6}y = -1$

47. $0.1x + 0.2y = 50$
    $0.3x - 0.1y = 10$

48. $0.5x + 0.2y = 14$
    $-0.1x + 0.4y = 6$

49. $x - 2y = 5$
    $-2x + 4y = -2$

50. $3x + 4y = 5$
    $6x + 8y = 10$

Writing and Solving Equations

Exercises 51–58: Complete the following.
(a) Write a system of linear equations that models the situation.
(b) Solve the resulting system either graphically or numerically.

51. The sum of two numbers is 18, and their difference is 6. Find the two numbers.

52. Twice a number minus a second number equals 5. The sum of the two numbers is 16. Find the two numbers.

53. An athlete ran for part of an hour at 6 miles per hour and for the rest of the hour at 9 miles per hour. The total distance traveled was 8 miles. How long did the athlete run at each speed?

54. A car was driven for 2 hours, part of the time at 40 miles per hour and the rest of the time at 60 miles per hour. The total distance traveled was 90 miles. How long did the car travel at each speed?

55. The perimeter of a rectangle is 76 inches. The rectangle's length is 4 inches longer than its width. Find the dimensions of this rectangle.

56. An isosceles triangle has a perimeter of 101 inches. The triangle's longest side is 8 inches longer than either of the other two sides. Find the length of each side of the triangle. (*Hint:* An isosceles triangle has two sides with equal measure.)

57. The largest angle in an isosceles triangle is 60° larger than either of the other two angles. Find the measure of each angle. (*Hint:* An isosceles triangle has two angles with equal measure.)

58. If 2 boxes of popcorn and 3 soft drinks cost $7 and 3 boxes of popcorn and 2 soft drinks cost $8, find the price of a box of popcorn and the price of a soft drink.

APPLICATIONS

59. *Passports* In 1998 and 1999, a combined total of 12,834,867 U.S. passports were issued. In 1999, 244,861 more passports were issued than in 1998. Find the number of passports issued each year. (*Source:* State Department.)

60. *Home Runs* In 1998, Mark McGwire and Sammy Sosa hit a total of 136 home runs. McGwire hit 4 more home runs than Sosa. How many home runs did each player hit?

61. *Handguns* In 1996 and 1997, a combined total of 2,890,982 handguns were manufactured in the United States. From 1996 to 1997 the number of handguns manufactured decreased by 77,972. Find the number of handguns manufactured each year. (*Source:* Bureau of Alcohol, Tobacco and Firearms.)

62. *Student Loans* A student takes out two loans totaling $4000 to help pay for college expenses. One loan is at 6% simple interest, and the other is at 8% simple interest. The first-year interest is $296. Find the amount of each loan.

63. *Roof Trusses* Linear systems are used in the design of roof trusses for houses and other types of buildings (see the accompanying figures). One of the simplest types of roof trusses is an equilateral triangle. If a 200-pound force is applied to the peak of the truss, the weights W_1 and W_2 exerted on each rafter are determined by the following system of equations.

$$W_1 - W_2 = 0$$
$$\frac{\sqrt{3}}{2}(W_1 + W_2) = 200$$

(a) Solve this system graphically.
(b) Solve this system numerically.

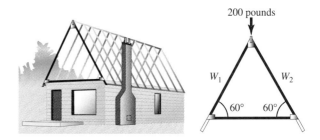

64. *Roof Trusses* (Refer to Exercise 63.) The weights W_1 and W_2 exerted on each rafter for the roof truss shown in the accompanying figure are determined by the following system of equations. Solve the system graphically and interpret the result.

$$W_1 + \sqrt{2}W_2 = 300$$
$$\sqrt{3}W_1 - \sqrt{2}W_2 = 0$$

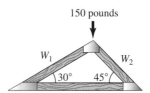

WRITING ABOUT MATHEMATICS

65. Discuss the types of linear systems with two variables. Explain how you can recognize each type graphically.

66. Given
$$ax + by = c$$
$$dx + ey = k,$$
explain how to solve this linear system graphically. Demonstrate your method with an example.

4.2 THE SUBSTITUTION AND ELIMINATION METHODS

The Substitution Method ~ The Elimination Method ~ Models and Applications

INTRODUCTION

The ability to solve systems of equations has resulted in the development of CAT scans, satellites, computers, compact discs, and accurate weather forecasts. In the preceding section we learned how to solve a system of two linear equations in two variables, using graphical and numerical methods. In this section we learn how to solve these systems symbolically. Small systems of equations can be solved with symbolic techniques. However, large systems of equations are usually solved with technology.

THE SUBSTITUTION METHOD

To apply the **substitution method**, we begin by solving an equation for one of its variables. Then we substitute the result into the other equation. For example, consider the following system of equations.

$$2x + y = 5$$
$$3x - 2y = 4$$

It is convenient to solve the first equation for y to obtain $y = 5 - 2x$. Now substitute $(5 - 2x)$ for y into the second equation,

$$3x - 2(y) = 4 \qquad \text{Second equation}$$
$$3x - 2(5 - 2x) = 4, \qquad \text{Substitute.}$$

to obtain a linear equation in one variable.

$$3x - 2(5 - 2x) = 4$$
$$3x - 10 + 4x = 4 \qquad \text{Distributive property}$$
$$7x - 10 = 4 \qquad \text{Combine like terms.}$$
$$7x = 14 \qquad \text{Add 10 to both sides.}$$
$$x = 2 \qquad \text{Divide both sides by 7.}$$

To determine y we substitute $x = 2$ into $y = 5 - 2x$ to obtain

$$y = 5 - 2(2) = 1.$$

The solution is (2, 1).

Note: When using substitution, we may begin by solving for either variable in either equation. The same result is obtained regardless of which equation is used.

EXAMPLE 1   Finding per capita income

In 1998, the average of the per capita (or per person) incomes for Massachusetts and Maine was $28,000. The per capita income in Massachusetts exceeded the per capita income in Maine by $10,000. Find the 1998 per capita income for each state.

Solution

Let x be the per capita income in Massachusetts and y be the per capita income in Maine. The following system of equations models the data.

$$\frac{x+y}{2} = 28{,}000 \qquad \text{Their average is \$28,000.}$$

$$x - y = 10{,}000 \qquad \text{Their difference is \$10,000.}$$

Begin by solving the second equation for x.

$$x = y + 10{,}000.$$

Substitute $(y + 10{,}000)$ into the first equation for x and solve for y.

$$\frac{(y + 10{,}000) + y}{2} = 28{,}000$$

$$\begin{aligned}
(y + 10{,}000) + y &= 56{,}000 &&\text{Multiply both sides by 2.} \\
2y + 10{,}000 &= 56{,}000 &&\text{Combine like terms.} \\
2y &= 46{,}000 &&\text{Subtract 10,000 from both sides.} \\
y &= 23{,}000 &&\text{Divide both sides by 2.}
\end{aligned}$$

Substituting for y in $x = y + 10{,}000$ yields $x = 33{,}000$. Thus, in 1998, the per capita income in Massachusetts was $33,000, and in Maine it was $23,000.

We can now solve a system of equations symbolically, graphically, and numerically as illustrated in the next example.

EXAMPLE 2   Finding fast-food sales

In 1998, McDonald's was number one in fast-food sales. McDonald's annual sales exceeded Burger King's by $10 billion. Their combined sales were $26 billion. Find the amount of business generated by each fast-food chain.   (*Source:* Technomic.)

Solution

Symbolic Solution   Let x be the sales for McDonald's and y be the sales for Burger King. The following system describes the data.

$$x - y = 10$$
$$x + y = 26$$

To use substitution, begin by solving the second equation for y.

$$y = 26 - x$$

Then substitute $(26 - x)$ into the first equation for y and solve for x.

$$x - (26 - x) = 10$$
$$x - 26 + x = 10 \quad \text{Distributive property}$$
$$2x - 26 = 10 \quad \text{Combine like terms.}$$
$$2x = 36 \quad \text{Add 26 to both sides.}$$
$$x = 18 \quad \text{Divide both sides by 2.}$$

Substituting $x = 18$ into $y = 26 - x$ gives $y = 8$, so the solution is $(18, 8)$. Thus McDonald's share of the market was $18 billion, and Burger King's was $8 billion.

Graphical and Numerical Solution To support this result graphically and numerically, solve each of the given equations for y. Then graph and construct the table for $Y_1 = X - 10$ and $Y_2 = 26 - X$. Figures 4.12 and 4.13 verify that the solution is $(18, 8)$.

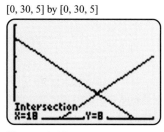

Figure 4.12   Figure 4.13

THE ELIMINATION METHOD

The **elimination** (or addition) **method** is a second way to solve linear systems symbolically. This method is based on the property that "equals added to equals are equal." That is, if

$$a = b \quad \text{and} \quad c = d,$$

then

$$a + c = b + d.$$

EXAMPLE 3   *Applying the elimination method*

Solve the following system by using elimination.

$$2x - y = 4$$
$$x + y = 1$$

Solution

Note that adding the two equations eliminates the variable y.

$$2x - y = 4$$
$$\underline{x + y = 1}$$
$$3x \quad\quad = 5 \quad \text{or} \quad x = \frac{5}{3} \quad \text{Add the two equations and solve for } x.$$

Substituting $x = \frac{5}{3}$ into the second equation gives

$$\frac{5}{3} + y = 1 \quad \text{or} \quad y = -\frac{2}{3}.$$

The solution is $\left(\frac{5}{3}, -\frac{2}{3}\right)$.

In the next example, we use multiplication before we add the two equations.

EXAMPLE 4 *Multiplying before applying elimination*

Solve the following system by using elimination.

$$3x - 2y = 11$$
$$2x + 3y = 3$$

Solution

If we add these equations, neither variable will be eliminated. However, if we multiply the first equation by 3 and multiply the second equation by 2, we can eliminate y.

$$9x - 6y = 33 \quad \text{Multiply by 3.}$$
$$\underline{4x + 6y = 6} \quad \text{Multiply by 2.}$$
$$13x = 39 \quad \text{or} \quad x = 3 \quad \text{Add the two equations.}$$

Substituting $x = 3$ into $3x - 2y = 11$ gives

$$3(3) - 2y = 11 \quad \text{or} \quad y = -1.$$

The solution is $(3, -1)$.

Note: In Example 4 we could have multiplied the first equation by 2 and the second equation by -3. Adding the resulting equations would have eliminated the variable x.

EXAMPLE 5 *Recognizing an inconsistent system*

If possible, use elimination to solve the following system.

$$3x - 4y = 5$$
$$-6x + 8y = 9$$

Solution

If we multiply the first equation by 2 and add, we obtain the following result.

$$6x - 8y = 10 \quad \text{Multiply by 2.}$$
$$\underline{-6x + 8y = 9}$$
$$0 = 19 \quad \text{Adding the two equations gives a false result.}$$

The statement $0 = 19$ is always false, which tells us that the system has no solution. If we solve each equation for y and graph, we obtain two parallel lines with slope $\frac{3}{4}$ that never intersect. This result is shown in Figure 4.14, where $Y_1 = (3X - 5)/4$ and $Y_2 = (6X + 9)/8$ have been graphed.

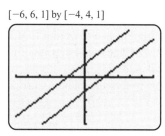

Figure 4.14

EXAMPLE 6 *Recognizing a dependent system*

Use elimination to solve the following system.

$$3x - 6y = 3$$
$$x - 2y = 1$$

Solution

If we multiply the second equation by -3 and add, we obtain the following.

$$\begin{array}{rl} 3x - 6y = & 3 \\ -3x + 6y = & -3 \quad \text{Multiply by } -3. \\ \hline 0 = & 0 \quad \text{Adding the two equations gives a true result.} \end{array}$$

The statement $0 = 0$ is always true, so the two equations are equivalent. If an ordered pair (x, y) satisfies the first equation, it also satisfies the second equation. Thus the solution set may be expressed as $\{(x, y) \mid x - 2y = 1\}$. For example, $(1, 0)$ and $(5, 2)$ are both solutions.

■ **MAKING CONNECTIONS**

Graphical Solutions and Elimination

1. *Unique Solution*   A graphical solution results in two lines intersecting at a unique point. Elimination gives unique values for x and y.
2. *Inconsistent Linear System*   A graphical solution results in two parallel lines. Elimination results in an equation that is *always false*, such as $0 = 1$.
3. *Dependent Linear System*   A graphical solution results in two identical lines. Elimination results in an equation that is *always* true, such as $0 = 0$.

MODELS AND APPLICATIONS

Linear systems are often used in applications. In the next example we use a linear system to determine the cost of tuition.

EXAMPLE 7 *Modeling tuition*

A student is attempting to graduate on schedule by taking 14 credits of day classes at one college and 4 credits of night classes at a different college. Credits for day classes cost $20 more than credits for night classes. If the student's total tuition was $2440, how much did each type of credit cost?

Solution

Let x represent the cost of a credit for day classes and y represent the cost of a credit for night classes. Then the data can be represented by the following system of equations.

$$x - y = 20 \quad \text{Credits are \$20 more during the day than at night.}$$

$$14x + 4y = 2440 \quad \text{The total cost is \$2440 for 14 credits during the day and 4 credits at night.}$$

To solve this equation we multiply the first equation by 4 and then add.

$$\begin{aligned} 4x - 4y &= 80 \quad &\text{Multiply by 4.} \\ 14x + 4y &= 2440 \\ \hline 18x &= 2520 \quad \text{or} \quad x = 140 \quad &\text{Add and solve for } x. \end{aligned}$$

Since $x = 140$ and $x - y = 20$, it follows that $y = 120$. Thus a credit for day classes costs $140 and a credit for night classes costs $120.

In the next example we determine a linear function for modeling rising funeral costs.

EXAMPLE 8 *Modeling real data with a linear function*

The average cost for an adult funeral rose from $1800 in 1980 to $5000 in 1998. This increase in cost may be modeled by $f(x) = ax + b$, where a and b are constants. (*Source:* National Funeral Directors Association.)

(a) Set up a linear system of equations whose solution gives values for a and b so that $f(x)$ models these data.
(b) Find a and b, using elimination.
(c) Use $f(x)$ to estimate funeral costs in 1991. Compare it to the actual cost of $3742.

Solution

(a) We want to find a linear function whose graph passes through the points (1980, 1800) and (1998, 5000), as illustrated in Figure 4.15.

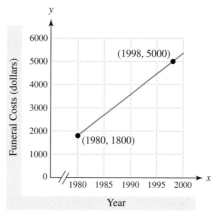

Figure 4.15 Increase in Funeral Costs

If $f(x) = ax + b$ is to model the data, then $f(1980) = 1800$ and $f(1998) = 5000$.

$$f(1980) = a(1980) + b = 1800$$
$$f(1998) = a(1998) + b = 5000$$

We can write the system of equations as follows.

$$1980a + b = 1800$$
$$1998a + b = 5000$$

(b) We eliminate the variable b by subtracting the two equations.

(*Note:* This step is equivalent to multiplying the second equation by -1 and adding.)

$$\begin{array}{rl} 1980a + b = & 1800 \\ \underline{1998a + b = } & \underline{5000} \\ -18a = & -3200 \end{array} \quad \text{or} \quad a = \frac{1600}{9} \qquad \text{Subtract and solve.}$$

We find b by solving the first equation for b and then substituting for a.

$$1980a + b = 1800 \qquad \text{First equation}$$
$$b = 1800 - 1980a \qquad \text{Solve for } b.$$
$$b = 1800 - 1980\left(\frac{1600}{9}\right) \qquad \text{Let } a = \frac{1600}{9}.$$
$$b = -350{,}200 \qquad \text{Simplify.}$$

Thus $f(x) = \dfrac{1600}{9}x - 350{,}200$.

(c) To estimate funeral costs in 1991 we use a calculator to evaluate $f(1991)$.

$$f(1991) = \frac{1600}{9}(1991) - 350{,}200 \approx \$3756$$

This estimate compares favorably with the actual value of $3742.

Critical Thinking

Suppose that the sum of two numbers is 20 and that their product is 84. If you use a system of equations to find the two numbers, is the system linear? Explain.

4.2 Putting It All Together

Systems of linear equations can be solved graphically, numerically, and symbolically. Two symbolic methods for solving a system of linear equations are substitution and elimination. The following table presents a summary of each method by means of an example.

Concept	Explanation
Substitution Method	1. Solve for a convenient variable such as y in the first equation. $$x + y = 3 \quad \text{or} \quad y = 3 - x$$ $$2x - y = 0$$ 2. Substitute $(3 - x)$ into the second equation for y. Solve the equation for x. $$2x - (3 - x) = 0 \quad \text{or} \quad x = 1.$$ 3. Substitute $x = 1$ into one of the given equations and find y. $$1 + y = 3 \quad \text{or} \quad y = 2.$$ 4. The solution is $(1, 2)$.
Elimination Method	1. Multiply the first equation by -2 so that the coefficients of x are additive inverses. $$x + 2y = 1 \quad \text{or} \quad -2x - 4y = -2$$ $$2x - 3y = 9 \qquad\qquad\quad 2x - 3y = 9$$ 2. Eliminate x by adding the two equations. $$-2x - 4y = -2$$ $$\underline{2x - 3y = 9}$$ $$-7y = 7 \quad \text{or} \quad y = -1$$

Concept	Explanation
Elimination Method *(continued)*	3. Substitute $y = -1$ into one of the given equations and solve for x. $x + 2(-1) = 1$ or $x = 3$. 4. The solution is $(3, -1)$.

4.2 EXERCISES

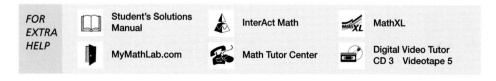

CONCEPTS

1. Name two methods for solving a linear system symbolically.

2. Are different solutions obtained when the same system of equations is solved symbolically, graphically, and numerically? Explain.

3. When using elimination, how can you recognize an inconsistent system of equations?

4. When using elimination, how can you recognize a dependent system of equations?

5. When you use substitution to solve
$$3x + y = 4$$
$$5x - 7y = -2,$$
what is a good first step?

6. When you are using elimination to solve
$$2x - 5y = 4$$
$$x + 3y = 11,$$
what is a good first step?

SUBSTITUTION METHOD

Exercises 7–10: Solve the system of linear equations by using substitution.

7. $y = 2x$
   $3x + y = 5$

8. $y = x + 1$
   $x + 2y = 8$

9. $x = 2y - 1$
   $x + 5y = 20$

10. $x = 3y$
    $-x + 2y = 4$

Exercises 11–12: Use substitution to solve the system of equations. Then use the graph to support your answer.

11. $x - 2y = 0$
    $3x + y = 7$

12. $2x + 3y = 8$
    $3x - 2y = -14$

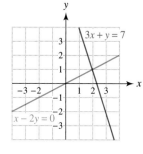
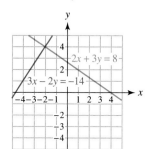

Exercises 13–16: Solve the system of equations by using substitution. Then solve the system graphically.

13. $2x - y = -\dfrac{1}{4}$
    $3x + 4y = \dfrac{15}{4}$

14. $x - 5y = 3$
    $2x + 3y = -\dfrac{1}{2}$

15. $2x - 3y = -3$
$6x + 4y = 4$

16. $10x + 25y = 5$
$15x - 10y = 150$

Exercises 17–24: Solve the system of equations by using substitution.

17. $2x - 5y = -1$
$-x + 3y = 0$

18. $5x + 10y = -5$
$-10x + 5y = 60$

19. $x - y = 1$
$2x + 6y = -2$

20. $4x - y = -4$
$-2x + 5y = 29$

21. $\frac{1}{2}x + \frac{2}{3}y = -2$
$\frac{1}{4}x - \frac{1}{3}y = 3$

22. $\frac{1}{5}x - \frac{1}{10}y = \frac{7}{40}$
$\frac{1}{4}x - \frac{1}{5}y = \frac{11}{40}$

23. $0.1x + 0.4y = 1.3$
$0.3x - 0.2y = 1.1$

24. $1.5x - 4.1y = -1.6$
$2.7x - 0.1y = 0.76$

25. The following system is dependent.

$x - y = 5$
$2x - 2y = 10$

Solve this system by using substitution. Explain how you can recognize a dependent system when you are using substitution.

26. The following system is inconsistent.

$-x + 2y = 5$
$2x - 4y = 10$

Solve this system by using substitution. Explain how you can recognize an inconsistent system when you are using substitution.

ELIMINATION METHOD

Exercises 27–28: Use elimination to solve the system of equations. Then use the graph to support your answer.

27. $x - y = 5$
$x + y = 9$

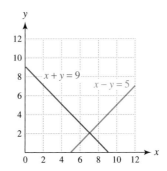

28. $3x - 2y = 9$
$5x + 2y = 7$

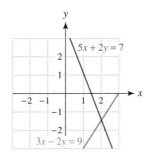

Exercises 29–40: Use elimination to solve the system.

29. $2x + y = 4$
$2x - y = -2$

30. $2x + 3y = -6$
$2x - y = 4$

31. $6x - 4y = 12$
$3x + 5y = -6$

32. $-5x - 4y = -8$
$x - 4y = 10$

33. $2x - 4y = 5$
$-x + 2y = 9$

34. $x - 3y = 5$
$2x - 6y = 1$

35. $2x + y = 2$
$4x + 2y = 4$

36. $\frac{1}{2}x + \frac{1}{7}y = 2$
$7x + 2y = 28$

37. $2x + 4y = -22$
$75x + 15y = -120$

38. $-3x + 20y = 67$
$2x + 5y = 47$

39. $0.3x + 0.2y = 0.8$
$0.4x + 0.3y = 1.1$

40. $1.2x + 4.3y = 1.7$
$2.4x - 1.5y = 1.38$

Using More Than One Method

Exercises 41–46: Solve the system of equations graphically, numerically, and symbolically.

41. $x - y = 4$
    $x + y = 6$

42. $2x - y = 7$
    $3x + y = 3$

43. $5x + 2y = 9$
    $3x - y = 1$

44. $-x + 4y = 17$
    $3x + 6y = 21$

45. $3x - 2y = 12$
    $2x + 4y = -8$

46. $-x - 3y = 0$
    $2x - 5y = 11$

Applications

47. *Burning Fat Calories* Two athletes engage in a strenuous 40-minute workout on a stair climber. The heavier athlete burns 58 more fat calories than the lighter athlete. Together they burn a total of 290 fat calories. How many fat calories did each athlete burn? If 1 fat gram equals 9 fat calories, how many fat grams did each athlete burn? **(Source: Runner's World.)**

48. *Burning Calories* During strenuous exercise, an athlete can burn 10 calories per minute on a rowing machine, whereas on a stair climber the athlete can burn 11.5 calories per minute. In a 60-minute workout an athlete burns 633 calories by using both exercise machines. How many minutes did the athlete spend on each type of workout equipment? **(Source: Runner's World.)**

49. *Mixing Antifreeze* A mixture of water and antifreeze in a car is 10% antifreeze. In colder climates this mixture should contain 50% antifreeze. If the radiator contains 4 gallons of fluid, how many gallons of radiator fluid should be drained and replaced with a mixture containing 80% antifreeze?

50. *Mixing Acid* Determine the milliliters of 10% sulfuric acid and the milliliters of 25% sulfuric acid that should be mixed to obtain 20 milliliters of 18% sulfuric acid.

51. *Phone Bills* (Refer to Example 8.) The average phone bill in 1990 was $80, but in 1998 this average fell to $39. **(Source: Telecommunications Industry Association.)**

   (a) Use a linear system to find values for a and b so that the graph of $f(x) = ax + b$ passes through (1990, 80) and (1998, 39).
   (b) Use $f(x)$ to predict the average phone bill in 1995.

52. *Federal Regulations* (Refer to Example 8.) The cost to operate federal agencies that regulate social and economic programs has increased dramatically. In 1960, it was $1.9 billion and in 2000 it was $16 billion. Both amounts have been adjusted to 1992 dollars. **(Source: Center for the Study of American Business.)**

   (a) Use a linear system to find values for a and b so that the graph of $f(x) = ax + b$ passes through (1960, 1.9) and (2000, 16).
   (b) Use f to estimate the value in 1990. Compare it to the actual value of $12.3 billion.

53. *Airplane Speed* A plane flies 2400 miles from Orlando, Florida, to Los Angeles, California, against the jet stream in 4 hours and 10 minutes. Then the plane flies back to Orlando with the jet stream in 3 hours and 45 minutes. Find the average air speed of the plane and the average speed of the jet stream.

54. *River Current* A tugboat can push a barge 165 miles upstream in 33 hours. The same tugboat and barge can make the return trip downstream in 15 hours. Determine the speed of the current and the speed of the tugboat when there is no current.

55. *Student Loans* A student takes out two loans to help pay for college. One loan is at 8% simple interest, and the other is at 9% simple interest. The total amount borrowed is $3500, and the interest after 1 year for both loans is $294. Find the amount of each loan.

56. *Computer Sales* Sales of personal computers are expected to increase in the twenty-first century. In 2000 and 2001, a combined total of 264 million personal computers were sold worldwide, with 12.9% more computers sold in 2001 than in 2000. How many personal computers were sold in 2000, and how many were sold in 2001? Round your answers to the nearest million. **(Source: International Data Corporation.)**

57. *Geometry* A basketball court has a perimeter of 296 feet. Its length is 44 feet greater than its width. Set up a system of two linear equations whose solution gives the length and width of the court. Solve the system
(a) symbolically,
(b) graphically, and
(c) numerically.

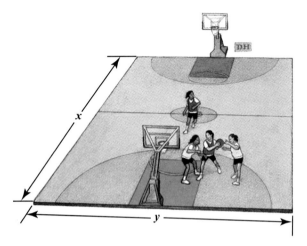

58. *Concert Tickets* Tickets for a concert sold for $55 and $40. If 2000 tickets were sold for total receipts of $90,500, find how many of each type of ticket were sold
(a) symbolically,
(b) graphically, and
(c) numerically.

WRITING ABOUT MATHEMATICS

59. Explain what it means for a linear system to be inconsistent. Give an example.

60. Describe two possible ways that you could use elimination to solve the following system of linear equations. (Do not actually solve the system.)

$$ax - y = c$$
$$x + dy = e$$

CHECKING BASIC CONCEPTS FOR SECTIONS 4.1 AND 4.2

1. Solve the system of linear equations
$$2x - y = 5$$
$$-x + 3y = 0$$
   (a) graphically and
   (b) numerically.
   Do your answers agree?

2. Use substitution to solve the system of equations.
$$4x - 3y = -1$$
$$x + 4y = 14$$

3. Use elimination to solve the system of equations. Is the system either dependent or inconsistent?
$$4x - 3y = -17$$
$$-6x + 2y = 23$$

4.3 SYSTEMS OF LINEAR INEQUALITIES

Basic Concepts ~ Solving Systems of Linear Inequalities

INTRODUCTION

People often walk or jog in an effort to increase their heart rates and get in better shape. During strenuous exercise, older people should maintain lower heart rates than younger people. A person cannot maintain precisely one heart rate, so a range of heart rates is recommended by health professionals. For aerobic fitness, a 50-year-old's heart rate should be between 120 and 140 beats per minute, whereas a 20-year-old's heart rate should be between 140 and 160. Systems of linear inequalities can be used to model these situations.

BASIC CONCEPTS

Suppose that candy costs $2 per pound and that peanuts cost $1 per pound. The total cost C of buying x pounds of candy and y pounds of peanuts is given by

$$C = 2x + y.$$

If we only have $5 to spend, the inequality

$$2x + y \leq 5$$

must be satisfied. To determine the different weight combinations of candy and peanuts that could be bought, we could use the graph shown in Figure 4.16. Points located on the line $2x + y = 5$ represent weight combinations resulting in a $5 purchase. For example, the point (2, 1) satisfies the equation $2x + y = 5$ and represents buying 2 pounds of candy and 1 pound of peanuts for $5. Ordered pairs (x, y) located in the shaded region below the line represent purchases of less than $5. The point (1, 2) lies in the shaded region and represents buying 1 pound of candy and 2 pounds of peanuts for $4. Note that, if we substitute $x = 1$ and $y = 2$ into the inequality, we obtain

$$2(1) + (2) \leq 5,$$

which is a true statement.

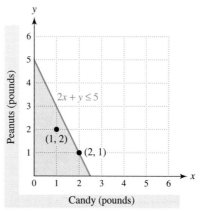

Figure 4.16 Purchases for Candy and Peanuts

When the equals sign in a linear equation of two variables is replaced with $<, \leq, >$, or $\geq$, a **linear inequality in two variables** results. Examples include

$$2x + y \leq 5, \quad y \geq x - 5, \quad \text{and} \quad \frac{1}{2}x - \frac{3}{5}y < 8.$$

EXAMPLE 1 *Solving a linear inequality graphically*

Shade the solution set for the linear inequality

$$4x - 3y < 12.$$

Solution

Begin by solving the inequality for y.

$$4x - 3y < 12 \qquad \text{Given inequality}$$
$$-3y < -4x + 12 \qquad \text{Subtract } 4x.$$
$$y > \frac{4}{3}x - 4 \qquad \text{Divide by } -3; \text{reverse inequality.}$$

The solution set can be illustrated graphically by shading all points above the line $y = \frac{4}{3}x - 4$, as shown in Figure 4.17. The line is dashed because points on the line do not satisfy the inequality.

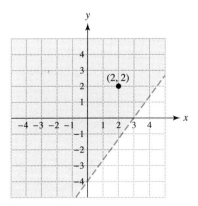

Figure 4.17

If we are uncertain about which region of the xy-plane to shade, we can select a **test point** in the region that we think should be shaded. For example, the point $(2, 2)$ lies in the shaded region shown in Figure 4.17 and it satisfies the inequality $4x - 3y < 12$. That is,

$$4(2) - 3(2) < 12$$

is a true statement. Thus we have shaded the correct region. If we substitute a test point into an inequality and it results in a false statement, the region containing the test point should *not* be shaded.

SOLVING SYSTEMS OF LINEAR INEQUALITIES

A **system of linear inequalities** results when the equals signs in a system of linear equations are replaced with $<, \leq, >,$ or $\geq$. The system of linear equations

$$x + y = 4$$
$$y = x$$

becomes a system of linear inequalities when it is written as

$$x + y \leq 4$$
$$y \geq x.$$

A solution to a system of inequalities must satisfy *both* inequalities. The ordered pair (1, 2) is a solution to this system because substituting $x = 1$ and $y = 2$ makes both inequalities true.

$$1 + 2 \leq 4 \quad \text{True}$$
$$2 \geq 1 \quad \text{True}$$

To graph the solution set for $x + y \leq 4$, we solve for y to obtain $y \leq -x + 4$. Its graph consists of points lying on the line $y = -x + 4$ and all points below the line. This region is shaded in Figure 4.18. Similarly, the solutions to $y \geq x$ include the line $y = x$ and all points above. This region is shaded in Figure 4.19.

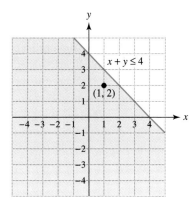

Figure 4.18

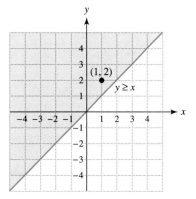

Figure 4.19

For a point (x, y) to represent a solution to the *system* of linear inequalities, it must be located in both shaded regions shown in Figures 4.18 and 4.19. The *intersection* of the shaded regions is shown in Figure 4.20. Note that the point (1, 2) is located in each shaded region shown in Figures 4.18 and 4.19. Therefore the point (1, 2) is located in the shaded region shown in Figure 4.20 and is a solution of the system of linear inequalities.

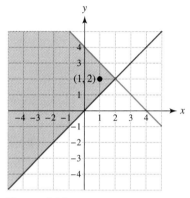

Figure 4.20

In the next example we demonstrate how systems of inequalities can model the real-life situation discussed in the Introduction.

EXAMPLE 2   *Modeling target heart rates*

When exercising, people often try to maintain target heart rates that are percentages of their maximum heart rates. A person's maximum heart rate (MHR) is MHR = 220 − A, where A represents age and the MHR is in beats per minute. The shaded region shown in Figure 4.21 represents target heart rates for aerobic fitness for various ages. (*Source:* Hebb Industries, Inc.)

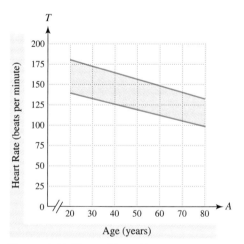

Figure 4.21 Target Heart Rates

(a) Estimate the range R of heart rates that are acceptable for someone 40 years old.

(b) Equations for the two lines shown in Figure 4.21 are

$$T = -0.8A + 196 \quad \text{Upper line}$$

and

$$T = -0.7A + 154, \quad \text{Lower line}$$

where A represents age and T represents the target heart rate. Write a system of inequalities whose solution set is the shaded region, including the two lines.

(c) Use Figure 4.21 to determine whether (30, 150) is a solution. Then verify your answer using the system of inequalities.

Solution

(a) When $A = 40$, target heart rates T are approximately $125 \leq T \leq 165$ beats per minute.

(b) The region lies below the upper line, above the lower line, and includes both lines. Therefore this region is modeled by

$$T \leq -0.8A + 196$$
$$T \geq -0.7A + 154.$$

(c) Figure 4.21 shows that a 30-year-old person with a heart rate of 150 beats per minute lies in the shaded region, so (30, 150) is a solution. This result can be verified by substituting $A = 30$ and $T = 150$ into the system.

$$150 \leq -0.8(30) + 196 = 172 \quad \text{True}$$
$$150 \geq -0.7(30) + 154 = 133 \quad \text{True}$$

Critical Thinking

Graph the solution set to the following system and discuss your results.

$$3x + y \geq 6$$
$$3x + y \leq 2$$

EXAMPLE 3 *Solving a linear inequality graphically*

Shade the solution set for the system of linear inequalities.

$$x + 3y < 3$$
$$2x - y < 2$$

Solution

Begin by solving the first inequality for y.

$$x + 3y < 3 \quad \text{First inequality}$$
$$3y < -x + 3 \quad \text{Subtract } x.$$
$$y < -\frac{1}{3}x + 1 \quad \text{Divide by 3.}$$

Similarly, solve the second inequality for *y*.

$$2x - y < 2 \quad \text{Second inequality}$$
$$-y < -2x + 2 \quad \text{Subtract } 2x.$$
$$y > 2x - 2 \quad \text{Multiply by } -1; \text{ reverse inequality.}$$

In Figure 4.22 the solution set for $y < -\frac{1}{3}x + 1$ is shaded, and in Figure 4.23 the solution set for $y > 2x - 2$ is shaded. The solution set for the system includes the *intersection* of these two shaded regions, as shown in Figure 4.24. This intersection lies below the line $y = -\frac{1}{3}x + 1$ and above the line $y = 2x - 2$.

For a test point we can use (0, 0), which is located in the shaded region.

$$0 + 3(0) < 3 \quad \text{True}$$
$$2(0) - 0 < 2 \quad \text{True}$$

The test point satisfies both equations, so the correct region in the *xy*-plane has been shaded.

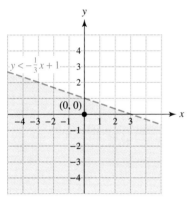

Figure 4.22

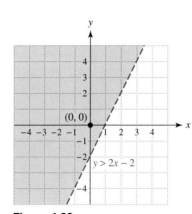

Figure 4.23

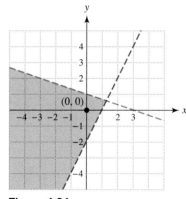

Figure 4.24

EXAMPLE 4 *Solving a system of linear inequalities with technology*

 Shade the solution set for the system of inequalities, using a graphing calculator.

$$2x + y \leq 5$$
$$-2x + y \geq 1$$

Solution

Begin by solving each inequality for *y* to obtain $y \leq 5 - 2x$ and $y \geq 2x + 1$. Graph $Y_1 = 5 - 2X$ and $Y_2 = 2X + 1$, as shown in Figure 4.25. Solutions lie below the line Y_1 and above the line Y_2. This area can be shaded by using a shading utility, as shown in Figures 4.26 and 4.27.

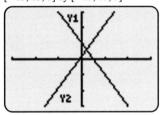
[−15, 15, 5] by [−15, 15, 5]

Figure 4.25

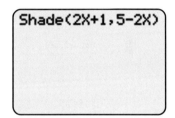

Figure 4.26

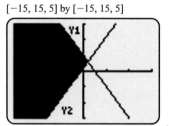
[−15, 15, 5] by [−15, 15, 5]

Figure 4.27

Technology Note *Shading of Linear Inequalities*

When shading the solution set for a linear inequality, graphing calculators often show solid lines even if a line should be dashed. For example, the graphs for $y < 5 - 2x$ and $y \leq 5 - 2x$ are identical on some graphing calculators.

4.3 Putting It All Together

In this section we presented systems of linear inequalities in two variables. These systems usually have infinitely many solutions and can be represented by a shaded region in the xy-plane. The following table shows an example.

Concept	Explanation
Systems of Linear Inequalities	To solve a system of linear inequalities graphically, start by solving each inequality for y. For example, $$x + y \leq 5 \quad \text{or} \quad y \leq 5 - x,$$ and $$2x - y > 3 \quad \text{or} \quad y < 2x - 3.$$ Solutions lie below both lines given by $y = 5 - x$ and $y = 2x - 3$. Note that a dashed line is used for $y < 2x - 3$ because the points lying on the line $y = 2x - 3$ are not solutions.

4.3 EXERCISES

CONCEPTS

1. If a system of inequalities contains two inequalities, how many of the inequalities must a solution satisfy?

2. Can a system of linear inequalities have infinitely many solutions? Explain.

3. Does (3, 1) satisfy $5x - 2y > 8$?

4. Does (3, 4) satisfy the following system of inequalities?
$$x - 2y \geq -8$$
$$2x - y < 1$$

SYSTEMS OF INEQUALITIES

Exercises 5–10: Shade the solution set in the xy-plane. Use a test point to check your graph.

5. $y \geq 3x$
6. $y \leq 1 - x$
7. $y < 3x - 2$
8. $x + y > 3$
9. $2x - y \leq 4$
10. $4x + 2y < 8$

Exercises 11–20: Shade the solution set in the xy-plane. Use a test point to check your graph.

11. $x \geq y$
    $y \leq 3$

12. $y > x + 1$
    $y < 2$

13. $\frac{1}{2}x + y \leq 1$
    $x - \frac{1}{3}y < 1$

14. $2x - y > 3$
    $y \leq x + 2$

15. $x \geq 1$
    $y \leq 1$

16. $4x + y \leq 4$
    $x - 3y \leq 3$

17. $\frac{1}{2}x - y \geq 1$
    $x - \frac{1}{3}y \geq 1$

18. $2x + y > 4$
    $x - y > 1$

19. $3x < 6 - 2y$
    $2y \geq 2 - 2x$

20. $-4x + 6y > 12$
    $2x - 4y < 8$

Exercises 21–24: Match the inequality or system of inequalities with its graph (a.–d.). Each graph uses the standard viewing rectangle.

21. $x + y \geq 2$

22. $y \leq x + 1$
    $y \geq x - 3$

23. $y \leq \frac{1}{2}x$

24. $y \geq x + 1$
    $y \leq 5$

a.
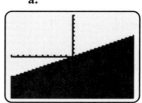

b.
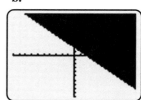

c.
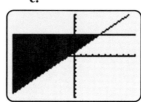

d.
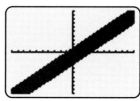

Exercises 25–28: Use the graph to write the inequality or system of inequalities.

25.
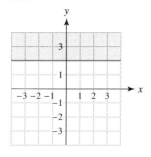

26.
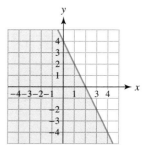

27. **28.**

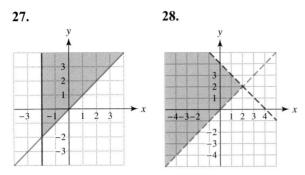

Exercises 29–34: (Refer to Example 4.) Use a graphing calculator to shade the solution set for the system of inequalities.

29. $y \leq 5$
    $y \geq -3$
30. $y \geq -x$
    $y \leq x + 1$
31. $x + 2y \geq 8$
    $6x - 3y \geq 10$
32. $y \geq 2.1x - 3.5$
    $y \leq 2.1x - 1.7$
33. $0.9x + 1.7y \leq 3.2$
    $1.9x - 0.7y \leq 1.3$
34. $21x \geq 31y - 51$
    $5x - 17y \leq 18$

APPLICATIONS

35. *Target Heart Rate* (Refer to Example 2.) The following graph shows target heart rates for general health and weight loss for a person x years old. These heart rates should be maintained for longer periods of time than the times specified for aerobic fitness.
    (a) What range of heart rates is recommended for someone 30 years old?
    (b) The upper line has a slope of -0.6 and passes through the point $(20, 140)$, and the lower line has a slope of -0.5 and passes through the point $(20, 110)$. Write a system of inequalities whose solution set lies in the shaded region in the graph. (*Hint:* The equation of a line with slope m passing through (h, k) is $y = m(x - h) + k$.)

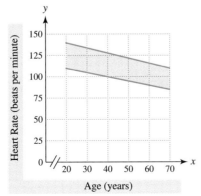

36. *Deserts, Grasslands, and Forests* Two factors that have a critical effect on plant growth are temperature and precipitation. If a region has too little precipitation, it will be a desert. Forests tend to grow in regions where trees can exist at relatively low temperatures and rainfall is sufficient. At other levels of precipitation and temperature, grasslands may prevail. The following figure illustrates the relationship between forests, grasslands, and deserts as suggested by average annual temperature T in degrees Fahrenheit and precipitation P in inches. (*Source:* A. Miller and J. Thompson, *Elements of Meteorology.*)
    (a) Determine a system of linear inequalities that describe where grasslands are likely to occur.
    (b) Bismarck, North Dakota, has an average annual temperature of 40°F and precipitation of 15 inches. According to the graph, what type of plant growth would you expect near Bismarck? Do these values satisfy the system of inequalities from part (a)?

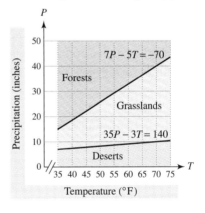

Exercises 37–40: Weight and Height The following graph shows a weight and height chart. The weight w is listed in pounds, and the height h in inches. The shaded area is the recommended region. (*Source:* Department of Agriculture.)

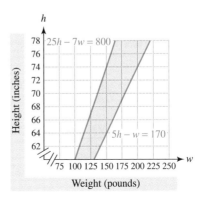

37. What does this chart indicate about an individual weighing 125 pounds with a height of 70 inches?

38. For a person 74 inches tall, use the graph to estimate the recommended weight range.

39. Use the graph to find a system of linear inequalities that describes the recommended region.

40. Explain why inequalities, rather than equalities, are more appropriate for describing recommended weight and height combinations.

WRITING ABOUT MATHEMATICS

41. A student changes the following system of linear inequalities

$$2x - y \geq 6$$
$$x - y \leq 4$$

to

$$y \geq 2x - 6$$
$$y \leq x - 4.$$

The student tests the point $(-2, -8)$, which satisfies the second system but not the first. Explain the student's error.

42. Explain how a system of linear inequalities with two variables can have no solutions. Give an example to illustrate your answer.

GROUP ACTIVITY
Working with Real Data

Directions: Form a group of 2 to 4 people. Select someone to record the group's responses for this activity. All members of the group should work cooperatively to answer the questions. If your instructor asks for your results, each member of the group should be prepared to respond. If the group is asked to turn in its work, be sure to include each person's name on the paper.

1. *CD Sales* In 1987, compact disc (CD) sales accounted for 12% of all recorded music sales. By 1997, such sales had increased to 70%.
(*Source:* Recording Industry Association of America.)
   (a) Set up a linear system to find values for a and b so that the graph of $y_1 = ax + b$ passes through the points (1987, 12) and (1997, 70).
   (b) Solve the system, using any method.

2. *Cassette Sales* In 1987, cassette tape sales accounted for 63% of all recorded music sales. By 1997, such sales had decreased to 18%.
(*Source:* Recording Industry Association of America.)
   (a) Set up a linear system to find values for c and d so that the graph of $y_2 = cx + d$ passes through the points (1987, 63) and (1997, 18).
   (b) Solve the system, using any method.

3. *Recorded Music Sales* Use graphing to estimate the year when CDs and cassette tapes had equal market shares. When were CD sales greater than cassette tape sales?

4.4 SYSTEMS OF EQUATIONS IN THREE VARIABLES

Basic Concepts ~ Solving Linear Systems with Substitution and Elimination ~ Modeling Data

INTRODUCTION

In sections 4.1 and 4.2, we described how to solve systems of linear equations in two variables. In real-world applications linear systems commonly have many variables.

Large linear systems are used in the design of electrical circuits, bridges, and ships. They also are used in business, economics, and psychology. Because of the enormous amount of work needed to solve large systems, technology is usually used to find approximate solutions. In this section we discuss symbolic methods for finding solutions of linear systems with three variables. These methods provide the basis for understanding how technology is able to solve large linear systems.

BASIC CONCEPTS

When we solve a linear system in two variables, we can express a solution as an ordered pair (x, y). A linear equation in two variables can be represented graphically by a line. If a solution is unique, we can represent a system of two linear equations with two variables graphically by two lines intersecting at a point, as shown in Figure 4.28.

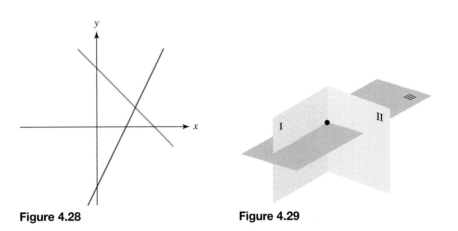

Figure 4.28　　　　**Figure 4.29**

When solving linear systems in three variables, we often use the variables x, y, and z. A solution is expressed as an **ordered triple** (x, y, z), rather than an ordered pair (x, y). For example, if the ordered triple $(1, 2, 3)$ is a solution, $x = 1$, $y = 2$, and $z = 3$ satisfy each equation. A linear equation in three variables can be represented by a flat plane in space. If the solution is unique, we can represent a linear system of three equations in three variables graphically by three planes intersecting a single point, as illustrated in Figure 4.29.

Critical Thinking

The following figure of three planes in space represents a system of three linear equations in three variables. How many solutions are there? Explain.

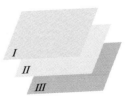

CHAPTER 4 Systems of Linear Equations

The next example shows how to determine whether an ordered triple represents a solution.

EXAMPLE 1 Checking solutions

Determine whether $(4, 2, -1)$ or $(-1, 0, 3)$ is a solution.

$$2x - 3y + z = 1$$
$$x - 2y + 2z = 5$$
$$2y + z = 3$$

Solution

To check $(4, 2, -1)$, substitute $x = 4$, $y = 2$, and $z = -1$ into each equation.

$2(4) - 3(2) + (-1) \stackrel{?}{=} 1$ True
$4 - 2(2) + 2(-1) \stackrel{?}{=} 5$ False
$2(2) + (-1) \stackrel{?}{=} 3$ True

The ordered triple $(4, 2, -1)$ does not satisfy all three equations, so it is not a solution. Next, substitute $x = -1$, $y = 0$, and $z = 3$.

$2(-1) - 3(0) + 3 \stackrel{?}{=} 1$ True
$-1 - 2(0) + 2(3) \stackrel{?}{=} 5$ True
$2(0) + 3 \stackrel{?}{=} 3$ True

The ordered triple $(-1, 0, 3)$ satisfies all three equations, so it is a solution.

In the next example we show how a linear system involving three equations and three variables can be used to model a real-world situation. We solve this system of equations in Example 6.

EXAMPLE 2 Modeling real data with linear systems

The Bureau of Land Management studies antelope populations in Wyoming. It monitors the number of adult antelope, the number of fawns each spring, and the severity of the winter. The first two columns of Table 4.1 contain counts of fawns and adults for three representative winters. The third column shows the severity of each winter. The severity of the winter is measured from 1 to 5, with 1 being mild and 5 being severe.

TABLE 4.1

Fawns (F)	Adults (A)	Winter (W)
405	870	3
414	848	2
272	684	5
?	750	4

4.4 Systems of Equations in Three Variables

We want to use the data in the first three rows of the table to estimate the number of fawns F when the number of adults is 750 and the severity of the winter is 4. To do so, we use the formula

$$F = a + bA + cW,$$

where a, b, and c are constants. Write a system of linear equations whose solution gives appropriate values for a, b, and c.

Solution

From the first row in the table, we know that, when $F = 405$, $A = 870$, and $W = 3$, the formula

$$F = a + bA + cW$$

becomes

$$405 = a + b(870) + c(3).$$

Similarly, $F = 414$, $A = 848$, and $W = 2$ gives

$$414 = a + b(848) + c(2),$$

and $F = 272$, $A = 684$, and $W = 5$ yields

$$272 = a + b(684) + c(5).$$

To find values for a, b, and c we can solve the following system of linear equations.

$$405 = a + 870b + 3c$$
$$414 = a + 848b + 2c$$
$$272 = a + 684b + 5c$$

We can also write these equations as a linear system in the following form.

$$a + 870b + 3c = 405$$
$$a + 848b + 2c = 414$$
$$a + 684b + 5c = 272$$

Finding values for a, b, and c will allow us to use the formula $F = a + bA + cW$ to predict the number of fawns F when the number of adults A is 750 and the severity of the winter W is 4 (see Example 6).

Critical Thinking

The model presented in Example 2 considers only two variables that might affect the fawn population. What other factors might be included in the model? If we included these factors, what would happen to the number of variables in the model?

Linear systems of two equations can have no solutions, one solution, or infinitely many solutions. The same is true for larger linear systems. In this section we focus on linear systems having one solution.

Solving Linear Systems with Substitution and Elimination

When solving systems of linear equations with more than two variables, we usually use both substitution and elimination. However, in the next example we use only substitution to solve a particular type of linear system in three variables.

EXAMPLE 3 *Using substitution to solve a linear system of equations*

Solve the following system.

$$2x - y + z = 7$$
$$3y - z = 1$$
$$z = 2$$

Solution

Note that the last equation gives us the value of z immediately. We can substitute $z = 2$ into the second equation and determine y.

$$3y - z = 1 \quad \text{Second equation}$$
$$3y - 2 = 1 \quad \text{Substitute } z = 2.$$
$$3y = 3 \quad \text{Add 2 to both sides.}$$
$$y = 1 \quad \text{Divide both sides by 3.}$$

Knowing that $y = 1$ and $z = 2$ allows us to find x by using the first equation.

$$2x - y + z = 7 \quad \text{First equation}$$
$$2x - 1 + 2 = 7 \quad \text{Let } y = 1 \text{ and } z = 2.$$
$$2x = 6 \quad \text{Simplify and subtract 1.}$$
$$x = 3 \quad \text{Divide both sides by 2.}$$

Thus $x = 3$, $y = 1$, and $z = 2$ and the solution is $(3, 1, 2)$.

In the next example we use elimination and substitution to solve a system of linear equations.

EXAMPLE 4 *Solving a linear system in three variables*

Solve the following system.

$$x - y + 2z = 6$$
$$2x + y - 2z = -3$$
$$-x - 2y + 3z = 7$$

Solution

STEP 1 We begin by eliminating the variable x from the second and third equations. To eliminate x from the second equation we multiply the first equation by -2

and then add it to the second equation. To eliminate x from the third equation we add the first and third equations.

$-2x + 2y - 4z = -12$	First equation times -2	$x - y + 2z = 6$ First equation
$2x + y - 2z = -3$	Second equation	$-x - 2y + 3z = 7$ Third equation
$3y - 6z = -15$	Add.	$-3y + 5z = 13$ Add.

STEP 2 Take the two resulting equations from Step 1 and eliminate either variable. Here we add the two equations to eliminate the variable y.

$$3y - 6z = -15$$
$$-3y + 5z = 13$$
$$-z = -2 \quad \text{Add the equations.}$$
$$z = 2 \quad \text{Multiply by } -1.$$

STEP 3 Now we can use substitution to find the values of x and y. We let $z = 2$ in either equation used in Step 2 to find y.

$$3y - 6z = -15$$
$$3y - 6(2) = -15 \quad \text{Substitute } z = 2.$$
$$3y - 12 = -15 \quad \text{Multiply.}$$
$$3y = -3 \quad \text{Add 12.}$$
$$y = -1 \quad \text{Divide by 3.}$$

STEP 4 Finally, we substitute $y = -1$ and $z = 2$ into any of the original equations to find x.

$$x - y + 2z = 6 \quad \text{First equation}$$
$$x - (-1) + 2(2) = 6 \quad \text{Let } y = -1 \text{ and } z = 2.$$
$$x + 1 + 4 = 6 \quad \text{Simplify.}$$
$$x = 1 \quad \text{Subtract 5.}$$

The solution is $(1, -1, 2)$.

In the next example we determine the price of tickets at a play.

EXAMPLE 5   *Finding ticket prices*

One thousand tickets were sold for a play, which generated $3800 in revenue. The prices of the tickets were $3 for children, $4 for students, and $5 for adults. There were 100 less student tickets sold than adult tickets. Find the number of each type of ticket sold.

Solution

Let x be the number of tickets sold to children, y be the number of tickets sold to students, and z be the number of tickets sold to adults. The total number of tickets sold was 1000, so

$$x + y + z = 1000.$$

Each child's ticket costs $3, so the revenue generated from selling x tickets would be $3x$. Similarly, the revenue generated from students would be $4y$, and the revenue from adults would be $5z$. Total ticket sales were $3800, so

$$3x + 4y + 5z = 3800.$$

The equation $y - z = -100$ must also be satisfied, as 100 fewer tickets were sold to students than adults.

To find the price of a ticket we need to solve the following system of linear equations.

$$\begin{aligned} x + y + z &= 1000 \\ 3x + 4y + 5z &= 3800 \\ y - z &= -100 \end{aligned}$$

STEP 1 We begin by eliminating the variable x from the second equation. To do so, we multiply the first equation by 3 and subtract the second equation.

$$\begin{array}{ll} 3x + 3y + 3z = 3000 & \text{First equation times 3} \\ \underline{3x + 4y + 5z = 3800} & \text{Second equation} \\ -y - 2z = -800 & \text{Subtract.} \end{array}$$

STEP 2 We then use the resulting equation from Step 1 and the third equation to eliminate y.

$$\begin{array}{ll} -y - 2z = -800 & \text{Equation from Step 1} \\ \underline{y - z = -100} & \text{Third equation} \\ -3z = -900 & \text{Add the equations.} \\ z = 300 & \text{Divide by } -3. \end{array}$$

STEP 3 To find y we can substitute $z = 300$ into the third equation.

$$\begin{array}{ll} y - z = -100 & \\ y - 300 = -100 & \text{Let } z = 300. \\ y = 200 & \text{Add 300.} \end{array}$$

STEP 4 Finally, we substitute $y = 200$ and $z = 300$ into the first equation.

$$\begin{array}{ll} x + y + z = 1000 & \text{First equation} \\ x + 200 + 300 = 1000 & \text{Let } y = 200 \text{ and } z = 300. \\ x = 500 & \text{Subtract 500.} \end{array}$$

Thus 500 tickets were sold to children, 200 to students, and 300 to adults.

Modeling Data

In the next example we solve the system of equations that we developed to model the data in Example 2.

EXAMPLE 6 *Predicting fawns in the spring*

Solve the following linear system for a, b, and c. Then use the formula $F = a + bA + cW$ to predict the number of fawns when the number of adults is 750 and the severity of the winter is 4.

$$a + 870b + 3c = 405$$
$$a + 848b + 2c = 414$$
$$a + 684b + 5c = 272$$

Solution

STEP 1 We begin by eliminating the variable a from the second and third equations. To do so, we subtract the second and third equations from the first equation.

$a + 870b + 3c = 405$	First equation	$a + 870b + 3c = 405$	First equation
$a + 848b + 2c = 414$	Second equation	$a + 684b + 5c = 272$	Third equation
$22b + c = -9$	Subtract.	$186b - 2c = 133$	Subtract.

STEP 2 We use the two resulting equations from Step 1 to eliminate c. To do so we multiply $22b + c = -9$ by 2 and add it to the other equation to eliminate c.

$$44b + 2c = -18 \quad \text{Twice } (22b + c = -9)$$
$$\underline{186b - 2c = 133}$$
$$230b = 115 \quad \text{Add the equations.}$$
$$b = 0.5 \quad \text{Divide by 230.}$$

STEP 3 To find c we substitute $b = 0.5$ into either equation used in Step 2.

$$44b + 2c = -18$$
$$44(0.5) + 2c = -18 \quad \text{Let } b = 0.5.$$
$$22 + 2c = -18 \quad \text{Multiply.}$$
$$2c = -40 \quad \text{Subtract 22.}$$
$$c = -20 \quad \text{Divide by 2.}$$

STEP 4 Finally, we substitute $b = 0.5$ and $c = -20$ into any of the original equations to find a.

$$a + 870b + 3c = 405 \quad \text{First equation}$$
$$a + 870(0.5) + 3(-20) = 405 \quad \text{Let } b = 0.5 \text{ and } c = -20.$$
$$a + 435 - 60 = 405 \quad \text{Multiply.}$$
$$a = 30 \quad \text{Solve for } a.$$

The solution is $a = 30$, $b = 0.5$, and $c = -20$. Thus we may write

$$F = a + bA + cW$$
$$= 30 + 0.5A - 20W$$

If there are 750 adults and the winter has a severity of 4, this model predicts

$$F = 30 + 0.5(750) - 20(4)$$
$$= 325 \text{ fawns.}$$

Critical Thinking

Give reasons why the coefficient for A is positive and the coefficient for W is negative in the formula

$$F = 30 + 0.5A - 20W.$$

4.4 Putting It All Together

In this section we discussed how to solve a system of three linear equations in three variables. Systems of linear equations can have no solutions, one solution, or infinitely many solutions. Our discussion focused on systems that have only one solution. The following table summarizes some of the important concepts presented in this section.

Concept	Explanation
System of Linear Equations	The following is a system of three linear equations with three variables. $$x - 2y + z = 0$$ $$-x + y + z = 4$$ $$-y + 4z = 10$$
Solution of a Linear System	The solution to a linear system in three variables is an ordered triple, expressed as (x, y, z). The solution to the preceding system is $(1, 2, 3)$ because substituting $x = 1, y = 2,$ and $z = 3$ into each equation results in a true statement. $$(1) - 2(2) + 3 = 0 \quad \text{True}$$ $$-(1) + (2) + (3) = 4 \quad \text{True}$$ $$-(2) + 4(3) = 10 \quad \text{True}$$
Solving a Linear System with Substitution and Elimination	Refer to Example 4. **STEP 1** Eliminate one variable, such as x, from two of the equations. **STEP 2** Use the two resulting equations in two variables to eliminate one of the variables, such as y. Solve for the remaining variable z. **STEP 3** Substitute z into one of the two equations from Step 2. Solve for the unknown variable y. **STEP 4** Substitute values for y and z into one of the original equations and find x. The solution is (x, y, z).

4.4 EXERCISES

CONCEPTS

1. Can a linear system of three linear equations and three variables have two solutions? Explain.

2. Give an example of a system of three linear equations with three variables.

3. Does the ordered triple $(1, 2, 3)$ satisfy the equation $x + y + z = 6$?

4. Does $(3, 4)$ represent a solution to the equation $x + y + z = 7$? Explain.

5. To solve for two variables, how many equations do you usually need?

6. To solve for three variables, how many equations do you usually need?

SOLVING LINEAR SYSTEMS

Exercises 7–10: Determine which ordered triple is a solution to the linear system.

7. $(1, 2, 3), (0, 2, 4)$
$$x + y + z = 6$$
$$x - y - z = -4$$
$$-x - y + z = 0$$

8. $(-1, 0, 2), (0, 4, 4)$
$$2x + y - 3z = -8$$
$$x - 3y + 2z = -4$$
$$3x - 2y + z = -4$$

9. $(1, 0, 3), (-1, 1, 2)$
$$3x - 2y + z = -3$$
$$-x + 3y - 2z = 0$$
$$x + 4y + 2z = 7$$

10. $\left(\frac{1}{2}, \frac{3}{2}, -\frac{1}{2}\right), (-1, 0, -2)$
$$x + 3y - 4z = 7$$
$$-x + 5y + 3z = \frac{11}{2}$$
$$3x - 2y - 7z = 2$$

Exercises 11–16: (Refer to Example 3.) Use substitution to solve the system of linear equations. Check your solution.

11. $x + y - z = 1$
$2y + z = -1$
$z = 1$

12. $2x + y - 3z = 1$
$y + 4z = 0$
$z = -1$

13. $-x - 3y + z = -2$
$2y + 3z = 3$
$z = 2$

14. $3x + 2y - 3z = -4$
$-y + 2z = 4$
$z = 0$

15. $a - b + 2c = 3$
$-3b + c = 4$
$c = -2$

16. $5a + 2b - 3c = 10$
$5b - 2c = -4$
$c = 3$

Exercises 17–28: (Refer to Example 4.) Use elimination and substitution to solve the system of linear equations.

17. $x + y - z = 11$
$-x + 2y + 3z = -1$
$2z = 4$

18. $x + 2y - 3z = -7$
$-2x + y + z = -1$
$3z = 9$

19. $x + y - z = -2$
$-x + z = 1$
$y + 2z = 3$

20. $x + y - 3z = 11$
$-2x + y + 2z = 1$
$-3y + 3z = -21$

21. $x + y - 2z = -7$
    $y + z = -1$
    $-y + 3z = 9$

22. $2x + 3y + z = 5$
    $y + 2z = 4$
    $-2y + z = 2$

23. $x + 2y + 2z = 1$
    $x + y + z = 0$
    $-x - 2y + 3z = -11$

24. $x + y - z = 0$
    $x - 3y + z = -2$
    $x - y + 3z = 8$

25. $2x + 6y - 2z = 47$
    $2x + y + 3z = -28$
    $-x + y + z = -\dfrac{7}{2}$

26. $x + y + 2z = 23$
    $3x - y + 3z = 8$
    $2x + 2y + z = 13$

27. $x + 3y - 4z = \dfrac{13}{2}$
    $-2x + 3y - z = \dfrac{1}{2}$
    $3x + z = 4$

28. $x - 2y + z = \dfrac{9}{2}$
    $4x - y + 3z = 9$
    $x + 2y = -\dfrac{3}{2}$

APPLICATIONS

29. *Finding Costs* The accompanying table shows the costs of purchasing different combinations of hamburgers, fries, and soft drinks.

Hamburgers	Fries	Soft Drinks	Total Cost
1	2	4	$10
1	4	6	$15
0	3	2	$6

(a) Let *x* be the cost of a hamburger, *y* the cost of fries, and *z* the cost of a soft drink. Write a system of three linear equations that represents the data in the table.
(b) Solve the system of linear equations and interpret your answer.

30. *Cost of CDs* The accompanying table shows the total cost of purchasing various combinations of differently priced CDs. The types of CDs are labeled A, B, and C.

A	B	C	Total cost
1	1	1	$37
3	2	1	$69
1	1	4	$82

(a) Let *x* be the cost of a CD of type A, *y* the cost of a CD of type B, and *z* the cost of a CD of type C. Write a system of three linear equations that represents the data in the table.
(b) Solve this linear system of equations and interpret your answer.

31. *Geometry* The largest angle in a triangle is 55° more than the smallest angle. The sum of the measures of the two smaller angles is 10° more than the measure of the largest angle.
(a) Let *x*, *y*, and *z* represent the measures of the three angles from largest to smallest. Write a system of three linear equations whose solution gives the measure of each angle. (*Hint:* The sum of the measures of the angles equals 180°.)
(b) Solve the system by using elimination and substitution.
(c) Check your solution.

32. *Geometry* The perimeter of a triangle is 90 inches. The longest side is 20 inches longer than the shortest side and 10 inches longer than the remaining side.
(a) Let *x*, *y*, and *z* represent the lengths of the three sides from largest to smallest. Write a system of three linear equations whose solution gives the lengths of each side.

(b) Solve the system by using elimination and substitution.
(c) Check your solution.

33. *Predicting Fawns* (Refer to Examples 2 and 6.) The accompanying table shows counts for fawns and adult deer and the severity of the winter. These data may be modeled by the equation $F = a + bA + cW$.

Fawns (F)	Adults (A)	Winter (W)
525	600	4
365	400	2
805	900	5

(a) Write a system of three linear equations in three variables whose solution gives values for a, b, and c.
(b) Solve this system of linear equations.
(c) Predict the number of fawns when there are 500 adults and the winter has severity 3.

34. *Predicting Home Prices* Selling prices of homes can depend on several factors such as size and age. The accompanying table shows the selling price for three homes. In this table price P is given in thousands of dollars, age A in years, and home size S in thousands of square feet. These data may be modeled by the equation $P = a + bA + cS$.

Price (P)	Age (A)	Size (S)
190	20	2
320	5	3
50	40	1

(a) Write a system of linear equations whose solution gives a, b, and c.
(b) Solve this system of linear equations.
(c) Predict the price of a home that is 10 years old and has 2500 square feet.

35. *Investment Mixture* A sum of $30,000 was invested in three mutual funds. In one year the first fund grew by 8%, the second by 10%, and the third by 15%. Total earnings were $3550. The amount invested in the third fund was $2000 less than the total amount invested in the other two funds combined. Use a linear system of equations to determine the amount invested in each fund.

36. *Football Tickets* A total of 2500 tickets were sold at a football game. Prices were $2 for children, $3 for students, and $5 for adults. Twice as many tickets were sold to students as children and ticket revenues were $7250. Use a system of linear equations to determine how many of each type of ticket were sold.

WRITING ABOUT MATHEMATICS

37. In the previous section we solved problems with two variables; to obtain a unique solution we needed two linear equations. In this section we solved problems with three variables; to obtain a unique solution we needed three equations. Try to generalize these results. In the design of aircraft, problems commonly involve 100,000 variables. How many equations are required to solve such problems? Can such problems be solved by hand? If not, how are such problems solved? Explain your answers.

38. In Exercise 34 the price of a home was estimated by its age and size. What other factors might affect the price of a home? Explain how these factors might affect the number of variables and equations in the linear system.

CHECKING BASIC CONCEPTS FOR SECTIONS 4.3 AND 4.4

1. Write an inequality that describes the shaded region in the graph.

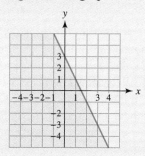

2. Graph the solution set for the linear system of inequalities. Use a test point to check your graph.

$$2x + 3y \leq 6$$
$$x + 2y \geq 4$$

3. Determine which ordered pair is the solution. $(5, -4, 0), (1, 3, -1)$

$$x - y + 7z = -9$$
$$2x - 2y + 5z = -9$$
$$-x + 3y - 2z = 10$$

4. Solve the system of equations by using elimination and substitution,

$$x - y + z = 2$$
$$2x - 3y + z = -1$$
$$-x + y + z = 4$$

4.5 MATRIX SOLUTIONS OF LINEAR SYSTEMS

Representing Systems of Linear Equations with Matrices ~ Gaussian Elimination ~ Using Technology to Solve Systems of Linear Equations

INTRODUCTION

Suppose that the size of a bear's head and its overall length are known. Can its weight be estimated from these variables? Can a bear's weight be estimated if its neck size and chest size are known? In this section we show that systems of linear equations can be used to make such estimates.

In the previous section we solved systems of three linear equations in three variables by using elimination and substitution. In real life, systems of equations often contain thousands of variables. To solve a large system of equations, we need an efficient method. Long before the invention of the computer, Carl Fredrich Gauss (1777–1855) developed a method called *Gaussian elimination* to solve systems of linear equations. Even though it was developed more than 150 years ago, it is still used today in modern computers and calculators. In this section we introduce this method.

REPRESENTING SYSTEMS OF LINEAR EQUATIONS WITH MATRICES

Arrays of numbers are used frequently in many different real-world situations. Spreadsheets often make use of arrays. A **matrix** is a rectangular array of numbers.

Each number in a matrix is called an **element**. The following are examples of *matrices* (plural of matrix), with their dimensions written below them.

$$\begin{bmatrix} 2 & 0 \\ 3 & 1 \end{bmatrix} \quad \begin{bmatrix} -1.2 & 5 & 0 \\ 1 & 0 & 1 \\ 4 & -5 & 7 \end{bmatrix} \quad \begin{bmatrix} 3 & -6 & 0 & \sqrt{3} \\ 1 & 4 & 0 & 9 \\ -3 & 1 & 1 & 18 \\ -10 & -4 & 5 & -1 \end{bmatrix} \quad \begin{bmatrix} 4 & 2 \\ 0 & 1 \\ 1 & 0 \end{bmatrix} \quad \begin{bmatrix} 1 & 5 & -1 \\ 3 & 4 & 2 \end{bmatrix}$$

2×2 3×3 4×4 3×2 2×3

rows × columns

The dimension of a matrix is stated much like the dimensions of a rectangular room. We might say that a room is n feet long and m feet wide. Similarly, the **dimension of a matrix** is $n \times m$ (n by m), if it has n rows and m columns. For example, the last matrix in the preceding group has a dimension of 2×3 because it has 2 rows and 3 columns. If the number of rows and columns are equal, the matrix is a **square matrix**. The first three matrices in that group are square matrices.

Matrices can be used to represent a system of linear equations. For example, if we have the system of equations

$$\begin{aligned} 3x - y + 2z &= 7 \\ x - 2y + z &= 0 \\ 2x + 5y - 7z &= -9, \end{aligned}$$

we can represent the system with the following **augmented matrix**. Note how the coefficients of the variables were placed in the matrix. A vertical line is positioned in the matrix where the equals signs occur in the system of equations. The rows and columns are labeled, and the **main diagonal** of the augmented matrix is circled. The matrix has dimension 3×4.

Main diagonal

$$\begin{bmatrix} 3 & -1 & 2 & | & 7 \\ 1 & -2 & 1 & | & 0 \\ 2 & 5 & -7 & | & -9 \end{bmatrix} \begin{matrix} \text{Row 1} \\ \text{Row 2} \\ \text{Row 3} \end{matrix}$$

Column 1   Column 2   Column 3   Column 4

EXAMPLE 1   *Representing a linear system*

Represent each linear system with an augmented matrix. State the dimension of the matrix.

(a) $x - 2y = 9$
      $6x + 7y = 16$

(b) $x - 3y + 7z = 4$
      $2x + 5y - z = 15$
      $2x + y = 8$

Solution

(a) This system can be represented by the following 2 × 3 matrix.

$$\begin{bmatrix} 1 & -2 & | & 9 \\ 6 & 7 & | & 16 \end{bmatrix}$$

(b) This system has three equations in three variables and can be represented by the following 3 × 4 matrix.

$$\begin{bmatrix} 1 & -3 & 7 & | & 4 \\ 2 & 5 & -1 & | & 15 \\ 2 & 1 & 0 & | & 8 \end{bmatrix}$$

GAUSSIAN ELIMINATION

A convenient matrix form for representing a system of linear equations is **reduced row–echelon form.** The following matrices are examples of reduced row–echelon form. Note that there are 1s on the main diagonal with 0s above and below each 1.

$$\begin{bmatrix} 1 & 0 & 3 \\ 0 & 1 & -2 \end{bmatrix} \quad \begin{bmatrix} 1 & 0 & 0 & 3 \\ 0 & 1 & 0 & 1 \\ 0 & 0 & 1 & -1 \end{bmatrix} \quad \begin{bmatrix} 1 & 0 & 0 & 8 \\ 0 & 1 & 0 & 2 \\ 0 & 0 & 1 & 3 \end{bmatrix}$$

If an augmented matrix representing a linear system is in reduced row–echelon form, we can usually determine the solution easily.

EXAMPLE 2 *Determining a solution from reduced row–echelon form*

Each matrix represents a system of linear equations. Find the solution.

(a) $\begin{bmatrix} 1 & 0 & 0 & | & 2 \\ 0 & 1 & 0 & | & -3 \\ 0 & 0 & 1 & | & 5 \end{bmatrix}$ (b) $\begin{bmatrix} 1 & 0 & | & 10 \\ 0 & 1 & | & -4 \end{bmatrix}$

Solution

(a) The top row represents $1x + 0y + 0z = 2$ or $x = 2$. Using similar reasoning for the next two rows, we have $y = -3$ and $z = 5$. The solution is $(2, -3, 5)$.
(b) The system involves two equations and two unknowns. The solution is $(10, -4)$.

We can use a numerical method called **Gaussian elimination** to solve a linear system. It makes use of the following matrix row transformations.

Matrix Row Transformations

For any augmented matrix representing a system of linear equations, the following row transformations result in an equivalent system of linear equations.

1. Any two rows may be interchanged.
2. The elements of any row may be multiplied by a nonzero constant.
3. Any row may be changed by adding to (or subtracting from) its elements a multiple of the corresponding elements of another row.

4.5 Matrix Solutions of Linear Systems

Gaussian elimination can be used to transform an augmented matrix into reduced row–echelon form. Its objective is to use these matrix row transformations to obtain a matrix that has the following reduced row–echelon form, where (a, b) represents the solution.

$$\begin{bmatrix} 1 & 0 & | & a \\ 0 & 1 & | & b \end{bmatrix}$$

This method is illustrated in the next example.

EXAMPLE 3 *Transforming a matrix into reduced row–echelon form*

Use Gaussian elimination to transform the augmented matrix of the linear system into reduced row–echelon form. Find the solution.

$$x + y = 5$$
$$-x + y = 1$$

Solution

Both the linear system and the augmented matrix are shown.

Linear System **Augmented Matrix**
$$x + y = 5 \qquad \begin{bmatrix} 1 & 1 & | & 5 \\ -1 & 1 & | & 1 \end{bmatrix}$$
$$-x + y = 1$$

First, we want to obtain a 0 in the second row, where the -1 is highlighted. To do so, we add row 1 to row 2 and place the result in row 2. This step is denoted $R_2 + R_1$ and eliminates the x-variable from the second equation.

$$x + y = 5 \qquad\qquad \begin{bmatrix} 1 & 1 & | & 5 \\ 0 & 2 & | & 6 \end{bmatrix}$$
$$2y = 6 \qquad R_2 + R_1 \rightarrow$$

To obtain a 1 where the 2 in the second row is located, we divide the second row by 2, denoted $\frac{R_2}{2}$.

$$x + y = 5 \qquad\qquad \begin{bmatrix} 1 & 1 & | & 5 \\ 0 & 1 & | & 3 \end{bmatrix}$$
$$y = 3 \qquad \frac{R_2}{2} \rightarrow$$

Next, we need to obtain a 0 where the 1 is highlighted. We do so by subtracting row 2 from row 1 and placing the result in row 1, denoted $R_1 - R_2$.

$$x = 2 \qquad R_1 - R_2 \rightarrow \begin{bmatrix} 1 & 0 & | & 2 \\ 0 & 1 & | & 3 \end{bmatrix}$$
$$y = 3$$

This matrix is in reduced row–echelon form. The solution is (2, 3).

In the next example, we use Gaussian elimination to solve a system with three linear equations and three variables. To do so we transform the matrix into the following reduced row–echelon form, where (a, b, c) represents the solution.

$$\begin{bmatrix} 1 & 0 & 0 & | & a \\ 0 & 1 & 0 & | & b \\ 0 & 0 & 1 & | & c \end{bmatrix}$$

Several calculations are involved in transforming the system to reduced row–echelon form. Later we use technology to perform the same calculations more efficiently.

EXAMPLE 4 *Transforming a matrix to reduced row–echelon form*

Use Gaussian elimination to transform the augmented matrix of the linear system into reduced row–echelon form. Find the solution.

$$x + y + 2z = 1$$
$$-x \phantom{{}+y} + z = -2$$
$$2x + y + 5z = -1$$

Solution

The linear system and the augmented matrix are both shown.

Linear System

$$x + y + 2z = 1$$
$$-x \phantom{{}+y} + z = -2$$
$$2x + y + 5z = -1$$

Augmented Matrix

$$\begin{bmatrix} 1 & 1 & 2 & | & 1 \\ -1 & 0 & 1 & | & -2 \\ 2 & 1 & 5 & | & -1 \end{bmatrix}$$

First, we want to put 0s into the second and third rows, where the -1 and 2 are highlighted. To obtain a 0 in the first position of the second equation we add row 2 and row 1 and place the result in row 2, denoted $R_2 + R_1$. To obtain a 0 in the first position of the third row we subtract 2 times row 1 from row 3 and place the result in row 3, denoted $R_3 - 2R_1$. Row 1 does not change. These steps eliminate the x-variable from the second and third equations.

$$x + y + 2z = 1$$
$$\phantom{x+{}}y + 3z = -1$$
$$\phantom{x+{}}-y + z = -3$$

$R_2 + R_1 \rightarrow$
$R_3 - 2R_1 \rightarrow$

$$\begin{bmatrix} 1 & 1 & 2 & | & 1 \\ 0 & 1 & 3 & | & -1 \\ 0 & -1 & 1 & | & -3 \end{bmatrix}$$

To eliminate the y-variable in row 1, we subtract row 2 from row 1. To eliminate the y-variable from row 3, we add row 2 to row 3.

$$x \phantom{{}+y} - z = 2$$
$$\phantom{x+{}}y + 3z = -1$$
$$\phantom{x+y+{}}4z = -4$$

$R_1 - R_2 \rightarrow$
$R_3 + R_2 \rightarrow$

$$\begin{bmatrix} 1 & 0 & -1 & | & 2 \\ 0 & 1 & 3 & | & -1 \\ 0 & 0 & 4 & | & -4 \end{bmatrix}$$

To obtain a 1 in row 3, where the highlighted 4 is located, we divide row 3 by 4.

$$x \phantom{{}+y} - z = 2$$
$$\phantom{x+{}}y + 3z = -1$$
$$\phantom{x+y+3{}}z = -1$$

$\dfrac{R_3}{4} \rightarrow$

$$\begin{bmatrix} 1 & 0 & -1 & | & 2 \\ 0 & 1 & 3 & | & -1 \\ 0 & 0 & 1 & | & -1 \end{bmatrix}$$

To transform the matrix into reduced row–echelon form, we need to put 0s into the highlighted locations. To do so we first add row 1 and row 3 and then subtract 3 times row 3 from row 2.

$$\begin{array}{l} x = 1 \\ y = 2 \\ z = -1 \end{array} \qquad \begin{array}{l} R_1 + R_3 \to \\ R_2 - 3R_3 \to \end{array} \left[\begin{array}{ccc|c} 1 & 0 & 0 & 1 \\ 0 & 1 & 0 & 2 \\ 0 & 0 & 1 & -1 \end{array}\right]$$

This matrix is now in reduced row–echelon form. The solution is $(1, 2, -1)$.

USING TECHNOLOGY TO SOLVE SYSTEMS OF LINEAR EQUATIONS

Example 4 involves a lot of arithmetic, which can easily lead to mistakes. Trying to solve a large system of equations by hand is an enormous—if not impossible—task. In the real world, people use technology to solve large systems. Many types of graphing calculators are programmed to solve systems of linear equations with the method discussed in Example 4. The calculator does all the arithmetic.

In the next example we solve the linear systems from Examples 3 and 4 with a graphing calculator.

EXAMPLE 5 *Using technology*

 Use a graphing calculator to solve the following systems of equations.

(a) $\quad x + y = 5$
$\quad -x + y = 1$

(b) $\quad x + y + 2z = 1$
$\quad -x + z = -2$
$\quad 2x + y + 5z = -1$

Solution

(a) The system can be written as an augmented matrix.

$$\left[\begin{array}{cc|c} 1 & 1 & 5 \\ -1 & 1 & 1 \end{array}\right]$$

Enter the 2×3 matrix into a graphing calculator and then transform the matrix into reduced row–echelon form (rref), as shown in Figures 4.30 and 4.31. Figure 4.31 shows that the solution is $(2, 3)$.

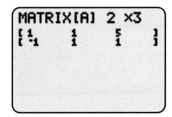

Figure 4.30

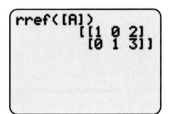

Figure 4.31

(b) The system can be written as an augmented matrix.

$$\begin{bmatrix} 1 & 1 & 2 & | & 1 \\ -1 & 0 & 1 & | & -2 \\ 2 & 1 & 5 & | & -1 \end{bmatrix}$$

Enter the 3 × 4 matrix into a graphing calculator and then transform the matrix into reduced row–echelon form (rref), as shown in Figures 4.32 and 4.33. (The fourth column of A can be seen by scrolling right.) Figure 4.33 shows that the solution is $(1, 2, -1)$.

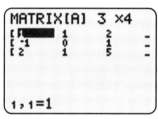

Figure 4.32

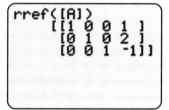

Figure 4.33

Technology Note *Reduced Row–Echelon Form*

On some graphing calculators a matrix can be entered from the MATRIX EDIT menu. Reduced row–echelon form can be accessed from the MATRIX CALC menu. You may want to consult your owner's manual for details.

In the next example we use technology to solve a real-world application.

EXAMPLE 6 *Modeling the weight of male bears*

The data shown in Table 4.2 give the weight W, head length H, and overall length L of three bears. These data can be modeled with the equation $W = a + bH + cL$, where a, b, and c are constants that we need to determine. **(Sources: M. Triola, *Elementary Statistics*; Minitab, Inc.)**

TABLE 4.2

W (pound)	H (inches)	L (inches)
362	16	72
300	14	68
147	11	52

(a) Set up a system of equations whose solution gives values for constants a, b, and c.
(b) Solve the system by using a graphing calculator.
(c) Predict the weight of a bear with head length $H = 13$ inches and overall length $L = 65$ inches.

Solution

(a) Substitute each row of the data into the equation $W = a + bH + cL$.

$$362 = a + b(16) + c(72)$$
$$300 = a + b(14) + c(68)$$
$$147 = a + b(11) + c(52)$$

Rewrite this system as

$$a + 16b + 72c = 362$$
$$a + 14b + 68c = 300$$
$$a + 11b + 52c = 147$$

and represent it as the augmented matrix

$$A = \begin{bmatrix} 1 & 16 & 72 & | & 362 \\ 1 & 14 & 68 & | & 300 \\ 1 & 11 & 52 & | & 147 \end{bmatrix}.$$

(b) Enter A in a graphing calculator and put it in reduced row–echelon form, as shown in Figures 4.34 and 4.35. The solution is $a = -374$, $b = 19$, and $c = 6$.

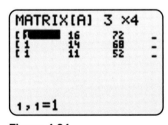

Figure 4.34

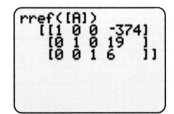

Figure 4.35

(c) For $W = a + bH + cL$, use

$$W = -374 + 19H + 6L$$

to predict the weight of a bear with head length $H = 13$ and overall length $L = 65$. Thus

$$W = -374 + 19(13) + 6(65)$$
$$= 263 \text{ pounds.}$$

In the next example we find the amounts invested in three different mutual funds.

EXAMPLE 7 Determining investment amounts

A total of $8000 was invested in three mutual funds that grew at a rate of 5%, 10%, and 20% over 1 year. After 1 year, the combined value of the three funds had grown by $1200. Five times as much money was invested at 20% as at 10%. Find the amount invested in each fund.

Solution

Let x be the amount invested at 5%, y the amount invested at 10%, and z the amount invested at 20%. The total amount invested was $8000, so

$$x + y + z = 8000.$$

The growth in the first mutual fund, paying 5% of x, is given by $0.05x$. Similarly, the growths in the other mutual funds are given by $0.10y$ and $0.20z$. As the total growth was $1200, we can write

$$0.05x + 0.10y + 0.20z = 1200.$$

Five times as much was invested at 20% as at 10%, so $z = 5y$, which can be written as $5y - z = 0$. These three equations can be written as a system of linear equations.

$$\begin{aligned} x + y + z &= 8000 \\ 0.05x + 0.10y + 0.20z &= 1200 \\ 5y - z &= 0 \end{aligned}$$

Entering the system into a graphing calculator and then solving, results in $x = 2000$, $y = 1000$, and $z = 5000$, as shown in Figures 4.36 and 4.37. Thus $2000 was invested at 5%, $1000 at 10%, and $5000 at 20%.

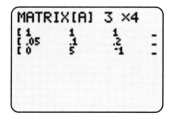

Figure 4.36

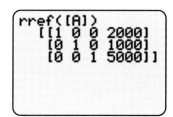

Figure 4.37

4.5 Putting It All Together

A matrix is a rectangular array of numbers. An augmented matrix may be used to represent a system of linear equations. One common method for solving a system of linear equations is Gaussian elimination. Matrix row operations may be used to transform an augmented matrix to reduced row–echelon form. Technology can be used to solve systems of linear equations efficiently. The following table summarizes augmented matrices and reduced row–echelon form.

Concept	Explanation
Augmented Matrix	A linear system can be represented using an augmented matrix. The following matrix has dimension 3×4. **Linear System** $\qquad$ **Augmented Matrix** $$\begin{aligned} x + 2y - z &= 6 \\ -2x + y - z &= 7 \\ 2x + 3z &= -11 \end{aligned} \qquad \begin{bmatrix} 1 & 2 & -1 & 6 \\ -2 & 1 & -1 & 7 \\ 2 & 0 & 3 & -11 \end{bmatrix}$$
Reduced Row–Echelon Form	The following augmented matrix is in reduced row–echelon form, which results from transforming the preceding system to reduced row–echelon form. There are 1s along the main diagonal and 0s elsewhere in the first three columns. The solution to the linear system is $(-1, 2, -3)$. $$\begin{bmatrix} 1 & 0 & 0 & -1 \\ 0 & 1 & 0 & 2 \\ 0 & 0 & 1 & -3 \end{bmatrix}$$

4.5 EXERCISES

FOR EXTRA HELP: Student's Solutions Manual, MyMathLab.com, InterAct Math, Math Tutor Center, MathXL, Digital Video Tutor CD 3 Videotape 6

CONCEPTS

1. What is a matrix?

2. Give an example of a matrix and state its dimension.

3. Give an example of an augmented matrix and state its dimension.

4. If an augmented matrix is used to solve a system of three linear equations in three variables, what will be its dimension?

5. Give an example of a matrix in reduced row–echelon form.

6. Identify the elements on the main diagonal in the augmented matrix
$$\begin{bmatrix} 4 & -6 & -1 & 3 \\ 6 & 2 & -2 & 9 \\ 7 & 5 & -3 & 1 \end{bmatrix}.$$

DIMENSIONS OF MATRICES AND AUGMENTED MATRICES

Exercises 7–10: State the dimension of the matrix.

7. $\begin{bmatrix} 3 & -3 & 7 \\ 2 & 6 & -2 \\ 4 & 2 & 5 \end{bmatrix}$
8. $\begin{bmatrix} -2 & 3 & 0 \\ 1 & -8 & 4 \end{bmatrix}$

9. $\begin{bmatrix} 1 & 7 \\ 0 & 2 \\ 2 & -5 \end{bmatrix}$
10. $\begin{bmatrix} 4 & 2 & -3 & -1 \\ 4 & -3 & 2 & -7 \\ 14 & 6 & 4 & 0 \end{bmatrix}$

Exercises 11–14: Represent the linear system as an augmented matrix.

11. $\quad x - 3y = 1$
    $\quad -x + 3y = -1$

12. $\quad 4x + 2y = -5$
    $\quad 5x + 8y = 2$

13. $\begin{aligned} 2x - y + 2z &= -4 \\ x - 2y &= 2 \\ -x + y - 2z &= -6 \end{aligned}$

14. $\begin{aligned} 3x - 2y + z &= 5 \\ -x + 2z &= -4 \\ x - 2y + z &= -1 \end{aligned}$

Exercises 15–20: Write the system of linear equations that the augmented matrix represents. Use the variables x, y, and z.

15. $\begin{bmatrix} 1 & 2 & | & -6 \\ 5 & -1 & | & 4 \end{bmatrix}$
16. $\begin{bmatrix} 1 & -5 & | & 7 \\ 0 & -3 & | & 6 \end{bmatrix}$

17. $\begin{bmatrix} 1 & -1 & 2 & | & 6 \\ 2 & 1 & -2 & | & 1 \\ -1 & 2 & -1 & | & 3 \end{bmatrix}$
18. $\begin{bmatrix} 3 & -1 & 2 & | & -1 \\ 2 & -2 & 2 & | & 4 \\ 1 & 7 & -2 & | & 2 \end{bmatrix}$

19. $\begin{bmatrix} 1 & 0 & 0 & | & 4 \\ 0 & 1 & 0 & | & -2 \\ 0 & 0 & 1 & | & 7 \end{bmatrix}$
20. $\begin{bmatrix} 1 & 0 & 0 & | & 6 \\ 0 & 1 & 0 & | & -2 \\ 0 & 0 & 1 & | & 4 \end{bmatrix}$

GAUSSIAN ELIMINATION

Exercises 21–30: Write the system of linear equations as an augmented matrix. Use Gaussian elimination to find the solution. Write the solution as an ordered pair or ordered triple and check the solution.

21. $\begin{aligned} x - y &= 5 \\ x + 3y &= -1 \end{aligned}$
22. $\begin{aligned} x + 4y &= 1 \\ 3x - 2y &= 10 \end{aligned}$

23. $\begin{aligned} 4x - 8y &= -10 \\ x + y &= 2 \end{aligned}$
24. $\begin{aligned} x - 7y &= -16 \\ 4x + 10y &= 50 \end{aligned}$

25. $\begin{aligned} x + y + z &= 0 \\ 2x + y + 2z &= -1 \\ x + y &= 0 \end{aligned}$

26. $\begin{aligned} x + y - 2z &= 5 \\ x + 2y - 2z &= 4 \\ -x - y + z &= -4 \end{aligned}$

27. $\begin{aligned} x + y + z &= 3 \\ -x - z &= -2 \\ x + y + 2z &= 4 \end{aligned}$
28. $\begin{aligned} x + 2y - z &= 3 \\ -x - y + z &= 0 \\ x + 2y &= 5 \end{aligned}$

29. $\begin{aligned} x + 2y + z &= 3 \\ 2x + y - z &= -6 \\ -x - y + 2z &= 5 \end{aligned}$

30. $\begin{aligned} x + y + z &= -3 \\ x - y - z &= -1 \\ -2x + y + 4z &= 4 \end{aligned}$

Exercises 31–40: **Technology** *Use a graphing calculator to solve the system of linear equations.*

31. $\begin{aligned} x + 4y &= 13 \\ 5x - 3y &= -50 \end{aligned}$
32. $\begin{aligned} 9x - 11y &= 7 \\ 5x + 6y &= 16 \end{aligned}$

33. $\begin{aligned} 2x - y + 3z &= 9 \\ -4x + 5y + 2z &= 12 \\ 2x + 7z &= 23 \end{aligned}$

34. $\begin{aligned} 3x - 2y + 4z &= 29 \\ 2x + 3y - 7z &= -14 \\ 5x - y + 11z &= 59 \end{aligned}$

35. $\begin{aligned} 6x + 2y + z &= 4 \\ -2x + 4y + z &= -3 \\ 2x - 8y &= -2 \end{aligned}$

36. $\begin{aligned} -x - 9y + 2z &= -28.5 \\ 2x - y + 4z &= -17 \\ x - y + 8z &= -9 \end{aligned}$

37. $\begin{aligned} 4x + 3y + 12z &= -9.25 \\ 15y + 8z &= -4.75 + x \\ 7z &= -5.5 - 6y \end{aligned}$

38. $\begin{aligned} 5x + 4y &= 13.3 + z \\ 7y + 9z &= 16.9 - x \\ x - 3y + 4z &= -4.1 \end{aligned}$

39. $\begin{aligned} 1.2x - 0.9y + 2.7z &= 5.37 \\ 3.1x - 5.1y + 7.2z &= 14.81 \\ 1.8y + 6.38 &= 3.6z - 0.2x \end{aligned}$

40. $\begin{aligned} 11x + 13y - 17z &= 380 \\ 5x - 14y - 19z &= 24 \\ -21y + 46z &= -676 + 7x \end{aligned}$

APPLICATIONS

41. *Weight of a Bear* Use the results of Example 6 to estimate the weight of a bear with a head length of 12 inches and an overall length of 60 inches.

42. *Weight of a Bear* (Refer to Example 6.) Head length and overall length are not the only variables that can be used to estimate the weight of a bear. The data in the accompanying table lists the weight W, neck size N, and chest size C of three bears. These data can be modeled with the equation $W = a + bN + cC$. (*Source:* M. Triola, *Elementary Statistics*; Minitab, Inc.)

W (pounds)	N (inches)	C (inches)
80	16	26
344	28	45
416	31	54

(a) Set up a system of equations whose solution gives values for the constants a, b, and c.
(b) Solve this system with a graphing calculator. Round each value to the nearest tenth.
(c) Predict the weight of a bear with neck size $N = 22$ inches and chest size $C = 38$ inches.

43. *Garbage and Household Size* A larger household produces more garbage, on average, than a smaller household. If we know the amount of metal M and plastic P waste produced each week, we can estimate the household size H from the equation $H = a + bM + cP$. The following table contains representative data for three households. (*Source:* M. Triola, *Elementary Statistics*.)

H (people)	M (pound)	P (pound)
3	2.00	1.40
2	1.50	0.65
6	4.00	3.40

(a) Set up a system of equations whose solution gives values for the constants a, b, and c.
(b) Solve this system with a graphing calculator.
(c) Predict the household size if it produces 3 pounds of metal waste and 2 pounds of plastic waste each week.

44. *Old Faithful Geyser* In Yellowstone National Park, Old Faithful Geyser has been a favorite attraction for decades. Although this geyser erupts about every 80 minutes, this time interval varies, as do the duration and height of the eruptions. The accompanying table shows the height H, duration D, and time interval T for three eruptions. (*Source:* National Park Service.)

H (feet)	D (seconds)	T (minutes)
160	276	94
125	203	84
140	245	79

(a) Assume that these data may be modeled by $H = a + bD + cT$. Set up a system of equations whose solution gives values for the constants a, b, and c.
(b) Solve this system by using a graphing calculator. Round each value to the nearest thousandth.
(c) Use this equation to estimate H when $D = 220$ and $T = 81$.

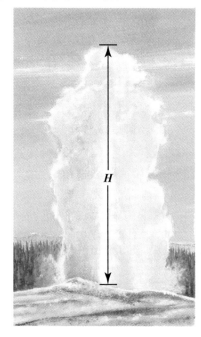

45. *Jogging Speeds* A runner in preparation for a marathon jogs at 5, 6, and 8 miles per hour. The runner travels a total distance of 12.5 miles in 2 hours and jogs the same length of time at 5 miles per hour and at 8 miles per hour. How long did the runner jog at each speed?

46. *Mixture Problem* Three different types of candy that cost $2, $3, and $4 per pound are to be mixed to produce a 5-pound bag of candy that costs $14.50. If there are to be equal amounts of the $3 per pound candy and the $4 per pound candy, how much of each type of candy should be included in the mixture?

47. *Interest and Investments* (Refer to Example 7.) A total of $3000 is invested at 5%, 8%, and 12% simple interest. The interest earned after 1 year equals $285. The amount invested at 12% is triple the amount invested at 5%. Find the amount invested at each interest rate.

48. *Geometry* The measure of the largest angle in a triangle is twice the measure of the smallest angle. The remaining angle is 10° less than the largest angle. Find the measure of each angle.

WRITING ABOUT MATHEMATICS

49. Explain what the dimension of a matrix means. What is the difference between a matrix that has dimension 3×4 and one that has dimension 4×3?

50. Discuss the advantages of using technology to transform an augmented matrix to reduced row–echelon form. Are there any disadvantages? Explain.

GROUP ACTIVITY
Working with Real Data

Directions: Form a group of 2 to 4 people. Select someone to record the group's responses for this activity. All members of the group should work cooperatively to answer the questions. If your instructor asks for the results, each member of the group should be prepared to respond. If the group is asked to turn in its work, be sure to include each person's name on the paper.

Charitable Giving In 1998, average charitable giving varied according to income. The average amount given for three different incomes is listed in the table.

Income	$10,000	$40,000	$100,000
Giving	$412	$843	$2550

Source: Gallup for Independent Sector.

(a) Let x represent income and y represent charitable giving. Set up an augmented matrix whose solution gives values for a, b, and c so that the graph of $y = ax^2 + bx + c$ passes through the points (10,000, 412), (40,000, 843), and (100,000, 2550). (*Hint:* Because the graph passes through the point (10,000, 412) the equation

$$412 = a(10,000)^2 + b(10,000) + c$$

must be satisfied.)

(b) Use a graphing calculator to solve this system of linear equations.

(c) Estimate the average charitable giving for someone earning $20,000. Compare your answer to the actual value of $525.

4.6 DETERMINANTS

Calculation of Determinants ~ Area of Regions ~ Cramer's Rule

INTRODUCTION

Surveyors commonly calculate the areas of parcels of land. To do so they frequently divide the land into triangular regions. When the coordinates of the vertices of a triangle are known, determinants may be used to find the area of the triangle. A determinant is a real number that can be calculated for any square matrix. In this section we use determinants to find areas and to solve systems with two linear equations in two variables.

CALCULATION OF DETERMINANTS

The concept of determinants originated with the Japanese mathematician Seki Kowa (1642–1708), who used them to solve systems of linear equations. Later, Gottfried Leibniz (1646–1716) formally described determinants and also used them to solve systems of linear equations. **(Source: Historical Topics for the Mathematical Classroom, NCTM.)**

We begin by defining a determinant for a 2×2 matrix.

Determinant for a 2×2 matrix

The **determinant** of

$$A = \begin{bmatrix} a & b \\ c & d \end{bmatrix}$$

is a *real number* defined by

$$\det A = ad - cb.$$

EXAMPLE 1 *Calculating a determinant*

Find det A for each 2×2 matrix.

(a) $A = \begin{bmatrix} 1 & 2 \\ 3 & 4 \end{bmatrix}$ \qquad **(b)** $A = \begin{bmatrix} -1 & -3 \\ 2 & -8 \end{bmatrix}$

Solution

(a) The determinant is calculated as follows.

$$\det A = \det \begin{bmatrix} 1 & 2 \\ 3 & 4 \end{bmatrix} = (1)(4) - (3)(2) = -2$$

(b) Similarly,

$$\det A = \det \begin{bmatrix} -1 & -3 \\ 2 & -8 \end{bmatrix} = (-1)(-8) - (2)(-3) = 14.$$

We can use determinants of 2 × 2 matrices to find determinants of 3 × 3 matrices. This method is called **expansion of a determinant by minors**.

Determinant of a 3 × 3 Matrix

$$\det A = \det \begin{bmatrix} a_1 & b_1 & c_1 \\ a_2 & b_2 & c_2 \\ a_3 & b_3 & c_3 \end{bmatrix} =$$

$$a_1 \cdot \det \begin{bmatrix} b_2 & c_2 \\ b_3 & c_3 \end{bmatrix} - a_2 \cdot \det \begin{bmatrix} b_1 & c_1 \\ b_3 & c_3 \end{bmatrix} + a_3 \cdot \det \begin{bmatrix} b_1 & c_1 \\ b_2 & c_2 \end{bmatrix}$$

The 2 × 2 matrices in this equation are called **minors**.

EXAMPLE 2 *Calculating a 3 × 3 determinant*

Evaluate det A.

(a) $A = \begin{bmatrix} 2 & 1 & -1 \\ -1 & 3 & 2 \\ 4 & -3 & -5 \end{bmatrix}$ (b) $A = \begin{bmatrix} 5 & -2 & 4 \\ 0 & 2 & 1 \\ -1 & 4 & -4 \end{bmatrix}$

Solution

(a) We evaluate the determinant as follows.

$$\det \begin{bmatrix} 2 & 1 & -1 \\ -1 & 3 & 2 \\ 4 & -3 & -5 \end{bmatrix} = 2 \cdot \det \begin{bmatrix} 3 & 2 \\ -3 & -5 \end{bmatrix} - (-1) \cdot \det \begin{bmatrix} 1 & -1 \\ -3 & -5 \end{bmatrix}$$
$$+ 4 \cdot \det \begin{bmatrix} 1 & -1 \\ 3 & 2 \end{bmatrix}$$
$$= 2(-9) + 1(-8) + 4(5)$$
$$= -6$$

(b) We evaluate the determinant as follows.

$$\det \begin{bmatrix} 5 & -2 & 4 \\ 0 & 2 & 1 \\ -1 & 4 & -4 \end{bmatrix} = 5 \cdot \det \begin{bmatrix} 2 & 1 \\ 4 & -4 \end{bmatrix} - (0) \cdot \det \begin{bmatrix} -2 & 4 \\ 4 & -4 \end{bmatrix}$$
$$+ (-1) \cdot \det \begin{bmatrix} -2 & 4 \\ 2 & 1 \end{bmatrix}$$
$$= 5(-12) - 0(-8) + (-1)(-10)$$
$$= -50$$

Many graphing calculators can evaluate the determinant of a matrix, as illustrated in the next example, where we evaluate the determinants from Example 2.

EXAMPLE 3 *Using technology to find a determinant*

Find the determinant of A, using a graphing calculator.

(a) $A = \begin{bmatrix} 2 & 1 & -1 \\ -1 & 3 & 2 \\ 4 & -3 & -5 \end{bmatrix}$ (b) $A = \begin{bmatrix} 5 & -2 & 4 \\ 0 & 2 & 1 \\ -1 & 4 & -4 \end{bmatrix}$

Solution

(a) Begin by entering the matrix and then evaluate the determinant, as shown in Figures 4.38 and 4.39. The result is det $A = -6$, which agrees with our earlier calculation.

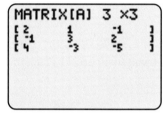

Figure 4.38   Figure 4.39

(b) The determinant of A evaluates to -50 (see Figures 4.40 and 4.41).

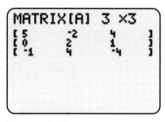

Figure 4.40   Figure 4.41

AREA OF REGIONS

A determinant may be used to find the area of a triangle. For example, if a triangle has vertices (a_1, a_2), (b_1, b_2), and (c_1, c_2), its area equals the absolute value of D, where

$$D = \frac{1}{2} \det \begin{bmatrix} a_1 & b_1 & c_1 \\ a_2 & b_2 & c_2 \\ 1 & 1 & 1 \end{bmatrix}.$$

If the vertices are entered in the columns of D counterclockwise, D will be positive.
(*Source:* W. Taylor, *The Geometry of Computer Graphics.*)

EXAMPLE 4 *Computing the area of a triangular parcel of land*

A triangular parcel of land is shown in Figure 4.42. If all units are miles, find the area of the parcel of land by using a determinant.

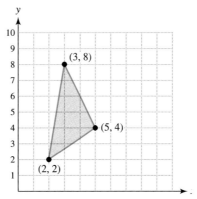

Figure 4.42

Solution

The vertices of the triangular parcel of land are (2, 2), (5, 4), and (3, 8). The area of the triangle is

$$D = \frac{1}{2}\det\begin{bmatrix} 2 & 5 & 3 \\ 2 & 4 & 8 \\ 1 & 1 & 1 \end{bmatrix} = \frac{1}{2} \cdot 16 = 8.$$

The area of the triangle is 8 square miles. This value, as obtained on a graphing calculator, is shown in Figure 4.43.

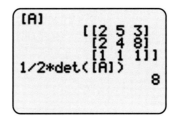

Figure 4.43

Critical Thinking

Suppose that you are given three distinct vertices and $D = 0$. What must be true about the three points?

CRAMER'S RULE

Determinants were developed independently by Gabriel Cramer (1704–1752). His work was published in 1750 and provided a method called **Cramer's rule** for solving systems of linear equations.

Cramer's Rule for Linear Systems in Two Variables

The solution to the linear system

$$a_1x + b_1y = c_1$$
$$a_2x + b_2y = c_2$$

is given by $x = \dfrac{E}{D}$ and $y = \dfrac{F}{D}$, where

$$E = \det\begin{bmatrix} c_1 & b_1 \\ c_2 & b_2 \end{bmatrix}, \quad F = \det\begin{bmatrix} a_1 & c_1 \\ a_2 & c_2 \end{bmatrix}, \quad \text{and} \quad D = \det\begin{bmatrix} a_1 & b_1 \\ a_2 & b_2 \end{bmatrix} \neq 0.$$

Note: If $D = 0$, the system has either no solutions or infinitely many solutions.

EXAMPLE 5 Using Cramer's rule

Use Cramer's rule to solve the following linear systems.

(a) $3x - 4y = 18$
 $7x + 5y = -1$

(b) $-4x + 9y = -24$
 $6x + 17y = -25$

Solution

(a) $E = \det\begin{bmatrix} c_1 & b_1 \\ c_2 & b_2 \end{bmatrix} = \det\begin{bmatrix} 18 & -4 \\ -1 & 5 \end{bmatrix} = (18)(5) - (-1)(-4) = 86$

$F = \det\begin{bmatrix} a_1 & c_1 \\ a_2 & c_2 \end{bmatrix} = \det\begin{bmatrix} 3 & 18 \\ 7 & -1 \end{bmatrix} = (3)(-1) - (7)(18) = -129$

$D = \det\begin{bmatrix} a_1 & b_1 \\ a_2 & b_2 \end{bmatrix} = \det\begin{bmatrix} 3 & -4 \\ 7 & 5 \end{bmatrix} = (3)(5) - (7)(-4) = 43$

The solution is $x = \dfrac{E}{D} = \dfrac{86}{43} = 2$ and $y = \dfrac{F}{D} = \dfrac{-129}{43} = -3$.

(b) $E = \det\begin{bmatrix} c_1 & b_1 \\ c_2 & b_2 \end{bmatrix} = \det\begin{bmatrix} -24 & 9 \\ -25 & 17 \end{bmatrix} = (-24)(17) - (-25)(9) = -183$

$F = \det\begin{bmatrix} a_1 & c_1 \\ a_2 & c_2 \end{bmatrix} = \det\begin{bmatrix} -4 & -24 \\ 6 & -25 \end{bmatrix} = (-4)(-25) - (6)(-24) = 244$

$D = \det\begin{bmatrix} a_1 & b_1 \\ a_2 & b_2 \end{bmatrix} = \det\begin{bmatrix} -4 & 9 \\ 6 & 17 \end{bmatrix} = (-4)(17) - (6)(9) = -122$

The solution is $x = \dfrac{E}{D} = \dfrac{-183}{-122} = 1.5$ and $y = \dfrac{F}{D} = \dfrac{244}{-122} = -2$.

USING CRAMER'S RULE IN REAL-WORLD APPLICATIONS. In real-world applications equations with hundreds of variables are routinely solved. Such systems of equations could be solved with Cramer's rule. However, using Cramer's rule and the expansion of a determinant with minors to solve a linear system of equation with n variables requires at least

$$1 \cdot 2 \cdot 3 \cdot 4 \cdot \cdots \cdot n \cdot (n+1)$$

multiplication operations. To solve a linear system involving only 25 variables would require about

$$1 \cdot 2 \cdot 3 \cdot 4 \cdot \cdots \cdot 25 \cdot 26 \approx 4 \times 10^{26}$$

multiplication operations. The fastest supercomputers can perform about 1 trillion (1×10^{12}) multiplication operations per second. This would take about

$$\frac{4 \times 10^{26}}{1 \times 10^{12}} = 4 \times 10^{14} \text{ seconds.}$$

With $60 \times 60 \times 24 \times 365 = 31{,}536{,}000$ seconds in a year, 4×10^{14} seconds equals

$$\frac{4 \times 10^{14}}{31{,}536{,}000} \approx 12{,}700{,}000 \text{ years!}$$

Modern software packages do *not* use Cramer's rule for three or more variables.

4.6 Putting It All Together

The determinant of a square matrix A is a real number, denoted det A. The computation of a determinant frequently involves a large number of arithmetic operations that can be done on a graphing calculator. Cramer's rule is a method that uses determinants to solve systems of linear equations. The following table summarizes important topics from this section.

Concept	Explanation
Determinant of a 2×2 Matrix	The determinant of a 2×2 matrix A is given by $$\det A = \det \begin{bmatrix} a & b \\ c & d \end{bmatrix} = ad - cb.$$
Determinant of a 3×3 Matrix	The determinant of a 3×3 matrix A is given by $$\det A = \det \begin{bmatrix} a_1 & b_1 & c_1 \\ a_2 & b_2 & c_2 \\ a_3 & b_3 & c_3 \end{bmatrix}$$ $$= a_1 \cdot \det \begin{bmatrix} b_2 & c_2 \\ b_3 & c_3 \end{bmatrix} - a_2 \cdot \det \begin{bmatrix} b_1 & c_1 \\ b_3 & c_3 \end{bmatrix}$$ $$+ a_3 \cdot \det \begin{bmatrix} b_1 & c_1 \\ b_2 & c_2 \end{bmatrix}$$

4.6 Determinants

Concept	Explanation
Area of a Triangle	If a triangle has vertices (a_1, a_2), (b_1, b_2), and (c_1, c_2), its area equals the absolute value of D, where $$D = \frac{1}{2}\det\begin{bmatrix} a_1 & b_1 & c_1 \\ a_2 & b_2 & c_2 \\ 1 & 1 & 1 \end{bmatrix}.$$
Cramer's Rule for Linear Systems in Two Variables	The solution to the linear system $$a_1 x + b_1 y = c_1$$ $$a_2 x + b_2 y = c_2$$ is given by $x = \dfrac{E}{D}$ and $y = \dfrac{F}{D}$, where $$E = \det\begin{bmatrix} c_1 & b_1 \\ c_2 & b_2 \end{bmatrix}, \quad F = \det\begin{bmatrix} a_1 & c_1 \\ a_2 & c_2 \end{bmatrix}, \text{ and}$$ $$D = \det\begin{bmatrix} a_1 & b_1 \\ a_2 & b_2 \end{bmatrix} \neq 0.$$ *Note:* If $D = 0$ then the system has either no solutions or infinitely many solutions.

4.6 EXERCISES

CALCULATING DETERMINANTS

Exercises 1–16: Evaluate det A by hand.

1. $A = \begin{bmatrix} 1 & -2 \\ 3 & -8 \end{bmatrix}$

2. $A = \begin{bmatrix} 5 & -1 \\ 3 & 7 \end{bmatrix}$

3. $A = \begin{bmatrix} -3 & 7 \\ 8 & -1 \end{bmatrix}$

4. $A = \begin{bmatrix} 0 & -7 \\ -3 & 1 \end{bmatrix}$

5. $A = \begin{bmatrix} 23 & 4 \\ 6 & -13 \end{bmatrix}$

6. $A = \begin{bmatrix} 44 & -51 \\ -9 & 32 \end{bmatrix}$

7. $A = \begin{bmatrix} 1 & -1 & 2 \\ 0 & 1 & -3 \\ 0 & -4 & 7 \end{bmatrix}$

8. $A = \begin{bmatrix} 2 & -1 & -5 \\ -1 & 4 & -2 \\ 0 & 1 & 4 \end{bmatrix}$

9. $A = \begin{bmatrix} 2 & -1 & 0 \\ 1 & -2 & 6 \\ 0 & 1 & 8 \end{bmatrix}$

10. $A = \begin{bmatrix} 0 & 1 & -4 \\ 3 & -6 & 10 \\ 4 & -2 & 7 \end{bmatrix}$

11. $A = \begin{bmatrix} -1 & 3 & 5 \\ 3 & -3 & 5 \\ 2 & -3 & 7 \end{bmatrix}$

12. $A = \begin{bmatrix} 6 & -1 & 9 \\ 7 & 0 & -3 \\ 2 & 5 & -1 \end{bmatrix}$

13. $A = \begin{bmatrix} 5 & 0 & 0 \\ 0 & -2 & 0 \\ 0 & 0 & 5 \end{bmatrix}$

14. $A = \begin{bmatrix} 1 & 2 & 3 \\ 2 & 4 & 6 \\ 3 & 6 & 9 \end{bmatrix}$

15. $A = \begin{bmatrix} 0 & 2 & -3 \\ 0 & 3 & -9 \\ 0 & 5 & 9 \end{bmatrix}$

16. $A = \begin{bmatrix} 3 & -1 & 2 \\ 0 & 5 & 7 \\ 0 & 0 & -1 \end{bmatrix}$

Exercises 17–20: Use technology to calculate det A.

17. $A = \begin{bmatrix} 2 & -5 & 13 \\ 10 & 15 & -10 \\ 17 & -19 & 22 \end{bmatrix}$

18. $A = \begin{bmatrix} 1.6 & 3.1 & 5.7 \\ 2.1 & 6.7 & 8.1 \\ -0.4 & -0.8 & -3.1 \end{bmatrix}$

19. $A = \begin{bmatrix} 17 & 0 & 4 \\ -9 & 14 & 1.5 \\ 13 & 67 & -11 \end{bmatrix}$

20. $A = \begin{bmatrix} 121 & 45 & -56 \\ -45 & 87 & 32 \\ -14 & -34 & 67 \end{bmatrix}$

CALCULATING AREA

Exercises 21–26: Find the area of the figure by using a determinant. Assume that units are feet.

21.
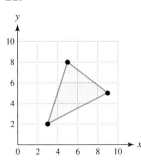

22.
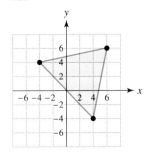

23.
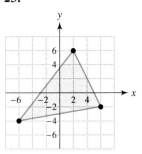

24.
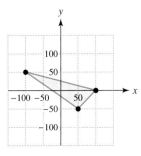

25.
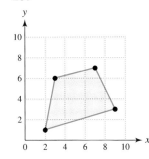

26.
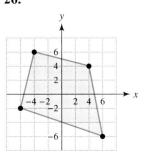

CRAMER'S RULE

Exercises 27–32: Solve the system of equations by using Cramer's rule.

27. $5x + 3y = 4$
    $6x - 4y = 20$

28. $-5x + 4y = -5$
    $4x + 4y = -32$

29. $7x - 5y = -3$
    $-4x + 6y = -8$

30. $-4x - 9y = -17$
    $8x + 4y = 9$

31. $8x = 3y - 61$
    $-x = 4y - 23$

32. $15y = -188 - 22x$
    $23y = -173 - 16x$

WRITING ABOUT MATHEMATICS

33. Suppose that the first column of a 3×3 matrix A is all 0s. What is the value of det A? Give an example and explain your answer.

34. Estimate how long it might take a supercomputer to solve a linear system of equations with 15 variables, using Cramer's rule and expansion of determinant by minors. Explain your calculations. (*Hint:* See the discussion at the end of this section.)

CHECKING BASIC CONCEPTS FOR SECTIONS 4.5 AND 4.6

1. Use an augmented matrix to represent the system of equations. Solve the system using
   (a) Gaussian elimination and
   (b) technology.

   $$x + 2y + z = 1$$
   $$x + y + z = -1$$
   $$y + z = 1$$

2. Evaluate each determinant by hand. Check your result with a graphing calculator.

   (a) $A = \begin{bmatrix} -3 & 4 \\ -2 & 3 \end{bmatrix}$

   (b) $A = \begin{bmatrix} 1 & -2 & 3 \\ 5 & 1 & 1 \\ 0 & 2 & -1 \end{bmatrix}$

3. Use Cramer's rule to solve the system of equations.

   $$2x - y = -14$$
   $$3x - 4y = -36$$

4. Find the area of a triangle with vertices $(-1, 2)$, $(5, 6)$, and $(2, -3)$.

Chapter 4 Summary

Section 4.1 Systems of Linear Equations in Two Variables

The following linear system has two equations and two variables.

$$x - 2y = -7$$
$$5x - y = 1$$

Its solution is $(1, 4)$ because, if we substitute $x = 1$ and $y = 4$, both equations are true.

A linear system of equations can have no solutions (inconsistent), one solution (unique solution), or infinitely many solutions (dependent). These situations are illustrated graphically in the accompanying figures.

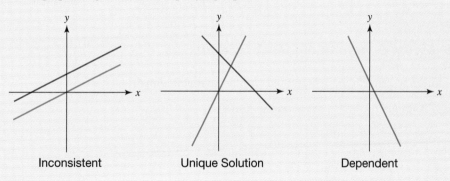

Inconsistent   Unique Solution   Dependent

Section 4.2 The Substitution and Elimination Methods

Systems with two linear equations and two variables may be solved graphically, numerically, or symbolically. Two symbolic methods are *substitution* and *elimination*.

Substitution

$x + y = 5$ or $y = 5 - x$
$x - y = -3$
Substituting into the second equation,
$x - (5 - x) = -3$, yields $x = 1$
and $y = 4$.

Elimination

$x + y = 5$
$x - y = -3$
$\overline{2x = 2}$ Add the equations.
Thus $x = 1$ and $y = 4$.

Section 4.3 Systems of Linear Inequalities

Linear inequalities involving two variables can be solved graphically in the xy-plane. Any point (x, y) in a shaded region must satisfy the given inequalities. The test point $(0, 0)$ lies in the shaded region and satisfies the inequality $x + y \leq 4$ because $0 + 0 \leq 4$.

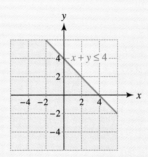

Section 4.4 Systems of Equations in Three Variables

Systems of linear equations with three equations in three variables can also be solved by substitution and elimination.

$$x - y + 2z = 3$$
$$2x - y + z = 5$$
$$x + y + z = 6$$

Its solution is $(3, 2, 1)$ because these values for (x, y, z) satisfy all three equations.

Section 4.5 Matrix Solutions of Linear Systems

Any linear system can be represented with an augmented matrix.

Linear System

$4x - 3y = 5$
$x + 2y = 4$

Augmented Matrix

$$\begin{bmatrix} 4 & -3 & | & 5 \\ 1 & 2 & | & 4 \end{bmatrix}$$

An efficient way to solve linear systems is to use Gaussian elimination. In Gaussian elimination, matrix row operations are used to transform an augmented matrix to reduced row–echelon form.

Section 4.6 Determinants

A determinant is a *real number*. The determinant of a 2 × 2 matrix is

$$\det A = \det \begin{bmatrix} a & b \\ c & d \end{bmatrix} = ad - cb.$$

Determinants of larger square matrices can be found either by hand or with a graphing calculator. Cramer's rule uses determinants to solve linear systems of equations. Determinants can also be used to find areas of triangles.

Chapter 4

Review Exercises

Section 4.1

Exercises 1 and 2: Decide which of the following is a solution.

1. $(2, -1), (3, 2)$
$$3x - 2y = 5$$
$$-2x + 4y = 2$$

2. $(4, -3), (-1, 2)$
$$x - 5y = 19$$
$$4x + 3y = 7$$

3. A system of linear equations has been solved graphically. Use the graph to find any solutions.

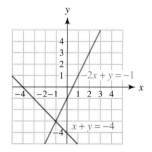

4. A system of linear equations has been solved numerically. Use the table to solve the system.
$$Y_1 = 3X - 2, Y_2 = X + 5$$

X	Y₁	Y₂
1	1	6
1.5	2.5	6.5
2	4	7
2.5	5.5	7.5
3	7	8
3.5	8.5	8.5
4	10	9

X=1

Exercises 5–10: Determine graphically whether the system is dependent, inconsistent, or has a unique solution. If possible, give any solutions.

5. $x + y = 6$
$x - y = -4$

6. $x - y = -2$
$-2x + 2y = 4$

7. $4x + 2y = 1$
$2x + y = 5$

8. $x - 3y = 5$
$x + 5y = -3$

9. $4x + 2y = 0$
$-x + 3y = 3.5$

10. $4x - 7y = -15$
$6x - 5y = -6$

Exercises 11 and 12: Solve the system of linear equations numerically.

11. $3x + y = 7$
    $6x + y = 16$

12. $4x + 2y = -6$
    $3x - y = -7$

Exercises 13 and 14: Complete the following.
(a) Write a system that models the situation.
(b) Solve the resulting system graphically or numerically.

13. The sum of two numbers is 25 and their difference is 10. Find the two numbers.

14. Three times a number minus two times another number equals 19. The sum of the two numbers is 18. Find the two numbers.

Section 4.2

Exercises 15 and 16: Use substitution to solve the system of equations.

15. $2x + 5y = -1$
    $x + 2y = -1$

16. $3x + y = 6$
    $4x + 5y = 8$

Exercises 17–20: Use elimination to solve the system of equations, if possible.

17. $3x + y = 4$
    $2x - y = -2$

18. $2x + 3y = -13$
    $3x - 2y = 0$

19. $3x - y = 5$
    $-6x + 2y = -10$

20. $8x - 6y = 7$
    $-4x + 3y = 11$

Section 4.3

Exercises 21–28: Shade the solution set in the xy-plane. Use a test point to check your graph.

21. $y \geq 2$

22. $y < 2x - 3$

23. $2x - y \leq 4$

24. $-x + 3y > 3$

25. $y - x \geq 1$
    $y \leq 2$

26. $-x + y \leq 3$
    $3x + 2y \geq 6$

27. $y > x - 1$
    $y < 4 - 3x$

28. $x + y \geq 5$
    $2x - 3 < 6$

Exercises 29 and 30: Use the graph to write the system of inequalities.

29.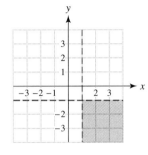

30.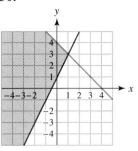

Section 4.4

31. Is $(3, -4, 5)$ a solution for $x + y + z = 4$?

32. Decide which is a solution: $(1, -1, 2), (1, 0, 5)$.
    $2x - 3y + z = 7$
    $-x - y + 3z = 6$
    $3x - 2y + z = 7$

Exercises 33–36: Use elimination and substitution to solve the system of linear equations.

33. $x - y - 2z = -11$
    $-x + 2y + 3z = 16$
    $3z = 6$

34. $x + y = 4$
    $-2x + y + 3z = -2$
    $x - 2y + 5z = -26$

35. $2x - y = -5$
    $x + 2y + z = 7$
    $-2x + y + z = 7$

36. $2x + 3y + z = 6$
    $-x + 2y + 2z = 3$
    $x + y + 2z = 4$

Section 4.5

Exercises 37–40: Write the system of linear equations as an augmented matrix. Then use Gaussian elimination to solve the system, writing the solution as an ordered triple. Check your solution.

37. $x + y + z = -6$
$x + 2y + z = -8$
$y + z = -5$

38. $x + y + z = -3$
$-x + y = 5$
$y + z = -1$

39. $x + 2y - z = 1$
$-x + y - 2z = 5$
$2y + z = 10$

40. $2x + 2y - 2z = -14$
$-2x - 3y + 2z = 12$
$x + y - 4z = -22$

Exercises 41–44: **Technology** *Use a graphing calculator to solve the system of linear equations.*

41. $3x - 2y + 6z = -17$
$-2x - y + 5z = 20$
$4y + 7z = 30$

42. $2x + 4y + 3z = 9$
$x + 8y - 6z = -9$
$7x - 2y + 7z = -3$

43. $4x + 8z = y - 5$
$2x - 3y = 2z - 5$
$2x + 5y + 4z = 36$

44. $19x - 13y - 7z = 7.4$
$22x + 33y - 8z = 110.5$
$10x - 56y + 9z = 23.7$

Section 4.6

Exercises 45–48: Evaluate det A.

45. $A = \begin{bmatrix} 6 & -5 \\ -4 & 2 \end{bmatrix}$

46. $A = \begin{bmatrix} 0 & -6 \\ 5 & 9 \end{bmatrix}$

47. $A = \begin{bmatrix} 3 & -5 & -3 \\ 1 & 4 & 7 \\ 0 & -3 & 1 \end{bmatrix}$

48. $A = \begin{bmatrix} -2 & -1 & -7 \\ 2 & 1 & -3 \\ 3 & -5 & 8 \end{bmatrix}$

Exercises 49 and 50: Use technology to calculate det A.

49. $A = \begin{bmatrix} 22 & -45 & 3 \\ 15 & -12 & -93 \\ 5 & 81 & -21 \end{bmatrix}$

50. $A = \begin{bmatrix} 0.5 & -7.3 & 9.6 \\ 0.1 & 3.1 & 9.2 \\ -0.5 & -1.9 & 5.4 \end{bmatrix}$

Exercises 51 and 52: Find the area of the triangle by using a determinant. Assume that the units are feet.

51.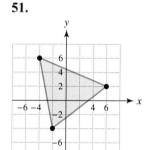

52.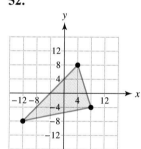

Exercises 53–56: Use Cramer's rule to solve each system.

53. $7x + 6y = 8$
$5x - 8y = 18$

54. $-2x + 5y = 25$
$3x + 4y = -3$

55. $3x - 6y = 1.5$
$7x - 5y = 8$

56. $-5x + 4y = -47$
$6x - 7y = 63$

Applications

57. *Pedestrian Fatalities* Forty-seven percent of pedestrian fatalities occur on Friday and Saturday nights. The combined total of pedestrian fatalities in 1988 and 1998 was 12,090. There were 1650 more fatalities in 1988 than in 1998. Find the number of pedestrian fatalities during each year. (*Source:* National Highway Traffic Safety Administration.)

58. Burning Calories During strenuous exercise, an athlete burns 11.5 calories per minute on a stair climber and 9 calories per minute on a stationary bicycle. In a 30-minute workout the athlete burns 290 calories. How many minutes were spent on each type of workout equipment? *(Source: Runner's World.)*

59. Pagers Today's pagers can send and receive e-mail, news, and other information. Because of their versatility, the number of paging subscribers has grown. In 1996 there were 42 million subscribers and in 1999 there were 58 million. *(Source: Strategis Group for Personal Communications Association.)*
(a) Use a linear system to find values for a and b so that the graph of $f(x) = ax + b$ passes through (1996, 42) and (1999, 58).
(b) Use $f(x)$ to estimate the number of subscribers in 1998. Compare it to the actual value of 53 million.

60. Mixing Antifreeze A car radiator should contain 4 gallons of fluid that is 40% antifreeze. An auto mechanic has a 30% solution of antifreeze and a 55% solution of antifreeze. If the car radiator is empty, how many gallons of each solution should be added?

61. Boat Speed A boat travels 18 miles down a river in 1 hour. The return trip against the current takes 1.5 hours. Find the average speed of the boat and the average speed of the current.

62. Tickets Tickets for a football game sold for $8 and $12. If 480 tickets were sold for total receipts of $4620, how many of each type of ticket were sold? Find the answer
(a) symbolically,
(b) graphically, and
(c) numerically.

63. Determining Costs The accompanying table shows the costs for purchasing different combinations of malts, cones, and ice cream bars.

Malts	Cones	Bars	Total Cost
1	3	5	$14
1	2	4	$11
0	1	3	$5

(a) Let x be the cost of a malt, y the cost of a cone, and z the cost of an ice cream bar. Write a system of three linear equations that represents the data in the table.
(b) Solve this system.

64. Geometry The largest angle in a triangle is 20° more than the sum of the two smaller angles. The measure of the largest angle is 85° more than the smallest angle. Find the measure of each angle in the triangle.

65. Mixture Problem Three different types of candy that cost $1.50, $2.00, and $2.50 per pound are to be mixed to produce 12 pounds of candy worth $26.00. If there is to be 2 pounds more of the $2.50 candy than the $2.00 candy, how much of each type of candy should be used in the mixture?

66. Estimating the Chest Size of a Bear The accompanying table shows the chest size C, weight W, and overall length L of three bears. These data can be modeled with the formula $C = a + bW + cL$. *(Source: M. Triola, Elementary Statistics; Minitab, Inc.)*

C (inches)	W (pounds)	L (inches)
40	202	63
50	365	70
55	446	77

(a) Set up a system of linear equations whose solution gives values for the constants a, b, and c.
(b) Solve this system with a graphing calculator. Round each value to the nearest thousandth.
(c) Predict the chest size of a bear weighing 300 pounds and having a length of 68 inches.

Chapter 4
Test

Exercises 1 and 2: Determine graphically whether the system is dependent, inconsistent, or has a unique solution. If possible, give any solutions.

1. $2x + y = \frac{3}{2}$
   $3x - 2y = \frac{15}{2}$

2. $8x - 4y = 3$
   $-4x + 2y = 6$

3. Solve the system of equations using substitution. Then support your answer numerically.
   $2x + 5y = -1$
   $3x + 2y = -7$

4. The difference of two numbers is 34. The first number is twice the second number.
   (a) Write a system of linear equations that models the situation.
   (b) Solve the system using elimination.
   (c) Support your results in part (b) graphically.

5. Shade the solution set in the xy-plane. Use a test point to check your graph.
   $-2x + y \geq 3$
   $x - 2y < -3$

6. Use elimination and substitution to solve the system of linear equations.
   $x + 3y = 2$
   $-2x + y + z = 5$
   $y + z = -3$

7. Use the graph to write the system of inequalities.

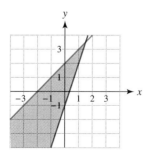

8. Consider the system of linear equations.
   $x + y + z = 2$
   $x - y - z = 3$
   $2x + 2y + z = 6$

   (a) Write the system of linear equations as an augmented matrix.
   (b) Use Gaussian elimination to solve the system, writing the solution as an ordered triple.
   (c) Check your solution with a graphing calculator.

9. Evaluate det A if $A = \begin{bmatrix} 3 & 2 & -1 \\ 6 & 2 & -6 \\ 0 & 8 & -3 \end{bmatrix}$.

10. Solve the system of linear equations with Cramer's rule.
    $5x - 3y = 7$
    $-4x + 2y = 11$

11. *College Tuition* In 1998, the average cost of private tuition, including room and board, was $11,500 more than comparable public tuition. Private tuition was 2.5 times higher than public tuition. (*Source:* **Department of Education.**)
    (a) Write a system of linear equations that models the situation.
    (b) Solve the system and interpret the solution.

12. *Burning Calories* An athlete burns 12 calories per minute while running and 9 calories per minute on a rowing machine. During a one-hour workout, the athlete burns 669 calories. How many minutes were spent on each type of exercise?

13. *Airplane Speed* An airplane travels 600 miles into the wind in 2.5 hours. The return trip with the wind takes 2 hours. Find the average ground speed of the airplane and the average wind speed.

14. *Geometry* The largest angle in a triangle is 50° more than the smallest angle. The sum of the measures of the smaller two angles is 10° more than the largest angle. Find the measure of each angle in the triangle.

Chapter 4

Extended and Discovery Exercises

Analysis of Traffic Flow

Exercises 1 and 2: Mathematics is frequently used to analyze traffic for timing of traffic lights. The following figure shows three one-way streets with intersections A, B, and C. The average number of cars traveling on certain portions of the streets is given in cars per minute. The variables x, y, and z represent unknown traffic flows that need to be determined. The number of vehicles entering an intersection must equal the number of vehicles exiting an intersection.

(a) Verify that the accompanying system of linear equations describes the traffic flow at each of three intersections.

(b) Express the system using an augmented matrix.

(c) Solve the system and interpret your solution.

1. A: $8 + 12 = y + x$
   B: $z + 4 = 8 + 5$
   C: $y + 6 = z + 9$

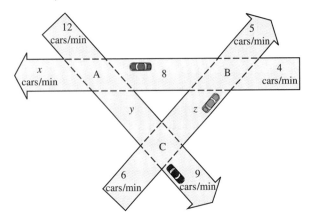

2. A: $6 + 3 = y + x$
   B: $z + 10 = 6 + 9$
   C: $y + 7 = z + 5$

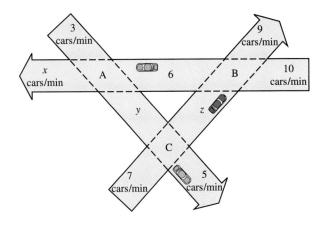

Matrices and Road Maps

Exercises 3–6: Adjacency Matrix A matrix A can be used to represent a map showing distances between cities. Let a_{ij} denote the number in row i and column j of a matrix A. Now consider the following map illustrating freeway distances in miles between four cities. Each city has been assigned a number. For example, there is a direct route from Denver, Colorado (city 1) to Colorado Springs, Colorado (city 2) of approximately 60 miles. Therefore $a_{12} = 60$ in the accompanying matrix A. (Note that a_{12} is the number in row 1 and column 2.) The distance from Colorado Springs to Denver is also 60 miles, so $a_{21} = 60$. As there is no direct freeway connection between Las Vegas, Nevada (city 4) and Colorado Springs (city 2), we let $a_{24} = a_{42} = *$. The matrix A is called an **adjacency matrix**. (Source: Baase, S., *Computer Algorithms Introduction to Design and Analysis*.)

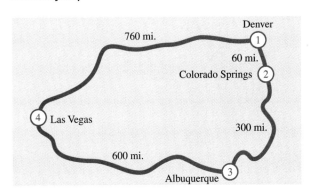

$$A = \begin{bmatrix} 0 & 60 & * & 760 \\ 60 & 0 & 300 & * \\ * & 300 & 0 & 600 \\ 760 & * & 600 & 0 \end{bmatrix}$$

3. Explain how to use A to find the freeway distance from Denver to Las Vegas.

4. Explain how to use A to find the freeway distance from Denver to Albuquerque.

5. If a map shows 20 cities, what would be the dimension of the adjacency matrix? How many elements would there be in this matrix?

6. Why are there only zeros on the main diagonal of A?

Exercises 7 and 8: (Refer to Exercises 3–6.) Determine an adjacency matrix A for the road map.

7.

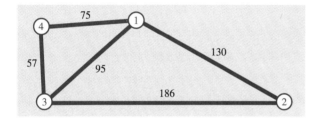

8.

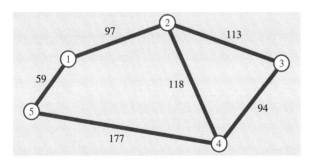

Exercises 9 and 10: (Refer to Exercises 3–6.) Sketch a road map represented by the adjacency matrix A. Is your answer unique?

9. $A = \begin{bmatrix} 0 & 30 & 20 & 5 \\ 30 & 0 & 15 & * \\ 20 & 15 & 0 & 25 \\ 5 & * & 25 & 0 \end{bmatrix}$

10. $A = \begin{bmatrix} 0 & 5 & * & 13 & 20 \\ 5 & 0 & 5 & * & * \\ * & 5 & 0 & 13 & * \\ 13 & * & 13 & 0 & 10 \\ 20 & * & * & 10 & 0 \end{bmatrix}$

SOLVING AN EQUATION IN FOUR VARIABLES

11. *Weight of a Bear* In Section 4.5 we estimated the weight of a bear by using two variables. We may be able to make more accurate estimates by using three variables. The accompanying table shows the weight W, neck size N, overall length L, and chest size C for four bears. (**Source:** M. Triola, *Elementary Statistics*; Minitab, Inc.)

W (pounds)	N (inches)	L (inches)	C (inches)
125	19	57.5	32
316	26	65	42
436	30	72	48
514	30.5	75	54

(a) Model these data with the equation $W = a + bN + cL + dC$, where a, b, c, and d are constants. To do so, represent a system of linear equations by a 4×5 augmented matrix whose solution gives values for a, b, c, and d.

(b) Solve the system with a graphing calculator. Round each value to the nearest thousandth.

(c) Predict the weight of a bear with $N = 24$, $L = 63$, and $C = 39$. Interpret the result.

CHAPTER 5
POLYNOMIAL EXPRESSIONS AND FUNCTIONS

Many times when data are plotted they do not lie on a line and, as a result, cannot be modeled with a linear function. For example, the following table of data and line graph show the pulse rate in beats per minute of an athlete x minutes after exercise has stopped. Note that the pulse rate decreases faster at first and then begins to level off, indicating that the data are nonlinear. To model these data accurately we can use a polynomial function that is nonlinear.

Time (minutes)	0	2	4	6	8
Pulse (beats per minute)	200	150	115	90	80

Source: V. Thomas, *Science and Sport*.

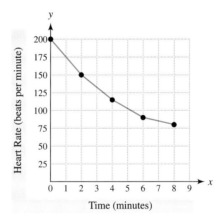

Generally, nonlinear functions are more complicated than linear functions. However, technology will help make our work easier.

The struggle is what teaches us.
—Sue Grafton

5.1 POLYNOMIAL EXPRESSIONS

Basic Concepts ~ **Addition and Subtraction of Monomials** ~
Addition and Subtraction of Polynomials ~
Evaluating Polynomial Expressions

INTRODUCTION

Many quantities in real life cannot be modeled with linear functions and equations. If data points do not lie on a (straight) line, we say the data are *nonlinear*. For example, a scatterplot of the cumulative number of AIDS deaths from 1981 through 1997 is shown in Figure 5.1. Polynomials are often used to model nonlinear data such as these. Before modeling nonlinear data, we must first discuss basic concepts of monomials and polynomials.

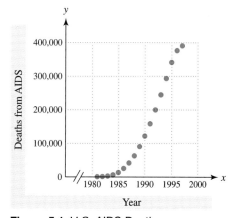

Figure 5.1 U.S. AIDS Deaths

BASIC CONCEPTS

A **term** is a number, a variable, or a *product* of numbers and variables raised to powers. Examples of terms include

$$-15, \quad y, \quad x^4, \quad 3x^3z, \quad x^{-1/2}y^{-2}, \quad \text{and} \quad 6x^{-1}y^3.$$

If the variables in a term have only *nonnegative integer* exponents, the term is called a **monomial.** Examples of monomials include

$$-4, \quad 5y, \quad x^2, \quad 5x^2z^6, \quad -xy^7, \quad \text{and} \quad 6xy^3.$$

EXAMPLE 1 *Identifying monomials*

Determine whether the expression is a monomial.

(a) $-8x^3y^5$ **(b)** xy^{-1} **(c)** $9 + x^2$

Solution

(a) The expression $-8x^3y^5$ represents a monomial because it is a product of the number -8 and the variables x and y, both of which have nonnegative integer exponents.

(b) The expression xy^{-1} is not a monomial because y^{-1} has a negative exponent.

(c) The expression $9 + x^2$ is not a monomial because it is the sum of two monomials, 9 and x^2.

Monomials occur in applications involving geometry and digital pictures, as illustrated in the next two examples.

EXAMPLE 2 *Writing monomials*

Write a monomial that represents the volume of a cube with sides of length x. Make a sketch of your result.

Solution

The volume of a rectangular box equals the product of its length, width, and height. For a cube all sides have the same length. If this length is x, its volume is given by $x \cdot x \cdot x = x^3$. See Figure 5.2.

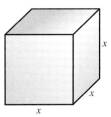

Figure 5.2 Volume of Cube: x^3

Critical Thinking

Write a monomial that represents the volume of four identical cubes with sides of length y.

Digital cameras are a new technology that are rapidly replacing cameras that rely on film. Digital pictures have been used by NASA for decades because they can be easily transmitted over large distances without loss of clarity. Today, digital pictures are ideal for use in computers and on the Internet.

EXAMPLE 3 *Counting pixels in digital pictures*

Digital pictures are made up of tiny rectangles called *pixels*. Individual pixels can be seen on a graphing calculator screen. The number of pixels used in making a picture varies by make and model of digital cameras. Pictures taken by a digital camera are often described in terms of pixels.

(a) Write a monomial that gives the number of pixels in an image that is x pixels wide and y pixels high.

(b) The image of an astronaut working on the Hubble telescope shown in Figure 5.3 is 356 by 480 pixels. How many pixels make up this image?

Figure 5.3 Digital Picture (*Source:* NASA)

Solution

(a) The digital picture of the letter F shown in Figure 5.4 is 4 pixels wide and 5 pixels high. It contains a total of $4 \times 5 = 20$ pixels. Therefore a digital picture that is x pixels wide and y pixels high contains xy pixels.

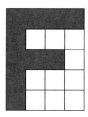

Figure 5.4
Digital Letter F

(b) The image shown in Figure 5.3 is $x = 356$ pixels wide and $y = 480$ pixels high. Therefore the image contains $xy = 356 \cdot 480 = 170{,}880$ pixels.

ADDITION AND SUBTRACTION OF MONOMIALS

Suppose that we have 3 rectangles of the same dimension having length x and width y, as shown in Figure 5.5 on the next page. The total area is given by

$$xy + xy + xy.$$

This area is equivalent to 3 times xy, which can be expressed as $3xy$. In symbols we write

$$xy + xy + xy = 3xy.$$

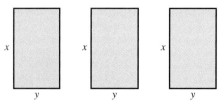

Figure 5.5 Total Area: 3xy

We can add these three terms because they are like terms. If two terms contain the same variables raised to the same powers, we call them **like terms.** We can add or subtract *like* terms, but not *unlike* terms. For example, if one cube has sides of length x, and another cube has sides of length y, their respective volumes are x^3 and y^3. The total volume of the two cubes equals

$$x^3 + y^3,$$

but we cannot combine these terms into one term because they are unlike terms. However,

$$x^3 + 3x^3 = (1 + 3)x^3 = 4x^3$$

because x^3 and $3x^3$ are like terms. To add or subtract monomials we simply combine like terms, as illustrated in the next example.

EXAMPLE 4 *Adding and subtracting monomials*

Simplify each of the following expressions by combining like terms.

(a) $8x^2 - 4x^2 + x^3$ **(b)** $9x - 6xy^2 + 2xy^2 + 4x$

Solution

(a) The terms $8x^2$ and $-4x^2$ are like terms, so they may be combined.

$$8x^2 - 4x^2 + x^3 = (8 - 4)x^2 + x^3 \qquad \text{Combine like terms.}$$
$$= 4x^2 + x^3 \qquad \text{Subtract.}$$

However, $4x^2$ and x^3 are unlike terms and cannot be combined.

(b) The terms $9x$ and $4x$ may be combined, as can $-6xy^2$ and $2xy^2$.

$$9x - 6xy^2 + 2xy^2 + 4x = 9x + 4x - 6xy^2 + 2xy^2 \qquad \text{Commutative property}$$
$$= (9 + 4)x + (-6 + 2)xy^2 \qquad \text{Combine like terms.}$$
$$= 13x - 4xy^2 \qquad \text{Add.}$$

The **degree of a monomial** equals the sum of the exponents of the variables. A constant term has degree 0, unless the term is 0 (which has no degree). The numeric constant in a monomial is called its **coefficient**. Table 5.1 shows the degree and coefficient of several monomials.

TABLE 5.1

Monomial	64	$4x^2y^3$	$-5x^2$	xy^3
Degree	0	5	2	4
Coefficient	64	4	−5	1

ADDITION AND SUBTRACTION OF POLYNOMIALS

A **polynomial** is either a monomial or a sum of monomials. Examples of polynomials include

$5x^4z^2$, $9x^4 - 5$, $4x^2 + 5xy - y^2$,   and $4 - y^2 + 5y^4 + y^5$

1 term   2 terms   3 terms   4 terms

Polynomials containing one variable are called **polynomials of one variable.** The second and fourth polynomials shown are examples of polynomials of one variable. The **leading coefficient** of a polynomial of one variable is the coefficient of the monomial with highest degree. The **degree of a polynomial** equals the degree of the monomial with highest degree. Table 5.2 shows several polynomials of one variable along with their degrees and leading coefficients. A polynomial of degree 1 is a **linear polynomial,** a polynomial of degree 2 is a **quadratic polynomial,** and a polynomial of degree 3 is a **cubic polynomial.**

TABLE 5.2

Polynomial	Degree	Leading Coefficient
-98	0	-98
$2x - 7$	1	2
$-5z + 9z^2 + 7$	2	9
$-2x^3 + 4x^2 + x - 1$	3	-2
$7 - x + 4x^2 + x^5$	5	1

To add two polynomials we combine like terms.

EXAMPLE 5 Adding polynomials

Simplify each expression.

(a) $(2x^2 - 3x + 7) + (3x^2 + 4x - 2)$ **(b)** $(z^3 + 4z + 8) + (4z^2 - z + 6)$

Solution

(a) $(2x^2 - 3x + 7) + (3x^2 + 4x - 2) = 2x^2 + 3x^2 - 3x + 4x + 7 - 2$
$= (2 + 3)x^2 + (-3 + 4)x + (7 - 2)$
$= 5x^2 + x + 5$

(b) $(z^3 + 4z + 8) + (4z^2 - z + 6) = z^3 + 4z^2 + 4z - z + 8 + 6$
$= z^3 + 4z^2 + (4 - 1)z + (8 + 6)$
$= z^3 + 4z^2 + 3z + 14$

To subtract integers we add the first integer with the *additive inverse* or *opposite* of the second integer. For example, to evaluate $3 - 5$ we perform the following operations.

$3 - 5 = 3 + (-5)$ Add the opposite.
$\quad\quad\ = -2$ Simplify.

CHAPTER 5 Polynomial Expressions and Functions

Similarly, to subtract two polynomials we add the first polynomial and the opposite of the second polynomial. To find the **opposite of a polynomial,** we simply negate each term. Table 5.3 shows three polynomials and their opposites.

TABLE 5.3

Polynomial	Opposite
$9 - x$	$-9 + x$
$5x^2 + 4x - 1$	$-5x^2 - 4x + 1$
$-x^4 + 5x^3 - x^2 + 5x - 1$	$x^4 - 5x^3 + x^2 - 5x + 1$

EXAMPLE 6 *Subtracting polynomials*

Simplify.

(a) $(y^5 + 3y^3) - (-y^4 + 2y^3)$ **(b)** $(5x^3 + 9x^2 - 6) - (5x^3 - 4x^2 - 7)$

Solution

(a) The opposite of $(-y^4 + 2y^3)$ is $(y^4 - 2y^3)$.

$$(y^5 + 3y^3) - (-y^4 + 2y^3) = (y^5 + 3y^3) + (y^4 - 2y^3) \quad \text{Add opposite.}$$
$$= y^5 + y^4 + (3 - 2)y^3 \quad \text{Combine like terms.}$$
$$= y^5 + y^4 + y^3 \quad \text{Simplify.}$$

(b) The opposite of $(5x^3 - 4x^2 - 7)$ is $(-5x^3 + 4x^2 + 7)$

Add opposite. $\quad (5x^3 + 9x^2 - 6) - (5x^3 - 4x^2 - 7) = (5x^3 + 9x^2 - 6) + (-5x^3 + 4x^2 + 7)$
Combine like terms. $\quad = (5 - 5)x^3 + (9 + 4)x^2 + (-6 + 7)$
Simplify. $\quad = 0x^3 + 13x^2 + 1$
$0x^3 = 0 \quad = 13x^2 + 1$

Critical Thinking

Is the sum of two quadratic polynomials always a quadratic polynomial? Explain.

EVALUATING POLYNOMIAL EXPRESSIONS

Frequently, monomials and polynomials represent formulas that may be evaluated. This situation is illustrated in the next two examples.

EXAMPLE 7 *Writing and evaluating a polynomial*

Write a polynomial that represents the total volume of two identical boxes having square bases. Make a sketch to illustrate your formula. Find the total volume of the boxes if each base is 11 inches on a side and the height of each box is 5 inches.

Solution

Let x represent the length of the base and y the height of a side. The volume of one box is x^2y, as illustrated in Figure 5.6, and the volume of two boxes is $2x^2y$. To calculate their total volumes let $x = 11$ and $y = 5$ in the expression $2x^2y$. Then

$$2 \cdot 11^2 \cdot 5 = 1210 \text{ cubic inches.}$$

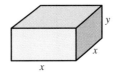

Figure 5.6 Volume: x^2y

Technology may be used to evaluate polynomials with more complicated formulas. This approach is demonstrated in the next example.

EXAMPLE 8 *Modeling growth of cellular phone use*

In the early years of cellular phone technology—from 1985 through 1991—the number of subscribers in thousands could be modeled by the polynomial $163x^2 - 146x + 205$, where $x = 0$ corresponds to 1985, $x = 1$ to 1986, and so on until $x = 6$ represents 1991. See Figure 5.7.

(a) Use the formula to find the number of subscribers in 1986.
(b) Construct a table for the formula starting at $x = 0$ and incrementing by 1 until $x = 6$.

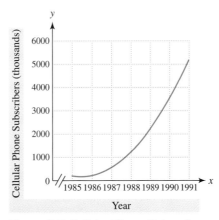

Figure 5.7 Cellular Phone Subscribers

Solution

(a) Because 1986 corresponds to $x = 1$, we substitute $x = 1$ into the polynomial, which gives

$$163(1)^2 - 146(1) + 205 = 17 + 205 = 222 \text{ thousand.}$$

(b) We construct the table for $Y_1 = 163X^2 - 146X + 205$, starting at $x = 0$ and incrementing by 1, as shown in Figures 5.8–5.10. The number of subscribers increased from 205 thousand in 1985 to 5197 thousand (or 5,197,000) in 1991.

Figure 5.8

Figure 5.9

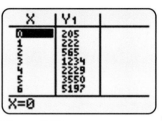

Figure 5.10

5.1 Putting It All Together

In this section we introduced monomials and polynomials. A monomial is a number or a product of numbers and variables, where each variable has a *nonnegative integer* exponent. A polynomial is a monomial or a sum of monomials. To add polynomials we combine like terms. To subtract polynomials we add the first polynomial with the opposite of the second polynomial. Examples are given in the following table.

Addition of Polynomials	$(-3x^2 + 2x - 7) + (4x^2 - x + 1) = -3x^2 + 4x^2 + 2x - x - 7 + 1$ $= (-3 + 4)x^2 + (2 - 1)x + (-7 + 1)$ $= x^2 + x - 6$ $(x^4 + 8x^3 - 7x) + (5x^4 - 3x) = x^4 + 5x^4 + 8x^3 - 7x - 3x$ $= (1 + 5)x^4 + 8x^3 + (-7 - 3)x$ $= 6x^4 + 8x^3 - 10x$
Opposite of a Polynomial	$-(5x^3 + 2x^2 - 3x + 6) = -5x^3 - 2x^2 + 3x - 6$ $-(-6x^6 + 4x^4 - 8x^2 - 17) = 6x^6 - 4x^4 + 8x^2 + 17$
Subtraction of Polynomials	$(5x^2 - 6x + 1) - (-5x^2 + 3x - 5) = (5x^2 - 6x + 1) + (5x^2 - 3x + 5)$ $= (5 + 5)x^2 + (-6 - 3)x + (1 + 5)$ $= 10x^2 - 9x + 6$ $(x^4 - 6x^2 + 5x) - (x^4 - 5x + 7) = (x^4 - 6x^2 + 5x) + (-x^4 + 5x - 7)$ $= (1 - 1)x^4 - 6x^2 + (5 + 5)x - 7$ $= -6x^2 + 10x - 7$
Evaluating a Polynomial	To evaluate the polynomial $-4x^2 + 3x - 1$ at $x = 2$, substitute 2 for x. $-4(2)^2 + 3(2) - 1 = -16 + 6 - 1 = -11$

5.1 EXERCISES

CONCEPTS

1. Give an example of a term that is a monomial.

2. Give an example of a term that is not a monomial.

3. What are the degree and leading coefficient of $3x^2 - x^3 + 1$?

4. Are $-5x^3y$ and $6xy^3$ like terms? Explain.

5. Give an example of a polynomial that has 3 terms and is degree 4.

6. Give an example of an expression that is not a polynomial.

7. Does the opposite of $x^2 + 1$ equal $-x^2 + 1$? Explain.

8. Evaluate $4x^3y$ when $x = 2$ and $y = 3$.

MONOMIALS

Exercises 9–16: Determine whether the expression is a monomial.

9. x^4
10. x^{-4}
11. $2x^2y + y^2$
12. $5 - x^{-1/2}$
13. $-4x^3y^3$
14. xy
15. $5x^{0.5}y^{-3}$
16. $\pi x^4 y^2 z$

Exercises 17–22: Write a monomial that represents the described quantity.

17. The area of a square with sides equal to x

18. The circumference of a circle with diameter d

19. The area of a circle with radius r

20. The area of three congruent triangles with base b and height h

21. The number of members in a marching band that has x rows with y people in each row

22. The revenue from selling w items for z dollars each

Exercises 23–28: Identify the degree and coefficient of the monomial.

23. $3x^7$
24. $-5y^3$
25. $-3x^2y^5$
26. xy^5
27. $-x^3y^3$
28. $\sqrt{2}xy$

Exercises 29–40: Combine like terms whenever possible.

29. $x^2 + 4x^2$
30. $-3z + 5z$
31. $6y^4 - 3y^4$
32. $9xy - 7xy$
33. $5x^2y + 8xy^2$
34. $5x + 4y$
35. $9x^2 - x + 4x - 6x^2$
36. $-xy^2 - \frac{1}{2}xy^2$
37. $x^2 + 9xy - y^2 + 4x^2 + y^2$
38. $6xy + 4x - 6xy$
39. $4x + 7x^3y^7 - \frac{1}{2}x^3y^7 + 9x - \frac{3}{2}x^3y^7$
40. $19x^3 + x^2 - 3x^3 + x - 4x^2 + 1$

POLYNOMIALS

Exercises 41–46: Identify the degree and leading coefficient of the polynomial.

41. $5x^2 - 4x + \frac{3}{4}$
42. $-9y^4 + y^2 + 5$
43. $5 - x + 3x^2 - \frac{2}{5}x^3$
44. $7x + 4x^4 - \frac{4}{3}x^3$
45. $8x^4 + 3x^3 - 4x + x^5$
46. $5x^2 - x^3 + 7x^4 + 10$

Exercises 47–54: Add the polynomials.

47. $(3x + 1) + (-x + 1)$

48. $(5y^3 + y) + (12y^3 - 5y)$

49. $(x^2 - 2x + 15) + (-3x^2 + 5x - 7)$

50. $(3x^3 - 4x + 3) + (5x^2 + 4x + 12)$

51. $(4x) + (1 - 4.5x)$

52. $(y^5 + y) + \left(5 - y + \frac{1}{3}y^2\right)$

53. $(x^4 - 3x^2 - 4) + \left(-8x^4 + x^2 - \frac{1}{2}\right)$

54. $(3z + z^4 + 2) + (-3z^4 - 5 + z^2)$

Exercises 55–60: Find the opposite of the polynomial.

55. $6x^5$

56. $-5y^7$

57. $19x^5 - 5x^3 + 3x$

58. $-x^2 - x - 5$

59. $-7z^4 + z^2 - 8$

60. $6 - 4x + 5x^2 - \frac{1}{10}x^3$

Exercises 61–68: Subtract the polynomials.

61. $(5x - 3) - (2x + 4)$

62. $(10x + 5) - (-6x - 4)$

63. $(x^2 - 3x + 1) - (-5x^2 + 2x - 4)$

64. $(-x^2 + x - 5) - (x^2 - x + 5)$

65. $(4x^4 + 2x^2 - 9) - (x^4 - 2x^2 - 5)$

66. $(8x^3 + 5x^2 - 3x + 1) - (-5x^3 + 6x - 11)$

67. $(x^4 - 1) - (4x^4 + 3x + 7)$

68. $(5x^4 - 6x^3 + x^2 + 5) - (x^3 + 11x^2 + 9x - 3)$

Exercises 69–72: Apply the distributive property and combine like terms.

69. $5x + 2(x - 5)$

70. $10x + 3(-2x + 2)$

71. $5x - 5(3x + 1)$

72. $5 - x - 4(-3x + 1)$

APPLICATIONS

73. *Women on the Run* The number of women participating in the New York marathon has increased dramatically in recent years. The polynomial $8.87x^2 + 232x + 769$ models the number of women running each year from 1978 through 1998, where $x = 0$ corresponds to 1978, $x = 10$ to 1988, and $x = 20$ to 1998. See the accompanying figure. (*Source: Runner's World.*)

(a) Use the graph to estimate the number of women running in 1980 and 1995.

(b) Answer part (a), using the polynomial. How do your answers compare with your answers in part (a)?

(c) Use the polynomial to find the increase in women runners from 1978 to 1998.

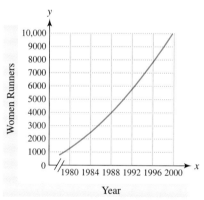

74. *Cars and Horsepower* Americans have traditionally favored big cars over small cars. The horsepower per 100 pounds of car weight from 1978 through 1998 is modeled by the polynomial $0.00258242x^2 - 10.19687x + 10,069.4$, where x represents the year. See the accompanying figure. (*Source: National Highway Traffic Safety Administration.*)

(a) Evaluate this polynomial at $x = 1980$ and 1996.

(b) Interpret your answers in part (a).

5.1 Polynomial Expressions

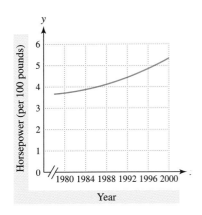

75. *A PC for All?* Worldwide sales of personal computers have climbed as prices continue to drop. The polynomial $0.7868x^2 + 12x + 79.5$ models the number of computers sold and expected to be sold in millions during year x, where $x = 0$ corresponds to 1997, $x = 1$ to 1998, and so on. Estimate the number of personal computers that will be sold in 2003, using both the graph and the polynomial. *(Source: International Data Corporation.)*

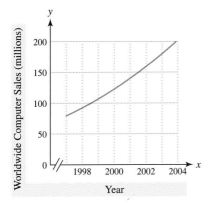

76. *Digital Cameras* (Refer to Example 3.)

(a) How many pixels are there in 5 digital pictures, each y pixels by z pixels?

(b) A DC-280 digital Kodak camera creates pictures that are 1760 by 1168 pixels. Why is this camera called a 2 megapixel camera?

(c) How many pixels are contained in 5 pictures from this camera?

77. *World Population* The following table lists the estimated world population between 1960 and 2013. Construct tables for both of the polynomials given and decide which models the data best. Let x represent the year.

$$0.054x - 102.1 \quad \text{or} \quad 0.0763x - 146.6$$

Year	1960	1974	1987	1999	2013
Population (billions)	3	4	5	6	7

Source: United Nations Population Division.

78. *Geometry* A house has 8-foot-high ceilings. Three rooms are x feet by y feet and two rooms are x feet by z feet. Write a polynomial that gives the total volume of the five rooms. Find the total volume of these rooms if $x = 10$ feet, $y = 12$ feet, and $z = 7$ feet.

79. *Squares and Circles* Write a polynomial that gives the sum of the areas of a square with sides of length x and a circle with radius x. Find the combined area when $x = 10$ inches.

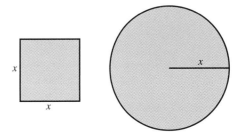

80. *Spheres* Write a monomial that gives the volume of 9 spheres with radius y. Find the combined volumes when $y = 3$ feet. (*Hint:* The volume of a sphere with radius r is $\frac{4}{3}\pi r^3$.)

WRITING ABOUT MATHEMATICS

81. Discuss how to subtract two polynomials. Demonstrate your method with an example.

82. Explain the difference between a monomial and a polynomial. Give examples of each and an example that is neither a monomial nor a polynomial.

GROUP ACTIVITY
Working with Real Data

Directions: Form a group of 2 to 4 people. Select someone to record the group's responses for this activity. All members of the group should work cooperatively to answer the questions. If your instructor asks for the results, each member of the group should be prepared to respond. If the group is asked to turn in its work, be sure to include each person's name on the paper.

Meals Eaten Out The following table shows the average number of meals per person purchased at restaurants annually from 1990 to 1999.

Year	1990	1991	1992	1993	1994
Meals	121	123	124	126	128

Year	1995	1996	1997	1998	1999
Meals	130	132	134	137	139

Source: The NPD Group, *Eating Patterns in America*.

(a) Make a scatterplot of the data in the viewing rectangle [1989, 2000, 1] by [115, 145, 5].
(b) Find a polynomial that models the data. What is the degree of your polynomial? Explain how you found your polynomial.
(c) Assuming that the trend continues, estimate the average number of meals eaten out per person in 2003.

5.2 POLYNOMIAL FUNCTIONS AND MODELS

Polynomial Functions ~ Evaluating Polynomial Functions ~ Models Involving Polynomial Functions

INTRODUCTION

Data can be either linear or nonlinear. A scatterplot of linear data lies on a line, whereas nonlinear data do not lie on a line. Figures 5.11 and 5.12 show scatterplots of linear and nonlinear data being modeled by graphs of functions. In Chapter 2 we modeled linear data with linear functions. In this section we model nonlinear data with polynomial functions of degree 2 or higher. Polynomial functions can be used to model the number of AIDS cases, ocean temperatures, and a person's heart rate after exercising.

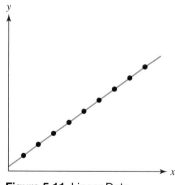

Figure 5.11 Linear Data

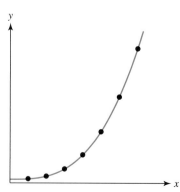

Figure 5.12 Nonlinear Data

5.2 Polynomial Functions and Models

POLYNOMIAL FUNCTIONS

In Section 5.1 we explained that the following expressions are examples of polynomials of one variable.

$$1 - 5x, \quad 3x^2 - 5x + 1, \quad x^3 + 5$$

As a result, we say that the following are *symbolic representations* of polynomial functions of one variable.

$$f(x) = 1 - 5x, \quad g(x) = 3x^2 - 5x + 1, \quad h(x) = x^3 + 5$$

Function f is a *linear function* with degree 1, function g is a *quadratic function* with degree 2, and function h is a *cubic function* with degree 3.

■ **MAKING CONNECTIONS**

Linear Functions and Polynomial Functions

Every linear function can be written as $f(x) = ax + b$ and is an example of a polynomial function. However, polynomial functions of degree 2 or higher are nonlinear functions. To model nonlinear data we use polynomial functions of degree 2 or higher.

EXAMPLE 1 *Identifying polynomial functions*

Determine whether $f(x)$ represents a polynomial function.

(a) $f(x) = 5x^3 - x + 10$ **(b)** $f(x) = x^{-2.5} + 1$ **(c)** $f(x) = 1 - 2x$

Solution

(a) The expression $5x^3 - x + 10$ is a polynomial, so $f(x)$ represents a polynomial function.
(b) $f(x)$ does not represent a polynomial function because the variables in a polynomial must have *nonnegative integer* exponents.
(c) $f(x) = 1 - 2x$ represents a polynomial function that is linear.

EVALUATING POLYNOMIAL FUNCTIONS

To make estimations and predictions with polynomial functions, we must be able to evaluate polynomial functions. For example, the polynomial function $f(x) = x^3$ calculates the volume of a cube with sides of length x. Thus a cube with 4-inch sides has a volume of

$$f(4) = 4^3 = 4 \cdot 4 \cdot 4 = 64 \text{ cubic inches.}$$

We can evaluate polynomial functions symbolically, graphically, and numerically, as shown in the next two examples.

EXAMPLE 2 *Evaluating a polynomial function*

Evaluate $f(x) = x^2 - 5x + 4$ at $x = 2$ symbolically, numerically, and graphically.

Solution

Symbolic Evaluation To evaluate $f(2)$ symbolically we let $x = 2$ in the formula.

$$f(2) = 2^2 - 5 \cdot 2 + 4 = -2.$$

Numerical Evaluation To evaluate $f(2)$ numerically we let $Y_1 = X^\wedge 2 - 5X + 4$. A table of Y_1 is shown in Figure 5.13. Note that, when $x = 2$, $Y_1 = -2$.

Graphical Evaluation In Figure 5.14 a graph of Y_1 is shown. We can use the trace feature to see that $Y_1 = -2$ when $x = 2$.

[−4.7, 4.7, 1] by [−3.1, 3.1, 1]

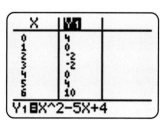
Figure 5.13

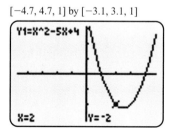
Figure 5.14

EXAMPLE 3 *Evaluating a polynomial function graphically*

GCLM A graph of $f(x) = 4x - x^3$ is shown in Figure 5.15. Evaluate $f(-1)$ graphically and check your result symbolically.

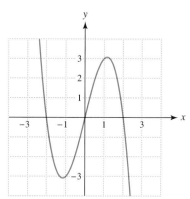
Figure 5.15

Solution

Graphical Evaluation To calculate $f(-1)$ graphically find -1 on the x-axis and move down until the graph of f is reached. Then move horizontally to the y-axis, as shown in Figure 5.16. Thus when $x = -1$, $y = -3$ and $f(-1) = -3$.

Symbolic Evaluation Symbolic evaluation is performed as follows.

$$f(-1) = 4(-1) - (-1)^3$$
$$= -4 - (-1)$$
$$= -3$$

5.2 Polynomial Functions and Models

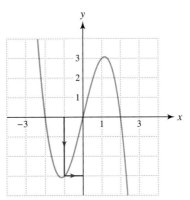

Figure 5.16

Technology Note *Evaluating a Function*

In addition to using the TRACE feature, you can evaluate a function graphically on some types of calculators with a VALUE feature. For example, if $Y_1 = 4X - X^3$, you can evaluate Y_1 at $x = -1$, as illustrated in Figures 5.17–5.19.

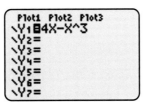

Figure 5.17

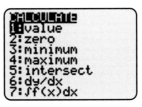

Figure 5.18

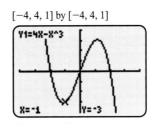

Figure 5.19

EXAMPLE 4 *Evaluating a polynomial function symbolically*

Evaluate $f(x)$ at the given value of x.
(a) $f(x) = -3x^4 - 2, x = 2$ (b) $f(x) = -2x^3 - 4x^2 + 5, x = -3$

Solution

(a) $f(2) = -3(2)^4 - 2 = -3 \cdot 16 - 2 = -50$
(b) $f(-3) = -2(-3)^3 - 4(-3)^2 + 5 = -2(-27) - 4(9) + 5 = 23$

MODELS INVOLVING POLYNOMIAL FUNCTIONS

Polynomials may be used to model a wide variety of data. A scatterplot of the cumulative number of reported AIDS cases in thousands from 1984 to 1994 is shown in

Figure 5.20. In this graph $x = 4$ corresponds to 1984, $x = 5$ to 1985, and so on until $x = 14$ represents 1994. These data can be modeled with a quadratic function, as shown in Figure 5.21, where $f(x) = 4.1x^2 - 25x + 46$ is graphed with the data. Note that $f(x)$ was found by using a graphing calculator. (*Source:* **Department of Health and Human Services.**)

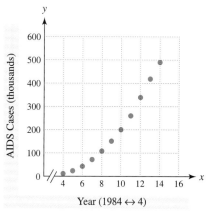

Figure 5.20 AIDS Cases (1984–1994)

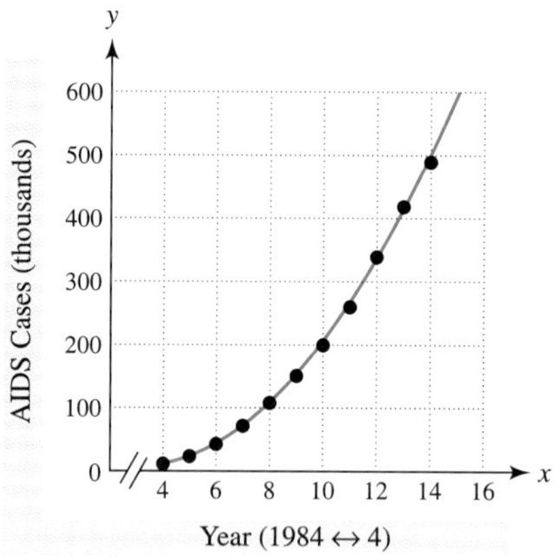

Figure 5.21 Modeling AIDS Cases

EXAMPLE 5 *Modeling AIDS cases in the United States*

Use $f(x) = 4.1x^2 - 25x + 46$ to model the number of AIDS cases.

(a) Estimate the number of AIDS cases reported by 1987. Compare it to the actual value of 71.4 thousand.

(b) By 1997, the total number of AIDS cases reported was 631 thousand. What estimate does $f(x)$ give? Discuss your result.

Solution

(a) The value $x = 7$ corresponds to 1987, so we evaluate $f(7)$, which gives

$$f(7) = 4.1(7)^2 - 25(7) + 46 = 71.9.$$

This model estimates that a cumulative total of 71.9 thousand AIDS cases were reported by 1987. This result compares favorably to the actual value of 71.4 thousand cases.

(b) To estimate the number in 1997, we evaluate $f(17)$ because $x = 17$ corresponds to 1997, obtaining

$$f(17) = 4.1(17)^2 - 25(17) + 46 = 805.9.$$

This result is considerably more than the actual value of 631 thousand. The reason is that f models data only from 1984 through 1994. After 1994, f gives estimates that are too large because the growth in AIDS cases has slowed in recent years.

A well-conditioned athlete's heart rate can reach 200 beats per minute during strenuous physical activity. Upon quitting, a typical heart rate decreases rapidly at first and then more gradually after a few minutes, as illustrated in the next example.

5.2 Polynomial Functions and Models

EXAMPLE 6 *Modeling heart rate of an athlete*

The polynomial $f(x) = 1.875x^2 - 30x + 200$ models a typical athlete's heart rate x minutes after exercise has stopped, where $0 \le x \le 8$. (**Source: V. Thomas,** *Science and Sport.*)

(a) What is the initial heart rate when the athlete stops exercising?
(b) Construct the table for f, starting at $x = 0$ and incrementing by 1. What is the heart rate after 4 minutes?
(c) Graph f in [0, 8, 1] by [0, 200, 25] and interpret the graph.

Solution

(a) To find the initial heart rate evaluate $f(x)$ at $x = 0$, or
$$f(0) = 1.875(0)^2 - 30(0) + 200 = 200.$$
When the athlete stops exercising, the heart rate is 200 beats per minute.

(b) Construct the table for $Y_1 = 1.875X^2 - 30X + 200$, as shown in Figure 5.22. After 4 minutes the heart rate is 110 beats per minute.

(c) Graph Y_1, as shown in Figure 5.23. The heart rate drops rapidly at first and then begins to level off.

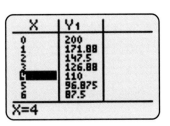

Figure 5.22

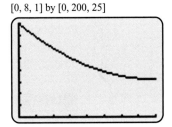

[0, 8, 1] by [0, 200, 25]

Figure 5.23

EXAMPLE 7 *Modeling ocean temperature*

The polynomial $f(x) = -0.07x^3 + 0.61x^2 + 3.2x + 57$ models the average ocean temperature in degrees Fahrenheit at St. Petersburg, Florida. In this formula $x = 1$ corresponds to January, $x = 2$ to February, and so on. (**Source: J. Williams,** *The Weather Almanac, 1995.*)

(a) What is the average ocean temperature in May?
(b) Construct the table for f, starting at $x = 1$ and incrementing by 1. During what month is the average temperature greatest?
(c) Graph f in [1, 12, 1] by [50, 90, 10] and interpret the graph.

Solution

(a) The value $x = 5$ corresponds to May, so evaluate $f(5)$ with a calculator.
$$f(5) = -0.07(5)^3 + 0.61(5)^2 + 3.2(5) + 57 = 79.5$$
The ocean temperature in May is about 79.5°F.

(b) Construct the table for $Y_1 = -.07X^3 + 0.61X^2 + 3.2X + 57$, as shown in Figure 5.24 on the next page. The highest temperature occurs in August ($x = 8$) when it reaches about 85.8°F.

(c) Graph Y_1 as shown in Figure 5.25. From January through August the ocean temperature increases gradually. After August the temperature decreases.

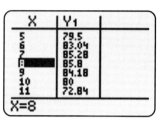
Figure 5.24

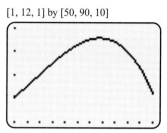
[1, 12, 1] by [50, 90, 10]
Figure 5.25

Critical Thinking

Make a rough sketch of the monthly average ocean temperature at Sydney, Australia, for a year.

5.2 Putting It All Together

Polynomial functions are important because they are frequently used to model real-world data. The following table summarizes some important concepts involving polynomial functions.

Polynomial Functions	The following are examples of polynomial functions. $f(x) = 3$ Degree = 0 Constant $f(x) = 5x - 3$ Degree = 1 Linear $f(x) = x^2 - 2x - 1$ Degree = 2 Quadratic $f(x) = 3x^3 + 2x^2 - 6$ Degree = 3 Cubic
Evaluating a Polynomial Function	Let $f(x) = x^2 + 2x - 1$. We can evaluate $f(1)$ symbolically, graphically, and numerically. **Symbolically** $f(1) = 1^2 + 2(1) - 1 = 2$ **Graphically** **Numerically** [−4.7, 4.7, 1] by [−3.1, 3.1, 1]

5.2 EXERCISES

CONCEPTS

1. How can you distinguish nonlinear data from linear data?

2. What types of functions are sometimes used to model nonlinear data?

3. When you evaluate a polynomial function for a value of x, can you calculate two answers? Explain.

4. Could the graph of a polynomial function be a line? Explain.

5. How can the degree of a polynomial function be determined?

6. Name three ways to evaluate a polynomial function.

Exercises 7–14: Determine whether $f(x)$ represents a polynomial function.

7. $f(x) = x^4 + 5x^2 - 6$
8. $f(x) = 5x^3$
9. $f(x) = x^{-2}$
10. $f(x) = |x|$
11. $f(x) = \dfrac{1}{x^2 + 1}$
12. $f(x) = 5x - 7$
13. $f(x) = \dfrac{1}{3} + 3x - x^2$
14. $f(x) = x^{1.5} + 3x^{0.5} + 6$

EVALUATING POLYNOMIAL FUNCTIONS

Exercises 15–20: Evaluate $f(x)$ at the given value of x by hand.

15. $f(x) = 3x^2$ $x = -2$
16. $f(x) = x^2 - 2x$ $x = 3$
17. $f(x) = 5 - 4x$ $x = -\dfrac{1}{2}$
18. $f(x) = x^2 + 4x - 5$ $x = -3$
19. $f(x) = 0.5x^4 - 0.3x^3 + 5$ $x = -1$
20. $f(x) = 6 - 2x + x^3$ $x = 0$

Exercises 21–22: Use the graph to evaluate $f(x)$ at the given value of x.

21. $x = -1$ 22. $x = 1.8$

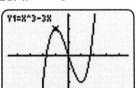
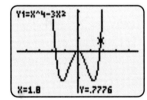

Exercises 23–26: Use the graph of f to evaluate both expressions.

23. $f(1)$ and $f(-2)$ 24. $f(0)$ and $f(2)$

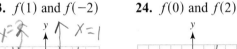
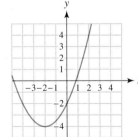

25. $f(-1)$ and $f(2)$ 26. $f(-1)$ and $f(0)$

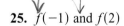
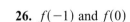
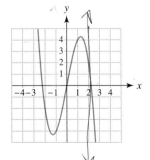
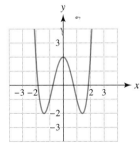

Exercises 27–28: Use the table to evaluate $f(x)$ at the given values of x.

27. $x = 2, x = 5$

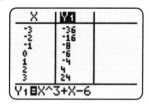

28. $x = -3, x = 1$

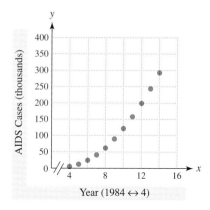

Exercises 29–36: (Refer to Example 2.) Evaluate $f(x)$ symbolically at the given x-value. Give graphical and numerical support.

29. $f(x) = -x^3$ $x = -1$

30. $f(x) = 3 - 2x$ $x = \dfrac{3}{10}$

31. $f(x) = -x^2 - 3x$ $x = -3$

32. $f(x) = 2x^3 - 4x + 1$ $x = 1$

33. $f(x) = 1 - 2x + x^2$ $x = 2.4$

34. $f(x) = 4x^2 - 20x + 25$ $x = 1.8$

35. $f(x) = x^5 - 5$ $x = -1$

36. $f(x) = 1.2x^4 - 5.7x + 3$ $x = \dfrac{3}{2}$

APPLICATIONS

37. *U.S. AIDS Cases* (Refer to Example 5.) Use $f(x)$ to estimate the cumulative number of AIDS cases diagnosed in 1990. Compare it to the true value of 199,608.

38. *U.S. AIDS Deaths* The scatterplot shows the cumulative number of reported AIDS deaths. In this graph $x = 4$ corresponds to 1984, $x = 5$ to 1985, and so on until $x = 14$ corresponds to 1994. The data may be modeled by $f(x) = 2.4x^2 - 14x + 23$, where the output is in thousands of deaths.

(a) Use $f(x)$ to estimate the cumulative total of AIDS deaths in 1990. Compare it with the actual value of 121.6 thousand.

(b) In 1997 the cumulative number of AIDS deaths was 390 thousand. What estimate does $f(x)$ give? Discuss your result.

39. *Heart Rate* (Refer to Example 6.) Make a table of $f(x) = 1.875x^2 - 30x + 200$, starting at $x = 0$ and incrementing by 1. When was the athlete's heart rate between 80 and 110, inclusive?

40. *Ocean Temperatures* The polynomial function

$$f(x) = -0.064x^3 + 0.56x^2 + 2.9x + 61$$

models the ocean temperature in degrees Fahrenheit at Naples, Florida. In this formula $x = 1$ corresponds to January, $x = 2$ to February, and so on. (*Source: J. Williams.*)

(a) What is the average ocean temperature in April?

(b) Make a table of f, starting at $x = 1$ and incrementing by 1. During what month is the average temperature least?

(c) Graph f in [1, 12, 1] by [50, 90, 10]. Estimate when the maximum ocean temperature occurs. What is this maximum?

41. *Manatee Deaths* In the United States, run-ins with boats kill more manatees than any other cause. A graph of a polynomial function that models the total number of deaths each year is shown in the accompanying figure. Use the graph to discuss the trend in manatee deaths from 1988 through 1998. (*Source:* Florida Department of Environmental Protection.)

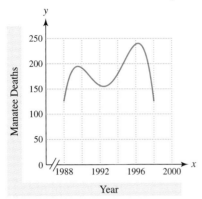

42. *Cigarette Butts* About 98% of cigarettes sold domestically in 1997 were filtered. As a result, millions of dollars are spent each year picking up cigarette butts and disposing of them. The number of filtered cigarettes sold in billions is modeled by

    $f(x) = -0.619608x^2 + 2457.58x - 2,436,356,$

    where x is the year. (*Source:* Federal Trade Commission.)
    (a) Make a table of f starting at $x = 1965$ and incrementing by 5. How many filtered cigarettes were sold in 1965 and in 1995?
    (b) Graph f in [1960, 2000, 5] by [200, 600, 100]. Discuss how domestic sales of filtered cigarettes have changed over this time period.

43. *Average Baseball Salaries* Average salaries for major league baseball players have risen dramatically since 1976. Average salaries in millions of dollars may be modeled by

    $f(x) = 0.002188x^2 - 0.012x + 0.047,$

    where $x = 6$ corresponds to 1976, $x = 7$ to 1977, and so on until $x = 30$ corresponds to 2000. (*Source:* Major League Baseball.)
    (a) Make a table of f starting at $x = 6$ and incrementing by 1. What was the average major league baseball player's salary in 1976 and in 1997?
    (b) Graph f in [6, 29, 1] by [0, 2, 0.5]. Use the trace feature on your calculator to estimate the year when average baseball salaries first reached $400 thousand.

44. *Student Aid* The following table lists aid to college students in constant dollars for selected academic years.

Year	1988	1994	1998
Aid ($ billions)	35	50	64

    Source: The College Board.

    Which of the following polynomials models the data best, where x is the year? Explain how you made your decision.
    (i) $f(x) = 0.1x^2 - 395.7x + 391,472.2$
    (ii) $g(x) = 0.375x^2 - 1490.75x + 1,481,587$

45. *Microsoft Stocks* If a person had bought $2500 of Microsoft stock in 1986, it would have been worth about $1.2 million in 1999. The following table shows the cost of this stock (adjusted for splits) in various years.

Year	1986	1994	1999
Cost ($)	0.19	20	90

    Source: USA Today.

    Which of the following polynomials models the data best, where $x = 6$ corresponds to 1986, $x = 14$ to 1994, and $x = 19$ to 1999. Explain how you made your decision.
    (i) $f(x) = 0.886x^2 - 15.25x + 59.79$
    (ii) $g(x) = 0.882x^2 - 15.4x + 60.15$

46. *SUV Sales* Sales of sport utility vehicles (SUVs) increased dramatically between 1991 and 1999. In 1991 about 1 million SUVs were sold, and in 1999 3.2 million were sold. Growth in sales could be modeled by a linear function f. (*Source:* Autodata Corporation.)
    (a) Find values for a and b so that $f(x) = ax + b$ models the data. (*Hint:* The graph of f should pass through the points (1991, 1) and (1999, 3.2).)
    (b) Use f to estimate the number of SUVs sold in 1995.

Writing About Mathematics

47. If a polynomial function has degree 1, what can be said about its graph? If a polynomial function has degree greater than 1, what can be said about its graph? If you model linear data with a polynomial function, what degree polynomial should you use? If you model nonlinear data with a polynomial function, what degree polynomial should you use?

48. Explain how to evaluate a function graphically. Use your method to evaluate $f(x) = x^2$ at $x = -2$.

CHECKING BASIC CONCEPTS FOR SECTIONS 5.1 AND 5.2

1. Combine like terms.
   (a) $8x^2 + 4x - 5x^2 + 3x$
   (b) $(5x^2 - 3x + 2) - (3x^2 - 5x^3 + 1)$

2. Write a polynomial that represents the revenue from ticket sales if $x + 120$ tickets are sold for x dollars each. Evaluate the polynomial for $x = 10$ and interpret the result.

3. Evaluate
$$f(x) = x^3 - 4x^2 + x - 1$$
at $x = -2$ graphically. Check your result symbolically.

4. An athlete's heart rate x minutes after exercise is stopped is modeled by
$$f(x) = 2x^2 - 25x + 160,$$
where $0 \leq x \leq 6$.
   (a) What was the initial heart rate when exercise stopped?
   (b) What was the heart rate after 4 minutes of rest?
   (c) Graph $f(x)$ in the viewing rectangle [0, 6, 1] by [0, 180, 20]. Discuss the information that can be obtained from the graph.

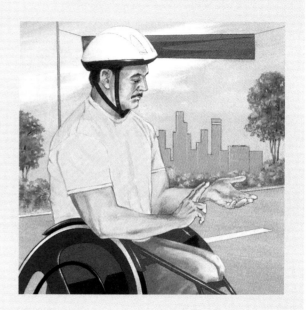

5.3 MULTIPLICATION OF POLYNOMIALS

Review of Basic Properties ~ Multiplying Polynomials ~ Some Special Products

INTRODUCTION

The study of polynomials dates back to Babylonian civilization in about 1800–1600 B.C. Much later, Gottfried Leibniz (1646–1716) was the first to generalize polynomial functions of degree n. Many eighteenth century mathematicians devoted their entire careers to the study of polynomials. This study included multiplying and factoring polynomials. Both skills are used to solve equations. In this section we discuss the basics of polynomial multiplication. (*Sources: Historical Topics for the Mathematics Classroom, Thirty-first Yearbook,* NCTM; L. Motz and J. Weaver, *The Story of Mathematics.*)

5.3 Multiplication of Polynomials

REVIEW OF BASIC PROPERTIES

Distributive properties are used frequently in the multiplication of polynomials. For all real numbers a, b, and c

$$a(b + c) = ab + ac \quad \text{and}$$
$$a(b - c) = ab - ac.$$

In the next example we use these distributive properties to multiply expressions.

EXAMPLE 1 *Using distributive properties*

Multiply.

(a) $4(5 + x)$ **(b)** $-3(x - 4y)$ **(c)** $(2x - 5)(6)$

Solution

(a) $4(5 + x) = 4 \cdot 5 + 4 \cdot x = 20 + 4x$

(b) $-3(x - 4y) = -3 \cdot x - (-3) \cdot (4y) = -3x + 12y$

(c) $(2x - 5)(6) = 2x \cdot 6 - 5 \cdot 6 = 12x - 30$

You can visualize the solution in part (a) of the preceding example by using areas of rectangles. If a rectangle has width 4 and length $5 + x$, its area is $20 + 4x$, as shown in Figure 5.26.

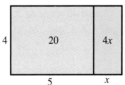

Figure 5.26 Area: $20 + 4x$

In Section 1.3 you learned that *when you multiply powers of a variable, you add exponents.* That is,

$$a^n \cdot a^m = a^{n+m},$$

where a is a nonzero real number and m and n are integers. This property is applied in the next example.

EXAMPLE 2 *Multiplying powers of variables*

Multiply each expression.

(a) $-2x^3 \cdot 4x^5$ **(b)** $3xy^3 \cdot 4x^2y^2$ **(c)** $6y^3(2y - y^2)$

Solution

(a) $-2x^3 \cdot 4x^5 = -2 \cdot 4 \cdot x^3 \cdot x^5 = -8x^{3+5} = -8x^8$

(b) $3xy^3 \cdot 4x^2y^2 = 3 \cdot 4 \cdot x \cdot x^2 \cdot y^3 \cdot y^2 = 12 \cdot x^{1+2} \cdot y^{3+2} = 12x^3y^5$

Note: We cannot simplify $12x^3y^5$ further because x^3 and y^5 have different bases.

(c) $6y^3(2y - y^2) = 6y^3 \cdot 2y - 6y^3 \cdot y^2 = 12y^4 - 6y^5$

Multiplying Polynomials

A polynomial with one term is a **monomial,** with two terms a **binomial,** and with three terms a **trinomial.** Examples are shown in Table 5.4.

TABLE 5.4

Monomials	$2x^2$	$-3x^4y$	9
Binomials	$3x - 1$	$2x^3 - x$	$x^2 + 5$
Trinomials	$x^2 - 3x + 5$	$5x^2 - 2x + 10$	$2x^3 - x^2 - 2$

In the next example we multiply two binomials, using both geometric and symbolic techniques.

EXAMPLE 3 *Multiplying binomials*

 Multiply $(x + 1)(x + 3)$

(a) geometrically and
(b) symbolically.

Solution

(a) To multiply $(x + 1)(x + 3)$ geometrically, draw a rectangle $x + 1$ wide and $x + 3$ long, as shown in Figure 5.27. The area of the rectangle equals the product $(x + 1)(x + 3)$. This large rectangle can be divided into four smaller rectangles as shown in Figure 5.28. The sum of the areas of these four smaller rectangles equals the area of the large rectangle. The smaller rectangles have areas of $x^2, x, 3x,$ and 3. Thus

$$(x + 1)(x + 3) = x^2 + x + 3x + 3$$
$$= x^2 + 4x + 3.$$

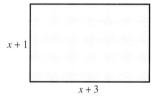
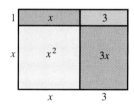

Figure 5.27 Area: $(x + 1)(x + 3)$

Figure 5.28 Area: $x^2 + 4x + 3$

(b) To multiply $(x + 1)(x + 3)$ symbolically we apply the distributive property.

$$(x + 1)(x + 3) = (x + 1)(x) + (x + 1)(3)$$
$$= x \cdot x + 1 \cdot x + x \cdot 3 + 1 \cdot 3$$
$$= x^2 + x + 3x + 3$$
$$= x^2 + 4x + 3$$

We can give graphical and numerical support to our result in the preceding example by letting $Y_1 = (X + 1)(X + 3)$ and $Y_2 = X^2 + 4X + 3$. The graphs of Y_1 and Y_2 appear to be identical in Figures 5.29 and 5.30. Figure 5.31 shows that $Y_1 = Y_2$ for each value of x.

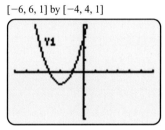
Figure 5.29

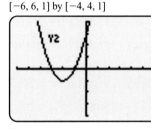
Figure 5.30

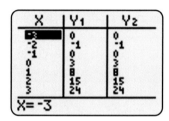
Figure 5.31

To multiply $(x + 1)$ by $(x + 3)$ we multiplied every term in $x + 1$ by every term in $x + 3$. That is,

$$(x + 1)(x + 3) = x^2 + 3x + x + 3$$
$$= x^2 + 4x + 3.$$

Note: This process of multiplying binomials is called *FOIL*. You may use it to remind yourself to multiply the first terms (*F*), outside terms (*O*), inside terms (*I*), and last terms (*L*).

Multiply the *First terms* to obtain x^2. $(x + 1)(x + 3)$

Multiply the *Outside terms* to obtain $3x$. $(x + 1)(x + 3)$

Multiply the *Inside terms* to obtain x. $(x + 1)(x + 3)$

Multiply the *Last terms* to obtain 3. $(x + 1)(x + 3)$

The following method summarizes how to multiply two polynomials in general.

Multiplication of Polynomials

The product of two polynomials may be found by multiplying every term in the first polynomial by every term in the second polynomial.

EXAMPLE 4 *Multiplying polynomials*

Multiply each binomial symbolically.

(a) $(2x - 1)(x + 2)$ **(b)** $(1 - 3x)(2 - 4x)$ **(c)** $(x^2 + 1)(5x - 3)$

Solution

(a) $(2x - 1)(x + 2) = 2x \cdot x + 2x \cdot 2 - 1 \cdot x - 1 \cdot 2$
$$= 2x^2 + 4x - x - 2$$
$$= 2x^2 + 3x - 2$$

(b) $(1 - 3x)(2 - 4x) = 1 \cdot 2 - 1 \cdot 4x - 3x \cdot 2 + 3x \cdot 4x$
$$= 2 - 4x - 6x + 12x^2$$
$$= 2 - 10x + 12x^2$$

(c) $(x^2 + 1)(5x - 3) = x^2 \cdot 5x - x^2 \cdot 3 + 1 \cdot 5x - 1 \cdot 3$
$$= 5x^3 - 3x^2 + 5x - 3$$

EXAMPLE 5 *Multiplying polynomials*

Multiply each expression symbolically.

(a) $3x(x^2 + 5x - 4)$ (b) $-x^2(x^4 - 2x + 5)$ (c) $(x + 2)(x^2 + 4x - 3)$

Solution

(a) $3x(x^2 + 5x - 4) = 3x \cdot x^2 + 3x \cdot 5x - 3x \cdot 4$
$= 3x^3 + 15x^2 - 12x$

(b) $-x^2(x^4 - 2x + 5) = -x^2 \cdot x^4 + x^2 \cdot 2x - x^2 \cdot 5$
$= -x^6 + 2x^3 - 5x^2$

(c) $(x + 2)(x^2 + 4x - 3) = x \cdot x^2 + x \cdot 4x - x \cdot 3 + 2 \cdot x^2 + 2 \cdot 4x - 2 \cdot 3$
$= x^3 + 4x^2 - 3x + 2x^2 + 8x - 6$
$= x^3 + 6x^2 + 5x - 6$

SOME SPECIAL PRODUCTS

We begin our discussion by reviewing two properties of exponents that we introduced in Section 1.3.

Properties of Exponents

For any nonzero numbers a and b and integers m and n,

$$(a^m)^n = a^{mn}$$

and

$$(ab)^n = a^n b^n.$$

The first property states that, to raise a power to a power, we multiply the exponents. The second property states that, to raise a product to a power, we raise each factor to the power.

EXAMPLE 6 *Using properties of exponents*

Simplify.

(a) $(x^2)^5$ (b) $(2x)^3$ (c) $(5x^3)^2$

Solution

(a) $(x^2)^5 = x^{2 \cdot 5} = x^{10}$ (b) $(2x)^3 = 2^3 x^3 = 8x^3$
(c) $(5x^3)^2 = 5^2(x^3)^2 = 25x^6$

The following special product often occurs in mathematics.

$$(a - b)(a + b) = a \cdot a + a \cdot b - b \cdot a - b \cdot b$$
$$= a^2 + ab - ba - b^2$$
$$= a^2 - b^2$$

That is, the product of a sum and difference equals the difference of their squares.

EXAMPLE 7 *Finding the product of a sum and difference*

Multiply.

(a) $(x - 3)(x + 3)$ **(b)** $(5 - 4x^2)(5 + 4x^2)$

Solution

(a) If we let $a = x$ and $b = 3$, we can apply the rule
$$(a - b)(a + b) = a^2 - b^2.$$
Thus
$$(x - 3)(x + 3) = (x)^2 - (3)^2$$
$$= x^2 - 9.$$

(b) Similarly, we can multiply as follows.
$$(5 - 4x^2)(5 + 4x^2) = (5)^2 - (4x^2)^2$$
$$= 25 - 16x^4$$

Two other special products involve *squaring a binomial:*
$$(a + b)^2 = (a + b)(a + b)$$
$$= a^2 + ab + ba + b^2$$
$$= a^2 + 2ab + b^2$$

and
$$(a - b)^2 = (a - b)(a - b)$$
$$= a^2 - ab - ba + b^2$$
$$= a^2 - 2ab + b^2.$$

The first product is illustrated geometrically in Figure 5.32, where each side of a square has length $(a + b)$. The area of the square is
$$(a + b)(a + b) = (a + b)^2.$$

This area can also be computed by adding the area of the four small rectangles.
$$a^2 + ab + ba + b^2 = a^2 + 2ab + b^2$$

Thus $(a + b)^2 = a^2 + 2ab + b^2$. Note that, to obtain the middle term, we multiply the two terms in the binomial and double the result.

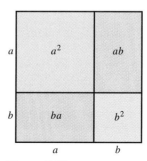

Figure 5.32
$(a + b)^2 = a^2 + 2ab + b^2$

EXAMPLE 8   Squaring a binomial

Multiply.

(a) $(x + 5)^2$ (b) $(3 - 2x)^2$

Solution

(a) If we let $a = x$ and $b = 5$, we can apply the formula
$$(a + b)^2 = a^2 + 2ab + b^2.$$
Thus
$$(x + 5)^2 = (x)^2 + 2(x)(5) + (5)^2 \quad \text{To find the middle term, multiply}$$
$$= x^2 + 10x + 25 \quad \text{a and b and double the result.}$$

(b) Applying the formula $(a - b)^2 = a^2 - 2ab + b^2$, we find
$$(3 - 2x)^2 = (3)^2 - 2(3)(2x) + (2x)^2$$
$$= 9 - 12x + 4x^2$$

Note: If you forget these special products, you can still use techniques learned earlier to multiply the polynomials in Examples 7 and 8. For example,
$$(3 - 2x)^2 = (3 - 2x)(3 - 2x)$$
$$= 3 \cdot 3 - 3 \cdot 2x - 2x \cdot 3 + 2x \cdot 2x$$
$$= 9 - 6x - 6x + 4x^2$$
$$= 9 - 12x + 4x^2.$$

Critical Thinking

Suppose that a student is convinced that the expressions
$$(x + 3)^2 \quad \text{and} \quad x^2 + 9$$
are equivalent. How could you use a graphing calculator to convince the student that $(x + 3)^2 \neq x^2 + 9$? Explain your answer.

5.3 Putting It All Together

The following table summarizes some important concepts in this section.

Concept	Explanation	Examples
Distributive Properties	For all real numbers a, b, and c, $a(b + c) = ab + ac$ $a(b - c) = ab - ac$.	$4(3 + a) = 12 + 4a$ $5(x - 1) = 5x - 5$

Concept	Explanation	Examples
Multiplying Polynomials	The product of two polynomials may be found by multiplying every term in the first polynomial by every term in the second polynomial.	$(2x + 3)(x - 7) = 2x \cdot x - 2x \cdot 7 + 3 \cdot x - 3 \cdot 7$ $= 2x^2 - 14x + 3x - 21$ $= 2x^2 - 11x - 21$
Properties of Exponents	For nonzero numbers a and b and integers m and n $a^m \cdot a^n = a^{m+n}$, $(a^m)^n = a^{mn}$, and $(ab)^n = a^n b^n$.	$5^2 \cdot 5^6 = 5^8$ $(2^3)^2 = 2^6$ $(5y^4)^2 = 25y^8$
Special Products of Binomials	Product of a sum and difference $(a - b)(a + b) = a^2 - b^2$ Squares of binomials $(a + b)^2 = a^2 + 2ab + b^2$ $(a - b)^2 = a^2 - 2ab + b^2$	$(x - 2)(x + 2) = x^2 - 4$ $(x + 4)^2 = x^2 + 8x + 16$ $(x - 4)^2 = x^2 - 8x + 16$

5.3 EXERCISES

CONCEPTS

1. The equation $5(x - 4) = 5x - 20$ illustrates what property?

2. Give an example of a monomial, a binomial, and a trinomial.

3. Simplify the expression $x^3 \cdot x^5$.

4. Simplify $(2x)^3$ and $(x^2)^3$.

5. $(a - b)(a + b) =$ _____.

6. $(a + b)^2 =$ _____.

Exercises 7–14: Multiply.

7. $5(y + 2)$
8. $4(x - 7)$
9. $-2(5x + 9)$
10. $-3x(5 + x)$
11. $-6y(y - 3)$
12. $(2x - 5)8x^3$
13. $(9 - 4x)3x$
14. $-(5 - x^2)$

Exercises 15–22: Multiply the monomials.

15. $x^4 \cdot x^8$
16. $2x \cdot 4x^3$
17. $-5y^7 \cdot 4y$
18. $3xy^2 \cdot 6x^3 y^2$
19. $(-xy)(4x^3 y^5)$
20. $(4z^3)(-5z^2)$
21. $(5y^2 z)(4x^2 yz^5)$
22. $x^2(-xy^2)$

MULTIPLYING POLYNOMIALS

Exercises 23–26: Use the figure to write the product. Find the area of the rectangle if $x = 5$ inches.

23. $(x + 1)(x + 2)$

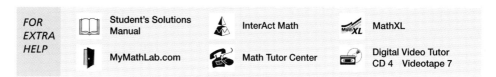

24. $(x + 3)(x + 4)$

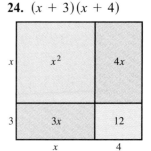

25. $(2x+1)(x+1)$ **26.** $(2x+4)(3x+2)$

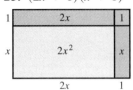 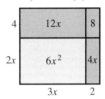

Exercises 27–30: (Refer to Example 3.) Multiply geometrically and symbolically.

27. $(x+5)(x+6)$ **28.** $(x+1)(x+4)$

29. $(2x+1)(2x+1)$ **30.** $(x+3)(2x+4)$

Exercises 31–36: Determine symbolically whether the equation is an identity. Support your conclusion graphically and numerically. (Hint: If it is an identity, the equation must be true for all x-values.)

31. $(x-2)^2 = x^2 - 4$

32. $(x+4)^2 = x^2 + 16$

33. $(x+3)(x-2) = x^2 + x - 6$

34. $(5x-1)^2 = 25x^2 - 10x + 1$

35. $2x(3x+1) = 6x^2 + 1$

36. $-(x^2+4x) = -x^2 + 4x$

Exercises 37–42: Multiply the binomials symbolically. Support your result either graphically or numerically.

37. $(x+5)(x+10)$ **38.** $(x-5)(x-10)$

39. $(x-3)(x-4)$ **40.** $(x+3)(x+4)$

41. $(2x-1)(x+2)$ **42.** $(2x+1)(x-2)$

Exercises 43–54: Multiply the binomials symbolically.

43. $(y+3)(y-4)$ **44.** $(2x+1)(5x+1)$

45. $(4x-3)(4-9x)$ **46.** $(1-x)(1+2x)$

47. $(-2x+3)(x-2)$ **48.** $(z-2)(4z+3)$

49. $\left(x-\frac{1}{2}\right)\left(x+\frac{1}{4}\right)$ **50.** $\left(z-\frac{1}{3}\right)\left(z-\frac{1}{6}\right)$

51. $(x^2+1)(2x^2-1)$ **52.** $(x^2-2)(x^2+4)$

53. $(x+y)(x-2y)$ **54.** $(x^2+y^2)(x-y)$

Exercises 55–62: Multiply the polynomials.

55. $4x(x^2-2x-3)$ **56.** $2x(3-x+x^2)$

57. $-x(x^4-3x^2+1)$

58. $-3x^2(4x^3+x^2-2x)$

59. $(2x^2-4x+1)(3x^2)$

60. $(x-y+5)(xy)$

61. $(x+1)(x^2+2x-3)$

62. $(2x-1)(3x^2-x+6)$

Exercises 63–70: Simplify the expression, using properties of exponents.

63. $(2^3)^2$ **64.** $(x^3)^5$

65. $2(z^3)^6$ **66.** $(5y)^3$

67. $(-5x)^2$ **68.** $(2y)^4$

69. $(-2xy^2)^3$ **70.** $(3x^2y^3)^4$

Exercises 71–86: Multiply the expressions.

71. $(x-3)(x+3)$ **72.** $(x+5)(x-5)$

73. $(2x-3)(2x+3)$ **74.** $(5x-4)(5x+4)$

75. $(x-y)(x+y)$

76. $(2x+2y)(2x-2y)$

77. $(x+2)^2$ **78.** $(y+5)^2$

79. $(2x+1)^2$ **80.** $(3x+5)^2$

81. $(x-1)^2$ **82.** $(x-7)^2$

83. $(3x-2)^2$ **84.** $(6x-5)^2$

85. $3x(x+1)(x-1)$ **86.** $-4x(3x-5)^2$

AREA

Exercises 87–90: Express the shaded area in terms of x. Find the area if $x = 20$ feet.

87.  **88.**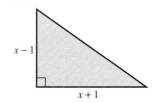

89.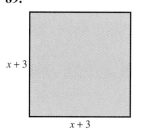

90.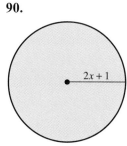

APPLICATIONS

91. *Interest* If the interest rate r is expressed as a decimal and N dollars are deposited into an account, then the amount in the account after two years is $N(1 + r)^2$.
(a) Multiply this expression.
(b) Let $r = 0.10$ and $N = 200$. Use both the given expression and the expression you obtained in part (a) to find the amount of money in the account after 2 years. Do your answers agree?

92. *Probability* Suppose that the likelihood, or chance, that a volleyball player's serve will be out of bounds is x percent. Then the likelihood that two consecutive serves are in bounds is given by $\left(1 - \dfrac{x}{100}\right)^2$.
(a) Multiply this expression.
(b) Let $x = 25$. Use both the given expression and the expression you obtained in part (a) to find the likelihood that two consecutive serves are in bounds.

93. Write a polynomial that represents the product of two consecutive integers, where x is the smaller integer.

94. Write a polynomial that represents the revenue received from selling $2y + 1$ items at a price of x dollars each.

95. A rectangular pen has a perimeter of 100 feet. If x represents its width, write a polynomial that gives the area of the pen in terms of x.

96. Find an expression that represents the product of the sum and difference of two numbers. Let the two numbers be x and y.

WRITING ABOUT MATHEMATICS

97. Suppose that a student insists that
$$(x + 5)(x + 5)$$
equals $x^2 + 25$. Explain how you could convince the student otherwise.

98. Explain how to multiply two polynomials. Give an example of how your method works. Does your method give the correct result for any two polynomials?

5.4 FACTORING POLYNOMIALS

Common Factors ~ Factoring and Equations ~ Factoring By Grouping

INTRODUCTION

If a baseball is hit upward with a velocity of 60 miles per hour, its height h in feet above the ground after t seconds is modeled by $h(t) = 88t - 16t^2$. To estimate when the ball strikes the ground, we can solve the *polynomial* equation
$$88t - 16t^2 = 0.$$
One method for solving this equation is by factoring (see Example 4). In this section we discuss factoring and how it is used to solve equations.

COMMON FACTORS

When factoring a polynomial, we first look for factors that are common to each term in an expression. By applying a distributive property, we can write a polynomial as two factors. For example, each term in the polynomial $2x^2 + 4x$ contains a factor of $2x$.

$$2x^2 + 4x = 2x(x + 2) \quad \text{Distributive property}$$

Thus the product $2x(x + 2)$ equals $2x^2 + 4x$. We can check this result by multiplying the two factors.

$$2x(x + 2) = 2x \cdot x + 2x \cdot 2$$
$$= 2x^2 + 4x \quad \text{It checks.}$$

EXAMPLE 1   *Finding common factors*

Factor.

(a) $4x^2 + 5x$ **(b)** $12x^3 - 4x^2$ **(c)** $6z^3 - 2z^2 + 4z$ **(d)** $4x^3y^2 + x^2y^3$

Solution

(a) Both $4x^2$ and $5x$ contain a common factor of x.

$$4x^2 + 5x = x(4x + 5)$$

(b) Both $12x^3$ and $4x^2$ contain a common factor of $4x^2$.

$$12x^3 - 4x^2 = 4x^2(3x - 1)$$

(c) The terms $6z^3$, $2z^2$, and $4z$ all contain a common factor of $2z$.

$$6z^3 - 2z^2 + 4z = 2z(3z^2 - z + 2)$$

(d) The terms $4x^3y^2$ and x^2y^3 contain a common factor of x^2y^2.

$$4x^3y^2 + x^2y^3 = x^2y^2(4x + y)$$

▪ **MAKING CONNECTIONS**

Checking Common Factors with Multiplication

When factoring, you can check your work by multiplying the result. For example, if you are uncertain whether the equation

$$12x^3 - 4x^2 = 4x^2(3x - 1)$$

is factored correctly, you can multiply the right side.

$$4x^2(3x - 1) = 4x^2 \cdot 3x - 4x^2 \cdot 1$$
$$= 12x^3 - 4x^2 \quad \text{It checks.}$$

In most situations we factor out the **greatest common factor** (GCF). For example, the polynomial $15x^4 - 5x^2$ has a common factor of $5x$. We could write this polynomial as

$$15x^4 - 5x^2 = 5x(3x^3 - x).$$

However, we can also factor out $5x^2$ to obtain

$$15x^4 - 5x^2 = 5x^2(3x^2 - 1).$$

Because no obvious common factors are left in the expression $(3x^2 - 1)$, we say that $5x^2$ is the greatest common factor of $15x^4 - 5x^2$.

FACTORING AND EQUATIONS

To solve equations using factoring, we use the **zero-product property.** It states that, if the product of two numbers is 0, then at least one of the numbers must equal 0.

Zero-Product Property

For all real numbers a and b, if $ab = 0$, then $a = 0$ or $b = 0$ (or both).

Note: The zero-product property works only for 0. If $ab = 1$, then it does *not* follow that $a = 1$ or $b = 1$. For example, $a = \frac{1}{3}$ and $b = 3$ also satisfy the equation $ab = 1$.

After factoring an expression, we can use the zero-product property to solve a polynomial equation. The left side of the equation

$$2x^2 + 4x = 0$$

may be factored to obtain

$$2x(x + 2) = 0.$$

Note that $2x$ times $(x + 2)$ equals 0. By the zero-product property, we must have either

$$2x = 0 \quad \text{or} \quad x + 2 = 0.$$

Solving each equation for x gives

$$x = 0 \quad \text{or} \quad x = -2.$$

The x-values of 0 and -2 are called **zeros** of the polynomial $2x^2 + 4x$ because, when they are substituted into $2x^2 + 4x$, the result is 0.

EXAMPLE 2 | *Solving a polynomial equation with factoring*

Solve each polynomial equation.

(a) $x^2 - x = 0$ **(b)** $4x^2 = 16x$

Solution

(a) We begin by factoring out the greatest common factor.

$\quad\quad x^2 - x = 0 \quad\quad$ Given equation
$\quad\quad x(x - 1) = 0 \quad\quad$ Factor out x.
$\quad x = 0 \quad \text{or} \quad x - 1 = 0 \quad\quad$ Zero-product property
$\quad x = 0 \quad \text{or} \quad x = 1 \quad\quad$ Solve.

(b) Write the equation so that there is a 0 on its right side.

$$4x^2 = 16x \qquad \text{Given equation}$$
$$4x^2 - 16x = 0 \qquad \text{Subtract } 16x \text{ from both sides.}$$
$$4x(x - 4) = 0 \qquad \text{Factor out } 4x.$$
$$4x = 0 \quad \text{or} \quad x - 4 = 0 \qquad \text{Zero-product property}$$
$$x = 0 \quad \text{or} \quad x = 4 \qquad \text{Solve.}$$

In the next example we solve these same equations graphically and numerically.

EXAMPLE 3 *Finding graphical and numerical solutions*

Solve each polynomial equation graphically and numerically.

(a) $x^2 - x = 0$ **(b)** $4x^2 = 16x$

Solution

(a) Begin by letting $Y_1 = X^2 - X$. Then a solution to $Y_1 = 0$ corresponds to the x-value at which the graph of Y_1 intersects the x-axis. In Figures 5.33 and 5.34 the solutions are $x = 0$ and $x = 1$. A numerical solution is shown in Figure 5.35, where $Y_1 = 0$ when $x = 0$ or $x = 1$.

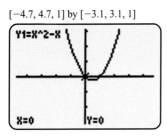

Figure 5.33

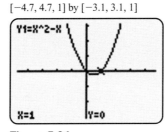

Figure 5.34

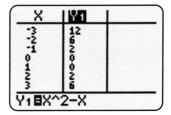

Figure 5.35

(b) Write the equation as $4x^2 - 16x = 0$ and let $Y_1 = 4X^2 - 16X$. Then a solution to $Y_1 = 0$ corresponds to the x-value at which the graph of Y_1 intersects the x-axis. In Figures 5.36 and 5.37 the solutions are $x = 0$ and $x = 4$. A numerical solution is shown in Figure 5.38, where $Y_1 = 0$ when $x = 0$ or $x = 4$.

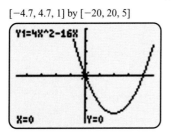

Figure 5.36

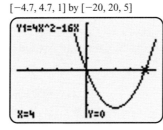

Figure 5.37

Figure 5.38

5.4 Factoring Polynomials

Technology Note *Locating x-intercepts*

Many calculators have the capability to locate an *x*-intercept or zero without using the trace feature. These calculators often ask for a left bound, a right bound, and a guess. This method is illustrated in Figures 5.39–5.43, where the solution $x = 2$ to the equation $x^2 - 4 = 0$ is found.

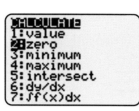

Figure 5.39

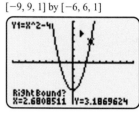

Figure 5.40   **Figure 5.41**

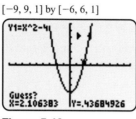 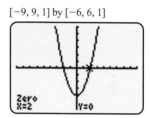

Figure 5.42   **Figure 5.43**

In the next example we use factoring to estimate when a baseball strikes the ground.

EXAMPLE 4   *Modeling the flight of a baseball*

If a baseball is hit upward at 60 miles per hour, its height h in feet after t seconds is modeled by $h(t) = 88t - 16t^2$.

(a) Use factoring to determine when the baseball strikes the ground.
(b) Solve part (a) graphically and numerically.

Solution

(a) *Symbolic Solution*   The baseball strikes the ground when its height is 0.

$$88t - 16t^2 = 0 \qquad h(t) = 0$$
$$8t(11 - 2t) = 0 \qquad \text{Factor out } 8t.$$
$$8t = 0 \quad \text{or} \quad 11 - 2t = 0 \qquad \text{Zero-product property}$$
$$t = 0 \quad \text{or} \quad t = \frac{11}{2} \qquad \text{Solve for } t.$$

The ball strikes the ground after $\frac{11}{2}$, or 5.5 seconds. The solution of $t = 0$ is not used in this problem.

(b) *Graphical and Numerical Solutions* Graph $Y_1 = 88X - 16X^2$. Using a graphing calculator to find an *x*-intercept (or zero) yields the solution, 5.5. See Figure 5.44. Numerical support is shown in Figure 5.45, where $Y_1 = 0$ when $x = 5.5$.

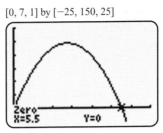

Figure 5.44

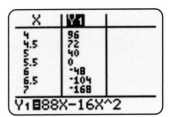

Figure 5.45

FACTORING BY GROUPING

Factoring by grouping is a technique that makes use of the distributive property. Consider the polynomial

$$3t^3 + 6t^2 + 2t + 4.$$

We can factor it as follows.

$(3t^3 + 6t^2) + (2t + 4)$	Associative property
$3t^2(t + 2) + 2(t + 2)$	Factor out common factors.
$(3t^2 + 2)(t + 2)$	Distributive property

EXAMPLE 5 *Factoring by grouping*

Factor each polynomial.

(a) $x^3 - 2x^2 + 3x - 6$ **(b)** $2x - 2y - ax + ay$

Solution

(a)
$x^3 - 2x^2 + 3x - 6$	Given expression
$(x^3 - 2x^2) + (3x - 6)$	Associative property
$x^2(x - 2) + 3(x - 2)$	Factor out common factors.
$(x^2 + 3)(x - 2)$	Distributive property

(b)
$2x - 2y - ax + ay$	Given expression
$(2x - 2y) - (ax - ay)$	Associative and distributive properties
$2(x - y) - a(x - y)$	Factor out common factors.
$(2 - a)(x - y)$	Distributive property

Critical Thinking

Solve the following equation, using factoring.

$$xy - 4x - 3y + 12 = 0$$

5.4 Putting It All Together

There are several different ways of solving an equation, including symbolic, graphical, and numerical techniques. Factoring may be used to solve an equation symbolically. The following table summarizes some important topics covered in this section.

Concept	Explanation	Examples
Greatest Common Factor (GCF)	We can sometimes factor a greatest common factor out of a polynomial.	$3x^4 - 9x^3 + 12x^2 = 3x^2(x^2 - 3x + 4)$ The terms in $(x^2 - 3x + 4)$ have no obvious common factor, so $3x^2$ is called the *greatest common factor* of $3x^4 - 9x^3 + 12x^2$.
Zero-Product Property	If the product of two numbers is 0, then at least one of the numbers equals zero.	$xy = 0$ implies that $x = 0$ or $y = 0$. $x(2x + 1) = 0$ implies that $x = 0$ or $2x + 1 = 0$.
Factoring and Equations	Factoring may be used to solve equations.	$6x^2 - 9x = 0$ $3x(2x - 3) = 0$ $3x = 0$ or $2x - 3 = 0$ $x = 0$ or $x = \frac{3}{2}$
Factoring by Grouping	Factoring by grouping is a method that can be used to factor four terms into a product of two binomials. It involves the use of the associative and distributive properties.	$4x^3 + 6x^2 + 10x + 15 = (4x^3 + 6x^2) + (10x + 15)$ $= 2x^2(2x + 3) + 5(2x + 3)$ $= (2x^2 + 5)(2x + 3)$

5.4 EXERCISES

CONCEPTS

1. Give one reason for factoring expressions in mathematics.

2. If $x(x - 3) = 0$, what can be said about x or $(x - 3)$? What property did you use?

3. Is $2x$ a common factor of $(4x^3 - 12x^2)$? Explain.

4. Is $2x$ the greatest common factor (GCF) of $(4x^3 - 12x^2)$? Explain.

5. If $ab = 2$, does it follow that either $a = 2$ or $b = 2$? Explain.

6. If $xy = 0$, does it follow that either $x = 0$ or $y = 0$? Explain.

Factoring and Equations

Exercises 7–16: Factor out the greatest common factor.

7. $10x - 15$
8. $32 - 16x$
9. $2x^3 - 5x$
10. $3y - 9y^2$
11. $8x^3 - 4x^2 + 16x$
12. $-5x^3 + x^2 - 4x$
13. $5x^2y^2 - 15x^2y^3$
14. $21xy + 14x^3y^3$
15. $15x^2y + 10xy - 25x^2y^2$
16. $14a^3b^2 - 21a^2b^2 + 35a^2b$

Exercises 17–20: (Refer to Examples 2 and 3.) Use factoring to solve the polynomial equation. Support your result graphically and numerically.

17. $x^2 - 2x = 0$
18. $15x - 5x^2 = 0$
19. $12x^2 = 3x$
20. $x^2 = 16x$

Exercises 21–32: Use factoring to solve the polynomial equation.

21. $x^2 - x = 0$
22. $4x - 2x^2 = 0$
23. $5x^2 - x = 0$
24. $4x^2 + 3x = 0$
25. $10x^2 = 25x + 2x^2$
26. $34x^2 = 51x$
27. $32x^4 - 16x^3 = 0$
28. $5x^3 = 6x^2 - 4x^3$
29. $2x(x + 2) + 4(x + 2) = 0$
30. $5(x - 1) - 2x(x - 1) = 0$
31. $(x - 5)x^2 - (x - 5)x = 0$
32. $7x(x - 1) - 3(x - 1) = 0$

Factoring by Grouping

Exercises 33–40: Use grouping to factor the polynomial.

33. $x^3 + 3x^2 + 2x + 6$
34. $4x^3 + 3x^2 + 8x + 6$
35. $6x^3 - 4x^2 + 9x - 6$
36. $x^3 - 3x^2 - 5x + 15$
37. $3x^3 - 15x^2 + 5x - 25$
38. $2x^4 - x^3 + 4x - 2$
39. $xy + x + 3y + 3$
40. $ax + bx - ay - by$

Applications

41. *Flight of a Baseball* (Refer to Example 4.) If a baseball is hit upward at a velocity of 128 feet per second (about 87 mph), its height h above the ground can be modeled by
$$h(t) = -16t^2 + 128t,$$
where t is in seconds.
    (a) Use factoring to determine when the baseball strikes the ground.
    (b) Graph h in $[0, 9.4, 1]$ by $[-10, 300, 100]$. Use the trace feature to determine whether the graph supports your answer in part (a).
    (c) Make a table of h, starting at $t = 3$ and incrementing by 1. Does the table support your answer in part (a)?
    (d) Use the table and graph of h to determine the maximum height of the baseball. At what value of t did this occur?

42. *U.S. AIDS Cases* The cumulative number of AIDS cases in thousands can be modeled by $f(x) = 4.1x^2 - 25x + 46$, where $x = 4$ corresponds to 1984, $x = 5$ to 1985, and so on. Estimate either graphically or numerically when the cumulative number of AIDS cases first reached 500 thousand.

43. *Space Shuttle* Approximate heights y in feet of the space shuttle *Endeavour*, t seconds after liftoff, are shown in the table.

t (seconds)	20	30	40
y (feet)	4500	10,000	18,000

t (seconds)	50	60
y (feet)	28,000	40,000

Source: NASA.

(a) The function $f(t) = 11t^2 + 6t$ models the data, where $y = f(t)$. Make a table of f, starting at $t = 20$ and incrementing by 10. Comment on how well f models the data.
(b) Predict the height of the shuttle 70 seconds after liftoff. Round your answer to the nearest thousand feet.
(c) Solve the equation $11t^2 + 6t = 0$. Do your results have any physical meaning?

44. *Average Temperature* Suppose that the monthly average temperatures T in degrees Fahrenheit for a city are given by $T(x) = 24x - 2x^2$, where $x = 1$ corresponds to January, $x = 2$ to February, and so on.
(a) What is the monthly average temperature in October?
(b) Use factoring to solve the equation $24x - 2x^2 = 0$. Do your results have any physical meaning?
(c) Support your results in part (b) either graphically or numerically.

45. *Area* Suppose that the area of a rectangular room is 216 square feet. Is it possible to determine the dimensions of the room from this information? Explain.

46. *Area* Suppose that the area of a rectangular room is $4x^2 + 8x$ square feet. If the length of the room is $4x$, is it possible to determine the width of the room in terms of x? Explain.

WRITING ABOUT MATHEMATICS

47. A student solves the following equation *incorrectly* by factoring.
$$x^2 - 5x \stackrel{?}{=} 1$$
$$x(x - 5) \stackrel{?}{=} 1$$
$$x \stackrel{?}{=} 1 \quad \text{or} \quad x - 5 \stackrel{?}{=} 1$$
$$x \stackrel{?}{=} 1 \quad \text{or} \quad x \stackrel{?}{=} 6$$
What is the student's mistake? Explain.

48. Explain how you would solve the equation $ax^2 + bx = 0$ symbolically and graphically.

CHECKING BASIC CONCEPTS FOR SECTIONS 5.3 AND 5.4

1. Multiply the expressions.
   (a) $-5(x - 6)$
   (b) $4x^3(3x^2 - 5x)$
   (c) $(2x - 1)(x + 3)$

2. Multiply the following special products.
   (a) $(5x - 6)(5x + 6)$
   (b) $(3x - 4)^2$

3. Make a sketch of a rectangle whose area illustrates the fact that
$$(x + 2)(x + 4) = x^2 + 6x + 8.$$

4. Factor out the greatest common factor.
   (a) $3x^2 - 6x$
   (b) $16x^3 - 8x^2 + 4x$

5. Use factoring to solve each equation. Give either graphical or numerical support for your answer.
   (a) $x^2 - 2x = 0$
   (b) $9x^2 = 81x$

5.5 FACTORING TRINOMIALS

Factoring Trinomials ~ Factoring Trinomials with Technology

INTRODUCTION

In business, an item usually sells more at a lower price. At a higher price fewer items are sold, but more money is made on each item. For example, suppose that, if concert tickets are priced at $100 no one will buy a ticket, but for each $1 reduction in price

100 additional tickets will be sold. If the promoters of the concert need to gross $240,000 from ticket sales, what ticket price accomplishes this goal? To solve this problem we need to set up and solve a polynomial equation. In this section we describe how to solve polynomial equations symbolically, graphically, and numerically.

FACTORING TRINOMIALS

A *trinomial* is a polynomial that has three terms. For now we focus on factoring trinomials of degree 2. We begin by reviewing multiplication of binomials, which was introduced in Section 5.3.

$$(x + 1)(x + 2) = x \cdot x + x \cdot 2 + 1 \cdot x + 1 \cdot 2$$
$$= x^2 + 3x + 2$$

Note that x^2 results from multiplying the *first* terms of each binomial. The middle term of $3x$ results from adding the product of the *outside* terms to the product of the *inside* terms. Finally, the last term of 2 results from multiplying the *last* terms of each binomial. Recall that this method of multiplying binomials is abbreviated *FOIL* and is applied as follows.

$$(x + 1)(x + 2) = x^2 + 3x + 2$$

$$\begin{array}{c} 1x \\ +2x \\ \hline 3x \end{array}$$

Now suppose that we want to factor $x^2 + 3x + 2$. We can apply this process in reverse. The factors of x^2 are x and x, so we do the following.

$$x^2 + 3x + 2 \stackrel{?}{=} (x + \underline{})(x + \underline{})$$

Next we consider the factors of 2. Two integer factors are 1 and 2, so we write them in the blanks.

$$x^2 + 3x + 2 \stackrel{?}{=} (x + 1)(x + 2)$$

For the factorization to be valid *we must add the product of the outside terms to the product of the inside terms*. The outside terms give $2x$, and the inside terms give $1x$. Their sum is $2x + 1x = 3x$, which is the correct middle term.

EXAMPLE 1 *Factoring the form $x^2 + bx + c$*

Factor each trinomial.

(a) $x^2 + 8x + 12$ **(b)** $x^2 - 7x + 12$ **(c)** $x^2 - 4x - 12$

Solution

(a) The factors of x^2 are x and x, so we begin by writing

$$x^2 + 8x + 12 \stackrel{?}{=} (x + \underline{})(x + \underline{}).$$

There are several factors for 12, such as 1 and 12, 2 and 6, or 3 and 4. If we try 3 and 4, then

$$(x + 3)(x + 4) = x^2 + 4x + 3x + 12$$
$$= x^2 + 7x + 12.$$

5.5 Factoring Trinomials

The middle term, $7x$, is incorrect because the sum of these factors is $3 + 4 = 7$ and a middle term of $8x$ is needed. We try the factors 2 and 6 because their sum is 8.

$$x^2 + 8x + 12 = (x + 6)(x + 2)$$

$$\underbrace{6x}_{} $$
$$\underline{+2x}$$
$$8x \quad \text{It checks.}$$

Thus $x^2 + 8x + 12 = (x + 2)(x + 6)$.

(b) To obtain a middle term of $-7x$ and a last term of *positive* 12, we need the factors of 12 to be both negative and sum to -7. We try the factors -3 and -4.

$$(x - 3)(x - 4) = x^2 - 4x - 3x + 12$$
$$= x^2 - 7x + 12 \quad \text{It checks.}$$

Thus $x^2 - 7x + 12 = (x - 3)(x - 4)$.

(c) For the last term to be -12, we need one factor of 12 to be positive and one to be negative. The middle term is $-4x$, so we need the sum of the two factors to be -4. We try the factors 2 and -6.

$$(x + 2)(x - 6) = x^2 - 6x + 2x - 12$$
$$= x^2 - 4x - 12 \quad \text{It checks.}$$

Thus $x^2 - 4x - 12 = (x + 2)(x - 6)$.

Factoring may be used to solve the problem presented in the introduction to this section. To do so we let x represent the number of dollars that the price of a ticket is reduced below $100. Then $100 - x$ represents the price of a ticket. For each $1 reduction x in price, 100 additional tickets will be sold, so the number of tickets sold is given by $100x$. Because the number of tickets sold times the price of each ticket equals the total sales, we need to solve the equation

$$100x(100 - x) = 240{,}000$$

to determine when gross sales will reach $240,000. Note that we cannot apply the zero-product property to this equation because the product of $(100x)$ and $(100 - x)$ does not equal 0.

EXAMPLE 2 *Determining ticket cost*

Solve the equation $100x(100 - x) = 240{,}000$ symbolically. What should be the price of the tickets and how many tickets will be sold? Support your answer graphically and numerically.

Solution

Symbolic Solution Begin by applying the distributive property.

$100x(100 - x) = 240{,}000$	Given equation
$10{,}000x - 100x^2 = 240{,}000$	Distributive property
$-100x^2 + 10{,}000x - 240{,}000 = 0$	Subtract 240,000.
$x^2 - 100x + 2400 = 0$	Divide each term by -100.
$(x - 40)(x - 60) = 0$	Factor.
$x = 40 \quad \text{or} \quad x = 60$	Solve.

If $x = 40$, the price is $100 - 40 = \$60$ per ticket. If $x = 60$, the price is $100 - 60 = \$40$. When $x = 40$, then $100 \cdot 40 = 4000$ tickets are sold at $60 each. When $x = 60$, then $100 \cdot 60 = 6000$ tickets are sold at $40 each. In either case, tickets sales are $240,000.

Graphical and Numerical Solution For graphical support, let $Y_1 = X^2 - 100X + 2400$. The graph of Y_1 intersects the x-axis at $x = 40$ and $x = 60$, as shown in Figures 5.46 and 5.47. Numerical support is shown in Figure 5.48, where $Y_1 = 0$ when $x = 40$ and $x = 60$.

[0, 80, 10] by [−500, 500, 100]

[0, 80, 10] by [−500, 500, 100]

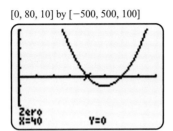

Figure 5.46

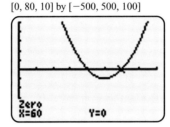

Figure 5.47

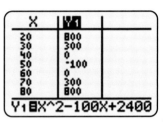

Figure 5.48

Critical Thinking

In Example 2 what price will maximize the revenue received from ticket sales? (*Hint:* Graph $y = 100x(100 - x)$.)

In the next example we factor expressions of the form $ax^2 + bx + c$, where $a \neq 1$. In this situation, we may need to *guess and check* or use *trial and error* a few times before finding the correct factors.

EXAMPLE 3 *Factoring the form $ax^2 + bx + c$*

Factor each trinomial.

(a) $2x^2 + 9x + 4$ (b) $6x^2 - x - 2$ (c) $4x^3 - 14x^2 + 6x$

Solution

(a) The factors of $2x^2$ are $2x$ and x, so we begin by writing

$$2x^2 + 9x + 4 \stackrel{?}{=} (2x + \underline{})(x + \underline{}).$$

The factors of the last term, 4, are either 1 and 4 or 2 and 2. Selecting the factors 2 and 2 results in a middle term of $6x$ rather than $9x$.

$$(2x + 2)(x + 2) = 2x^2 + 6x + 4$$

Next we try the factors 1 and 4.

$$(2x + 4)(x + 1) = 2x^2 + 6x + 4$$

Again we obtain the wrong middle term. By interchanging the 1 and 4, we find the correct factorization.

$$(2x + 1)(x + 4) = 2x^2 + 9x + 4$$

with middle terms $1x + 8x = 9x$

(b) The factors of $6x^2$ are either $2x$ and $3x$ or $6x$ and x. The factors of -2 are either -1 and 2 or 1 and -2. To obtain a middle term of $-x$ we use the following factors.

$$(3x - 2)(2x + 1) = 6x^2 + 3x - 4x - 2$$
$$= 6x^2 - x - 2 \qquad \text{It checks.}$$

To find the correct factorization we may need to guess and check a few times.

(c) Each term contains a common factor of $2x$, so we do the following step first.

$$4x^3 - 14x^2 + 6x = 2x(2x^2 - 7x + 3)$$

Next we factor $2x^2 - 7x + 3$. The factors of $2x^2$ are $2x$ and x. Because the middle term is negative, we use -1 and -3 for factors of 3.

$$4x^3 - 14x^2 + 6x = 2x(2x^2 - 7x + 3)$$
$$= 2x(2x - 1)(x - 3)$$

Again we may need to guess and check more than once before obtaining the correct factors.

EXAMPLE 4 *Estimating passing distance*

A car traveling 48 miles per hour accelerates at a constant rate to pass a car in front of it. A no-passing zone begins 2000 feet away. The distance traveled in feet by the car after t seconds is modeled by $d(t) = 3t^2 + 70t$. How long does it take for the car to reach the no-passing zone?

Solution

We must determine the time when $d(t) = 2000$.

$3t^2 + 70t = 2000$	Equation to be solved
$3t^2 + 70t - 2000 = 0$	Subtract 2000.
$(3t - 50)(t + 40) = 0$	Factor.
$3t - 50 = 0 \quad \text{or} \quad t + 40 = 0$	Zero-product property
$t = \dfrac{50}{3} \quad \text{or} \quad t = -40$	Solve.

The car reaches the no-passing zone after $\dfrac{50}{3} \approx 16.7$ seconds. (The solution $t = -40$ has no physical meaning in this problem.)

FACTORING TRINOMIALS WITH TECHNOLOGY

Technology may be used to factor trinomials and other types of polynomials. One way to do so is first to find the zeros of a polynomial. A number p is a zero if a value of 0 results when p is substituted into the polynomial. For example, both 2 and -2 are zeros of the polynomial $x^2 - 4$ because

$$2^2 - 4 = 0 \quad \text{and} \quad (-2)^2 - 4 = 0.$$

Now consider the trinomial $x^2 - 3x + 2$. If we graph $Y_1 = X^2 - 3X + 2$, its zeros (or x-intercepts) are 1 and 2, as shown in Figures 5.49 and 5.50. Note that this trinomial factors as

$$x^2 - 3x + 2 = (x - 1)(x - 2).$$

In general, if p is a zero of a trinomial, $(x - p)$ is a factor, as summarized in the following technology note.

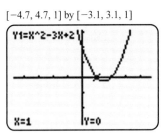

Figure 5.49   Figure 5.50

Technology Note *Factoring a Trinomial with Technology*

To factor the trinomial $x^2 + bx + c$ either graph or make a table of $y = x^2 + bx + c$. If the zeros of the trinomial are p and q, then the trinomial may be factored as

$$x^2 + bx + c = (x - p)(x - q).$$

If the trinomial $ax^2 + bx + c$ has zeros p and q, then it may be factored as

$$ax^2 + bx + c = a(x - p)(x - q).$$

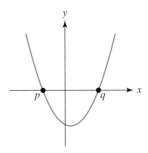

We demonstrate this method in the next example.

EXAMPLE 5   *Factoring with technology*

Factor each trinomial graphically or numerically.

(a) $x^2 - 2x - 24$ **(b)** $2x^2 - 51x + 220$

5.5 Factoring Trinomials

Solution

(a) Graph $Y_1 = X^2 - 2X - 24$. The zeros of the trinomial are -4 and 6, as shown in Figures 5.51 and 5.52. Thus the trinomial factors as follows.

$$x^2 - 2x - 24 = (x - (-4))(x - 6)$$
$$= (x + 4)(x - 6)$$

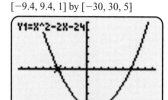

Figure 5.51 Figure 5.52

(b) Construct a table for $Y_1 = 2X^2 - 51X + 220$. Figures 5.53 and 5.54 reveal that the zeros are 5.5 and 20. The leading coefficient of $2x^2 - 51x + 220$ is 2, so we factor this expression as follows.

$$2x^2 - 51x + 220 = 2(x - 5.5)(x - 20)$$
$$= (2x - 11)(x - 20)$$

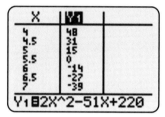

Figure 5.53 Figure 5.54

5.5 Putting It All Together

Equations of the form $ax^2 + bx + c = 0$ occur in applications. One way to solve such equations is to factor the trinomial $ax^2 + bx + c$. This factorization may be done symbolically, graphically, or numerically. The following table includes an explanation and examples of each technique.

Technique	Explanation	Example
Symbolic Factoring	To factor $x^2 + bx + c$ find two factors of c whose sum is b.	To factor $x^2 - 3x - 4$ find two factors of -4 whose sum is -3. These factors are 1 and -4. $$x^2 - 3x - 4 = (x + 1)(x - 4).$$ It checks.

continued on next page

Technique	Explanation	Example
Symbolic Factoring *continued*	To factor $ax^2 + bx + c$ find factors of ax^2 and c so that the middle term is bx.	To factor $2x^2 + 3x - 2$ find factors of $2x^2$ and -2 so that the middle term is $3x$. $$2x^2 + 3x - 2 = (2x - 1)(x + 2).$$ $\begin{array}{r} -x \\ +4x \\ \hline 3x \end{array}$ It checks.
Graphical Factoring	If p and q are zeros of $ax^2 + bx + c$ then the expression factors as $a(x - p)(x - q)$.	To factor $x^2 - 3x - 4 (a = 1)$, graph $Y_1 = X^2 - 3X - 4$. The zeros are -1 and 4 and it follows that $(x - (-1))$ and $(x - 4)$ are factors. Thus, $$x^2 - 3x - 4 = (x + 1)(x - 4)$$
Numerical Factoring	If p and q are zeros of $ax^2 + bx + c$ then the expression factors as $a(x - p)(x - q)$.	To factor $2x^2 + 3x - 2$, make a table for $Y_1 = 2X^2 + 3X - 2$. The zeros are 0.5 and -2, so $(x - 0.5)$ and $(x + 2)$ are factors. Thus $$2x^2 + 3x - 2 = 2(x - 0.5)(x + 2)$$ $$= (2x - 1)(x + 2).$$

5.5 EXERCISES

CONCEPTS

1. What is a trinomial? Give an example.

2. What is the method of *FOIL* used for?

3. What does each letter in *FOIL* stand for?

4. If you factor $x^2 + 3x + 2$, do you obtain $(x + 1)(x + 2)$? Explain.

5. If you factor $x^2 - 7x + 6$, do you obtain $(x - 2)(x - 3)$? Explain.

5.5 Factoring Trinomials

6. Suppose that you graph $y = 6x^2 - 15x + 6$ and find the x-intercepts to be $\frac{1}{2}$ and 2. What are the solutions to the equation
$$6x^2 - 15x + 6 = 0?$$

FACTORING TRINOMIALS

Exercises 7–14: Determine whether the factors of the trinomial are the given binomials.

7. $x^2 + 5x + 6$ $(x + 2)(x + 3)$
8. $x^2 - 13x + 30$ $(x - 10)(x - 3)$
9. $x^2 - x - 20$ $(x + 5)(x - 4)$
10. $x^2 + 3x - 28$ $(x + 4)(x - 7)$
11. $6x^2 - 11x + 4$ $(2x - 1)(3x - 4)$
12. $4x^2 - 19x + 12$ $(2x - 4)(2x - 3)$
13. $10x^2 - 21x + 10$ $(5x + 2)(2x - 5)$
14. $12x^2 + 5x - 2$ $(3x + 2)(4x - 1)$

Exercises 15–34: Factor the expression.

15. $x^2 + 7x + 10$
16. $x^2 + 3x - 10$
17. $x^2 + 8x + 12$
18. $x^2 - 8x + 12$
19. $x^2 - 13x + 36$
20. $x^2 + 11x + 24$
21. $x^2 - 7x - 8$
22. $x^2 - 21x - 100$
23. $2x^2 + 7x + 3$
24. $2x^2 - 5x - 3$
25. $6x^2 - x - 2$
26. $10x^2 + 3x - 1$
27. $1 + x - 2x^2$
28. $3 - 5x - 2x^2$
29. $20 + 7x - 6x^2$
30. $4 + 13x - 12x^2$
31. $5x^3 + x^2 - 6x$
32. $2x^3 + 8x^2 - 24x$
33. $6x^3 + 21x^2 + 9x$
34. $12x^3 - 8x^2 - 20x$

FACTORING WITH TECHNOLOGY

Exercises 35–38: Use the table to factor the expression. Check your answer by multiplying.

35. $x^2 - 3x - 4$
36. $x^2 - 40x + 300$

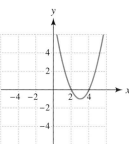
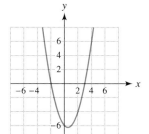

37. $2x^2 - 2x - 4$
38. $3x^2 - 18x + 24$

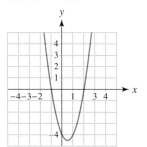
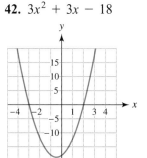

Exercises 39–42: Factor the expression by using the graph of $y = f(x)$, where $f(x)$ is the given expression. Check your answer by multiplying.

39. $x^2 - 6x + 8$
40. $x^2 - x - 6$

41. $2x^2 - 2x - 4$
42. $3x^2 + 3x - 18$

Exercises 43–52: Factor the expression graphically or numerically.

43. $x^2 + 3x - 10$
44. $x^2 + 7x + 12$
45. $x^2 - 3x - 28$
46. $x^2 - 25x + 100$
47. $2x^2 - 14x + 20$
48. $12x^2 - 6x - 6$
49. $5x^2 - 30x - 200$
50. $12x^2 - 30x + 12$
51. $8x^2 - 44x + 20$
52. $20x^2 + 3x - 9$

APPLICATIONS

53. *Ticket Prices* (Refer to Example 2.) If tickets are sold for $35, 1200 tickets are expected to be sold. For each $1 reduction in ticket price, an additional 100 tickets will be sold.
    (a) Write an expression that gives the revenue from ticket sales when ticket prices are reduced by x dollars.
    (b) Determine symbolically the ticket prices that result in sales of $54,000.
    (c) Solve part (b) either graphically or numerically.
    (d) Find the ticket price that gives maximum revenue.

54. *Passing Distances* (Refer to Example 4.) The distance traveled in feet by a car passing another car after t seconds is given by $2t^2 + 88t$.
    (a) Use factoring to determine how long it takes for the car to travel 600 feet.
    (b) Solve part (a) either graphically or numerically.

WRITING ABOUT MATHEMATICS

55. A student factors each trinomial into the given pair of binomials.
    (i) $x^2 - 5x - 6$ $(x - 1)(x + 6)$
    (ii) $10x^2 - 11x - 6$ $(5x - 3)(2x + 2)$
    (iii) $12x^2 + 13x + 3$ $(6x + 1)(2x + 3)$
    Explain the error that the student made.

56. Suppose that a graph of $y = x^2 + bx + c$ intersects the x-axis twice at p and q. Explain how you could use the graph to factor $x^2 + bx + c$.

GROUP ACTIVITY
Working with Real Data

Directions: Form a group of 2 to 4 people. Select someone to record the group's responses for this activity. All members of the group should work cooperatively to answer the questions. If your instructor asks for the results, each member of the group should be prepared to respond. If the group is asked to turn in its work, be sure to include each person's name on the paper.

Women in College The following table shows the number of women attending 4-year institutions of higher education.

Year	1983	1993	2003
Women (millions)	3.8	4.7	5.2

Source: National Center for Education Statistics.

(a) Verify that these data are modeled by $f(x) = -0.002x^2 + 8.042x - 8078.908$.
(b) Make a table of $f(x)$, starting at $x = 2004$ and incrementing by 1. What does this model predict will happen in about 2010? Is this model accurate that far into the future? Explain.
(c) Would $f(x)$ be more accurate predicting the number of women attending college in 2001? Explain.
(d) The zeros of $f(x)$ are approximately 1958 and 2062. Do these zeros have meaning in this model? Explain.

5.6 SPECIAL TYPES OF FACTORING

**Difference of Two Squares ~ Perfect Square Trinomials ~
Sum and Difference of Two Cubes ~ Solving Polynomial Equations**

INTRODUCTION

Some polynomials can be factored with special methods. These factoring techniques are used in other mathematics courses to solve equations and simplify expressions. In this section we discuss these methods and their use in solving polynomial equations.

DIFFERENCE OF TWO SQUARES

In Section 5.3 we learned that

$$(a - b)(a + b) = a^2 - b^2.$$

We can use this equation to factor a difference of two squares. For example, if we want to factor $x^2 - 25$, we can think of it being in the form $a^2 - b^2$, where $a = x$ and $b = 5$. Thus substituting into

$$a^2 - b^2 = (a - b)(a + b)$$

gives

$$x^2 - 25 = (x - 5)(x + 5).$$

EXAMPLE 1 *Factoring the difference of two squares*

Factor each polynomial.

(a) $9x^2 - 64$ **(b)** $4x^2 - 9y^2$

Solution

(a) Note that $9x^2 = (3x)^2$ and $64 = 8^2$.
$$9x^2 - 64 = (3x)^2 - (8)^2$$
$$= (3x - 8)(3x + 8)$$

(b) $4x^2 - 9y^2 = (2x)^2 - (3y)^2$
$$= (2x - 3y)(2x + 3y)$$

PERFECT SQUARE TRINOMIALS

In Section 5.3 we also learned how to expand $(a + b)^2$ and $(a - b)^2$.

$$(a + b)^2 = a^2 + 2ab + b^2$$
$$(a - b)^2 = a^2 - 2ab + b^2$$

The expressions $a^2 + 2ab + b^2$ and $a^2 - 2ab + b^2$ are called **perfect square trinomials.** If we can recognize a perfect square trinomial, we can use these formulas to factor it, as demonstrated in the next example.

Polynomials are used to model real-world data. However, using factoring to solve a polynomial equation may not be convenient. In that case, technology is commonly used to solve the equation graphically or numerically. This approach is demonstrated in the next example, where a quadratic polynomial (degree 2) is used to model a proposed highway.

EXAMPLE 5 *Modeling a hill on a highway*

When designing a highway, engineers commonly use quadratic polynomials to model hills and valleys along a proposed route. Suppose that the elevation E in feet of a highway x horizontal feet along a hill is modeled by

$$E = -0.0001x^2 + 0.2x + 1000,$$

where $0 \leq x \leq 2000$. See Figure 5.58. (*Source:* F. Manning and W. Kilareski, *Principles of Highway Engineering and Traffic Analysis.*)

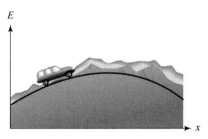

Figure 5.58

(a) Make a table of the elevations for $x = 0, 200, 400, \ldots, 2000$ and interpret the values in the table.
(b) Use graphing to determine where the elevation of the highway is 1070 feet.

Solution

(a) Construct the table for $Y_1 = -.0001X^2 + .2X + 1000$, starting at $x = 0$ and incrementing by 200, as shown in Figures 5.59 and 5.60. The elevation increases from 1000 feet when $x = 0$ to a maximum of 1100 feet when $x = 1000$ feet. It then decreases to 1000 feet at a distance of 2000 horizontal feet along the proposed route.

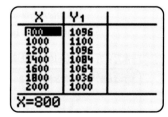

Figure 5.59   **Figure 5.60**

(b) Solve the equation

$$-0.0001x^2 + 0.2x + 1000 = 1070.$$

Begin by subtracting 1070 from both sides.
$$-0.0001x^2 + 0.2x - 70 = 0$$

Graph $Y_1 = -.0001X^2 + 0.2X - 70$. The zeros of Y_1 are shown in Figures 5.61 and 5.62 as $x \approx 452$ and $x \approx 1548$. Thus the elevation is 1070 feet, either 452 feet or 1548 feet (horizontally) along the hill.

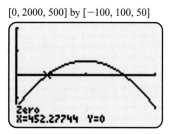

Figure 5.61   Figure 5.62

5.6 Putting It All Together

Factoring is often used to solve equations. In this section we studied three special types of factoring, which are summarized in the following table.

Type of Factoring	Description	Example
The difference of two squares	To factor the difference of two squares use the following. $a^2 - b^2 = (a - b)(a + b)$ **Note:** $a^2 + b^2$ cannot be factored.	$25x^2 - 16 = (5x - 4)(5x + 4)$ $(a = 5x, b = 4)$
A perfect square trinomial	To factor a perfect square trinomial use one of the following. $a^2 + 2ab + b^2 = (a + b)^2$ $a^2 - 2ab + b^2 = (a - b)^2$	$x^2 + 4x + 4 = (x + 2)^2 \quad (a = x, b = 2)$ $16x^2 - 24x + 9 = (4x - 3)^2 \quad (a = 4x, b = 3)$
The sum and difference of two cubes	The sum and difference of two cubes can be factored as follows: $a^3 + b^3 = (a + b)$ $\cdot (a^2 - ab + b^2)$ $a^3 - b^3 = (a - b)$ $\cdot (a^2 + ab + b^2)$	$x^3 + 8y^3 = (x + 2y)(x^2 - x \cdot 2y + (2y)^2)$ $= (x + 2y)(x^2 - 2xy + 4y^2)$ $(a = x, b = 2y)$ $125x^3 - 64 = (5x - 4)((5x)^2 + 5x \cdot 4 + 4^2)$ $= (5x - 4)(25x^2 + 20x + 16)$ $(a = 5x, b = 4)$

5.6 EXERCISES

CONCEPTS

1. Give an example of a difference of two squares.
2. Give an example of a trinomial perfect square.
3. Give an example of the sum of two cubes.
4. Factor $a^2 - b^2$.
5. Factor $a^2 + 2ab + b^2$.
6. Factor $a^3 - b^3$.

DIFFERENCE OF TWO SQUARES

Exercises 7–10: Determine whether the expression represents a difference of two squares. Factor the expression.

7. $x^2 - 25$
8. $16x^2 - 100$
9. $x^3 + y^3$
10. $9x^2 + 36y^2$

Exercises 11–20: Factor the expression completely.

11. $x^2 - 36$
12. $y^2 - 144$
13. $4x^2 - 25$
14. $36 - y^2$
15. $36x^2 - 100$
16. $9x^2 - 4y^2$
17. $64z^2 - 25z^4$
18. $100x^3 - x$
19. $16x^4 - y^4$
20. $x^4 - 9y^2$

PERFECT SQUARE TRINOMIALS

Exercises 21–28: Determine whether the trinomial is a perfect square. If it is, factor it.

21. $x^2 + x + 16$
22. $4x^2 - 2x + 25$
23. $x^2 + 8x + 16$
24. $x^2 - 4x + 4$
25. $4z^2 - 4z + 1$
26. $9z^2 + 6z + 4$
27. $16t^2 - 12t + 9$
28. $4t^2 + 12t + 9$

Exercises 29–36: Factor the expression.

29. $x^2 + 2x + 1$
30. $x^2 - 6x + 9$
31. $4x^2 + 20x + 25$
32. $x^2 + 10x + 25$
33. $x^2 - 12x + 36$
34. $16z^4 - 24z^3 + 9z^2$
35. $9z^3 - 6z^2 + z$
36. $49y^2 + 42y + 9$

SUM AND DIFFERENCE OF TWO CUBES

Exercises 37–46: Factor the expression.

37. $x^3 - 8$
38. $x^3 + 8$
39. $y^3 + z^3$
40. $y^3 - z^3$
41. $27x^3 - 8$
42. $64 - y^3$
43. $8x^4 + 125x$
44. $x^3y^2 + 8y^5$
45. $27y - 8x^3y$
46. $x^3y^3 + 1$

GENERAL FACTORING

Exercises 47–58: Factor the expression completely.

47. $25x^2 - 64$
48. $25x^2 - 30x + 9$
49. $x^3 + 27$
50. $4 - 16y^2$
51. $64x^2 + 16x + 1$
52. $2x^2 - 5x + 3$
53. $3x^2 + 14x + 8$
54. $125x^3 - 1$
55. $x^4 + 8x$
56. $2x^3 - 12x^2 + 18x$
57. $64x^3 + 8y^3$
58. $54 - 16x^3$

SOLVING POLYNOMIAL EQUATIONS

Exercises 59–62: Use the graph to solve the equation. Then solve the equation by factoring.

59. $x^2 + x - 6 = 0$ **60.** $x^2 - 2x - 3 = 0$

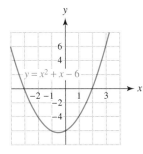 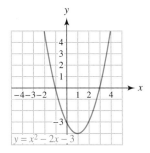

61. $2x^2 - 4x - 16 = 0$ **62.** $2x^2 + 3x - 2 = 0$

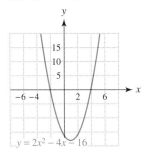 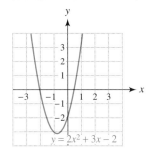

Exercises 63–70: (Refer to Example 4.) Solve the equation
  (a) *symbolically*,
  (b) *graphically, and*
  (c) *numerically.*

63. $x^2 - 2x - 15 = 0$

64. $x^2 + 7x + 10 = 0$

65. $2x^2 - 3x = 2$

66. $4t^2 + 25 = 20t$

67. $3t^2 + 18t + 15 = 0$

68. $4z^2 = 16$

69. $4x^4 + 16x^2 = 16x^3$

70. $2x^3 + 12x^2 + 18x = 0$

APPLICATIONS

71. *Modeling a Highway* (Refer to Example 5.) In highway design quadratic polynomials are used not only to model hills but also to model valleys. Valleys or sags in the road are sometimes referred to as *sag curves*. Suppose that the elevation E in feet of the sag curve x (horizontal) feet along a proposed route is modeled by

$$E = 0.0002x^2 - 0.3x + 500,$$

where $0 \leq x \leq 1500$. See the accompanying figure. (*Source:* F. Mannering.)

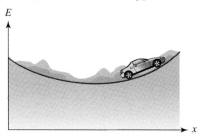

  (a) Make a table of the elevations for $x = 0$, 300, 600, ..., 1500 and interpret the values in the table.
  (b) Determine where the elevation is 400 feet graphically.

72. *Biology* Some types of worms have a remarkable ability to live without moisture. The following table from one study shows the number of worms y surviving after x days.

x (days)	0	20	40
y (worms)	50	48	45
x (days)	80	120	160
y (worms)	36	20	3

Source: D. Brown and P. Rothery, Models in Biology.

  (a) Plot the data and graph

$$y = -0.00138x^2 - 0.076x + 50.1$$

  in the same viewing rectangle.
  (b) Discuss how well the equation models the data.
  (c) Estimate the day x when 10 worms remained.

WRITING ABOUT MATHEMATICS

73. Explain why the graph of $y = ax^2 + bx + c$ is used to model a dip or valley in a highway rather than the graph of $y = |ax + b|$. Make a sketch to support your reasoning.

74. Explain the basic steps that you would use to solve the equation $ax^2 + bx + c = 0$ graphically and numerically. Assume that the equation has two solutions.

CHECKING BASIC CONCEPTS FOR SECTIONS 5.5 AND 5.6

1. Factor each trinomial.
   (a) $x^2 + 3x - 10$
   (b) $x^2 - 3x - 10$
   (c) $8x^2 + 14x + 3$

2. Factor each expression.
   (a) $25x^2 - 16$
   (b) $x^2 + 12x + 36$
   (c) $x^3 - 27$
   (d) $x^3 - 3x^2 + x - 3$ (*Hint:* Factor by grouping.)

3. Solve $x^2 + 3x + 2 = 0$ symbolically. Support your answer either graphically or numerically.

4. A baseball is thrown upward at approximately 44 miles per hour. Its height above the ground after t seconds is given by
$$h(t) = -16t^2 + 64t + 8.$$
Determine symbolically when the baseball is 56 feet above the ground. Support your answer either graphically or numerically.

Chapter 5 Summary

Section 5.1 Polynomial Expressions

A monomial is a term in which each variable has only nonnegative integer exponents.

Examples: -5, $8x^2$, and $5xy^3z^2$

When adding or subtracting monomials, combine like terms. For example,
$$3x^3 - 5x^3 = -2x^3,$$
whereas the sum
$$x^2 + 4x^3$$
cannot be combined. A polynomial is either a monomial or a sum of monomials.

Example: $5x^3 + 6x - 7$ has one variable, degree 3, and leading coefficient 5.

Section 5.2 Polynomial Functions and Models

If the formula for a function is a polynomial, it is a polynomial function. The following are examples of polynomial functions.

$f(x) = 2x - 7$	Linear	Degree 1
$f(x) = 4x^2 - 5x + 1$	Quadratic	Degree 2
$f(x) = 9x^3 + 6x$	Cubic	Degree 3

Linear data can be modeled with a degree 1 polynomial function because its graph is a line. Nonlinear data can be modeled by polynomial functions of degree 2 or higher because these functions have graphs that are curved.

Section 5.3 Multiplication of Polynomials

The product of two polynomials may be found by multiplying every term in the first polynomial by every term in the second polynomial.

Example: $(x + 4)(2x - 7) = x \cdot 2x - x \cdot 7 + 4 \cdot 2x - 4 \cdot 7$
$= 2x^2 + x - 28$

The following special products are often used.

$$(x - y)(x + y) = x^2 - y^2$$
$$(x + y)^2 = x^2 + 2xy + y^2$$
$$(x - y)^2 = x^2 - 2xy + y^2$$

Section 5.4 Factoring Polynomials

Factoring is often used to solve polynomial equations. Different methods can be used to factor polynomials. One method is to factor out any common factors.

$$5x^3 + 10x = 5x(x^2 + 2)$$

A second method is factoring by grouping.

$$5x^3 - 10x^2 + 2x - 4 = (5x^3 - 10x^2) + (2x - 4)$$
$$= 5x^2(x - 2) + 2(x - 2)$$
$$= (5x^2 + 2)(x - 2)$$

Section 5.5 Factoring Trinomials

Correctly factoring a trinomial requires checking the middle term.

$$2x^2 + 3x - 14 = (2x + 7)(x - 2)$$

$\lfloor 7x \rfloor$
$-4x$
$3x$ It checks.

Trinomials can also be factored either graphically or numerically. For example, the x-intercepts on the graph of $f(x) = 2x^2 + x - 1$ are $x = 0.5$ and $x = -1$. Thus $f(x)$ can be factored as

$$2x^2 + x - 1 = 2(x - 0.5)(x + 1)$$
$$= (2x - 1)(x + 1).$$

Section 5.6 Special Types of Factoring

The following are some special types of factoring.

Difference of Two Squares $\quad x^2 - y^2 = (x - y)(x + y)$

Perfect Square Trinomial $\quad x^2 + 2xy + y^2 = (x + y)^2$
$\qquad\qquad\qquad\qquad\qquad x^2 - 2xy + y^2 = (x - y)^2$

Sum of Two Cubes $\quad x^3 + y^3 = (x + y)(x^2 - xy + y^2)$

Difference of Two Cubes $\quad x^3 - y^3 = (x - y)(x^2 + xy + y^2)$

CHAPTER 5

REVIEW EXERCISES

SECTION 5.1

1. Give an example of a binomial and an example of a trinomial.

2. Write a monomial that represents the volume of 10 identical boxes x inches tall with square bases y inches on a side. What is the total volume of the boxes if $x = 5$ and $y = 8$?

Exercises 3–6: Identify the degree and coefficient of the monomial.

3. $-4x^5$
4. y^3
5. $5xy^6$
6. $-9y^{10}$

Exercises 7–10: Combine like terms.

7. $5x - 4x + 10x$
8. $-3y + 6y$
9. $9x^3 - 5x^3 + x^2$
10. $6x^3y - 4x^3 + 8x^3y + 5x^3$

Exercises 11–14: Identify the degree and leading coefficient of the polynomial.

11. $5x^2 - 3x + 5$
12. $-9x^3 - 4x^2 + 6x + 1$
13. $1 - 5x + 8x^2$
14. $x^3 - 5x + 5 + 2x^4$

Exercises 15–18: Combine the polynomials.

15. $(3x^2 - x + 7) + (5x^2 + 4x - 8)$
16. $(6z^3 + z) + (17z^3 - 4z^2)$
17. $(-4x^2 - 6x + 1) - (-3x^2 - 7x + 1)$
18. $(3x^3 - 5x + 7) - (8x^3 + x^2 - 2x + 1)$

SECTION 5.2

Exercises 19–22: Evaluate $f(x)$ at the given value of x.

19. $f(x) = -4x^3 \qquad x = 2$
20. $f(x) = 4x - x^2 \qquad x = -3$
21. $f(x) = 2x^2 - 3x + 2 \qquad x = -1$
22. $f(x) = 1 - x - 4x^3 \qquad x = 4$

23. Use the graph to evaluate $f(2)$.

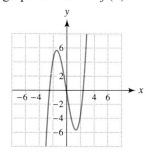

24. Use the table to evaluate $f(x) = 5x - 3x^2$ at $x = 1.5$.

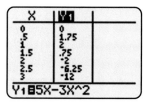

Section 5.3

Exercises 25 and 26: Multiply.

25. $5(3x - 4)$

26. $-2x(1 + x - 4x^2)$

Exercises 27–30: Multiply the monomials.

27. $x^3 \cdot x^5$

28. $-2x^3 \cdot 3x$

29. $-7xy^7 \cdot 6xy$

30. $12xy^4 \cdot 5x^2y$

Exercises 31–42: Multiply the expressions.

31. $(x + 4)(x + 5)$

32. $(x - 7)(x - 8)$

33. $(6x + 3)(2x - 9)$

34. $(2x + 3)(x - 4)$

35. $\left(x - \frac{1}{3}\right)\left(x + \frac{1}{3}\right)$

36. $\left(y - \frac{2}{5}\right)\left(y - \frac{3}{5}\right)$

37. $4x^2(2x^2 - 3x - 1)$

38. $-x(4 + 5x - 7x^2)$

39. $(4x + y)(4x - y)$

40. $(x + 3)^2$

41. $(2y - 5)^2$

42. $(a - b)(a^2 + ab + b^2)$

Section 5.4

Exercises 43–46: Factor out the greatest common factor.

43. $25x^2 - 30x$

44. $18x^3 + 6x$

45. $5y + 15y^2$

46. $12x^3 + 8x^2 - 16x$

Exercises 47–52: Use factoring to solve the polynomial equation. Support your answer graphically or numerically.

47. $x^2 + 3x = 0$

48. $10x + 5x^2 = 0$

49. $7x^4 = 28x^2$

50. $10x^4 - 30x^3 = 0$

51. $2t^2 - 3t + 1 = 0$

52. $4z(z - 3) + 4(z - 3) = 0$

Exercises 53–56: Use grouping to factor the polynomial.

53. $2x^3 + 2x^2 - 3x - 3$

54. $2x^3 + 3x^2 + 6x + 9$

55. $z^3 + z^2 + z + 1$

56. $ax - bx + ay - by$

Sections 5.5 and 5.6

Exercises 57–68: Factor completely.

57. $x^2 + 8x + 12$

58. $x^2 - 5x - 50$

59. $9x^2 + 25x - 6$

60. $4x^2 - 22x + 10$

61. $t^2 - 49$

62. $4y^2 - 9x^2$

63. $x^2 + 4x + 4$

64. $16x^2 - 8x + 1$

65. $x^3 - 27$

66. $8y^3 + 1$

67. $125a^3 - 64$

68. $64x^3 + 27y^3$

Exercises 69 and 70: Use the table to factor the expression. Check your answer by multiplying.

69. $x^2 - 2x - 15$

70. $x^2 - 24x + 143$

71. Use the graph to factor $x^2 + 3x - 28$. Check your answer by multiplying.

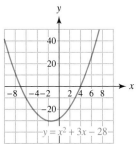

72. Use the graph to solve $x^2 - 21x + 104 = 0$. Check your solutions.

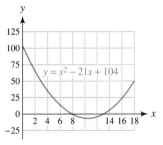

APPLICATIONS

73. *Thanksgiving Travel* The number of people traveling at Thanksgiving in the 1990s increased to an all-time high in 1999. The accompanying graph shows the number of people in millions who traveled 100 miles or more during the Thanksgiving holiday. The function

$$f(x) = -0.061x^2 + 1.52x + 25.2$$

models these data, where $x = 0$ corresponds to 1990, $x = 1$ to 1991, and so on until $x = 9$ corresponds to 1999. (*Source:* **American Automobile Association.**)

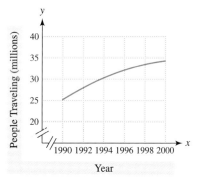

(a) Use $f(x)$ to determine how many people traveled in 1994.
(b) Use the graph to estimate when the number of travelers reached 32 million.
(c) Make a table of $f(x)$ starting at $x = 0$ and incrementing by 1. In what year were there about 28 million travelers?

74. *Cigar Sales* The sales of imported premium cigars increased dramatically from 1994 through 1997 and then fell, as shown in the accompanying line graph. Sales are shown in millions of cigars sold. (*Source:* **Consolidated Cigar Corporation.**)

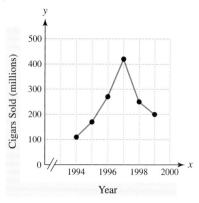

(a) A student proposes that

$$f(x) = 22.5x^2 - 144.5x + 328.5$$

models the data in the graph, where $x = 4$ corresponds to 1994, $x = 5$ to 1995, and so on until $x = 9$ represents 1999. Use $f(x)$ to estimate sales in 1995 and 1997. Compare these values with the sales shown in the graph for 1995 and 1997.
(b) Make a table of $f(x)$, starting at $x = 4$ and incrementing by 1. Use $f(x)$ to determine cigar sales in 1999. Now use the graph to determine how many were sold in 1999. Does $f(x)$ model the data in the graph accurately? Explain.

75. *Air Temperatures* The formula

$$f(x) = -1.466x^2 + 20.25x + 9$$

models the monthly average high temperatures in degrees Fahrenheit at Columbus, Ohio. In this formula $x = 1$ corresponds to January, $x = 2$ to February, and so on. (*Source:* **J. Williams, The Weather Almanac, 1995.**)

(a) What is the average high temperature in May?
(b) Make a table of $f(x)$, starting at $x = 1$ and incrementing by 1. During what month is the average temperature the greatest?
(c) Graph f in [1, 12, 1] by [30, 90, 10] and interpret the graph.

76. *Probability* Suppose that the likelihood, or chance, that a pitch in softball will be a strike is x percent. Then the likelihood that two consecutive pitches will not be strikes is given by $\left(1 - \dfrac{x}{100}\right)^2$.

(a) Multiply this expression.
(b) Let $x = 70$. Use both the given expression and the expression you found in part (a) to obtain the likelihood that two consecutive pitches are not strikes.

77. Write a polynomial that represents the product of three consecutive integers, where x is the largest integer.

(a) Discuss how the number of adult flies varied during this time period.
(b) Discuss how the number of eggs varied during this time period.
(c) How do the peaks in the graph of the adult flies compare to the peaks in the graph of the eggs?
(d) Explain the relationship between the two graphs.

MODELING NONLINEAR DATA

Exercises 3 and 4: Real-world data often do not lie on a line when they are plotted. Such data are called nonlinear data, and nonlinear functions are used to model this type of data. In the next four exercises you will be asked to find your own functions that model real-world data.

3. *Planetary Orbits* Johannes Kepler (1571–1630) was the first person to find a formula that models the relationship between a planet's average distance from the sun and the time it takes to orbit the sun. The following table lists a planet's average distance x from the sun and the number of years y it takes to orbit the sun. In this table distances have been normalized so that Earth's distance is exactly 1. For example, Saturn is 9.54 times farther from the sun than Earth is and requires 29.5 years to orbit the sun.

Planet	x (distance)	y (period)
Mercury	0.387	0.241
Venus	0.723	0.615
Earth	1.00	1.00
Mars	1.52	1.88
Jupiter	5.20	11.9
Saturn	9.54	29.5

Source: C. Ronan, *The National History of the Universe.*

(a) Make a scatterplot of the data. Are the data linear or nonlinear?
(b) Kepler discovered a formula having the form $y = x^k$, which models the data in the table. Graph $Y_1 = X^{2.5}$ and the data in the same viewing rectangle. Does Y_1 model the data accurately? Explain.
(c) Try to discover a value for k so that $Y_1 = X^K$ models the data.
(d) The average distances of Neptune and Pluto from the sun are 30.1 and 39.4, respectively. Estimate how long it takes for each planet to orbit the sun. Check a reference to see whether your answers are accurate.

4. *Women in the Work Force* The number of women gainfully employed in the workforce has changed significantly since 1900. The table lists these numbers in millions at 10-year intervals.

Year	1900	1910	1920	1930	1940
Work Force (millions)	5.3	7.4	8.6	10.8	12.8

Year	1950	1960	1970	1980	1990
Work Force (millions)	18.4	23.2	31.5	45.5	56.6

Source: Department of Labor.

(a) Make a scatterplot of the data in [1890, 2000, 20] by [0, 60, 10]. Are the data linear or nonlinear? Explain your answer.
(b) Use trial and error to find a value for k so that

$$Y_1 = K(X - 1900)^2 + 5.3$$

models the data.
(c) Use Y_1 to predict the number of women that will be in the workforce in 2005.

5. *Aging of America* Americans are living longer. The following table shows the number of Americans expected to be more than 100 years old for various years.

Year	1994	1996	1998
Number (thousands)	50	56	65

Year	2000	2002	2004
Number (thousands)	75	94	110

Source: Bureau of the Census.

(a) Make a scatterplot of the data. Choose an appropriate viewing rectangle.
(b) Use trial and error to find a value for k so that
$$Y_1 = K(X - 1994)^2 + 50$$
models the data.
(c) Use Y_1 to predict the number of Americans that will be more than 100 years old in 2006.

6. *Highway Design* In order to allow enough distance for cars to pass on two-lane highways, engineers calculate minimum sight distances for curves and for hills. See the accompanying figure.

The following table shows the minimum sight distance y in feet for a car traveling x miles per hour.

(a) Make a scatterplot of the data. As x increases, describe how y changes. Are the data linear or nonlinear?
(b) Find a formula that models the data.
(c) Use your formula to predict the minimum sight distance for a car traveling at 56 miles per hour.

x miles per hour	20	30	40	50
y feet	810	1090	1480	1840

x miles per hour	60	65	70
y feet	2140	2310	2490

Source: Haefner, L., *Introduction to Transportation Systems.*

Chapter 6: Quadratic Functions and Equations

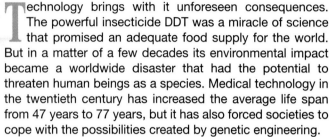

Technology brings with it unforeseen consequences. The powerful insecticide DDT was a miracle of science that promised an adequate food supply for the world. But in a matter of a few decades its environmental impact became a worldwide disaster that had the potential to threaten human beings as a species. Medical technology in the twentieth century has increased the average life span from 47 years to 77 years, but it has also forced societies to cope with the possibilities created by genetic engineering.

In this course we use technology to help teach important concepts in mathematics and to solve applications. Without technology many modern inventions would be impossible. However, technology is not a replacement for mathematical understanding and judgment because fundamental concepts change little over time. On the one hand, the human mind is capable of mathematical insight and decision making, but it is not particularly proficient at performing long, tedious calculations. On the other hand, computers and calculators are incapable of possessing genuine mathematical insight, but they are excellent at performing arithmetic and other routine computation. In this way, technology can complement the human mind in the study and application of mathematics.

People are still the most extraordinary computers of all.
—*John F. Kennedy*

Source: F. Allen, "Technology at the End of the Century," *Invention and Technology,* Winter 2000.

6.1 QUADRATIC FUNCTIONS AND THEIR GRAPHS

Representations of Quadratic Functions ~ Translations of Parabolas ~ Modeling with Quadratic Functions

INTRODUCTION

A taxiway used by an airplane to exit a runway often contains curves. A curve that is too sharp for the speed of the plane is a safety hazard. The scatterplot in Figure 6.1 shows an appropriate radius R of a curve designed for an airplane taxiing x miles per hour. The data are nonlinear because they do not lie on a line. In this section we learn how a quadratic function may be used to model such data. First, we discuss representations of quadratic functions.

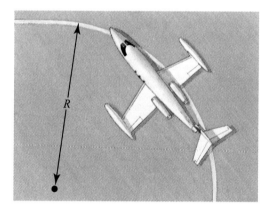

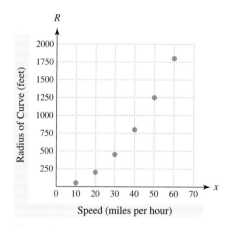

Figure 6.1

REPRESENTATIONS OF QUADRATIC FUNCTIONS

In Chapter 5 you learned that a quadratic function may be represented by a polynomial of degree 2. We now give an alternative definition of a quadratic function.

6.1 Quadratic Functions and Their Graphs

Quadratic Function

A quadratic function may be written in the form

$$f(x) = ax^2 + bx + c,$$

where a, b, and c are real numbers with $a \neq 0$.

A simple quadratic function is $f(x) = x^2$, and its graph is shown in Figure 6.2. The graph of any quadratic function is a **parabola**. A parabola is a U-shaped graph that either opens upward or downward. The graph of $y = x^2$ is a parabola that opens upward, with its *vertex* located at the origin. The **vertex** is the lowest point on the graph of a parabola that opens upward. A parabola opening downward is shown in Figure 6.3. Its vertex is the point $(0, 2)$ and is the highest point on a graph. If we were to fold either graph along the y-axis, the left and right sides of the graph would match. That is, both parts of the graph are symmetric with respect to the y-axis. In this case the y-axis is the **axis of symmetry** for the graph. Figure 6.4 shows a parabola that opens upward and has vertex $(2, -1)$ and axis of symmetry $x = 2$.

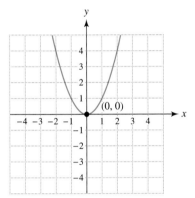

Figure 6.2 Vertex $(0, 0)$

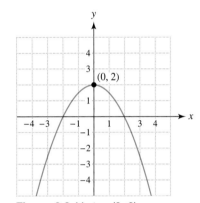

Figure 6.3 Vertex $(0, 2)$

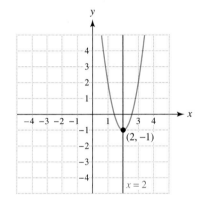

Figure 6.4 Vertex $(2, -1)$

EXAMPLE 1 *Graphing quadratic functions*

Graph each quadratic function. Make a table of function values. Identify the vertex and the axis of symmetry.

(a) $f(x) = 3 - 2x^2$ **(b)** $f(x) = x^2 + 4x + 3$

Solution

(a) Graph $Y_1 = 3 - 2X^2$, as shown in Figure 6.5 on the next page. The parabola opens downward, and the point $(0, 3)$ is the vertex because it is the highest point on the graph. The axis of symmetry corresponds to the y-axis, which may be expressed as $x = 0$. A table of values, or *numerical representation*, is shown in Figure 6.6. Note that the greatest value of Y_1 occurs when $x = 0$.

342 CHAPTER 6 Quadratic Functions and Equations

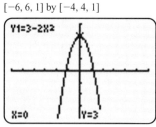
Figure 6.5

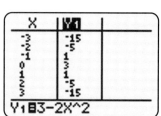
Figure 6.6

(b) Graph and construct a table for $Y_1 = X^2 + 4X + 3$, as shown in Figures 6.7 and 6.8. The vertex is $(-2, -1)$ and is the lowest point on the graph. The axis of symmetry is $x = -2$ because, if the graph is folded on the vertical line $x = -2$, the left and right sides of the parabola match. See Figure 6.9.

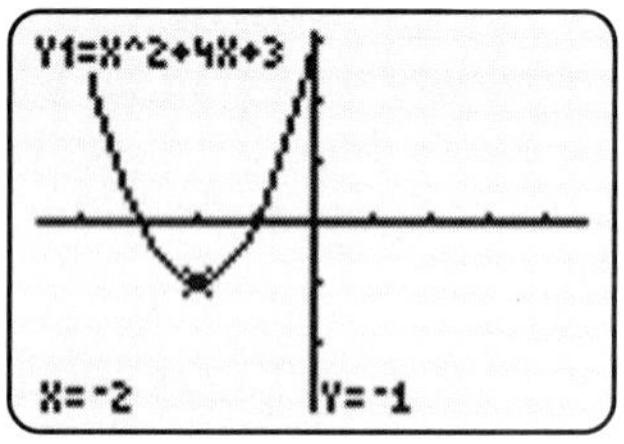

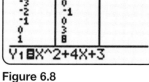

Figure 6.7

Figure 6.8

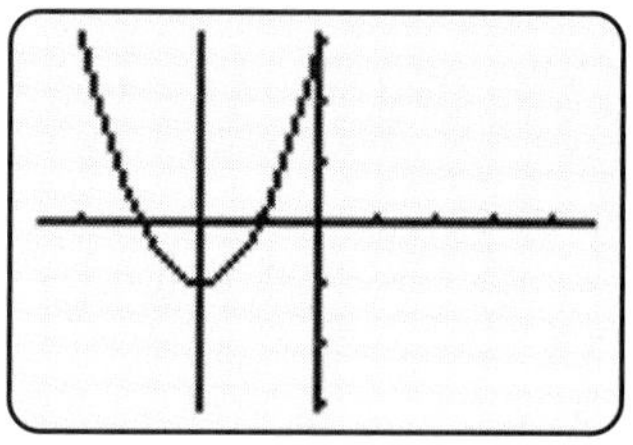

Figure 6.9 Axis of Symmetry

The vertex of a parabola can be found symbolically with the following formula.

Vertex formula

The x-coordinate of the vertex of the graph of $y = ax^2 + bx + c$, $a \neq 0$, is given by

$$x = -\frac{b}{2a}.$$

To find the y-coordinate of the vertex, substitute this x-value into the equation.

We demonstrate this technique in the next example.

EXAMPLE 2 *Finding the vertex of a parabola*

Find the vertex of the graph of $y = 2x^2 - 4x + 1$ symbolically. Support your answer graphically.

Solution

For $y = 2x^2 - 4x + 1$, $a = 2$ and $b = -4$. The x-value of the vertex is

$$x = -\frac{b}{2a} = -\frac{(-4)}{2(2)} = 1.$$

To find the y-value of the vertex, substitute $x = 1$ into the given equation.
$$y = 2(1)^2 - 4(1) + 1 = -1.$$
Thus the vertex is located at $(1, -1)$, which is supported by Figure 6.10.

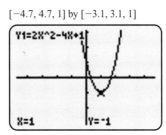

Figure 6.10

Sometimes when a quadratic function f is used to model real-world data, the vertex provides important information. The reason is that the y-coordinate of the vertex gives either the maximum value of $f(x)$ or the minimum value of $f(x)$. We demonstrate this feature in the next example.

EXAMPLE 3 *Maximizing revenue*

A hotel is considering giving the following group-discount on room rates. The regular price for a room is $80, but for each room rented the price will decrease by $2.
(a) Write a formula $f(x)$ that gives the revenue from renting x rooms at the group rate.
(b) Graph $y = f(x)$ in [0, 40, 10] by [0, 1000, 100] and interpret the graph.
(c) What is the maximum revenue? How many rooms should be rented to obtain this revenue?
(d) Solve part (c) symbolically.

Solution

(a) If x rooms are rented, the price for each room is $80 - 2x$. The revenue equals the number of rooms rented times the price of each room. Thus $f(x) = x(80 - 2x)$.
(b) The graph of $Y_1 = X(80 - 2X)$ is shown in Figure 6.11. The revenue increases at first, reaches a maximum, which corresponds to the vertex, and then decreases.

Figure 6.11   **Figure 6.12**

(c) Figure 6.12 reveals that the vertex is $(20, 800)$. Thus the maximum revenue of $800 occurs when 20 rooms are rented.

(d) First, multiply $x(80 - 2x)$ to obtain $80x - 2x^2$ and then let $f(x) = -2x^2 + 80x$. The x-coordinate of the vertex is

$$x = -\frac{b}{2a} = -\frac{80}{2(-2)} = 20.$$

The y-coordinate is

$$y = f(20) = -2(20)^2 + 80(20) = 800.$$

Technology Note *Locating a Vertex*

Some graphing calculators can locate a vertex on a parabola with either the MAXIMUM or MINIMUM utility. For example, the coordinates of the vertex in Example 3 are found on a graphing calculator for a slightly different viewing rectangle in Figures 6.13–6.17.

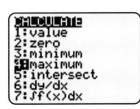

Figure 6.13

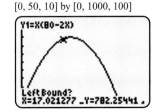

Figure 6.14

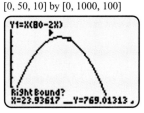

Figure 6.15

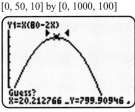

Figure 6.16

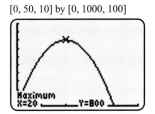

Figure 6.17

In the introduction to this section we discussed airport taxiway curves designed for airplanes. The data shown in Figure 6.1 are listed in Table 6.1.

TABLE 6.1

x (mph)	10	20	30	40	50	60
R (ft)	50	200	450	800	1250	1800

Source: Federal Aviation Administration.

A second scatterplot of the data is shown in Figure 6.18. The data may be modeled by $R(x) = ax^2$ for some value a. To illustrate this relation graph R for different values of a. In Figures 6.19–6.21, R has been graphed for $a = 2, -1,$ and $\frac{1}{2}$, respectively. Note that when $a > 0$ the parabola opens upward and when $a < 0$ the parabola opens downward. Larger values of $|a|$ make a parabola narrower, whereas smaller values of $|a|$ make the parabola wider. Through trial and error, $a = \frac{1}{2}$ gives a good fit to the data, so $R(x) = \frac{1}{2}x^2$ models the data.

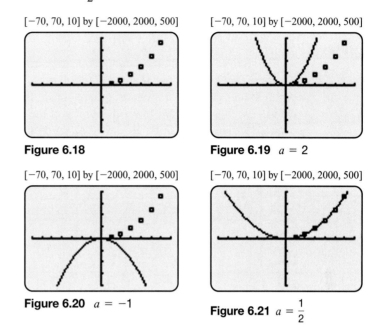

Figure 6.18

Figure 6.19 $a = 2$

Figure 6.20 $a = -1$

Figure 6.21 $a = \frac{1}{2}$

We can also find this value of a symbolically, as demonstrated in the next example.

EXAMPLE 4 *Modeling Safe Taxiway Speed*

Find a value for the constant a symbolically so that $R(x) = ax^2$ models the data in Table 6.1. Check your result by making a table of values for $R(x)$.

Solution

Table 6.1 shows that, when $x = 10$ miles per hour, the curve radius is $R(x) = 50$ feet. Therefore

$$R(10) = 50 \quad \text{or} \quad a(10)^2 = 50.$$

Solving for a gives

$$a = \frac{50}{10^2} = \frac{1}{2}.$$

To be sure that $R(x) = \frac{1}{2}x^2$ is correct, construct a table for $Y_1 = .5X^2$, as shown in Figure 6.22. Its values agree with those in Table 6.1.

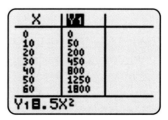

Figure 6.22

Critical Thinking

If the speed of a taxiing airplane doubles, what should happen to the radius of a safe taxiway curve?

TRANSLATIONS OF PARABOLAS

The graph of $y = x^2$ is a parabola opening upward with vertex $(0, 0)$, as shown in Figure 6.23. Now suppose that we graph $y = x^2 + 1$ and $y = x^2 - 2$, as shown in Figures 6.24 and 6.25. These graphs have the same shape as $y = x^2$; however, $y = x^2 + 1$ is shifted upward 1 unit and the graph of $y = x^2 - 2$ is shifted downward 2 units. Such shifts are called *translations*.

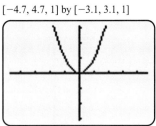

Figure 6.23 $y = x^2$

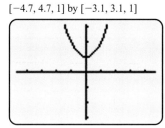

Figure 6.24 $y = x^2 + 1$

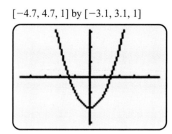

Figure 6.25 $y = x^2 - 2$

Next, suppose that we graph $y = (x - 1)^2$, $y = (x + 2)^2$, and $y = (x - 2)^2$, as illustrated in Figures 6.26–6.28. The graph of $y = (x - 1)^2$ is the same shape as $y = x^2$, but it has been translated *right* 1 unit. Similarly, the graph of $y = (x + 2)^2$ has been translated *left* 2 units, and the graph of $y = (x - 2)^2$ has been translated *right* 2 units.

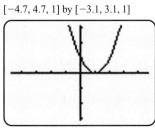

Figure 6.26 $y = (x - 1)^2$

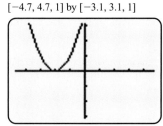

Figure 6.27 $y = (x + 2)^2$

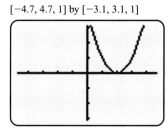

Figure 6.28 $y = (x - 2)^2$

These results are summarized as follows.

Vertical and Horizontal Translations of Parabolas

Let k be a positive number.

To graph	shift the graph of $y = x^2$ by k units
$y = x^2 + k$	upward.
$y = x^2 - k$	downward.
$y = (x - k)^2$	right.
$y = (x + k)^2$	left.

6.1 Quadratic Functions and Their Graphs

EXAMPLE 5 *Translating the graph $y = x^2$*

Sketch the graph of $y = x^2$ by hand and identify the vertex.

(a) $y = x^2 + 2$ (b) $y = (x + 3)^2$ (c) $y = (x - 2)^2 - 3$

Solution

(a) The graph of $y = x^2 + 2$ is similar to the graph of $y = x^2$ except that it has been translated upward 2 units, as shown in Figure 6.29. The vertex is $(0, 2)$.

(b) The graph of $y = (x + 3)^2$ is similar to the graph of $y = x^2$ except that it has been translated *left* 3 units, as shown in Figure 6.30. The vertex is $(-3, 0)$.

Note: If you are thinking that the graph should be shifted right (instead of left) 3 units, try graphing $Y_1 = (X + 3)^{\wedge}2$ on your graphing calculator.

(c) The graph of $y = (x - 2)^2 - 3$ is similar to the graph of $y = x^2$ except that it has been translated downward 3 units *and* right 2 units, as shown in Figure 6.31. The vertex is $(2, -3)$.

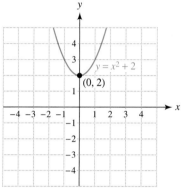

Figure 6.29 Vertex $(0, 2)$

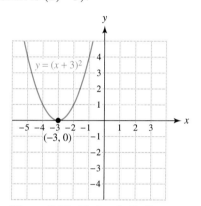

Figure 6.30 Vertex $(-3, 0)$

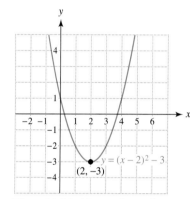

Figure 6.31 Vertex $(2, -3)$

The equation of any parabola may be written in a vertex form. The vertex form is equivalent to $y = ax^2 + bx + c$, but the vertex is easier to identify in vertex form.

Vertex Form of a Parabola

The **vertex form of a parabola** with vertex (h, k) is

$$y = a(x - h)^2 + k,$$

where $a \neq 0$ is a constant. If $a > 0$, the parabola opens upward; if $a < 0$, the parabola opens downward.

EXAMPLE 6 *Graphing a parabola*

GCLM Compare the graphs of $y = f(x)$ to the graph of $y = x^2$. Graph both equations in the standard viewing rectangle to support your answer.

(a) $f(x) = \dfrac{1}{2}(x - 5)^2 + 2$ (b) $f(x) = -3(x + 5)^2 - 1$

Solution

(a) Both graphs are parabolas. However, compared to the graph of $y = x^2$, the graph of $y = f(x)$ is translated 5 units right and 2 units upward. The vertex for $f(x)$ is (5, 2), whereas the vertex of $y = x^2$ is (0, 0). Because $a = \frac{1}{2}$, the graph of $y = f(x)$ opens wider than the graph of $y = x^2$. This difference is shown in Figure 6.32, where $Y_1 = X\wedge 2$ and $Y_2 = .5(X - 5)\wedge 2 + 2$.

[−10, 10, 1] by [−10, 10, 1]

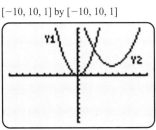

Figure 6.32

[−10, 10, 1] by [−10, 10, 1]

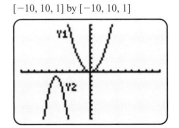

Figure 6.33

(b) Compared to the graph of $y = x^2$, the graph of $y = f(x)$ is translated 5 units left and 1 unit downward. The vertex for $f(x)$ is $(-5, -1)$. Because $a = -3$, the graph of $y = f(x)$ opens downward and is narrower than the graph of $y = x^2$. This difference is shown in Figure 6.33, where $Y_1 = X\wedge 2$ and $Y_2 = -3(X + 5)\wedge 2 - 1$.

■ MAKING CONNECTIONS ─────────────────

Parabolas and the value of a

The value of a affects the overall shape of a parabola expressed either as $y = ax^2 + bx + c$ or $y = a(x - h)^2 + k$. If $a > 0$, the parabola opens upward; if $a < 0$, the parabola opens downward. Larger values of $|a|$ cause the parabola to be narrower, and smaller values of $|a|$ cause the parabola to be wider. In Figures 6.34 and 6.35, the parabolas $y = ax^2$ for $a = \pm\frac{1}{3}$ and $a = \pm 2$ have been graphed to illustrate the effect of a.

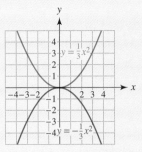

Figure 6.34 $a = \pm\frac{1}{3}$

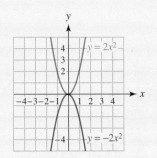

Figure 6.35 $a = \pm 2$

EXAMPLE 7 | Finding equations of parabolas

Write the vertex form of a parabola with $a = 2$ and vertex $(-2, 1)$. Then express this equation in the form $y = ax^2 + bx + c$.

Solution

The vertex form of a parabola is

$$y = a(x - h)^2 + k,$$

where the vertex is (h, k). For $a = 2$, $h = -2$, and $k = 1$, the equation becomes

$$y = 2(x - (-2))^2 + 1 \quad \text{or} \quad y = 2(x + 2)^2 + 1.$$

To write this equation in the form $y = ax^2 + bx + c$, do the following.

$y = 2(x^2 + 4x + 4) + 1$ Multiply $(x + 2)^2$.
$y = (2x^2 + 8x + 8) + 1$ Distributive property
$y = 2x^2 + 8x + 9$ Add.

The equivalent equation is $y = 2x^2 + 8x + 9$.

MODELING WITH QUADRATIC FUNCTIONS

In 1981, the first cases of AIDS were reported in the United States. Since then, AIDS has become one of the most devastating diseases of recent times. Table 6.2 lists the *cumulative* number of AIDS cases in the United States for various years. For example, between 1981 and 1990, a total of 199,608 AIDS cases were reported.

TABLE 6.2

Year	1981	1984	1987	1990	1993	1996
AIDS Cases	425	11,106	71,414	199,608	417,835	609,933

Source: Department of Health and Human Services.

A scatterplot of these data is shown in Figure 6.36. To model these nonlinear data, we want to find (the right half of) a parabola with the shape illustrated in Figure 6.37.

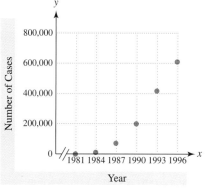

Figure 6.36 AIDS Cases

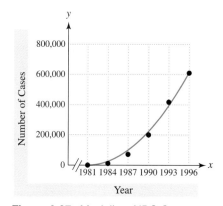

Figure 6.37 Modeling AIDS Cases

EXAMPLE 8 Modeling AIDS cases

GCLM Use the data in Table 6.2 to complete the following.

(a) Make a scatterplot of the data in [1980, 1997, 2] by [−10,000, 800,000, 100,000].
(b) The lowest data point in Table 6.2 is (1981, 425). Let this point be the vertex of a parabola. Graph $y = a(x - 1981)^2 + 425$ together with the data by first letting $a = 1000$.
(c) Use trial and error to adjust the value of a until the graph models the data.
(d) Use your final equation to estimate the number of AIDS cases in 1992. Compare it to the known value of 338,786.

Solution

(a) A scatterplot of the data in Table 6.2 is shown in Figure 6.38.

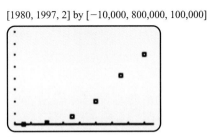

Figure 6.38

(b) A graph of $y = 1000(x - 1981)^2 + 425$ is shown in Figure 6.39. To have a better fit of the data, a larger value for a is needed.

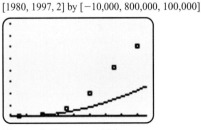

Figure 6.39 $a = 1000$

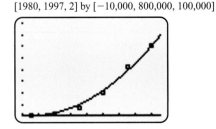

Figure 6.40 $a = 2700$

(c) Figure 6.40 shows the effect of adjusting the value of a to 2700. This value provides a reasonably good fit. (Note that you may decide on a slightly different value for a.)
(d) If $a = 2700$, the modeling equation becomes

$$y = 2700(x - 1981)^2 + 425.$$

To estimate the number of AIDS cases in 1992, substitute $x = 1992$ to obtain

$$y = 2700(1992 - 1981)^2 + 425 = 327{,}125.$$

This number is about 12,000 less than the known value of 338,786.

6.1 Putting It All Together

A quadratic function is an example of a relatively simple nonlinear function that may be written either as $f(x) = ax^2 + bx + c$ or $f(x) = a(x - h)^2 + k$. The graph of a quadratic function is a parabola. A parabola is a U-shaped graph that can open upward or downward. The following table summarizes some results from this section.

Feature	Description	Example
Vertex of a parabola	The x-coordinate of the vertex for $y = ax^2 + bx + c$ with $a \neq 0$ is given by $$x = -\frac{b}{2a}.$$ The y-coordinate of the vertex is found by substituting this x-value into the equation for y.	If $y = -2x^2 + 8x - 7$, then $$x = -\frac{8}{2(-2)} = 2$$ and $$y = -2(2)^2 + 8(2) - 7 = 1.$$ The vertex is $(2, 1)$. Its graph opens downward because $a = -2 < 0$.
Vertex form of a parabola	The vertex form of a parabola with vertex (h, k) is $$y = a(x - h)^2 + k,$$ where $a \neq 0$ is a constant. If $a > 0$, the parabola opens upward; if $a < 0$, the parabola opens downward.	The graph of $y = 3(x + 2)^2 - 7$ has a vertex of $(-2, -7)$ and opens upward because $a = 3 > 0$.

6.1 EXERCISES

Concepts

1. The graph of a quadratic function is called a _____.

2. If a parabola opens upward, what is the lowest point on the parabola called?

3. If a parabola is symmetric with respect to the y-axis, the y-axis is called the _____.

4. A quadratic function may be written either in the form _____ or _____.

5. The vertex of $y = x^2$ is _____.

6. Sketch a parabola that opens downward with a vertex of $(1, 2)$.

7. If $y = ax^2 + bx + c$, the x-coordinate of the vertex is given by $x = $ _____.

8. The vertex form of a parabola is given by _____ and its vertex is _____.

Exercises 9–16: The accompanying figure shows the graph of a polynomial function. Decide whether the graph represents a linear function, a quadratic function, or neither. If the graph represents a quadratic function, identify the vertex, the axis of symmetry, and whether it opens upward or downward.

9.
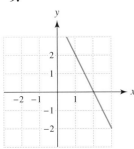

10.
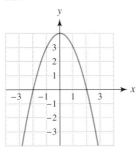

11.
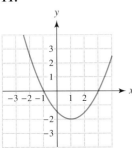

12.
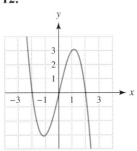

13.
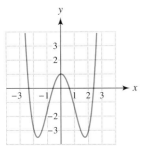

14.
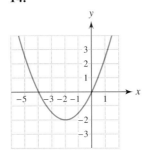

15.
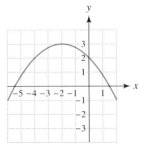

16.
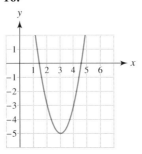

REPRESENTATIONS OF QUADRATIC FUNCTIONS

Exercises 17–24: Complete the following for the quadratic function.
- **(a)** Graph $y = f(x)$.
- **(b)** Use the graph to identify the vertex and axis of symmetry.
- **(c)** Make a table of values using $x = -3, -2, -1, \ldots, 3$.
- **(d)** Use the table to evaluate $f(-2)$ and $f(3)$.

17. $f(x) = x^2 - 2$

18. $f(x) = 1 - 2x^2$

19. $f(x) = 2x - x^2$

20. $f(x) = x^2 + 2x - 8$

21. $f(x) = -2x^2 + 4x - 1$

22. $f(x) = -\dfrac{1}{2}x^2 + 2x - 3$

23. $f(x) = \dfrac{1}{4}x^2 - x + 5$

24. $f(x) = 3 - 6x - 4x^2$

Exercises 25–28: Use the graph of f to evaluate the expressions.

25. $f(-2)$ and $f(0)$
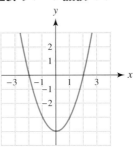

26. $f(-2)$ and $f(2)$
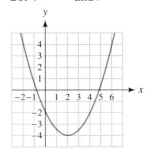

27. $f(-3)$ and $f(1)$
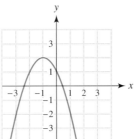

28. $f(-1)$ and $f(2)$
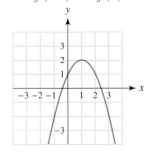

Exercises 29–36: Find the vertex of the parabola symbolically. Support your result graphically.

29. $f(x) = x^2 - 4x - 2$

30. $f(x) = 2x^2 + 6x - 3$

31. $f(x) = -\frac{1}{3}x^2 - 2x + 1$

32. $f(x) = 5 - 4x + x^2$

33. $f(x) = 3 - 2x^2$

34. $f(x) = \frac{1}{4}x^2 - 3x - 2$

35. $f(x) = -0.3x^2 + 0.6x + 1.1$

36. $f(x) = 25 - 10x + 20x^2$

Exercises 37–40: Find a value for the constant a so that $f(x) = ax^2$ models the data. If you are uncertain about your value for a, check it by making a table of values for $f(x)$.

37.
x	1	2	3
y	2	8	18

38.
x	-2	0	2
y	6	0	6

39.
x	2	4	6	8
y	1.2	4.8	10.8	19.2

40.
x	5	10	15	20
y	17.5	70	157.5	280

TRANSLATIONS OF PARABOLAS

Exercises 41–50: Complete the following.

(a) Graph $y = f(x)$ in the standard viewing rectangle.

(b) Compare the graph of f with the graph of $y = x^2$.

41. $f(x) = x^2 - 4$

42. $f(x) = 2x^2 + 1$

43. $f(x) = (x - 3)^2$

44. $f(x) = (x + 1)^2$

45. $f(x) = -x^2$

46. $f(x) = -(x + 2)^2$

47. $f(x) = (x + 1)^2 - 2$

48. $f(x) = (x - 3)^2 + 1$

49. $f(x) = (x - 1)^2 + 2$

50. $f(x) = \frac{1}{2}(x + 3)^2 - 3$

Exercises 51–56: Sketch the graph of f by hand and identify the vertex.

51. $f(x) = x^2$

52. $f(x) = x^2 - 3$

53. $f(x) = 2 - x^2$

54. $f(x) = (x - 1)^2$

55. $f(x) = (x + 2)^2$

56. $f(x) = (x - 2)^2 - 3$

Exercises 57–60: (Refer to Example 7.) Write the vertex form of a parabola that satisfies the conditions given. Then write the equation in the form $y = ax^2 + bx + c$.

57. Vertex $(3, 4)$ and $a = 3$

58. Vertex $(-1, 3)$ and $a = -5$

59. Vertex $(5, -2)$ and $a = -\frac{1}{2}$

60. Vertex $(-2, -6)$ and $a = \frac{3}{4}$

Exercises 61–64: Write the vertex form of a parabola that satisfies the conditions given. Assume that $a = \pm 1$.

61. Opens upward, vertex $(1, 2)$

62. Opens downward, vertex $(-1, -2)$

63. Opens downward, vertex $(0, -3)$

64. Opens upward, vertex $(5, -4)$

Exercises 65–68: Write the vertex form of the parabola shown in the graph. Assume that $a = \pm 1$.

65.
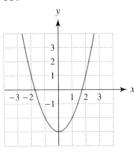

66.
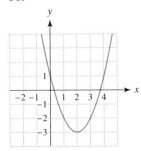

67.
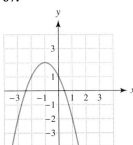

68.
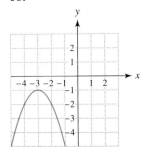

APPLICATIONS

Exercises 69–72: Quadratic Models Match the physical situation with the graph (a.-d.) that models it best.

69. The height y of a stone thrown from ground level after x seconds.

70. The number of people attending a popular movie x weeks after its opening.

71. The temperature after x hours in a house when the furnace quits and a repair person fixes it.

72. The population of the United States from 1800 to the present.

a.

b.

c.

d.
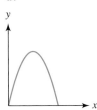

73. *Maximizing Area* A rectangular pen being constructed for a pet requires 60 feet of fence.
  (a) Write a formula $f(x)$ that gives the area of the pen if one side of the pen has length x.
  (b) Find the dimensions of the pen that give the largest area. What is the largest area?

74. *Seedling Growth* The effect of temperature on the growth of melon seedlings was studied. In this study the seedlings were grown at different temperatures, and their heights were measured after a fixed period of time. The findings of this study can be modeled by

$$f(x) = -0.095x^2 + 5.4x - 52.2,$$

where x is the temperature in degrees Celsius and the output $f(x)$ gives the resulting average height in centimeters. (**Source:** R. Pearl, "The growth of *Cucumis melo* seedlings at different temperatures.")
  (a) Graph f in [20, 40, 5] by [0, 30, 5].
  (b) Estimate graphically the temperature that resulted in the greatest height for the melon seedlings.
  (c) Solve part (b) symbolically.

75. *Height of a Baseball* Suppose that a baseball is thrown upward with an initial velocity of 66 feet per second (45 miles per hour) and it is released 6 feet above the ground. Its height h after t seconds may be modeled by

$$h(t) = -16t^2 + 66t + 6.$$

After how many seconds does the baseball reach a maximum height? Estimate this height.

76. *Throwing a Baseball on the Moon* (Refer to Exercise 75.) If the same baseball were thrown the same way on the moon, its height h above the moon's surface after t seconds would be

$$h(t) = -5.1t^2 + 66t + 6.$$

Does the baseball go higher on the moon or on Earth? What is the difference in these two heights?

77. *Maximizing Revenue* (Refer to Example 3.) The regular price for a round-trip ticket to Las Vegas, Nevada, charged by an airline charter company is $300. For a group rate the company will reduce the price of a ticket by $1.50 for every passenger on the flight.
(a) Write a formula $f(x)$ that gives the revenue from selling x tickets.
(b) Determine how many tickets should be sold to maximize the revenue. What is the maximum revenue?

78. *Budget for National Parks* The annual budget in billions of dollars for national parks from 1994 through 1998 may be modeled by

$$f(x) = 0.04(x - 1995.7)^2 + 1.35,$$

where x represents the year. (*Source:* National Park Service.)

(a) Graph f in the window [1994, 1998, 1] by [1.3, 1.7, 0.1].
(b) Describe how the budget has changed over this time period.

Exercises 79–82: Modeling Quadratic Data (Refer to Example 8.) Find a quadratic function expressed in vertex form that models the data in the given table.

79.

x	1	2	3	4
y	−3	−1	5	15

80.

x	−2	−1	0	1	2
y	5	2	−7	−22	−43

81.

x	1980	1990	2000	2010
y	6	55	210	450

82.

x	1990	1995	2000	2005
y	10	60	205	470

WRITING ABOUT MATHEMATICS

83. If a quadratic function is represented by $f(x) = ax^2 + bx + c$, explain how the values of a and c affect the graph of f.

84. If a quadratic function is represented by $f(x) = a(x - h)^2 + k$, explain how the values of a, h, and k affect the graph of f.

GROUP ACTIVITY
Working with Real Data

Directions: Form a group of 2 to 4 people. Select someone to record the group's responses for this activity. All members of the group should work cooperatively to answer the questions. If your instructor asks for the results, each member of the group should be prepared to respond. If the group is asked to turn in its work, be sure to include each person's name on the paper.

Minimum Wage The following table shows the minimum wage for three different years.

Year	1940	1968	1997
Wage	$0.25	$1.60	$5.45

Source: Bureau of Labor Statistics.

(a) Make a scatterplot of the data in the window [1930, 2010, 10] by [0, 6,1].

(b) Find a quadratic function given by $f(x) = a(x - h)^2 + k$ that models the data.

(c) Estimate the minimum wage in 1976 and compare it to the actual value of $2.30.

(d) If current trends continue, predict the minimum wage in 2005.

6.2 QUADRATIC EQUATIONS

Basics of Quadratic Equations ~ The Square Root Property ~ Completing the Square ~ Applications of Quadratic Equations

INTRODUCTION

In Section 6.1 we modeled curves on airport taxiways by using $R(x) = \frac{1}{2}x^2$. In this formula x represented the airplane's speed in miles per hour, and R represented the radius of the curve in feet. This formula may be used to determine the speed limit for a curve with a radius of 650 feet by solving the *quadratic equation*

$$\frac{1}{2}x^2 = 650.$$

In this section we demonstrate techniques for solving quadratic equations.

BASICS OF QUADRATIC EQUATIONS

Any quadratic function f can be represented by $f(x) = ax^2 + bx + c, a \neq 0$. Examples of quadratic functions include

$$f(x) = 2x^2 - 1, \quad g(x) = -\frac{1}{3}x^2 + 2x, \quad \text{and} \quad h(x) = x^2 + 2x - 1.$$

Quadratic functions can be used to write quadratic equations. Examples of quadratic equations include

$$2x^2 - 1 = 0, \quad -\frac{1}{3}x^2 + 2x = 0, \quad \text{and} \quad x^2 + 2x - 1 = 3.$$

Quadratic Equation

A **quadratic equation** is an equation that can be written as

$$ax^2 + bx + c = 0,$$

where a, b, and c are real numbers, with $a \neq 0$.

6.2 Quadratic Equations

Solutions to the quadratic equation $ax^2 + bx + c = 0$ correspond to *x*-intercepts of the graph of $y = ax^2 + bx + c$. Because the graph of a quadratic function is U-shaped, it can intersect the *x*-axis zero, one, or two times, as illustrated in Figures 6.41–6.43. Hence a quadratic equation can have zero, one, or two solutions.

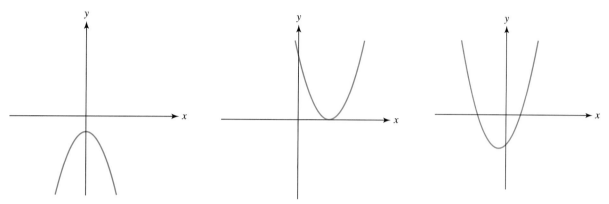

Figure 6.41 No *x*-Intercepts   **Figure 6.42** One *x*-Intercept   **Figure 6.43** Two *x*-Intercepts

We have already solved quadratic equations by factoring, graphing, and constructing tables. We review these techniques in the next example.

EXAMPLE 1 Solving quadratic equations

Solve each quadratic equation. Support your results graphically and numerically.

(a) $2x^2 + 1 = 0$ (No solutions)
(b) $x^2 + 4 = 4x$ (One solution)
(c) $x^2 - 6x + 8 = 0$ (Two solutions)

Solution

(a) *Symbolic Solution*

$2x^2 + 1 = 0$ Given equation
$2x^2 = -1$ Subtract 1.
$x^2 = -\dfrac{1}{2}$ Divide by 2.

This equation has no real number solutions because $x^2 \geq 0$ for all real numbers *x*.

Graphical and Numerical Solutions For graphical and numerical support, let $Y_1 = 2X^2 + 1$. Figure 6.44 shows that the graph of Y_1 has no *x*-intercepts. Values for Y_1 are always greater than 0. See Figure 6.45.

[−6, 6, 1] by [−4, 4, 1]

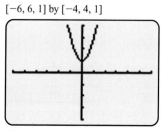
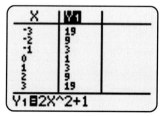

Figure 6.44 No Solutions   **Figure 6.45** No Solutions

(b) *Symbolic Solution*

$$x^2 + 4 = 4x \quad \text{Given equation}$$
$$x^2 - 4x + 4 = 0 \quad \text{Subtract } 4x \text{ from both sides.}$$
$$(x - 2)(x - 2) = 0 \quad \text{Factor.}$$
$$x - 2 = 0 \quad \text{or} \quad x - 2 = 0 \quad \text{Zero-product property}$$
$$x = 2 \quad \text{There is one solution.}$$

Graphical and Numerical Solutions For graphical and numerical support, graph and construct the table for $Y_1 = X^2 - 4X + 4$. Figure 6.46 shows that the graph of Y_1 has one x-intercept, 2. Figure 6.47 also shows that $Y_1 = 0$ when $x = 2$.

[−6, 6, 1] by [−4, 4, 1]

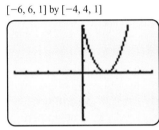
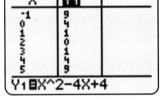

Figure 6.46 One Solution   **Figure 6.47** One Solution

(c) *Symbolic Solution*

$$x^2 - 6x + 8 = 0 \quad \text{Given equation}$$
$$(x - 2)(x - 4) = 0 \quad \text{Factor.}$$
$$x - 2 = 0 \quad \text{or} \quad x - 4 = 0 \quad \text{Zero-product property}$$
$$x = 2 \quad \text{or} \quad x = 4 \quad \text{There are two solutions.}$$

Graphical and Numerical Solutions For graphical and numerical support, graph and construct a table for $Y_1 = X^2 - 6X + 8$. Figure 6.48 shows that the graph of Y_1 has two x-intercepts, 2 and 4. Figure 6.49 shows that $Y_1 = 0$ when $x = 2$ or $x = 4$.

[−6, 6, 1] by [−4, 4, 1]

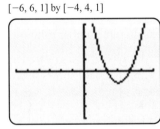
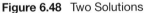

Figure 6.48 Two Solutions   **Figure 6.49** Two Solutions

THE SQUARE ROOT PROPERTY

The *square root property* is a symbolic method used to solve quadratic equations that are missing x-terms. The following is an example of the square root property.

$$x^2 = 25 \quad \text{is equivalent to} \quad x = \pm 5$$

6.2 Quadratic Equations

The expression $x = \pm 5$ (read "x equals plus or minus 5") indicates that either $x = 5$ or $x = -5$. Each value is a solution because $(5)^2 = 25$ and $(-5)^2 = 25$.

> **Square Root Property**
>
> Let k be a nonnegative number. Then the solutions to the equation
> $$x^2 = k$$
> are $x = \pm\sqrt{k}$. If $k < 0$, then this equation has no real solutions.

Before applying the square root property in the next two examples, we review a basic property of square roots. If a and b are positive numbers, then

$$\sqrt{\frac{a}{b}} = \frac{\sqrt{a}}{\sqrt{b}}.$$

For example,

$$\sqrt{\frac{25}{36}} = \frac{\sqrt{25}}{\sqrt{36}} = \frac{5}{6}.$$

EXAMPLE 2 *Using the square root property*

Solve each equation.

(a) $x^2 = 7$ (b) $16x^2 - 9 = 0$ (c) $(x - 4)^2 = 25$

Solution

(a) $x^2 = 7$ is equivalent to $x = \pm\sqrt{7}$ by the square root property.

(b) $16x^2 - 9 = 0$ Given equation
$16x^2 = 9$ Add 9.
$x^2 = \dfrac{9}{16}$ Divide by 16.
$x = \pm\sqrt{\dfrac{9}{16}}$ Square root property
$x = \pm\dfrac{3}{4}$ Simplify.

(c) $(x - 4)^2 = 25$ Given equation.
$(x - 4) = \pm\sqrt{25}$ Square root property
$x - 4 = \pm 5$ Simplify.
$x = 4 \pm 5$ Add 4.
$x = 9 \text{ or } x = -1$ Evaluate $4 + 5$ and $4 - 5$.

If an object is dropped from a height of h feet, its distance above the ground after t seconds is given by

$$d(t) = h - 16t^2.$$

This formula can be used to estimate the time it takes for a falling object to hit the ground.

EXAMPLE 3 *Modeling a falling object*

A toy falls from a window 30 feet above the ground. How long does it take the toy to hit the ground?

Solution

The height of the window above the ground is 30 feet so let $d(t) = 30 - 16t^2$. The toy strikes the ground when the distance d above the ground equals 0.

$$30 - 16t^2 = 0 \qquad \text{Equation to solve}$$
$$-16t^2 = -30 \qquad \text{Subtract 30.}$$
$$t^2 = \frac{30}{16} \qquad \text{Divide by } -16.$$
$$t = \pm\sqrt{\frac{30}{16}} \qquad \text{Square root property}$$
$$t = \pm\frac{\sqrt{30}}{4} \qquad \text{Simplify.}$$

Time cannot be negative in this problem, so the appropriate solution is $t = \frac{\sqrt{30}}{4} \approx 1.4$. The toy hits the ground after about 1.4 seconds.

COMPLETING THE SQUARE

An important technique in mathematics is **completing the square.** Although this technique can be used in several different types of situations, we use it to solve quadratic equations. If a quadratic equation is written in the form $x^2 + kx = d$, where k and d are constants, the equation can be solved by completing the square if we add $\left(\frac{k}{2}\right)^2$ to both sides.

In the equation $x^2 + 6x = 7$ we have $k = 6$, so we add $\left(\frac{6}{2}\right)^2 = 9$ to both sides.

$$x^2 + 6x = 7 \qquad \text{Given equation}$$
$$x^2 + 6x + 9 = 7 + 9 \qquad \text{Add 9 to both sides.}$$
$$(x + 3)^2 = 16 \qquad \text{Perfect square trinomial}$$
$$x + 3 = \pm 4 \qquad \text{Square root property}$$
$$x = -3 \pm 4 \qquad \text{Add } -3 \text{ to both sides.}$$
$$x = 1 \quad \text{or} \quad x = -7 \qquad \text{Simplify.}$$

Note that, if the square is completed correctly, the left side of the equation will be a perfect square trinomial.

EXAMPLE 4 *Creating a perfect square trinomial*

Find the term that should be added to $x^2 - 10x$ to form a perfect square trinomial.

Solution

The coefficient of the x-term is -10, so we let $k = -10$. To complete the square we divide k by 2 and then square the result.

$$\left(\frac{k}{2}\right)^2 = \left(\frac{-10}{2}\right)^2 = 25$$

If we add 25, a perfect square trinomial is formed.

$$x^2 - 10x + 25 = (x - 5)(x - 5)$$
$$= (x - 5)^2$$

Completing the square can be used to solve quadratic equations when a trinomial does not factor easily.

EXAMPLE 5 *Completing the square when the leading coefficient is 1*

Solve the equation $x^2 - 4x + 2 = 0$.

Solution

Start by writing the equation in the form $x^2 + kx = d$.

$x^2 - 4x + 2 = 0$	Given equation
$x^2 - 4x = -2$	Subtract 2.
$x^2 - 4x + 4 = -2 + 4$	Add $\left(\frac{k}{2}\right)^2 = \left(\frac{-4}{2}\right)^2 = 4$.
$(x - 2)^2 = 2$	Perfect square trinomial
$x - 2 = \pm\sqrt{2}$	Square root property
$x = 2 \pm \sqrt{2}$	Solve.

The solutions are $x = 2 + \sqrt{2} \approx 3.41$ and $x = 2 - \sqrt{2} \approx 0.59$.

EXAMPLE 6 *Completing the square when the leading coefficient is not 1*

Solve the equation $2x^2 + 4x - 5 = 0$.

Solution

Start by writing the equation in the form $x^2 + kx = d$. That is, add 5 to each side and then divide the equation by 2 so that the leading coefficient of the x^2-term becomes a 1.

$2x^2 + 4x - 5 = 0$	Given equation
$2x^2 + 4x = 5$	Add 5.
$x^2 + 2x = \frac{5}{2}$	Divide by 2.
$x^2 + 2x + 1 = \frac{5}{2} + 1$	Add $\left(\frac{k}{2}\right)^2 = \left(\frac{2}{2}\right)^2 = 1$.
$(x + 1)^2 = \frac{7}{2}$	Perfect square trinomial
$x + 1 = \pm\sqrt{\frac{7}{2}}$	Square root property
$x = -1 \pm \sqrt{\frac{7}{2}}$	Add -1.

The solutions are $x = -1 + \sqrt{\frac{7}{2}} \approx 0.87$ and $x = -1 - \sqrt{\frac{7}{2}} \approx -2.87$.

Critical Thinking

What happens if you try to solve

$$2x^2 - 13 = 1$$

by completing the square? What method should you use to solve this problem?

APPLICATIONS OF QUADRATIC EQUATIONS

In the introduction to this section we discussed how the solution to the equation

$$\frac{1}{2}x^2 = 650$$

would give a safe speed limit for a curve with a radius of 650 feet on an airport taxiway. We solve this problem in the next example.

EXAMPLE 7 *Finding a safe speed limit*

Solve the equation $\frac{1}{2}x^2 = 650$ and interpret any solutions.

Solution

Use the square root property to solve this problem.

$$\frac{1}{2}x^2 = 650 \quad \text{Given equation}$$
$$x^2 = 1300 \quad \text{Multiply by 2.}$$
$$x = \pm\sqrt{1300} \quad \text{Square root property}$$

The solutions are $x = \sqrt{1300} \approx 36$ and $x = -\sqrt{1300} \approx -36$. The solution of $x \approx 36$ indicates that a safe speed limit for a curve with a radius of 650 feet should be 36 miles per hour. (The negative solution has no physical meaning in this problem.)

In real-world applications, solving a quadratic equation either graphically or numerically is often easier than solving it symbolically. This is done in the next example.

EXAMPLE 8 *Modeling Internet users*

Use of the Internet in Western Europe is expected to increase dramatically by 2002. Figure 6.50 shows a scatterplot of online users in Western Europe, together with a graph of a function f that models the data. The function f is given by $f(x) = 0.976x^2 - 4.643x + 0.238$, where the output is in millions of users. In this formula $x = 6$ corresponds to 1996, $x = 7$ to 1997, and so on, until $x = 12$ represents 2002. (*Source:* Nortel Networks.)

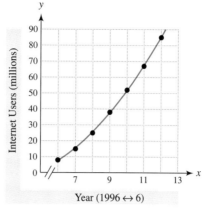

Figure 6.50 Internet Usage in Western Europe

(a) Evaluate $f(10)$ and interpret the result.
(b) Graph f and estimate the year when the number of Internet users may reach 85 million. Compare your result with the given graph.
(c) Solve part (b) numerically.

Solution

(a) Substituting $x = 10$ into the formula yields

$$f(10) = 0.976(10)^2 - 4.643(10) + 0.238 \approx 51.4.$$

Because $x = 10$ corresponds to 2000, some 51.4 million Internet users in Western Europe are expected by 2000.

(b) Graph $Y_1 = .976X^2 - 4.643X + .238$ and $Y_2 = 85$, as shown in Figure 6.51. Their graphs intersect near $x \approx 12$, which corresponds to 2002, which agrees with Figure 6.50.

(c) Construct the table for Y_1, as shown in Figure 6.52, which reveals $Y_1 \approx 85$ when $x = 12$.

[5, 13, 1] by [0, 100, 10]

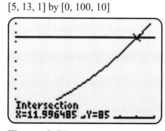

Figure 6.51

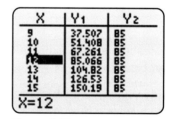

Figure 6.52

6.2 Putting It All Together

Quadratic equations can be expressed in the form $ax^2 + bx + c = 0, a \neq 0$. They can have zero, one, or two solutions and may be solved symbolically, graphically, and numerically. Symbolic techniques for solving quadratic equations include factoring, the square root property, and completing the square. We discussed factoring extensively in Chapter 5, so the following table summarizes only the square root property and completing the square techniques.

Technique	Description	Example
Square root property	If $k \geq 0$, the solutions to the equation $x^2 = k$ are $x = \pm\sqrt{k}$.	$x^2 = 100$ is equivalent to $x = \pm 10$ $x^2 = 13$ is equivalent to $x = \pm\sqrt{13}$
Completing the square	To solve an equation in the form $x^2 + kx = d$, add $\left(\dfrac{k}{2}\right)^2$ to both sides of the equation. Factor the resulting perfect square trinomial and solve for x by applying the square root property.	To solve $x^2 + 8x - 3 = 0$, begin by adding 3 to both sides. $$x^2 + 8x = 3$$ Because $k = 8$, add $\left(\dfrac{8}{2}\right)^2 = 16$ to both sides. $x^2 + 8x + 16 = 3 + 16$ $(x + 4)^2 = 19$ Perfect square trinomial $x + 4 = \pm\sqrt{19}$ Square root property $x = -4 \pm \sqrt{19}$ Add -4. $x \approx 0.36, -8.36$ Approximate

6.2 EXERCISES

CONCEPTS

1. Give an example of a quadratic equation. How many real solutions can a quadratic equation have?

2. Is a quadratic equation a linear equation or a nonlinear equation?

3. Name three symbolic methods that can be used to solve a quadratic equation.

4. Sketch a graph of a quadratic function that has two x-intercepts and opens downward.

5. Sketch a graph of a quadratic function that has no x-intercepts and opens upward.

6. If the graph of $y = ax^2 + bx + c$ intersects the x-axis twice, how many solutions does the equation $ax^2 + bx + c = 0$ have? Explain.

7. Solve $x^2 = 64$. What property did you use?

8. To solve $x^2 + kx = 6$ by completing the square, what value should be added to both sides of the equation?

Exercises 9–16: Determine whether the equation is quadratic.

9. $x^2 - 3x + 1 = 0$ 10. $2x^2 - 3 = 0$

11. $3x + 1 = 0$

12. $x^3 - 3x^2 + x = 0$

13. $-3x^2 + x = 16$

14. $x^2 - 1 = 4x$

15. $x^2 = \sqrt{x} + 1$ 16. $\dfrac{1}{x - 1} = 5$

Solving Quadratic Equations Graphically and Numerically

Exercises 17–20: A graph of $y = ax^2 + bx + c$ is given. Use this graph to solve $ax^2 + bx + c = 0$, if possible.

17.

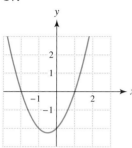

18.

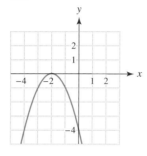

19.

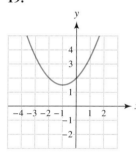

20.

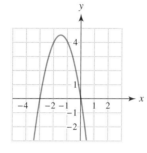

Exercises 21–24: A table of $y = ax^2 + bx + c$ is given. Use this table to solve $ax^2 + bx + c = 0$.

21.
22.
23.
24.

Exercises 25–32: Solve the quadratic equation
(a) *graphically and*
(b) *numerically.*

25. $x^2 - 4x - 5 = 0$ **26.** $x^2 - x - 6 = 0$

27. $2x^2 - 2x - \dfrac{3}{2} = 0$ **28.** $\dfrac{1}{4}x^2 + x - \dfrac{5}{4} = 0$

29. $x^2 = 9$ **30.** $2x^2 + x = 1$

31. $\dfrac{1}{2}x^2 + x = -\dfrac{3}{2}$ **32.** $23x = 10x^2 + 12$

Solving Quadratic Equations Symbolically

Exercises 33–40: Solve by factoring.

33. $x^2 + 2x - 35 = 0$ **34.** $2x^2 - 7x + 3 = 0$

35. $6x^2 - x - 1 = 0$

36. $x^2 + 4x + 6 = -3x$

37. $2x^2 + 3 = 5x$ **38.** $6x^2 - x = 15$

39. $10x^2 + 18 = 27x$ **40.** $15(3x^2 + x) = 10$

Exercises 41–50: Use the square root property to solve the equation.

41. $x^2 = 144$ **42.** $4x^2 - 5 = 0$

43. $5x^2 - 64 = 0$ **44.** $3x^2 = -7$

45. $(x + 1)^2 = 25$ **46.** $(x + 4)^2 = 9$

47. $(x - 1)^2 = -64$ **48.** $(x - 3)^2 = 0$

49. $(2x - 1)^2 = 5$ **50.** $(5x + 3)^2 = 7$

Exercises 51–54: To solve the equation by completing the square, what value should you add to both sides of the equation?

51. $x^2 + 4x = -3$ **52.** $x^2 - 6x = 4$

53. $x^2 - 5x = 4$ **54.** $x^2 + 3x = 1$

Exercises 55–58: (Refer to Example 4.) Find the term that should be added to the expression to form a perfect square trinomial. Write the resulting perfect square trinomial in factored form.

55. $x^2 - 8x$ **56.** $x^2 - 5x$

57. $x^2 + 9x$ **58.** $x^2 + x$

Exercises 59–66: Solve the equation by completing the square.

59. $x^2 - 2x = 24$ **60.** $x^2 - 2x + \frac{1}{2} = 0$

61. $x^2 + 6x - 2 = 0$ **62.** $x^2 - 16x = 5$

63. $2x^2 - 3x = 4$

64. $3x^2 + 6x - 5 = 0$

65. $4x^2 - 8x - 7 = 0$

66. $25x^2 - 20x - 1 = 0$

SOLVING EQUATIONS BY MORE THAN ONE METHOD

Exercises 67–72: Solve the quadratic equation
(a) symbolically,
(b) graphically, and
(c) numerically.

67. $x^2 - 3x - 18 = 0$

68. $\frac{1}{2}x^2 + 2x - 6 = 0$

69. $x^2 - 8x + 15 = 0$

70. $2x^2 + 3 = 7x$

71. $4x^2 + 140 = 48x$

72. $4x(2 - x) = -5$

APPLICATIONS

73. *Safe Curve Speed* (Refer to Example 7.) Find a safe speed limit x for an airport taxiway curve with the given radius R.
(a) $R = 450$ feet
(b) $R = 800$ feet

74. *Modeling Internet Users* (Refer to Example 8.) Estimate either graphically or numerically when the number of Internet users in Western Europe is expected to reach 150 million.

75. *Seedling Growth* (Refer to Exercise 74, Section 6.1.) The height of melon seedlings grown at different temperatures is shown in the following graph. At what temperatures was the height of the seedling about 22 centimeters? (**Source: R. Pearl, "The growth of *Cucumis melo* seedlings at different temperatures."**)

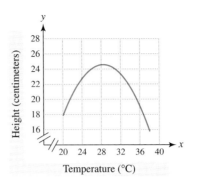

76. *Construction* A rectangular plot of land has an area of 520 square feet and is 6 feet longer than it is wide.
(a) Write a quadratic equation in the form $ax^2 + bx + c = 0$, whose solution gives the width of the plot of land.
(b) Solve the equation either graphically or numerically.
(c) Solve the equation symbolically.

77. *U.S. Population* The accompanying table shows the population of the United States from 1800 through 2000 at 20-year intervals.

Year	1800	1820	1840	1860
Population (millions)	5	10	17	31

Year	1880	1900	1920	1940
Population (millions)	50	76	106	132

Year	1960	1980	2000
Population (millions)	179	226	269

Source: Bureau of the Census.

(a) Without plotting the data, how do you know that the data are nonlinear?
(b) These data are modeled by
$$f(x) = 0.0066(x - 1800)^2 + 5.$$
Find the vertex of the graph of f and interpret the result.
(c) Estimate graphically when the U.S. population was 85 million.
(d) Solve part (c) symbolically.

78. *Trade Deficit* The U.S. trade deficit in billions of dollars for the years 1997, 1998, and 1999 can be computed by
$$f(x) = 16x^2 - 63,861x + 63,722,378,$$
where x is the year. In which year was the trade deficit $164 billion? (*Source: Department of Commerce.*)

79. *Braking Distance* The braking distance y in feet that it takes for a car to stop on wet, level pavement can be estimated by $y = \frac{1}{9}x^2$, where x is the speed of the car in miles per hour. Find the speed associated with each stopping distance. (*Source: L. Haefner, Introduction to Transportation Systems.*)
(a) 25 feet
(b) 361 feet
(c) 784 feet

80. *Modeling Motion* The height h of a tennis ball above the ground after t seconds is shown in the graph. Estimate when the ball was 25 feet above the ground.

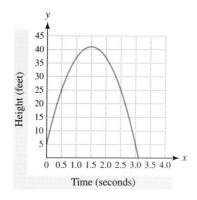

81. *Falling Object* (Refer to Example 3.) How long does it take for a toy to hit the ground if it is dropped out of a window 60 feet above the ground? Does it take twice as long as it takes to fall from a window 30 feet above the ground?

82. *Falling Object* If a metal ball is thrown downward with an initial velocity of 22 feet per second (15 mph) from a 100-foot water tower, its height h in feet above the ground after t seconds is modeled by
$$h(t) = -16t^2 - 22t + 100.$$
(a) Determine symbolically when the height of the ball is 62 feet.
(b) Support your result either graphically or numerically.

WRITING ABOUT MATHEMATICS

83. Suppose that you are asked to solve
$$ax^2 + bx + c = 0.$$
Explain how the graph of $y = ax^2 + bx + c$ can be used to find any solutions to the equation.

84. Explain why a quadratic equation could not have more than 2 solutions. (*Hint:* Consider the graph of $y = ax^2 + bx + c$.)

CHECKING BASIC CONCEPTS FOR SECTIONS 6.1 AND 6.2

1. Give symbolic, graphical, and numerical representations for the function that squares the input x and then subtracts 1. For the numerical representation let $x = -3, -2, -1, \ldots, 3$.

2. Find the vertex of $f(x) = 2x^2 - 4x$ symbolically and graphically.

3. Write an equation for a parabola that is similar to the graph of $y = x^2$ except that its vertex is $(-2, 1)$.

4. Solve the quadratic equation
$$2x^2 - 7x + 3 = 0$$
   (a) symbolically,
   (b) graphically, and
   (c) numerically.

5. Solve the equation symbolically.
   (a) $x^2 = 5$
   (b) $x^2 - 4x + 1 = 0$

6.3 THE QUADRATIC FORMULA

Solving Quadratic Equations ~ The Discriminant ~ Derivation of the Quadratic Formula

INTRODUCTION

To model the stopping distance of a car, highway engineers compute two quantities. The first quantity is the *reaction distance*, which is the distance a car travels between the time a driver first recognizes a hazard and applies the brakes. The second quantity is *braking distance*, which is the distance a car travels after a driver applies the brakes. *Stopping distance* equals the sum of the reaction distance and the braking distance. If a car is traveling x miles per hour, highway engineers estimate the reaction distance as $\frac{11}{3}x$ and the braking distance as $\frac{1}{9}x^2$. To estimate the total stopping distance d in feet, they add the two expressions to obtain

$$d(x) = \frac{1}{9}x^2 + \frac{11}{3}x.$$

If a car's headlights do not illuminate the road beyond 500 feet, a safe speed x for the car can be determined by solving the quadratic equation

$$\frac{1}{9}x^2 + \frac{11}{3}x = 500.$$

In this section we learn how to solve this equation with the quadratic formula.
(Source: L. Haefner, *Introduction to Transportation Systems*.)

SOLVING QUADRATIC EQUATIONS

Recall that any quadratic equation can be written in the form

$$ax^2 + bx + c = 0.$$

If we solve this equation for x in terms of a, b, and c by completing the square, we obtain the **quadratic formula**. We give its derivation at the end of this section.

6.3 The Quadratic Formula

Quadratic Formula

The solutions of the quadratic equation $ax^2 + bx + c = 0$, $a \neq 0$, are

$$x = \frac{-b \pm \sqrt{b^2 - 4ac}}{2a}.$$

EXAMPLE 1 *Solving a quadratic equation with two solutions*

Solve the equation $2x^2 - 3x - 1 = 0$ symbolically. Support your results graphically.

Solution

Symbolic Solution Let $a = 2$, $b = -3$, and $c = -1$.

$$x = \frac{-b \pm \sqrt{b^2 - 4ac}}{2a} \qquad \text{Quadratic formula}$$

$$x = \frac{-(-3) \pm \sqrt{(-3)^2 - 4(2)(-1)}}{2(2)} \qquad \text{Substitute for } a, b, \text{ and } c.$$

$$x = \frac{3 \pm \sqrt{17}}{4} \qquad \text{Simplify.}$$

The solutions are $x = \dfrac{3 + \sqrt{17}}{4} \approx 1.78$ and $x = \dfrac{3 - \sqrt{17}}{4} \approx -0.28$.

Graphical Solution Graph $Y_1 = 2X^2 - 3X - 1$. Figures 6.53 and 6.54 show that the x-intercepts are approximately 1.78 and -0.28.

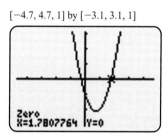

Figure 6.53

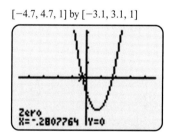

Figure 6.54

Critical Thinking

Use the results of Example 1 to help evaluate each expression mentally.

$$2\left(\frac{3 + \sqrt{17}}{4}\right)^2 - 3\left(\frac{3 + \sqrt{17}}{4}\right) - 1$$

$$2\left(\frac{3 - \sqrt{17}}{4}\right)^2 - 3\left(\frac{3 - \sqrt{17}}{4}\right) - 1$$

EXAMPLE 2 *Solving a quadratic equation with one solution*

Solve the equation $25x^2 + 20x + 4 = 0$ symbolically. Support your result graphically and numerically.

Solution

Symbolic Solution   Let $a = 25$, $b = 20$, and $c = 4$.

$$x = \frac{-b \pm \sqrt{b^2 - 4ac}}{2a} \quad \text{Quadratic formula}$$

$$x = \frac{-20 \pm \sqrt{20^2 - 4(25)(4)}}{2(25)} \quad \text{Substitute for } a, b, \text{ and } c.$$

$$x = \frac{-20 \pm \sqrt{0}}{50} \quad \text{Simplify.}$$

$$x = \frac{-20}{50} = -0.4 \quad \sqrt{0} = 0$$

There is one solution, $x = -0.4$.

Graphical Solution   Graph $Y_1 = 25X^2 + 20X + 4$. From Figure 6.55, there is one x-intercept, $x = -0.4$.

Numerical Solution   Figure 6.56 shows that $Y_1 = 0$ when $x = -0.4$.

[−3.7, 1, 0.2] by [−10, 10, 1]

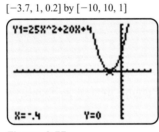

Figure 6.55

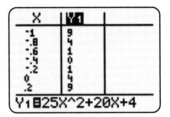

Figure 6.56

EXAMPLE 3 *Recognizing a quadratic equation with no real solutions*

Solve the equation $5x^2 - x + 3 = 0$ symbolically. Support your result graphically.

Solution

Symbolic Solution   Let $a = 5$, $b = -1$, and $c = 3$.

$$x = \frac{-b \pm \sqrt{b^2 - 4ac}}{2a} \quad \text{Quadratic formula}$$

$$x = \frac{-(-1) \pm \sqrt{(-1)^2 - 4(5)(3)}}{2(5)} \quad \text{Substitute for } a, b, \text{ and } c.$$

$$x = \frac{1 \pm \sqrt{-59}}{10} \quad \text{Simplify.}$$

There are no real solutions to this equation because $\sqrt{-59}$ is not defined. (In Chapter 8 you will learn how to evaluate the square root of a negative number by using complex numbers.)

Graphical Solution Graph $Y_1 = 5X^2 - X + 3$. Figure 6.57 reveals that there are no x-intercepts, so there are no real solutions.

[−5, 5, 1] by [−20, 20, 5]

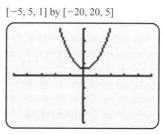

Figure 6.57

In the introduction to this section we discussed how engineers estimate safe stopping distances for automobiles. In the next example we solve the problem presented in the introduction.

EXAMPLE 4 *Modeling stopping distance*

If a car's headlights do not illuminate the road beyond 500 feet, estimate a safe speed x for the car by solving $\frac{1}{9}x^2 + \frac{11}{3}x = 500$.

Solution

Begin by subtracting 500 from both sides of the equation.

$$\frac{1}{9}x^2 + \frac{11}{3}x - 500 = 0.$$

To eliminate fractions multiply both sides by 9. (This step is not necessary, but it makes the problem easier to work.)

$$x^2 + 33x - 4500 = 0.$$

Now let $a = 1$, $b = 33$, and $c = -4500$.

$$x = \frac{-b \pm \sqrt{b^2 - 4ac}}{2a} \qquad \text{Quadratic formula}$$

$$x = \frac{-33 \pm \sqrt{(33)^2 - 4(1)(-4500)}}{2(1)} \qquad \text{Substitute for } a, b, \text{ and } c.$$

$$x = \frac{-33 \pm \sqrt{19{,}089}}{2} \qquad \text{Simplify.}$$

The solutions are

$$\frac{-33 + \sqrt{19{,}089}}{2} \approx 52.6 \quad \text{and} \quad \frac{-33 - \sqrt{19{,}089}}{2} \approx -85.6.$$

The negative solution has no physical meaning because negative speeds are not possible. The other solution is 52.6, so an appropriate speed might be 50 miles per hour.

THE DISCRIMINANT

The expression $b^2 - 4ac$ in the quadratic formula is called the **discriminant.** It provides information about the number of solutions to a quadratic equation.

> **The Discriminant and Quadratic Equations**
>
> To determine the number of solutions to $ax^2 + bx + c = 0$, evaluate the discriminant $b^2 - 4ac$.
>
> 1. If $b^2 - 4ac > 0$, there are two real solutions.
> 2. If $b^2 - 4ac = 0$, there is one real solution.
> 3. If $b^2 - 4ac < 0$, there are no real solutions.
>
> *Note:* When $b^2 - 4ac < 0$, the solutions to a quadratic equation may be expressed as complex numbers (discussed in Chapter 8).

EXAMPLE 5 *Using the discriminant*

 Use the discriminant to determine the number of solutions to $4x^2 + 25 = 20x$. Then solve the equation, using the quadratic formula. Support your result graphically.

Solution

Write the equation as $4x^2 - 20x + 25 = 0$ so that $a = 4$, $b = -20$, and $c = 25$. The discriminant evaluates to

$$b^2 - 4ac = (-20)^2 - 4(4)(25) = 0.$$

Thus there is one solution.

$$\begin{aligned}
x &= \frac{-b \pm \sqrt{b^2 - 4ac}}{2a} &&\text{Quadratic formula} \\
&= \frac{-(-20) \pm \sqrt{0}}{2(4)} &&\text{Substitute.} \\
&= \frac{20}{8} = 2.5 &&\text{Simplify.}
\end{aligned}$$

The only solution is 2.5. A graph of $Y_1 = 4X^2 - 20X + 25$ is shown in Figure 6.58. Note that the graph has one *x*-intercept, 2.5.

6.3 The Quadratic Formula

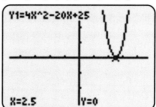

Figure 6.58

Technology Note: *Evaluating the Discriminant*

In Example 5 it is important to enter the discriminant in a calculator as
$$(-20)^2 - 4(4)(25)$$
rather than as
$$-20^2 - 4(4)(25)$$
because $(-20)^2 = 400$, whereas $-20^2 = -400$. See Figure 6.59.

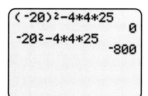

Figure 6.59

EXAMPLE 6 *Analyzing graphs of quadratic functions*

A graph of $f(x) = ax^2 + bx + c$ is shown in Figure 6.60.

(a) State whether $a > 0$ or $a < 0$.
(b) Solve the equation $ax^2 + bx + c = 0$.
(c) Determine whether the discriminant is positive, negative, or zero.

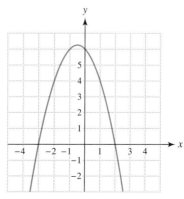

Figure 6.60

Solution

(a) The parabola opens downward, so $a < 0$.
(b) The solutions correspond to the x-intercepts, -3 and 2.
(c) There are two solutions, so the discriminant is positive.

DERIVATION OF THE QUADRATIC FORMULA

Consider the quadratic equation

$$ax^2 + bx + c = 0, \quad a > 0.$$

We can solve this equation for x in terms of a, b, and c to obtain the quadratic formula. We do so by completing the square.

$ax^2 + bx = -c$	Subtract c.
$x^2 + \dfrac{b}{a}x = -\dfrac{c}{a}$	Divide by a.
$x^2 + \dfrac{b}{a}x + \dfrac{b^2}{4a^2} = -\dfrac{c}{a} + \dfrac{b^2}{4a^2}$	Add $\left(\dfrac{b/a}{2}\right)^2 = \dfrac{b^2}{4a^2}$.
$\left(x + \dfrac{b}{2a}\right)^2 = -\dfrac{c}{a} + \dfrac{b^2}{4a^2}$	Perfect square trinomial
$\left(x + \dfrac{b}{2a}\right)^2 = -\dfrac{c \cdot 4a}{a \cdot 4a} + \dfrac{b^2}{4a^2}$	Multiply $-\dfrac{c}{a}$ by $\dfrac{4a}{4a}$.
$\left(x + \dfrac{b}{2a}\right)^2 = -\dfrac{4ac}{4a^2} + \dfrac{b^2}{4a^2}$	Simplify.
$\left(x + \dfrac{b}{2a}\right)^2 = \dfrac{-4ac + b^2}{4a^2}$	Add fractions.
$\left(x + \dfrac{b}{2a}\right)^2 = \dfrac{b^2 - 4ac}{4a^2}$	Rewrite.
$x + \dfrac{b}{2a} = \pm\sqrt{\dfrac{b^2 - 4ac}{4a^2}}$	Square root property
$x = -\dfrac{b}{2a} \pm \sqrt{\dfrac{b^2 - 4ac}{4a^2}}$	Add $-\dfrac{b}{2a}$.
$x = -\dfrac{b}{2a} \pm \dfrac{\sqrt{b^2 - 4ac}}{2a}$	Property of square roots
$x = \dfrac{-b \pm \sqrt{b^2 - 4ac}}{2a}$	Combine fractions.

■ MAKING CONNECTIONS

Completing the Square and the Quadratic Formula

The quadratic formula results from completing the square for the equation $ax^2 + bx + c = 0$. When you use the quadratic formula, the work of completing the square has already been done for you.

6.3 Putting It All Together

Quadratic equations can be solved symbolically using factoring, the square root property, completing the square, and the quadratic formula. Graphical and numerical methods can also be used to solve quadratic equations. In this section we discussed the quadratic formula and its discriminant, which we summarize in the following table.

Method	Explanation	Example
Quadratic Formula	The quadratic formula can be used to solve any quadratic equation written as $ax^2 + bx + c = 0$. The solutions are given by $$x = \frac{-b \pm \sqrt{b^2 - 4ac}}{2a}.$$	For the equation $$2x^2 - 3x + 1 = 0$$ and the values $$a = 2, \quad b = -3, \quad \text{and} \quad c = 1,$$ $$\frac{-(-3) \pm \sqrt{(-3)^2 - 4(2)(1)}}{2(2)} = \frac{3 \pm \sqrt{1}}{4} = 1, \frac{1}{2}.$$
The Discriminant	The expression $b^2 - 4ac$ is called the discriminant. The discriminant may be used to determine the number of solutions to $$ax^2 + bx + c = 0.$$ $b^2 - 4ac > 0$ indicates two real solutions. $b^2 - 4ac = 0$ indicates one real solution. $b^2 - 4ac < 0$ indicates no real solutions.	For the equation $$x^2 + 4x - 1 = 0$$ and the values $$a = 1, \quad b = 4, \quad \text{and} \quad c = -1,$$ $$b^2 - 4ac = 4^2 - 4(1)(-1) = 20 > 0,$$ indicating two real solutions.

6.3 EXERCISES

FOR EXTRA HELP: Student's Solutions Manual, MyMathLab.com, InterAct Math, Math Tutor Center, MathXL, Digital Video Tutor CD 4 Videotape 10

CONCEPTS

1. What is the quadratic formula used for?

2. What basic algebraic technique is used to derive the quadratic formula?

3. What is the discriminant?

4. If the discriminant evaluates to 0, what does that indicate about the quadratic equation?

5. Name four symbolic techniques for solving a quadratic equation.

6. Does every quadratic equation have at least one real solution? Explain.

THE QUADRATIC FORMULA

Exercises 7–10: Use the quadratic formula to solve the equation. Support your result graphically and numerically.

7. $2x^2 + 11x - 6 = 0$

8. $x^2 + 2x - 24 = 0$

9. $-x^2 + 2x - 1 = 0$ 10. $3x^2 - x + 1 = 0$

Exercises 11–24: Solve, using the quadratic formula.

11. $x^2 - 6x - 16 = 0$ 12. $2x^2 - 9x + 7 = 0$

13. $4x^2 - x - 1 = 0$ 14. $-x^2 + 2x + 1 = 0$

15. $-3x^2 + 2x - 1 = 0$

16. $x^2 + x + 3 = 0$

17. $36x^2 - 36x + 9 = 0$

18. $4x^2 - 5.6x + 1.96 = 0$

19. $2x(x - 3) = 2$

20. $x(x + 1) + x = 5$

21. $(x - 1)(x + 1) + 2 = 4x$

22. $\frac{1}{2}(x - 6) = x^2 + 1$

23. $\frac{1}{2}x(x + 1) = 2x^2 - \frac{3}{2}$

24. $\frac{1}{2}x^2 - \frac{1}{4}x + \frac{1}{2} = x$

THE DISCRIMINANT

Exercises 25–30: A graph of $y = ax^2 + bx + c$ is shown.
  (a) State whether $a > 0$ or $a < 0$.
  (b) Solve $ax^2 + bx + c = 0$, if possible.
  (c) Determine whether the discriminant is positive, negative, or zero.

25.
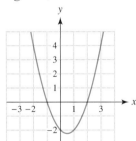

26.
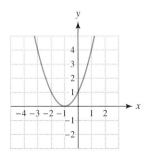

27.
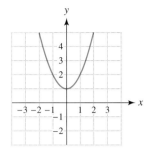

28.
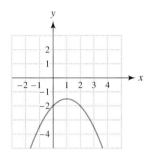

29.
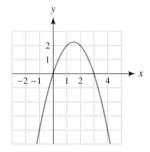

30.
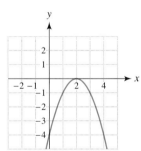

Exercises 31–38: Complete the following for the given equation.
  (a) Evaluate the discriminant.
  (b) How many real solutions are there?
  (c) Support your answer for part (b) by using graphing.

31. $3x^2 + x - 2 = 0$ 32. $5x^2 - 13x + 6 = 0$

33. $x^2 - 4x + 4 = 0$ 34. $\frac{1}{4}x^2 + 4 = 2x$

35. $\frac{1}{2}x^2 + \frac{3}{2}x + 2 = 0$ 36. $x - 3 = 2x^2$

37. $x(x + 3) = 3$

38. $(4x - 1)(x - 3) = -25$

YOU DECIDE THE METHOD

Exercises 39–44: Find exact solutions to the quadratic equation, using a method of your choice. Explain why you chose the method you did.

39. $x^2 - 3x + 2 = 0$ **40.** $x^2 + 2x + 1 = 0$

41. $0.5x^2 - 1.75x - 1 = 0$

42. $\frac{3}{5}x^2 + \frac{9}{10}x - \frac{3}{5} = 0$

43. $x^2 - 5x + 2 = 0$ **44.** $2x^2 - x - 4 = 0$

APPLICATIONS

Exercises 45–49: Modeling Stopping Distance (Refer to Example 4.) Find a safe speed for the following stopping distances.

45. 42 feet   **46.** 152 feet

47. 390 feet   **48.** 726 feet

49. *Modeling U.S. AIDS Deaths* The cumulative numbers in thousands of AIDS deaths from 1984 through 1994 may be modeled by
$$f(x) = 2.39x^2 + 5.04x + 5.1,$$
where $x = 0$ corresponds to 1984, $x = 1$ to 1985, and so on until $x = 10$ corresponds to 1994. See the accompanying graph. Use the formula for $f(x)$ to estimate symbolically the year when the total number of AIDS deaths reached 200 thousand. Compare your result with that shown in the graph.

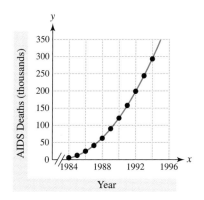

50. *Screen Dimensions* The width of a rectangular computer screen is 3 inches more than its height. If the area of the screen is 154 square inches, find its dimensions
  (a) graphically,
  (b) numerically, and
  (c) symbolically.

51. *Modeling Water Flow* When water runs out of a hole in a cylindrical container, the height of the water in the container can often be modeled by a quadratic function. The data in the following table show the height y of water at 30-second intervals in a metal can that had a small hole in it.

Time (seconds)	0	30	60	90
Height (centimeters)	16	11.9	8.4	5.3

Time (seconds)	120	150	180
Height (centimeters)	3.1	1.4	0.5

These data are modeled by
$$f(x) = 0.0004x^2 - 0.15x + 16.$$

(a) Explain why a linear function would not be appropriate for modeling this data.
(b) Use the table to estimate the time at which the height was 7 centimeters.
(c) Use $f(x)$ and the quadratic formula to solve part (b).

52. *Hospitals* The general trend in the number of hospitals in the United States from 1945 through 1995 is modeled by
$$f(x) = -1.38x^2 + 84x + 5865,$$
where $x = 5$ corresponds to 1945, $x = 10$ to 1950, and so on until $x = 55$ represents 1995. See the scatterplot and accompanying graph.

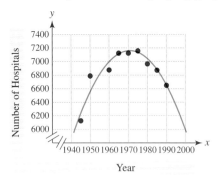

(a) Describe any trends in the numbers of hospitals from 1945 to 1995.

(b) What information does the vertex of the graph of f give?

(c) Use the formula for $f(x)$ to estimate the number of hospitals in 1970 ($x = 30$). Compare your result with that shown in the graph.

(d) Use the formula for $f(x)$ to estimate the year (or years) when there were 6500 hospitals. Compare your result with that shown in the graph.

Writing About Mathematics

53. Explain how the discriminant can be used to determine the number of real solutions to a quadratic equation.

54. Let $f(x) = ax^2 + bx + c$. If you know the value of $b^2 - 4ac$, what information does this give you about the graph of f? Explain your answer.

6.4 Quadratic Inequalities

Basic Concepts ~ Graphical and Numerical Solutions ~ Symbolic Solutions

Introduction

Many new and amazing inventions continue to flood the market. Genetic engineering, medical marvels, space craft, and digital cameras were made possible, in part, because of people's ability to solve nonlinear equations and inequalities. Before modern technology, many equations were simply too difficult to solve by hand. Now these equations can be solved with high-performance computers. In fact, today's computers are even used to design tomorrow's super computers. (*Reference:* F. Allen, "Technology at the End of the Century," *American Heritage of Invention & Technology,* Winter 2000.)

Quadratic inequalities are nonlinear inequalities. These inequalities are simple enough to be solved by hand. In this section we discuss how to solve them graphically, numerically, and symbolically.

Basic Concepts

If the equals sign in a quadratic equation is replaced with $>$, $\geq$, $<$, or $\leq$, a **quadratic inequality** results. Examples of quadratic inequalities include
$$x^2 + 4x - 3 < 0, \quad 5x^2 \geq 5, \quad \text{and} \quad 1 - z \leq z^2.$$

Solution

(a) Begin by solving $6x^2 - 7x - 5 = 0$.

$$6x^2 - 7x - 5 = 0$$
$$(2x + 1)(3x - 5) = 0 \quad \text{Factor.}$$
$$2x + 1 = 0 \quad \text{or} \quad 3x - 5 = 0 \quad \text{Zero-product property}$$
$$x = -\frac{1}{2} \quad \text{or} \quad x = \frac{5}{3} \quad \text{Solve.}$$

Therefore the solutions to $6x^2 - 7x - 5 \geq 0$ lie "outside" these two values and satisfy $x \leq -\frac{1}{2}$ or $x \geq \frac{5}{3}$.

(b) First, rewrite the inequality as follows.

$$x(3 - x) > -18$$
$$3x - x^2 > -18 \quad \text{Distributive property}$$
$$3x - x^2 + 18 > 0 \quad \text{Add 18.}$$
$$-x^2 + 3x + 18 > 0 \quad \text{Rewrite.}$$
$$x^2 - 3x - 18 < 0 \quad \text{Multiply by } -1 \text{ and reverse the inequality sign.}$$

Next, solve $x^2 - 3x - 18 = 0$.

$$(x + 3)(x - 6) = 0 \quad \text{Factor.}$$
$$x = -3 \quad \text{or} \quad x = 6 \quad \text{Solve.}$$

Solutions to $x^2 - 3x - 18 < 0$ lie between these two values and satisfy $-3 < x < 6$.

Critical Thinking ───

Graph $y = x^2 + 1$ and solve the following inequalities.

i. $x^2 + 1 > 0$ **ii.** $x^2 + 1 < 0$

Now graph $y = (x - 1)^2$ and solve the following inequalities.

iii. $(x - 1)^2 \geq 0$ **iv.** $(x - 1)^2 \leq 0$

EXAMPLE 5   *Finding the dimensions of a building*

A rectangular building needs to be 7 feet longer than it is wide, as illustrated in Figure 6.72. The area of the building must be at least 450 square feet. What widths x are possible for the building?

Figure 6.72

Solution

Symbolic Solution   If x is the width of the building, $x + 7$ is the length of the building and its area is $x(x + 7)$. The area must be at least 450 square feet, so the inequality $x(x + 7) \geq 450$ must be satisfied. First solve the following quadratic equation.

$x(x + 7) = 450$	Quadratic equation
$x^2 + 7x = 450$	Distributive property
$x^2 + 7x - 450 = 0$	Subtract 450.
$x = \dfrac{-7 \pm \sqrt{7^2 - 4(1)(-450)}}{2(1)}$	Quadratic formula
$= \dfrac{-7 \pm \sqrt{1849}}{2}$	Simplify.
$= \dfrac{-7 \pm 43}{2}$	$\sqrt{1849} = 43$
$= 18, -25$	Evaluate.

Thus the solution to $x(x + 7) \geq 450$ is $x \leq -25$ or $x \geq 18$. The width is positive, so the building width must be 18 feet or more.

Numerical Solution   Numerical support is shown in Figure 6.73, where $Y_1 = X(X + 7)$ equals 450 when $x = 18$. For $x \geq 18$ the area is at least 450 square feet.

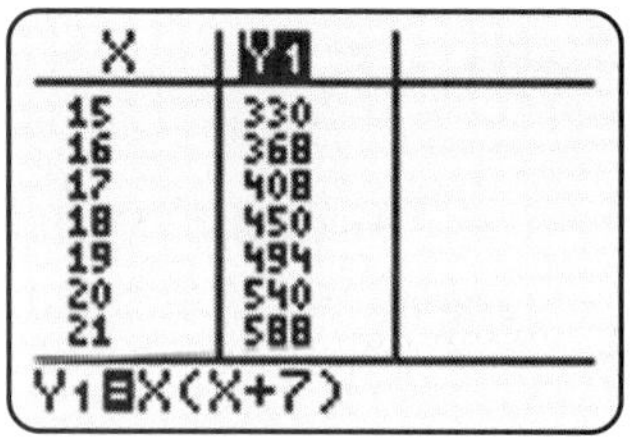

Figure 6.73

6.4 Putting It All Together

The following table summarizes solutions of quadratic inequalities containing the symbols $<$ or $>$. Cases involving $\leq$ or $\geq$ are solved similarly.

Method	Explanation
Solving a quadratic inequality symbolically	Let $ax^2 + bx + c = 0$, $a > 0$, have two solutions p and q, where $p < q$. $ax^2 + bx + c < 0$ is equivalent to $p < x < q$ $ax^2 + bx + c > 0$ is equivalent to $x < p$ or $x > q$

Method	Explanation
Solving a quadratic inequality symbolically (continued)	*Example:* The solutions to $x^2 - 3x + 2 = 0$ are $x = 1$ or $x = 2$. The solutions to $x^2 - 3x + 2 < 0$ are $1 < x < 2$. The solutions to $x^2 - 3x + 2 > 0$ are $x < 1$ or $x > 2$.
Solving a quadratic inequality graphically	Given $ax^2 + bx + c < 0, a > 0$, graph $y = ax^2 + bx + c$ and locate any x-intercepts. If there are two x-intercepts, then solutions correspond to x-values between the x-intercepts. Solutions to $ax^2 + bx + c > 0$ correspond to x-values "outside" the x-intercepts.
Solving a quadratic inequality numerically	If a quadratic inequality is expressed as $ax^2 + bx + c < 0, a > 0$, solve $y = ax^2 + bx + c = 0$ with a table. If there are two solutions, then the solutions to the given inequality lie between these values. Solutions to $ax^2 + bx + c > 0$ lie "outside" these values.

6.4 EXERCISES

CONCEPTS

1. How is a quadratic inequality different from a quadratic equation?

2. Do quadratic inequalities typically have two solutions? Explain.

3. Is $x = 3$ a solution to $x^2 < 7$?

4. Is $x = 5$ a solution to $x^2 \geq 25$?

5. The solutions to $x^2 - 2x - 8 = 0$ are -2 and 4. What are the solutions to $x^2 - 2x - 8 < 0$?

6. The solutions to $x^2 + 2x - 3 = 0$ are -3 and 1. What are the solutions to $x^2 + 2x - 3 > 0$?

Exercises 7–12: Determine whether the inequality is quadratic.

7. $x^2 + 4x + 5 < 0$ 8. $x > x^3 - 5$

9. $x^2 > 19$

10. $x(x - 1) - 2 \geq 0$

11. $4x > 1 - x$

12. $2x(x^2 + 3) < 0$

Exercises 13–18: Determine whether the given value of x is a solution.

13. $2x^2 + x - 1 > 0,$ $x = 3$

14. $x^2 - 3x + 2 \leq 0,$ $x = 2$

15. $x^2 + 2 \leq 0,$ $x = 0$

16. $2x(x - 3) \geq 0,$ $x = 1$

17. $x^2 - 3x \leq 1,$ $x = -3$

18. $4x^2 - 5x + 1 > 30,$ $x = -2$

Solving Quadratic Inequalities

Exercises 19–24: Use the graph of
$$y = ax^2 + bx + c$$
to solve each quadratic equation or inequality.
(a) $ax^2 + bx + c = 0$
(b) $ax^2 + bx + c < 0$
(c) $ax^2 + bx + c > 0$

19.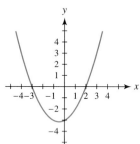

20.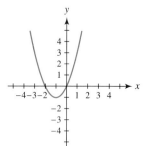

21.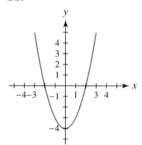

22.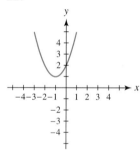

23.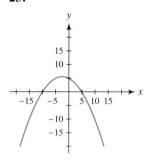

24.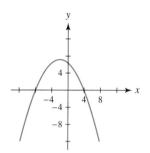

25.

X	Y1
-3	5
-2	0
-1	-3
0	-4
1	-3
2	0
3	5

Y1■X^2−4

26.

X	Y1
-3	10
-2	4
-1	0
0	-2
1	-2
2	0
3	4

Y1■X^2−X−2

27.

X	Y1
-5	5
-4	0
-3	-3
-2	-4
-1	-3
0	0
1	5

Y1■X^2+4X

28.

X	Y1
-2	-2.5
-1.5	0
-1	1.5
-.5	2
0	1.5
.5	0
1	-2.5

Y1■−2X^2−2X+1.5

Exercises 29–38: Solve the quadratic inequality
(a) *graphically and*
(b) *numerically.*

29. $x^2 + 4x + 3 < 0$

30. $x^2 + x - 2 \leq 0$

31. $2x^2 - x - 15 \geq 0$

32. $3x^2 - 3x - 6 > 0$

33. $2x^2 \leq 8$

34. $x^2 < 9$

35. $x^2 > -5$

36. $-x^2 \geq 1$

37. $-x^2 + 3x > 0$

38. $-8x^2 - 2x + 1 \leq 0$

Exercises 25–28: Use the table for
$$y = ax^2 + bx + c$$
to solve each quadratic equation or inequality.
(a) $ax^2 + bx + c = 0$
(b) $ax^2 + bx + c < 0$
(c) $ax^2 + bx + c > 0$

Exercises 39–42: Solve the quadratic equation in part (a) symbolically. Use the results to solve the inequalities in parts (b) and (c).

39. (a) $x^2 - 4 = 0$
    (b) $x^2 - 4 < 0$
    (c) $x^2 - 4 > 0$

40. (a) $x^2 - 5 = 0$
    (b) $x^2 - 5 \leq 0$
    (c) $x^2 - 5 \geq 0$

41. (a) $x^2 + 4x - 5 = 0$
    (b) $x^2 + 4x - 5 \leq 0$
    (c) $x^2 + 4x - 5 \geq 0$

42. (a) $x^2 + x - 1 = 0$
    (b) $x^2 + x - 1 < 0$
    (c) $x^2 + x - 1 > 0$

Exercises 43–52: Solve the quadratic inequality symbolically.

43. $x^2 + 10x + 21 \leq 0$

44. $x^2 - 7x - 18 < 0$

45. $3x^2 - 9x + 6 > 0$

46. $7x^2 + 34x - 5 \geq 0$

47. $x^2 < 10$

48. $x^2 \geq 64$

49. $x(x - 6) > 0$

50. $1 - x^2 \leq 0$

51. $x(4 - x) \leq 2$

52. $2x(1 - x) \geq 2$

APPLICATIONS

53. *Highway Design* (Refer to Example 3 and Figure 6.67.) The elevation E of a sag curve in feet is given by
$$E(x) = 0.0000375x^2 - 0.175x + 1000,$$
where $0 \leq x \leq 4000$.
    (a) Estimate graphically the x-values for which the elevation is 850 feet or less. (*Hint:* Use [0, 4000, 1000] by [500, 1200, 100] as a viewing rectangle.)
    (b) For what x-values is the elevation 850 feet or more?

54. *Early Cellular Phone Use* Our society is in transition from an industrial to an informational society. Cellular communication has played an increasingly large role in this transition. The number of cellular subscribers in the United States in thousands from 1985 to 1991 can be modeled by
$$f(x) = 163x^2 - 146x + 205,$$
where x is the year and $x = 0$ corresponds to 1985, $x = 1$ to 1986, and so on. (*Source:* Paetsch, M., *Mobile Communication in the U.S. and Europe.*)
    (a) Write a quadratic inequality whose solution set represents the years when there were 2 million subscribers or more.
    (b) Solve this inequality.

55. *Heart Disease Death Rates* From 1960 to 1995, age-adjusted heart disease rates decreased dramatically. The number of deaths per 100,000 people can be modeled by
$$f(x) = -0.014777x^2 + 54.14x - 49,060,$$
where x is the year, as illustrated in the accompanying figure. (*Source:* **Department of Health and Human Services.**)

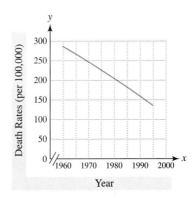

    (a) Evaluate $f(1985)$, using both the formula and the graph. How do your results compare?
    (b) Use the graph to estimate the years when this death rate was 250 or less.
    (c) Solve part (b) by graphing $y = f(x)$ and $y = 250$ in [1960, 1995, 10] by [0, 300, 50].

56. *Accidental Deaths* From 1910 to 1996 the number of accidental deaths per 100,000 people generally decreased and can be modeled by

$$f(x) = -0.001918x^2 + 6.93x - 6156,$$

where x is the year, as shown in the accompanying figure. *(Source: Department of Health and Human Services.)*

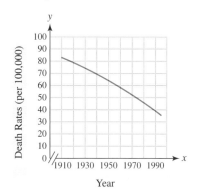

(a) Evaluate $f(1955)$, using both the formula and the graph. How do your results compare?

(b) Use the graph to estimate when this death rate was 60 or more.

(c) Solve part (b) by graphing $y = f(x)$ and $y = 60$ in [1910, 1996, 10] by [0, 100, 10].

57. *Dimensions of a Pen* A rectangular pen for a pet is 5 feet longer than it is wide. Give possible values for the width x of the pen if its area must be between 176 and 500 square feet, inclusively.

58. *Dimensions of a Cylinder* The volume of a cylindrical can is given by $V = \pi r^2 h$, where r is its radius and h is its height. See the accompanying figure. If $h = 6$ inches and the volume of the can must be 50 cubic inches or more, estimate to the nearest tenth of an inch possible values for r.

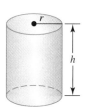

WRITING ABOUT MATHEMATICS

59. Consider an inequality of the form $x^2 < 0$. Discuss the solutions to this equation and explain your reasoning.

60. Explain how the graph of $y = ax^2 + bx + c$ can be used to solve the inequality

$$ax^2 + bx + c > 0$$

when $a < 0$. Assume that the x-intercepts of the graph are p and q with $p < q$.

CHECKING BASIC CONCEPTS FOR SECTIONS 6.3 AND 6.4

1. Use the quadratic formula to solve each equation.
   (a) $2x^2 = 3x + 1$
   (b) $9x^2 - 24x + 16 = 0$

2. Calculate the discriminant for each equation and give the number of real solutions.
   (a) $x^2 - 5x + 5 = 0$
   (b) $2x^2 - 5x + 4 = 0$
   (c) $49x^2 - 56x + 16 = 0$

3. Solve the quadratic inequality

$$x^2 - x - 6 > 0.$$

4. Solve the quadratic inequality

$$3x^2 + 5x + 2 \leq 0.$$

Chapter 6 Summary

Section 6.1 Quadratic Functions and Their Graphs

Any quadratic function f can be written as
$$f(x) = ax^2 + bx + c,$$
$a \neq 0$. Its graph is U-shaped and called a parabola. If $a > 0$, the parabola opens upward; if $a < 0$, the parabola opens downward. The vertex of a parabola is either the lowest point on a parabola that opens upward or the highest point on a parabola that opens downward. The x-coordinate of the vertex is given by $x = -\dfrac{b}{2a}$.

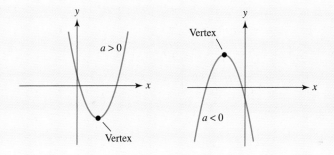

A quadratic function can also be expressed in vertex form as
$$f(x) = a(x - h)^2 + k.$$
In this form the point (h, k) is the vertex.

Section 6.2 Quadratic Equations

Any quadratic equation can be written in the form
$$ax^2 + bx + c = 0,$$
$a \neq 0$, and can have zero, one, or two solutions. These solutions correspond to the x-intercepts on the graph of $y = ax^2 + bx + c$.

The square root property may be used to solve quadratic equations with missing x-terms such as $x^2 - 4 = 0$ or $3x^2 = 6$. For example,
$$x^2 = 11 \quad \text{is equivalent to} \quad x = \pm\sqrt{11}.$$

Quadratic equations can also be solved symbolically by factoring, completing the square, and using the quadratic formula. To complete the square for an equation in the form

$$x^2 + kx = d,$$

add $\left(\dfrac{k}{2}\right)^2$ to both sides of the equation.

Section 6.3 *The Quadratic Formula*

The quadratic formula can be used to solve any quadratic equation in the form

$$ax^2 + bx + c = 0,$$

$a \neq 0$. The solutions are given by

$$x = \frac{-b \pm \sqrt{b^2 - 4ac}}{2a}.$$

The expression $b^2 - 4ac$ is called the discriminant. If $b^2 - 4ac < 0$ there are no real solutions, if $b^2 - 4ac = 0$ there is one solution, and if $b^2 - 4ac > 0$ there are two real solutions.

For example, to solve $2x^2 + 3x - 1 = 0$, let $a = 2$, $b = 3$, and $c = -1$.

$$x = \frac{-3 \pm \sqrt{3^2 - 4(2)(-1)}}{2(2)} = \frac{-3 \pm \sqrt{17}}{4} \approx 0.28, -1.78$$

Note that $b^2 - 4ac = 3^2 - 4(2)(-1) = 17 > 0$ and that there are two solutions to the quadratic equation.

Section 6.4 *Quadratic Inequalities*

When the equals sign in a quadratic equation is replaced with $<, >, \leq,$ or $\geq$, a quadratic inequality results. For example,

$$3x^2 - x + 1 = 0$$

is a quadratic equation and

$$3x^2 - x + 1 > 0$$

is a quadratic inequality. Like quadratic equations, quadratic inequalities can be solved symbolically, graphically, and numerically. An important first step in solving a quadratic inequality is to solve the corresponding quadratic equation.

Example: The solutions to $x^2 - 5x - 6 = 0$ are $x = -1, 6$.

The solutions to $x^2 - 5x - 6 < 0$ satisfy $-1 < x < 6$.

The solutions to $x^2 - 5x - 6 > 0$ satisfy $x < -1$ or $x > 6$.

Chapter 6
Review Exercises

Section 6.1

Exercises 1–4: State whether the graph represents a linear function, a quadratic function, or neither. If the function is quadratic, identify the vertex, the axis of symmetry, and if it opens upward or downward.

1.
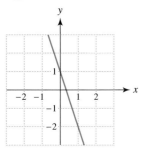

2.
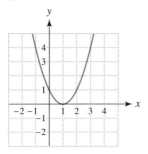

3.
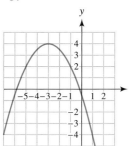

4.
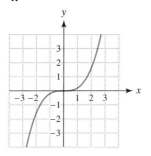

Exercises 5–8: Complete the following.
   (a) Graph f in $[-4.7, 4.7, 1]$ by $[-3.1, 3.1, 1]$.
   (b) Use the graph to identify the vertex and axis of symmetry.
   (c) Use the graph to evaluate $f(x)$ at the given value of x. Check your result symbolically.

5. $f(x) = x^2 - 2,$ $\quad x = -1$

6. $f(x) = -x^2 + 4x - 3,$ $\quad x = 3$

7. $f(x) = -\frac{1}{2}x^2 + x + \frac{3}{2},$ $\quad x = -2$

8. $f(x) = 2x^2 + 8x + 5,$ $\quad x = -3$

Exercises 9–12: Find the vertex of the parabola symbolically. Check your answer graphically.

9. $f(x) = x^2 - 4x - 2$

10. $f(x) = 5 - x^2$

11. $f(x) = -\frac{1}{4}x^2 + x + 1$

12. $f(x) = 2 + 2x + x^2$

Exercises 13 and 14: Find a value for the constant a so that $f(x) = ax^2 - 1$ models the data.

13.
x	1	2	3
$f(x)$	2	11	26

14.
x	-1	0	1
$f(x)$	$-3/4$	-1	$-3/4$

Exercises 15–20: Complete the following.
   (a) Graph f in the standard viewing rectangle.
   (b) Compare the graph of f with the graph of $y = x^2$.

15. $f(x) = x^2 + 2$

16. $f(x) = 3x^2$

17. $f(x) = (x - 2)^2$

18. $f(x) = (x + 1)^2 - 3$

19. $f(x) = \frac{1}{2}(x + 1)^2 + 2$

20. $f(x) = 2(x - 1)^2 - 3$

Section 6.2

Exercises 21–24: Use the graph of
$$y = ax^2 + bx + c$$
to solve $ax^2 + bx + c = 0$.

21.

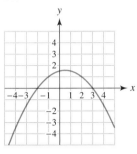

22.

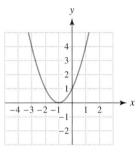

23.

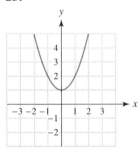

24.

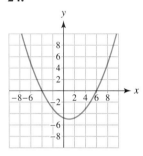

Exercises 25 and 26: Use the table of
$$y = ax^2 + bx + c$$
to solve $ax^2 + bx + c = 0$.

25.

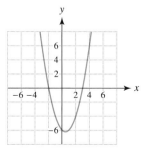

26.

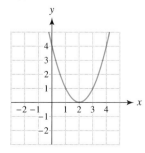

Exercises 27–30: Solve the quadratic equation
(a) graphically and
(b) numerically.

27. $x^2 - 5x - 50 = 0$ 28. $\frac{1}{2}x^2 + x - \frac{3}{2} = 0$

29. $\frac{1}{4}x^2 + \frac{1}{2}x = 2$ 30. $\frac{1}{2}x + \frac{3}{4} = \frac{1}{4}x^2$

Exercises 31–34: Solve the quadratic equation by factoring.

31. $x^2 + x - 20 = 0$ 32. $x^2 + 11x + 24 = 0$

33. $15x^2 - 4x - 4 = 0$

34. $7x^2 - 25x + 12 = 0$

Exercises 35–38: Use the square root property to solve the equation.

35. $x^2 = 100$ 36. $3x^2 = \frac{1}{3}$

37. $4x^2 - 6 = 0$ 38. $5x^2 = x^2 - 4$

Exercises 39–42: Solve the equation by completing the square.

39. $x^2 + 6x = -2$ 40. $x^2 - 4x = 6$

41. $x^2 - 2x - 5 = 0$ 42. $2x^2 + 6x - 1 = 0$

Section 6.3

Exercises 43–48: Solve the equation, using the quadratic formula.

43. $x^2 - 9x + 18 = 0$

44. $x^2 - 24x + 143 = 0$

45. $6x^2 + x = 1$ 46. $5x^2 + 1 = 5x$

47. $x(x - 8) = 5$

48. $2x(2 - x) = 3 - 2x$

Exercises 49–52: A graph of $y = ax^2 + bx + c$ is shown.
(a) State whether $a > 0$ or $a < 0$.
(b) Solve $ax^2 + bx + c = 0$.
(c) Determine whether the discriminant is positive, negative, or zero.

49.

50.

51. 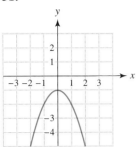 **52.**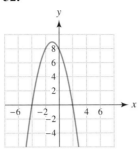

59. 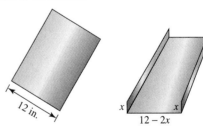 **60.**

Exercises 53–56: Complete the following for the given equation.
(a) Evaluate the discriminant.
(b) How many real solutions are there?
(c) Support your answer for part (b) graphically.

53. $2x^2 - 3x + 1 = 0$ **54.** $7x^2 + 2x - 5 = 0$

55. $3x^2 + x + 2 = 0$

56. $4.41x^2 - 12.6x + 9 = 0$

SECTION 6.4

Exercises 57 and 58: Use the graph of
$$y = ax^2 + bx + c$$
to solve each quadratic equation or inequality.
(a) $ax^2 + bx + c = 0$
(b) $ax^2 + bx + c < 0$
(c) $ax^2 + bx + c > 0$

57. 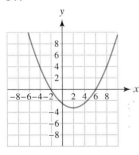 **58.**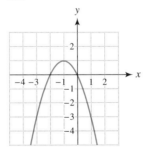

Exercises 59–60: Use the table of
$$y = ax^2 + bx + c$$
to solve each quadratic equation or inequality.
(a) $ax^2 + bx + c = 0$
(b) $ax^2 + bx + c < 0$
(c) $ax^2 + bx + c > 0$

Exercises 61 and 62: Solve the quadratic equation in part (a) symbolically. Use the results to solve the inequalities in parts (b) and (c).

61. (a) $x^2 - 2x - 3 = 0$
     (b) $x^2 - 2x - 3 < 0$
     (c) $x^2 - 2x - 3 > 0$

62. (a) $2x^2 - 7x - 15 = 0$
     (b) $2x^2 - 7x - 15 \leq 0$
     (c) $2x^2 - 7x - 15 \geq 0$

Exercises 63–66: Solve the quadratic inequality.

63. $x^2 + 4x + 3 \leq 0$ **64.** $5x^2 - 16x + 3 < 0$

65. $6x^2 - 13x + 2 > 0$ **66.** $x^2 \geq 5$

APPLICATIONS

67. *Construction* A rain gutter is being fabricated from a flat sheet of metal so that the cross section of the gutter is a rectangle, as shown in the accompanying figure. The width of the metal sheet is 12 inches.

(a) Write a formula $f(x)$ that gives the area of the cross section.
(b) Graph f in [0, 7, 1] by [0, 20, 5].
(c) To hold the greatest amount of rainwater, the cross section of the gutter should have maximum area. Use the graph to find the dimensions that result in the greatest cross-sectional area.
(d) Solve part (c) symbolically.

68. *Height of a Stone* Suppose that a stone is thrown upward with an initial velocity of 44 feet per second (30 miles per hour) and is released 4 feet above the ground. Its height h in feet after t seconds is modeled by
$$h(t) = -16t^2 + 44t + 4.$$
(a) Graph h in [0, 3, 1] by [0, 40, 10].
(b) When does the stone reach a height of 32 feet?
(c) After how many seconds does the stone reach maximum height? Estimate this height.
(d) Solve part (c) symbolically.

69. *Maximizing Revenue* A hotel is planning a group room rate. Rooms normally cost $90 per night. However, for a group rate the management is considering reducing the cost of a room by $3 for every room rented.
(a) Write a formula $f(x)$ that gives the revenue from renting x rooms at the group rate.
(b) Graph f in [0, 30, 5] by [0, 800, 100].
(c) How many rooms should be rented to yield revenue of $600?
(d) Use the graph to estimate how many rooms should be rented to maximize revenue.
(e) Solve part (d) symbolically.

70. *Airline Complaints* In 2000, major U.S. airlines promised better customer service. From 1997 through 1999, the number of complaints per 100,000 passengers can be modeled by
$$f(x) = 0.4(x - 1997)^2 + 0.8,$$
where x represents the year. (*Source:* Department of Transportation.)
(a) Evaluate $f(1999)$. Interpret the result.
(b) Graph f in [1997, 1999, 1] by [0.5, 3, 0.5]. Discuss how complaints changed over this time period.

71. *Braking Distance* On dry pavement a safe braking distance d in feet for a car traveling x miles per hour is $d = \dfrac{x^2}{12}$. For each distance d, find x. (F. Mannering, *Principles of Highway Engineering and Traffic Control.*)
(a) $d = 144$ feet (b) $d = 300$ feet

72. *Numbers* The product of two numbers is 143. One number is 2 more than the other.
(a) Write an equation whose solution gives the smaller number x.
(b) Solve the equation graphically.
(c) Solve the equation numerically.
(d) Solve the equation symbolically.

73. *Educational Attainment* From 1940 through 1991, the percentage of people with a high school diploma increased dramatically, as shown in the table.

Year	1940	1950	1960
H. S. Diploma (%)	25	34	44

Year	1970	1980	1991
H. S. Diploma (%)	55	69	78

Source: Bureau of the Census.

(a) Plot the data.
(b) Would it be reasonable to model this data with a linear function rather than a quadratic function? Explain your reasoning.
(c) Find a function that models the data.

74. *U. S. Energy Consumption* From 1950 to 1970 per capita consumption of energy in millions of Btu can be modeled by
$$f(x) = \frac{1}{4}(x - 1950)^2 + 220,$$
where x is the year. (*Source:* Department of Energy.)
(a) Find and interpret the vertex.
(b) Graph f in [1950, 1970, 5] by [200, 350, 25]. What happened to energy consumption during this time period?
(c) Use f to predict the consumption in 1996. Actual consumption was 354. Did f provide a good model for 1996? Explain.

75. *Screen Dimensions* A square computer screen has an area of 123 square inches. Approximate its dimensions.

76. *Flying a Kite* A kite is being flown, as illustrated in the accompanying figure. If 130 feet of string have been let out, find the value of x.

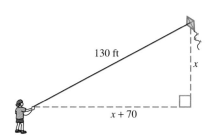

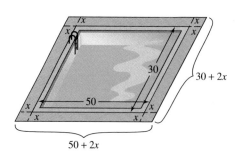

77. *Area* A uniform strip of grass is to be planted around a rectangular swimming pool, as illustrated in the figure at top right. The swimming pool is 30 feet wide and 50 feet long. If there is only enough grass seed to cover 250 square feet, estimate the width x that the strip of grass should be.

78. *Dimensions of a Cone* The volume V of a cone is given by $V = \frac{1}{3}\pi r^2 h$, where r is its base radius and h is its height. See the accompanying figure. If $h = 20$ inches and the volume of the cone must be between 750 and 1700 cubic inches, inclusively, estimate to the nearest tenth of an inch possible values for r.

Chapter 6

Test

1. Find the vertex and axis of symmetry for the graph of $f(x) = -\frac{1}{2}x^2 + x + 1$ symbolically. Check your answer graphically.

2. Find the exact value for the constant a so that $f(x) = ax^2 + 2$ models the data in the table.

x	−2	0	2	4
$f(x)$	0	2	0	−6

3. Graph $f(x) = \frac{1}{2}(x - 3)^2 + 2$ in the standard viewing rectangle and compare the graph of f to the graph of $y = x^2$.

4. Use the graph of
$$y = ax^2 + bx + c$$
to solve $ax^2 + bx + c = 0$.

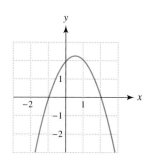

Exercises 5 and 6: Solve the quadratic equation symbolically. Check your answer either graphically or numerically.

5. $3x^2 + 11x - 4 = 0$ 6. $2x^2 = 2 - 6x^2$

7. Solve $x^2 - 5x = 1$ by completing the square.

8. Solve $x(-2x + 3) = -1$, using the quadratic formula.

9. A graph of $y = ax^2 + bx + c$ is shown.
   (a) State whether $a > 0$ or $a < 0$.
   (b) Solve $ax^2 + bx + c = 0$.
   (c) Determine whether the discriminant is positive, negative, or zero.

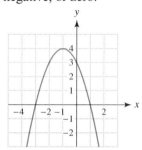

10. Complete the following for $-3x^2 + 4x - 5 = 0$.
    (a) Evaluate the discriminant.
    (b) How many real solutions are there?
    (c) Support your answer for part (b) graphically.

Exercises 11 and 12: Use the graph of
$$y = ax^2 + bx + c$$
to solve each equation or inequality.
   (a) $ax^2 + bx + c = 0$
   (b) $ax^2 + bx + c < 0$
   (c) $ax^2 + bx + c > 0$

11.
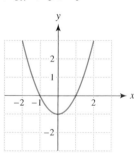

12.
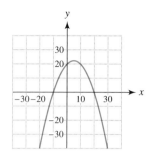

13. Solve the quadratic equation in part (a) symbolically. Use the result to solve the inequalities in parts (b) and (c).
    (a) $8x^2 - 2x - 3 = 0$
    (b) $8x^2 - 2x - 3 \leq 0$
    (c) $8x^2 - 2x - 3 \geq 0$

14. *Braking Distance* On wet pavement a safe braking distance d in feet for a car traveling x miles per hour is $d = \dfrac{x^2}{9}$. What speed corresponds to a stopping distance of 250 feet?
    (F. Mannering, *Principles of Highway Engineering and Traffic Control*.)

15. *Construction* A fence is being constructed along a 20-foot building, as shown in the accompanying figure. No fencing is used along the building.
    (a) If 200 feet of fence are available, find a formula $f(x)$ that gives the area enclosed by the fence and building.
    (b) Graph $y = f(x)$.
    (c) What value of x gives the greatest area?

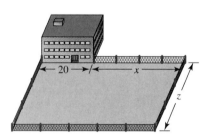

16. *Height of a Stone* Suppose that a stone is thrown upward with an initial velocity of 88 feet per second (60 miles per hour) and is released 8 feet above the ground. Its height h in feet after t seconds is modeled by
$$h(t) = -16t^2 + 88t + 8.$$

(a) Graph h in [0, 6, 1] by [0, 150, 50].
(b) When does the stone strike the ground?
(c) After how many seconds does the stone reach maximum height? Estimate this height.

CHAPTER 6
EXTENDED AND DISCOVERY EXERCISES

MODELING DATA WITH A QUADRATIC FUNCTION

1. *Survival Rate of Birds* The survival rate of sparrowhawks varies according to their age. The following table summarizes the results of one study by listing the age in years and the percentage of birds that survived the previous year. For example, 52% of sparrowhawks that reached age 6 lived to be 7 years old. (*Source:* D. Brown and P. Rothery, *Models in Biology: Mathematics, Statistics and Computing.*)

Age	1	2	3	4	5
Percent (%)	45	60	71	67	67

Age	6	7	8	9
Percent (%)	61	52	30	25

(a) Try to explain the relationship between age and the likelihood of surviving the next year.
(b) Make a scatterplot of the data. What type of function might model the data? Explain your reasoning.
(c) Graph each function. Which of the following functions models the data better?
$$f_1(x) = -3.57x + 71.1$$
$$f_2(x) = -2.07x^2 + 17.1x + 33$$
(d) Use one of these functions to estimate the likelihood of a 5.5-year-old sparrowhawk surviving for 1 more year.

2. *Photosynthesis and Temperature* Photosynthesis is the process by which plants turn sunlight into energy. At very cold temperatures photosynthesis may halt even though the sun is shining. In one study the efficiency of photosynthesis for an Antarctic species of grass was investigated. The following table lists results for various temperatures. The temperature x is in degrees Celsius, and the efficiency y is given as a percent. The purpose of the research was to determine the temperature at which photosynthesis is most efficient. (*Source:* D. Brown.)

x (°C)	−1.5	0	2.5	5	7	10	12
y (%)	33	46	55	80	87	93	95

x (°C)	15	17	20	22	25	27	30
y (%)	91	89	77	72	54	46	34

(a) Plot the data.
(b) What type of function might model these data? Explain your reasoning.
(c) Find a function f that models the data.
(d) Use f to estimate the temperature at which photosynthesis is most efficient in this type of grass.

Translations of Parabolas in Computer Graphics

Exercises 3 and 4: In video games with two-dimensional graphics, the background is often translated to give the illusion that a character in the game is moving. The simple scene on the left shows a mountain and an airplane. To make it appear that the airplane is flying, the mountain can be translated to the left, as shown in the figure on the right. **(Reference: C. Pokorny and C. Gerald, *Computer Graphics*.)**

3. *Video Games* Suppose that the mountain in the figure on the left can be modeled by $f(x) = -0.4x^2 + 4$ and that the airplane is located at the point (1, 5).
   (a) Graph f in $[-4, 4, 1]$ by $[0, 6, 1]$, where the units are kilometers. Plot the point (1, 5) to show the airplane.
   (b) Assume that the airplane is moving horizontally to the right at 0.2 kilometer per second. To give a video player the illusion that the airplane is moving, graph the image of the mountain and the position of the airplane after 10 seconds.

4. *Video Games* (Refer to Exercise 3.) Discuss how you could create the illusion of the airplane moving to the left and gaining altitude as it passes over the mountain. Try to perform a translation of this type. Explain your reasoning.

Chapter 7: Rational Expressions and Functions

Railroads, airports, and highways have been essential in the development of the United States. Without them much of the country's economic and technological growth would have been impossible. Today people are able to travel not only throughout the United States but also throughout the world by airplane, car, train, bus, or boat.

As transportation systems have become more complex, mathematics has played an increasingly important role in their design and operation. Highway designers use mathematics and cannot depend on trial and error for solutions. For example, the following table shows braking distances D for a car traveling downhill on wet pavement at 30 miles per hour for different grades G of the hill. (Grade corresponds to the slope of the road: A larger grade indicates a steeper hill.)

G	0.00	0.05	0.10	0.15	0.20	0.25
D (feet)	86	100	120	150	200	300

In this chapter we show how a rational function can model these data. We also discuss railroad track design, time spent waiting in a line, the ozone layer, skin cancer, probability, population growth, electricity, aerial photography, and memory requirements for storing music on a compact disc.

The future belongs to those who believe in the beauty of their dreams.
—*Eleanor Roosevelt*

Source: N. Garber and L. Hoel, *Traffic and Highway Engineering.*

7.1 RATIONAL FUNCTIONS AND EQUATIONS

Rational Expressions and Functions ~
Introduction to Rational Equations

INTRODUCTION

Suppose that a parking ramp attendant can wait on 5 cars per minute and that cars are randomly leaving the parking ramp at a rate of 4 cars per minute. Is it possible to predict how long the average driver will wait in line? Using rational functions, we can answer this question (see Exercise 69). Rational functions also occur in other types of applications, such as in the design of highways and train tracks. In this section we introduce rational functions and equations.

RATIONAL EXPRESSIONS AND FUNCTIONS

In Chapter 5 we discussed polynomials. Examples of polynomials include

$$4, \quad x, \quad 2x - 5, \quad 3x^2 - 6x + 1, \quad \text{and} \quad 3x^4 - 7.$$

Rational expressions result when a polynomial is divided by a nonzero polynomial. Three examples of rational expressions are

$$\frac{4}{x}, \quad \frac{x}{2x - 5}, \quad \text{and} \quad \frac{3x^2 - 6x + 1}{3x^4 - 7}.$$

EXAMPLE 1 *Recognizing a rational expression.*

Determine whether each expression is rational.

(a) $\dfrac{4 - x}{5x^2 + 3}$ (b) $\dfrac{x^2}{\sqrt{x} + 2x}$ (c) $\dfrac{6}{x - 1}$

Solution

(a) This expression is rational because both $4 - x$ and $5x^2 + 3$ are polynomials.
(b) This expression is not rational because $\sqrt{x} + 2x$ is not a polynomial.
(c) This expression is rational because both 6 and $x - 1$ are polynomials.

Rational expressions are used to define *rational functions*. For example, $\dfrac{4}{x + 1}$ is a rational expression and $f(x) = \dfrac{4}{x + 1}$ represents a rational function.

Rational Function

Let $p(x)$ and $q(x)$ be polynomials. Then a **rational function** is given by

$$f(x) = \frac{p(x)}{q(x)}.$$

The domain of f includes all x-values such that $q(x) \neq 0$.

Like other types of functions, rational functions may be represented symbolically, numerically, and graphically, as demonstrated in the next example.

EXAMPLE 2 Representing a rational function

Let f be the rational function that computes the reciprocal of x. Represent f symbolically, graphically, and numerically.

Solution

Symbolic Representation The reciprocal of x is computed by $f(x) = \frac{1}{x}$.

Graphical Representation Graph $Y_1 = 1/X$, as shown in Figure 7.1. Note that the expression $\frac{1}{x}$ is undefined when $x = 0$. Therefore no point on the graph of f has an x-coordinate of 0. Input 0 is not in the domain of f.

Numerical Representation Construct the table for $Y_1 = 1/X$, starting at $x = -3$ and incrementing by 1, as shown in Figure 7.2. When $x = 0$, an error occurs because 0 is not a valid input for $f(x)$. Note that, when $x = 2, f(x) = \frac{1}{2} = 0.5$ and that, when $x = 3, f(x) = \frac{1}{3} \approx 0.3333$.

[−4.7, 4.7, 1] by [−3.1, 3.1, 1]

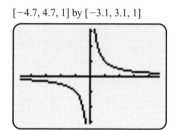
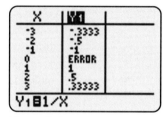

Figure 7.1   Figure 7.2

In the next example we evaluate a rational function symbolically, graphically, and numerically.

EXAMPLE 3 Evaluating a rational function

A rational function f is represented numerically, symbolically, and graphically. Use Table 7.1, $f(x)$, and Figure 7.3 on the next page to evaluate $f(-1), f(1)$, and $f(2)$.

(a) TABLE 7.1

x	−3	−2	−1	0	1	2	3
$f(x)$	3/2	4/3	1	0	Error	4	3

(b) $f(x) = \dfrac{2x}{x - 1}$

(c)

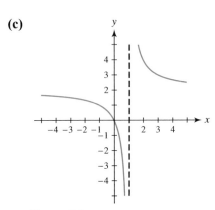

Figure 7.3

Solution

(a) *Numerical Evaluation* Table 7.1 shows that
$$f(-1) = 1, \quad f(1) = \text{Error}, \quad \text{and} \quad f(2) = 4.$$

(b) *Symbolic Evaluation*
$$f(-1) = \frac{2(-1)}{-1-1} = 1$$
$$f(1) = \frac{2(1)}{1-1} = \frac{2}{0}, \text{ which is undefined. Input 1 is not in the domain of } f.$$
$$f(2) = \frac{2(2)}{2-1} = 4$$

(c) *Graphical Evaluation* To evaluate $f(-1)$ graphically, find $x = -1$ on the x-axis and move upward to the graph of f. The y-value is 1 at the point of intersection, so $f(-1) = 1$, as shown in Figure 7.4. In Figure 7.5 the vertical line $x = 1$ is called a *vertical asymptote*. Because the graph of f does not intersect this line, $f(1)$ is undefined. Figure 7.6 reveals that $f(2) = 4$.

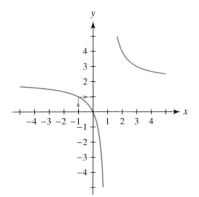

Figure 7.4

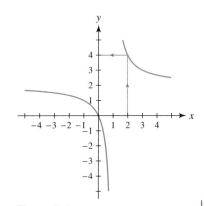

Figure 7.5 **Figure 7.6**

A **vertical asymptote** occurs in the graph of a rational function when the denominator of the rational expression is 0 but the numerator is not 0. The graph of a rational function *never* crosses a vertical asymptote. In Figure 7.1, the vertical asymptote is the y-axis or $x = 0$.

Applications involving rational functions are numerous. One instance is in the design of curves for train tracks, which we discuss in the next example.

EXAMPLE 4 Modeling a train track curve

When curves are designed for train tracks, sometimes the outer rail is elevated, or banked, so that a locomotive and cars can safely negotiate the curve at a higher speed than if the tracks were level. Suppose that a circular curve with a radius of r feet is being designed for a train traveling 60 miles per hour. Then $f(r) = \dfrac{2540}{r}$ calculates the proper elevation y in inches for the outer rail, where $y = f(r)$. See Figure 7.7.
(*Source:* L. Haefner, *Introduction to Transportation Systems.*)

(a) Evaluate $f(300)$ and interpret the result.
(b) Graph f in [0, 600, 100] by [0, 50, 10].
(c) Discuss how the elevation of the outer rail changes as the radius r increases.

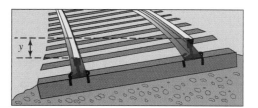

Figure 7.7

Solution

(a) $f(300) = \dfrac{2540}{300} \approx 8.5$. Thus the outer rail on a curve with a radius of 300 feet should be elevated about 8.5 inches for a train to safely travel through it at 60 miles per hour.

(b) A graph of $Y_1 = 2540/X$ is shown in Figure 7.8.

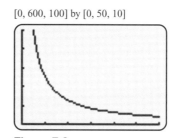

[0, 600, 100] by [0, 50, 10]

Figure 7.8

(c) Figure 7.8 shows that, as the radius increases (and the curve becomes less sharp), the outer rail needs to be elevated less.

Critical Thinking

Refer to Example 4. If the radius of a circular curve becomes large, what happens to the curve? According to $f(x)$, what should happen to the elevation of the outer rail when the radius becomes very large? Does this result agree with your intuition? Explain.

Introduction to Rational Equations

When working with rational functions, we commonly encounter rational equations. In Example 4, we used $f(r) = \dfrac{2540}{r}$ to calculate the elevation of the outer rail in the train track curve. Suppose that the outer rail for a curve is elevated 6 inches. What should be the radius of the curve? To answer this question, we need to solve the **rational equation**

$$\dfrac{2540}{r} = 6.$$

We solve this equation in the next example.

EXAMPLE 5 *Determining the proper radius for a train track curve*

Solve the rational equation $\dfrac{2540}{r} = 6$ and interpret the result.

Solution

We begin by multiplying both sides of the equation by r.

$$r \cdot \dfrac{2540}{r} = 6 \cdot r \quad \text{Multiply by } r.$$

$$2540 = 6r \quad \text{Simplify.}$$

$$r = \dfrac{2540}{6} \quad \text{Solve for } r.$$

$$r = 423\dfrac{1}{3} \quad \text{Rewrite.}$$

A train track curve designed for 60 miles per hour and banked 6 inches should have a radius of about 423 feet.

Multiplication is often used as a first step in solving rational equations. To do so we apply the following property of rational expressions, with $D \neq 0$.

$$D \cdot \dfrac{C}{D} = C$$

Examples of this property include,

$$5 \cdot \dfrac{7}{5} = 7 \quad \text{and} \quad (x-1) \cdot \dfrac{4x}{x-1} = 4x.$$

EXAMPLE 6 *Solving rational equations*

Solve each rational equation and check your answer.

(a) $\dfrac{3x}{2x-1} = 3$ **(b)** $\dfrac{x+1}{x-2} = \dfrac{3}{x-2}$

Solution

(a) Begin by multiplying both sides of the equation by $2x - 1$.

$$(2x - 1) \cdot \frac{3x}{2x - 1} = 3 \cdot (2x - 1) \quad \text{Multiply by } (2x - 1).$$

$$3x = 3(2x - 1) \quad \text{Simplify.}$$

$$3x = 6x - 3 \quad \text{Distributive property}$$

$$3 = 3x \quad \text{Simplify.}$$

$$x = 1 \quad \text{Solve for } x.$$

To check your answer substitute $x = 1$ into the given equation.

$$\frac{3(1)}{2(1) - 1} = 3 \quad \text{The answer checks.}$$

(b) Begin by multiplying both sides of the equation by $x - 2$.

$$(x - 2) \cdot \frac{x + 1}{x - 2} = \frac{3}{x - 2} \cdot (x - 2) \quad \text{Multiply by } x - 2.$$

$$x + 1 = 3 \quad \text{Simplify.}$$

$$x = 2 \quad \text{Solve for } x.$$

Substituting $x = 2$ results in both sides of the equation being undefined because the denominators are 0. In this case $x = 2$ is not a valid solution. Instead it is called an **extraneous solution,** which cannot be used. This result illustrates why it is *important* to check your answers.

The *grade* x of a hill is a measure of its steepness and corresponds to the slope of the road. For example, if a road rises 10 feet for every 100 feet of horizontal distance, it has an uphill grade of $x = \frac{10}{100}$, or 10%, as illustrated in Figure 7.9. The braking distance D in feet for a car traveling 60 miles per hour on a wet, uphill grade is given by

$$D(x) = \frac{3600}{30x + 9}.$$

In the next example we use this formula to determine the grade associated with a given braking distance. **(Source:** N. Garber and L. Hoel, *Traffic and Highway Engineering.***)**

Figure 7.9

EXAMPLE 7 *Solving a rational equation graphically, numerically, and symbolically*

The braking distance for a car traveling at 60 miles per hour on a wet, uphill grade is 250 feet. Find the grade of the hill

(a) graphically, **(b)** numerically, and **(c)** symbolically.

Solution

(a) *Graphical Solution* We must find an *x*-value that satisfies the rational equation

$$\frac{3600}{30x + 9} = 250.$$

To solve this equation graphically, we let $Y_1 = 3600/(30X + 9)$ and $Y_2 = 250$. Their graphs intersect at (0.18, 250), as shown in Figure 7.10. Thus the highway grade is 18%.

[0, 0.3, 0.1] by [0, 500, 100]

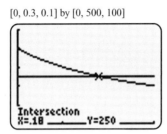
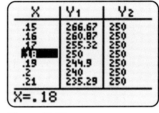

Figure 7.10 **Figure 7.11**

(b) *Numerical Solution* The table shown in Figure 7.11 indicates that $Y_1 = 250$ when $x = 0.18$.

(c) *Symbolic Solution* To solve the equation $\frac{3600}{30x + 9} = 250$ symbolically, we begin by multiplying both sides by $30x + 9$.

$$\frac{3600}{30x + 9} = 250 \qquad \text{Equation to be solved}$$

$$(30x + 9) \cdot \frac{3600}{30x + 9} = 250 \cdot (30x + 9) \qquad \text{Multiply by } (30x + 9).$$

$$3600 = 250(30x + 9) \qquad \text{Simplify.}$$

$$3600 = 7500x + 2250 \qquad \text{Distributive property}$$

$$1350 = 7500x \qquad \text{Subtract 2250.}$$

$$x = \frac{1350}{7500} \qquad \text{Solve for } x.$$

$$x = 0.18 \qquad \text{Write } x \text{ in decimal form.}$$

Technology Note *Asymptotes, Dot Mode, and Decimal Windows*

When rational functions are graphed on graphing calculators, pseudo-asymptotes often occur because the calculator is simply connecting dots to draw a graph. For example, Figures 7.12–7.14 show the graph of $y = \frac{2}{x - 2}$ in connected mode, dot mode, and with a *decimal,* or *friendly, window*. In dot mode, pixels in the calculator screen are not connected. With dot mode (and sometimes with a decimal window) pseudo-asymptotes do not appear. To learn more about these features consult your owner's manual.

[−6, 6, 1] by [−4, 4, 1]   [−6, 6, 1] by [−4, 4, 1]

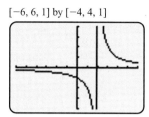
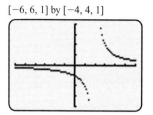

Figure 7.12 Connected Mode   **Figure 7.13** Dot Mode

[−4.7, 4.7, 1] by [−3.1, 3.1, 1]

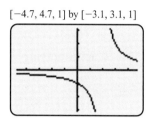

Figure 7.14 Decimal Window

7.1 Putting It All Together

A rational expression results when a polynomial is divided by a nonzero polynomial. Rational expressions may be used to define rational functions. The following table summarizes basic concepts about rational functions and equations.

Term	Explanation	Example
Rational function	Let $p(x)$ and $q(x)$ be polynomials with $q(x) \neq 0$. Then a *rational function* is given by $$f(x) = \frac{p(x)}{q(x)}.$$	$$f(x) = \frac{3x^2}{x + 4},$$ The domain of f includes all real numbers except $x = -4$. A vertical asymptote occurs at $x = -4$.
Rational equation	An equation that contains rational expressions is a *rational equation*. When solving a rational equation, multiplication is often a good first step.	To solve $\frac{3x}{x + 1} = 6$ begin by multiplying both sides by $x + 1$. $$(x + 1) \cdot \frac{3x}{x + 1} = 6 \cdot (x + 1)$$ $$3x = 6x + 6$$ $$-6 = 3x$$ $$x = -2$$ Be sure to check your results.

7.1 EXERCISES

CONCEPTS

1. What is a rational expression? Give an example.

2. If $f(x) = \dfrac{6}{x-4}$, for what input x is $f(x)$ undefined?

3. What would be a good first step in solving $\dfrac{3}{x+7} = x$?

4. Is $x = 5$ a solution to the equation
$$\dfrac{x+5}{x-5} = \dfrac{10}{x-5}?$$
Explain.

5. Does $x \cdot \dfrac{5+x}{x}$ simplify to $5x$? Explain.

6. Name three ways to represent a rational function.

Exercises 7–12: Determine whether the expression is rational.

7. $\dfrac{2}{x}$

8. $\dfrac{5-2x}{x^2}$

9. $\dfrac{3x-5}{2x+1}$

10. $|x| + \dfrac{4}{|x|}$

11. $\dfrac{\sqrt{x}}{x^2 - 1}$

12. $\dfrac{4x^2 - 3x + 1}{x^3 - 4x}$

Exercises 13–16: (Refer to Example 2.) Use the verbal representation of a rational function f to find symbolic, graphical, and numerical representations.

13. Divide x by the quantity x plus 1.

14. Add 2 to x and then divide the result by the quantity x plus 5.

15. Divide x squared by the quantity x minus 2.

16. Compute the reciprocal of twice x.

Exercises 17–22: Use your graphing calculator as an aid to sketch a graph of $y = f(x)$ by hand. Show vertical asymptotes with dashed lines. Give x-values not in the domain of f.

17. $f(x) = \dfrac{1}{x-1}$

18. $f(x) = \dfrac{1}{x+3}$

19. $f(x) = \dfrac{x}{x-1}$

20. $f(x) = \dfrac{2x}{x-2}$

21. $f(x) = \dfrac{4}{x^2 + 1}$

22. $f(x) = \dfrac{4}{x^2 - 4}$

Exercises 23–28: Evaluate $f(x)$ at the given value of x by hand.

23. $f(x) = \dfrac{1}{x-1}$ $x = -2$

24. $f(x) = \dfrac{3x}{x^2 - 1}$ $x = 2$

25. $f(x) = \dfrac{x+1}{x-1}$ $x = -3$

26. $f(x) = \dfrac{2x+1}{3x-1}$ $x = 0$

27. $f(x) = \dfrac{x^2 - 3x + 5}{x^2 + 1}$ $x = -2$

28. $f(x) = \dfrac{5}{x^2 - x}$ $x = -1$

7.1 Rational Functions and Equations

Exercises 29–32: Use the graph to evaluate each expression. Give the equation of any vertical asymptotes.

29. $f(-3)$ and $f(1)$ **30.** $f(-3)$ and $f(2)$

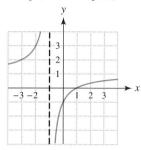
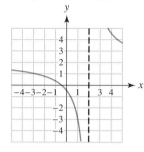

31. $f(-1)$ and $f(2)$ **32.** $f(-1)$ and $f(1)$

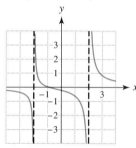
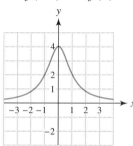

Exercises 33–36: Give a numerical representation of $f(x)$ for $x = -3, -2, -1, \ldots, 3$. Then evaluate $f(x)$ for the given values of x.

33. $f(x) = \dfrac{2}{2x - 1}$, $x = -2, x = 1$

34. $f(x) = \dfrac{x}{x^2 + 1}$, $x = -3, x = 2$

35. $f(x) = \dfrac{4x}{x^2 - 9}$, $x = -3, x = 0$

36. $f(x) = \dfrac{x^2 + 2x}{x^2 + x + 1}$, $x = -2, x = -1$

RATIONAL EQUATIONS

Exercises 37–50: Solve the rational equation symbolically. Check your result.

37. $\dfrac{3}{x} = 5$

38. $\dfrac{5}{x} - 2 = 5$

39. $\dfrac{1}{x - 2} = -1$

40. $\dfrac{-5}{x + 4} = -5$

41. $\dfrac{x}{x + 1} - 1 = 1$

42. $\dfrac{3x}{x + 3} = -6$

43. $\dfrac{2x + 1}{3x - 2} = 1$

44. $\dfrac{x + 3}{4x + 1} = \dfrac{3}{5}$

45. $\dfrac{1}{x^2 - 1} = -1$

46. $\dfrac{5x}{x^2 - 5} = \dfrac{x + 2}{x^2 - 5}$

47. $\dfrac{x}{x - 5} = \dfrac{2x - 5}{x - 5}$

48. $\dfrac{x}{x + 1} = \dfrac{2x + 1}{x + 1}$

49. $\dfrac{2x}{x + 2} = \dfrac{x - 4}{x + 2}$

50. $\dfrac{x + 3}{x + 1} = \dfrac{3x + 4}{x + 1}$

Exercises 51–56: Solve the rational equation either graphically or numerically.

51. $\dfrac{4 + x}{2x} = -0.5$

52. $\dfrac{2x - 1}{x - 5} = -\dfrac{5}{2}$

53. $\dfrac{2x}{x^2 - 4} = -\dfrac{2}{3}$

54. $\dfrac{x - 1}{x + 2} = x$

55. $\dfrac{1}{x - 1} = x - 1$

56. $\dfrac{2}{x + 2} = x + 1$

USING MORE THAN ONE METHOD

Exercises 57–60: Solve the equation (a) symbolically, (b) graphically, and (c) numerically.

57. $\dfrac{1}{x + 2} = 1$

58. $\dfrac{3}{x - 1} = 2$

59. $\dfrac{x}{2x + 1} = \dfrac{2}{5}$

60. $\dfrac{4x - 2}{x + 1} = \dfrac{3}{2}$

APPLICATIONS

Exercises 61–64: **Rational Models** *Match the physical situation with the graph of the rational function on the following page that models it best.*

61. A population of fish that increases and then levels off

62. An insect population that dies out

63. The length of a ticket line as the rate at which people arrive in line increases

64. The wind speed during a day that begins calm, becomes windy, and then is calm again

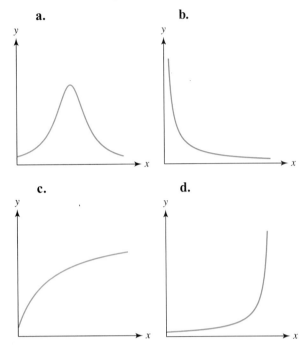

a.
b.
c.
d.

65. *Train Track Curves* (Refer to Example 4.) Let $f(r) = \dfrac{2540}{r}$ compute the elevation of the outer rail in inches for a curve designed for 60 miles per hour with a radius of r feet. (*Source:* L. Haefner.)
   (a) Evaluate $f(400)$ and interpret the result.
   (b) Construct a table of f, starting at $r = 100$ and incrementing by 50.
   (c) If the radius of the curve doubles, what happens to the elevation of the outer rail?

66. *Highway Curves* Engineers need to calculate a minimum safe radius for highway curves. If a curve is too sharp for a given speed, it can be dangerous. To make a sharp curve safer, the road can be banked, or elevated, as illustrated in the accompanying figure. If a curve is designed for a speed of 40 miles per hour and is banked with slope m, then a minimum radius R is computed by

$$R(m) = \dfrac{1600}{15m + 2}.$$

(*Source:* N. Garber.)

(a) Evaluate $R(0.1)$ and interpret the result.
(b) Graph R in $[0, 1, 0.1]$ by $[0, 500, 100]$.
(c) Describe what happens to the radius of the curve as the slope of the banking increases.
(d) If the curve has a radius of 320 feet, what should m be?

67. *Uphill Highway Grade* (Refer to Example 7.) The braking distance for a car traveling 30 miles per hour on a wet *uphill* grade x is given by

$$D(x) = \dfrac{900}{10.5 + 30x}.$$

(*Source:* N. Garber.)
(a) Evaluate $D(0.05)$ and interpret the result.
(b) If the braking distance for this car is 60 feet, find the uphill grade x.

68. *Downhill Highway Grade* (See Exercise 67 and the Chapter Introduction.) The braking distance for a car traveling 30 miles per hour on a wet, *downhill* grade x is given by

$$S(x) = \dfrac{900}{10.5 - 30x}.$$

(*Source:* N. Garber.)
(a) Evaluate $S(0.05)$ and interpret the result.
(b) Construct a table for $D(x)$ from Exercise 67 and $S(x)$, starting at $x = 0$ and incrementing by 0.05.
(c) How do the braking distances for uphill and downhill grades compare? Does this result agree with your driving experience?

69. *Time Spent in Line* Suppose that a parking ramp attendant can wait on 5 vehicles per minute and vehicles are leaving the ramp randomly at an average rate of x vehicles per minute. Then the average time T in minutes spent waiting in line and paying the attendant is given by

$$T(x) = \dfrac{1}{5 - x},$$

where $x < 5$. (*Source:* N. Garber.)

(a) Evaluate $T(4)$ and interpret the result.
(b) Graph T in [0, 5, 1] by [0, 5, 1].
(c) Interpret the graph as x increases from 0 to 5. Does this result agree with your intuition?
(d) Find x if the waiting time is 3 minutes.

70. *People Waiting in Line* At a post office, workers can wait on 50 people per hour. If people arrive randomly at an average rate of x per hour, then the average number of people N waiting in line is given by
$$N(x) = \frac{x^2}{2500 - 50x},$$
where $x < 50$. (Source: N. Garber.)
(a) Evaluate $N(30)$ and interpret the result.
(b) Graph N in [0, 50, 10] by [0, 20, 5].
(c) Interpret the graph as x increases from 0 to 50. Does this result agree with your intuition?
(d) Find x if the average number of people waiting in line is 8.

71. *Probability* A jar contains x balls. Each ball has a unique number written on it and only one ball has the winning number. The likelihood, or probability P, of *not* drawing the winning ball is given by
$$P(x) = \frac{x-1}{x},$$
where $x > 0$. For example, if $P = 0.99$, then there is a 99% chance of not drawing the winning ball.
(a) Evaluate $P(1)$ and $P(50)$ and interpret the result.
(b) Graph P in [0, 100, 10] by [0, 1, 0.1].
(c) What happens to the probability of not winning as the number of balls increases? Does this result agree with your intuition? Explain.
(d) How many balls are in the jar if the probability of not winning is 0.975?

72. *Insect Population* Suppose that an insect population in thousands per acre is modeled by
$$P(x) = \frac{5x+2}{x+1},$$
where $x \geq 0$ is time in months.
(a) Evaluate $P(10)$ and interpret the result.
(b) Graph P in [0, 50, 10] by [0, 6, 1].
(c) What happens to the insect population after several years?
(d) After how many months is the insect population 4.8 thousand per acre?

WRITING ABOUT MATHEMATICS

73. Is every polynomial function a rational function? Explain your answer.

74. The domain of a polynomial function includes all real numbers. Does the domain of a rational function include all real numbers? Explain your answer and give an example.

GROUP ACTIVITY
Working with Real Data

Directions: Form a group of 2 to 4 people. Select someone to record the group's responses for this activity. All members of the group should work cooperatively to answer the questions. If your instructor asks for your results, each member of the group should be prepared to respond. If the group is asked to turn in its work, be sure to include each person's name on the paper.

Slippery Roads If a car is moving on a level highway, its braking distance depends on road conditions. If the road is slippery, stopping may take longer. A measure of the slipperiness of a road is the coefficient of friction x between the tire and the road, where x satisfies $0 < x \leq 1$. A smaller value for x indicates that the road is slipperier. The stopping distance D in feet for a car traveling at 50 miles per hour on a road with a coefficient of friction x is given by

$$D(x) = \frac{250}{3x}.$$

(a) Graph D in the viewing rectangle $[0, 1, 0.2]$ by $[0, 800, 100]$. Identify any vertical asymptotes in the graph of D.

(b) What happens to the stopping distance as x increases to 1? Explain.
(c) What happens to the stopping distance as x decreases to 0? Explain.
(d) In reality, could $x = 0$? Explain. What would happen if a road could have $x = 0$ and a car tried to stop?

7.2 MULTIPLICATION AND DIVISION OF RATIONAL EXPRESSIONS

Review of Multiplication and Division of Fractions ~
Multiplication and Division of Rational Expressions ~
Rational Equations

INTRODUCTION

In previous chapters we demonstrated how to add, subtract, and multiply real numbers and polynomials. In this section we show you how to multiply and divide rational expressions and in the next section we discuss addition and subtraction of rational expressions. We also discuss applications involving rational equations.

REVIEW OF MULTIPLICATION AND DIVISION OF FRACTIONS

Recall that to multiply two fractions we use the property

$$\frac{a}{b} \cdot \frac{c}{d} = \frac{ac}{bd}.$$

For example,

$$\frac{2}{5} \cdot \frac{3}{7} = \frac{2 \cdot 3}{5 \cdot 7} = \frac{6}{35}.$$

Sometimes we can simplify or *reduce* a fraction by using the following property.

$$\frac{ac}{bc} = \frac{a}{b}$$

This property holds because $\frac{c}{c} = 1$ and $\frac{a}{b} \cdot 1 = \frac{a}{b}$. It can be used to simplify the product of two fractions, as in

$$\frac{3}{4} \cdot \frac{2}{11} = \frac{6}{44} = \frac{3 \cdot 2}{22 \cdot 2} = \frac{3}{22}.$$

7.2 Multiplication and Division of Rational Expressions

EXAMPLE 1 *Multiplying fractions*

Multiply and simplify the product. Verify your results with a calculator.

(a) $\dfrac{4}{9} \cdot \dfrac{3}{8}$ (b) $\dfrac{2}{3} \cdot \dfrac{3}{4} \cdot \dfrac{5}{6}$

Solution

(a) $\dfrac{4}{9} \cdot \dfrac{3}{8} = \dfrac{12}{72} = \dfrac{1 \cdot 12}{6 \cdot 12} = \dfrac{1}{6}$ (b) $\dfrac{2}{3} \cdot \dfrac{3}{4} \cdot \dfrac{5}{6} = \dfrac{6 \cdot 5}{12 \cdot 6} = \dfrac{5}{12}$

These results are verified in Figure 7.15.

Figure 7.15

Recall that to divide two fractions we "invert and multiply." That is, we change a division problem to a multiplication problem, using the property

$$\dfrac{a}{b} \div \dfrac{c}{d} = \dfrac{a}{b} \cdot \dfrac{d}{c}.$$

For example,

$$\dfrac{3}{4} \div \dfrac{5}{4} = \dfrac{3}{4} \cdot \dfrac{4}{5} = \dfrac{3 \cdot 4}{5 \cdot 4} = \dfrac{3}{5}.$$

MULTIPLICATION AND DIVISION OF RATIONAL EXPRESSIONS

Multiplying and dividing rational expressions is similar to multiplying and dividing fractions.

Products and Quotients of Rational Expressions

To multiply two rational expressions, multiply numerators and multiply denominators.

$$\dfrac{A}{B} \cdot \dfrac{C}{D} = \dfrac{AC}{BD} \qquad B \text{ and } D \text{ are nonzero.}$$

To divide two rational expressions, multiply by the reciprocal of the divisor.

$$\dfrac{A}{B} \div \dfrac{C}{D} = \dfrac{A}{B} \cdot \dfrac{D}{C} \qquad B, C, \text{ and } D \text{ are nonzero.}$$

CHAPTER 7 Rational Expressions and Functions

EXAMPLE 2 *Multiplying rational expressions*

Multiply.

(a) $\dfrac{1}{x} \cdot \dfrac{x+1}{2x}$ (b) $\dfrac{x-1}{x} \cdot \dfrac{x-1}{x+2}$

Solution

(a) $\dfrac{1}{x} \cdot \dfrac{x+1}{2x} = \dfrac{1 \cdot (x+1)}{x \cdot 2x} = \dfrac{x+1}{2x^2}$

(b) $\dfrac{x-1}{x} \cdot \dfrac{x-1}{x+2} = \dfrac{(x-1)(x-1)}{x(x+2)}$

■ **MAKING CONNECTIONS**

Reducing Fractions and Rational Expressions

We often simplify, or reduce, fractions. For example,

$$\dfrac{10}{15} = \dfrac{2 \cdot 5}{3 \cdot 5} = \dfrac{2}{3}.$$

To reduce rational expressions we often factor the numerator and denominator. For example,

$$\dfrac{x^2}{x^2 + 5x} = \dfrac{x \cdot x}{x(x+5)} = \dfrac{x}{x+5}.$$

Note: We factor *first* and then reduce.

EXAMPLE 3 *Simplifying rational expressions*

Simplify.

(a) $\dfrac{(x-1)(x+2)}{(x-3)(x-1)}$ (b) $\dfrac{x^2 - 4}{2x^2 - 3x - 2}$

Solution

(a) $\dfrac{(x-1)(x+2)}{(x-3)(x-1)} = \dfrac{(x+2)(x-1)}{(x-3)(x-1)}$ Commutative property

$= \dfrac{x+2}{x-3}$ Reduce.

(b) $\dfrac{x^2 - 4}{2x^2 - 3x - 2} = \dfrac{(x+2)(x-2)}{(2x+1)(x-2)}$ Factor.

$= \dfrac{x+2}{2x+1}$ Reduce.

In the next example we divide and simplify rational expressions.

7.2 Multiplication and Division of Rational Expressions

EXAMPLE 4 *Dividing two rational expressions*

Divide and simplify.

(a) $\dfrac{2}{x} \div \dfrac{2x - 1}{4x}$

(b) $\dfrac{x^2 - 1}{x^2 + x - 6} \div \dfrac{x - 1}{x + 3}$

Solution

(a) $\dfrac{2}{x} \div \dfrac{2x - 1}{4x} = \dfrac{2}{x} \cdot \dfrac{4x}{2x - 1}$ "Invert and multiply."

$= \dfrac{8x}{x(2x - 1)}$ Multiply.

$= \dfrac{8}{2x - 1}$ Reduce.

(b) $\dfrac{x^2 - 1}{x^2 + x - 6} \div \dfrac{x - 1}{x + 3} = \dfrac{x^2 - 1}{x^2 + x - 6} \cdot \dfrac{x + 3}{x - 1}$ "Invert and multiply."

$= \dfrac{(x + 1)(x - 1)}{(x - 2)(x + 3)} \cdot \dfrac{x + 3}{x - 1}$ Factor.

$= \dfrac{(x + 1)(x - 1)(x + 3)}{(x - 2)(x - 1)(x + 3)}$ Commutative property

$= \dfrac{(x + 1)}{(x - 2)}$ Reduce.

RATIONAL EQUATIONS

One way to solve a rational equation is to *clear fractions* by multiplying both sides of an equation by a common denominator. For example, to solve the rational equation

$$\dfrac{1}{x} + \dfrac{1}{x^2} = \dfrac{3}{4}$$

we can multiply by $4x^2$.

$4x^2 \cdot \left(\dfrac{1}{x} + \dfrac{1}{x^2} \right) = \dfrac{3}{4} \cdot 4x^2$ Multiply by $4x^2$.

$\dfrac{4x^2}{x} + \dfrac{4x^2}{x^2} = \dfrac{12x^2}{4}$ Distributive property

$4x + 4 = 3x^2$ Reduce.

$0 = 3x^2 - 4x - 4$ Subtract $4x + 4$.

$0 = (3x + 2)(x - 2)$ Factor.

$x = -\dfrac{2}{3}$ or $x = 2$ Solve.

If each of the denominators is a factor of an expression, that expression is a **common denominator**. For example, $4x^2$ is a common denominator of x, x^2, and 4 because all are factors of $4x^2$.

$$4x^2 = x \cdot 4x \qquad x \text{ is a factor.}$$
$$4x^2 = x^2 \cdot 4 \qquad x^2 \text{ is a factor.}$$
$$4x^2 = 4 \cdot x^2 \qquad 4 \text{ is a factor.}$$

The expression $4x^3$ is also a common denominator of x, x^2, and 4. However, we say that $4x^2$ is the **least common denominator** (LCD) because it is the common denominator with fewest factors.

Note: A common denominator can always be found by taking the product of the denominators. However, it may not be the *least* common denominator.

EXAMPLE 5 *Finding a least common denominator*

Find the LCD for the given expressions.

(a) $\dfrac{1}{x}, \dfrac{1}{x-1}$ (b) $\dfrac{1}{x+2}, \dfrac{2x}{x^2-4}, \dfrac{1}{3}$

Solution

(a) The LCD of x and $x - 1$ is their product, $x(x - 1)$.
(b) A common denominator of $x + 2$, $x^2 - 4$, and 3 would be their product. However, as

$$x^2 - 4 = (x + 2)(x - 2),$$

the LCD is

$$(x + 2)(x - 2)(3).$$

EXAMPLE 6 *Solving rational equations*

GCLM Solve each equation. Support your results either graphically or numerically.

(a) $\dfrac{1}{2} + \dfrac{x}{3} = \dfrac{x}{5}$ (b) $\dfrac{3}{x-2} = \dfrac{5}{x+2}$

Solution

(a) The LCD for 2, 3, and 5 is their product, 30.

$$30 \cdot \left(\dfrac{1}{2} + \dfrac{x}{3}\right) = \dfrac{x}{5} \cdot 30 \qquad \text{Multiply by the LCD.}$$

$$\dfrac{30}{2} + \dfrac{30x}{3} = \dfrac{30x}{5} \qquad \text{Distributive property}$$

$$15 + 10x = 6x \qquad \text{Reduce.}$$

$$4x = -15 \qquad \text{Subtract } 6x \text{ and } 15.$$

$$x = -\dfrac{15}{4} \qquad \text{Solve.}$$

This result can be supported graphically by letting $Y_1 = 1/2 + X/3$ and $Y_2 = X/5$. Their graphs intersect at the point $(-3.75, -0.75)$, as shown in Figure 7.16.

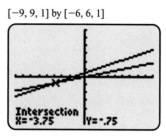

Figure 7.16

(b) The LCD for $x - 2$ and $x + 2$ is their product. Multiply both sides of the equation by $(x - 2)(x + 2)$.

$$(x - 2)(x + 2) \cdot \frac{3}{x - 2} = \frac{5}{x + 2} \cdot (x - 2)(x + 2)$$

$$\frac{3(x - 2)(x + 2)}{(x - 2)} = \frac{5(x - 2)(x + 2)}{x + 2} \quad \text{Multiply expressions.}$$

$$3(x + 2) = 5(x - 2) \quad \text{Reduce.}$$

$$3x + 6 = 5x - 10 \quad \text{Distributive property}$$

$$16 = 2x \quad \text{Add 10 and subtract } 3x.$$

$$x = 8 \quad \text{Solve.}$$

To support this result numerically let $Y_1 = 3/(X - 2)$ and $Y_2 = 5/(X + 2)$. Figure 7.17 shows that $Y_1 = Y_2 = 0.5$ when $x = 8$.

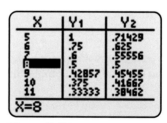

Figure 7.17

Critical Thinking

A bicyclist rides uphill at 6 miles per hour for one mile and then rides downhill at 12 miles per hour for 1 mile. What was the average speed of the bicyclist?

Rational equations sometimes occur in problems involving time and rate, as demonstrated in the next example.

EXAMPLE 7 Determining the time required to empty a pool

A pump can empty a swimming pool in 50 hours. To speed up the process a second pump is used that can empty the pool in 80 hours. How long will it take for both pumps working together to empty the pool?

Solution

Because the first pump can empty the entire pool in 50 hours, it can empty $\frac{1}{50}$ of the pool in 1 hour, $\frac{2}{50}$ of the pool in 2 hours, $\frac{3}{50}$ of the pool in 3 hours, and in general, it can empty $\frac{t}{50}$ of the pool in t hours. The second pump can empty the pool in 80 hours, so (using similar reasoning) it can empty $\frac{t}{80}$ of the pool in t hours.

Together the pumps can empty

$$\frac{t}{50} + \frac{t}{80}$$

of the pool in t hours. The job will be complete when the fraction of the pool that is empty equals 1. Thus we must solve the equation

$$\frac{t}{50} + \frac{t}{80} = 1.$$

To solve we multiply both sides by $(50)(80)$.

$$(50)(80)\left(\frac{t}{50} + \frac{t}{80}\right) = 1(50)(80)$$

$$\frac{50 \cdot 80 \cdot t}{50} + \frac{50 \cdot 80 \cdot t}{80} = 4000 \qquad \text{Distributive property}$$

$$80t + 50t = 4000 \qquad \text{Reduce.}$$

$$130t = 4000 \qquad \text{Combine like terms.}$$

$$t = \frac{4000}{130} \approx 30.8 \text{ hours} \qquad \text{Solve.}$$

The two pumps can empty the pool in about 30.8 hours.

If a person drives a car 60 miles per hour for 4 hours, the total distance d traveled is

$$d = 60 \cdot 4 = 240 \text{ miles.}$$

That is, distance equals the product of the rate (speed) and the elapsed time. This may be written as $d = rt$ and expressed verbally as "distance equals rate times time." This formula is used in the next example.

EXAMPLE 8   Solving an application graphically and numerically

Suppose that the winner of a 3-mile cross country race finishes 3 minutes ahead of another runner. If the winner runs 2 miles per hour faster than the slower runner, find the average speed of each runner.

Solution

Let x represent the speed of the slower runner. Then $x + 2$ represents the speed of the winner. To determine the time for each runner to finish the race divide both sides of $d = rt$ by r to obtain

$$t = \frac{d}{r}.$$

The slower runner ran 3 miles at x miles per hour, so the time is $\frac{3}{x}$; the winner ran 3 miles at $x + 2$ miles per hour, so the winning time is $\frac{3}{x + 2}$. If you add 3 minutes (or equivalently $\frac{3}{60} = \frac{1}{20}$ hour) to the winner's time, it equals the slower runner's time, or

$$\frac{3}{x + 2} + \frac{1}{20} = \frac{3}{x}.$$

Note: The runners' speeds are in miles per *hour*, so you need to keep all times in hours.

Solve this equation graphically and numerically by letting

$$Y_1 = 3/(X + 2) + 1/20 \text{ and } Y_2 = 3/X.$$

As shown in Figure 7.18 on the next page, their graphs intersect at $(10, 0.3)$. Thus the slower runner's speed is 10 miles per hour. The winner runs 2 miles per hour faster, or

12 miles per hour. The result is supported numerically in Figure 7.19, where $Y_1 = Y_2 = 0.3$ when $x = 10$.

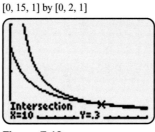

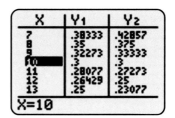

Figure 7.18

Figure 7.19

Technology Note *Entering Rational Expressions*

When entering a rational expression, use parentheses around the numerator and denominator. The following are examples. Enter

$$y = \frac{3}{x + 2} \quad \text{as} \quad Y_1 = 3/(X + 2).$$

Enter

$$y = \frac{x + 1}{2x - 1} \quad \text{as} \quad Y_1 = (X + 1)/(2X - 1).$$

7.2 Putting It All Together

The table summarizes some important concepts found in this section.

Concept	Explanation	Example
Multiplication and division of rational expressions	To multiply two rational expressions, multiply numerators and multiply denominators. $$\frac{A}{B} \cdot \frac{C}{D} = \frac{AC}{BD} \quad \text{B and D are nonzero.}$$ To divide two rational expressions, multiply by the reciprocal of the divisor. ("Invert and multiply.") $$\frac{A}{B} \div \frac{C}{D} = \frac{A}{B} \cdot \frac{D}{C} \quad \text{B, C, and D are nonzero.}$$	$$\frac{x}{x+1} \div \frac{x+2}{x+1} = \frac{x}{x+1} \cdot \frac{x+1}{x+2}$$ $$= \frac{x(x+1)}{(x+2)(x+1)}$$ $$= \frac{x}{x+2}$$
Least common denominator	A common denominator has each of the given denominators as factors. The least common denominator (LCD) has the fewest factors.	A common denominator for $\frac{1}{x^2}$ and $\frac{1}{x(x+1)}$ is $x^3(x+1)$. However, the LCD is $x^2(x+1)$.

7.2 Multiplication and Division of Rational Expressions

Concept	Explanation	Example
Rational equations	To solve a rational equation, multiply both sides of the equation by the LCD.	Multiply the given equation by $4x^2$. $$\frac{1}{4x} + \frac{2}{x^2} = \frac{1}{x}$$ $$4x^2\left(\frac{1}{4x} + \frac{2}{x^2}\right) = \frac{1}{x} \cdot 4x^2$$ $$\frac{4x^2}{4x} + \frac{8x^2}{x^2} = \frac{4x^2}{x}$$ $$x + 8 = 4x$$ $$8 = 3x$$ $$x = \frac{8}{3}$$

7.2 EXERCISES

FOR EXTRA HELP: Student's Solutions Manual, MyMathLab.com, InterAct Math, Math Tutor Center, MathXL, Digital Video Tutor CD 5 Videotape 11

CONCEPTS

1. Simplify $\dfrac{2x + 3}{2x + 3}$.

2. Find a common denominator for $\dfrac{1}{x}, \dfrac{1}{2},$ and $\dfrac{1}{4}$ that is *not* the LCD. What is the LCD?

3. Is $\dfrac{x + 1}{x}$ equal to 1? Is it equal to 2? Explain.

4. Is $\dfrac{3}{3 + x}$ equal to $\dfrac{1}{x}$? Is it equal to $1 + \dfrac{3}{x}$?

5. To divide $\dfrac{2}{3}$ by $\dfrac{5}{7}$ multiply _____ by _____.

6. $\dfrac{a}{b} \cdot \dfrac{c}{d} =$ _____.

7. $\dfrac{a}{b} \div \dfrac{c}{d} =$ _____.

8. $\dfrac{ac}{bc} =$ _____.

REVIEW OF FRACTIONS

Exercises 9–16: Simplify.

9. $\dfrac{1}{2} \cdot \dfrac{4}{5}$

10. $\dfrac{5}{6} \cdot \dfrac{3}{10}$

11. $\dfrac{7}{8} \cdot \dfrac{4}{3} \cdot (-3)$

12. $4 \cdot \dfrac{7}{4} \cdot \dfrac{1}{2}$

13. $\dfrac{5}{7} \div \dfrac{15}{14}$

14. $-\dfrac{2}{3} \div 2$

15. $6 \div \left(-\dfrac{1}{3}\right)$

16. $\dfrac{10}{9} \div \dfrac{5}{3}$

SIMPLIFYING RATIONAL EXPRESSIONS

Exercises 17–20: Simplify the expression.

17. $\dfrac{5x}{x^2}$

18. $\dfrac{18t^3}{6t}$

19. $\dfrac{(x - 1)(x + 1)}{x - 1}$

20. $(x + 2) \cdot \dfrac{x - 6}{x + 2}$

Exercises 21–24: Simplify the expression.

21. $\dfrac{x^2 - 4}{x + 2}$

22. $\dfrac{(x + 2)(x - 5)}{(x + 10)(x + 2)}$

23. $\dfrac{x+5}{x^2+2x-15}$

24. $\dfrac{x^2-9}{x^2+6x+9} \cdot (x+3)$

MULTIPLICATION AND DIVISION OF RATIONAL EXPRESSIONS

Exercises 25–40: Simplify the expression.

25. $\dfrac{1}{2x} \cdot \dfrac{4x}{2}$

26. $\dfrac{5a^2}{7} \cdot \dfrac{7}{10a}$

27. $\dfrac{3x}{2} \div \dfrac{2x}{5}$

28. $\dfrac{x^2+x}{2x+6} \div \dfrac{x}{x+3}$

29. $\dfrac{x+1}{2x-5} \cdot \dfrac{2x-5}{x}$

30. $\dfrac{x+1}{x} \cdot \dfrac{x}{x+2}$

31. $\dfrac{(x-5)(x+3)}{3x-1} \cdot \dfrac{x(3x-1)}{(x-5)}$

32. $\dfrac{b^2+1}{b^2-1} \cdot \dfrac{b-1}{b+1}$

33. $\dfrac{x^2-2x-35}{2x^3-3x^2} \cdot \dfrac{x^3-x^2}{2x-14}$

34. $\dfrac{2x+4}{x+1} \cdot \dfrac{x^2+3x+2}{4x+2}$

35. $\dfrac{6b}{b+2} \div \dfrac{3b^4}{2b+4}$

36. $\dfrac{5x^5}{x-2} \div \dfrac{10x^3}{5x-10}$

37. $\dfrac{3a+1}{a^7} \div \dfrac{a+1}{3a^8}$

38. $\dfrac{x^2-16}{x+3} \div \dfrac{x+4}{x^2-9}$

39. $\dfrac{x+5}{x^3-x} \div \dfrac{x^2-25}{x^3}$

40. $\dfrac{x^2+x-12}{2x^2-9x-5} \div \dfrac{x^2+7x+12}{2x^2-7x-4}$

SOLVING RATIONAL EQUATIONS

Exercises 41–46: Find the LCD.

41. $\dfrac{1}{x}, \dfrac{1}{5}$

42. $\dfrac{1}{x-1}, \dfrac{1}{x+5}$

43. $\dfrac{1}{x-1}, \dfrac{1}{x^2-1}$

44. $\dfrac{1}{x^2-x}, \dfrac{1}{2x}$

45. $\dfrac{3}{2}, \dfrac{x}{2x+1}, \dfrac{x}{2x-4}$

46. $\dfrac{1}{x}, \dfrac{1}{x^2-4x}, \dfrac{1}{2x}$

Exercises 47–56: Solve the rational equation. Check your result.

47. $\dfrac{x}{3} + \dfrac{1}{2} = \dfrac{5}{6}$

48. $\dfrac{7}{8} - \dfrac{x}{4} = -\dfrac{11}{8}$

49. $\dfrac{2}{x} - \dfrac{7}{3} = -\dfrac{29}{15}$

50. $\dfrac{1}{3x} + \dfrac{1}{x} = \dfrac{4}{15}$

51. $\dfrac{3}{x-1} = \dfrac{6}{x+4}$

52. $\dfrac{1}{x+5} = \dfrac{2}{2x+1}$

53. $\dfrac{1}{x} + \dfrac{1}{x^2} = 2$

54. $\dfrac{1}{x} - \dfrac{1}{x+1} = \dfrac{1}{56}$

55. $\dfrac{x}{x+2} = \dfrac{4}{x-3}$

56. $\dfrac{2x}{1-x} - \dfrac{1}{x} = 0$

Exercises 57–62: Solve the rational equation either graphically or numerically. Approximate your answer to the nearest hundredth when appropriate.

57. $\dfrac{1}{x-2} = 2$

58. $\dfrac{2x}{x+1} = 3$

59. $\dfrac{1}{x} + \dfrac{1}{x^2} = \dfrac{15}{4}$

60. $\dfrac{1}{x} + \dfrac{1}{2x} = 2x$

61. $\dfrac{1}{x+2} - \dfrac{1}{x-2} = \dfrac{4}{3}$

62. $\dfrac{x}{2x+4} - \dfrac{2x}{x^2-4} = 2.7$

USING MORE THAN ONE METHOD

Exercises 63–66: Solve the equation (a) symbolically, (b) graphically, and (c) numerically.

63. $\dfrac{1}{x} + \dfrac{2}{3} = 1$

64. $\dfrac{1}{x} - \dfrac{1}{x^2} = -2$

65. $\dfrac{1}{x} + \dfrac{1}{x+2} = \dfrac{4}{3}$

66. $\dfrac{1}{x-2} + \dfrac{1}{x+2} = -\dfrac{2}{3}$

APPLICATIONS

67. *Mowing the Lawn* Suppose that a person with a push mower can mow a large lawn in 5 hours, whereas the lawn can be mowed with a riding mower in 2 hours.
   (a) Write an equation whose solution gives the time needed to mow the lawn if both mowers are used at the same time.
   (b) Solve the equation in part (a) symbolically.
   (c) Solve the equation in part (a) either graphically or numerically.

68. *Pumping Water* (Refer to Example 7.) Suppose that a large pump can empty a swimming pool in 40 hours and that a small pump can empty a pool in 70 hours.
   (a) Write an equation whose solution gives the time needed for both pumps to empty the pool.
   (b) Solve the equation in part (a) symbolically.
   (c) Solve the equation in part (a) either graphically or numerically.

69. *Running a Race* (Refer to Example 8.) The winner of a 5-mile race finishes 7.5 minutes ahead of the second-place runner. On average, the winner ran 2 miles per hour faster than the second-place runner. Find the average running speed for each runner.

70. *Filling a Water Tank* A large pump can fill a 10,000 gallon tank 5 hours faster than a small pump. The large pump outputs water 100 gallons per hour faster than the small pump. Find the number of gallons pumped in 1 hour by each pump.

71. *Aerial Photographs* An aerial photograph is being taken of an area of land. The scale S for the photograph is planned to be $S = \frac{1}{10,000}$. (This value for S means that a distance of 10,000 feet on land will be represented by 1 foot in the photograph.) If H represents the height of the airplane above the ground (see the accompanying figure), these quantities are related by the rational equation

$$S = \frac{0.625}{H}.$$

Determine the height that the plane must fly at for the photograph to have the correct scale.

(*Source:* N. Garber and L. Hoel, *Traffic and Highway Engineering.*)

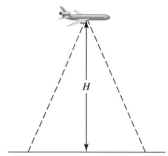

72. *Aerial Photographs* (Continuation of Exercise 71.) Suppose that the scale for the photograph is changed from $\frac{1}{10,000}$ to $\frac{1}{5000}$.
   (a) Should the plane fly higher or lower for this new scale? Explain your reasoning.
   (b) Test your conjecture in part (a) by letting $S = \frac{1}{5000}$ and finding H.

73. *Speed of a Boat* In still water a tugboat can travel 15 miles per hour. It travels 36 miles upstream and then 36 miles downstream in a total time of 5 hours. Find the speed of the current.

74. *Wind Speed* Without any wind an airplane flies at 200 miles per hour. The plane travels 800 miles into the wind and then returns with the wind in a total time of 8 hours and 20 minutes. Find the average speed of the wind.

75. *Speed of an Airplane* When there is a 50–mile per hour wind, an airplane can fly 675 miles with the wind in the same time that it can fly 450 miles against the wind. Find the speed of the plane when there is no wind.

76. *Waiting in Line* If a ticket agent can wait on 25 people per hour and people are arriving randomly at a rate of x people per hour, the average time T in hours for people waiting in line is given by

$$T(x) = \frac{1}{25 - x},$$

where $x < 25$. (*Source:* N. Garber.)
   (a) Solve the equation $T(x) = 5$ symbolically and interpret the result.
   (b) Solve part (a) either graphically or numerically.

Writing About Mathematics

77. A student does the following to simplify a rational expression. Is the work correct? Explain any errors and how you would correct them.

$$\frac{3x + x^2}{3x} \stackrel{?}{=} 1 + x^2$$

78. Explain in words how to multiply two rational expressions and how to divide two rational expressions.

CHECKING BASIC CONCEPTS FOR SECTIONS 7.1 AND 7.2

1. Let $f(x) = \dfrac{x}{x-1}$.
  (a) Evaluate $f(2)$.
  (b) Find any x-values not in the domain of f.
  (c) Graph f. Identify any vertical asymptotes.

2. Solve $\dfrac{6}{x} - \dfrac{1}{2} = 1$
  (a) symbolically,
  (b) graphically, and
  (c) numerically.

3. Simplify the expression.
  (a) $\dfrac{2x^2}{x^2 - 1} \cdot \dfrac{x + 1}{4x}$
  (b) $\dfrac{1}{x - 2} \div \dfrac{3}{(x-2)(x+3)}$

4. Solve $\dfrac{3}{x - 4} = \dfrac{1}{x + 1}$ using any method.

7.3 ADDITION AND SUBTRACTION OF RATIONAL EXPRESSIONS

Review of Addition and Subtraction of Fractions ~
Addition and Subtraction of Rational Expressions ~
Simplifying Complex Fractions

INTRODUCTION

In this section we demonstrate how to add and subtract rational expressions. These techniques are similar to the techniques used to add and subtract fractions. We begin with a brief review of addition and subtraction of fractions.

REVIEW OF ADDITION AND SUBTRACTION OF FRACTIONS

Recall that to add two fractions we use the property

$$\frac{a}{c} + \frac{b}{c} = \frac{a + b}{c}.$$

This property requires that the fractions have like denominators. For example,

$$\frac{1}{5} + \frac{3}{5} = \frac{1 + 3}{5} = \frac{4}{5}.$$

7.3 Addition and Subtraction of Rational Expressions

When the denominators are not alike, we must find a common denominator. Before adding two fractions, such as $\frac{2}{3}$ and $\frac{1}{4}$, we write them with 12 as their common denominators.

$$\frac{2}{3} = \frac{2}{3} \cdot \frac{4}{4} = \frac{8}{12}$$

$$\frac{1}{4} = \frac{1}{4} \cdot \frac{3}{3} = \frac{3}{12}$$

Once the fractions have a common denominator, we can add them, as in

$$\frac{2}{3} + \frac{1}{4} = \frac{8}{12} + \frac{3}{12} = \frac{11}{12}.$$

EXAMPLE 1 *Adding fractions*

Find the sum. Support your work with a calculator.

(a) $\frac{3}{4} + \frac{1}{8}$ **(b)** $\frac{3}{5} + \frac{2}{7}$

Solution

(a) The LCD is 8.
$$\frac{3}{4} + \frac{1}{8} = \frac{3}{4} \cdot \frac{2}{2} + \frac{1}{8} = \frac{6}{8} + \frac{1}{8} = \frac{7}{8}$$

(b) The LCD is 35.
$$\frac{3}{5} + \frac{2}{7} = \frac{3}{5} \cdot \frac{7}{7} + \frac{2}{7} \cdot \frac{5}{5} = \frac{21}{35} + \frac{10}{35} = \frac{31}{35}$$

These calculations are supported in Figure 7.20.

```
3/4+1/8▶Frac
            7/8
3/5+2/7▶Frac
          31/35
```

Figure 7.20

Recall that subtraction is similar. To subtract two fractions with like denominators we use the property

$$\frac{a}{c} - \frac{b}{c} = \frac{a - b}{c}.$$

For example,

$$\frac{3}{11} - \frac{7}{11} = \frac{3 - 7}{11} = -\frac{4}{11}.$$

EXAMPLE 2 *Subtracting fractions*

Find the difference. Support your work with a calculator.

(a) $\dfrac{3}{10} - \dfrac{2}{15}$ (b) $\dfrac{3}{8} - \dfrac{5}{6}$

Solution

(a) The LCD is 30.
$$\frac{3}{10} - \frac{2}{15} = \frac{3}{10} \cdot \frac{3}{3} - \frac{2}{15} \cdot \frac{2}{2} = \frac{9}{30} - \frac{4}{30} = \frac{5}{30} = \frac{1}{6}$$

(b) The LCD is 24.
$$\frac{3}{8} - \frac{5}{6} = \frac{3}{8} \cdot \frac{3}{3} - \frac{5}{6} \cdot \frac{4}{4} = \frac{9}{24} - \frac{20}{24} = -\frac{11}{24}$$

These calculations are supported in Figure 7.21.

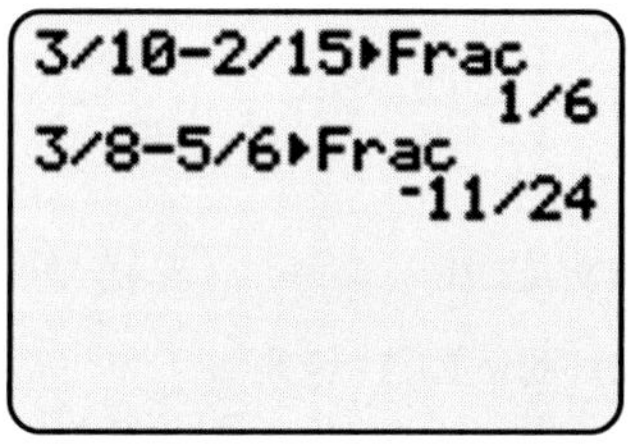

```
3/10-2/15▶Frac
              1/6
3/8-5/6▶Frac
            -11/24
```

Figure 7.21

ADDITION AND SUBTRACTION OF RATIONAL EXPRESSIONS

Addition and subtraction of rational expressions with like denominators are performed in the following manner.

Sums and Differences of Rational Expressions

To add (or subtract) two rational expressions with like denominators, add (or subtract) their numerators. The denominator does not change.

$$\frac{A}{C} + \frac{B}{C} = \frac{A + B}{C}$$

$$\frac{A}{C} - \frac{B}{C} = \frac{A - B}{C}, \qquad C \neq 0$$

Note: If the denominators are not alike, begin by writing each rational expression, using a common denominator. Then add or subtract the numerators.

EXAMPLE 3 *Adding rational expressions*

Add and simplify.

(a) $\dfrac{x}{x + 2} + \dfrac{3x + 1}{x + 2}$ (b) $\dfrac{1}{x} + \dfrac{2}{x^2}$ (c) $\dfrac{1}{x - 1} + \dfrac{2x}{x + 1}$

7.3 Addition and Subtraction of Rational Expressions

Solution

(a) The denominators are alike, so we add the numerators and keep the same denominator.

$$\frac{x}{x+2} + \frac{3x+1}{x+2} = \frac{x+3x+1}{x+2} \quad \text{Add numerators.}$$

$$= \frac{4x+1}{x+2} \quad \text{Combine like terms.}$$

(b) The LCD for x and x^2 is x^2.

$$\frac{1}{x} + \frac{2}{x^2} = \frac{1}{x} \cdot \frac{x}{x} + \frac{2}{x^2} \quad \text{Change to a common denominator.}$$

$$= \frac{x}{x^2} + \frac{2}{x^2} \quad \text{Multiply.}$$

$$= \frac{x+2}{x^2} \quad \text{Add numerators.}$$

(c) The LCD for $x - 1$ and $x + 1$ is their product, $(x-1)(x+1)$.

$$\frac{1}{x-1} + \frac{2x}{x+1} = \frac{1}{x-1} \cdot \frac{x+1}{x+1} + \frac{2x}{x+1} \cdot \frac{x-1}{x-1} \quad \text{Change to a common denominator.}$$

$$= \frac{x+1}{(x-1)(x+1)} + \frac{2x(x-1)}{(x+1)(x-1)} \quad \text{Multiply.}$$

$$= \frac{x+1+2x^2-2x}{(x-1)(x+1)} \quad \text{Add numerators; distributive property}$$

$$= \frac{2x^2-x+1}{(x-1)(x+1)} \quad \text{Combine like terms.}$$

Subtraction of rational expressions is similar.

EXAMPLE 4 Subtracting rational expressions

Subtract and simplify.

(a) $\dfrac{3}{x^2} - \dfrac{x+3}{x^2}$ **(b)** $\dfrac{x-1}{x} - \dfrac{5}{x+5}$ **(c)** $\dfrac{1}{x^2-3x+2} - \dfrac{1}{x^2-x-2}$

Solution

(a) The denominators are alike, so we subtract the numerators and keep the same denominator.

$$\frac{3}{x^2} - \frac{x+3}{x^2} = \frac{3-(x+3)}{x^2} \quad \text{Subtract numerators.}$$

$$= \frac{3-x-3}{x^2} \quad \text{Distributive property}$$

$$= \frac{-x}{x^2} \quad \text{Simplify numerator.}$$

$$= -\frac{1}{x} \quad \text{Reduce.}$$

(b) The LCD is $x(x + 5)$.

$$\frac{x-1}{x} - \frac{5}{x+5} = \frac{x-1}{x} \cdot \frac{x+5}{x+5} - \frac{5}{x+5} \cdot \frac{x}{x} \quad \text{Change to a common denominator.}$$

$$= \frac{(x-1)(x+5)}{x(x+5)} - \frac{5x}{x(x+5)} \quad \text{Multiply.}$$

$$= \frac{(x-1)(x+5) - 5x}{x(x+5)} \quad \text{Subtract numerators.}$$

$$= \frac{x^2 + 4x - 5 - 5x}{x(x+5)} \quad \text{Multiply binomials.}$$

$$= \frac{x^2 - x - 5}{x(x+5)} \quad \text{Combine like terms.}$$

(c) Because $x^2 - 3x + 2 = (x - 2)(x - 1)$ and $x^2 - x - 2 = (x - 2)(x + 1)$, the LCD is

$$(x - 1)(x + 1)(x - 2).$$

$$\frac{1}{x^2 - 3x + 2} - \frac{1}{x^2 - x - 2}$$

$$= \frac{1}{(x-2)(x-1)} \cdot \frac{(x+1)}{(x+1)} - \frac{1}{(x-2)(x+1)} \cdot \frac{(x-1)}{(x-1)} \quad \text{Change to a common denominator.}$$

$$= \frac{(x+1)}{(x-2)(x-1)(x+1)} - \frac{(x-1)}{(x-2)(x+1)(x-1)} \quad \text{Multiply.}$$

$$= \frac{(x+1) - (x-1)}{(x-2)(x-1)(x+1)} \quad \text{Subtract numerators.}$$

$$= \frac{x + 1 - x + 1}{(x-2)(x-1)(x+1)} \quad \text{Distributive property}$$

$$= \frac{2}{(x-2)(x-1)(x+1)} \quad \text{Simplify numerator.}$$

Technology Note *Graphical and Numerical Support*

We can give graphical and numerical support to our work in Example 4(a) by letting $Y_1 = 3/X^2 - (X + 3)/X^2$, the given expression, and $Y_2 = -1/X$, the simplified expression. In Figures 7.22 and 7.23 the graphs of Y_1 and Y_2 appear to be identical. In Figure 7.24 numerical support is given, where $Y_1 = Y_2$ for each value of x.

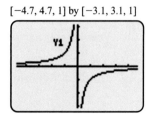

Figure 7.22

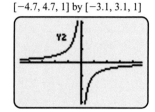

Figure 7.23

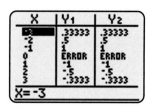

Figure 7.24

AN APPLICATION FROM ELECTRICITY Sums of rational expressions occur in applications such as electrical circuits. The flow of electricity through a wire can be compared to the flow of water through a hose. Voltage is the force "pushing" the electricity and corresponds to water pressure in a hose. Resistance is the opposition to the flow of electricity, and more resistance results in less flow of electricity. Resistance corresponds to the diameter of a hose; if the diameter is smaller, less water flows. An ordinary light bulb is an example of a resistor in an electrical circuit.

Suppose that two light bulbs are wired in parallel so that electricity can flow through either light bulb, as depicted in Figure 7.25. If their resistances are R_1 and R_2, their combined resistance R can be computed by the equation

$$\frac{1}{R} = \frac{1}{R_1} + \frac{1}{R_2}.$$

Resistance is often measured in a unit called *ohms*. A standard 60-watt light bulb might have a resistance of about 200 ohms. **(Source: R. Weidner and R. Sells, *Elementary Classical Physics, Vol. 2*.)**

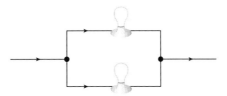

Figure 7.25

EXAMPLE 5 *Modeling electrical resistance*

A 100-watt light bulb with a resistance of $R_1 = 120$ ohms and a 75-watt light bulb with a resistance of $R_2 = 160$ ohms are placed in an electrical circuit, as shown in Figure 7.25. Find their combined resistance.

Solution

Let $R_1 = 120$ and $R_2 = 160$ in the given equation and solve for R.

$$\frac{1}{R} = \frac{1}{R_1} + \frac{1}{R_2} \quad \text{Given equation}$$

$$= \frac{1}{120} + \frac{1}{160} \quad \text{Substitute for } R_1 \text{ and } R_2.$$

$$= \frac{1}{120} \cdot \frac{4}{4} + \frac{1}{160} \cdot \frac{3}{3} \quad \text{LCD is 480.}$$

$$= \frac{4}{480} + \frac{3}{480} \quad \text{Multiply.}$$

$$= \frac{7}{480} \quad \text{Add.}$$

See Figure 7.26 on the following page. Because $\frac{1}{R} = \frac{7}{480}$, $R = \frac{480}{7} \approx 69$ ohms.

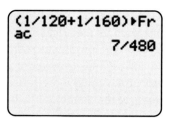

Figure 7.26

Critical Thinking

If $\dfrac{1}{R} = \dfrac{1}{R_1} + \dfrac{1}{R_2}$, does $R = R_1 + R_2$? Explain your answer.

SIMPLIFYING COMPLEX FRACTIONS

A **complex fraction** is a rational expression that contains fractions in its numerator, denominator, or both. Examples of complex fractions include

$$\dfrac{1 + \dfrac{1}{x}}{1 - \dfrac{1}{x}}, \qquad \dfrac{2x}{\dfrac{4}{x} + \dfrac{3}{x}}, \qquad \text{and} \qquad \dfrac{\dfrac{a}{3} + \dfrac{a}{4}}{a - \dfrac{1}{a-1}}.$$

One strategy for simplifying a complex fraction is to multiply the numerator and denominator by the LCD of the fractions in the numerator and denominator. For example, the LCD for the complex fraction

$$\dfrac{1 - \dfrac{1}{x}}{1 + \dfrac{1}{2x}}$$

is $2x$. To simplify, multiply the complex fraction by 1, expressed in the form $\dfrac{2x}{2x}$.

$$\dfrac{\left(1 - \dfrac{1}{x}\right) \cdot 2x}{\left(1 + \dfrac{1}{2x}\right) \cdot 2x} = \dfrac{2x - \dfrac{2x}{x}}{2x + \dfrac{2x}{2x}} \qquad \text{Distributive property}$$

$$= \dfrac{2x - 2}{2x + 1} \qquad \text{Reduce.}$$

In the next example we simplify other complex fractions.

EXAMPLE 6 *Simplifying complex fractions*

Simplify.

(a) $\dfrac{\dfrac{1}{x} - \dfrac{1}{y}}{x - y}$ (b) $\dfrac{\dfrac{3}{x-1} - \dfrac{2}{x}}{\dfrac{1}{x-1} + \dfrac{3}{x}}$

Solution

(a) The LCD of x and y is their product, xy. Multiply the expression by $\dfrac{xy}{xy}$.

$$\dfrac{\left(\dfrac{1}{x}-\dfrac{1}{y}\right)\cdot xy}{(x-y)\cdot xy} = \dfrac{\dfrac{xy}{x}-\dfrac{xy}{y}}{xy(x-y)} \quad \text{Distributive and commutative properties}$$

$$= \dfrac{y-x}{xy(x-y)} \quad \text{Reduce.}$$

$$= \dfrac{-1(x-y)}{xy(x-y)} \quad \text{Factor out } -1.$$

$$= -\dfrac{1}{xy} \quad \text{Reduce.}$$

(b) The LCD of x and $x-1$ is their product, $x(x-1)$. Multiply the expression by $\dfrac{x(x-1)}{x(x-1)}$.

$$\dfrac{\left(\dfrac{3}{x-1}-\dfrac{2}{x}\right)}{\left(\dfrac{1}{x-1}+\dfrac{3}{x}\right)} \cdot \dfrac{x(x-1)}{x(x-1)} = \dfrac{\dfrac{3x(x-1)}{x-1}-\dfrac{2x(x-1)}{x}}{\dfrac{x(x-1)}{x-1}+\dfrac{3x(x-1)}{x}} \quad \text{Distributive property}$$

$$= \dfrac{3x-2(x-1)}{x+3(x-1)} \quad \text{Reduce.}$$

$$= \dfrac{3x-2x+2}{x+3x-3} \quad \text{Distributive property}$$

$$= \dfrac{x+2}{4x-3} \quad \text{Combine like terms.}$$

7.3 Putting It All Together

The following table summarizes some important concepts from this section.

Concept	Explanation	Example
Addition and subtraction of rational expressions	To add (or subtract) two rational expressions with like denominators, add (or subtract) their numerators. The denominator does not change. $$\dfrac{A}{C}+\dfrac{B}{C}=\dfrac{A+B}{C}$$ $$\dfrac{A}{C}-\dfrac{B}{C}=\dfrac{A-B}{C}, \quad C\neq 0$$	$\dfrac{x}{x+1}+\dfrac{3x}{x+1}=\dfrac{4x}{x+1}$ If the denominators are not alike, find the LCD first. $$\dfrac{1}{x}-\dfrac{2x}{x+1}=\dfrac{1}{x}\cdot\dfrac{x+1}{x+1}-\dfrac{2x}{x+1}\cdot\dfrac{x}{x}$$ $$=\dfrac{x+1}{x(x+1)}-\dfrac{2x^2}{x(x+1)}$$ $$=\dfrac{-2x^2+x+1}{x(x+1)}$$

continued on next page

continued from previous page

Concept	Explanation	Example
Simplifying complex fractions	To simplify a complex fraction start by multiplying the numerator and denominator by the LCD of the fractions in the numerator and in the denominator.	The LCD in the following complex fraction is x. $$\frac{2 - \frac{1}{x}}{2 + \frac{1}{x}} = \frac{\left(2 - \frac{1}{x}\right) \cdot x}{\left(2 + \frac{1}{x}\right) \cdot x} = \frac{2x - 1}{2x + 1}$$

7.3 EXERCISES

FOR EXTRA HELP: Student's Solutions Manual, MyMathLab.com, InterAct Math, Math Tutor Center, MathXL, Digital Video Tutor CD 5 Videotape 11

CONCEPTS

1. What do you need to find before you can add $\frac{1}{4}$ and $\frac{1}{3}$?

2. What do you need to find before you can add $\frac{2}{x}$ and $\frac{1}{2x-1}$?

3. $\frac{a}{c} + \frac{b}{c} = $ _____ .

4. $\frac{a}{c} - \frac{b}{c} = $ _____ .

5. What is the LCD for $\frac{1}{6}$ and $\frac{1}{9}$?

6. What is the LCD for $\frac{1}{x^2 - 25}$ and $\frac{1}{x + 5}$?

7. What is a good first step when simplifying
$$\frac{2 + \frac{1}{x-1}}{2 - \frac{1}{x-1}}?$$

8. Explain what a complex fraction is.

REVIEW OF FRACTIONS

Exercises 9–16: Simplify. Support your result with a calculator.

9. $\frac{1}{7} + \frac{4}{7}$

10. $\frac{2}{5} + \frac{1}{2}$

11. $\frac{2}{3} + \frac{5}{6} + \frac{1}{4}$

12. $\frac{3}{11} + \frac{1}{2} + \frac{1}{6}$

13. $\frac{1}{10} - \frac{3}{10}$

14. $\frac{2}{9} - \frac{1}{11}$

15. $\frac{3}{2} - \frac{1}{8}$

16. $\frac{3}{12} - \frac{5}{16}$

ADDITION AND SUBTRACTION OF RATIONAL EXPRESSIONS

Exercises 17–22: Simplify. Support your result either graphically or numerically.

17. $\frac{1}{x} + \frac{3}{x}$

18. $\frac{2}{x-1} + \frac{x}{x-1}$

19. $\frac{2}{x^2 - 4} - \frac{x+1}{x^2 - 4}$

20. $\frac{2x-1}{x^2 + 6} - \frac{2x+1}{x^2 + 6}$

21. $\frac{x}{x+4} - \frac{x+1}{x}$

22. $\frac{4x}{x+2} + \frac{x-5}{x-2}$

7.3 Addition and Subtraction of Rational Expressions

Exercises 23–38: Simplify.

23. $\dfrac{2}{x^2} - \dfrac{4x-1}{x}$

24. $\dfrac{2x}{x-5} - \dfrac{x}{x+5}$

25. $\dfrac{x+3}{x-5} + \dfrac{5}{x-3}$

26. $\dfrac{x}{2x-1} + \dfrac{1-x}{3x}$

27. $\dfrac{3}{x-5} - \dfrac{1}{x-3} - \dfrac{2x}{x-5}$

28. $\dfrac{2x+1}{x-1} - \dfrac{3}{x+1} + \dfrac{x}{x-1}$

29. $\dfrac{x}{x^2-9} + \dfrac{5x}{x-3}$

30. $\dfrac{a^2+1}{a^2-1} + \dfrac{a}{1-a^2}$

31. $\dfrac{b}{2b-4} - \dfrac{b-1}{b-2}$

32. $\dfrac{y^2}{2-y} - \dfrac{y}{y^2-4}$

33. $\dfrac{2x}{x-5} + \dfrac{2x-1}{3x^2-16x+5}$

34. $\dfrac{x+3}{2x-1} + \dfrac{3}{10x^2-5x}$

35. $\dfrac{4x}{x-y} - \dfrac{9}{x+y}$

36. $\dfrac{1}{a-b} - \dfrac{3a}{a^2-b^2}$

37. $\dfrac{3}{x^2-2x+1} + \dfrac{1}{x^2-3x+2}$

38. $\dfrac{x}{x^2-4} - \dfrac{1}{x^2+4x+4}$

COMPLEX FRACTIONS

Exercises 39–50: Simplify the expression.

39. $\dfrac{2+\dfrac{2}{3}}{2-\dfrac{1}{4}}$

40. $\dfrac{\dfrac{1}{2}+\dfrac{3}{4}}{\dfrac{1}{2}-\dfrac{3}{4}}$

41. $\dfrac{1+\dfrac{1}{x}}{x+1}$

42. $\dfrac{2-x}{\dfrac{1}{x}-\dfrac{1}{2}}$

43. $\dfrac{\dfrac{1}{x-3}}{\dfrac{1}{x}-\dfrac{3}{x-3}}$

44. $\dfrac{5+\dfrac{1}{x-1}}{\dfrac{1}{x-1}-\dfrac{1}{4}}$

45. $\dfrac{\dfrac{1}{x}+\dfrac{2}{x^2}}{\dfrac{3}{x}-\dfrac{1}{x^2}}$

46. $\dfrac{\dfrac{1}{x-1}+\dfrac{2}{x}}{2-\dfrac{1}{x}}$

47. $\dfrac{\dfrac{1}{x+3}+\dfrac{2}{x-3}}{2-\dfrac{1}{x-3}}$

48. $\dfrac{\dfrac{1}{x}+\dfrac{2}{x}}{\dfrac{1}{x-1}+\dfrac{x}{2}}$

49. $\dfrac{\dfrac{4}{x-5}}{\dfrac{1}{x+5}+\dfrac{1}{x}}$

50. $\dfrac{\dfrac{1}{x-4}+\dfrac{1}{x-4}}{1-\dfrac{1}{x+4}}$

APPLICATIONS

51. *Electrical Resistance* (Refer to Example 5.) A 150-watt light bulb with a resistance of 80 ohms and a 40-watt light bulb with a resistance of 300 ohms are wired in parallel. Find the combined resistance of the light bulbs.

52. *Electrical Resistance* (Continuation of Exercise 51.) Solve the formula
$$\dfrac{1}{R} = \dfrac{1}{R_1} + \dfrac{1}{R_2}$$
for R. Use your formula to solve Exercise 51. Did you find the same value for R?

53. *Photography* A lens in a camera has a focal length, which is important when focusing the camera. If an object is at distance D from a lens that has a focal length F, then to be in focus the distance S between the lens and the film should satisfy the equation
$$\dfrac{1}{S} = \dfrac{1}{F} - \dfrac{1}{D}.$$
(See the accompanying figure.) If the focal length is $F = 0.25$ foot and the object is $D = 10$ feet from the camera, find S.

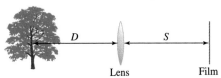

54. *Photography* (Continuation of Exercise 53.) Suppose that the object is closer to the lens than the focal length of the lens. That is, suppose $D < F$. According to the formula, is it possible to focus the image on the film? Explain your reasoning.

WRITING ABOUT MATHEMATICS

55. A student does the following to add two rational expressions. Explain the error that the student is making and how it can be corrected.

$$\frac{x}{x+1} + \frac{6x}{x} \stackrel{?}{=} \frac{7x}{2x+1}$$

56. Explain in words how to subtract two rational expressions with like denominators.

7.4 MODELING WITH PROPORTIONS AND VARIATION

Proportions ~ Direct Variation ~ Inverse Variation ~ Joint Variation

INTRODUCTION

Proportions are used frequently to solve problems in everyday life. The following are examples.

- If someone earns $100 per day, that person can earn $500 in 5 days.
- If a car goes 210 miles on 10 gallons of gas, the car can go 420 miles on 20 gallons of gas.
- If a person walks a mile in 16 minutes, that person can walk a half mile in 8 minutes.

Many applications involve proportions or variation. In this section we discuss some of them.

PROPORTIONS

A 650-megabyte compact disc (CD) can store about 74 minutes of music. Suppose that we have already recorded some music on the CD and 256 megabytes are still available. Using proportions we can determine how many more minutes of music could be recorded. A **proportion** is a statement that two ratios are equal. **(Source: Maxell Corporation.)**

Let x represent the number of minutes available on the CD. Then 74 minutes are to 650 megabytes as x minutes are to 256 megabytes, which can be written as the proportion

$$\frac{74}{650} = \frac{x}{256}.$$

Solving this equation for x gives

$$x = \frac{74 \cdot 256}{650} \approx 29.1 \text{ minutes.}$$

About 29 minutes are still available on the CD.

■ MAKING CONNECTIONS

Proportions and Fractional Parts

We could have solved the preceding problem by noting that the fraction of the CD still available for recording music is $\frac{256}{650}$. So $\frac{256}{650}$ of 74 minutes equals

$$\frac{256}{650} \cdot 74 \approx 29.1 \text{ minutes.}$$

■

Using the following property is often a convenient way to solve proportions:

$$\frac{a}{b} = \frac{c}{d} \quad \text{is equivalent to} \quad ad = bc,$$

provided $b \neq 0$ and $d \neq 0$. This is a result of multiplying both sides of the equation by a common denominator bd, and is sometimes referred to as **clearing fractions.**

$$bd \cdot \frac{a}{b} = \frac{c}{d} \cdot bd \qquad \text{Multiply by } bd.$$

$$\frac{bda}{b} = \frac{cbd}{d} \qquad \text{Property of multiplying fractions}$$

$$ad = bc \qquad \text{Reduce.}$$

For example, the proportion

$$\frac{6}{5} = \frac{8}{x}$$

is equivalent to

$$6x = 40 \quad \text{or} \quad x = \frac{40}{6} = \frac{20}{3}.$$

EXAMPLE 1 *Calculating the water content in snow*

Six inches of light, fluffy snow are equivalent to about half an inch of rain in terms of water content. If 21 inches of snow fall, estimate the water content.

Solution

Let x represent the equivalent amount of rain. Then 6 inches of snow is to $\frac{1}{2}$ inch of rain as 21 inches of snow is to x inches of rain, which can be written as the proportion

$$\frac{6}{1/2} = \frac{21}{x}.$$

Solving this equation gives

$$6x = \frac{21}{2} \quad \text{or} \quad x = \frac{21}{12} = 1.75.$$

Thus, 21 inches of light, fluffy snow is equivalent to about 1.75 inches of rain.

Proportions frequently occur in geometry when we work with similar figures. Two triangles are similar if their corresponding angles are equal, and corresponding sides of similar triangles are proportional. Figure 7.27 shows two right triangles, which are similar because each has angles of 30°, 60°, and 90°.

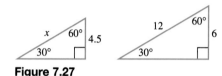

Figure 7.27

We can find the length of side x by using proportions. Side x is to 4.5 as 12 is to 6, which can be written as the proportion

$$\frac{x}{4.5} = \frac{12}{6}. \quad \begin{array}{l}\text{Hypotenuse}\\ \hline \text{Shorter leg}\end{array}$$

Solving yields the equation

$$6x = 4.5 \cdot 12 \quad \text{Clear fractions.}$$
$$x = 9. \quad \text{Divide by 6.}$$

EXAMPLE 2 *Calculating the height of a tree*

A 6–foot tall person casts a 4–foot long shadow. If a nearby tree casts a 44–foot long shadow, estimate the height of the tree. See Figure 7.28.

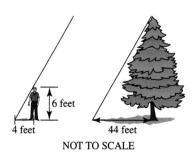

Figure 7.28

Solution

The triangles shown in Figure 7.29 are similar because corresponding angles are equal. Therefore their sides are proportional. Let h be the height of the tree.

$$\frac{h}{44} = \frac{6}{4} \qquad \frac{\text{Height}}{\text{Shadow length}}$$

$$4h = 6 \cdot 44 \qquad \text{Clear fractions.}$$

$$h = \frac{6 \cdot 44}{4} \qquad \text{Divide by 4.}$$

$$= 66 \qquad \text{Simplify.}$$

The tree is 66 feet tall.

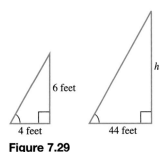

Figure 7.29

DIRECT VARIATION

If your wage is $9 per hour, the amount you earn is proportional to the number of hours that you work. If you worked H hours, your total pay P satisfies the equation

$$\frac{P}{H} = \frac{9}{1}, \qquad \frac{\text{Pay}}{\text{Hours}}$$

or, equivalently,

$$P = 9H.$$

We say that your pay P is *directly proportional* to the number of hours H worked. The constant of proportionality is 9.

> **Direct variation**
>
> Let x and y denote two quantities. Then y is **directly proportional** to x, or y **varies directly** with x, if there is a nonzero number k such that
>
> $$y = kx.$$
>
> The number k is called the **constant of proportionality,** or the **constant of variation.**

The graph of $y = kx$ is a line passing through the origin, as illustrated in Figure 7.30. Sometimes data in a scatterplot indicate that two quantities are directly proportional. The constant of proportionality k corresponds to the slope of the graph.

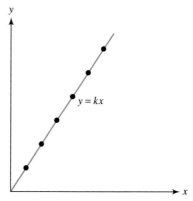

Figure 7.30 Direct Variation, $k > 0$

EXAMPLE 3 *Modeling college tuition*

 Table 7.2 lists the tuition for taking various numbers of credits.

TABLE 7.2

Number of Credits	3	5	8	11	17
Tuition	$189	$315	$504	$693	$1071

(a) Make a scatterplot of the data and discuss the graph. Are the data linear?
(b) Explain why tuition is directly proportional to the number of credits taken.
(c) Find the constant of proportionality. Interpret your result.
(d) Predict the cost of taking 15 credits.

Solution

(a) A scatterplot of the data in Table 7.2 is shown in Figure 7.31. (A scatterplot can also be done by hand.) The data are linear and suggest a line passing through the origin.
(b) Because the data can be modeled by a line passing through the origin, tuition is directly proportional to the number of credits taken. Hence doubling the credits will double the tuition and tripling the credits will triple the tuition.
(c) The slope of the line equals the constant of proportionality k. If we use the first and last data points (3, 189) and (17, 1071), the slope is

$$k = \frac{1071 - 189}{17 - 3} = 63.$$

That is, tuition is $63 per credit. If we graph the line $y = 63x$, it models the data as shown in Figure 7.32.
(d) If y represents tuition and x represents the credits taken, 15 credits would cost

$$y = 63 \cdot 15 = \$945.$$

[0, 20, 5] by [0, 1200, 400]   [0, 20, 5] by [0, 1200, 400]

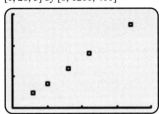
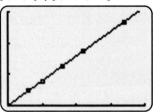

Figure 7.31   Figure 7.32

■ **MAKING CONNECTIONS**

Ratios and the Constant of Proportionality

The constant of proportionality in Example 3 can also be found by calculating the ratios $\frac{y}{x}$, where y is the tuition and x is the credits taken. Note that each ratio in Table 7.3 equals 63 because the equation $y = 63x$ is equivalent to the equation $\frac{y}{x} = 63$.

TABLE 7.3

x	3	5	8	11	17
y	189	315	504	693	1071
y/x	63	63	63	63	63

INVERSE VARIATION

When two quantities vary inversely, an increase in one quantity results in a decrease in the second quantity. For example, at 25 miles per hour a car travels 100 miles in 4 hours, whereas at 50 miles per hour the car travels 100 miles in 2 hours. Doubling the speed (or rate) decreases the travel time by half. Distance equals rate times time, so $d = rt$. Thus

$$100 = rt, \quad \text{or equivalently,} \quad t = \frac{100}{r}.$$

We say that the time t to travel 100 miles is *inversely proportional* to the speed or rate r. The constant of proportionality or constant of variation is 100.

Inverse variation

Let x and y denote two quantities. Then y is **inversely proportional** to x, or y **varies inversely** with x, if there is a nonzero number k such that

$$y = \frac{k}{x}.$$

Note: We assume that the constant k is positive.

The data shown in Figure 7.33 represent inverse variation and are modeled by $y = \dfrac{k}{x}$. Note that, as x increases, y decreases.

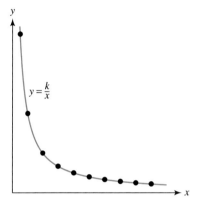

Figure 7.33 Inverse Variation, $k > 0$

A wrench is commonly used to loosen a nut on a bolt. See Figure 7.34. If the nut is difficult to loosen, a wrench with a longer handle is often helpful.

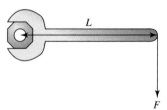

Figure 7.34

EXAMPLE 4 *Loosening a nut on a bolt*

Table 7.4 lists the force F necessary to loosen a particular nut with wrenches of different lengths L.

TABLE 7.4

L (inches)	6	10	12	15	20
F (pounds)	10	6	5	4	3

(a) Make a scatterplot of the data and discuss the graph. Are the data linear?
(b) Explain why the force F is inversely proportional to the handle length L. Find k so that $F = \dfrac{k}{L}$ models the data.
(c) Graph the data and the equation in the same viewing rectangle.
(d) Predict the force needed to loosen the nut with an 8-inch wrench.

7.4 Modeling with Proportions and Variation

Solution

(a) The scatterplot shown in Figure 7.35 reveals that the data are nonlinear. As the length L of the wrench increases, the force F necessary to loosen the nut decreases.

(b) If F is inversely proportional to L, then $F = \dfrac{k}{L}$, or $FL = k$. That is, the product of F and L equals the constant of proportionality. In Table 7.4, the product of F and L always equals 60 for each data point. Thus F is inversely proportional to L with constant of proportionality $k = 60$.

(c) Graph $Y_1 = 60/X$, as shown in Figure 7.36. The graph passes through each data point.

(d) If $L = 8$, then $F = \dfrac{60}{8} = 7.5$. A wrench with an 8-inch handle requires a force of 7.5 pounds to loosen the nut.

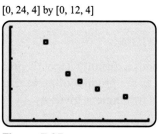

Figure 7.35   Figure 7.36

EXAMPLE 5 *Analyzing data*

 Determine whether the data in each table represent direct variation, inverse variation, or neither.

(a)
x	4	5	10	20
y	50	40	20	10

(b)
x	2	5	9	11
y	14	35	63	77

(c)
x	2	4	6	8
y	10	16	24	48

Solution

(a) As x increases, y decreases. Because $xy = 200$ for each data point, the equation $y = \dfrac{200}{x}$ models the data. The data represent inverse variation. The data and $Y_1 = 200/X$ are graphed in Figure 7.37 on the following page.

(b) As $\dfrac{y}{x} = 7$ for each data point in the table, the equation $y = 7x$ models the data.

These data represent direct variation. The data and $Y_1 = 7X$ are graphed in Figure 7.38.

(c) Neither the product xy nor the ratio $\dfrac{y}{x}$ are constant for the data in the table. Therefore these data represent neither direct variation nor inverse variation. The data are plotted in Figure 7.39. Note that the data values increase, and are nonlinear.

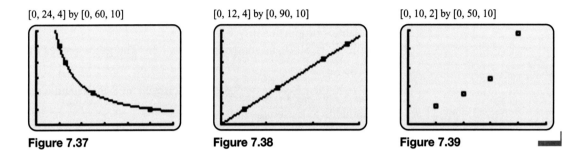

[0, 24, 4] by [0, 60, 10]   [0, 12, 4] by [0, 90, 10]   [0, 10, 2] by [0, 50, 10]

Figure 7.37   Figure 7.38   Figure 7.39

JOINT VARIATION

In many applications a quantity depends on more than one variable. In **joint variation** a quantity varies as the product of more than one variable. For example, the formula for the area A of a rectangle is given by

$$A = WL,$$

where W and L are the width and length, respectively. Thus the area of a rectangle varies jointly with the width and length.

> ### Joint variation
>
> Let x, y, and z denote three quantities. Then z **varies jointly** as x and y if there is a nonzero number k such that
>
> $$z = kxy.$$

Sometimes joint variation can involve a power of a variable. For example, the volume V of a cylinder is given by $V = \pi r^2 h$, where r is its radius and h is its height, as illustrated in Figure 7.40. In this case we say that the volume varies jointly with the height and the *square* of the radius.

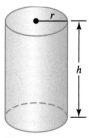

Figure 7.40

EXAMPLE 6 Strength of a rectangular beam

The strength S of a rectangular beam varies jointly as its width w and the square of its thickness t. See Figure 7.41. If a beam 3 inches wide and 5 inches thick supports 750 pounds, how much can a similar beam 2 inches wide and 6 inches thick support?

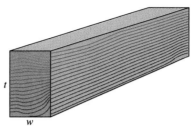

Figure 7.41

Solution

The strength of a beam is modeled by $S = kwt^2$, where k is a constant of variation. We can find k by substituting $S = 750$, $w = 3$, and $t = 5$ into the formula.

$$750 = k \cdot 3 \cdot 5^2$$

$$k = \frac{750}{3 \cdot 5^2} \qquad \text{Solve for } k.$$

$$= 10 \qquad \text{Simplify.}$$

Thus $S = 10wt^2$ models the strength of this type of beam. When $w = 2$ and $t = 6$, the beam can support

$$S = 10 \cdot 2 \cdot 6^2 = 720 \text{ pounds.}$$

7.4 Putting It All Together

A proportion is a statement that two ratios are equal. Proportions can be used to solve for unknown quantities. For example, if a pump can empty 50% of a water tank in 3 hours, then it can empty 80% of the tank in x hours, where x may be determined from the proportion

$$\frac{50}{3} = \frac{80}{x}.$$

If a quantity y is directly proportional to x, then y doubles whenever x doubles. If a quantity y is inversely proportional to x, then y decreases by half whenever x doubles. Joint variation occurs when a quantity varies as the *product* of more than one variable. The different types of variation are summarized in the following table.

Type of Variation	Explanation
Direct	Let x and y denote two quantities. Then y is directly proportional to x, or varies directly with x, if there is a nonzero number k such that $$y = kx.$$ The number k is called the constant of proportionality, or constant of variation. If data vary directly, the ratios $\frac{y}{x}$ are constant and equal k.
Inverse	Let x and y denote two quantities. Then y is inversely proportional to x, or varies inversely with x, if there is a nonzero number k such that $$y = \frac{k}{x}.$$ If data vary inversely, the products xy are constant and equal k.
Joint	Let x, y, and z denote three quantities. Then z varies jointly as x and y if there is a nonzero number k such that $$z = kxy.$$

7.4 EXERCISES

CONCEPTS

1. What is a proportion?

2. If 5 is to 6 as x is to 7, write a proportion that allows you to find x.

3. Suppose that y is directly proportional to x. If x doubles, what happens to y?

4. Suppose that y is inversely proportional to x. If x doubles, what happens to y?

5. If y varies inversely with x, then xy equals a _____.

6. If y varies directly with x, then $\frac{y}{x}$ equals a _____.

7. If z varies jointly with x and y, then $z =$ _____.

8. If z varies jointly with the square of x and the cube of y, then $z =$ _____.

9. Would a food bill B generally vary directly or inversely with the number of people N being fed? Explain your reasoning.

10. Would the number of people N painting a building vary directly or inversely with the time T needed to complete the job?

PROPORTIONS

Exercises 11–16: Solve the proportion.

11. $\dfrac{x}{14} = \dfrac{5}{7}$

12. $\dfrac{x}{5} = \dfrac{4}{9}$

13. $\dfrac{8}{x} = \dfrac{2}{3}$

14. $\dfrac{5}{11} = \dfrac{9}{x}$

15. $\dfrac{6}{13} = \dfrac{h}{156}$

16. $\dfrac{25}{a} = \dfrac{15}{8}$

Exercises 17–24: Complete the following.
 (a) *Write a proportion that models the situation described.*
 (b) *Solve the proportion for x.*

17. 7 is to 9, as 10 is to x

18. x is to 11, as 9 is to 2

19. A triangle has sides of 3, 4, and 6. In a similar triangle the shortest side is 5 and the longest side is x.

20. A rectangle has sides of 9 and 14. In a similar rectangle the longer side is 8 and the shorter side is x.

21. If you earn $78 in 6 hours, you can earn x dollars in 8 hours.

22. If 12 gallons of gasoline contain 1.2 gallons of ethanol, 18 gallons of gasoline contain x gallons of ethanol.

23. If 2 cassette tapes can record 90 minutes of music, 5 cassette tapes can record x minutes.

24. If a gas pump fills a 30-gallon tank in 8 minutes, it can fill a 17-gallon tank in x minutes.

VARIATION

Exercises 25–28: Suppose that y is directly proportional to x.
 (a) *Use the given information to find the constant of proportionality k.*
 (b) *Find y when x = 7.*

25. $y = 6$ when $x = 3$

26. $y = 7$ when $x = 14$

27. $y = 5$ when $x = 2$

28. $y = 11$ when $x = 22$

Exercises 29–32: Suppose that y is inversely proportional to x.
 (a) *Use the given information to find the constant of proportionality k.*
 (b) *Find y when x = 10.*

29. $y = 5$ when $x = 4$

30. $y = 2$ when $x = 30$

31. $y = 100$ when $x = \dfrac{1}{2}$

32. $y = \dfrac{1}{4}$ when $x = 40$

Exercises 33–36: Suppose that z varies jointly with x and y.
 (a) *Use the given information to find the constant of variation k.*
 (b) *Find z when x = 5 and y = 7.*

33. $z = 6$ when $x = 3$ and $y = 8$

34. $z = 135$ when $x = 2.5$ and $y = 9$

35. $z = 5775$ when $x = 25$ and $y = 21$

36. $z = 1530$ when $x = 22.5$ and $y = 4$

Exercises 37–42: (Refer to Example 5.)
 (a) *Determine whether the data represent direct variation, inverse variation, or neither.*
 (b) *If the data represent either direct or inverse variation find an equation that models the data.*
 (c) *Graph the equation and the data.*

37.
x	2	3	4	5
y	3	4.5	6	7.5

38.
x	3	6	9	12
y	12	6	4	3

39.
x	10	20	30	40
y	12	6	5	4

40.

x	2	6	10	14
y	105	35	21	15

41.

x	4	6	12	20
y	10	20	30	40

42.

x	1	5	9	15
y	6	30	54	90

Exercises 43–48: Use the graph to determine whether the data represent direct variation, inverse variation, or neither. Find the constant of variation whenever possible.

43.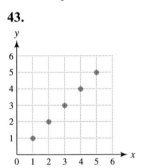

44.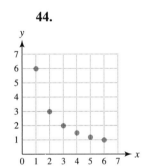

45.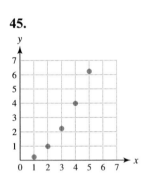

46.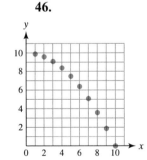

47.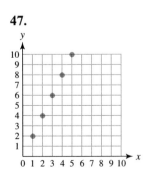

48.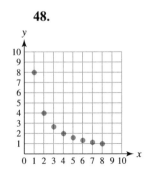

APPLICATIONS

49. *Recording Music* A 600-megabyte CD can record 68 minutes of music. How many minutes can be recorded on 360 megabytes?

50. *Height of a Tree* (Refer to Example 2.) A 6-foot person casts a 7-foot shadow, and a nearby tree casts a 27-foot shadow. Estimate the height of the tree.

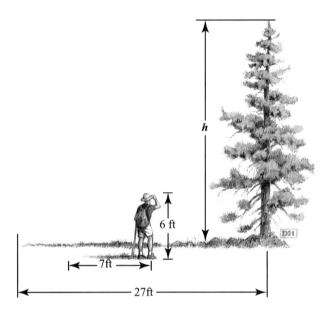

51. *Water Content in Snow* (Refer to Example 1.) Eight inches of heavy, wet snow are equivalent to an inch of rain. Estimate the water content in 11 inches of heavy, wet snow.

52. *Wages* If a person working for an hourly wage earns $143 in 11 hours, how much will that person earn in 17 hours?

53. *Rolling Resistance of Cars* If you were to try and push a car, you would experience *rolling resistance*. This resistance equals the force necessary to keep the car moving slowly in neutral gear. The following table shows the rolling resistance R for passenger cars of different gross weights W. (*Source:* N. Garber and L. Hoel, *Traffic and Highway Engineering*.)

W (pounds)	2000	2500	3000	3500
R (pounds)	24	30	36	42

(a) Do the data represent direct or inverse variation? Explain.
(b) Find an equation that models the data. Graph the equation with the data.
(c) Estimate the rolling resistance of a 3200-pound car.

54. *Transportation Costs* The use of a particular toll bridge varies inversely according to the toll. When the toll is $0.75, 6000 vehicles are using the bridge. Estimate the number of users if the toll is $0.40. (*Source:* N. Garber.)

55. *Flow of Water* The gallons of water G flowing in 1 minute through a hose with a cross-sectional area A are shown in the table.

A (square inch)	0.2	0.3	0.4	0.5
G (gallons)	5.4	8.1	10.8	13.5

(a) Do the data represent direct or inverse variation? Explain.
(b) Find an equation that models the data. Graph the equation with the data.
(c) Interpret the constant of variation k.

56. *Hooke's Law* The accompanying table shows the distance D that a spring stretches when a weight W is hung on it.
(a) Do the data represent direct or inverse variation? Explain.
(b) Find an equation that models the data.
(c) How far will the spring stretch if an 11-pound weight is hung on it, as depicted in the accompanying figure?

W (pounds)	2	6	9	15
D (inches)	1.5	4.5	6.75	11.25

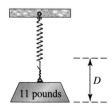

57. *Tightening Lug Nuts* (Refer to Example 4.) When a tire is mounted on a car, the lug nuts should not be over-tightened. The following table shows the maximum force used with wrenches of different lengths.
(a) Model the data, using the equation $F = \dfrac{k}{L}$.
(b) How much force should be used with a wrench 20 inches long?

L (inches)	8	10	16
F (pounds)	150	120	75

Source: Tires Plus.

58. *Cost of Tuition* (Refer to Example 3.) The cost of tuition is directly proportional to the number of credits taken. If 6 credits cost $435, find the cost of 11 credits. What does the constant of proportionality represent?

59. *Air Temperature and Altitude* In the first 6 miles of Earth's atmosphere, air cools as the altitude increases. The following graph shows the temperature change y in degrees Fahrenheit at an altitude of x miles. (*Source:* A. Miller and R. Anthes, *Meteorology.*)

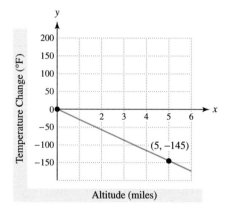

(a) Does this graph represent direct variation or inverse variation?
(b) Find an equation that models the data in the graph.
(c) Is the constant of proportionality k positive or negative? Interpret k.
(d) Estimate the change in air temperature 3.5 miles high.

60. *Ozone and UV Radiation* Ozone in the upper atmosphere filters out approximately 90% of the harmful ultraviolet (UV) rays from the sun. Depletion of the ozone layer has caused an increase in the amount of UV radiation reaching Earth's surface. An increase in UV radiation is associated with skin cancer. The following graph shows the percentage increase y in UV radiation for a decrease in the ozone layer of x percent.
(*Source:* R. Turner, D. Pearce, and I. Bateman, *Environmental Economics.*)

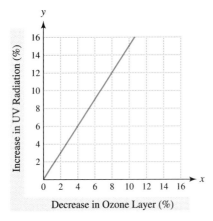

(a) Does this graph represent direct variation or inverse variation?
(b) Find an equation that models the data in the graph.
(c) Estimate the percentage increase in UV radiation if the ozone layer decreases by 7%.

61. *Electrical Resistance* The electrical resistance of a wire is directly proportional to its length. If a 35-foot long wire has a resistance of 2 ohms, find the resistance of a 25-foot long wire.

62. *Resistance and Current* The current that flows through an electrical circuit is inversely proportional to the resistance. When the resistance R is 150 ohms, the current I is 0.8 amp. Find the current when the resistance is 40 ohms.

63. *Joint Variation* The variable z varies jointly as the second power of x and the third power of y. Write a formula for z if $z = 31.9$ when $x = 2$ and $y = 2.5$.

64. *Wind Power* The electrical power generated by a windmill varies jointly with the square of the diameter of the area swept out by the blades and the cube of the wind velocity. If a windmill with an 8-foot diameter and a 10-mile per hour wind generates 2405 watts, how much power would be generated if the blades were 12 feet in diameter and the wind speed was 15 miles per hour?

65. *Strength of a Beam* (Refer to Example 6.) If a beam 5 inches wide and 3 inches thick supports 300 pounds, how much can a similar beam 5 inches wide and 2 inches thick support?

66. *Carpeting* The cost of carpet for a room varies jointly as its width and length. If a room 11 feet wide and 14 feet long costs $539 to carpet, find the cost to carpet a room 17 feet by 19 feet. Interpret the constant of variation k.

Writing About Mathematics

67. Explain in words what it means for a quantity y to be directly proportional to a quantity x.

68. Explain in words that it means for a quantity y to be inversely proportional to a quantity x.

CHECKING BASIC CONCEPTS FOR SECTIONS 7.3 AND 7.4

1. Simplify the expression.
   (a) $\dfrac{x}{x^2 - 1} + \dfrac{1}{x^2 - 1}$
   (b) $\dfrac{1}{x - 2} - \dfrac{3}{x}$

2. Simplify $\dfrac{3 - \dfrac{1}{x^2}}{3 + \dfrac{1}{x^2}}$.

3. If a gas pump can fill a 20-gallon tank in 4 minutes, how long will it take to fill a 42-gallon tank?

4. Determine whether the data represent direct variation or inverse variation. Find an equation that models the data.

   (a)
x	10	15	25	40
y	4	6	10	16

   (b)
x	5	10	15	20
y	24	12	8	6

7.5 DIVISION OF POLYNOMIALS

Division by a Monomial ~ Division by a Polynomial ~ Synthetic Division

INTRODUCTION

The study of polynomials has occupied the minds of mathematicians for centuries. During the sixteenth century, Italian mathematicians discovered how to solve higher degree polynomial equations. In this section we demonstrate symbolic methods for dividing polynomials. Division is often needed to factor higher degree polynomials and to solve polynomial equations. (Source: H. Eves, *An Introduction to the History of Mathematics*.)

DIVISION BY A MONOMIAL

To divide a polynomial by a monomial we use the two properties

$$\dfrac{a + b}{c} = \dfrac{a}{c} + \dfrac{b}{c} \quad \text{and} \quad \dfrac{a - b}{c} = \dfrac{a}{c} - \dfrac{b}{c}.$$

For example,

$$\dfrac{3x^2 + x}{x} = \dfrac{3x^2}{x} + \dfrac{x}{x}$$
$$= 3x + 1, \qquad x \neq 0.$$

Division by a Monomial

Exercises 7–10: Divide and check. Give numerical support for your result.

7. $\dfrac{4x-6}{2}$
8. $\dfrac{3x^4+x^2}{6x^2}$
9. $(4x^2-x+1)\div 2x^2$
10. $(5x^3-4x+2)\div 20x$

Exercises 11–16: Divide.

11. $\dfrac{9x^2-12x-3}{3}$
12. $\dfrac{50x^4+25x^2+100x}{25x}$
13. $(16x^3-24x)\div(12x)$
14. $(2y^4-4y^2+16)\div 2y^2$
15. $(a^2b^2-4ab+ab^2)\div ab$
16. $(10x^3y^2+5x^2y^3)\div(5x^2y^2)$

Division by a Polynomial

Exercises 17–20: Divide and check.

17. $\dfrac{3x^2-16x+21}{x-3}$
18. $\dfrac{4x^2-x-18}{x+2}$
19. $(6x^2-11x+4)\div(2x-3)$
20. $(10x^2-5)\div(x-1)$

Exercises 21–32: Divide.

21. $\dfrac{4x^3+8x^2-x-2}{x+2}$
22. $\dfrac{3x^3-4x^2+3x-2}{x-1}$
23. $\dfrac{x^3+3x-4}{x+4}$
24. $\dfrac{2x^3-x^2+5}{x-2}$
25. $(3x^3+8x^2-21x+7)\div(3x-1)$
26. $(14x^3+3x^2-9x+3)\div(7x-2)$
27. $(2a^4+5a^3-2a^2-5a)\div(2a+5)$
28. $(4b^4+10b^3-2b^2-3b+5)\div(2b+1)$
29. $\dfrac{3x^3+4x^2-12x-16}{x^2-4}$
30. $\dfrac{2x^4-x^3-5x^2+4x-12}{2x^2-x+3}$
31. $(2a^4-3a^3+14a^2-8a+10)\div(a^2-a+5)$
32. $(2z^3+3z^2-9z-12)\div(z^2-1)$

Synthetic Division

Exercises 33–44: Use synthetic division to divide.

33. $\dfrac{x^2+3x-1}{x-1}$
34. $\dfrac{2x^2+x-1}{x-3}$
35. $(3x^2-22x+7)\div(x-7)$
36. $(5x^2+29x-6)\div(x+6)$
37. $\dfrac{x^3+7x^2+14x+8}{x+4}$
38. $\dfrac{2x^3+3x^2+2x+4}{x+1}$
39. $\dfrac{2x^3+x^2-1}{x-2}$
40. $\dfrac{x^3+x-2}{x+3}$
41. $(x^3-2x^2-2x+4)\div(x-4)$
42. $(2x^4+3x^2-4)\div(x+2)$
43. $(b^4-1)\div(b-1)$
44. $(a^2+a)\div(a+2.5)$

Writing About Mathematics

45. A student dividing $2x^3+5x^2-13x+5$ by $2x-1$ does the following.

    $$\begin{array}{r|rrrr} 1 & 2 & 5 & -13 & 5 \\ & & 2 & 7 & -6 \\ \hline & 2 & 7 & -6 & -1 \end{array}$$

    What would you tell the student?

46. If you add, subtract, or multiply two polynomials is the result a polynomial? If you divide two polynomials is the result always a polynomial? Explain.

CHECKING BASIC CONCEPTS FOR SECTION 7.5

1. Divide and simplify.
   (a) $\dfrac{2x - x^2}{x^2}$
   (b) $\dfrac{6a^2 - 9a + 15}{3a}$

2. Divide $\dfrac{10x^2 - x + 4}{2x + 1}$.

3. Use synthetic division to divide
   $$2x^3 - 5x^2 - 1$$
   by each expression.
   (a) $x - 1$
   (b) $x + 2$

Chapter 7 Summary

Section 7.1 *Rational Functions and Equations*

Any rational function f may be represented symbolically by
$$f(x) = \frac{p(x)}{q(x)},$$
where $p(x)$ and $q(x)$ are polynomials. Examples include
$$f(x) = \frac{1}{x}, \qquad g(x) = \frac{3x - 5}{8x + 1}, \quad \text{and} \quad h(x) = \frac{x^2 + 5x - 7}{6x}.$$
Vertical asymptotes in the graph of a rational function f occur at x-values where the denominator of $f(x)$ is 0 but the numerator is nonzero. The graph of $f(x) = \dfrac{2x}{x - 2}$ is shown in the accompanying figure. It has a vertical asymptote at $x = 2$. The graph of a rational function never crosses a vertical asymptote.

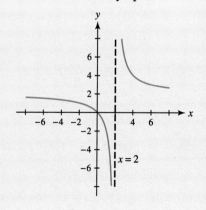

If an equation contains a rational expression, it is called a rational equation and can be solved by multiplying both sides by the LCD. Be sure to check all answers.

Example:

$$\frac{2}{x} + 1 = \frac{4-x}{x} \quad \text{Given equation}$$

$$x \cdot \left(\frac{2}{x} + 1\right) = \left(\frac{4-x}{x}\right) \cdot x \quad \text{Multiply by } x.$$

$$2 + x = 4 - x \quad \text{Clear fractions.}$$

$$2x = 2 \quad \text{Add } x \text{ and subtract 2.}$$

$$x = 1 \quad \text{Divide by 2.}$$

Section 7.2 Multiplication and Division of Rational Expressions

To multiply or divide two rational expressions we use the following two properties.

$$\frac{A}{B} \cdot \frac{C}{D} = \frac{A \cdot C}{B \cdot D} \quad \text{Multiplication of rational expressions}$$

$$\frac{A}{B} \div \frac{C}{D} = \frac{A}{B} \cdot \frac{D}{C} = \frac{A \cdot D}{B \cdot C} \quad \text{Division of rational expressions}$$

Examples:

$$\frac{x}{x-1} \cdot \frac{5x^2}{x+1} = \frac{5x^3}{x^2-1}$$

$$\frac{2}{x} \div \frac{x}{2x+3} = \frac{2}{x} \cdot \frac{2x+3}{x} = \frac{4x+6}{x^2}$$

Section 7.3 Addition and Subtraction of Rational Expressions

Before we can add two rational expressions they must have a common denominator. If the expressions have a common denominator, we can use the property

$$\frac{A}{C} + \frac{B}{C} = \frac{A+B}{C}.$$

Subtraction is performed similarly.

Examples:

$$\frac{2x}{2x+1} + \frac{3x}{2x+1} = \frac{5x}{2x+1}$$

$$\frac{2}{x+1} - \frac{1}{x-1} = \frac{2(x-1)}{(x+1)(x-1)} - \frac{(x+1)}{(x+1)(x-1)} = \frac{x-3}{(x+1)(x-1)}$$

Section 7.4 Modeling with Proportions and Variation

Two quantities x and y are directly proportional (or vary directly) if they can be modeled by $y = kx$, where k is the constant of proportionality. If x doubles, y also doubles.

Two quantities are inversely proportional (or vary inversely) if they can be modeled by $y = \frac{k}{x}$. In this case, if x doubles, y is halved.

A quantity z varies jointly with x and y if $z = kxy$ for some constant of variation k.

Section 7.5 Division of Polynomials

Division of polynomials may be performed similarly to division of natural numbers. Synthetic division is a fast way to divide a polynomial by an expression in the form $x - k$, where k is a fixed number.

Example: Divide $3x^3 + 4x^2 - 7x - 1$ by $x + 2$.

$$\begin{array}{r|rrrr} -2 & 3 & 4 & -7 & -1 \\ & & -6 & 4 & 6 \\ \hline & 3 & -2 & -3 & 5 \end{array}$$

The quotient is $3x^2 - 2x - 3$ with remainder 5, which can be written as

$$3x^2 - 2x - 3 + \frac{5}{x + 2}.$$

Chapter 7 Review Exercises

Section 7.1

Exercises 1 and 2: Use the verbal description of a rational function f to find symbolic, graphical, and numerical representations. For the numerical representation let $x = -3, -2, -1, \ldots, 3$.

1. Divide 1 by the quantity x minus 1.

2. Subtract 3 from x and then divide the result by x.

3. Use your graphing calculator as an aid to sketch a graph of $f(x) = \dfrac{1}{2x + 4}$. Show any vertical asymptotes as dashed lines.

4. Evaluate $f(x) = \dfrac{1}{x^2 - 1}$ at $x = 3$. Find any x-values that are not in the domain of f.

Exercises 5 and 6: Use the graph to evaluate the expressions. Give the equation of any vertical asymptotes.

5. $f(0)$ and $f(2)$

6. $f(-3)$ and $f(-2)$

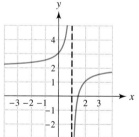
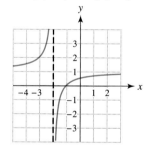

Exercises 7 and 8: Give a numerical representation of $f(x)$ by letting $x = -3, -2, -1, \ldots, 3$. Then evaluate $f(x)$ at the given values of x.

7. $f(x) = \dfrac{3}{x + 2}$, $x = -3, x = 2$

8. $f(x) = \dfrac{2x}{x^2 - 4}$, $x = -2, x = 3$

Section 7.2

Exercises 9–12: Simplify the expression.

9. $\dfrac{4a}{6a^4}$

10. $\dfrac{(x-3)(x+2)}{(x+1)(x-3)}$

11. $\dfrac{x^2-4}{x-2}$

12. $\dfrac{x^2-6x-7}{2x^2-x-3}$

Exercises 13–20: Simplify the expression.

13. $\dfrac{1}{2y} \cdot \dfrac{4y^2}{8}$

14. $\dfrac{x+2}{x-5} \cdot \dfrac{x-5}{x+1}$

15. $\dfrac{x^2+1}{x^2-1} \cdot \dfrac{x-1}{x+1}$

16. $\dfrac{x^2+2x+1}{x^2-9} \cdot \dfrac{x+3}{x+1}$

17. $\dfrac{1}{3y} \div \dfrac{1}{9y^4}$

18. $\dfrac{2x+2}{3x-3} \div \dfrac{x+1}{x-1}$

19. $\dfrac{x^2+2x}{x^2-25} \div \dfrac{x+2}{x+5}$

20. $\dfrac{x^2+2x-15}{x^2+4x+3} \div \dfrac{x-3}{x+1}$

Section 7.3

Exercises 21–26: Simplify.

21. $\dfrac{1}{x+4} + \dfrac{3}{x+4}$

22. $\dfrac{2}{x} + \dfrac{x-3}{x^2}$

23. $\dfrac{1}{x+1} - \dfrac{x}{(x+1)^2}$

24. $\dfrac{2x}{x^2-4} - \dfrac{2}{x-2}$

25. $\dfrac{5x}{x-2} + \dfrac{x-4}{x}$

26. $\dfrac{x}{x+6} - \dfrac{2x}{x+1} + \dfrac{1}{x+6}$

Exercises 27–32: Solve the rational equation symbolically. Support your results either graphically or numerically.

27. $\dfrac{3-x}{x} = 2$

28. $\dfrac{1}{x+3} = \dfrac{1}{5}$

29. $\dfrac{2}{x^2} - \dfrac{1}{x} = 1$

30. $\dfrac{1}{x+4} - \dfrac{1}{x} = 1$

31. $\dfrac{1}{x^2-1} - \dfrac{1}{x-1} = \dfrac{2}{3}$

32. $\dfrac{1}{x-3} + \dfrac{1}{x+3} = \dfrac{-5}{x^2-9}$

Exercises 33–36: Simplify the complex fraction.

33. $\dfrac{2 + \dfrac{3}{x}}{2 - \dfrac{3}{x}}$

34. $\dfrac{\dfrac{1}{x} + \dfrac{1}{2}}{\dfrac{1}{4} - \dfrac{2}{x}}$

35. $\dfrac{\dfrac{4}{x} + \dfrac{1}{x-1}}{\dfrac{1}{x} - \dfrac{2}{x-1}}$

36. $\dfrac{\dfrac{1}{x+3} - \dfrac{1}{x-3}}{\dfrac{4}{x+3} - \dfrac{2}{x-3}}$

Section 7.4

Exercises 37–40: Solve the proportion.

37. $\dfrac{x}{6} = \dfrac{6}{20}$

38. $\dfrac{11}{x} = \dfrac{5}{7}$

39. $\dfrac{x+1}{5} = \dfrac{x}{3}$

40. $\dfrac{3}{7} = \dfrac{4}{x-1}$

41. If $x+1$ is to 5 as 10 is to 15, use a proportion to find x.

42. A rectangle has sides with lengths 7 and 8. Find the longer side of a similar rectangle whose shorter side has length 11.

43. Suppose that y varies directly as x. If $y = 8$ when $x = 2$, find y when $x = 7$.

44. Suppose that y varies inversely as x. If $y = 5$ when $x = 10$, find y when $x = 25$.

45. Suppose that z varies jointly with x and y. If $z = 483$ when $x = 23$ and $y = 7$, find the constant of variation k.

46. Suppose that z varies jointly with x and the square of y. If $z = 891$ when $x = 22$ and $y = 3$, find z when $x = 10$ and $y = 4$.

CHAPTER 7 Review Exercises 461

Exercises 47 and 48: Use the table to determine whether the data represent direct or inverse variation. Find an equation that models the data.

47.
x	2	4	5	8
y	100	50	40	25

48.
x	3	7	8	11
y	9	21	24	33

Exercises 49–50: Determine whether the graph represents direct or inverse variation. Find the constant of variation k.

49.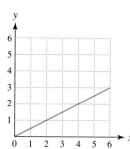

50.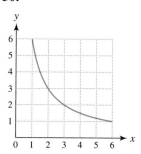

SECTION 7.5

Exercises 51–58: Divide and check.

51. $\dfrac{10x + 15}{5}$

52. $\dfrac{2x^2 + x}{x}$

53. $(4x^3 - x^2 + 2x) \div 2x^2$

54. $(4a^3b - 6ab^2) \div (2a^2b^2)$

55. $\dfrac{2x^2 - x - 2}{x + 1}$

56. $(x^3 + x - 1) \div (x + 1)$

57. $(6x^3 - 7x^2 + 5x - 1) \div (3x - 2)$

58. $\dfrac{2x^3 - 9x^2 + 21x - 21}{x^2 - 2}$

Exercises 59–60: Use synthetic division to divide.

59. $\dfrac{2x^2 - 11x + 13}{x - 3}$

60. $(3x^3 + 10x^2 - 4x + 19) \div (x + 4)$

APPLICATIONS

61. *Downhill Highway Grade* The braking distance D for a car traveling downhill at 40 miles per hour on a wet grade x is given by
$$D(x) = \dfrac{1600}{9.6 - 30x}.$$
(Source: N. Garber.)
(a) In this formula a level road is represented by $x = 0$ and a 10% downhill grade is represented by $x = 0.1$. Evaluate $D(0)$ and $D(0.1)$. How much does the downhill grade add to the braking distance compared to a level road?
(b) If the braking distance is 200 feet, estimate the downhill grade x.

62. *Time Spent in Line* Suppose that amusement park attendants can wait on 10 vehicles per minute and vehicles are arriving at the park randomly at an average rate of x vehicles per minute. Then the average time T in minutes spent waiting in line and paying the attendant is given by
$$T(x) = \dfrac{1}{10 - x},$$
where $x < 10$. (Source: N. Garber and L. Hoel, *Traffic and Highway Engineering*.)
(a) Construct a table of T, starting at $x = 9$ and incrementing by 0.1.
(b) What happens to the waiting time as x approaches 10? Interpret this result.

63. *Number of Cars Waiting* A car wash can clean 15 cars per hour. If cars are arriving randomly at an average rate of x per hour, the average number N of cars waiting in line is given by
$$N(x) = \dfrac{x^2}{225 - 15x},$$
where $x < 15$. (Source: N. Garber.)
(a) Estimate the average length of a line when 14 cars per hour are arriving.
(b) Graph N in [0, 15, 5] by [0, 10, 5].
(c) Interpret the graph as x increases. Does this agree with your intuition?

64. *Fish Population* Suppose that a fish population P in thousands found in a lake is modeled by

$$P(x) = \frac{5}{x^2 + 1},$$

where $x \geq 0$ is time in years.
(a) Evaluate $P(0)$ and interpret the result.
(b) Graph P in [0, 6, 1] by [0, 6, 1].
(c) What happens to the population over this 6-year period?
(d) When was the population 1000?

65. *Working Together* Suppose that two students are working collaboratively to solve a large number of quadratic equations with the quadratic formula. The first student can solve all the problems in 2 hours, whereas the second student can solve them in 3 hours.
(a) Write an equation whose solution gives the time needed to solve all the problems if the two students split the problems up and share answers.
(b) Solve the equation in part (a) symbolically.
(c) Solve the equation in part (a) either graphically or numerically.

66. *Speed of a Boat* In still water a riverboat can travel 12 miles per hour. It travels 48 miles upstream and then 48 miles downstream in a total time of 9 hours. Find the speed of the current.

67. *Recording Music* A 650-megabyte CD can record 74 minutes of music. Estimate the number of minutes that can be recorded on 387 megabytes.

68. *Height of a Building* A 5–foot tall person casts a 3–foot long shadow, while a nearby building casts a 26–foot long shadow, as illustrated in the accompanying figure. Find the height of the building.

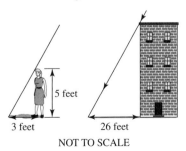

NOT TO SCALE

69. *Light Bulbs* The approximate resistances R for light bulbs of wattage W are measured and recorded in the following table. (*Source:* **D. Horn, Basic Electronics Theory.**)

W (watts)	50	100	200	250
R (ohms)	242	121	60.5	48.4

(a) Make a scatterplot of the data. Do the data represent direct or inverse variation?
(b) Find an equation that models the data. What is the constant of proportionality k?
(c) Find R for a 55-watt light bulb.

70. *Scales* The distance D that the spring in a produce scale stretches is directly proportional to the weight W of the fruits and vegetables placed in the pan.

(a) If 5 pounds of apples stretch the spring 0.3 inch, find an equation that relates D and W. What is the constant of proportionality k?
(b) How far will 7 pounds of oranges stretch the spring?

71. *Air Temperature and Altitude* In the first 30,000 feet of Earth's atmosphere the *change* in air temperature is directly proportional to the altitude. If the temperature is 80°F on the ground and 62°F 4000 feet above the ground, find the air temperature at 6000 feet. (*Source:* L. Battan, *Weather in Your Life.*)

72. *Skin Cancer and UV Radiation* Depletion of the ozone layer has caused an increase in the amount of UV radiation reaching Earth's surface. The following table shows the estimated percentage increase in skin cancer y for an x percent increase in the amount of UV radiation reaching Earth's surface. (*Source:* R. Turner, D. Pearce, and I. Bateman, *Environmental Economics.*)

x (%)	0	1	2	3	4
y (%)	0	3.5	7	10.5	14

(a) Do these data represent direct variation, inverse variation, or neither?
(b) Find an equation that models the data in the table.
(c) Estimate the percentage increase in skin cancer if UV radiation increases by 2.3%.

CHAPTER 7
TEST

1. Use the verbal description "Divide x by the quantity x plus 2" to find symbolic, graphical, and numerical representations of the rational function f. For the numerical representation let $x = -3, -2, -1, \ldots, 3$, and for the graphical representation use the viewing rectangle $[-4.7, 4.7, 1]$ by $[-4.7, 4.7, 1]$.

2. Use $f(x)$ in Exercise 1 to answer the following.
   (a) Find any x-values that are not in the domain of f.
   (b) Evaluate $f(-1)$.

3. Use your graphing calculator as an aid to sketch a graph of $f(x) = \dfrac{1}{x^2 - 1}$. Show any vertical asymptotes as dashed lines.

4. Use factoring to simplify $\dfrac{x^2 - 2x - 15}{2x^2 - x - 21}$.

Exercises 5–8: Simplify.

5. $\dfrac{x^2 + 4}{x^2 - 4} \cdot \dfrac{x - 2}{x + 2}$

6. $\dfrac{1}{4y^2} \div \dfrac{1}{8y^4}$

7. $\dfrac{1}{z + 4} - \dfrac{z}{(z + 4)^2}$

8. $\dfrac{\dfrac{1}{x - 2} + \dfrac{x}{x - 2}}{\dfrac{1}{3} - \dfrac{5}{x - 2}}$

Exercises 9–10: Solve the rational equation symbolically. Support your results either graphically or numerically.

9. $\dfrac{t}{5t + 1} = \dfrac{2}{7}$

10. $\dfrac{1}{x^2 - 4} - \dfrac{1}{x - 2} = \dfrac{1}{x + 2}$

11. A triangle has sides with lengths 12, 15, and 20. Find the longest side of a similar triangle with a shortest side of length 7.

12. Suppose that y varies directly as x. If $y = 8$ when $x = 23$, find y when $x = 10$.

13. Use the table to determine whether the data represent direct or inverse variation. Find an equation that models the data.

x	2	4	5	10
y	50	25	20	10

14. Determine whether the data represent direct or inverse variation. Find an equation that models the data.

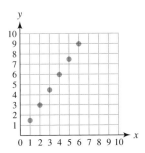

Exercises 15 and 16: Divide and check.

15. $\dfrac{4a^3 + 10a}{2a}$

16. $\dfrac{3x^3 + 5x^2 - 2}{x + 2}$

17. *Height of a Building* A person 73 inches tall casts a shadow 50 inches long while a nearby building casts a shadow 15 feet long. Find the height of the building.

18. *Time Spent in Line* Suppose that parking lot attendants can wait on 25 vehicles per minute and vehicles are arriving at the lot randomly at an average rate of x vehicles per minute. Then the average time T in hours spent waiting in line and paying the attendant is given by

$$T(x) = \dfrac{1}{25 - x},$$

where $x < 25$. (*Source:* N. Garber and L. Hoel, *Traffic and Highway Engineering.*)

(a) Graph T in $[0, 25, 5]$ by $[0, 2, 0.5]$. Identify any vertical asymptotes.

(b) If the wait is 1 minute, how many vehicles are arriving on average?

19. *Working Together* Suppose that one pump can empty a pool in 24 hours, and a second pump can empty the pool in 30 hours.

(a) Write an equation whose solution gives the time needed for the pumps working together to empty the pool.

(b) Solve the equation in part (a) symbolically.

20. *Dew Point and Altitude* In the first 30,000 feet of Earth's atmosphere the change in the dew point is directly proportional to the altitude. If the dew point is 50°F on the ground and 39°F 10,000 feet above the ground, find the dew point at 7500 feet. (*Source:* L. Battan, *Weather in Your Life.*)

CHAPTER 7

EXTENDED AND DISCOVERY EXERCISES

RATIONAL APPROXIMATIONS

1. Rational expressions are used to approximate other types of expressions in computer software. Graph each expression on the left for $1 \leq x \leq 15$ and try to match it with the graph of the rational expression on the right that approximates it best.

(a) $\sqrt{x}$

(b) $\sqrt{4x + 1}$

(c) $\sqrt[3]{x}$

(d) $\dfrac{1 - \sqrt{x}}{1 + \sqrt{x}}$

i. $\dfrac{2 - 2x^2}{3x^2 + 10x + 3}$

ii. $\dfrac{15x^2 + 75x + 33}{x^2 + 23x + 31}$

iii. $\dfrac{10x^2 + 80x + 32}{x^2 + 40x + 80}$

iv. $\dfrac{7x^3 + 42x^2 + 30x + 2}{2x^3 + 30x^2 + 42x + 7}$

OTHER TYPES OF DIRECT VARIATION

Exercises 2–6: Sometimes a quantity y varies directly as a power of x. For example, the area A of a circle varies directly as the second power of the radius r, since $A = \pi r^2$ and π is a constant. Let x and y

denote two quantities and n be a positive number. Then y is directly proportional to the nth power of x, or y varies directly as the nth power of x, if there exists a nonzero number k such that

$$y = kx^n.$$

2. Let y be directly proportional to the second power of x. If x doubles, what happens to y?

3. Let y be directly proportional to the $\frac{1}{2}$ power of x. If x quadruples, what happens to y?

4. *Modeling a pendulum* The time T required for a pendulum to swing back and forth once is called its period. See the accompanying figure. The length L of a pendulum is directly proportional to the nth power of T for some positive integer n. The accompanying table lists the period T for various lengths L.

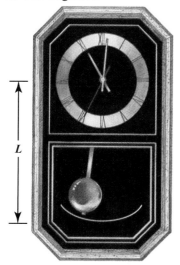

(a) Find the constant of proportionality k and the value of n. (*Hint:* If $L = kT^n$ for constants k and n, the ratio $\frac{L}{T^n} = k$.)

(b) Predict T for a pendulum having a length of 2.2 feet.

T (seconds)	1.11	1.36	1.57	1.76
L (feet)	1.0	1.5	2.0	2.5

T (seconds)	1.92	2.08	2.22
L (feet)	3.0	3.5	4.0

5. *Allometric Growth* If x is the weight of a fiddler crab and y is the weight of its claws, then y is directly proportional to the 1.25 power of x; that is, $y = kx^{1.25}$. Suppose that a typical crab with a body weight of 2.1 grams has claws weighing 1.125 grams. (*Source:* D. Brown and P. Rothery, *Models in Biology: Mathematics, Statistics, and Computing.*)

(a) Find the constant of proportionality k. Round your answer to the nearest thousandth.

(b) Estimate the weight of a fiddler crab with claws weighing 0.8 gram.

6. *Volume* The volume V of a cylinder is directly proportional to the square of its radius r. If a cylinder with a radius of 5 inches has a volume of 150 cubic inches, what is the volume of a cylinder with the same height and a radius of 7 inches?

OTHER TYPES OF INVERSE VARIATION

Exercises 7–10: Sometimes a quantity varies inversely to a power of x. Let x and y denote two quantities and n be a positive number. Then y is inversely proportional to the nth power of x, or y varies inversely as the nth power of x, if there exists a nonzero number k such that

$$y = \frac{k}{x^n}.$$

7. Let y be inversely proportional to the second power of x. If x doubles, what happens to y?

8. Let y be inversely proportional to the third power of x. If x doubles, what happens to y?

9. *Earth's Gravity* The weight W of an object can be modeled by $W = \frac{k}{d^2}$, where d is the distance that the object is from Earth's center and k is a constant. Earth's radius is about 4000 miles.

(a) Find k for a person who weighs 200 pounds on Earth's surface.

(b) Graph W in a convenient viewing rectangle. At what distance from Earth's center is this person's weight 50 pounds?

(c) How far from the center of Earth would an object be if its weight were 1% of its weight on the surface of Earth?

10. *Modeling Brightness* Inverse variation occurs when the intensity of a light is measured. If you increase your distance from a light bulb, the intensity of the light decreases. Intensity I is inversely proportional to the second power of the distance d. The equation $I = \dfrac{k}{d^2}$ models this phenomenon. The following table gives the intensity of a 100-watt light bulb at various distances. (*Source:* R. Weidner and R. Sells, *Elementary Classical Physics, Vol. 2.*)

d (meters)	0.5	2	3	4
I (watts/ square meter)	31.68	1.98	0.88	0.495

(a) Find the constant of proportionality k.
(b) Graph I in [0, 5, 1] by [0, 30, 5]. What happens to the intensity as the distances increase?
(c) If the distance from the light bulb doubles, what happens to the intensity?
(d) Determine d when $I = 1$ watt per square meter.

CHAPTER 8
RADICAL EXPRESSIONS AND FUNCTIONS

Mathematics can be both applied and theoretical. A common misconception is that theoretical mathematics is unimportant. Many new ideas that have great practical importance in today's world were first developed as abstract concepts. For example, in 1854 George Boole published *Laws of Thought*, which outlined the basis for Boolean algebra. That was 85 years before the invention of the first digital computer. Boolean algebra became the basis for operation of modern computer hardware and, without it, we would not have digital computers.

In this chapter we describe a new number system called *complex numbers*, which involve square roots of negative numbers. The Italian mathematician Cardano (1501–1576) was one of the first mathematicians to work with complex numbers and called them useless. Rene Descartes (1596–1650) originated the term *imaginary number*, which is associated with complex numbers. However, today complex numbers are used in many applications, such as electricity, fiber optics, and the design of airplanes. We are privileged to read in a period of hours what took people centuries to discover.

> *Bear in mind that the wonderful things you learn in schools are the work of many generations, produced by enthusiastic effort and infinite labor in every country.*
> —Albert Einstein

Source: *Historical Topics for the Mathematics Classroom, Thirty-first Yearbook*, NCTM.

8.1 RADICAL EXPRESSIONS AND RATIONAL EXPONENTS

Radical Notation ~ Rational Exponents ~
Properties of Rational Exponents

INTRODUCTION

Cellular phone technology has become a part of everyday life. In order to have cellular phone coverage transmission towers, or cellular sites, are spread throughout a region. To estimate the minimum broadcasting distance for each cellular site, radical expressions are needed. In this section we discuss radical expressions and rational exponents and show how to manipulate them symbolically. (*Source:* **C. Smith,** *Practical Cellular & PCS Design.*)

RADICAL NOTATION

Recall the definition of the square root of a number a.

Square Root

The number b is a *square root* of a if $b^2 = a$.

EXAMPLE 1 *Finding square roots*

Find the square roots of 100.

Solution

The square roots of 100 are 10 and -10 because $10^2 = 100$ and $(-10)^2 = 100$.

Every positive number a has two square roots, one positive and one negative. Recall that the *positive* square root is called the *principal square root* and is denoted $\sqrt{a}$. The **negative square root** is denoted $-\sqrt{a}$. To identify both square roots we write $\pm\sqrt{a}$. The symbol $\sqrt{}$ is called the **radical sign.** The expression under the radical sign is called the **radicand,** and an expression containing a radical sign is called a **radical expression.** Examples of radical expressions include

$$\sqrt{6}, \quad 5 + \sqrt{x + 1}, \quad \text{and} \quad \sqrt{\frac{3x}{2x - 1}}.$$

EXAMPLE 2 *Finding principal square roots*

Find the principal square roots of each expression.

(a) 25
(b) 17
(c) $c^2, \quad c > 0$

8.1 Radical Expressions and Rational Exponents

Solution

(a) The principal, or positive, square root of 25 is $\sqrt{25} = 5$.
(b) The principal square root of 17 is $\sqrt{17}$. This value is not an integer, but we can approximate it. Figure 8.1 shows that $\sqrt{17} \approx 4.12$. Note that calculators do not give exact answers when approximating many radical expressions.
(c) The principal square root of c^2 is $\sqrt{c^2} = c$, as c is positive.

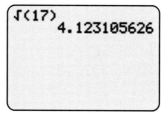

Figure 8.1

Technology Note *Square Roots*

The square root of a negative number is not a real number. Many calculators display ERROR when square roots of negative numbers are input for calculation. Figures 8.2 and 8.3 illustrate this case. Other calculators give a complex number for the answer. We discuss complex numbers in Section 8.5.

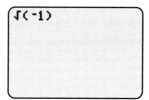

Figure 8.2

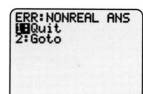

Figure 8.3

Another common radical expression is the cube root of a number a, denoted $\sqrt[3]{a}$.

Cube Root

The number b is a *cube root* of a if $b^3 = a$.

Although the square root of a negative number is not a real number, the cube root of a negative number is a negative real number. Every real number has one cube root.

EXAMPLE 3 *Finding cube roots*

GCLM Find the cube root of each expression.
(a) 8   (b) -27 (c) 16   (d) d^6

Solution

(a) $\sqrt[3]{8} = 2$ because $2^3 = 2 \cdot 2 \cdot 2 = 8$.
(b) $\sqrt[3]{-27} = -3$ because $(-3)^3 = (-3) \cdot (-3) \cdot (-3) = -27$.
(c) $\sqrt[3]{16}$ is not an integer. Figure 8.4 shows that $\sqrt[3]{16} \approx 2.52$.
(d) $\sqrt[3]{d^6} = d^2$ because $(d^2)^3 = d^2 \cdot d^2 \cdot d^2 = d^{2+2+2} = d^6$.

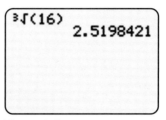

Figure 8.4

In the next example we use the principal square root to estimate the minimum transmission distance for a cellular site.

EXAMPLE 4 *Estimating cellular phone transmission distance*

If the ground is level, a cellular transmission tower will broadcast its signal in roughly a circular pattern, whose radius can be altered by changing the strength of its signal. See Figure 8.5. Suppose that a city has an area of 50 square miles and 10 transmission towers that are spread evenly throughout the city. Estimate a *minimum* transmission radius R for each tower. (Note that a larger distance would probably be necessary to adequately cover the city.) (*Source:* C. Smith.)

Figure 8.5

Solution

The circular area A covered by one transmission tower is $A = \pi R^2$. The total area covered by 10 towers is $10\pi R^2$, which must equal at least 50 square miles.

$$10\pi R^2 = 50$$
$$\pi R^2 = 5 \qquad \text{Divide by 10.}$$
$$R^2 = \frac{5}{\pi} \qquad \text{Divide by } \pi.$$
$$R = \sqrt{\frac{5}{\pi}} \approx 1.26 \qquad \text{Take principal square root.}$$

Each transmission tower must broadcast with a minimum radius of approximately 1.26 miles.

We can generalize square roots and cube roots to include the *n*th root of a number *a*. The number *b* is an **nth root** of *a* if $b^n = a$, where *n* is a positive integer, and is denoted $\sqrt[n]{a} = b$. The number *n* is called the **index**. For the square root the index is 2, although we usually write $\sqrt{a}$ rather than $\sqrt[2]{a}$. When *n* is odd, we are finding an **odd root**, and when *n* is even, we are finding an **even root**. The square root $\sqrt{a}$ is an example of an even root, and the cube root $\sqrt[3]{a}$ is an example of an odd root. An odd root of a negative number is a negative number, but the even root of a negative number is *not* a real number.

EXAMPLE 5 *Finding nth roots*

Find each root, if possible.

(a) $\sqrt[4]{16}$
(b) $\sqrt[5]{-32}$
(c) $\sqrt[4]{-81}$
(d) $\sqrt{(x-2)^2}, \quad x - 2 > 0$

Solution

(a) $\sqrt[4]{16} = 2$ because $2^4 = 2 \cdot 2 \cdot 2 \cdot 2 = 16$.
(b) $\sqrt[5]{-32} = -2$ because $(-2)^5 = (-2) \cdot (-2) \cdot (-2) \cdot (-2) \cdot (-2) = -32$.
(c) The even root of a negative number is not a real number.
(d) $\sqrt{(x-2)^2} = x - 2$ because $x - 2 > 0$.

Consider the calculations

$$\sqrt{3^2} = \sqrt{9} = 3, \qquad \sqrt{(-4)^2} = \sqrt{16} = 4, \quad \text{and} \quad \sqrt{(-6)^2} = \sqrt{36} = 6.$$

In general, the expression $\sqrt{x^2}$ equals $|x|$. Graphical support is shown in Figures 8.6 and 8.7, where the graphs of $Y_1 = \sqrt{(X^\wedge 2)}$ and $Y_2 = \text{abs}(X)$ appear to be identical. Numerical support is shown in Figure 8.8, where $Y_1 = Y_2$ for every value of *x*.

[−6, 6, 1] by [−4, 4, 1]

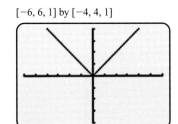

Figure 8.6 $y_1 = \sqrt{x^2}$

[−6, 6, 1] by [−4, 4, 1]

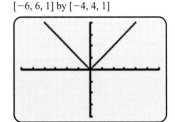

Figure 8.7 $y_2 = |x|$

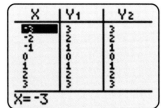

Figure 8.8 $y_1 = y_2$

> ### The Expression $\sqrt{x^2}$
> For every real number *x*, $\sqrt{x^2} = |x|$.

Rational Exponents

We now define values for expressions having fractional exponents, such as $3^{1/2}$. When m and n are integers, we know that

$$a^n \cdot a^m = a^{n+m}. \quad \text{Add exponents.}$$

If we extend this rule to include fractions,

$$3^{1/2} \cdot 3^{1/2} = 3^{1/2+1/2} = 3^1 = 3.$$

Since $\sqrt{3} \cdot \sqrt{3} = 3$, it suggests that $3^{1/2} = \sqrt{3}$ and motivates the following definition.

The Expression $a^{1/n}$

If n is an integer greater than 1, then

$$a^{1/n} = \sqrt[n]{a}.$$

Note: If $a < 0$ and n is an even positive integer, then $a^{1/n}$ is not a real number.

EXAMPLE 6 *Interpreting rational exponents*

Write each expression in radical notation. Then evaluate the expression to the nearest hundredth when appropriate.

(a) $36^{1/2}$ **(b)** $23^{1/5}$

Solution

(a) The exponent $\frac{1}{2}$ indicates a square root. Thus $36^{1/2} = \sqrt{36}$, which evaluates to 6.

(b) The exponent $\frac{1}{5}$ indicates a fifth root. Thus $23^{1/5} = \sqrt[5]{23}$, which is not an integer.

Figure 8.9 shows this expression approximated in both exponential and radical notation. In either case $23^{1/5} \approx 1.87$.

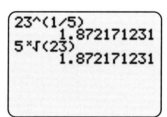

Figure 8.9

Suppose that we want to define the expression $8^{2/3}$. On the one hand, using properties of exponents we have

$$8^{1/3} \cdot 8^{1/3} = 8^{(1/3+1/3)} = 8^{2/3}.$$

On the other hand, we have
$$8^{1/3} \cdot 8^{1/3} = \sqrt[3]{8} \cdot \sqrt[3]{8} = 2 \cdot 2 = 4.$$
Thus $8^{2/3} = 4$, and that value is obtained whether we interpret $8^{2/3}$ as either
$$8^{2/3} = (8^{1/3})^2 = (\sqrt[3]{8})^2 = 2^2 = 4$$
or
$$8^{2/3} = (8^2)^{1/3} = \sqrt[3]{8^2} = \sqrt[3]{64} = 4.$$
This result leads to the following definition.

The Expression $a^{m/n}$

If m and n are positive integers with $\dfrac{m}{n}$ in lowest terms, then
$$a^{m/n} = \sqrt[n]{a^m} = (\sqrt[n]{a})^m.$$

Note: If $a < 0$ and n is an even integer, then $a^{m/n}$ is not a real number.

EXAMPLE 7 *Interpreting rational exponents*

Write each expression in radical notation. Then evaluate the expression to the nearest hundredth when appropriate.
(a) $(-27)^{2/3}$ **(b)** $12^{3/5}$

Solution

(a) The exponent $\dfrac{2}{3}$ indicates that we either take the cube root of -27 and then square it or that we square -27 and then take the cube root. In either case the result will be the same. Thus
$$(-27)^{2/3} = (\sqrt[3]{-27})^2 = (-3)^2 = 9 \quad \text{or} \quad (-27)^{2/3} = \sqrt[3]{(-27)^2} = \sqrt[3]{729} = 9,$$
as shown in the first two lines in Figure 8.10.

(b) The exponent $\dfrac{3}{5}$ indicates that we either take the fifth root of 12 and then cube it or that we cube 12 and then take the fifth root. Thus
$$12^{3/5} = \sqrt[5]{12^3} \quad \text{or} \quad 12^{3/5} = (\sqrt[5]{12})^3.$$
In either case we need a calculator to approximate the expression, $12^{3/5} \approx 4.44$, as shown in the last two lines of Figure 8.10.

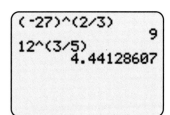

Figure 8.10

Technology Note Rational Exponents

When evaluating expressions with rational (fractional) exponents, be sure to put parentheses around the fraction. For example, for the term $8^{2/3}$ most calculators will evaluate 8^(2/3) and 8^2/3 differently. Figure 8.11 shows evaluation of the term input correctly, 8^(2/3), as 4 but shows evaluation of the term input incorrectly, 8^2/3, as $\frac{64}{3} \approx 21.\overline{3}$.

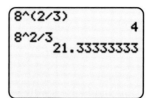

Figure 8.11

From properties of exponents we know that $a^{-n} = \frac{1}{a^n}$, where n is a positive integer. We now define this property for negative rational exponents.

The Expression $a^{-m/n}$

If m and n are positive integers with $\frac{m}{n}$ in lowest terms, then

$$a^{-m/n} = \frac{1}{a^{m/n}}, \qquad a \neq 0.$$

EXAMPLE 8   Interpreting negative rational exponents

Write each expression in radical notation and then evaluate.
(a) $(64)^{-1/3}$ (b) $(81)^{-3/4}$

Solution

(a) $(64)^{-1/3} = \frac{1}{64^{1/3}} = \frac{1}{\sqrt[3]{64}} = \frac{1}{4}$. See Figure 8.12.

(b) $(81)^{-3/4} = \frac{1}{81^{3/4}} = \frac{1}{(\sqrt[4]{81})^3} = \frac{1}{3^3} = \frac{1}{27}$. See Figure 8.12.

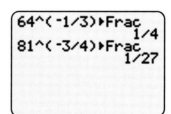

Figure 8.12

Properties of Rational Exponents

Any rational number can be written as a ratio of two integers. That is, if p is a rational number, then $p = \dfrac{m}{n}$, where m and n are integers. Properties for integer exponents also apply to rational exponents—with one exception. If n is even in the expression $a^{m/n}$, then a must be positive for the result to be a real number.

Properties of Exponents

Let p and q be rational numbers. For all real numbers a and b for which the expressions are real numbers the following properties hold.

1. $a^p \cdot a^q = a^{p+q}$ Product rule for exponents
2. $a^{-p} = \dfrac{1}{a^p}$ Negative exponents
3. $\left(\dfrac{a}{b}\right)^{-p} = \left(\dfrac{b}{a}\right)^{p}$ Negative exponents for quotients
4. $\dfrac{a^p}{a^q} = a^{p-q}$ Quotient rule for exponents
5. $(a^p)^q = a^{pq}$ Power rule for exponents
6. $(ab)^p = a^p b^p$ Power rule for products
7. $\left(\dfrac{a}{b}\right)^p = \dfrac{a^p}{b^p}$ Power rule for quotients

EXAMPLE 9   *Applying properties of exponents*

Write each expression with rational exponents and simplify. Write the answer with a positive exponent. Assume that all variables are positive numbers.

(a) $\sqrt{x} \cdot \sqrt[3]{x}$ (b) $\sqrt[3]{27x^2}$

(c) $\dfrac{\sqrt[4]{16x}}{\sqrt[3]{x}}$ (d) $\left(\dfrac{x^2}{81}\right)^{-1/2}$

Solution

(a) $\sqrt{x} \cdot \sqrt[3]{x} = x^{1/2} \cdot x^{1/3}$ Use rational exponents.
$\phantom{\sqrt{x} \cdot \sqrt[3]{x}} = x^{1/2 + 1/3}$ Product rule for exponents
$\phantom{\sqrt{x} \cdot \sqrt[3]{x}} = x^{5/6}$ Simplify.

(b) $\sqrt[3]{27x^2} = (27x^2)^{1/3}$ Use rational exponents.
$\phantom{\sqrt[3]{27x^2}} = 27^{1/3}(x^2)^{1/3}$ Power rule for products
$\phantom{\sqrt[3]{27x^2}} = 3x^{2/3}$ Power rule for exponents

(c) $\dfrac{\sqrt[4]{16x}}{\sqrt[3]{x}} = \dfrac{(16x)^{1/4}}{x^{1/3}}$ Use rational exponents.

$= \dfrac{16^{1/4} x^{1/4}}{x^{1/3}}$ Power rule for products

$= 16^{1/4} x^{(1/4 - 1/3)}$ Quotient rule for exponents

$= 2x^{-1/12}$ Subtract exponents.

$= \dfrac{2}{x^{1/12}}$ Negative exponents

(d) $\left(\dfrac{x^2}{81}\right)^{-1/2} = \left(\dfrac{81}{x^2}\right)^{1/2}$ Negative exponents for quotients

$= \dfrac{(81)^{1/2}}{(x^2)^{1/2}}$ Power rule for quotients

$= \dfrac{9}{x}$ Power rule for exponents

EXAMPLE 10 *Applying properties of exponents*

Simplify each expression. Assume that all variables are real numbers.

(a) $\sqrt{(x-y)^2}$ **(b)** $\sqrt[4]{(1-4x)^8}$

Solution

(a) To simplify this expression, we use the property $\sqrt{x^2} = |x|$.

$$\sqrt{(x-y)^2} = |x-y|$$

(b) The expression $(1-4x)^8$ is never negative, so its fourth root is always defined.

$$\sqrt[4]{(1-4x)^8} = ((1-4x)^8)^{1/4} = (1-4x)^2$$

8.1 Putting It All Together

Properties of radicals and rational exponents are summarized in the following table.

Feature	Property	Example
nth root of a real number	The nth root of real number a is b if $b^n = a$ and is denoted $\sqrt[n]{a} = b$. If $a < 0$ and n is even, $\sqrt[n]{a}$ is not a real number.	The square roots of 25 are 5 and -5. The principal square root is $\sqrt{25} = 5$. $\sqrt[3]{-125} = -5$ because $(-5)^3 = (-5)\cdot(-5)\cdot(-5) = -125$
Rational exponents	If m and n are positive integers with $\dfrac{m}{n}$ in lowest terms, $a^{m/n} = \sqrt[n]{a^m} = \left(\sqrt[n]{a}\right)^m$. *Note:* If $a < 0$ and n is an even integer, $a^{m/n}$ is not a real number.	$8^{4/3} = \left(\sqrt[3]{8}\right)^4 = 2^4 = 16$

8.1 Radical Expressions and Rational Exponents

Feature	Property	Example
Rational exponents *continued*	Let p and q be rational numbers. For all real numbers a and b for which the expressions are real numbers the following properties hold. 1. $a^p \cdot a^q = a^{p+q}$ 2. $a^{-p} = \dfrac{1}{a^p}$ 3. $\left(\dfrac{a}{b}\right)^{-p} = \left(\dfrac{b}{a}\right)^{p}$ 4. $\dfrac{a^p}{a^q} = a^{p-q}$ 5. $(a^p)^q = a^{pq}$ 6. $(ab)^p = a^p b^p$ 7. $\left(\dfrac{a}{b}\right)^p = \dfrac{a^p}{b^p}$	$2^{1/3} \cdot 2^{2/3} = 2^{(1/3+2/3)} = 2^1 = 2$ $2^{-1/2} = \dfrac{1}{2^{1/2}}$ $\left(\dfrac{3}{4}\right)^{-4/5} = \left(\dfrac{4}{3}\right)^{4/5}$ $\dfrac{7^{2/3}}{7^{1/3}} = 7^{(2/3-1/3)} = 7^{1/3}$ $(8^{2/3})^{1/2} = 8^{2/6} = 8^{1/3} = 2$ $(2x)^{1/3} = 2^{1/3} x^{1/3}$ $\left(\dfrac{x}{y}\right)^{1/6} = \dfrac{x^{1/6}}{y^{1/6}}$

8.1 EXERCISES

FOR EXTRA HELP: Student's Solutions Manual, InterAct Math, MathXL, MyMathLab.com, Math Tutor Center, Digital Video Tutor CD 5 Videotape 13

CONCEPTS

1. What are the square roots of 9?
2. What is the principal square root of 9?
3. What is the cube root of 8?
4. Does every real number have a cube root?
5. If $b^n = a$ and $b > 0$, then $\sqrt[n]{a} =$ _____.
6. Write $a^{1/n}$ in radical notation.
7. Write $a^{m/n}$ in radical notation.
8. Does $(a^{1/3})^2 = a^{2/3}$?

RADICAL EXPRESSIONS

Exercises 9–32: If possible, evaluate the expression by hand. If not, approximate the answer to the nearest hundredth.

9. $\sqrt{9}$
10. $\sqrt{121}$
11. $-\sqrt{5}$
12. $\sqrt{11}$
13. $\sqrt{z^2}$
14. $-\sqrt{(x+2)^2}$
15. $\sqrt[3]{27}$
16. $\sqrt[3]{64}$
17. $\sqrt[3]{-64}$
18. $-\sqrt[3]{-1}$
19. $\sqrt[3]{5}$
20. $\sqrt[3]{-13}$
21. $-\sqrt[3]{x^9}$
22. $\sqrt[3]{(x+1)^6}$
23. $\sqrt[3]{(2x)^6}$
24. $\sqrt[3]{9x^3}$
25. $\sqrt[4]{81}$
26. $\sqrt[5]{-1}$
27. $\sqrt[5]{-7}$
28. $\sqrt[4]{6}$
29. $\sqrt[4]{x^{12}}$
30. $\sqrt[6]{x^6}$
31. $\sqrt[4]{16x^8}$
32. $\sqrt[5]{32(x+4)^5}$

RATIONAL EXPONENTS

Exercises 33–38: Write each expression in radical notation.

33. $6^{1/2}$
34. $7^{1/3}$
35. $(xy)^{1/2}$
36. $x^{2/3}y^{1/5}$
37. $y^{-1/5}$
38. $\left(\dfrac{x}{y}\right)^{-2/7}$

Exercises 39–60: If possible, evaluate the expression by hand. If not, approximate the answer to the nearest hundredth.

39. $16^{1/2}$
40. $8^{1/3}$
41. $256^{1/4}$
42. $4^{3/2}$
43. $5^{2/3}$
44. $13^{3/4}$
45. $(-8)^{4/3}$
46. $(-1)^{3/5}$
47. $2^{1/2} \cdot 2^{2/3}$
48. $5^{3/5} \cdot 5^{1/10}$
49. $\left(\dfrac{4}{9}\right)^{1/2}$
50. $\left(\dfrac{27}{64}\right)^{1/3}$
51. $\dfrac{4^{2/3}}{4^{1/2}}$
52. $\dfrac{6^{1/5} \cdot 6^{3/5}}{6^{2/5}}$
53. $4^{-1/2}$
54. $9^{-3/2}$
55. $(-8)^{-1/3}$
56. $(49)^{-1/2}$
57. $\left(\dfrac{1}{16}\right)^{-1/4}$
58. $\left(\dfrac{16}{25}\right)^{-3/2}$
59. $(2^{1/2})^3$
60. $(5^{6/5})^{-1/2}$

Exercises 61–80: Simplify the expression. Assume that all variables are positive.

61. $(x^2)^{3/2}$
62. $(y^4)^{1/2}$
63. $(x^2 y^8)^{1/2}$
64. $(y^{10} z^4)^{1/4}$
65. $\sqrt[3]{x^3 y^6}$
66. $\sqrt{16x^4}$
67. $\sqrt{\dfrac{y^4}{x^2}}$
68. $\sqrt[3]{\dfrac{x^{12}}{z^6}}$
69. $\sqrt{y^3} \cdot \sqrt[3]{y^2}$
70. $\left(\dfrac{x^6}{81}\right)^{1/4}$
71. $\left(\dfrac{x^6}{27}\right)^{2/3}$
72. $\left(\dfrac{1}{x^8}\right)^{-1/4}$
73. $\left(\dfrac{x^2}{y^6}\right)^{-1/2}$
74. $\dfrac{\sqrt{x}}{\sqrt[3]{27x^6}}$
75. $\sqrt{\sqrt{y}}$
76. $\sqrt[3]{(3x)^2}$
77. $(a^{-1/2})^{4/3}$
78. $(x^{-3/2})^{2/3}$
79. $(a^3 b^6)^{1/3}$
80. $(64x^3 y^{18})^{1/6}$

Exercises 81–86: Simplify the expression. Assume that all variables are real numbers.

81. $\sqrt{y^2}$
82. $\sqrt{4x^2}$
83. $\sqrt{(x-5)^2}$
84. $\sqrt{(2x-1)^2}$
85. $\sqrt[4]{y^4}$
86. $\sqrt[4]{x^8 y^4}$

APPLICATIONS

87. *Cellular Phone Technology* (Refer to Example 4.) Suppose that a city has an area of 65 square miles and 15 cellular transmission towers spread evenly throughout it. Estimate a minimum radius R for each tower.

88. *Musical Notes* One octave on a piano contains 12 half notes, or keys on a piano (including both the black and white keys). The frequency of each successive half note increases by a factor of $2^{1/12}$. For example, middle C is two half notes below the first D above it. Therefore the frequency of this D is
$$2^{1/12} \cdot 2^{1/12} = 2^{1/6} \approx 1.12$$
times greater than middle C.
   (a) If notes are one octave apart, how do these frequencies compare?

(b) The A note below middle C has a frequency of 220 cycles per second. Middle C is 3 half notes above this A note. Estimate the frequency of middle C.

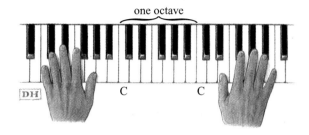

WRITING ABOUT MATHEMATICS

89. Try to calculate $\sqrt{-7}$, $\sqrt[4]{-56}$, and $\sqrt[6]{-10}$ with a calculator. Describe what happens when you evaluate an even root of a negative number. Does the same difficulty occur when you evaluate an odd root of a negative number? Try to evaluate $\sqrt[3]{-7}$, $\sqrt[5]{-56}$, and $\sqrt[7]{-10}$. Explain.

90. Explain the difference between a root and a power of a number. Give examples.

8.2 RADICAL FUNCTIONS

Root Functions ~ Power Functions ~ Modeling with Power Functions

INTRODUCTION

A good punter can kick a football so that it has a long *hang time*. Hang time is the length of time that the football is in the air. Long hang time gives the kicking team time to run down the field and stop the punt return. Using radicals, we can derive a function that calculates hang time.

ROOT FUNCTIONS

To derive a function that calculates the hang time of a football we need two facts from physics. First, when a ball is kicked into the air, half the hang time of the ball is spent going up and the other half is spent coming down. Second, the time t in seconds required for a ball to fall from a height of h feet is modeled by the equation

$$16t^2 = h.$$

Solving this equation for t gives half of the hang time.

$$16t^2 = h \qquad \text{Given equation}$$

$$t^2 = \frac{h}{16} \qquad \text{Divide by 16.}$$

$$t = \pm\sqrt{\frac{h}{16}} \qquad \text{Square root property}$$

$$t = \pm\frac{\sqrt{h}}{4} \qquad \text{Properties of square roots}$$

Because time is positive, we drop the negative solution. Half the hang time is $\frac{\sqrt{h}}{4}$, so the total hang time T in seconds is given by

$$T(h) = \frac{\sqrt{h}}{2},$$

where h is the maximum height of the ball.

EXAMPLE 1 *Calculating hang time*

If a football is kicked 50 feet into the air, estimate the hang time. Does the hang time double for a football kicked 100 feet into the air?

Solution

A football kicked 50 feet into the air has a hang time of

$$T(50) = \frac{\sqrt{50}}{2} \approx 3.5 \text{ seconds}.$$

If the football is kicked 100 feet into the air, the hang time is

$$T(100) = \frac{\sqrt{100}}{2} = 5 \text{ seconds}.$$

The hang time does not double.

Critical Thinking ———————————————————

How high would a football have to be kicked to have twice the hang time of a football kicked 50 feet into the air?

THE SQUARE ROOT FUNCTION. The square root function is given symbolically by $f(x) = \sqrt{x}$. The domain of the square root function is all nonnegative real numbers because we have not defined the square root of a negative number. A graph of $y = \sqrt{x}$ is shown in Figure 8.13. The table of values shown in Figure 8.14 contains some perfect square roots. Note also that, when the input x is negative, Y_1 displays ERROR.

[−6, 6, 1] by [−4, 4, 1]

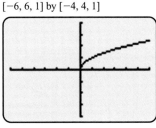

Figure 8.13

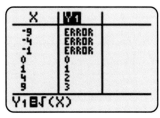

Figure 8.14

EXAMPLE 2 *Finding the domain of a function*

Find the domain of $f(x) = \sqrt{x - 1}$. Give graphical and numerical support.

Solution

The square root function is defined only for a nonnegative input. Therefore the domain of f includes all x-values that satisfy

$$x - 1 \geq 0 \quad \text{or} \quad x \geq 1.$$

Graphical support is given in Figure 8.15, where the graph $Y_1 = \sqrt{(X - 1)}$ appears only when $x \geq 1$. Numerical support is given in Figure 8.16. Note that when $x < 1$, ERROR appears.

[−4.7, 4.7, 1] by [−3.1, 3.1, 1]

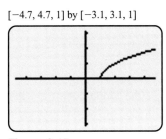

Figure 8.15

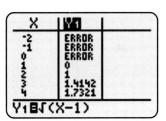

Figure 8.16

In the next example we use two functions to create the graph of an equation.

EXAMPLE 3 *Graphing $x = y^2$*

GCLM Graph the equation $x = y^2$ by solving for y. Does the graph represent a function?

Solution

Use the square root property to solve for y.

$$y^2 = x \qquad \text{Given equation}$$
$$y = \pm\sqrt{x} \qquad \text{Square root property}$$

Graph $Y_1 = \sqrt{(X)}$ and $Y_2 = -\sqrt{(X)}$, as shown in Figure 8.17 on the next page. The graph is a parabola that opens to the right rather than upward or downward. This parabola does not represent a function because it fails the vertical line test. It is created from the graphs of two functions: $f(x) = \sqrt{x}$ and $g(x) = -\sqrt{x}$. The graph of f generates the upper portion of the parabola, whereas the graph of g generates the lower portion, as illustrated in Figures 8.18 and 8.19.

[−6, 6, 1] by [−4, 4, 1]

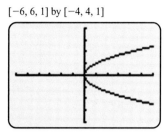

Figure 8.17 $x = y^2$

[−6, 6, 1] by [−4, 4, 1]

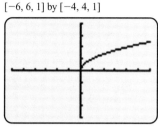

Figure 8.18 $f(x) = \sqrt{x}$

[−6, 6, 1] by [−4, 4, 1]

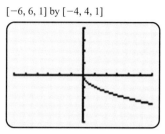

Figure 8.19 $g(x) = -\sqrt{x}$

THE CUBE ROOT FUNCTION. The cube root function is given symbolically by $f(x) = \sqrt[3]{x}$. Cube roots are defined for both positive and negative numbers, so the domain of the cube root function is all real numbers. A graph of $y = \sqrt[3]{x}$ is shown in Figure 8.20. The table of values shown in Figure 8.21 presents only perfect cube roots.

[−6, 6, 1] by [−4, 4, 1]

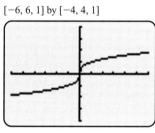

Figure 8.20

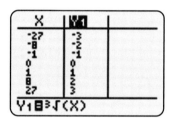

Figure 8.21

In some parts of the United States, wind power is used to generate electricity. Suppose that the diameter of the circular path created by the blades for a wind-powered generator is 8 feet. Then the wattage W generated by a wind velocity of v miles per hour is modeled by

$$W(v) = 2.4v^3.$$

If the wind blows at 10 miles per hour, the generator can produce about

$$W(10) = 2.4 \cdot 10^3 = 2400 \text{ watts.}$$

(*Source: Conquering the Sciences,* **Sharp Electronics.**)

EXAMPLE 4 Modeling a windmill

The formula $W(v) = 2.4v^3$ is used to calculate the watts generated when there is a wind velocity of v miles per hour.

(a) Find a function f that calculates the wind velocity when W watts are being produced.

(b) Graph f in [0, 30,000, 5000] by [0, 30, 10]. Is f a linear or nonlinear function?

(c) If the wattage doubles, has the wind velocity also doubled? Explain.

Solution

(a) Solve $W = 2.4v^3$ for v.

$$W = 2.4v^3 \qquad \text{Given formula}$$

$$\frac{W}{2.4} = v^3 \qquad \text{Divide by 2.4.}$$

$$v = \sqrt[3]{\frac{W}{2.4}} \qquad \text{Take the cube root of each side and rewrite equation.}$$

Thus $f(W) = \sqrt[3]{\frac{W}{2.4}}$.

(b) Graph either $Y_1 = \sqrt[3]{(X/2.4)}$ or $Y_1 = (X/2.4)\wedge(1/3)$, as shown in Figure 8.22. The graph is curved, and therefore f is nonlinear.

(c) A table of Y_1 is shown in Figure 8.23. When the power doubles from 1000 watts to 2000 watts, the wind velocity increases only from about 7.5 to 9.4 miles per hour. For the wattage to double, the wind velocity does not need to double.

[0, 30,000, 5000] by [0, 30, 10]

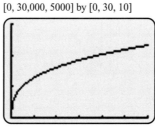
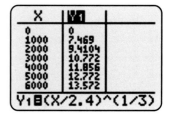

Figure 8.22 **Figure 8.23**

POWER FUNCTIONS

Power functions are a generalization of root functions. Examples of power functions include

$$f(x) = x^{1/2}, \qquad g(x) = x^{2/3}, \quad \text{and} \quad h(x) = x^{-3/5}.$$

The exponents for power functions are rational numbers. Any rational number can be written as $\frac{m}{n}$, where m and n are integers.

Power Function

If a function can be represented by

$$f(x) = x^p,$$

where p is a rational number, then it is a **power function**. If $p = \frac{1}{n}$, where $n \geq 2$ is an integer, then f is also a **root function**, which is given by

$$f(x) = \sqrt[n]{x}.$$

Note: If we let $p = n$, where n is a positive integer, then $f(x) = x^n$ represents both a power function and a polynomial function.

EXAMPLE 5 *Evaluating a power function*

Evaluate $f(x) = x^{3/4}$ at $x = 16$ and $x = 13$.

Solution

We can evaluate $f(16)$ by hand as
$$f(16) = 16^{3/4} = (16^{1/4})^3 = 2^3 = 8$$
because $16^{1/4} = \sqrt[4]{16} = 2$.

To evaluate $f(13)$ we use a calculator to obtain $13^{3/4} \approx 6.85$. See Figure 8.24.

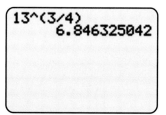

Figure 8.24

In the next example, we investigate the graph of $y = x^p$ for different values of p.

EXAMPLE 6 *Graphing power functions*

Graph the following power functions in the same viewing rectangle [0, 6, 1] by [0, 6, 1]. Then discuss how the value of p affects the graph of $y = x^p$.
$$f(x) = x^{1/3}, \qquad g(x) = x^{0.75}, \quad \text{and} \quad h(x) = x^{1.4}$$

Solution

First note that g and h can be written as $g(x) = x^{3/4}$ and $h(x) = x^{7/5}$. The graphs of f, g, and h are shown in Figure 8.25, where uppercase letters were used to distinguish the functions. As the exponent becomes larger, the graph of the power function increases faster for $x \geq 1$.

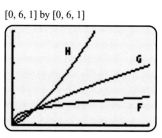

Figure 8.25

MODELING WITH POWER FUNCTIONS

Allometry is the study of the relative sizes of different characteristics of an organism. For example, the weight of a bird is related to the surface area of its wings: Heavier birds tend to have larger wings. Allometric relations are often modeled with $f(x) = kx^p$, where k and p are constants. *(Source: C. Pennycuick, Newton Rules Biology.)*

EXAMPLE 7 *Modeling surface area of wings*

The surface area A of a bird's wings with weight x is shown in Table 8.1.

TABLE 8.1

x (kilograms)	0.5	2.0	3.5	5.0
$A(x)$ (square meters)	0.069	0.175	0.254	0.325

(a) Make a scatterplot of the data. Discuss any trends in the data.
(b) Biologists modeled the data with $A(x) = kx^{2/3}$, where k is a constant. Find k.
(c) Graph A and the data in the same viewing rectangle.
(d) Estimate the area of the wings of a 3-kilogram bird.

Solution

(a) A scatterplot of the data is shown in Figure 8.26. As the weight of a bird increases so does the surface area of its wings.

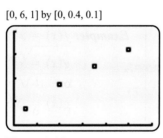
[0, 6, 1] by [0, 0.4, 0.1]

Figure 8.26

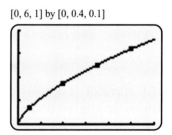
[0, 6, 1] by [0, 0.4, 0.1]

Figure 8.27

(b) To determine k, substitute one of the data points into $A(x)$.

$$A(x) = kx^{2/3} \quad \text{Given formula}$$
$$0.175 = k(2)^{2/3} \quad \text{Let } x = 2 \text{ and } A(x) = 0.175 \text{ (any data point could be used).}$$
$$k = \frac{0.175}{2^{2/3}} \quad \text{Solve for } k.$$
$$k \approx 0.11 \quad \text{Approximate } k.$$

Thus $A(x) = 0.11x^{2/3}$.

(c) The data and graph of $Y_1 = 0.11X^{\wedge}(2/3)$ are shown in Figure 8.27. Note that the graph appears to pass through each data point.

(d) $A(3) = 0.11(3)^{2/3} \approx 0.23$ square meter

These examples suggest the following rule.

> **Product Rule for Radical Expressions**
>
> Let a and b be real numbers, where $\sqrt[n]{a}$ and $\sqrt[n]{b}$ are both defined. Then
> $$\sqrt[n]{a} \cdot \sqrt[n]{b} = \sqrt[n]{a \cdot b}.$$
> To multiply radical expressions with the same index, multiply the radicands.

EXAMPLE 1 *Multiplying radical expressions*

Multiply each pair of radical expressions. Check your results with a calculator.
(a) $\sqrt{5} \cdot \sqrt{20}$ (b) $\sqrt[3]{3} \cdot \sqrt[3]{9}$

Solution
(a) $\sqrt{5} \cdot \sqrt{20} = \sqrt{5 \cdot 20} = \sqrt{100} = 10$. See Figure 8.30.
(b) $\sqrt[3]{3} \cdot \sqrt[3]{9} = \sqrt[3]{3 \cdot 9} = \sqrt[3]{27} = 3$. See Figure 8.30.

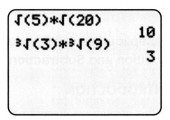

Figure 8.30

EXAMPLE 2 *Multiplying radical expressions containing variables*

Multiply each pair of radical expressions. Assume that all variables are positive.
(a) $\sqrt{x} \cdot \sqrt{x^3}$ (b) $\sqrt[3]{2a} \cdot \sqrt[3]{5a}$

Solution
(a) $\sqrt{x} \cdot \sqrt{x^3} = \sqrt{x \cdot x^3} = \sqrt{x^4} = x^2$
(b) $\sqrt[3]{2a} \cdot \sqrt[3]{5a} = \sqrt[3]{2a \cdot 5a} = \sqrt[3]{10a^2}$

Critical Thinking

Can you use the product rule for radical expressions to simplify $\sqrt{2} \cdot \sqrt[3]{3}$? Explain.

Consider the following example of dividing radical expressions.

$$\sqrt{\frac{4}{9}} = \sqrt{\frac{2}{3} \cdot \frac{2}{3}} = \frac{2}{3} \quad \text{and} \quad \frac{\sqrt{4}}{\sqrt{9}} = \frac{2}{3}$$

implies that

$$\sqrt{\frac{4}{9}} = \frac{\sqrt{4}}{\sqrt{9}}. \text{ See Figure 8.31.}$$

8.3 Operations on Radical Expressions

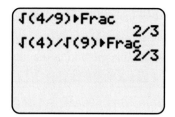

Figure 8.31

The preceding example suggests the following rule.

> **Quotient Rule for Radical Expressions**
>
> Let a and b be real numbers, where $\sqrt[n]{a}$ and $\sqrt[n]{b}$ are both defined and $b \neq 0$.
>
> $$\sqrt[n]{\frac{a}{b}} = \frac{\sqrt[n]{a}}{\sqrt[n]{b}}$$

EXAMPLE 3 *Simplifying quotients*

Simplify each radical expression. Assume that all variables are positive.

(a) $\sqrt[3]{\dfrac{5}{8}}$ (b) $\sqrt[4]{\dfrac{x}{16}}$ (c) $\sqrt{\dfrac{16}{y^2}}$

Solution

(a) $\sqrt[3]{\dfrac{5}{8}} = \dfrac{\sqrt[3]{5}}{\sqrt[3]{8}} = \dfrac{\sqrt[3]{5}}{2}$

(b) $\sqrt[4]{\dfrac{x}{16}} = \dfrac{\sqrt[4]{x}}{\sqrt[4]{16}} = \dfrac{\sqrt[4]{x}}{2}$

(c) $\sqrt{\dfrac{16}{y^2}} = \dfrac{\sqrt{16}}{\sqrt{y^2}} = \dfrac{4}{y}$

■ **MAKING CONNECTIONS**

> **Rules for Radical Expressions and Rational Exponents**
>
> The rules for radical expressions are a result of the properties of rational exponents.
>
> $\sqrt[n]{a \cdot b} = \sqrt[n]{a} \cdot \sqrt[n]{b}$ is equivalent to $(a \cdot b)^{1/n} = a^{1/n} \cdot b^{1/n}$.
>
> $\sqrt[n]{\dfrac{a}{b}} = \dfrac{\sqrt[n]{a}}{\sqrt[n]{b}}$ is equivalent to $\left(\dfrac{a}{b}\right)^{1/n} = \dfrac{a^{1/n}}{b^{1/n}}$.

Simplifying Radical Expressions

The product and quotient rules can be used to help simplify radical expressions. We illustrate this method in the next two examples.

EXAMPLE 4 *Simplifying radical expressions*

Simplify each expression. Assume that all variables are positive.

(a) $\sqrt{25x^4}$ (b) $\sqrt[3]{2a} \cdot \sqrt[3]{4a^2b}$

Solution

(a) $\sqrt{25x^4} = \sqrt{25} \cdot \sqrt{x^4} = 5x^2$

(b) $\sqrt[3]{2a} \cdot \sqrt[3]{4a^2b} = \sqrt[3]{2a \cdot 4a^2b} = \sqrt[3]{8a^3b} = \sqrt[3]{8} \cdot \sqrt[3]{a^3} \cdot \sqrt[3]{b} = 2a\sqrt[3]{b}$

EXAMPLE 5 *Simplifying radical expressions*

Simplify each expression. Assume that all variables are positive.

(a) $\dfrac{\sqrt{40}}{\sqrt{10}}$ (b) $\sqrt[4]{\dfrac{16x^3}{y^4}}$

Solution

(a) $\dfrac{\sqrt{40}}{\sqrt{10}} = \sqrt{\dfrac{40}{10}} = \sqrt{4} = 2$ (b) $\sqrt[4]{\dfrac{16x^3}{y^4}} = \dfrac{\sqrt[4]{16x^3}}{\sqrt[4]{y^4}} = \dfrac{\sqrt[4]{16} \cdot \sqrt[4]{x^3}}{\sqrt[4]{y^4}} = \dfrac{2\sqrt[4]{x^3}}{y}$

Addition and Subtraction

We can add $2x^2$ and $5x^2$ to obtain $7x^2$ because they are *like* terms. That is,

$$2x^2 + 5x^2 = (2 + 5)x^2 = 7x^2.$$

We can add and subtract **like radicals**, which have the same index and the same radicand. For example, we can add $3\sqrt{2}$ and $5\sqrt{2}$ because they are like radicals.

$$3\sqrt{2} + 5\sqrt{2} = (3 + 5)\sqrt{2} = 8\sqrt{2}$$

Sometimes two radical expressions that are not alike can be added by changing them to like radicals. For example, $\sqrt{20}$ and $\sqrt{5}$ are unlike radicals. However,

$$\sqrt{20} = \sqrt{4 \cdot 5} = \sqrt{4} \cdot \sqrt{5} = 2\sqrt{5},$$

so

$$\sqrt{20} + \sqrt{5} = 2\sqrt{5} + \sqrt{5} = 3\sqrt{5}.$$

We cannot combine $x + x^2$ because they are unlike terms. Similarly, we cannot combine $\sqrt{2} + \sqrt{5}$ because they are unlike radicals.

EXAMPLE 6 *Finding like radicals*

Write each pair of terms as like radicals, if possible.

(a) $\sqrt{45}, \sqrt{20}$ (b) $\sqrt{27}, \sqrt{5}$ (c) $5\sqrt[3]{16}, 4\sqrt[3]{54}$

Solution

(a) The expressions $\sqrt{45}$ and $\sqrt{20}$ are unlike radicals. However, they can be changed to like radicals as follows.

$$\sqrt{45} = \sqrt{9 \cdot 5} = \sqrt{9} \cdot \sqrt{5} = 3\sqrt{5}$$
$$\sqrt{20} = \sqrt{4 \cdot 5} = \sqrt{4} \cdot \sqrt{5} = 2\sqrt{5}$$

The expressions $3\sqrt{5}$ and $2\sqrt{5}$ are like radicals.

(b) The radical expressions $\sqrt{27} = 3\sqrt{3}$ and $\sqrt{5}$ are unlike radicals and cannot be written as like radicals.

(c) $5\sqrt[3]{16} = 5\sqrt[3]{8 \cdot 2} = 5\sqrt[3]{8} \cdot \sqrt[3]{2} = 5 \cdot 2 \cdot \sqrt[3]{2} = 10\sqrt[3]{2}$
$4\sqrt[3]{54} = 4\sqrt[3]{27 \cdot 2} = 4\sqrt[3]{27} \cdot \sqrt[3]{2} = 4 \cdot 3 \cdot \sqrt[3]{2} = 12\sqrt[3]{2}$

The expressions $10\sqrt[3]{2}$ and $12\sqrt[3]{2}$ are like radicals.

We use these techniques to add radical expressions in the next example.

EXAMPLE 7 *Adding radical expressions*

Add the expressions and simplify.

(a) $10\sqrt{11} + 4\sqrt{11}$ (b) $5\sqrt[3]{6} + \sqrt[3]{6}$

(c) $\sqrt{12} + 7\sqrt{3}$ (d) $3\sqrt{2} + \sqrt{8} + \sqrt{18}$

Solution

(a) $10\sqrt{11} + 4\sqrt{11} = (10 + 4)\sqrt{11} = 14\sqrt{11}$

(b) $5\sqrt[3]{6} + \sqrt[3]{6} = (5 + 1)\sqrt[3]{6} = 6\sqrt[3]{6}$

(c) $\sqrt{12} + 7\sqrt{3} = \sqrt{4 \cdot 3} + 7\sqrt{3}$
$= \sqrt{4} \cdot \sqrt{3} + 7\sqrt{3}$
$= 2\sqrt{3} + 7\sqrt{3}$
$= 9\sqrt{3}$

(d) $3\sqrt{2} + \sqrt{8} + \sqrt{18} = 3\sqrt{2} + \sqrt{4 \cdot 2} + \sqrt{9 \cdot 2}$
$= 3\sqrt{2} + \sqrt{4} \cdot \sqrt{2} + \sqrt{9} \cdot \sqrt{2}$
$= 3\sqrt{2} + 2\sqrt{2} + 3\sqrt{2}$
$= 8\sqrt{2}$

Caution: $\sqrt{9 + 4} \neq \sqrt{9} + \sqrt{4} = 3 + 2 = 5$. Rather, $\sqrt{9 + 4} = \sqrt{13} \approx 3.61$.

Subtraction of radical expressions is similar, as illustrated in the next example.

EXAMPLE 8 *Subtracting radical expressions*

Subtract each pair of expressions and simplify. Assume that all variables are positive.

(a) $5\sqrt{7} - 3\sqrt{7}$ (b) $3\sqrt[3]{xy^2} - 2\sqrt[3]{xy^2}$

(c) $\sqrt{16x^3} - \sqrt{x^3}$ (d) $\sqrt[3]{\dfrac{5x}{27}} - \dfrac{\sqrt[3]{5x}}{6}$

Solution

(a) $5\sqrt{7} - 3\sqrt{7} = (5 - 3)\sqrt{7} = 2\sqrt{7}$

(b) $3\sqrt[3]{xy^2} - 2\sqrt[3]{xy^2} = (3 - 2)\sqrt[3]{xy^2} = \sqrt[3]{xy^2}$

(c) $\sqrt{16x^3} - \sqrt{x^3} = \sqrt{16} \cdot \sqrt{x^3} - \sqrt{x^3}$
$= 4\sqrt{x^3} - \sqrt{x^3}$
$= 3\sqrt{x^3}$

(d) $\sqrt[3]{\dfrac{5x}{27}} - \dfrac{\sqrt[3]{5x}}{6} = \dfrac{\sqrt[3]{5x}}{\sqrt[3]{27}} - \dfrac{\sqrt[3]{5x}}{6}$ Quotient rule for radical expressions

$= \dfrac{\sqrt[3]{5x}}{3} - \dfrac{\sqrt[3]{5x}}{6}$ Evaluate $\sqrt[3]{27} = 3$.

$= \dfrac{2\sqrt[3]{5x}}{6} - \dfrac{\sqrt[3]{5x}}{6}$ Find a common denominator.

$= \dfrac{2\sqrt[3]{5x} - \sqrt[3]{5x}}{6}$ Subtract numerators.

$= \dfrac{\sqrt[3]{5x}}{6}$ Simplify.

8.3 Putting It All Together

In this section we discussed how to simplify and perform arithmetic with radical expressions. Results are summarized in the following table.

Procedure	Explanation	Examples
Product rule for radical expressions	Let a and b be real numbers, where $\sqrt[n]{a}$ and $\sqrt[n]{b}$ are both defined. $\sqrt[n]{a} \cdot \sqrt[n]{b} = \sqrt[n]{a \cdot b}$	$\sqrt{2} \cdot \sqrt{32} = \sqrt{64} = 8$
Quotient rule for radical expressions	Let a and b be real numbers, where $\sqrt[n]{a}$ and $\sqrt[n]{b}$ are both defined and $b \neq 0$. $\sqrt[n]{\dfrac{a}{b}} = \dfrac{\sqrt[n]{a}}{\sqrt[n]{b}}$	$\dfrac{\sqrt{60}}{\sqrt{15}} = \sqrt{\dfrac{60}{15}} = \sqrt{4} = 2$ $\sqrt[3]{\dfrac{x^2}{27}} = \dfrac{\sqrt[3]{x^2}}{\sqrt[3]{27}} = \dfrac{\sqrt[3]{x^2}}{3}$
Like radicals	Like radicals have the same index and the same radicand.	$7\sqrt{5}$ and $3\sqrt{5}$ are like radicals. $5\sqrt[3]{ab}$ and $\sqrt[3]{ab}$ are like radicals. $\sqrt[3]{5}$ and $\sqrt[3]{4}$ are unlike radicals. $\sqrt[3]{7}$ and $\sqrt{7}$ are unlike radicals.

8.3 Operations on Radical Expressions

Procedure	Explanation	Examples
Adding and subtracting radical expressions	Combine like radicals when adding or subtracting. We cannot combine unlike radicals such as $\sqrt{2}$ and $\sqrt{5}$.	$6\sqrt{13} + \sqrt{13} = (6+1)\sqrt{13} = 7\sqrt{13}$ $\sqrt{40} - \sqrt{10} = \sqrt{4} \cdot \sqrt{10} - \sqrt{10}$ $= 2\sqrt{10} - \sqrt{10}$ $= \sqrt{10}$

8.3 EXERCISES

FOR EXTRA HELP: Student's Solutions Manual, MyMathLab.com, InterAct Math, Math Tutor Center, MathXL, Digital Video Tutor CD 6 Videotape 13

CONCEPTS

1. Does $\sqrt{2} \cdot \sqrt{3}$ equal $\sqrt{6}$?
2. Does $\sqrt{5} \cdot \sqrt[3]{5}$ equal 5?
3. $\sqrt[3]{a} \cdot \sqrt[3]{b} = $ _____
4. $\dfrac{\sqrt{a}}{\sqrt{b}} = \sqrt{?}$
5. $\dfrac{\sqrt[n]{a}}{\sqrt[n]{b}} = \sqrt[n]{?}$
6. $\sqrt{a} + \sqrt{a} = $ _____
7. We cannot simplify $\sqrt[3]{4} + \sqrt[3]{7}$ because they are not _____ radicals.
8. Can we simplify $4\sqrt{15} - 3\sqrt{15}$? Explain.

MULTIPLYING AND DIVIDING

Exercises 9–36: Simplify the expression. Assume that all variables are positive.

9. $\sqrt{3} \cdot \sqrt{3}$
10. $\sqrt{2} \cdot \sqrt{18}$
11. $\sqrt{2} \cdot \sqrt{50}$
12. $\sqrt[3]{2} \cdot \sqrt[3]{4}$
13. $\sqrt[3]{4} \cdot \sqrt[3]{16}$
14. $\sqrt[3]{x} \cdot \sqrt[3]{x^2}$
15. $\sqrt{\dfrac{9}{25}}$
16. $\sqrt[3]{\dfrac{x}{8}}$
17. $\sqrt{\dfrac{1}{2}} \cdot \sqrt{\dfrac{1}{8}}$
18. $\sqrt{\dfrac{5}{3}} \cdot \sqrt{\dfrac{1}{3}}$
19. $\sqrt{\dfrac{x}{2}} \cdot \sqrt{\dfrac{x}{8}}$
20. $\sqrt{\dfrac{4}{y}} \cdot \sqrt{\dfrac{y}{5}}$
21. $\dfrac{\sqrt{45}}{\sqrt{5}}$
22. $\dfrac{\sqrt{7}}{\sqrt{28}}$
23. $\dfrac{\sqrt{a^2 b}}{\sqrt{b}}$
24. $\dfrac{\sqrt{4xy^2}}{\sqrt{x}}$
25. $\dfrac{\sqrt[3]{54}}{\sqrt[3]{2}}$
26. $\dfrac{\sqrt[3]{x^3 y^7}}{\sqrt[3]{y^4}}$
27. $\sqrt{4x^4}$
28. $\sqrt[3]{8y^3}$
29. $\sqrt[3]{5a^6}$
30. $\sqrt{9x^2 y}$
31. $\sqrt[4]{16x^4 y}$
32. $\sqrt[3]{8xy^3}$
33. $\sqrt{3x} \cdot \sqrt{12x}$
34. $\sqrt{6x^5} \cdot \sqrt{6x}$
35. $\sqrt[3]{8x^6 y^3 z^9}$
36. $\sqrt{16x^4 y^6}$

Usually these equations can be solved symbolically, graphically, and numerically and they may have more than one solution. We begin by discussing graphical and numerical solutions.

EXAMPLE 1 *Solving a radical equation*

Solve $\sqrt{2x - 1} = 3$ graphically and numerically.

Solution

Graphical Solution Graph $Y_1 = \sqrt{(2X - 1)}$ and $Y_2 = 3$. Their graphs intersect at $(5, 3)$, as shown in Figure 8.32. Thus $x = 5$ is the solution.

Numerical Solution Construct a table for Y_1 and Y_2. Figure 8.33 reveals that $Y_1 = Y_2 = 3$ when $x = 5$.

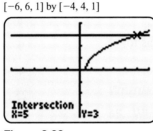
Figure 8.32

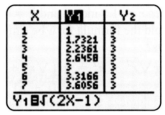
Figure 8.33

In the next example we solve an equation containing a rational exponent.

EXAMPLE 2 *Solving an equation with a rational exponent*

Solve $x^{2/3} = 3 - x^2$ graphically.

Solution

Graph $Y_1 = X^{\wedge}(2/3)$ and $Y_2 = 3 - X^{\wedge}2$. Their graphs intersect near $(-1.34, 1.21)$ and $(1.34, 1.21)$, as shown in Figures 8.34 and 8.35. Thus the solutions are $x \approx \pm 1.34$.

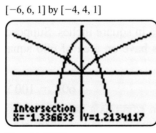
Figure 8.34

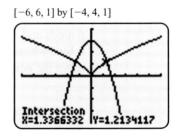
Figure 8.35

SYMBOLIC METHODS

One strategy for solving an equation containing a square root is to square each side of the equation. In the next example, we solve the equation from Example 1 symbolically.

EXAMPLE 3   Solving a radical equation symbolically

Solve $\sqrt{2x-1} = 3$ symbolically. Check your solution.

Solution

Begin by squaring both sides of the equation.

$\sqrt{2x-1} = 3$ Given equation
$(\sqrt{2x-1})^2 = 3^2$ Square both sides.
$2x - 1 = 9$ Simplify.
$2x = 10$ Add 1.
$x = 5$ Divide by 2.

To check our work we substitute $x = 5$ into the original equation.

$\sqrt{2(5)-1} \stackrel{?}{=} \sqrt{9}$
$3 = 3$ It checks.

In the next example we solve the equation presented in the section introduction.

EXAMPLE 4   Finding the weight of a bird

Solve the equation $600 = 100\sqrt[3]{W^2}$ symbolically and graphically to determine the weight of a bird having wings with an area of 600 square inches.

Solution

Symbolic Solution   Begin by dividing both sides of the equation by 100.

$\dfrac{600}{100} = \sqrt[3]{W^2}$ Divide both sides by 100.
$(6)^3 = (\sqrt[3]{W^2})^3$ Cube both sides.
$216 = W^2$ Simplify.
$W = \sqrt{216}$ Take principal square root, $W > 0$.
$W \approx 14.7$ Approximate.

The weight of the bird is approximately 14.7 pounds.

Graphical Solution   The graphs of $Y_1 = 600$ and $Y_2 = 100W^{\wedge}(2/3)$ intersect near (14.7, 600), as shown in Figure 8.36. (Negative variables are not valid in this application, so only the first quadrant is shown.)

[0, 20, 5] by [0, 800, 100]

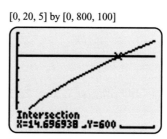

Figure 8.36

Technology Note *Graphing radical expressions*

On some graphing calculators it is more convenient to use rational exponents to enter radical expressions. In Example 4 we used

$$\sqrt[3]{W^2} = (W^2)^{1/3} = W^{2/3}.$$

Thus we let $Y_2 = X\wedge(2/3)$. Be sure to include parentheses around the exponent, as most calculators will interpret X^2/3 as $\dfrac{x^2}{3} \neq x^{2/3}$.

In the next example we show why it is important to check an answer when squaring both sides of an equation.

EXAMPLE 5 *Solving a radical equation*

Solve $\sqrt{3x + 3} = 2x - 1$ symbolically. Check your results and then solve the equation graphically.

Solution

Symbolic Solution Begin by squaring both sides of the equation.

$$\sqrt{3x + 3} = 2x - 1 \qquad \text{Given equation}$$
$$(\sqrt{3x + 3})^2 = (2x - 1)^2 \qquad \text{Square both sides.}$$
$$3x + 3 = 4x^2 - 4x + 1 \qquad \text{Expand.}$$
$$0 = 4x^2 - 7x - 2 \qquad \text{Subtract } 3x + 3.$$
$$0 = (4x + 1)(x - 2) \qquad \text{Factor.}$$
$$x = -\frac{1}{4} \quad \text{or} \quad x = 2 \qquad \text{Solve for } x.$$

To check these values substitute $x = -\frac{1}{4}$ and $x = 2$ into the given equation.

$$\sqrt{3\left(-\frac{1}{4}\right) + 3} \stackrel{?}{=} 2\left(-\frac{1}{4}\right) - 1$$
$$\sqrt{2.25} \stackrel{?}{=} -1.5$$
$$1.5 \neq -1.5 \qquad \text{It does not check.}$$

Thus $x = -\frac{1}{4}$ is *not* a solution and is called an *extraneous solution*. Next substitute $x = 2$ into the given equation.

$$\sqrt{3 \cdot 2 + 3} \stackrel{?}{=} 2 \cdot 2 - 1$$
$$\sqrt{9} \stackrel{?}{=} 3$$
$$3 = 3 \qquad \text{It checks.}$$

The only solution is $x = 2$.

Graphical Solution The solution $x = 2$ is supported graphically in Figure 8.37, where the graphs of $Y_1 = \sqrt{(3X + 3)}$ and $Y_2 = 2X - 1$ intersect at the point $(2, 3)$.

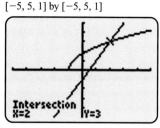

[−5, 5, 1] by [−5, 5, 1]

Figure 8.37

The preceding example demonstrates that checking solutions is essential when you are squaring both sides of an equation. Squaring may introduce *extraneous solutions*, which are solutions to the new equation but are not solutions to the given equation.

> **Power Rule for Solving Equations**
>
> If each side of an equation is raised to the same integer power, then any solutions to the original equations are among the solutions to the new equation. That is, the solutions to the equation $a = b$ are among the solutions to $a^n = b^n$.

For example, consider the equation $2x = 6$. If we square both sides, we obtain $4x^2 = 36$. Solving this new equation gives $x^2 = 9$, or $x = \pm 3$. Here, $x = 3$ is a solution to both equations, but $x = -3$ is an extraneous solution that satisfies the second equation but not the given equation.

When an equation contains two or more terms with square roots, it may be necessary to square each side of the equation more than once. In these situations, we isolate one of the square roots and then square each side of the equation. If a radical term remains after simplifying, we repeat these steps. We apply this technique in the following example.

EXAMPLE 6 *Squaring twice*

Solve $\sqrt{2x} - 1 = \sqrt{x + 1}$ symbolically. Check your results graphically.

Solution

Begin by squaring both sides of the equation.

$$\sqrt{2x} - 1 = \sqrt{x + 1} \qquad \text{Given equation}$$
$$(\sqrt{2x} - 1)^2 = (\sqrt{x + 1})^2 \qquad \text{Square both sides.}$$
$$(\sqrt{2x})^2 - 2(\sqrt{2x})(1) + 1^2 = x + 1 \qquad (a - b)^2 = a^2 - 2ab + b^2$$
$$2x - 2\sqrt{2x} + 1 = x + 1 \qquad \text{Simplify.}$$
$$2x - 2\sqrt{2x} = x \qquad \text{Subtract 1.}$$
$$x = 2\sqrt{2x} \qquad \text{Subtract } x \text{ and add } 2\sqrt{2x}.$$
$$x^2 = 4(2x) \qquad \text{Square both sides.}$$
$$x^2 - 8x = 0 \qquad \text{Subtract } 8x.$$
$$x(x - 8) = 0 \qquad \text{Factor.}$$
$$x = 0 \quad \text{or} \quad x = 8 \qquad \text{Solve.}$$

To check your solutions substitute $x = 0$ into the original equation.

$$\sqrt{2 \cdot 0 - 1} \stackrel{?}{=} \sqrt{0 + 1}$$

$$-1 \neq 1 \qquad \text{It does not check.}$$

Now substitute $x = 8$.

$$\sqrt{2 \cdot 8 - 1} \stackrel{?}{=} \sqrt{8 + 1}$$

$$3 = 3 \qquad \text{It checks.}$$

The only solution is $x = 8$. This solution is supported graphically in Figure 8.38, where the graphs of $Y_1 = \sqrt{(2X) - 1}$ and $Y_2 = \sqrt{(X + 1)}$ intersect at one point, $(8, 3)$.

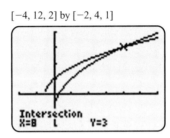

Figure 8.38

THE DISTANCE FORMULA

One of the most famous theorems in mathematics is the **Pythagorean theorem.** It states that, if a right triangle has legs a and b with hypotenuse c (see Figure 8.39), then

$$a^2 + b^2 = c^2.$$

For example, if the legs of a right triangle are $a = 3$ and $b = 4$, the hypotenuse is $c = 5$ because $3^2 + 4^2 = 5^2$.

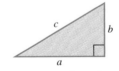

Figure 8.39 $a^2 + b^2 = c^2$

EXAMPLE 7 *Applying the Pythagorean theorem*

A rectangular television screen has a width of 20 inches and a height of 15 inches, as depicted in Figure 8.40. Find the diagonal of the television. Why is it called a 25-inch television?

Figure 8.40

8.4 Equations Involving Radical Expressions

Solution

Let $a = 20$ and $b = 15$. Then the diagonal of the television corresponds to the hypotenuse of a right triangle with legs of 20 inches and 15 inches.

$$c^2 = a^2 + b^2 \qquad \text{Pythagorean theorem}$$
$$c = \sqrt{a^2 + b^2} \qquad \text{Take the principal square root, } c > 0.$$
$$c = \sqrt{20^2 + 15^2} \qquad \text{Substitute } a = 20 \text{ and } b = 15.$$
$$c = 25 \qquad \text{Simplify.}$$

A 25-inch television has a diagonal of 25 inches.

The Pythagorean theorem can be used to determine the distance between two points. Suppose that a line segment has endpoints (x_1, y_1) and (x_2, y_2), as illustrated in Figure 8.41. The lengths of the legs of the right triangle are $x_2 - x_1$ and $y_2 - y_1$. The distance d is the hypotenuse of a right triangle. Applying the Pythagorean theorem, we have

$$d^2 = (x_2 - x_1)^2 + (y_2 - y_1)^2.$$

Distance is nonnegative, so we let d be the principal square root and obtain

$$d = \sqrt{(x_2 - x_1)^2 + (y_2 - y_1)^2}.$$

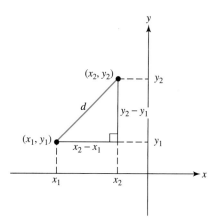

Figure 8.41

Distance Formula

The **distance** d between the points (x_1, y_1) and (x_2, y_2) in the xy-plane is

$$d = \sqrt{(x_2 - x_1)^2 + (y_2 - y_1)^2}.$$

EXAMPLE 8 *Finding distance between points*

Find the distance between the points $(-2, 3)$ and $(1, -4)$.

Solution

Start by letting $(x_1, y_1) = (-2, 3)$ and $(x_2, y_2) = (1, -4)$. Then substitute these values into the distance formula.

$$d = \sqrt{(x_2 - x_1)^2 + (y_2 - y_1)^2} \quad \text{Distance formula}$$
$$= \sqrt{(1 - (-2))^2 + (-4 - 3)^2} \quad \text{Substitute.}$$
$$= \sqrt{9 + 49} \quad \text{Simplify.}$$
$$= \sqrt{58} \quad \text{Take the square root.}$$
$$\approx 7.62 \quad \text{Approximate.}$$

The distance between the points, as shown in Figure 8.42, is exactly $\sqrt{58}$ units, or about 7.62 units.

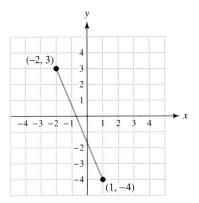

Figure 8.42

EXAMPLE 9   Designing a highway curve

Figure 8.43 shows a circular highway curve joining a straight section of road. A surveyor is trying to locate the *x*-coordinate of the *point of curvature PC* where the two sections of the highway meet. The distance between the surveyor and *PC* should be 400 feet. Estimate the *x*-coordinate of *PC* if *x* is positive.

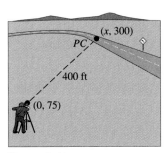

Figure 8.43

Solution

The distance between the points $(0, 75)$ and $(x, 300)$ is 400 feet. We can apply the distance formula and solve for *x*.

$$d = \sqrt{(x_2 - x_1)^2 + (y_2 - y_1)^2} \quad \text{Distance formula}$$
$$400 = \sqrt{(x - 0)^2 + (300 - 75)^2} \quad \text{Substitute.}$$
$$400^2 = (x - 0)^2 + (300 - 75)^2 \quad \text{Square both sides.}$$
$$160{,}000 = x^2 + 50{,}625 \quad \text{Simplify.}$$

$$109{,}375 = x^2 \qquad \text{Subtract } 50{,}625.$$
$$x = \sqrt{109{,}375} \qquad \text{Take the principal square root, } x > 0.$$
$$x \approx 330.7 \qquad \text{Approximate.}$$

The x-coordinate is about 330.7 feet.

8.4 Putting It All Together

In this section we focused on equations that contain either radical expressions or rational exponents. A good strategy for solving an equation containing radical expressions *symbolically*, is to raise both sides of the equation to the same integer power. However, checking answers is important to eliminate any extraneous solutions. Note that these extraneous solutions do not occur when an equation is being solved graphically or numerically.

Solution Method	Description	Example
Power rule for solving equations	If both sides of an equation are raised to the same power, any solutions to the original equation are among the solutions to the new equation.	$\sqrt{2x} = x$ $2x = x^2$ — Square both sides. $x^2 - 2x = 0$ — Rewrite equation. $x = 0$ or $x = 2$ — Factor and solve. Be sure to check any solutions.
Pythagorean theorem	If c is the hypotenuse of a right triangle and a and b are its legs, then $a^2 + b^2 = c^2$.	The sides of the right triangle are $a = 5$, $b = 12$, and $c = 13$. They satisfy the equation $$a^2 + b^2 = c^2$$ because $$5^2 + 12^2 = 13^2.$$
Distance formula	The distance d between the points (x_1, y_1) and (x_2, y_2) is $$d = \sqrt{(x_2 - x_1)^2 + (y_2 - y_1)^2}.$$	The distance between $(2, 3)$ and $(-3, 4)$ is $$d = \sqrt{(-3 - 2)^2 + (4 - 3)^2}$$ $$= \sqrt{(-5)^2 + (1)^2} = \sqrt{26}.$$

8.4 EXERCISES

Concepts

1. What is a good first step to use for solving $\sqrt{4x - 1} = 5$?

2. What is a good first step to use for solving $\sqrt[3]{x + 1} = 6$?

3. Can an equation involving rational exponents have more than one solution?

4. When you square each side of an equation to solve for an unknown, what must you do with any solutions?

5. What is the Pythagorean theorem used for?

6. If the legs of a right triangle are 3 and 4, what is the length of the hypotenuse?

7. What formula can you use to find the distance between the points $(3, -4)$ and $(1, 2)$?

8. Write the equation $\sqrt{x} + \sqrt[4]{x^3} = 2$ using rational exponents.

Graphical and Numerical Solutions

Exercises 9–14: Solve the equation graphically and numerically.

9. $\sqrt{3x} = 9$

10. $\sqrt{2x + 2} = 2$

11. $\sqrt{5z - 1} = \sqrt{z + 1}$

12. $y = \sqrt{y + 1} + 1$

13. $\sqrt[3]{x - 1} = \sqrt[3]{x}$

14. $\sqrt[3]{1 - x} = 1 - x$

Exercises 15–24: Solve the equation graphically. Approximate solutions to the nearest hundredth when appropriate.

15. $\sqrt[3]{x + 5} = 2$

16. $\sqrt[3]{x} + \sqrt{x} = 3.43$

17. $\sqrt{2x - 3} = \sqrt{x} - \frac{1}{2}$

18. $x^{4/3} - 1 = 2$

19. $x^{5/3} = 2 - 3x^2$

20. $x^{3/2} = \sqrt{x + 2} - 2$

21. $z^{1/3} - 1 = 2 - z$

22. $z^{3/2} - 2z^{1/2} - 1 = 0$

23. $\sqrt{y + 2} + \sqrt{3y + 2} = 2$

24. $\sqrt{x + 1} - \sqrt{x - 1} = 4$

Symbolic Solutions

Exercises 25–40: Solve the equation symbolically. Check your results.

25. $\sqrt{x} = 8$

26. $\sqrt{3z} = 6$

27. $\sqrt[4]{x} = 3$

28. $\sqrt[3]{x - 4} = 2$

29. $\sqrt{2t + 4} = 4$

30. $\sqrt{y + 4} = 3$

31. $\sqrt{x + 6} = x$

32. $\sqrt{z + 6} = z$

33. $\sqrt[3]{x} = 3$

34. $\sqrt[3]{x + 10} = 4$

35. $\sqrt{1 - 2x} = x + 7$

36. $\sqrt{4 - y} = y - 2$

37. $\sqrt{x} = \sqrt{x - 5} + 1$

38. $\sqrt{x - 1} = \sqrt{x + 4} - 1$

39. $\sqrt{2t - 2} + \sqrt{t} = 7$

40. $\sqrt{x + 1} - \sqrt{x - 6} = 1$

Using More Than One Method

Exercises 41–44: Solve the equation
   (a) symbolically,
   (b) graphically, and
   (c) numerically.

41. $2\sqrt{x} = 8$

42. $\sqrt[3]{5 - x} = 2$

43. $\sqrt{6z - 2} = 8$ 44. $\sqrt{y + 4} = \dfrac{y}{3}$

PYTHAGOREAN THEOREM

Exercises 45–52: If the sides of a triangle are a, b, and c and they satisfy $a^2 + b^2 = c^2$, the triangle is a right triangle. Determine whether the triangle with the given sides is a right triangle.

45. $a = 6$ $b = 8$ $c = 10$

46. $a = 5$ $b = 12$ $c = 13$

47. $a = 4$ $b = 5$ $c = 7$

48. $a = \sqrt{5}$ $b = \sqrt{9}$ $c = \sqrt{14}$

49. $a = 7$ $b = 24$ $c = 25$

50. $a = 1$ $b = \sqrt{3}$ $c = 2$

51. $a = 8$ $b = 8$ $c = 16$

52. $a = 11$ $b = 60$ $c = 61$

Exercises 53–56: Find the length of the missing side in the right triangle.

53.

54.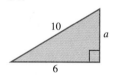

55.

56.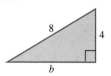

Exercises 57–62: A right triangle has legs a and b with hypotenuse c. Find the length of the missing side.

57. $a = 3, b = 4$

58. $a = 4, b = 7$

59. $a = \sqrt{3}, c = 8$

60. $a = \sqrt{6}, c = \sqrt{10}$

61. $b = 48, c = 50$

62. $b = 10, c = 26$

DISTANCE FORMULA

Exercises 63–66: Find the length of the line segment in the figure.

63.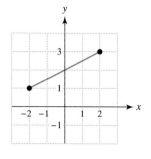

64.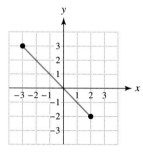

65.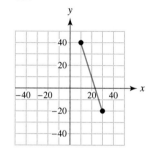

66.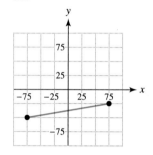

Exercises 67–70: Find the distance between the points.

67. $(-1, 2), (4, 10)$ 68. $(5, -40), (-6, 20)$

69. $(0, -3), (4, 0)$ 70. $(3, 9), (-4, 2)$

Exercises 71–74: (Refer to Example 9.) Find x if the distance between the given points is d. Assume that x is positive.

71. $(x, 3), (0, 6)$ $d = 5$

72. $(x, -1), (6, 11)$ $d = 13$

73. $(x, -5), (62, 6)$ $d = 61$

74. $(x, 3), (12, -4)$ $d = 25$

APPLICATIONS

Exercises 75 and 76: **Weight of a Bird** *(Refer to Example 4.) Estimate the weight of a bird having wings of area A.*

75. $A = 400$ square inches

76. $A = 1000$ square inches

Exercises 77–80: *Distance to the Horizon* Because of Earth's curvature, a person can see a limited distance to the horizon. The higher the location of the person, the farther that person can see. The distance D in miles to the horizon can be estimated by $D(h) = 1.22\sqrt{h}$, where h is the height of the person above the ground in feet.

77. Find D for a 6-foot tall person standing on level ground.

78. Find D for a person on top of Mount Everest with a height of 29,028 feet.

79. How high does a person need to be to see 20 miles?

80. How high does a plane need to fly for the pilot to be able to see 100 miles?

81. *Diagonal of a Television* (Refer to Example 7.) A rectangular television screen is 11.4 inches by 15.2 inches. Find the diagonal of the television set.

82. *Dimensions of a Television* The height of a television with a 13-inch diagonal is $\frac{3}{4}$ of its width. Find the width and height of the television set.

83. *DVD and Picture Dimensions* If the picture shown on a television set is h units high and w units wide, the *aspect ratio* of the picture is $\frac{w}{h}$ (see the accompanying figure). Digital video discs support the newer aspect ratio of $\frac{16}{9}$ rather than the older ratio of $\frac{4}{3}$. If the width of a picture with an aspect ratio of $\frac{16}{9}$ is 29 inches, approximate the height and diagonal of the rectangular picture. **(J. Taylor, *DVD Demystified*.)**

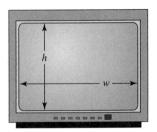

84. *Flood Control* The spillway capacity of a dam is important in flood control. Spillway capacity Q in cubic feet of water per second flowing over the spillway depends on the width W and the depth D of the spillway, as illustrated in the accompanying figure. If W and D are measured in feet, capacity can be modeled by $Q = 3.32WD^{3/2}$. **(Source: D. Callas, Project Director, *Snapshots of Applications in Mathematics*.)**

(a) Find the capacity of a spillway with $W = 20$ feet and $D = 5$ feet.

(b) A spillway with a width of 30 feet is to have a capacity of $Q = 2690$ cubic feet per second. Estimate the appropriate depth of the spillway.

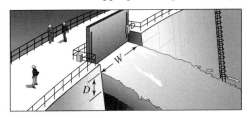

85. *Sky Diving* When sky divers initially fall from an airplane, their velocity v in miles per hour after free falling d feet can be approximated by $v = \frac{60}{11}\sqrt{d}$. Because of air resistance, they will eventually reach a terminal velocity and the formula will no longer be valid. How far do sky divers need to fall to attain the following velocities? (These values for d represent minimum distances.)

(a) 60 miles per hour
(b) 100 miles per hour

86. *Guy Wire* A guy wire attached to the top of a 30-foot long pole is anchored 10 feet from the base of the pole, as illustrated in the accompanying figure. Find the length of the guy wire to the nearest tenth of a foot.

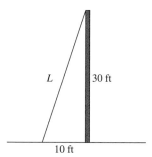

87. *Skid Marks* Vehicles involved in accidents often leave skid marks, which can be used to determine how fast a vehicle was traveling. To determine this speed, officials often use a test vehicle to compare skid marks on the same section of road. Suppose that a vehicle in a crash left skid marks D feet long and that a test vehicle traveling at v miles per hour leaves skid marks d feet long. Then the speed V of the vehicle involved in the crash is given by

$$V = v\sqrt{\frac{D}{d}}.$$

(Source: N. Garber and L. Hoel, *Traffic and Highway Engineering*.)

(a) Find V if $v = 30$ mph, $D = 285$ feet, and $d = 178$ feet. Interpret your result.
(b) A test vehicle traveling at 45 mph leaves skid marks that are 255 feet long. How long would the skid marks be for a vehicle traveling at 60 miles per hour?

88. *Highway Curves* If a circular curve without any banking has a radius of R feet, the speed limit L in miles per hour for the curve is given by $L = 1.5\sqrt{R}$. (Source: N. Garber.)

(a) Find the speed limit for a curve having a radius of 400 feet.
(b) If the radius of a curve doubles, what happens to the speed limit?
(c) A curve with a 40 mile per hour speed limit is being designed. What should be its radius?

89. *45°–45° Right Triangle* Suppose that the legs of a right triangle with angles of 45° and 45° both have length a, as depicted in the accompanying figure. Find the length of the hypotenuse.

90. *30°–60° Right Triangle* In a right triangle with angles of 30° and 60°, the shortest side is half the length of the hypotenuse (see the accompanying figure). If the hypotenuse has length c, find the length of the other two sides.

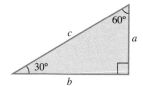

WRITING ABOUT MATHEMATICS

91. A student solves an equation *incorrectly* as follows.

$$\sqrt{3-x} = \sqrt{x} - 1$$
$$(\sqrt{3-x})^2 \stackrel{?}{=} (\sqrt{x})^2 - (1)^2$$
$$3 - x \stackrel{?}{=} x + 1$$
$$-2x \stackrel{?}{=} -2$$
$$x \stackrel{?}{=} 1$$

(a) How could you convince the student that the answer is wrong?
(b) Discuss where the error was made.

92. When both sides of an equation are squared, you must check your results. Explain why.

CHECKING BASIC CONCEPTS FOR SECTIONS 8.3 AND 8.4

1. Simplify each expression by hand.
   (a) $\sqrt{3} \cdot \sqrt{12}$
   (b) $\dfrac{\sqrt[3]{81}}{\sqrt[3]{3}}$
   (c) $\sqrt{36x^6},\ x > 0$

2. Simplify each expression.
   (a) $5\sqrt{6} + 2\sqrt{6} + \sqrt{7}$
   (b) $8\sqrt[3]{x} - 3\sqrt[3]{x}$
   (c) $\sqrt{4x} - \sqrt{x}$

3. Solve each equation symbolically. Check your results.
   (a) $x^{1/3} = 3$
   (b) $\sqrt{2x - 4} = 2$

4. Solve the equation symbolically and graphically.
$$\sqrt[3]{x - 1} = 3$$

5. A 16-inch diagonal television set has a rectangular picture with a width of 12.8 inches. Find the height of the picture.

8.5 COMPLEX NUMBERS

Basic Concepts ~ Addition, Subtraction, and Multiplication ~ Complex Conjugates and Division ~ Quadratic Equations

INTRODUCTION

Throughout history, people have created new numbers. Often these new numbers were met with resistance and regarded as being imaginary or unreal. The number 0 was not invented at the same time as the natural numbers. There was no Roman numeral for 0, which is one reason why our calendar started with A.D. 1 and, as a result, the twenty-first century began in 2001. No doubt there were skeptics during the time of the Roman Empire who questioned why anyone needed a number to represent nothing. Negative numbers also met strong resistance. After all, how could anyone possibly have -6 apples? The same was true for complex numbers, which relate to square roots of negative numbers. However, complex numbers are no more imaginary than any other number created by mathematics. Although the purpose of complex numbers was initially theoretical, today complex numbers are used in the design of electrical circuits, ships, and airplanes. Even the *fractal image* shown in Figure 8.44 would not have been discovered without complex numbers. (Source: D. Kincaid and W. Cheney, *Numerical Analysis*.)

8.5 Complex Numbers

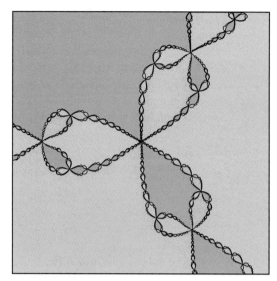

Figure 8.44 A Fractal: *The Cube Roots of Unity*

BASIC CONCEPTS

A graph of $y = x^2 + 1$ is shown in Figure 8.45. There are no *x*-intercepts, so the equation $x^2 + 1 = 0$ has no real number solutions.

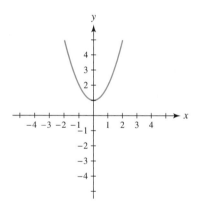

Figure 8.45

If we try to solve $x^2 + 1 = 0$ symbolically, it results in $x^2 = -1$. Because $x^2 \geq 0$ for any real number *x*, there are no real solutions. However, we can invent solutions.

$$x^2 = -1$$
$$x = \pm\sqrt{-1} \quad \text{Square root property}$$

We now define a new number called the **imaginary unit**, denoted *i*.

Properties of the Imaginary Unit *i*

$$i = \sqrt{-1} \quad \text{and} \quad i^2 = -1$$

By inventing the number *i*, the solutions to the equation $x^2 + 1 = 0$ are i and $-i$. Using the real numbers and the imaginary unit *i*, we can define a new set of numbers called the *complex numbers*. A **complex number** can be written in **standard form**, as $a + bi$, where *a* and *b* are real numbers. The **real part** is *a* and the **imaginary part** is *b*. Every real number *a* is also a complex number because it can be written $a + 0i$. A complex number $a + bi$ with $b \neq 0$ is an **imaginary number**. Table 8.2 lists several complex numbers with their real and imaginary parts.

TABLE 8.2

Complex Number: $a + bi$	$-3 + 2i$	5	$-3i$	$-1 + 7i$	$-5 - 2i$	$4 + 6i$
Real Part: a	-3	5	0	-1	-5	4
Imaginary Part: b	2	0	-3	7	-2	6

Figure 8.46 shows how different sets of numbers are related. Note that *the set of complex numbers contains the set of real numbers.*

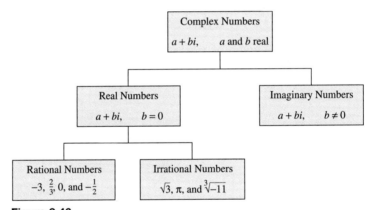

Figure 8.46

Using the imaginary unit *i*, we may write the square root of a negative number as a complex number. For example, $\sqrt{-2} = i\sqrt{2}$, and $\sqrt{-4} = i\sqrt{4} = 2i$. This method is summarized as follows.

The Expression $\sqrt{-a}$

If $a > 0$, then $\sqrt{-a} = i\sqrt{a}$.

EXAMPLE 1 *Writing the square root of a negative number*

Write each square root using the imaginary unit *i*.

(a) $\sqrt{-25}$ (b) $\sqrt{-7}$

Solution

(a) $\sqrt{-25} = i\sqrt{25} = 5i$ (b) $\sqrt{-7} = i\sqrt{7}$

ADDITION, SUBTRACTION, AND MULTIPLICATION

Arithmetic operations can be defined for complex numbers.

ADDITION AND SUBTRACTION. To add the complex numbers $(-3 + 2i)$ and $(2 - i)$ add the real parts and then add the imaginary parts.

$$(-3 + 2i) + (2 - i) = (-3 + 2) + (2i - i)$$
$$= (-3 + 2) + (2 - 1)i$$
$$= -1 + i$$

This same process works for subtraction.

$$(6 - 3i) - (2 + 5i) = (6 - 2) + (-3i - 5i)$$
$$= (6 - 2) + (-3 - 5)i$$
$$= 4 - 8i$$

This method is summarized as follows.

Sum or Difference of Complex Numbers

Let $a + bi$ and $c + di$ be two complex numbers. Then

$$(a + bi) + (c + di) = (a + c) + (b + d)i \quad \text{Sum}$$

and

$$(a + bi) - (c + di) = (a - c) + (b - d)i \quad \text{Difference}$$

EXAMPLE 2 *Adding and subtracting complex numbers*

Write each sum or difference in standard form.

(a) $(-7 + 2i) + (3 - 4i)$ **(b)** $3i - (5 - i)$

Solution

(a) $(-7 + 2i) + (3 - 4i) = (-7 + 3) + (2 - 4)i = -4 - 2i$
(b) $3i - (5 - i) = 3i - 5 + i = -5 + (3 + 1)i = -5 + 4i$

Technology Note *Complex Numbers* ─────────────

Many calculators can perform arithmetic with complex numbers. Figure 8.47 shows the graphing calculator display for completion of Example 2.

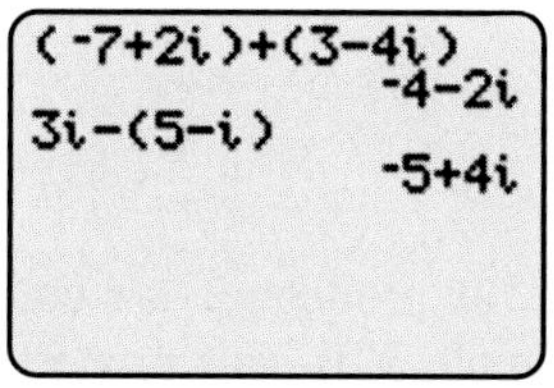

```
(-7+2i)+(3-4i)
              -4-2i
3i-(5-i)
              -5+4i
```

Figure 8.47

MULTIPLICATION. We multiply two complex numbers like binomials and apply the property $i^2 = -1$.

EXAMPLE 3   Multiplying complex numbers

Write each product in standard form.
(a) $(2 - 3i)(1 + 4i)$ (b) $(5 - 2i)(5 + 2i)$

Solution

(a) Multiply the complex numbers like binomials.
$$(2 - 3i)(1 + 4i) = (2)(1) + (2)(4i) - (3i)(1) - (3i)(4i)$$
$$= 2 + 8i - 3i - 12i^2$$
$$= 2 + 5i - 12(-1)$$
$$= 14 + 5i$$

(b) $(5 - 2i)(5 + 2i) = (5)(5) + (5)(2i) - (2i)(5) - (2i)(2i)$
$$= 25 + 10i - 10i - 4i^2$$
$$= 25 - 4(-1)$$
$$= 29$$

These results are supported in Figure 8.48.

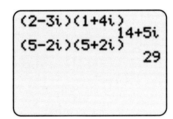

Figure 8.48

COMPLEX CONJUGATES AND DIVISION

The **complex conjugate** of $a + bi$ is $a - bi$. To find the conjugate, we change the sign of the imaginary part b. Table 8.3 contains examples of complex numbers and their conjugates.

TABLE 8.3   **Complex Conjugates**

Number	$2 + 5i$	$6 - 3i$	$-2 + 7i$	$-1 - i$	5	$-4i$
Conjugate	$2 - 5i$	$6 + 3i$	$-2 - 7i$	$-1 + i$	5	$4i$

The product of two complex conjugates is a real number, as we demonstrated in Example 3(b). This property is used to divide two complex numbers. To convert the quotient $\frac{2 + 3i}{3 - i}$ into standard form $a + bi$, we multiply the numerator and the denominator by the complex conjugate of the denominator, which is $3 + i$.

EXAMPLE 4 Dividing complex numbers

Write each quotient in standard form.

(a) $\dfrac{2 + 3i}{3 - i}$ (b) $\dfrac{4}{2i}$

Solution

(a) Multiply the numerator and denominator by $3 + i$.

$$\dfrac{2 + 3i}{3 - i} = \dfrac{(2 + 3i)(3 + i)}{(3 - i)(3 + i)} \quad \text{Multiply by the complex conjugate.}$$

$$= \dfrac{2(3) + (2)(i) + (3i)(3) + (3i)(i)}{(3)(3) + (3)(i) - (i)(3) - (i)(i)} \quad \text{Expand.}$$

$$= \dfrac{6 + 2i + 9i + 3i^2}{9 + 3i - 3i - i^2} \quad \text{Simplify.}$$

$$= \dfrac{6 + 11i + 3(-1)}{9 - (-1)} \quad i^2 = -1$$

$$= \dfrac{3 + 11i}{10} \quad \text{Simplify.}$$

$$= \dfrac{3}{10} + \dfrac{11}{10}i \quad \dfrac{a + bi}{c} = \dfrac{a}{c} + \dfrac{b}{c}i$$

(b) Multiply the numerator and denominator by $-2i$.

$$\dfrac{4}{2i} = \dfrac{(4)(-2i)}{(2i)(-2i)} \quad \text{Multiply by the complex conjugate.}$$

$$= \dfrac{-8i}{-4i^2} \quad \text{Simplify.}$$

$$= \dfrac{-8i}{-4(-1)} \quad i^2 = -1$$

$$= \dfrac{-8i}{4} \quad \text{Simplify.}$$

$$= -2i \quad \text{Divide.}$$

These results are supported in Figure 8.49.

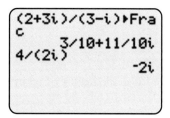

Figure 8.49

QUADRATIC EQUATIONS

In Chapter 6 we showed that, if the discriminant $b^2 - 4ac < 0$, there are no real solutions to a quadratic equation. However, using the imaginary unit i, we can obtain two imaginary solutions. For example, suppose that we solve the equation $x^2 + 1 = 0$.

Then $a = 1, b = 0, c = 1$, and $b^2 - 4ac = -4 < 0$.

$$x = \frac{-b \pm \sqrt{b^2 - 4ac}}{2a} \quad \text{Quadratic formula}$$

$$= \frac{-0 \pm \sqrt{0^2 - 4(1)(1)}}{2(1)} \quad \text{Substitute for } a, b, \text{ and } c.$$

$$= \frac{\pm \sqrt{-4}}{2} \quad \text{Simplify.}$$

$$= \pm \frac{2i}{2} \quad \sqrt{-4} = i\sqrt{4} = 2i$$

$$= \pm i \quad \text{Divide.}$$

The solutions to the equation are $\pm i$. We can check these solutions by substituting them into the given equation, $x^2 + 1 = 0$.

$$(i)^2 + 1 = -1 + 1 = 0 \quad \text{It checks.}$$
$$(-i)^2 + 1 = i^2 + 1 = -1 + 1 = 0 \quad \text{It checks.}$$

EXAMPLE 5 *Solving a quadratic equation having complex solutions*

Solve $2x^2 + x + 3 = 0$.

Solution

Let $a = 2, b = 1$, and $c = 3$. Note that $b^2 - 4ac = 1^2 - 4(2)(3) = -23 < 0$.

$$x = \frac{-b \pm \sqrt{b^2 - 4ac}}{2a} \quad \text{Quadratic formula}$$

$$= \frac{-1 \pm \sqrt{1^2 - 4(2)(3)}}{2(2)} \quad \text{Substitute for } a, b, \text{ and } c.$$

$$= \frac{-1 \pm \sqrt{-23}}{4} \quad \text{Simplify.}$$

$$= \frac{-1 \pm i\sqrt{23}}{4} \quad \sqrt{-23} = i\sqrt{23}$$

$$= -\frac{1}{4} \pm i\frac{\sqrt{23}}{4} \quad \frac{x + yi}{z} = \frac{x}{z} + i\frac{y}{z}$$

Critical Thinking

Use the results of Example 5 to simplify

$$2\left(-\frac{1}{4} + i\frac{\sqrt{23}}{4}\right)^2 + \left(-\frac{1}{4} + i\frac{\sqrt{23}}{4}\right) + 3$$

and

$$2\left(-\frac{1}{4} - i\frac{\sqrt{23}}{4}\right)^2 + \left(-\frac{1}{4} - i\frac{\sqrt{23}}{4}\right) + 3.$$

8.5 Putting It All Together

In this section we discussed complex numbers and how to perform arithmetic operations with them. Complex numbers allow us to solve equations that could not be solved only with real numbers. For example, quadratic equations that have no real solutions now have two imaginary solutions. The following table summarizes the important concepts in the section.

Concept	Explanation	Example
Complex numbers	A complex number can be expressed as $a + bi$, where a and b are real numbers. The imaginary unit i satisfies $i = \sqrt{-1}$ and $i^2 = -1$. As a result, we can write $\sqrt{-a} = i\sqrt{a}$ if $a > 0$.	$\sqrt{-13} = i\sqrt{13}$ and $\sqrt{-9} = 3i$
Addition, subtraction, and multiplication	To add (subtract) complex numbers, add (subtract) the real parts and then add (subtract) the imaginary parts. Multiply complex numbers in a similar manner to how *FOIL* is used to multiply binomials. Then apply the property $i^2 = -1$.	$(3 + 6i) + (-1 + 2i)$ Sum $= (3 + -1) + (6 + 2)i$ $= 2 + 8i$ $(2 - 5i) - (1 + 4i)$ Difference $= (2 - 1) + (-5 - 4)i$ $= 1 - 9i$ $(-1 + 2i)(3 + i)$ Product $= (-1)(3) + (-1)(i) + (2i)(3) + (2i)(i)$ $= -3 - i + 6i + 2i^2$ $= -3 + 5i + 2(-1)$ $= -5 + 5i$
Complex conjugates	The conjugate of $a + bi$ is $a - bi$.	The conjugate of $3 - 5i$ is $3 + 5i$.
Division	To simplify a quotient, multiply the numerator and denominator by the complex conjugate of the denominator. Then simplify the expression and write it in standard form as $a + bi$.	$\dfrac{10}{1 + 2i} = \dfrac{10(1 - 2i)}{(1 + 2i)(1 - 2i)}$ $= \dfrac{10 - 20i}{5}$ $= 2 - 4i$
Quadratic formula and complex solutions	If the discriminant is negative ($b^2 - 4ac < 0$), the solutions to a quadratic equation are imaginary numbers.	$2x^2 - x + 3 = 0$ $x = \dfrac{1 \pm \sqrt{(-1)^2 - 4(2)(3)}}{2(2)}$ $= \dfrac{1 \pm \sqrt{-23}}{4} = \dfrac{1}{4} \pm i\dfrac{\sqrt{23}}{4}$

8.5 EXERCISES

CONCEPTS

1. Give an example of a complex number that is not a real number.

2. Can you give an example of a real number that is not a complex number? Explain.

3. $\sqrt{-1} = $ _____

4. $i^2 = $ _____

5. $\sqrt{-a} = $ _____, if $a > 0$.

6. The complex conjugate of $10 + 7i$ is _____.

7. The standard form for a complex number is _____.

8. Write $\dfrac{2 + 4i}{2}$ in standard form.

9. The real part of $4 - 5i$ is _____.

10. The imaginary part of $4 - 5i$ is _____.

COMPLEX NUMBERS

Exercises 11–14: Use the imaginary unit to write the expression.

11. $\sqrt{-5}$
12. $\sqrt{-21}$
13. $\sqrt{-100}$
14. $\sqrt{-49}$

Exercises 15–26: Write the expression in standard form.

15. $(5 + 3i) + (-2 - 3i)$
16. $(1 - i) + (5 - 7i)$
17. $(2i) + (-8 + 5i)$
18. $(-3i) + (5i)$
19. $(2 - 7i) - (1 + 2i)$
20. $(1 + 8i) - (3 + 9i)$
21. $(5i) - (10 - 2i)$
22. $(1 + i) - (1 - i)$
23. $4(5 - 3i)$
24. $(1 + 2i)(-6 - i)$
25. $(-3 - 4i)(5 - 4i)$
26. $(3 + 5i)(3 - 5i)$

Exercises 27–30: Write the complex conjugate.

27. $3 + 4i$
28. $1 - 4i$
29. $-6i$
30. -10

Exercises 31–38: Write the expression in standard form.

31. $\dfrac{2}{1 + i}$
32. $\dfrac{-6}{2 - i}$
33. $\dfrac{3i}{5 - 2i}$
34. $\dfrac{-8}{2i}$
35. $\dfrac{8 + 9i}{5 + 2i}$
36. $\dfrac{3 - 2i}{1 + 4i}$
37. $\dfrac{5 + 7i}{1 - i}$
38. $\dfrac{-7 + 4i}{3 - 2i}$

QUADRATIC EQUATIONS

Exercises 39–48: Solve the equation. Write complex solutions in standard form.

39. $x^2 + 2 = 0$
40. $x^2 + 4 = 0$
41. $x^2 - x + 2 = 0$
42. $x^2 + 2x + 3 = 0$
43. $2x^2 + 3x + 4 = 0$
44. $3x^2 - x = 1$
45. $x^2 + 1 = 4x$
46. $3x^2 + 2 = x$
47. $x^2 + x = -2$
48. $x(x - 4) = -8$

APPLICATIONS

Exercises 49–50: Corrosion in Airplanes Corrosion in the metal surface of an airplane can be difficult to detect visually. One test used to locate it involves passing an alternating current through a small area on the plane's surface. If the current varies from one region to another, it may indicate that corrosion is occurring. The impedance Z (or opposition to the flow of electricity) of the metal is related to the voltage V and current I by the equation $Z = \dfrac{V}{I}$, where Z, V, and I are complex numbers. Calculate Z for the given values of V and I. **(Source: Society for Industrial and Applied Mathematics).**

49. $V = 40 + 70i, I = 2 + 3i$

50. $V = 10 + 20i, I = 3 + 7i$

WRITING ABOUT MATHEMATICS

51. A student multiplies $(2 + 3i)(4 - 5i)$ *incorrectly* to obtain $8 - 15i$. What was the student's mistake?

52. A student divides $\dfrac{6 - 10i}{3 + 2i}$ *incorrectly* to obtain $2 - 5i$. What was the student's mistake?

CHECKING BASIC CONCEPTS FOR SECTION 8.5

1. Use the imaginary unit i to write each expression.
   (a) $\sqrt{-64}$
   (b) $\sqrt{-17}$

2. Simplify each expression.
   (a) $(2 - 3i) + (1 - i)$
   (b) $4i - (2 + i)$
   (c) $(3 - 2i)(1 + i)$
   (d) $\dfrac{3}{2 - 2i}$

3. Graph $y = x^2 + 2x + 3$.
   (a) How many real solutions are there to $x^2 + 2x + 3 = 0$?
   (b) Solve the equation $x^2 + 2x + 3 = 0$.

CHAPTER 8 SUMMARY

Section 8.1 Radical Expressions and Rational Exponents

Square Root
The number b is a square root of a if $b^2 = a$. The square roots of 16 are 4 and -4. The principal square root of 16 is denoted $\sqrt{16} = 4$ and is always positive.

Cube Root
The number c is a cube root of a if $c^3 = a$ and is denoted $\sqrt[3]{a} = c$. Thus $\sqrt[3]{27} = 3$ because $3^3 = 3 \cdot 3 \cdot 3 = 27$.

nth Root
In general, d is an nth root of a if $d^n = a$ and is denoted $\sqrt[n]{a} = d$.

Rational Exponents
The nth root of a can be denoted by using a rational exponent $\frac{1}{n}$. The expression $a^{1/n}$ equals $\sqrt[n]{a}$. For example, $16^{1/4} = \sqrt[4]{16} = 2$ because $2 \cdot 2 \cdot 2 \cdot 2 = 16$.

The expressions $\sqrt[n]{a^m}$ or $(\sqrt[n]{a})^m$ may be represented by rational exponents as $a^{m/n}$, where $\frac{m}{n}$ is in lowest terms. For example, $8^{2/3} = (\sqrt[3]{8})^2 = 2^2 = 4$ and $8^{2/3} = \sqrt[3]{8^2} = \sqrt[3]{64} = 4$.

Properties of Exponents
Properties of rational exponents are similar to the properties of integer exponents. However, the even root of a negative number is not a real number. For example, $\sqrt{-8}$ does not represent a real number, whereas $\sqrt[3]{-8} = -2$ does.

Section 8.2 Radical Functions

Square Root Function
The square root function is given by $f(x) = \sqrt{x}$. Its domain is all nonnegative real numbers.

Cube Root Function
The cube root function is given by $f(x) = \sqrt[3]{x}$. Its domain is all real numbers.

Power Function
A power function may be written as $f(x) = x^p$, where p is a rational number. If $p = \frac{1}{n}$, where $n \geq 2$ is an integer, then f is also a root function. A root function can be represented by $f(x) = \sqrt[n]{x}$.

Section 8.3 Operations on Radical Expressions

To add or subtract radical expressions, we combine like radicals. Radical expressions can be simplified, added, subtracted, multiplied, and divided with the help of the properties

$$\sqrt[n]{a} \cdot \sqrt[n]{b} = \sqrt[n]{ab} \quad \text{and} \quad \sqrt[n]{\frac{a}{b}} = \frac{\sqrt[n]{a}}{\sqrt[n]{b}}.$$

Examples:
$5\sqrt{2} + 3\sqrt{2} = 8\sqrt{2}.$
$10\sqrt{12} - 5\sqrt{3} = 10\sqrt{4} \cdot \sqrt{3} - 5\sqrt{3} = 20\sqrt{3} - 5\sqrt{3} = 15\sqrt{3}$
$\sqrt[3]{4} \cdot \sqrt[3]{16} = \sqrt[3]{64} = 4$
$\frac{\sqrt{20}}{\sqrt{5}} = \sqrt{\frac{20}{5}} = \sqrt{4} = 2$

Section 8.4 *Equations Involving Radical Expressions*

One strategy for solving equations containing radical expressions is to use the power rule. To apply this rule raise each side of an equation to the same power. Be sure to check each solution.

Example:
$$\sqrt{2x + 2} = 2x \quad \text{Given equation}$$
$$(\sqrt{2x + 2})^2 = (2x)^2 \quad \text{Square each side.}$$
$$2x + 2 = 4x^2 \quad \text{Properties of exponents}$$
$$4x^2 - 2x - 2 = 0 \quad \text{Write as } ax^2 + bx + c = 0$$
$$(4x + 2)(x - 1) = 0 \quad \text{Factor.}$$
$$x = -\frac{1}{2} \quad \text{or} \quad x = 1 \quad \text{Solve.}$$

The solution of $x = 1$ checks, but $x = -\frac{1}{2}$ does not check.

If a right triangle has legs a and b with hypotenuse c, by the Pythagorean theorem $a^2 + b^2 = c^2$.

Example: If the legs of a triangle are 5 and 7 units, the hypotenuse is
$$c^2 = 5^2 + 7^2 = 74 \quad \text{or} \quad c = \sqrt{74}.$$

To find the distance d between two points (x_1, y_1) and (x_2, y_2) use the distance formula
$$d = \sqrt{(x_2 - x_1)^2 + (y_2 - y_1)^2}.$$

Example: The distance between $(1, -2)$ and $(-4, 3)$ is
$$d = \sqrt{(-4 - 1)^2 + (3 + 2)^2} = \sqrt{50}$$

Section 8.5 *Complex Numbers*

We defined a new number called the imaginary unit i, such that
$$i = \sqrt{-1} \quad \text{and} \quad i^2 = -1.$$

The introduction of i allowed creation of a new number system called the *complex numbers*. The standard form of a complex number is $a + bi$, where a and b are real numbers. When $b \neq 0$, $a + bi$ is an imaginary number. An example is $5 - 2i$, where the real part is $a = 5$, and the imaginary part is $b = -2$. We can add, subtract, multiply, and divide complex numbers.

Examples:
$$(2 + 3i) + (3 - i) = (2 + 3) + (3 - 1)i = 5 + 2i$$
$$(2 + 3i) - (3 - i) = (2 - 3) + (3 + 1)i = -1 + 4i$$
$$(2 + 3i)(3 - i) = (2)(3) - (2)(i) + (3i)(3) - (3i)(i) = 9 + 7i$$
$$\frac{2 + 3i}{3 - i} = \frac{(2 + 3i)(3 + i)}{(3 - i)(3 + i)} = \frac{3}{10} + \frac{11}{10}i$$

Complex numbers provide solutions for quadratic equations that have no real number solutions. For example, $x^2 = -4$ has no real number solutions. The use of complex numbers yields two solutions: $x = \pm 2i$. When $b^2 - 4ac < 0$, the quadratic equation $ax^2 + bx + c = 0$ has two imaginary solutions.

CHAPTER 8
REVIEW EXERCISES

SECTION 8.1

Exercises 1–12: Simplify the expression.

1. $\sqrt{4}$
2. $\sqrt{36}$
3. $\sqrt{9x^2}$
4. $\sqrt{(x-1)^2}$
5. $\sqrt[3]{64}$
6. $\sqrt[3]{-125}$
7. $\sqrt[3]{x^6}$
8. $\sqrt[3]{27x^3}$
9. $\sqrt[4]{16}$
10. $\sqrt[5]{-1}$
11. $\sqrt[4]{x^8}$
12. $\sqrt[5]{(x+1)^5}$

Exercises 13–16: Write the expression in radical notation.

13. $14^{1/2}$
14. $(-5)^{1/3}$
15. $\left(\dfrac{x}{y}\right)^{3/2}$
16. $(xy)^{-2/3}$

Exercises 17–20: Evaluate the expression without a calculator.

17. $(-27)^{2/3}$
18. $16^{1/4}$
19. $16^{3/2}$
20. $81^{3/4}$

Exercises 21–24: Simplify the expression. Assume that all variables are positive.

21. $(z^3)^{2/3}$
22. $(x^2 y^4)^{1/2}$
23. $\left(\dfrac{x^2}{y^6}\right)^{3/2}$
24. $\left(\dfrac{x^3}{y^6}\right)^{-1/3}$

SECTION 8.2

Exercises 25 and 26: Graph the equation by solving for y and using your graphing calculator.

25. $x = y^2 + 1$
26. $x = 4y^2$

Exercises 27–30: Graph the following equation. Compare the graph to either $y = \sqrt{x}$ or $y = \sqrt[3]{x}$.

27. $y = \sqrt{x} - 2$
28. $y = \sqrt{(x+3)}$
29. $y = \sqrt[3]{x} - 1$
30. $y = 2\sqrt[3]{x}$

31. Evaluate $f(x) = x^{-1/2}$ at $x = 16$.
32. Graph $f(x) = x^{2/3}$ in $[-6, 6, 1]$ by $[-6, 6, 1]$.

SECTION 8.3

Exercises 33–38: Simplify the expression. Assume that all variables are positive.

33. $\sqrt{2} \cdot \sqrt{32}$
34. $\sqrt[3]{-4} \cdot \sqrt[3]{2}$
35. $\sqrt[3]{x^4} \cdot \sqrt[3]{x^2}$
36. $\dfrac{\sqrt{80}}{\sqrt{20}}$
37. $\sqrt[3]{-\dfrac{x}{8}}$
38. $\sqrt{\dfrac{1}{3}} \cdot \sqrt{\dfrac{1}{3}}$

Exercises 39–46: Simplify the expression. Assume that all variables are positive.

39. $3\sqrt{3} + \sqrt{3}$
40. $\sqrt[3]{x} + 2\sqrt[3]{x}$
41. $3\sqrt[3]{5} - 6\sqrt[3]{5}$
42. $\sqrt[4]{y} - 2\sqrt[4]{y}$
43. $2\sqrt{12} + 7\sqrt{3}$
44. $3\sqrt{18} - 2\sqrt{2}$
45. $7\sqrt[3]{16} - \sqrt[3]{2}$
46. $\sqrt{4x + 4} + \sqrt{x + 1}$

SECTION 8.4

Exercises 47 and 48: Solve the equation graphically and numerically.

47. $\sqrt{x + 2} = x$
48. $\sqrt{2x - 1} = \sqrt{x + 3}$

Exercise 49 and 50: Solve the equation graphically. Approximate solutions to the nearest hundredth when appropriate.

49. $\sqrt[3]{2x - 1} = 2$
50. $x^{2/3} = 3 - x$

Exercises 51–54: Solve the equation symbolically. Check your results.

51. $\sqrt{x - 1} = 2$
52. $\sqrt[3]{3x} = 3$
53. $\sqrt{2x} = x - 4$
54. $\sqrt{x + 1} = \sqrt{x + 2}$

Exercises 55 and 56: A right triangle has legs a and b with hypotenuse c. Find the length of the missing side.

55. $a = 4, b = 7$ **56.** $a = 5, c = 8$

Exercises 57 and 58: Find the distance between the points.

57. $(-2, 3), (2, -2)$ **58.** $(2, -3), (-4, 1)$

Section 8.5

Exercises 59–62: Write the complex expression in standard form.

59. $(1 - 2i) + (-3 + 2i)$

60. $(1 + 3i) - (3 - i)$ **61.** $(1 - i)(2 + 3i)$

62. $\dfrac{3 + i}{1 - i}$

Exercises 63–66: Solve the equation. Write any complex solution in standard form.

63. $x^2 + 9 = 0$ **64.** $x^2 + x + 1 = 0$

65. $3x^2 + 1 = x$ **66.** $x(x - 1) = -2$

Applications

67. *Hang Time*   (Refer to Example 1, Section 8.2.) A football is punted and has a hang time of 4.6 seconds. Estimate the height that it was kicked to the nearest foot.

68. *Baseball Diamond*   The four bases of a baseball diamond form a square 90 feet on a side. Find the distance from home plate to second base.

69. *Falling Time*   The time T in seconds for an object to fall from a height of h feet is given by $T = \frac{1}{4}\sqrt{h}$. If a person steps off a 10-foot high board into a swimming pool, how long is the person in the air?

70. *Highway Curves*   If a circular highway curve is banked with a slope of $m = \dfrac{1}{10}$ (see the accompanying figure) and has a radius of R feet, then the speed limit L in miles per hour for the curve is given by
$$L = \sqrt{3.75R}.$$
(Source: N. Garber and L. Hoel, *Traffic and Highway Engineering*.)
(a) Find the speed limit if the curve has a radius of 500 feet.
(b) With no banking, the speed limit is given by $L = 1.5\sqrt{R}$. Find the speed limit for a curve with no banking and a radius of 500 feet. How does banking affect the speed limit? Does this result agree with your intuition?

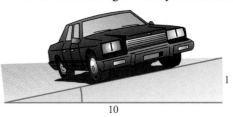

71. *Geometry*   Find the length of a side of a square if the square has an area of 7 square feet.

72. *Geometry*   A cube has sides of length $\sqrt{5}$.
(a) Find the area of one side of the cube.
(b) Find the volume of the cube.
(c) Find the length of the diagonal of one of the sides.
(d) Find the distance from A to B in the figure.

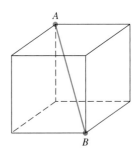

73. *Pendulum*   The time required for a pendulum to swing back and forth is called its *period* (see the accompanying figure). The period T of a pendulum does not depend on its weight, only on its length L and gravity. It is given by
$$T = 2\pi\sqrt{\dfrac{L}{32.2}},$$
where T is in seconds and L is in

feet. Estimate the length of a pendulum with a period of 1 second.

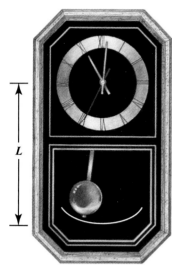

74. *Pendulum* (Refer to Exercise 73.) If a pendulum were on the moon, its period could be calculated by $T = 2\pi\sqrt{\dfrac{L}{5.1}}$. Estimate the length of a pendulum with a period of 1 second on the moon. Compare your answer to that for Exercise 73.

75. *Population Growth* In 1790 the population of the United States was 4 million, and by 1990 it had grown to 249 million. The annual percentage growth in the population r (expressed as a decimal) can be determined by the equation $249 = 4(1 + r)^{200}$. Solve this equation for r and interpret the result.

76. *Surface Area of a Cone* The surface area of a cone having radius r and height h is given by $S = \pi r\sqrt{r^2 + h^2}$. See the accompanying figure. Estimate the surface area if $r = 11$ inches and $h = 60$ inches.

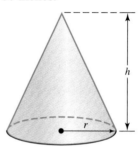

77. *Radioactive Carbon Dating* Living plants and animals have a constant amount of radioactive carbon in their cells, which comes from the carbon dioxide they breathe. When a plant or animal dies, the exchange of oxygen and carbon dioxide halts and the amount of radioactive carbon starts to decrease. The fraction of radioactive carbon remaining t years after death is given by $2^{-t/5700}$. Find the fraction left after each time period.
(a) 5700 years (b) 20,000 years

CHAPTER 8
TEST

Exercises 1 and 2: Simplify the expression.

1. $\sqrt{25x^4}$ 2. $\sqrt[3]{8z^6}$

Exercises 3 and 4: Write the expression in radical notation.

3. $7^{2/5}$ 4. $\left(\dfrac{x}{y}\right)^{-2/3}$

Exercises 5 and 6: Evaluate the expression without a calculator.

5. $(-8)^{4/3}$ 6. $36^{-3/2}$

7. Graph the equation $x = 2y^2 - 3$ by solving for y and using your graphing calculator.

8. Compare the graph of $f(x) = \sqrt[3]{x - 3} + 1$ to the graph of $y = \sqrt[3]{x}$.

Exercises 9–14: Simplify the expression. Assume that all variables are positive.

9. $(2z^{1/2})^3$ 10. $\left(\dfrac{y^2}{z^3}\right)^{-1/3}$

11. $\sqrt{3} \cdot \sqrt{27}$ 12. $\dfrac{\sqrt{y^3}}{\sqrt{4y}}$

13. $7\sqrt{7} - 3\sqrt{7} + \sqrt{5}$

14. $7\sqrt[3]{x} - \sqrt[3]{x}$

15. Solve the equation $\sqrt{2x + 2} = x - 11$ symbolically.

16. Solve the equation $\sqrt{3x + 1} = \sqrt{x + 4} + 1$ either graphically or numerically.

17. One leg of a right triangle has length 7 and the hypotenuse has length 13. Find the length of the third side.

18. Find the distance between the points $(-3, 5)$ and $(-1, 7)$.

Exercises 19–22: Write the complex expression in standard form.

19. $(-5 + i) + (7 - 20i)$

20. $(3i) - (6 - 5i)$

21. $\left(\frac{1}{2} - i\right)\left(\frac{1}{2} + i\right)$

22. $\frac{2i}{5 + 2i}$

23. Solve $2x^2 - x + 3 = 0$. Write the complex solutions in standard form.

24. *Volume of a Sphere* The volume V of a sphere is given by $V = \frac{4}{3}\pi r^3$, where r is its radius.
    (a) Solve the equation for r.
    (b) Find the radius of a sphere with a volume of 50 cubic inches.

25. *Windchill Factor* The windchill factor is a measure of the cooling effect that the wind has on a person's skin. It calculates the equivalent cooling temperature if there were no wind. The windchill factor W is based on two variables: the outside temperature T and the wind velocity V. The formula
$$W = 91.4 - (91.4 - T) \cdot (0.478 + 0.301\sqrt{V} - 0.02V)$$
can be used to calculate the windchill factor, where T is in degrees Fahrenheit and V is in miles per hour.
    (a) Find the windchill when $T = -20°F$ and $V = 20$ mph.
    (b) Substitute $T = 10$ into the formula. Then graph W in $[0, 50, 10]$ by $[-40, 40, 10]$. Describe what happens to the windchill as the wind velocity increases.

26. *Wing Span of a Bird* The wing span L of a bird with weight W can sometimes be modeled by $L = 27.4W^{1/3}$, where L is in inches and W is in pounds. Use this formula to estimate the weight of a bird that has a wing span of 30 inches. (*Source:* C. Pennycuick, *Newton Rules Biology.*)

CHAPTER 8

EXTENDED AND DISCOVERY EXERCISES

1. *Moons of Jupiter* The following table lists the orbital distances and periods of several moons of Jupiter. Let x represent the distance and y the period. These data can be modeled by a power function of the form $f(x) = 0.0002x^{m/n}$. Use trial and error to find the value of the fraction $\frac{m}{n}$. Graph f and the data in the same viewing rectangle.

Moon	Distance (10^3 kilometers)	Period (days)
Metis	128	0.29
Almathea	181	0.50
Thebe	222	0.67
Io	422	1.77
Europa	671	3.55
Ganymede	1070	7.16
Callisto	1883	16.69

Source: M. Zeilik, *Introductory Astronomy and Astrophysics.*

2. *Powers of i* Calculate the following powers of i.

$$i^2, \quad i^3, \quad i^4, \quad i^5, \quad i^6, \quad i^7, \quad i^8, \text{ and } i^9$$

For example, $i^3 = i \cdot i^2 = i \cdot (-1) = -i$. Continue calculating powers of i until you discover a pattern. If you are given i^n, for some natural number n, can you determine the value of i^n without actually calculating it? Explain. Try your method on i^{200} and i^{77}.

3. *Modeling Wood in a Tree* In forestry the volume of timber in a given area of forest is often estimated. To make such estimates formulas are developed to find the amount of wood contained in a tree with height h in feet and diameter d in inches. One study concluded that the volume V of wood in a tree could be modeled by the equation $V = kh^{1.12}d^{1.98}$, where k is a constant. The diameter is measured 4.5 feet above the ground. (Source: B. Ryan, B. Joiner, and T. Ryan, *Minitab Handbook*.)
   (a) A tree with 11-inch diameter and a 47-foot height has a volume of 11.4 cubic feet. Approximate the constant k.
   (b) Estimate the volume of wood in the same type of tree with $d = 20$ inches and $h = 105$ feet.

4. *Area of Skin* The surface area of the skin covering the human body is a function of more than one variable. Both height and weight influence the surface area of a person's body. Hence a taller person tends to have a larger surface area, as does a heavier person. A formula to determine the surface area of a person's body in square inches is computed by $S = 15.7(w^{0.425})(h^{0.725})$, where w is weight in pounds and h is height in inches. (Source: Lancaster, *Quantitative Methods in Biological and Medical Sciences*.)
   (a) Use S to estimate the surface area of a person who is 65 inches tall and weighs 154 pounds.
   (b) If a person's weight doubles, what happens to the area of the person's skin? Explain.
   (c) If a person's height doubles, what happens to the area of the person's skin? Explain.

5. *Minimizing Cost* A natural gas line running along a river is to be connected from point A to a cabin on the other bank located at point D, as illustrated in the accompanying figure. The width of the river is 500 feet, and the distance from point A to point C is 1000 feet. The cost of running the pipe along the shoreline is \$30 per foot, and the cost of running it underwater is \$50 per foot. The cost of connecting the gas line from A to D is to be minimized.

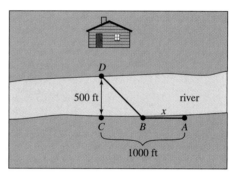

   (a) Write an expression that gives the cost of running the line from A to B if the distance between these points is x feet.
   (b) Find the distance from B to D in terms of x.
   (c) Write an expression that gives the cost of running the line from B to D.
   (d) Use your answer from parts (a) and (c) to write an expression that gives the cost of running the line from A to B to D.
   (e) Graph your expression from part (d) in [0, 1000, 100] by [40,000, 60,000, 5000] to determine the value of x that minimizes the cost of the line going from A to D. What is the minimum cost?

Chapter 9: Conic Sections

Throughout history people have been fascinated with the universe around them and compelled to understand its mysteries. Conic sections, which include parabolas, ellipses, and hyperbolas, have played an important role in gaining this understanding. Although conic sections were described and named by the Greek astronomer Apollonius in 200 B.C., not until much later were they used to model motion in the universe. In the sixteenth century Tycho Brahe, the greatest observational astronomer of the age, recorded precise data on planetary movement in the sky. Using Brahe's data in 1619, Johannes Kepler determined that planets move in elliptical orbits around the sun. In 1686 Newton used Kepler's work to show that elliptical orbits are the result of his famous theory of gravitation. We now know that all celestial objects—including planets, comets, asteroids, and satellites—travel in paths described by conic sections. Today scientists search the sky for information about the universe with enormous radio telescopes in the shape of parabolic dishes.

Conic sections have had a profound influence on people's understanding of their world and the cosmos. In this chapter we introduce you to these age-old curves.

> *The art of asking the right questions in mathematics is more important than the art of solving them.*
> —Georg Cantor

Source: *Historical Topics for the Mathematics Classroom, Thirty-first Yearbook*, NCTM.

9.1 Parabolas and Circles

Types of Conic Sections ~ Graphs of Parabolas with Horizontal Axes ~ Equations of Circles

Introduction

In this section we discuss two types of conic sections: parabolas and circles. Recall that we discussed parabolas with vertical axes of symmetry in Chapter 6. In this section we discuss parabolas with horizontal axes of symmetry, but first we introduce the three basic types of conic sections.

Types of Conic Sections

Conic sections are named after the different ways that a plane can intersect a cone. The three basic curves are parabolas, ellipses, and hyperbolas, as illustrated in Figure 9.1. A circle is a special case of an ellipse.

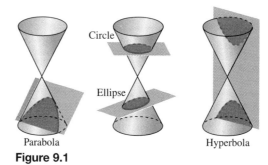

Figure 9.1

Figures 9.2–9.4 show examples of these three conic sections.

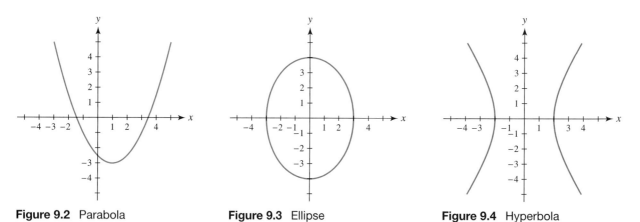

Figure 9.2 Parabola   **Figure 9.3** Ellipse   **Figure 9.4** Hyperbola

Graphs of Parabolas with Horizontal Axes

Recall that the *vertex form of a parabola* with a vertical axis of symmetry is

$$y = a(x - h)^2 + k,$$

where (h, k) is the vertex. If $a > 0$, the parabola opens upward; if $a < 0$, the parabola opens downward, as shown in Figures 9.5 and 9.6. The preceding equation can also be expressed in the form

$$y = ax^2 + bx + c.$$

In this form the x-coordinate of the vertex is $x = -\dfrac{b}{2a}$.

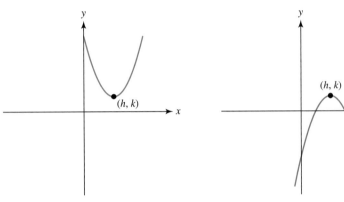

Figure 9.5 $y = a(x - h)^2 + k$, $a > 0$

Figure 9.6 $y = a(x - h)^2 + k$, $a < 0$

Interchanging the roles of x and y gives equations for parabolas that open to the right or the left. In this case, their axes of symmetry are horizontal.

Parabolas with Horizontal Axes of Symmetry

The graph of $x = a(y - k)^2 + h$ is a parabola that opens to the right if $a > 0$ and to the left if $a < 0$. The vertex of the parabola is located at (h, k).

The graph of $x = ay^2 + by + c$ is a parabola opening to the right if $a > 0$ and to the left if $a < 0$. The y-coordinate of its vertex is $y = -\dfrac{b}{2a}$.

These parabolas are illustrated in Figures 9.7 and 9.8 on the next page.

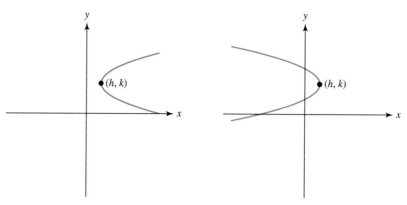

Figure 9.7 $x = a(y - k)^2 + h$, $a > 0$

Figure 9.8 $x = a(y - k)^2 + h$, $a < 0$

EXAMPLE 1 *Graphing a parabola*

Graph $x = -\frac{1}{2}y^2$. Find its vertex and axis of symmetry.

Solution

The equation can be written in vertex form as $x = -\frac{1}{2}(y - 0)^2 + 0$. The vertex is $(0, 0)$, and as $a = -\frac{1}{2} < 0$, the parabola opens to the left. We can make a table of values, as in Table 9.1, and plot a few points to help determine the location and shape of the graph. To obtain Table 9.1, first choose a *y*-value and then calculate an *x*-value.

TABLE 9.1

y	−2	−1	0	1	2
x	−2	−0.5	0	−0.5	−2

The graph is shown in Figure 9.9. Its axis of symmetry is the *x*-axis, or $y = 0$.

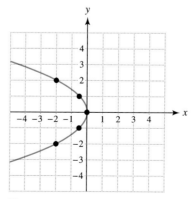

Figure 9.9

EXAMPLE 2 Graphing a parabola

Graph $x = (y - 3)^2 + 2$. Give its vertex and axis of symmetry.

Solution

The vertex is (2, 3), and because $a = 1 > 0$, the parabola opens to the right. This parabola has the same shape as $y = x^2$, except that it opens to the right. We can make a table of values and plot a few points to graph the parabola. To obtain Table 9.2 we first chose y-values and then calculated the x-values, using $x = (y - 3)^2 + 2$.

TABLE 9.2

y	1	2	3	4	5
x	6	3	2	3	6

These points and the graph of the parabola are shown in Figure 9.10. The axis of symmetry is $y = 3$ because, if we fold the graph on the horizontal line $y = 3$, the two sides match.

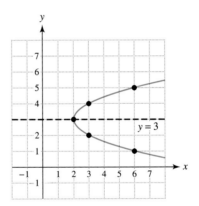

Figure 9.10

EXAMPLE 3 Graphing a parabola with a graphing calculator

 Graph $x = 2y^2 + 4y - 3$, using a graphing calculator, and estimate the coordinates of the vertex. Then find the vertex symbolically.

Solution

Some calculators have the capability to graph equations in the form $x = f(y)$ directly. Other calculators have an inverse draw feature that can be used to graph a parabola that opens to the right or to the left. To use this feature, first enter the equation $y = 2x^2 + 4x - 3$. Then from the home screen find DrawInv in the DRAW menu. Finally, use the VARS, Y-VARS, and FUNCTION menus to find the variable Y_1 and produce the graph (see Figures 9.11–9.13 on the next page). In Figure 9.13 the vertex appears to be near $(-5, -1)$. Note that if Y_1 is not "deselected," as shown in Figure 9.11, its graph will also appear.

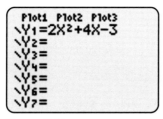
Figure 9.11

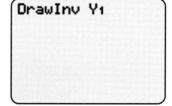
Figure 9.12

[−6, 6, 1] by [−4, 4, 1]

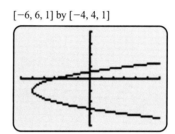
Figure 9.13

For $x = 2y^2 + 4y - 3$, $a = 2$ and $b = 4$. Then the y-coordinate of the vertex is

$$y = -\frac{b}{2a} = -\frac{4}{2(2)} = -1,$$

and the x-coordinate is

$$x = 2(-1)^2 + 4(-1) - 3 = -5.$$

The vertex is $(-5, -1)$, as suggested by the graph.

Equations of Circles

A **circle** consists of the set of points in a plane that are the same distance from a fixed point. The fixed distance is called the **radius,** and the fixed point is called the **center** of the circle. In Figure 9.14 all points lying on the circle are a distance of 2 units from the center (2, 1). Therefore the radius of the circle equals 2.

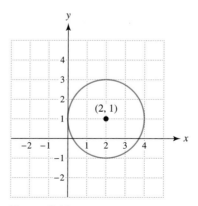
Figure 9.14

We can find the equation of the circle shown in Figure 9.14 by using the distance formula. If a point (x, y) lies on the circle, its distance from the center $(2, 1)$ is 2 and

$$\sqrt{(x - 2)^2 + (y - 1)^2} = 2.$$

Squaring both sides gives

$$(x - 2)^2 + (y - 1)^2 = 2^2.$$

This equation represents the standard equation for a circle with center $(2, 1)$ and radius 2.

Standard Equation of a Circle

The **standard equation of a circle** with center (h, k) and radius r is

$$(x - h)^2 + (y - k)^2 = r^2.$$

EXAMPLE 4 *Graphing a circle*

Graph $x^2 + y^2 = 9$. Find the radius and center.

Solution

The equation $x^2 + y^2 = 9$ can be written in standard form as

$$(x - 0)^2 + (y - 0)^2 = 3^2.$$

Therefore the center is $(0, 0)$ and the radius is 3. Its graph is shown in Figure 9.15.

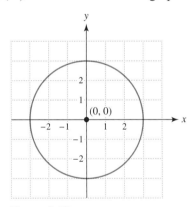

Figure 9.15

EXAMPLE 5 *Graphing a circle*

Graph $(x + 1)^2 + (y - 3)^2 = 4$. Find the radius and center.

Solution

Write the equation as

$$(x - (-1))^2 + (y - 3)^2 = 2^2.$$

The center is $(-1, 3)$, and the radius is 2. Its graph is shown in Figure 9.16 on the next page.

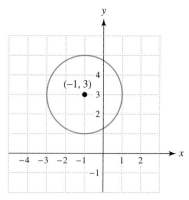

Figure 9.16

In the next example we use the technique of *completing the square* to find the center and radius of a circle. (To review completing the square, refer back to Section 6.2.)

EXAMPLE 6 *Finding the center of a circle*

Find the center and radius of the circle given by $x^2 + 4x + y^2 - 6y = 3$.

Solution

Begin by writing the equation as

$$(x^2 + 4x + \underline{}) + (y^2 - 6y + \underline{}) = 3.$$

To complete the square, add $\left(\dfrac{4}{2}\right)^2 = 4$ and $\left(\dfrac{-6}{2}\right)^2 = 9$ to both sides of the equation.

$$(x^2 + 4x + 4) + (y^2 - 6y + 9) = 3 + 4 + 9$$

Factoring the binomials as perfect trinomial squares yields

$$(x + 2)^2 + (y - 3)^2 = 4^2.$$

The center is $(-2, 3)$, and the radius is 4.

Critical Thinking

What is the center and radius of

$$x^2 + y^2 + 10y = -32?$$

The graph of a circle does not represent a function. One way to graph a circle with a graphing calculator is to solve the equation for y and obtain two equations. One equation gives the upper half of the circle, and the other equation gives the lower half. We illustrate this technique in the next example.

EXAMPLE 7 *Graphing a circle with a graphing calculator*

GCLM Graph $x^2 + y^2 = 4$ in the viewing rectangle $[-4.7, 4.7, 1]$ by $[-3.1, 3.1, 1]$.

Solution

Begin by solving for y.

$$x^2 + y^2 = 4 \qquad \text{Given equation}$$
$$y^2 = 4 - x^2 \qquad \text{Subtract } x^2.$$
$$y = \pm\sqrt{4 - x^2} \qquad \text{Square root property}$$

Graph $Y_1 = \sqrt{(4 - X^2)}$ and $Y_2 = -\sqrt{(4 - X^2)}$. The graph of Y_1 generates the upper half of the circle, and the graph of Y_2 generates the lower half of the circle, as shown in Figures 9.17–9.19.

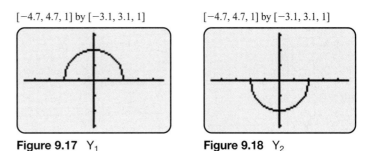

Figure 9.17 Y_1 **Figure 9.18** Y_2

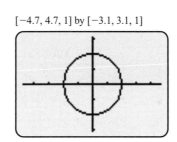

Figure 9.19 Y_1 and Y_2

Technology Note *Square Viewing Rectangles*

If a circle is not graphed in a *square viewing rectangle*, it will appear to be an oval rather than a circle. In a square viewing rectangle a circle will appear circular. Figure 9.20 shows the circle from Example 7 graphed in a viewing rectangle that is not square. Consult your owner's manual to learn more about square viewing rectangles for your calculator.

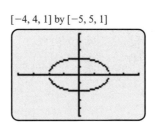

Figure 9.20

9.1 Putting It All Together

The following table summarizes some basic concepts about parabolas and circles.

Concept	Explanation	Example
Parabola with horizontal axis	Vertex form: $x = a(y - k)^2 + h$. If $a > 0$, it opens to the right; if $a < 0$, it opens to the left. The vertex is (h, k). [graph: $a > 0$, vertex (h, k), opens right] [graph: $a < 0$, vertex (h, k), opens left] These parabolas may also be expressed as $x = ay^2 + by + c$, where the y-coordinate of the vertex is $y = -\dfrac{b}{2a}$.	$x = 2(y - 1)^2 + 4$ opens to the right and its vertex is $(4, 1)$.
Standard equation of a circle	Standard equation: $(x - h)^2 + (y - k)^2 = r^2$. The radius is r and the center is (h, k).	$(x + 2)^2 + (y - 1)^2 = 16$ has center $(-2, 1)$ and radius 4.

9.1 EXERCISES

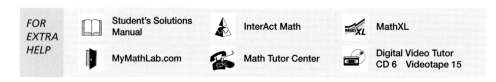

CONCEPTS

1. Name the three general types of conic sections.

2. What is the difference between the graphs of $y = ax^2 + bx + c$ and $x = ay^2 + by + c$?

3. If a parabola has a horizontal axis of symmetry, does it represent a function?

4. Sketch a graph of a parabola with a horizontal axis.

5. If a parabola has two y-intercepts, does it represent a function? Why or why not?

6. If $x = a(y - k)^2 + h$, what is the vertex?

7. The graph of $x = -y^2$ opens to the _____.

8. The graph of $x = 2y^2 + y - 1$ opens to the _____.

9. The graph of $(x - h)^2 + (y - k)^2 = r^2$ is a _____ with center _____.

10. The graph of $x^2 + y^2 = r^2$ is a circle with center _____ and radius _____.

PARABOLAS

Exercises 11–24: Graph the parabola by hand. Find the vertex and axis of symmetry.

11. $x = y^2$
12. $x = -y^2$
13. $x = y^2 + 1$
14. $x = y^2 - 1$
15. $x = 2y^2$
16. $x = \frac{1}{4}y^2$
17. $x = (y - 1)^2 + 2$
18. $x = (y - 2)^2 + 1$
19. $y = (x + 2)^2 + 1$
20. $y = (x - 4)^2 + 5$
21. $x = \frac{1}{2}(y + 1)^2 - 3$
22. $x = -2(y + 3)^2 + 1$
23. $x = -3(y - 1)^2$
24. $x = \frac{1}{4}(y + 2)^2 - 3$

Exercises 25–36: (Refer to Example 3.) Graph the parabola with a graphing calculator. Find the vertex and axis of symmetry.

25. $x = -3y^2$
26. $x = \frac{1}{2}y^2$
27. $x = 2y^2 - 3$
28. $x = y^2 + 2y - 1$
29. $x = y^2 + 3y + 2$
30. $x = y^2 - 5y + 1$
31. $y = 2x^2 - x + 1$
32. $y = -x^2 + 2x + 2$
33. $x = \frac{1}{2}y^2 + y - 1$
34. $x = -2y^2 + 3y + 2$
35. $x = 3y^2 + y$
36. $x = -\frac{3}{2}y^2 - 2y + 1$

Exercises 37–40: Use the graph to determine the equation of the parabola. Then check your answer with a graphing calculator. (Hint: Either $a = 1$ or $a = -1$.)

37.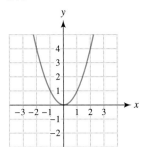

38.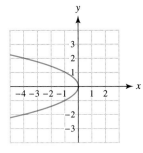

39.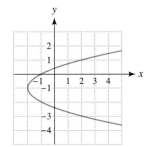

40.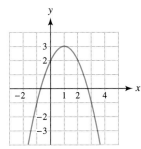

Exercises 41–44: Determine the direction that the parabola opens if it satisfies the given conditions.

41. Passing through $(2, 0)$, $(-2, 0)$, and $(0, -2)$

42. Passing through $(0, -3)$, $(0, 2)$, and $(1, 1)$

43. Vertex $(1, 2)$ passing through $(-1, -2)$ with a vertical axis

44. Vertex $(-1, 3)$ passing through $(0, 0)$ with a horizontal axis

45. What x-values are possible for the graph of $x = 2y^2$?

46. What y-values are possible for the graph of $x = 2y^2$?

47. How many y-intercepts does a parabola
$$x = a(y - k)^2 + h$$
have if $a > 0$ and $h < 0$?

48. Does the graph of $x = ay^2 + by + c$ always have a y-intercept? Explain.

49. What is the x-intercept for the graph of
$$x = 3y^2 - y + 1?$$

50. What are the y-intercepts for the graph of
$$x = y^2 - 3y + 2?$$

CIRCLES

Exercises 51–56: Write the equation of the circle with the given radius r and center C.

51. $r = 1$ $C = (0, 0)$
52. $r = 4$ $C = (2, 3)$
53. $r = 3$ $C = (-1, 5)$
54. $r = 5$ $C = (5, -3)$
55. $r = \sqrt{2}$ $C = (-4, -6)$
56. $r = \sqrt{6}$ $C = (0, 4)$

Exercises 57–60: Use the graph to find the equation of the circle.

57.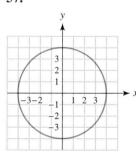

58.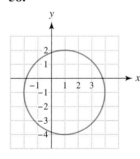

59.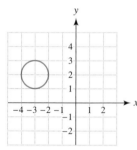

60.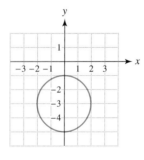

Exercises 61–70: Find the radius and center of the circle. Then graph the circle.

61. $x^2 + y^2 = 9$
62. $x^2 + y^2 = 1$
63. $(x - 1)^2 + (y - 3)^2 = 9$
64. $(x + 2)^2 + (y + 1)^2 = 4$
65. $(x + 5)^2 + (y - 5)^2 = 25$
66. $(x - 4)^2 + (y + 3)^2 = 16$
67. $x^2 + 6x + y^2 - 2y = -1$
68. $x^2 + y^2 + 12y + 32 = 0$
69. $x^2 + 6x + y^2 - 2y + 3 = 0$
70. $x^2 - 4x + y^2 + 4y = -3$

APPLICATIONS

71. *Radio Telescopes* The Parks radio telescope has the shape of a parabolic dish, as depicted in the accompanying figure. A cross section of this telescope can be modeled by $x = \dfrac{32}{11,025} y^2$, where $-105 \le y \le 105$; the units are feet.

(a) Graph the cross-sectional shape of the dish in $[-40, 40, 10]$ by $[-120, 120, 20]$.
(b) Find the depth d of the dish.

72. *Train Tracks* To make a curve safer for trains, parabolic curves are sometimes used instead of circular curves. See the accompanying figure. (*Source: F. Mannering and W. Kilareski, Principles of Highway Engineering and Traffic Analysis.*)
(a) Suppose that a curve must pass through the points $(-1, 0), (0, 2)$, and $(0, -2)$, where

the units are kilometers. Find an equation for the train tracks in the form

$$x = a(y - h)^2 + k.$$

(b) Find another point that lies on the train tracks.

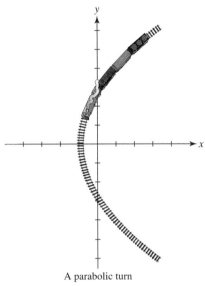

A parabolic turn

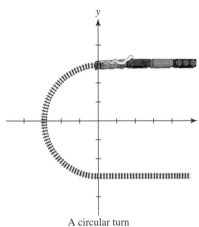

A circular turn

73. *Trajectories of Comets* Under certain circumstances, a comet can pass by the sun once and never return. In this situation the comet may travel in a parabolic path, as illustrated in the accompanying figure. Suppose that a comet's path is given by $x = -2.5y^2$, where the sun is located at $(-0.1, 0)$ and the units are astronomical units (A.U.). One astronomical unit equals 93 million miles. **(Source: W. Thomson, *Introduction to Space Dynamics*.)**

(a) Plot a point for the sun's location and then graph the path of the comet in the viewing rectangle $[-1.5, 1.5, 0.5]$ by $[-1, 1, 0.5]$.
(b) Find the distance from the sun to the comet when the comet is located at $(-2.5, 1)$.

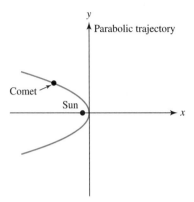

74. *Speed of a Comet* (Continuation of Exercise 73.) The velocity V in meters per second of a comet traveling in a parabolic trajectory about the sun is given by $V = \dfrac{k}{\sqrt{D}}$, where D is the distance from the sun in meters and $k = 1.15 \times 10^{10}$.

(a) How does the velocity of the comet change as its distance from the sun changes?
(b) Calculate the velocity of the comet when it is closest to the sun. (*Hint:* 1 mile $\approx$ 1609 meters.)

WRITING ABOUT MATHEMATICS

75. Suppose that you are given the equation
$$x = a(y - k)^2 + h.$$
(a) Explain how you can determine the direction that the parabola opens.
(b) Explain how to find the axis of symmetry and the vertex.
(c) If the points $(0, 4)$ and $(0, -2)$ lie on the graph of x, what is the axis of symmetry?
(d) Generalize part (c) if $(0, y_1)$ and $(0, y_2)$ lie on the graph of x.

76. Suppose that you are given the vertex of a parabola. Can you determine the axis of symmetry? Explain.

GROUP ACTIVITY
Working with Real Data

Directions: Form a group of 2 to 4 people. Select someone to record the group's responses for this activity. All members of the group should work cooperatively to answer the questions. If your instructor asks for your results, each member of the group should be prepared to respond. If the group is asked to turn in its work, be sure to include each person's name on the paper.

Radio Telescope The U.S. Naval Research Laboratory designed a giant radio telescope weighing 3450 tons. Its parabolic dish has a diameter of 300 feet and a depth of 44 feet, as shown in the accompanying figure. **(Source: J. Mar, *Structure Technology for Large Radio and Radar Telescope Systems*.)**

(a) Determine an equation that is of the form $x = ay^2$, $a > 0$, and that models a cross section of the dish.

(b) Graph your equation in an appropriate viewing rectangle.

9.2 ELLIPSES AND HYPERBOLAS

Equations of Ellipses ~ Equations of Hyperbolas

INTRODUCTION

Celestial objects in the sky travel in paths or trajectories determined by conic sections. For this reason, conic sections have been studied for centuries. In modern times physicists have learned that subatomic particles can also travel in trajectories determined by conic sections. Recall that the three main types of conic sections are parabolas, el-

9.2 Ellipses and Hyperbolas

lipses, and hyperbolas and that circles are a special type of ellipse. In Chapter 6 and Section 9.1, we discussed parabolas and circles. In this section we focus on ellipses and hyperbolas and some of their applications.

EQUATIONS OF ELLIPSES

One method used to sketch an ellipse is to tie the ends of a string to two nails driven into a flat board. If a pencil is placed against the string anywhere between the nails, as shown in Figure 9.21, and is used to draw a curve, the resulting curve is an ellipse. The sum of the distances d_1 and d_2 between the pencil and each of the nails is always fixed by the string. The location of the nails correspond to the foci of the ellipse. An **ellipse** is the set of points in a plane, the sum of whose distances from two fixed points is constant. Each fixed point is called a **focus** (plural *foci*) of the ellipse.

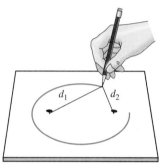

Figure 9.21

Critical Thinking

What happens to the shape of the ellipse shown in Figure 9.21 as the nails are moved farther apart? What happens to its shape as the nails are moved closer together? When would a circle be formed?

In Figures 9.22 and 9.23 the **major axis** and the **minor axis** are labeled for each ellipse. The major axis is the longer of the two axes. Figure 9.22 shows an ellipse with a *horizontal* major axis, and Figure 9.23 shows an ellipse with a *vertical* major axis. The **vertices**, V_1 and V_2, of each ellipse are located at the endpoints of the major axis.

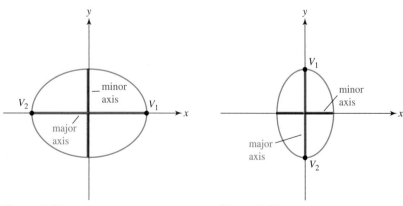

Figure 9.22   **Figure 9.23**

A vertical line can intersect the graph of an ellipse more than once, so an ellipse cannot be modeled by a function. However, some ellipses can be represented by the following equations.

> **Standard Equations for Ellipses Centered at (0, 0)**
>
> The ellipse with center at the origin, *horizontal* major axis, and equation
>
> $$\frac{x^2}{a^2} + \frac{y^2}{b^2} = 1, \quad a > b > 0,$$
>
> has vertices $(\pm a, 0)$ and endpoints of the minor axis $(0, \pm b)$.
>
> The ellipse with center at the origin, *vertical* major axis, and equation
>
> $$\frac{x^2}{b^2} + \frac{y^2}{a^2} = 1, \quad a > b > 0,$$
>
> has vertices $(0, \pm a)$ and endpoints of the minor axis $(\pm b, 0)$.

Figures 9.24 and 9.25 show two ellipses having a horizontal major axis and a vertical major axis, respectively. The coordinates of the vertices V_1 and V_2 and endpoints of the minor axis U_1 and U_2 are labeled.

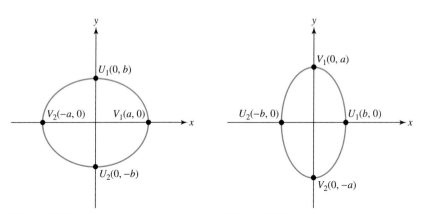

Figure 9.24 **Figure 9.25**

Critical Thinking

Suppose that $a = b$ for an ellipse centered at $(0, 0)$. What can be said about the ellipse? Explain.

EXAMPLE 1 *Sketching graphs of ellipses*

Sketch a graph of each ellipse. Label the vertices and endpoints of the minor axes.

(a) $\dfrac{x^2}{25} + \dfrac{y^2}{4} = 1$

(b) $9x^2 + 4y^2 = 36$

Solution

(a) The equation $\frac{x^2}{25} + \frac{y^2}{4} = 1$ describes an ellipse with $a^2 = 25$ and $b^2 = 4$. (When you are deciding whether 25 or 4 represents a^2, let a^2 be the larger of the two numbers.) Thus $a = 5$ and $b = 2$, so the ellipse has a horizontal major axis with vertices $(\pm 5, 0)$ and the endpoints of the minor axis are $(0, \pm 2)$. Plot these four points and then sketch the ellipse, as shown in Figure 9.26.

(b) The equation $9x^2 + 4y^2 = 36$ can be put into standard form by dividing both sides by 36.

$$9x^2 + 4y^2 = 36 \quad \text{Given equation}$$

$$\frac{9x^2}{36} + \frac{4y^2}{36} = \frac{36}{36} \quad \text{Divide by 36.}$$

$$\frac{x^2}{4} + \frac{y^2}{9} = 1 \quad \text{Simplify.}$$

This ellipse has a vertical major axis with $a = 3$ and $b = 2$. The vertices are $(0, \pm 3)$, and the endpoints of the minor axis are $(\pm 2, 0)$, as shown in Figure 9.27.

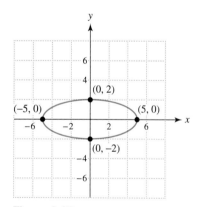

Figure 9.26

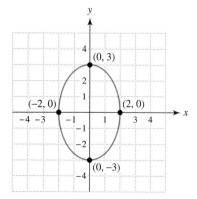

Figure 9.27

544 CHAPTER 9 Conic Sections

Planets travel around the sun in elliptical orbits. Astronomers have measured the values of *a* and *b* for each planet. Utilizing this information, we can find the equation of a planet's orbit.

EXAMPLE 2 *Modeling the orbit of Mercury*

 Except for Pluto, the planet Mercury has the least circular orbit of the nine planets. For Mercury $a = 0.387$ and $b = 0.379$. The units are astronomical units (A.U.), where 1 A.U. equals 93 million miles. Graph $\frac{x^2}{a^2} + \frac{y^2}{b^2} = 1$ to model the orbit of Mercury in $[-0.6, 0.6, 0.1]$ by $[-0.4, 0.4, 0.1]$. Then plot the sun at the point $(0.08, 0)$.
(*Source:* M. Zeilik, *Introductory Astronomy and Astrophysics.*)

Solution

The orbit of Mercury is given by

$$\frac{x^2}{0.387^2} + \frac{y^2}{0.379^2} = 1.$$

To graph an ellipse with some types of graphing calculators, we must solve the equation for *y*. Doing so results in two equations.

$$\frac{x^2}{0.387^2} + \frac{y^2}{0.379^2} = 1$$

$$\frac{y^2}{0.379^2} = 1 - \frac{x^2}{0.387^2}$$

$$\frac{y}{0.379} = \pm\sqrt{1 - \frac{x^2}{0.387^2}}$$

$$y = \pm 0.379\sqrt{1 - \frac{x^2}{0.387^2}}$$

The orbit of Mercury results from graphing these two equations. See Figures 9.28 and 9.29. The point $(0.08, 0)$ represents the position of the sun in Figure 9.29.

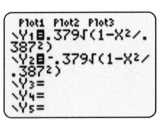

Figure 9.28

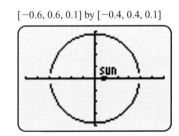

Figure 9.29

Critical Thinking

Use Figure 9.29 and the information in Example 2 to estimate the minimum and maximum distances that Mercury is from the sun.

EQUATIONS OF HYPERBOLAS

The third type of conic section is a **hyperbola**, which is the set of points in a plane, the difference of whose distances from two fixed points is constant. Each fixed point is called a **focus** of the hyperbola. Figure 9.30 shows a hyperbola whose equation is

$$\frac{x^2}{4} - \frac{y^2}{9} = 1.$$

This hyperbola is centered at the origin and has two **branches**, a *left branch* and a *right branch*. The **vertices** are $(-2, 0)$ and $(2, 0)$, and the line segment connecting the vertices is called the **transverse axis**.

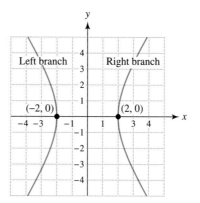

Figure 9.30

By the vertical line test, a hyperbola cannot be represented by a function, but many hyperbolas can be described by the following equations.

Standard Equations for Hyperbolas Centered at (0, 0)

The hyperbola with center at the origin, *horizontal* transverse axis, and equation

$$\frac{x^2}{a^2} - \frac{y^2}{b^2} = 1$$

has vertices $(\pm a, 0)$.

The hyperbola with center at the origin, *vertical* transverse axis, and equation

$$\frac{y^2}{a^2} - \frac{x^2}{b^2} = 1$$

has vertices $(0, \pm a)$.

Hyperbolas, along with the coordinates of their vertices, are shown in Figures 9.31 and 9.32 on the next page. The two parts of the hyperbola in Figure 9.31 are the *left branch* and *right branch*, whereas in Figure 9.32 the hyperbola has an *upper branch* and a *lower branch*. The dashed rectangle in each figure is called the **fundamental rectangle**, and its four vertices are determined by either $(\pm a, \pm b)$ or $(\pm b, \pm a)$. If its diagonals are extended, they correspond to the asymptotes of the hyperbola. The lines $y = \pm \frac{b}{a}x$ and $y = \pm \frac{a}{b}x$ are **asymptotes** for the hyperbolas, respectively, and may be used as an aid to graphing them.

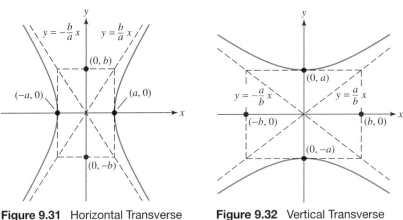

Figure 9.31 Horizontal Transverse Axis

Figure 9.32 Vertical Transverse Axis

Note: A hyperbola consists of two solid curves, or branches. The dashed lines and rectangles are not part of the actual graph but are used as an aid for sketching the graph.

One interpretation of an asymptote can be based on trajectories of comets as they approach the sun. Comets travel in parabolic, elliptic, or hyperbolic trajectories. If the speed of a comet is too slow, the gravitational pull of the sun captures the comet in an elliptic orbit (see Figure 9.33). If the speed of the comet is too fast, the sun's gravity is too weak to capture the comet and the comet passes by it in a hyperbolic trajectory. Near the sun the gravitational pull is stronger, and the comet's trajectory is curved. Farther from the sun, the gravitational pull becomes weaker, and the comet eventually returns to a straight-line trajectory determined by the *asymptote* of the hyperbola (see Figure 9.34). Finally, if the speed is neither too slow nor too fast, the comet will travel in a parabolic path (see Figure 9.35).

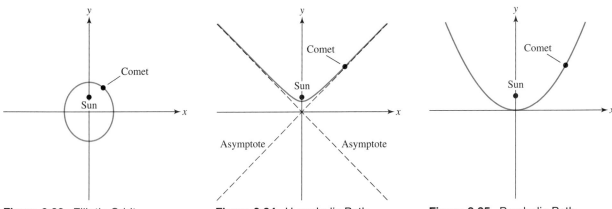

Figure 9.33 Elliptic Orbit

Figure 9.34 Hyperbolic Path

Figure 9.35 Parabolic Path

Critical Thinking

If a comet is observed at regular intervals, which type of conic section describes its path?

EXAMPLE 3 *Sketching the graph of a hyperbola*

Sketch a graph of $\frac{y^2}{4} - \frac{x^2}{9} = 1$. Label the vertices and show the asymptotes.

Solution

The equation is in standard form with $a^2 = 4$ and $b^2 = 9$, so $a = 2$ and $b = 3$. It has a vertical transverse axis with vertices $(0, -2)$ and $(0, 2)$. The vertices of the fundamental rectangle are $(\pm 3, \pm 2)$, that is, $(3, 2), (3, -2), (-3, 2),$ and $(-3, -2)$. The asymptotes are the diagonals of this rectangle and are given by $y = \pm \frac{a}{b}x$ or $y = \pm \frac{2}{3}x$. Figure 9.36 shows all these features.

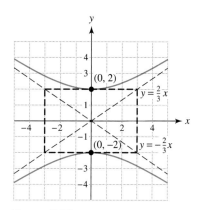

Figure 9.36

EXAMPLE 4 *Graphing a hyperbola with a graphing calculator*

Graph $\frac{y^2}{4} - \frac{x^2}{8} = 1$ with a graphing calculator.

Solution

Begin by solving the given equation for y.

$$\frac{y^2}{4} = 1 + \frac{x^2}{8} \qquad \text{Add } \frac{x^2}{8}.$$

$$y^2 = 4\left(1 + \frac{x^2}{8}\right) \qquad \text{Multiply by 4.}$$

$$y = \pm 2\sqrt{1 + \frac{x^2}{8}} \qquad \text{Square root property}$$

Graph $Y_1 = 2\sqrt{(1 + X^2/8)}$ and $Y_2 = -2\sqrt{(1 + X^2/8)}$. The results are shown in Figures 9.37 and 9.38 on the next page.

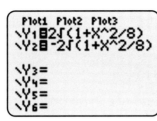
Figure 9.37

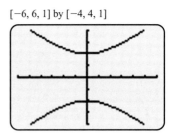
[−6, 6, 1] by [−4, 4, 1]
Figure 9.38

9.2 Putting It All Together

The following table summarizes some basic concepts about ellipses and hyperbolas.

Concept	Description
Ellipses centered at $(0, 0)$ with $a > b > 0$	**Horizontal Major Axis** Vertices: $(a, 0)$ and $(-a, 0)$ $$\frac{x^2}{a^2} + \frac{y^2}{b^2} = 1$$ **Vertical Major Axis** Vertices: $(0, a)$ and $(0, -a)$ $$\frac{y^2}{a^2} + \frac{x^2}{b^2} = 1$$
Hyperbolas centered at $(0, 0)$ with $a > 0$ and $b > 0$	**Horizontal Transverse Axis** Vertices: $(a, 0)$ and $(-a, 0)$ Asymptotes: $y = \pm \frac{b}{a} x$ $$\frac{x^2}{a^2} - \frac{y^2}{b^2} = 1$$ **Vertical Transverse Axis** Vertices: $(0, a)$ and $(0, -a)$ Asymptotes: $y = \pm \frac{a}{b} x$ $$\frac{y^2}{a^2} - \frac{x^2}{b^2} = 1$$

9.2 EXERCISES

CONCEPTS

1. Sketch an ellipse with a horizontal major axis.

2. Sketch a hyperbola with a vertical transverse axis.

3. The ellipse whose equation is $\frac{x^2}{a^2} + \frac{y^2}{b^2} = 1$, $a > b > 0$, has a _____ major axis.

4. The ellipse whose equation is $\frac{y^2}{a^2} + \frac{x^2}{b^2} = 1$, $a > b > 0$, has a _____ major axis.

5. What is the maximum number of times that a line can intersect an ellipse?

6. What is the maximum number of times that a parabola can intersect an ellipse?

7. The hyperbola whose equation is $\frac{x^2}{a^2} - \frac{y^2}{b^2} = 1$ has _____ and _____ branches.

8. The hyperbola whose equation is $\frac{y^2}{a^2} - \frac{x^2}{b^2} = 1$ has _____ and _____ branches.

9. How are the asymptotes of a hyperbola related to the fundamental rectangle?

10. Could an ellipse be centered at the origin and have vertices $(4, 0)$ and $(0, -5)$?

ELLIPSES

Exercises 11–22: Graph the ellipse by hand. Label the vertices and endpoints of the minor axis.

11. $\frac{x^2}{9} + \frac{y^2}{25} = 1$

12. $\frac{y^2}{9} + \frac{x^2}{25} = 1$

13. $\frac{x^2}{9} + \frac{y^2}{4} = 1$

14. $\frac{x^2}{3} + \frac{y^2}{9} = 1$

15. $x^2 + \frac{y^2}{4} = 1$

16. $\frac{x^2}{9} + y^2 = 1$

17. $\frac{y^2}{5} + \frac{x^2}{7} = 1$

18. $\frac{y^2}{11} + \frac{x^2}{6} = 1$

19. $36x^2 + 4y^2 = 144$

20. $25x^2 + 16y^2 = 400$

21. $6y^2 + 7x^2 = 42$

22. $9x^2 + 5y^2 = 45$

Exercises 23–26: Use a graphing calculator to graph the ellipse.

23. $\frac{x^2}{13} + \frac{y^2}{11} = 1$

24. $\frac{y^2}{4.1} + \frac{x^2}{6.2} = 1$

25. $2x^2 + 3y^2 = 6$

26. $\frac{x^2}{3.6} + \frac{y^2}{7.8} = 1$

Exercises 27–30: Use the graph to determine the equation of the ellipse.

27.
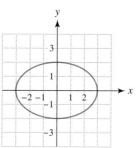

28.
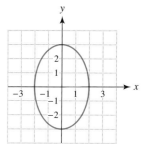

29.
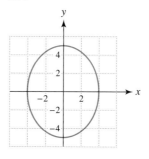

30.
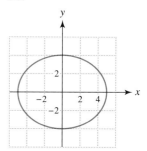

HYPERBOLAS

Exercises 31–42: Graph the hyperbola by hand. Show the asymptotes and vertices.

31. $\frac{x^2}{4} - \frac{y^2}{9} = 1$

32. $\frac{y^2}{4} - \frac{x^2}{9} = 1$

33. $\frac{x^2}{25} - \frac{y^2}{16} = 1$

34. $\frac{y^2}{25} - \frac{x^2}{16} = 1$

35. $x^2 - y^2 = 1$ 36. $y^2 - x^2 = 1$

37. $\dfrac{x^2}{3} - \dfrac{y^2}{4} = 1$ 38. $\dfrac{y^2}{5} - \dfrac{x^2}{8} = 1$

39. $9y^2 - 4x^2 = 36$ 40. $36x^2 - 25y^2 = 900$

41. $16x^2 - 4y^2 = 64$ 42. $y^2 - 9x^2 = 9$

Exercises 43–46: Use the graph to determine an equation of the hyperbola.

43.
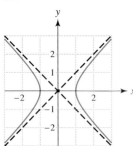

44.
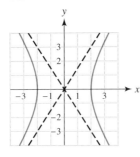

45.
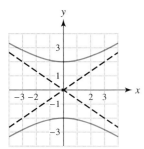

46.
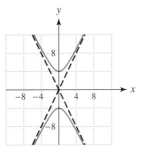

Exercises 47–50: Use a graphing calculator to graph the hyperbola.

47. $\dfrac{x^2}{13} - \dfrac{y^2}{17} = 1$ 48. $\dfrac{y^2}{7.2} - \dfrac{x^2}{4.6} = 1$

49. $2.1y^2 - 5.3x^2 = 7$

50. $4.8x^2 - 1.1y^2 = 5.1$

APPLICATIONS

51. *Geometry of an Ellipse* The area inside an ellipse is given by $A = \pi ab$, and its perimeter can be approximated by
$$P = 2\pi\sqrt{\dfrac{a^2 + b^2}{2}}.$$

Approximate A and P to the nearest hundredth for each ellipse.

(a) $\dfrac{x^2}{16} + \dfrac{y^2}{25} = 1$ (b) $\dfrac{x^2}{7} + \dfrac{y^2}{2} = 1$

52. *Geometry of an Ellipse* (Refer to Exercise 51.) If $a = b$ in the equation for an ellipse, the ellipse is a circle. Let $a = b$ in the formulas for the area and perimeter of an ellipse. Do the equations simplify to the area and perimeter for a circle? Explain.

53. *Planet Orbit* (Refer to Example 2.) The planet Pluto has the least circular orbit of any planet. For Pluto $a = 39.44$ and $b = 38.20$.
   (a) Graph the elliptic orbit of Pluto in the window $[-60, 60, 10]$ by $[-40, 40, 10]$. Plot the point $(9.82, 0)$ to show the position of the sun. Assume that the major axis is horizontal.
   (b) Use the information in Exercise 51 to determine how far Pluto travels in one orbit around the sun and approximate the area inside its orbit.

54. *Halley's Comet* (Refer to Example 2.) One of the most famous comets is Halley's comet. It travels in an elliptical orbit with $a = 17.95$ and $b = 4.44$ and passes by Earth roughly every 76 years. The most recent pass by Earth was in February 1986. (Source: M. Zeilik.)
   (a) Graph the orbit of Halley's comet in $[-21, 21, 5]$ by $[-14, 14, 5]$. Assume that the major axis is horizontal and that all units are in astronomical units. Plot a point at $(17.36, 0)$ to represent the position of the sun.
   (b) Use the formula in Exercise 51 to estimate how many miles Halley's comet travels in one orbit around the sun.
   (c) Estimate the average speed of Halley's comet in miles per hour.

55. *Satellite Orbit* The orbit of Explorer VII and the outline of Earth's surface are shown in the accompanying figure. This orbit is described by
$$\dfrac{x^2}{4464^2} + \dfrac{y^2}{4462^2} = 1,$$

and the surface of Earth is described by
$$\dfrac{(x - 164)^2}{3960^2} + \dfrac{y^2}{3960^2} = 1.$$

Find the maximum and minimum heights of the satellite above Earth's surface if all units are miles. (*Source:* **W. Thomson,** *Introduction to Space Dynamics.*)

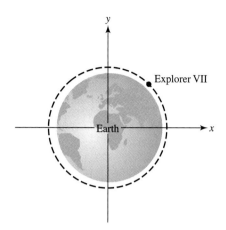

56. *Weight Machines* Elliptic shapes are used rather than circular shapes in modern weight machines. Suppose that the ellipse shown in the accompanying figure is represented by the equation

$$\frac{x^2}{16} + \frac{y^2}{100} = 1,$$

where the units are inches. Find r_1 and r_2.

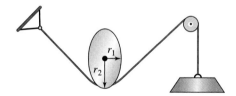

57. *Arch Bridge* The arch under a bridge is designed as the upper half of an ellipse as illustrated in the accompanying figure. Its equation is modeled by

$$400x^2 + 10{,}000y^2 = 4{,}000{,}000,$$

where the units are feet. Find the height and width of the arch.

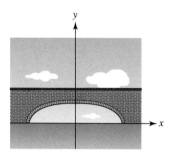

58. *Population Growth* Suppose that the population y of a country can be modeled by the upper right branch of the hyperbola

$$\frac{x^2}{a^2} - \frac{y^2}{b^2} = 1,$$

where x represents time in years. What happens to the population after a long period of time?

WRITING ABOUT MATHEMATICS

59. Explain how the values of a and b affect the graph of $\frac{x^2}{a^2} + \frac{y^2}{b^2} = 1$. Assume that $a > b > 0$.

60. Explain how the values of a and b affect the graph of $\frac{x^2}{a^2} - \frac{y^2}{b^2} = 1$. Assume that a and b are positive.

CHECKING BASIC CONCEPTS FOR SECTIONS 9.1 AND 9.2

1. Graph the parabola $x = (y - 2)^2 + 1$. Find the vertex and axis of symmetry.

2. Find the equation of the circle with center $(1, -2)$ and radius 2. Graph the circle.

3. Find the x- and y-intercepts on the graph of $\frac{x^2}{4} + \frac{y^2}{9} = 1$.

4. Graph the following. Label any vertices and state the type of conic section that it represents.
   (a) $x = y^2$
   (b) $\frac{x^2}{16} + \frac{y^2}{25} = 1$
   (c) $\frac{x^2}{4} - \frac{y^2}{9} = 1$
   (d) $(x - 1)^2 + (y + 2)^2 = 9$

9.3 NONLINEAR SYSTEMS OF EQUATIONS AND INEQUALITIES

Basic Concepts ~ Solving Nonlinear Systems of Equations ~ Solving Nonlinear Systems of Inequalities

INTRODUCTION

To describe characteristics of curved objects we often need to use nonlinear equations. For example, cylinders have a curved shape, as illustrated in Figure 9.39. If the radius of a cylinder is denoted r and its height denoted h, its volume V is given by the nonlinear equation

$$V = \pi r^2 h$$

and its side area A is given by the nonlinear equation

$$A = 2\pi rh.$$

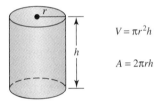

Figure 9.39 Cylindrical Container

If we want to manufacture a cylindrical container that holds 35 cubic inches and whose side area is 50 square inches, we need to solve the following **system of nonlinear equations**. (This system is solved in Example 4.)

$$\pi r^2 h = 35$$
$$2\pi rh = 50$$

In this section we solve systems of nonlinear equations and inequalities.

BASIC CONCEPTS

One way to locate the points at which the line $y = 2x$ intersects the circle $x^2 + y^2 = 5$, is to graph both equations (see Figure 9.40).

The equation describing the circle is nonlinear. Another way to locate the points of intersection is symbolically, by solving the nonlinear system of equations.

$$y = 2x$$
$$x^2 + y^2 = 5.$$

Linear systems of equations can have zero, one, or infinitely many solutions. It is possible for a system of nonlinear equations to have *any number* of solutions. Figure 9.40 shows that this nonlinear system of equations has two solutions: $(-1, -2)$ and $(1, 2)$.

9.3 Nonlinear Systems of Equations and Inequalities

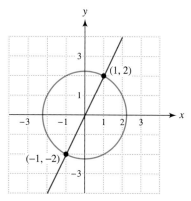

Figure 9.40

SOLVING NONLINEAR SYSTEMS OF EQUATIONS

Nonlinear systems of equations can sometimes be solved graphically, numerically, and symbolically. One symbolic technique is the **method of substitution**, which we demonstrate in the next example.

EXAMPLE 1 *Solving a nonlinear system of equations symbolically*

Solve

$$y = 2x$$
$$x^2 + y^2 = 5$$

symbolically. Check any solutions.

Solution

Substitute $y = 2x$ into the second equation and solve for x.

$x^2 + (2x)^2 = 5$	Let $y = 2x$ in the second equation.
$x^2 + 4x^2 = 5$	Properties of exponents
$5x^2 = 5$	Combine like terms.
$x^2 = 1$	Divide by 5.
$x = \pm 1$	Square root property

To determine corresponding y-values, substitute $x = \pm 1$ into $y = 2x$; the solutions are $(1, 2)$ and $(-1, -2)$. To check $(1, 2)$, substitute $x = 1$ and $y = 2$ into the given equations.

$$2 \stackrel{?}{=} 2(1) \quad \text{True}$$
$$(1)^2 + (2)^2 \stackrel{?}{=} 5 \quad \text{True}$$

To check $(-1, -2)$, substitute $x = -1$ and $y = -2$ into the given equations.

$$-2 \stackrel{?}{=} 2(-1) \quad \text{True}$$
$$(-1)^2 + (-2)^2 \stackrel{?}{=} 5 \quad \text{True}$$

EXAMPLE 2 Solving a nonlinear system of equations

Solve the nonlinear system of equations graphically, numerically, and symbolically.

$$x^2 - y = 2$$
$$x^2 + y = 4$$

Solution

Graphical Solution Begin by solving each equation for y.

$$y = x^2 - 2$$
$$y = 4 - x^2$$

Graph $Y_1 = X^2 - 2$ and $Y_2 = 4 - X^2$. The solutions are approximately $(-1.73, 1)$ and $(1.73, 1)$, as shown in Figures 9.41 and 9.42. The graphs consist of two parabolas intersecting at two points.

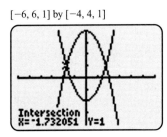
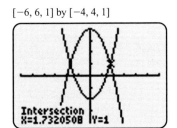

Figure 9.41   **Figure 9.42**

Numerical Solution Construct tables Y_1 and Y_2, as shown in Figures 9.43 and 9.44. Note that $Y_1 = Y_2 = 1$, when $x \approx 1.73$.

Figure 9.43   **Figure 9.44**

Symbolic Solution Solving the first equation for y gives $y = x^2 - 2$. Substitute this expression for y into the second equation and solve for x.

$$x^2 + y = 4 \qquad \text{Second equation}$$
$$x^2 + (x^2 - 2) = 4 \qquad \text{Substitute } y = x^2 - 2.$$
$$2x^2 = 6 \qquad \text{Combine like terms.}$$
$$x^2 = 3 \qquad \text{Divide by 2.}$$
$$x = \pm\sqrt{3} \qquad \text{Square root property}$$

To determine y, substitute $x = \pm\sqrt{3}$ into $y = x^2 - 2$.
$$y = (\sqrt{3})^2 - 2 = 3 - 2 = 1$$
$$y = (-\sqrt{3})^2 - 2 = 3 - 2 = 1$$
The *exact* solutions are $(\sqrt{3}, 1)$ and $(-\sqrt{3}, 1)$.

EXAMPLE 3 *Solving a nonlinear system of equations*

Solve the linear system of equations symbolically and graphically.
$$x^2 - y^2 = 3$$
$$x^2 + y^2 = 5$$

Solution

Symbolic Solution Instead of using substitution on this nonlinear system of equations, we use elimination. Note that, if we add the two equations, the y-variable will be eliminated.

$$\begin{aligned} x^2 - y^2 &= 3 \\ x^2 + y^2 &= 5 \\ \hline 2x^2 &= 8 \end{aligned}$$

Solving gives $x^2 = 4$, or $x = \pm 2$. To determine y, substitute $x^2 = 4$ into $x^2 + y^2 = 5$.
$$4 + y^2 = 5 \quad \text{or} \quad y^2 = 1$$
Because $y^2 = 1$, $y = \pm 1$. Thus there are four solutions: $(2, 1)$, $(2, -1)$, $(-2, 1)$, and $(-2, -1)$.

Graphical Solution The graph of the first equation is a hyperbola, and the graph of the second is a circle with radius $\sqrt{5}$. Begin by solving each equation for y.
$$y = \pm\sqrt{x^2 - 3} \quad \text{Hyperbola}$$
$$y = \pm\sqrt{5 - x^2} \quad \text{Circle}$$
Graph $Y_1 = \sqrt{(X^\wedge 2 - 3)}$, $Y_2 = -\sqrt{(X^\wedge 2 - 3)}$, $Y_3 = \sqrt{(5 - X^\wedge 2)}$, and $Y_4 = -\sqrt{(5 - X^\wedge 2)}$. The points of intersection are $(\pm 2, \pm 1)$. Figure 9.45 shows the solution located in the first quadrant.

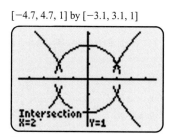

Figure 9.45

In the next example we use graphical and symbolic methods to solve the system of equations presented in the Introduction.

EXAMPLE 4 Modeling the dimensions of a can

Find the dimension of a can having a volume V of 35 cubic inches and a side area A of 50 square inches by solving

$$\pi r^2 h = 35$$
$$2\pi r h = 50$$

graphically and symbolically.

Solution

Graphical Solution Begin by solving each equation for h.

$$h = \frac{35}{\pi r^2}$$

$$h = \frac{50}{2\pi r}$$

Let x correspond to r and y correspond to h, and graph $Y_1 = 35/(\pi X^2)$ and $Y_2 = 50/(2\pi X)$. Their graphs intersect near the point (1.4, 5.68), as shown in Figure 9.46. A can having a volume of 35 square inches and a side area of 50 square inches has a radius of about 1.4 inches and a height of 5.68 inches.

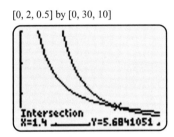

Figure 9.46

Symbolic Solution We can find r by solving the following equation.

$$\frac{50}{2\pi r} = \frac{35}{\pi r^2} \qquad h = \frac{50}{2\pi r} \text{ and } h = \frac{35}{\pi r^2}$$

$$50\pi r^2 = 70\pi r \qquad \text{Clear fractions.}$$

$$50\pi r^2 - 70\pi r = 0 \qquad \text{Subtract } 70\pi r.$$

$$10\pi r(5r - 7) = 0 \qquad \text{Factor out } 10\pi r.$$

$$10\pi r = 0 \quad \text{or} \quad 5r - 7 = 0 \qquad \text{Zero-product property}$$

$$r = 0 \quad \text{or} \quad r = \frac{7}{5} = 1.4 \qquad \text{Solve.}$$

Because $h = \dfrac{50}{2\pi r}$, $r = 0$ is not possible, but we can find h by substituting $r = 1.4$ into the formula.

$$h = \frac{50}{2\pi(1.4)} \approx 5.68$$

Note that these computations agree with the graphical solution.

Solving Nonlinear Systems of Inequalities

In Section 4.3 we solved systems of linear inequalities. A **system of nonlinear inequalities** can be solved similarly by using graphical techniques. For example, consider the system of nonlinear equalities

$$y \geq x^2 - 2$$
$$y \leq 4 - x^2.$$

The graph of $y = x^2 - 2$ is a parabola opening upward. The solution set for $y \geq x^2 - 2$ includes all points lying on or above this parabola. See Figure 9.47. Similarly, the graph of $y = 4 - x^2$ is a parabola opening downward. The solution set for $y \leq 4 - x^2$ includes all points lying on or below this parabola. See Figure 9.48.

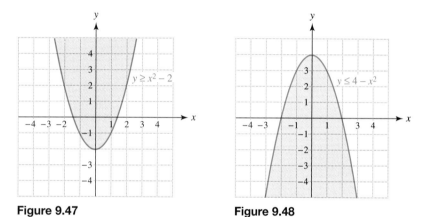

Figure 9.47   **Figure 9.48**

The solution set for the *system* of nonlinear inequalities includes all points (x, y) in both shaded regions. The *intersection* of the shaded regions is shown in Figure 9.49.

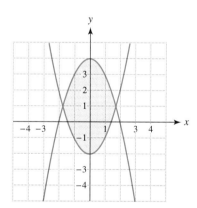

Figure 9.49

EXAMPLE 5   *Solving a system of inequalities graphically*

Shade the solution set for the system of inequalities.

$$\frac{x^2}{4} + \frac{y^2}{9} < 1$$
$$y > 1.$$

Solution

The solutions to $\frac{x^2}{4} + \frac{y^2}{9} < 1$ lie inside the ellipse $\frac{x^2}{4} + \frac{y^2}{9} = 1$. See Figure 9.50. Solutions to $y > 1$ lie above the line $y = 1$, as shown in Figure 9.51. The intersection of these two regions is shown in Figure 9.52. Note that a dashed curve and line are used when equality is not involved.

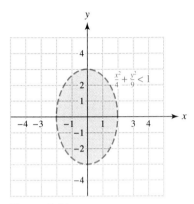

Figure 9.50

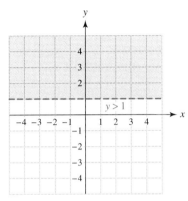

Figure 9.51

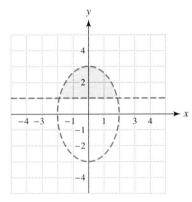

Figure 9.52

EXAMPLE 6 Solving a system of inequalities with a graphing calculator

 Shade the solution set for the following system of inequalities, using a graphing calculator.

$$x^2 + y \leq 4$$
$$-x + y \geq 2$$

Solution

Solve each inequality for y, obtaining $y \leq 4 - x^2$ and $y \geq x + 2$. Graph $Y_1 = 4 - X^{\wedge}2$ and $Y_2 = X + 2$, as shown in Figure 9.53. The solutions lie below the parabola Y_1 and above the line Y_2. This area can be shaded by using a graphing calculator and its shading utility. See Figures 9.54 and 9.55. The shade routine in the DRAW menu of many calculators may be used to shade a region between two functions.

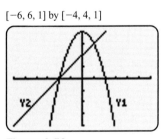

Figure 9.53

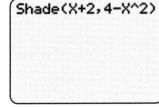

Figure 9.54

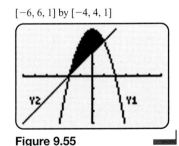

Figure 9.55

In the next example we use a graphing calculator to shade a region that lies above both graphs, using the "$Y_1 =$" menu. With it we can shade either above or below the graph of a function.

9.3 Nonlinear Systems of Equations and Inequalities

EXAMPLE 7 *Solving a system of inequalities with a graphing calculator*

 Shade the solution set for the following system of inequalities.

$$y \geq x^2 - 2$$
$$y \geq -1 - x$$

Solution

Enter $Y_1 = X^2 - 2$ and $Y_2 = -1 - X$, as shown in Figure 9.56. Note that the option to shade above the graphs of Y_1 and Y_2 was selected to the left of Y_1 and Y_2. Then the two equations were graphed in Figure 9.57. The solution set corresponds to where there are both vertical and horizontal lines.

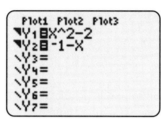

Figure 9.56

[−4.7, 4.7, 1] by [−3.1, 3.1, 1]

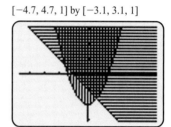

Figure 9.57

9.3 Putting It All Together

In this section we discussed nonlinear systems of equations in two variables. Unlike a linear system of equations, a nonlinear system of equations can have any number of solutions. Systems of nonlinear equations can be solved symbolically, graphically, and numerically. Systems of nonlinear inequalities involving two variables usually have infinitely many solutions, which can be represented by a shaded region in the xy-plane. The following table summarizes these concepts.

Concept	Explanation
Nonlinear systems of equations	To solve the following system of equations symbolically, using *substitution*, solve the first equation for y. $$x + y = 5 \quad \text{or} \quad y = 5 - x$$ $$x^2 - y = 1$$ Substitute $y = 5 - x$ into the second equation and solve the resulting quadratic equation. (*Elimination* can also be used on this system.) $$x^2 - (5 - x) = 1 \quad \text{or} \quad x^2 + x - 6 = 0$$ implies that $$x = -3, 2.$$

continued on next page

Concept	Explanation
Nonlinear systems of equations *continued*	Then $y = 5 - (-3) = 8$ and $y = 5 - 2 = 3$. The solutions are $(-3, 8)$ and $(2, 3)$. Graphical support is shown in the accompanying figures, where $Y_1 = 5 - X$ and $Y_2 = X^2 - 1$.
Nonlinear systems of inequalities	To solve the following system of inequalities graphically, solve each equation for y. $$x + y \leq 5 \quad \text{or} \quad y \leq 5 - x$$ $$x^2 - y \leq 1 \quad \text{or} \quad y \geq x^2 - 1$$ The solutions lie above the parabola and below the line, as shown in the following figure.

9.3 EXERCISES

CONCEPTS

1. How many solutions can a system of nonlinear equations have?

2. If a system of nonlinear equations has two equations, how many equations does a solution have to satisfy?

3. Determine mentally the number of solutions to the following system of equations. Explain your reasoning.

$$y = x$$
$$x^2 + y^2 = 4$$

4. Describe the solution set to $x^2 + y^2 \leq 1$.

9.3 Nonlinear Systems of Equations and Inequalities

5. Does $(-2, -1)$ satisfy $5x^2 - 2y^2 > 18$?

6. Does $(3, 4)$ satisfy $x^2 - 2y \geq 4$?

7. Sketch a parabola and ellipse with four points of intersection.

8. Sketch a line and a hyperbola with two points of intersection.

NONLINEAR SYSTEMS OF EQUATIONS

Exercises 9–12: Use the graph to estimate all solutions to the system of equations. Check each solution.

9. $x^2 + y^2 = 10, y = 3x$

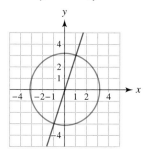

10. $x^2 + 3y^2 = 16, y = -x$

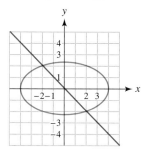

11. $y^2 - x^2 = 1, x^2 + 3y^2 = 3$

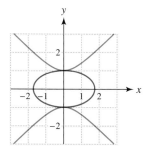

12. $y = 1 - x^2, x^2 + y^2 = 1$

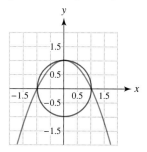

Exercises 13–18: Solve the system of equations symbolically. Check your solutions.

13. $y = 2x$
    $x^2 + y^2 = 45$

14. $y = x$
    $y^2 = 3 - 2x^2$

15. $x + y = 1$
    $x^2 - y^2 = 3$

16. $y - x = -1$
    $y = 2x^2$

17. $y - x^2 = 0$
    $x^2 + y^2 = 6$

18. $x^2 - y^2 = 4$
    $x^2 + y^2 = 4$

Exercises 19–24: Solve the system of equations graphically. Round answers to the nearest hundredth when appropriate.

19. $y = 2x$
    $x^2 + y = 5$

20. $y = x^2 - 3$
    $2x^2 - y = 1 - 3x$

21. $x + y = 2$
    $x - y^2 = 3$

22. $y - 2x = 3$
    $x - y^2 = -2$

23. $xy = 1$
    $y = x$

24. $y = (x + 1)^2$
    $x^2 + y^2 = 4$

Exercises 25–28: Solve the system of equations
  (a) symbolically,
  (b) graphically, and
  (c) numerically.

25. $\quad y = -2x$
    $x^2 + y = 3$

26. $4x - y = 0$
    $x^3 - y = 0$

27. $\quad xy = 1$
    $x - y = 0$

28. $x^2 + y^2 = 4$
    $\quad y - x = 2$

NONLINEAR SYSTEMS OF INEQUALITIES

Exercises 29–32: Shade the solution set in the xy-plane.

29. $y \geq x^2$

30. $y \leq x^2 - 1$

31. $\dfrac{x^2}{4} + \dfrac{y^2}{9} > 1$

32. $x^2 + y^2 \leq 1$

Exercises 33–40: Shade the solution set in the xy-plane. Then use the graph to select one solution.

33. $y > x^2 + 1$
    $y < 3$

34. $y > x^2$
    $y < x + 2$

35. $x^2 + y^2 \leq 1$
    $\quad\quad y < x$

36. $y > x^2 - 2$
    $y \leq 2 - x^2$

37. $\quad x^2 + y^2 \leq 1$
    $(x - 2)^2 + y^2 \leq 1$

38. $\quad x^2 - y \geq 2$
    $(x + 1)^2 + y^2 \leq 4$

39. $x^2 - y^2 \leq 4$
    $x^2 + y^2 \leq 9$

40. $3x + 2y < 6$
    $x^2 + y^2 \leq 16$

Exercises 41–42: Match the inequality or system of inequalities with its graph (a. or b.). The standard viewing rectangle was used to produce the graphs.

41. $y \leq \dfrac{1}{2}x^2$

42. $y \geq x^2 + 1$
    $y \leq 5$

a.

b.
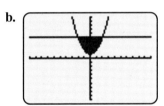

Exercises 43–44: Use the graph to write the inequality or system of inequalities.

43.
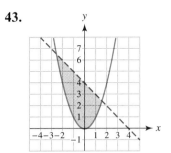

44.
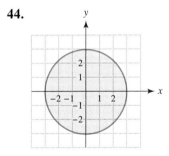

APPLICATIONS

45. *Dimensions of a Can* (Refer to Example 4.) Find the dimensions of a cylindrical container with volume of 40 cubic inches and side area of 50 square inches.
(a) graphically and (b) symbolically.

46. *Dimensions of a Can* (Refer to Example 4.) Is it possible to design an aluminum can with volume of 60 cubic inches and side area of 60 square inches? If so, find the dimensions of the can graphically.

47. *Dimensions of a Cone* The volume V of a cone is given by $V = \frac{1}{3}\pi r^2 h$, and the surface area S of its side is given by $S = \pi r \sqrt{r^2 + h^2}$, where h is the height and r is the radius of the base (see the accompanying figure).

(a) Solve each equation for h.
(b) Estimate r and h graphically for a cone with volume of 34 cubic feet and surface area of 52 square feet.

48. *Area and Perimeter* The area of a room is 143 feet, and its perimeter is 48 feet. Let x be the width and y the length of the room. See the accompanying figure.
(a) Write a system of equations that models this situation.
(b) Solve the system graphically.

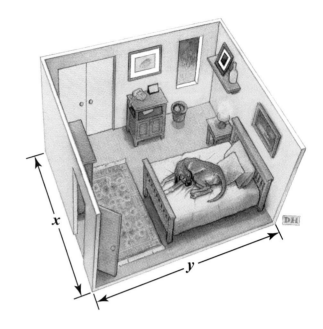

WRITING ABOUT MATHEMATICS

49. A student *incorrectly* changes the system of inequalities

$$x^2 - y \geq 6$$
$$2x - y \leq -3$$

to

$$y \geq x^2 - 6$$
$$y \leq 2x + 3.$$

The student discovers that the point $(1, 2)$ satisfied the second system but not the first. Explain the student's error.

50. Explain graphically how systems of nonlinear equations can have any number of solutions. Sketch graphs of different systems with 0, 1, 2, and 3 solutions.

CHECKING BASIC CONCEPTS FOR SECTION 9.3

1. Solve the following system of equations symbolically and graphically.

$$x^2 - y = 2x$$
$$2x - y = 3$$

2. Determine mentally the number of solutions to the following system of equations.

$$y = x^2 - 4$$
$$y = x$$

3. The solution set for a system of inequalities is shown in the following figure.

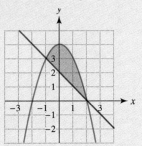

(a) Find one ordered pair (x, y) that is a solution and one that is not.

(b) Write the system of inequalities represented by the graph.

4. Shade the solution set for the following system of inequalities.

$$x^2 + y^2 \le 4$$
$$y < 1$$

CHAPTER 9 SUMMARY

Section 9.1 Parabolas and Circles

There are three basic types of conic sections: parabolas, ellipses, and hyperbolas. A parabola can have a vertical or a horizontal axis. Two forms of an equation for a parabola with a vertical axis are

$$y = ax^2 + bx + c \quad \text{and} \quad y = a(x - h)^2 + k.$$

If $a > 0$ the parabola opens upward, and if $a < 0$ it opens downward (see the figure on the left). The vertex is located at (h, k). Two forms of an equation for a parabola with a horizontal axis are

$$x = ay^2 + by + c \quad \text{and} \quad x = a(y - k)^2 + h.$$

If $a > 0$ the parabola opens to the right and if $a < 0$ it opens to the left (see the figure on the right). The vertex is located at (h, k).

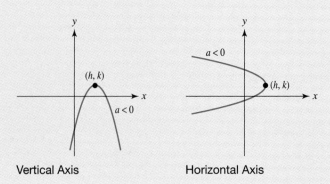

Vertical Axis　　　　Horizontal Axis

Section 9.1 Parabolas and Circles (continued)

The standard equation for a circle with center (h, k) and radius r is
$$(x - h)^2 + (y - k)^2 = r^2.$$

Section 9.2 Ellipses and Hyperbolas

Ellipses

The standard equation for an ellipse centered at the origin with a horizontal major axis is $\frac{x^2}{a^2} + \frac{y^2}{b^2} = 1$, $a > b > 0$, and the vertices are $(\pm a, 0)$, as shown in the figure on the left. The standard equation for an ellipse centered at the origin with a vertical major axis is $\frac{y^2}{a^2} + \frac{x^2}{b^2} = 1$, $a > b > 0$, and the vertices are $(0, \pm a)$, as shown in the figure on the right. Circles are a special type of ellipse, with the major and minor axes having equal lengths.

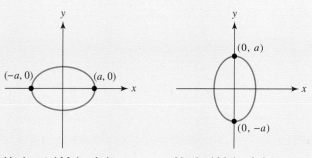

Horizontal Major Axis   Vertical Major Axis

Hyperbolas

The standard equation for a hyperbola centered at the origin with a horizontal transverse axis is $\frac{x^2}{a^2} - \frac{y^2}{b^2} = 1$, the asymptotes are given by $y = \pm \frac{b}{a}x$, and the vertices are $(\pm a, 0)$, as shown in the figure on the left. The standard equation for a hyperbola centered at the origin with a vertical transverse axis is $\frac{y^2}{a^2} - \frac{x^2}{b^2} = 1$, the asymptotes are $y = \pm \frac{a}{b}x$, and the vertices are $(0, \pm a)$, as shown in the figure on the right.

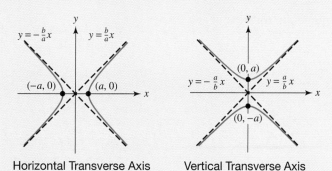

Horizontal Transverse Axis   Vertical Transverse Axis

Section 9.3 Nonlinear Systems of Equations and Inequalities

Systems of nonlinear equations can have any number of solutions. The methods of substitution or elimination can often be used to solve a system of nonlinear equations symbolically. Nonlinear systems can also be solved graphically or numerically. The solution set for a system of nonlinear inequalities with two equations and two variables is typically a region in the xy-plane. A solution is an ordered pair (x, y) that satisfies both inequalities. The solution set for $y \geq x^2 - 2$ and $y \leq 4 - \frac{1}{2}x^2$ is shaded in the accompanying figure.

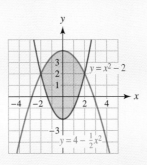

CHAPTER 9
REVIEW EXERCISES

SECTION 9.1

Exercises 1–6: Graph the parabola. Find the vertex and axis of symmetry.

1. $x = 2y^2$
2. $x = -(y + 1)^2$
3. $x = -2(y - 2)^2$
4. $x = (y + 2)^2 - 1$
5. $x = -3y^2 + 1$
6. $x = \frac{1}{2}y^2 + y - 3$

7. Use the graph to determine the equation of the parabola.

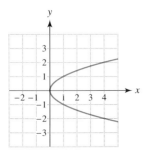

8. Use the graph to find the equation of the circle.

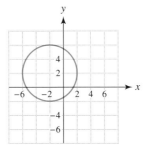

9. Write the equation of the circle with radius 1 and center $(0, 0)$.

10. Write the equation of the circle with radius 4 and center $(2, -3)$.

Exercises 11–14: Find the radius and center of the circle. Then graph the circle.

11. $x^2 + y^2 = 25$
12. $(x - 2)^2 + y^2 = 9$

13. $(x + 3)^2 + (y - 1)^2 = 5$

14. $x^2 - 2x + y^2 + 2y = 7$

Section 9.2

Exercises 15–18: Graph the ellipse by hand. Label the vertices and endpoints of the minor axis.

15. $\dfrac{x^2}{4} + \dfrac{y^2}{25} = 1$

16. $x^2 + \dfrac{y^2}{4} = 1$

17. $25x^2 + 20y^2 = 500$

18. $4x^2 + 9y^2 = 36$

19. Use the graph to determine the equation of the ellipse.

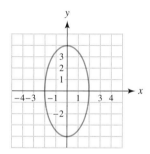

20. Use the graph to determine the equation of the hyperbola.

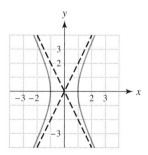

Exercises 21–24: Graph the hyperbola. Show the asymptotes.

21. $\dfrac{x^2}{9} - \dfrac{y^2}{4} = 1$

22. $\dfrac{y^2}{25} - \dfrac{x^2}{16} = 1$

23. $y^2 - x^2 = 1$ **24.** $25x^2 - 16y^2 = 400$

Exercises 25–30: Use a graphing calculator to graph the conic section in an appropriate viewing rectangle. Identify what type of conic section that each equation represents.

25. $x = 4.1y^2 - 3.7$

26. $x = -0.4y^2 - 1.1y + 2.4$

27. $\dfrac{x^2}{4.2} + \dfrac{y^2}{3.2} = 1$ **28.** $\dfrac{y^2}{17} + \dfrac{x^2}{31} = 1$

29. $\dfrac{y^2}{5} - \dfrac{x^2}{11} = 1$ **30.** $\dfrac{x^2}{3.1} - \dfrac{y^2}{6.3} = 1$

Section 9.3

Exercises 31–34: Use the graph to estimate all solutions to the system of equations. Check each solution.

31. $x^2 + y^2 = 9, \quad x + y = 3$

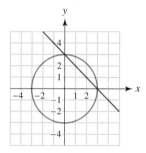

32. $xy = 2, \quad y = 2x$

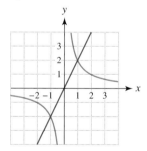

33. $x^2 - y = x, \quad y = x$

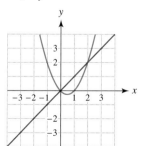

34. $x^2 + y^2 = 5, \quad x^2 - y^2 = 3$

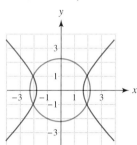

Exercises 35–38: Solve the system of equations symbolically. Check your solutions.

35. $\quad y = x$
$\quad\ \ x^2 + y^2 = 32$

36. $\quad x - y = 4$
$\qquad x^2 + y^2 = 16$

37. $\quad y = x^2$
$\quad\ \ 2x^2 + y = 3$

38. $\quad y = x^2 + 1$
$\quad\ \ 2x^2 - y = 3x - 3$

Exercises 39 and 40: Solve the system of equations graphically. Estimate solutions to the nearest hundredth when appropriate.

39. $2x - y = 4$
$\quad\ x^2 + y = 4$

40. $\quad x^2 + y = 4$
$\qquad x^2 + y^2 = 4$

Exercises 41 and 42: Solve the system of equations
  (a) symbolically,
  (b) graphically, and
  (c) numerically.

41. $\quad y = x$
$\quad\ \ x^2 + 2y = 8$

42. $\quad y = x^3$
$\qquad x^2 - y = 0$

Exercises 43–50: Shade the solution set in the xy-plane.

43. $y \geq 2x^2$

44. $y < 2x - 3$

45. $y < -x^2$

46. $\dfrac{x^2}{9} + \dfrac{y^2}{16} \leq 1$

47. $y - x^2 \geq 1$
$\quad\ \ y \leq 2$

48. $\quad x^2 + y \leq 4$
$\qquad 3x + 2y \geq 6$

49. $y > x^2$
$\quad\ y < 4 - x^2$

50. $\dfrac{x^2}{4} + \dfrac{y^2}{9} > 1$
$\quad\ \ x^2 + y^2 < 16$

Exercises 51 and 52: Use the graph to write the system of inequalities.

51.

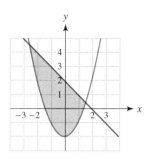

52.

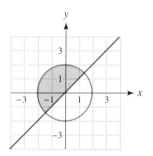

APPLICATIONS

53. *Area and Perimeter* The area of a desktop is 1000 square inches, and its perimeter is 130 inches. Let x be the width and y the length of the desktop.
  (a) Write a system of equations that models this situation.
  (b) Solve the system graphically.
  (c) Solve the system symbolically.

54. *Numbers* The product of two positive numbers is 60, and their difference is 7. Let x be the smaller number and y be the larger number.
    (a) Write a system of equations whose solution gives the two numbers.
    (b) Solve the system graphically.
    (c) Solve the system symbolically.

55. *Dimensions of a Container* The volume of a cylindrical container is $V = \pi r^2 h$, and its surface area, *excluding* the top and bottom is $A = 2\pi rh$. Graphically find the dimensions of a container with $A = 100$ square feet and $V = 50$ cubic feet. Is your answer unique?

56. *Dimensions of a Container* The volume of a cylindrical container is $V = \pi r^2 h$, and its surface area *including* the top and bottom, is $A = 2\pi rh + 2\pi r^2$. Graphically find the dimensions of a can with $A = 80$ square inches and $V = 35$ cubic inches. Is your answer unique?

57. *Geometry of an Ellipse* The area inside an ellipse is given by $A = \pi ab$, and its perimeter P can be approximated by

$$P = 2\pi \sqrt{\frac{a^2 + b^2}{2}}.$$

   (a) Graph $\dfrac{x^2}{5} + \dfrac{y^2}{12} = 1$.
   (b) Estimate its area and perimeter.

58. *Orbit of Mars* Mars has an elliptical orbit that is nearly circular, with $a = 1.524$ and $b = 1.517$, where the units are astronomical units (1 A.U. = 93 million miles). *(Source: M. Zeilik.)*
   (a) Graph the orbit of Mars in $[-3, 3, 1]$ by $[-2, 2, 1]$. Plot the point $(0.14, 0)$ to show the position of the sun. Assume that the major axis is horizontal.
   (b) Use the information in Exercise 57 to estimate how far Mars travels in one orbit around the sun. Approximate the area inside its orbit.

CHAPTER 9

TEST

1. Graph the parabola $x = (y - 4)^2 - 2$. Find the vertex and axis of symmetry.

2. Use the graph to determine the equation of the parabola.

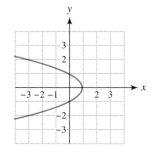

3. Use the graph to find the equation of the circle.

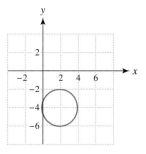

4. Write the equation of the circle with radius 10 and center $(-5, 2)$.

5. Find the radius and center of the circle given by
$$x^2 + 4x + y^2 - 6y = 3.$$
Then graph the circle.

6. Graph the ellipse $\dfrac{x^2}{16} + \dfrac{y^2}{49} = 1$ by hand. Label the vertices and endpoints of the minor axis.

7. Use the graph to determine the equation of the ellipse.

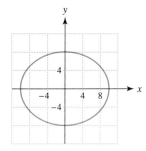

8. Graph the hyperbola $4x^2 - 9y^2 = 36$. Show the asymptotes.

9. Use a graphing calculator to graph
$$\dfrac{x^2}{5.7} + \dfrac{y^2}{1.8} = 1$$
in an appropriate viewing rectangle. Identify the type of conic section that the equation represents.

10. Use the graph to estimate all solutions to the system of equations. Check each solution by substitution in the system of equations.
$$x^2 + y^2 = 16 \quad \text{and} \quad x - y = 4$$

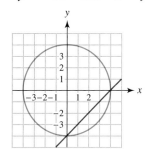

11. Solve the system of equations symbolically.
$$x - y = 3 \quad \text{and} \quad x^2 + y^2 = 17$$

12. Solve the system of equations graphically. Estimate solutions to the nearest hundredth when appropriate.
$$2x^2 - y = x \quad \text{and} \quad x^2 + 3y = 4$$

13. Shade the solution set in the *xy*-plane.
$$3x + y > 6 \quad \text{and} \quad x^2 + y^2 < 25$$

14. Use the graph to write the system of inequalities.

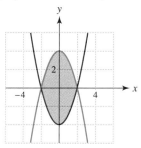

15. *Area and Perimeter* The area of a rectangular swimming pool is 5000 square feet, and its perimeter is 300 feet. Let x be the width and y the length of the pool.
   (a) Write a system of equations that models this situation.
   (b) Solve the system either graphically or numerically.
   (c) Solve the system symbolically.

16. *Dimensions of a Box* The volume of a rectangular box with a square bottom and open top is $V = x^2 y$, and its surface area is $A = x^2 + 4xy$, where x represents its width and length and y represents its height. Estimate graphically the dimensions of a box with $V = 1183$ cubic inches and $A = 702$ square inches. Is your answer unique? (*Hint:* Substitute appropriate values for V and A, and then solve each equation for y.)

17. *Orbit of a Uranus* The planet Uranus has an elliptical orbit that is nearly circular, with $a = 19.18$ and $b = 19.16$, where units are astronomical units (1 A.U. = 93 million miles). (*Source:* M. Zeilik.)
   (a) Graph the orbit of Uranus in $[-30, 30, 10]$ by $[-20, 20, 10]$. Plot the point $(0.9, 0)$ to show the position of the sun. Assume that the major axis is horizontal.
   (b) Find the minimum distance between Uranus and the sun.

Chapter 9

Extended and Discovery Exercises

Exercises 1–2: Foci of Parabolas The focus of a parabola is a point that has special significance. When a parabola is rotated about its axis, it sweeps out a shape called a **paraboloid,** as illustrated in the accompanying figure to the right. Paraboloids have an important reflective property. When incoming rays of light from the sun or distant stars strike the surface of a paraboloid, each ray is reflected toward the focus. If the rays are sunlight, intense heat is produced, which can be used to generate solar heat. Radio signals from distant space also concentrate at the focus, and scientists can measure these signals by placing a receiver there. The same reflective property of a paraboloid can be used in reverse. If a light source is placed at the focus, the light is reflected straight ahead. Searchlights, flashlights, and car headlights make use of this reflective property.

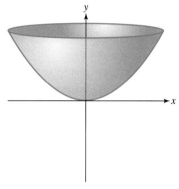

Paraboloid

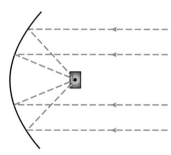

Reflective Property

The focus is always located inside a parabola, on its axis of symmetry. If the distance between the vertex and the focus is $|p|$, the following equations can be used to locate the focus. Note that the value of p may be either positive or negative.

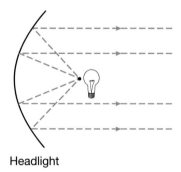

Headlight

Equation of a Parabola with Vertex (0, 0)

Vertical Axis

The parabola with a focus at $(0, p)$ has the equation $x^2 = 4py$.

Horizontal Axis

The parabola with a focus at $(p, 0)$ has the equation $y^2 = 4px$.

1. Sketch a graph of each parabola. Label the vertex and focus.
   (a) $x^2 = 4y$
   (b) $y^2 = -8x$
   (c) $x = 2y^2$

2. The reflective property of paraboloids is used in satellite dishes and radio telescopes. The U.S. Naval Research Laboratory designed a giant radio telescope weighing 3450 tons. Its parabolic dish has a diameter of 300 feet and a depth of 44 feet, as shown in the accompanying figures.
   (*Source: J. Mar, Structure Technology for Large Radio and Radar Telescope Systems.*)
   (a) Find an equation in the form $y = ax^2$ that describes a cross section of this dish.
   (b) If the receiver is located at the focus, how far should it be from the vertex?

Radio Telescope

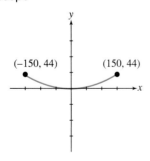

Exercises 3 and 4: Translations of Ellipses and Hyperbolas Ellipses and hyperbolas can be translated so that they are centered at a point (h, k), rather than at the origin. These techniques are the same as those used for parabolas and circles. To translate a conic section so that it is centered at (h, k) rather than (0, 0), replace x with (x − h) and replace y with (y − k). For example, to center $\frac{x^2}{9} + \frac{y^2}{4} = 1$ at $(-1, 2)$, change its equation to $\frac{(x+1)^2}{9} + \frac{(y-2)^2}{4} = 1$. See the accompanying figures.

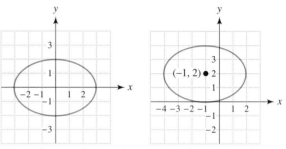

Ellipse Centered at (0, 0) Ellipse Centered at (−1, 2)

3. Graph each conic section. For hyperbolas give the equations of the asymptotes.
   (a) $\dfrac{(x-3)^2}{25} + \dfrac{(y-1)^2}{9} = 1$
   (b) $\dfrac{(x+1)^2}{4} + \dfrac{(y+2)^2}{16} = 1$
   (c) $\dfrac{(x+1)^2}{4} - \dfrac{(y-3)^2}{9} = 1$
   (d) $\dfrac{(y-4)^2}{16} - \dfrac{(x+1)^2}{4} = 1$

4. Write the equation of an ellipse having the following properties.
   (a) Horizontal major axis of length 8, minor axis of length 4, and centered at $(-3, 5)$.
   (b) Vertical major axis of length 10, minor axis of length 6, and centered at $(2, -3)$.

5. Determine the center of the conic section by completing the square.
   (a) $9x^2 - 18x + 4y^2 + 24y + 9 = 0$
   (b) $25x^2 + 150x - 16y^2 + 32y - 191 = 0$

CHAPTER 10
EXPONENTIAL AND LOGARITHMIC FUNCTIONS

In 1900, the Swedish scientist Svante Arrhenius first predicted a greenhouse effect resulting from emissions of carbon dioxide by the industrialized countries. His classic calculation made use of logarithms and predicted that a doubling of the carbon dioxide concentration in the atmosphere would raise the average global temperature by 7°F to 11°F. An increase in world population has resulted in higher emissions of greenhouse gases such as carbon dioxide, and these emissions have the potential to alter Earth's climate and destroy portions of the ozone layer.

In this chapter we use exponential and logarithmic functions to model a wide variety of phenomena, such as greenhouse gases, acid rain, the decline of the bluefin tuna, the demand for liver transplants, diversity of bird species, hurricanes, and earthquakes. Mathematics plays a key role in understanding, controlling, and predicting natural phenomena and people's effect on them.

Human history becomes more and more a race between education and catastrophe.
—H. G. Wells

Source: M. Kraljic, *The Greenhouse Effect.*

10.1 EXPONENTIAL FUNCTIONS

Basic Concepts ~ Graphs of Exponential Functions ~
Models Involving Exponential Functions ~
The Natural Exponential Function

INTRODUCTION

Many times the growth of a quantity depends on the amount or number present. The more money deposited in an account, the more interest the account earns; that is, the interest earned is proportional to the amount of money in an account. For example, if a person has $100 in an account and receives 10% annual interest, the interest accrued the first year will be $10, and the balance in the account will be $110 at the end of 1 year. Similarly, if a person begins with $1000 in an account, the balance in the account will be $1100 at the end of 1 year. This type of growth is called *exponential growth* and can be modeled by an *exponential function*.

BASIC CONCEPTS

Suppose that an insect population doubles each week. Table 10.1 shows the populations after x weeks. Note that, as the population of insects becomes larger, the *increase* in population each week becomes greater. The population is increasing by 100%, or doubling numerically, each week. When a quantity increases by a constant percentage (or constant factor) at regular intervals, its growth is exponential.

TABLE 10.1

Week	0	1	2	3	4	5
Population	100	200	400	800	1600	3200

We can model the data in Table 10.1 by using the exponential function

$$f(x) = 100(2)^x.$$

For example,

$$f(0) = 100(2)^0 = 100 \cdot 1 = 100,$$
$$f(1) = 100(2)^1 = 100 \cdot 2 = 200,$$
$$f(2) = 100(2)^2 = 100 \cdot 4 = 400,$$

and so on. Note that the exponential function f has a *variable as an exponent*.

Exponential Function

A function represented by

$$f(x) = Ca^x, \quad a > 0 \text{ and } a \neq 1,$$

is an **exponential function with base a.** (Unless stated otherwise, assume that $C > 0$.)

The set of valid inputs (domain) for an exponential function includes all real numbers. The set of corresponding outputs (range) includes all positive real numbers.

In the following example we evaluate some exponential functions both by hand and with a calculator. When evaluating an exponential function, we evaluate a^x before multiplying by C. This standard order of precedence is much like doing multiplication before addition. For example, $2(3)^2$ should be evaluated as

$$2(9) = 18 \quad \textit{not as} \quad (6)^2 = 36.$$

EXAMPLE 1 *Evaluating exponential functions*

Evaluate $f(x)$ for the given value of x by hand. Check your answers with a calculator.

(a) $f(x) = 10(3)^x \qquad x = 2$

(b) $f(x) = 5\left(\dfrac{1}{2}\right)^x \qquad x = 3$

(c) $f(x) = \dfrac{1}{3}(2)^x \qquad x = -1$

Solution

(a) $f(2) = 10(3)^2 = 10 \cdot 9 = 90$

(b) $f(3) = 5\left(\dfrac{1}{2}\right)^3 = 5 \cdot \dfrac{1}{8} = \dfrac{5}{8}$

(c) $f(-1) = \dfrac{1}{3}(2)^{-1} = \dfrac{1}{3} \cdot \dfrac{1}{2} = \dfrac{1}{6}$

These results are supported by Figure 10.1.

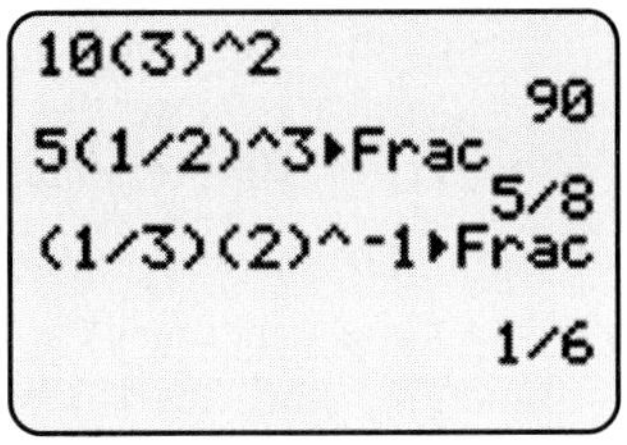

```
10(3)^2
                    90
5(1/2)^3▶Frac
                   5/8
(1/3)(2)^-1▶Frac
                   1/6
```

Figure 10.1

■ **MAKING CONNECTIONS**

The expressions a^{-x} and $\left(\dfrac{1}{a}\right)^x$

Using properties of exponents, we can write 2^{-x} as

$$2^{-x} = \dfrac{1}{2^x} = \left(\dfrac{1}{2}\right)^x.$$

In general, the expressions a^{-x} and $\left(\dfrac{1}{a}\right)^x$ are equal.

In the formula $f(x) = Ca^x$, a is called the **growth factor** when $a > 1$ and the **decay factor** when $0 < a < 1$. For an exponential function, each time x increases by 1 unit $f(x)$ increases by a *factor* of a when $a > 1$ and decreases by a factor of a when $0 < a < 1$. Moreover, since

$$f(0) = Ca^0 = C(1) = C,$$

the value of C equals the value of $f(x)$ when $x = 0$. If x represents time, C represents the initial value of f when time equals 0.

EXAMPLE 2 *Finding linear and exponential functions*

Use the table in each case to determine whether f is a linear or exponential function. Find a formula for f. Check your formula by making a table of values.

(a)
x	0	1	2	3	4
$f(x)$	16	8	4	2	1

(b)
x	0	1	2	3	4
$f(x)$	5	7	9	11	13

(c)
x	0	1	2	3	4
$f(x)$	1	3	9	27	81

Solution

(a) Each time x increases by 1 unit, $f(x)$ decreases by a factor of $\frac{1}{2}$. Therefore f is an exponential function with a decay factor of $\frac{1}{2}$. Because $f(0) = 16$, $C = 16$ and so $f(x) = 16\left(\frac{1}{2}\right)^x$. This expression can also be written as $f(x) = 16(2)^{-x}$.
Figure 10.2 shows a table for $Y_1 = 16(.5)^{\wedge}X$ that agrees with the given table.

(b) Each time x increases by 1 unit, $f(x)$ increases by 2 units. Therefore f is a linear function, and the slope of its graph equals 2. The y-intercept is 5, so $f(x) = 2x + 5$. Figure 10.3 shows a table for $Y_1 = 2X + 5$ that agrees with the given table.

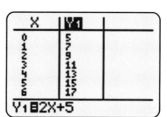

Figure 10.2   Figure 10.3

(c) Each time x increases by 1 unit, $f(x)$ increases by a factor of 3. Therefore f is an exponential function with a growth factor of 3. Because $f(0) = 1$, $C = 1$ and so

$f(x) = 1(3)^x$, or $f(x) = 3^x$. Figure 10.4 shows a table for $Y_1 = 3\wedge X$ that agrees with the given table.

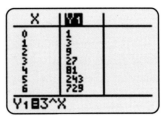

Figure 10.4

■ **MAKING CONNECTIONS**

Linear and Exponential Functions

For a *linear function*, given by $f(x) = ax + b$, each time x increases by 1 unit y increases (or decreases) by a units, where a equals the slope of the graph of f.

For an *exponential function*, given by $f(x) = Ca^x$, each time x increases by 1 unit y increases by a factor of a when $a > 1$ and decreases by a factor of a when $0 < a < 1$. The constant a equals either the growth factor or the decay factor.

If \$100 is deposited in a savings account paying 10% annual interest, the interest earned after 1 year equals $\$100 \times 0.10 = \$10$. The total amount of money in the account after 1 year is $100(1 + 0.10) = \$110$. Each year the money in the account increases by a factor of 1.10, so after x years there will be $100(1.10)^x$ dollars in the account. Thus compound interest is an example of exponential growth.

Compound Interest

If C dollars are deposited in an account paying an annual rate of interest r, expressed in decimal form, then after x years the account will contain A dollars, where

$$A = C(1 + r)^x.$$

The growth factor is $(1 + r)$.

Note: The compound interest formula takes the form of an exponential function with $a = 1 + r$.

EXAMPLE 3 *Calculating compound interest*

A 20-year-old worker deposits \$2000 in a retirement account that pays 13% annual interest. How much money will be in the account when the worker is 65 years old? What is the growth factor?

Solution

Here, $C = 2000$, $r = 0.13$, and $x = 45$. The amount in the account after 45 years equals

$$A = 2000(1 + 0.13)^{45} \approx \$489{,}282.80,$$

which is supported by Figure 10.5. In this dramatic example of exponential growth, $2000 grows to nearly half a million dollars in 45 years. Each year the amount of money on deposit is multiplied by a factor of $(1 + 0.13)$, so the growth factor is 1.13.

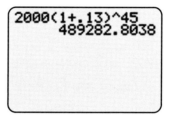

Figure 10.5

GRAPHS OF EXPONENTIAL FUNCTIONS

We can graph $f(x) = 2^x$ by first plotting some points, as in Table 10.2.

TABLE 10.2

x	-2	-1	0	1	2
2^x	$\dfrac{1}{4}$	$\dfrac{1}{2}$	1	2	4

If we plot these points and sketch the graph, we obtain Figure 10.6. Note that, for negative values of x, $0 < 2^x < 1$ and that, for positive values of x, $2^x > 1$. The graph of $y = 2^x$ never intersects the x-axis, and $y = 2^x$ is always positive.

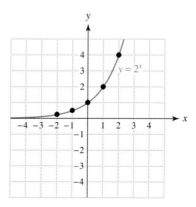

Figure 10.6

We can investigate the graphs of exponential functions further by graphing $y = 1.3^x$, $y = 1.7^x$, and $y = 2.5^x$ (see Figure 10.7). For $a > 1$ the graph of $y = a^x$

increases at a faster rate for larger values of *a*. Now graph $y = 0.7^x$, $y = 0.5^x$, and $y = 0.15^x$ (see Figure 10.8). Note that, if $0 < a < 1$, the graph of $y = a^x$ *decreases more rapidly for smaller values of a.* The graph of $y = a^x$ is increasing when $a > 1$ and decreasing when $0 < a < 1$ (from left to right).

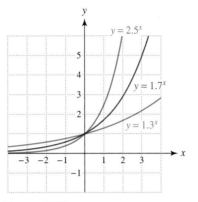

Figure 10.7   **Figure 10.8**

Critical Thinking

Every graph of $y = a^x$ passes through what point? Why?

EXAMPLE 4   *Comparing exponential and linear functions*

Compare $f(x) = 3^x$ and $g(x) = 3x$ graphically and numerically for $x \geq 0$.

Solution

Graphical Comparison   The graphs of $Y_1 = 3\text{\textasciicircum}X$ and $Y_2 = 3X$ are shown in Figure 10.9. The graph of the exponential function Y_1 increases faster than the graph of the linear function Y_2.

Numerical Comparison   The tables of $Y_1 = 3\text{\textasciicircum}X$ and $Y_2 = 3X$ are shown in Figure 10.10. The values for Y_1 increase faster than the values for Y_2.

[0, 5, 1] by [0, 120, 20]

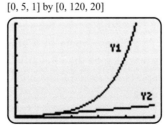
Figure 10.9

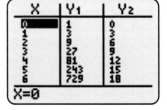
Figure 10.10

The results of Example 4 are true in general: For large enough inputs, exponential functions with $a > 1$ grow faster than any linear function.

■ MAKING CONNECTIONS

Exponential and Polynomial Functions

The function $f(x) = 2^x$ is an exponential function. The base 2 is a constant and the exponent x is a variable, so $f(3) = 2^3 = 8$.

The function $g(x) = x^2$ is a polynomial function. The base x is a variable and the exponent 2 is a constant, so $g(3) = 3^2 = 9$.

Figure 10.11 clearly shows that the exponential function $Y_1 = 2\string^X$ grows much faster than the polynomial function $Y_2 = X\string^2$.

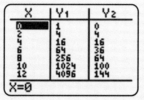

Figure 10.11

MODELS INVOLVING EXPONENTIAL FUNCTIONS

Traffic flow on highways can be modeled with exponential functions whenever traffic patterns occur randomly. In the next example we model traffic at an intersection by using an exponential function.

EXAMPLE 5   Modeling traffic flow

On average, a particular intersection has 360 vehicles arriving randomly each hour. Highway engineers use $f(x) = (0.905)^x$ to estimate the likelihood, or probability, that no vehicle will enter the intersection within a period of x seconds.   (*Source*: **F. Mannering and W. Kilareski,** *Principles of Highway Engineering and Traffic Analysis.*)

(a) Compute $f(5)$ and interpret the results.
(b) Graph f in [0, 10, 1] by [0, 1.5, 0.5] and discuss the graph.
(c) Is this function an example of exponential growth or decay?

Solution

(a) The result $f(5) = (0.905)^5 \approx 0.61$ indicates that there is a 61% chance that no vehicle will enter the intersection during any particular 5-second interval.
(b) Graph $Y_1 = 0.905\string^X$, as shown in Figure 10.12. As the number of seconds increases, there is less chance that no car will enter the intersection.
(c) Because the graph is decreasing and $a = 0.905 < 1$, this function is an example of exponential decay.

[0, 10, 1] by [0, 1.5, 0.5]

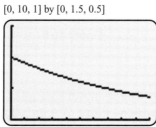

Figure 10.12

10.1 Exponential Functions

In the next example, we use an exponential function to model how trees grow in a forest.

EXAMPLE 6 *Modeling tree density in a forest*

Ecologists studied the spacing of individual trees in a British Columbia forest. This pine forest was 40 to 50 years old and contained approximately 1600 randomly spaced trees per acre. The probability or likelihood that no tree is located within a circle of radius x feet can be estimated by $P(x) = (0.892)^x$. For example, $P(4) \approx 0.63$ means that, if a person picks a point at random in the forest, there is a 63% chance that no tree will be located within 4 feet of the person. See Figure 10.13. *(Source: Pielou, E., Populations and Community Ecology.)*

(a) Evaluate $P(8)$ and interpret the result.
(b) Graph P in $[0, 20, 5]$ by $[0, 1, 0.1]$ and discuss the graph.

Figure 10.13

Solution

(a) The probability $P(8) = (0.892)^8 \approx 0.40$ means that there is a 40% chance that no tree is growing within a circle of radius 8 feet. See Figure 10.14.
(b) The graph of $Y_1 = 0.892\text{\textasciicircum}X$, as shown in Figure 10.15, indicates that the larger the circle, the less the likelihood is of no tree being inside the circle.

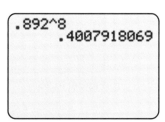

Figure 10.14

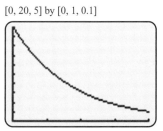

Figure 10.15

The Natural Exponential Function

A special type of exponential function is called the *natural exponential function*, $f(x) = e^x$. The base e is a special number in mathematics similar to π. The number π is approximately 3.14, whereas the number e is approximately 2.72. The natural exponential function has e as its base. Most calculators have a special key that can be used to compute the natural exponential function.

> **Natural Exponential Function**
>
> The function represented by
>
> $$f(x) = e^x,$$
>
> is the **natural exponential function,** where $e \approx 2.71828$.

This function is frequently used to model **continuous growth.** For example, the fact that births and deaths occur throughout the year, not just at one time during the year, must be recognized when population growth is being modeled. If a population P is growing continuously at r percent per year, expressed as a decimal, we can model this population after x years by

$$P = Ce^{rx}.$$

To evaluate natural exponential functions, we use a calculator.

EXAMPLE 7 *Modeling population*

In 1994 Florida's population was 14 million people and was growing at a continuous rate of 1.53%. This population x years after 1994 could be modeled by

$$f(x) = 14e^{0.0153x}.$$

Estimate the population in 2002.

Solution

As 2002 is 8 years after 1994, we evaluate $f(8)$ and obtain

$$f(8) = 14e^{0.0153(8)} \approx 15.8,$$

which is supported by Figure 10.16. (Be sure to include parentheses around the exponent of e.) This model estimates that the population of Florida will be about 15.8 million in 2002.

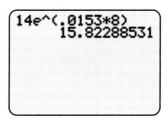

Figure 10.16

Critical Thinking

Graph $y = 2^x$ and $y = 3^x$ in the same viewing rectangle, $[-3, 3, 1]$ by $[0, 4, 1]$. Conjecture how the graph of $y = e^x$ will appear. Test your conjecture.

10.1 Putting It All Together

The following table summarizes some important concepts about exponential functions and compound interest.

Topic	Explanation	Example
Exponential function	An exponential function can be written as $f(x) = Ca^x$, where $a > 0$ and $a \neq 1$. If $a > 1$, the function models exponential growth, and if $0 < a < 1$, the function models exponential decay. The natural exponential function has $C = 1$ and $a = e \approx 2.71828$; that is, $f(x) = e^x$. Exponential Growth Exponential Decay	$f(x) = 3(2)^x$ models exponential growth and $g(x) = 2\left(\frac{1}{3}\right)^x$ models exponential decay.
Compound interest	If C dollars are deposited in an account paying an annual rate of interest r, expressed as a decimal, then after x years the account will contain A dollars, where $$A = C(1 + r)^x.$$ The growth factor is $(1 + r)$.	If $1000 is deposited in an account paying 5% annual interest, then after 6 years the amount A in the account is $$A = 1000(1 + 0.05)^6 \approx \$1340.10.$$

10.1 EXERCISES

CONCEPTS

1. Give a general formula for an exponential function f.

2. Sketch a graph of an exponential function that illustrates exponential decay.

3. Does the graph of $f(x) = a^x$, $a > 1$, illustrate exponential growth or decay?

4. Evaluate the expressions 2^x and x^2 for $x = 5$.

5. Give an approximate value for e to the nearest thousandth.

6. Evaluate e^2 and π^2 using your calculator.

7. If a quantity y grows exponentially, then for each unit increase in x, y increases by a constant _____.

8. If $f(x) = 1.5^x$ what is the growth factor?

EVALUATING AND GRAPHING EXPONENTIAL FUNCTIONS

Exercises 9–20: Evaluate the exponential function for the given values of x by hand when possible. Approximate answers to the nearest hundredth when appropriate.

9. $f(x) = 3^x$ $x = -2, x = 2$

10. $f(x) = 5^x$ $x = -1, x = 3$

11. $f(x) = 5(2^x)$ $x = 0, x = 5$

12. $f(x) = 3(7^x)$ $x = -2, x = 0$

13. $f(x) = \left(\frac{1}{2}\right)^x$ $x = -2, x = 3$

14. $f(x) = \left(\frac{1}{4}\right)^x$ $x = 0, x = 2$

15. $f(x) = 5(3)^{-x}$ $x = -1, x = 2$

16. $f(x) = 4\left(\frac{3}{7}\right)^x$ $x = 1, x = 4$

17. $f(x) = 1.8^x$ $x = -3, x = 1.5$

18. $f(x) = 0.91^x$ $x = 5.1, x = 10$

19. $f(x) = 3(0.6)^x$ $x = -1, x = 2$

20. $f(x) = 5(4.5)^{-x}$ $x = -2.1, x = 5.9$

Exercises 21–28: Graph f in $[-4, 4, 1]$ by $[0, 8, 1]$. State whether the graph illustrates exponential growth or exponential decay.

21. $f(x) = 1.5^x$

22. $f(x) = 3^x$

23. $f(x) = \left(\frac{1}{4}\right)^x$

24. $f(x) = \left(\frac{1}{2}\right)^x$

25. $f(x) = 2^{-x}$

26. $f(x) = 3^{-x}$

27. $f(x) = 0.5(1.7)^x$

28. $f(x) = 3(0.85)^x$

Exercises 29–34: (Refer to Example 2.) A table for a function f is given.
    (a) *Determine whether f represents exponential growth, exponential decay, or linear growth.*
    (b) *Find a formula for f.*

29.

x	0	1	2	3	4
$f(x)$	64	16	4	1	$\frac{1}{4}$

30.

x	0	1	2	3	4
$f(x)$	$\frac{1}{2}$	1	2	4	8

31.

x	0	1	2	3	4
$f(x)$	8	11	14	17	20

32.

x	-2	-1	0	1	2
$f(x)$	4	2	1	$\frac{1}{2}$	$\frac{1}{4}$

33.

x	−2	−1	0	1	2
f(x)	2.56	3.2	4	5	6.25

34.

x	−2	−1	0	1	2
f(x)	−6	−2	2	6	10

Exercises 35–38: Use the graph of $y = Ca^x$ to determine C and a.

35.
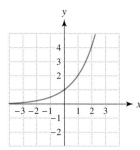

36.
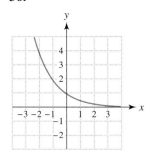

37.
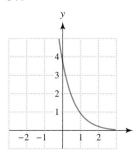

38.
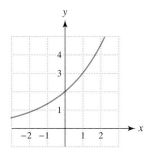

Exercises 39–42: Match the symbolic representation with its graph (a.–d.). Do not use a calculator.

39. $f(x) = 1.5^x$

40. $f(x) = \frac{1}{4}(2^x)$

41. $f(x) = 4\left(\frac{1}{2}\right)^x$

42. $f(x) = \left(\frac{1}{3}\right)^x$

a.
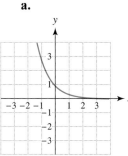

b.
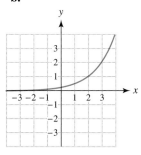

c.
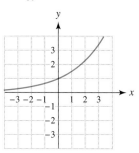

d.
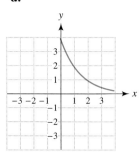

Compound Interest

Exercises 43–48: (Refer to Example 3.) If C dollars are deposited in an account paying r percent annual interest, approximate the amount in the account after x years.

43. $C = \$1500$ $r = 9\%$ $x = 10$ years

44. $C = \$1500$ $r = 15\%$ $x = 10$ years

45. $C = \$200$ $r = 20\%$ $x = 50$ years

46. $C = \$5000$ $r = 8.4\%$ $x = 7$ years

47. $C = \$560$ $r = 1.4\%$ $x = 25$ years

48. $C = \$750$ $r = 10\%$ $x = 13$ years

49. *Interest* Suppose that $1000 is deposited in an account paying 8% interest for 10 years. If $2000 had been deposited instead of $1000, would there be twice the money in the account after 10 years? Explain.

50. *Interest* Suppose that $500 is deposited in an account paying 5% interest for 10 years. If the interest rate had been 10% instead of 5%, would the total interest earned after 10 years be twice as much? Explain.

51. *Federal Debt* In fiscal 1996 the federal budget deficit was $107.3 billion. At the same time, 30-year treasury bonds were paying 6.83% interest. Suppose that U.S. taxpayers loaned $107.3 billion to the federal government at 6.83%. If the federal government waited 30 years to pay the entire amount back, including the interest, how much would it be? (*Source:* **U.S. Treasury Department.**)

52. Federal Debt Repeat Exercise 51 but suppose that the interest rate is 2% higher, or 8.83%. How much would the federal government owe after 30 years? Is the national debt sensitive to interest rates?

THE NATURAL EXPONENTIAL FUNCTION

Exercises 53–56: Evaluate $f(x)$ for the given value of x. Approximate answers to the nearest hundredth.

53. $f(x) = e^x$ $x = 1.2$

54. $f(x) = 2e^x$ $x = 2$

55. $f(x) = 1 - e^x$ $x = -2$

56. $f(x) = 4e^{-x}$ $x = 1.5$

Exercises 57–60: Graph $f(x)$ in $[-4, 4, 1]$ by $[0, 8, 1]$. State whether the graph illustrates exponential growth or exponential decay.

57. $f(x) = e^{0.5x}$ 58. $f(x) = e^x + 1$

59. $f(x) = 1.5e^{-0.32x}$ 60. $f(x) = 2e^{-x} + 1$

APPLICATIONS

61. Modeling Population (Refer to Example 7.) In 1997 the population of Arizona was 4.56 million and growing continuously at a rate of 3.1%.
(a) Write a function f that models the population of Arizona x years after 1997.
(b) Graph f in $[0, 10, 1]$ by $[4, 7, 1]$.
(c) Estimate the population of Arizona in 2003.

62. Dating Artifacts Radioactive carbon-14 is found in all living things and is used to date objects containing organic material. Suppose that an object initially contains C grams of carbon-14. After x years it will contain A grams, where
$$A = C(0.99988)^x.$$
(a) Let $C = 10$ and graph A over a 20,000-year period. Is this function an example of exponential growth or decay?
(b) How many grams are left after 5700 years? What fraction of the carbon-14 is left?

63. E. coli Bacteria A strain of bacteria that inhabits the intestines of animals is named *Escherichia coli* (*E. coli*). These bacteria are capable of rapid growth and can be dangerous to humans—particularly children. The accompanying table shows the results of one study of the growth of *E. coli* bacteria, where concentrations are listed in thousands of bacteria per milliliter.

t (minutes)	0	50	100
Concentration	500	1000	2000

t (minutes)	150	200
Concentration	4000	8000

Source: G. S. Stent, *Molecular Biology of Bacterial Viruses.*

(a) Find C and a so that $f(t) = Ca^{t/50}$ models the data.
(b) Use $f(t)$ to estimate the concentration of bacteria after 170 minutes.
(c) Graph f in $[0, 300, 50]$ by $[0, 35{,}000, 5000]$. Discuss the growth of this strain of bacteria over a 300-minute time period.

64. Internet Use Internet use in Western Europe is expected to expand. The following table shows the number of Internet users y in millions during year x, where $x = 0$ corresponds to 2000, $x = 1$ to 2001, and $x = 2$ to 2002.

x (year)	0	1	2
y (millions)	52	67	85

Source: Nortel Networks.

(a) Approximate C and a so that $f(x) = Ca^x$ models the data. (*Hint:* To find a, estimate the factor by which y increases each year.)
(b) Use $f(x)$ to estimate the number of users in 2004 ($x = 4$).
(c) How long is this type of growth likely to continue?

65. Cellular Phone Use In 1985, there were about 203,000 cellular phone subscribers in the United States. This number increased to about 84 million users in 2000, as illustrated in the accompanying figure. The rapid growth in cellular phone subscribers in millions can be modeled by $f(x) = 0.0272(1.495)^{x-1980}$, where x is the year.
(*Source:* Cellular Telecommunications Industry Association.)

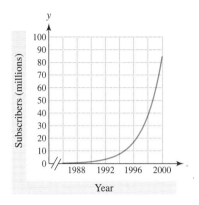

(a) Evaluate $f(1995)$ and interpret the result.
(b) What is the growth factor for $f(x)$? Explain what the growth factor indicates about cellular phone subscribers from 1985 to 2000.

66. *Swimming Pool Maintenance* Chlorine is frequently used to disinfect swimming pools. The chlorine concentration should remain between 1.5 and 2.5 parts per million (ppm). After a warm, sunny day only 80% of the chlorine may remain in the water, with the other 20% dissipating into the air or combining with other chemicals in the water. (*Source:* **D. Thomas, Swimming Pool Operators Handbook.**)
    (a) Let $f(x) = 3(0.8)^x$ model the concentration of chlorine in parts per million after x days. What is the initial concentration of chlorine in the pool?
    (b) If no more chlorine is added, estimate graphically when the chlorine level drops below 1.5 parts per million.

67. *Modeling Traffic Flow* (Refer to Example 5.) Construct a table of $f(x) = (0.905)^x$, starting at $x = 0$ and incrementing by 10.
    (a) Evaluate $f(0)$ and interpret the result.
    (b) After how many seconds is there only a 5% chance that no cars have entered the intersection?

68. *Modeling Tree Density* (Refer to Example 6.)
    (a) Evaluate $P(10), P(20),$ and $P(30)$. Interpret the results.
    (b) What happens to $P(x)$ as x becomes large? Explain how this probability relates to the spacing of trees in a forest.

WRITING ABOUT MATHEMATICS

69. A student evaluates $f(x) = 4(2)^x$ at $x = 3$ and obtains 512. Did the student evaluate the function correctly? What was the student's error?

70. For a set of data, how can you distinguish between linear growth and exponential growth? Give an example of each type of data.

GROUP ACTIVITY
Working with Real Data

Directions: Form a group of 2 to 4 people. Select someone to record the group's responses for this activity. All members of the group should work cooperatively to answer the questions. If your instructor asks for your results, each member of the group should be prepared to respond. If the group is asked to turn in its work, be sure to include each person's name on the paper.

Greenhouse Gases Carbon dioxide (CO_2) is a greenhouse gas in the atmosphere that may raise average temperatures on Earth. The burning of fossil fuels could be responsible for the increased levels of carbon dioxide in the atmosphere. If current trends continue, future concentrations of atmospheric carbon dioxide in parts per million (ppm) are expected to reach the levels shown in the accompanying table.

The CO_2 concentration in the year 2000 is greater than it has been at any time in the previous 160,000 years.

Year	2000	2050	2100	2150	2200
CO_2 (ppm)	364	467	600	769	987

Source: R. Turner, *Environmental Economics*.

(a) Let x represent the year, where $x = 0$ corresponds to 2000, $x = 1$ to 2001, and so on. Find values for C and a so that $f(x) = Ca^x$ models the data.

(b) Graph f and the data in the same viewing rectangle.

(c) Use $f(x)$ to estimate graphically the year when the carbon dioxide concentration will be double the preindustrial level of 280 ppm.

10.2 LOGARITHMIC FUNCTIONS

The Common Logarithmic Function ~ Logarithms with Other Bases

INTRODUCTION

Logarithmic functions are used in many applications. For example, if one airplane weighs twice as much as another, does the heavier airplane typically need a runway that is twice as long? Using a logarithmic function, we can answer that question. Logarithmic functions are also used to measure the intensity of sound. In this section we discuss logarithmic functions and several of their applications.

THE COMMON LOGARITHMIC FUNCTION

In applications, measurements can vary greatly in size. Table 10.3 lists some examples of objects, with the approximate distances in meters across each.

TABLE 10.3

Object	Distance (meters)
Atom	10^{-9}
Protozoan	10^{-4}
Small Asteroid	10^2
Earth	10^7
Universe	10^{26}

Source: Ronan, C., *The Natural History of the Universe*.

Each distance is listed in the form 10^k for some k. The value of k distinguishes one measurement from another. The *common logarithmic function* or *base-10 logarithmic function* denoted log outputs k if the input x can be expressed as 10^k for some real number k. For example, $\log 10^{-9} = -9$, $\log 10^2 = 2$, and $\log 10^{1.43} = 1.43$. For any real number k, $\log 10^k = k$. Some values for $f(x) = \log x$ are given in Table 10.4.

TABLE 10.4

x	10^{-4}	10^{-3}	10^{-2}	10^{-1}	10^0	10^1	10^2	10^3	10^4
$\log x$	-4	-3	-2	-1	0	1	2	3	4

We use this information to define the common logarithm.

Common Logarithm

The **common logarithm of a positive number** x, denoted $\log x$, may be calculated as follows. If x is expressed as $x = 10^k$, then

$$\log x = k,$$

where k is a real number. That is, $\log 10^k = k$.

The function given by

$$f(x) = \log x,$$

is called the **common logarithmic function.**

The common logarithmic function outputs an exponent k, which may be positive, negative, or zero. However, a valid input must be positive. The expression $\log x$ equals the exponent k on base 10 that equals the number x. For example, $\log 1000 = 3$ because $1000 = 10^3$.

EXAMPLE 1 *Evaluating common logarithms*

 Simplify each common logarithm.

(a) $\log 100$ **(b)** $\log \dfrac{1}{10}$ **(c)** $\log \sqrt{1000}$ **(d)** $\log 45$

Solution

(a) $100 = 10^2$, so $\log 100 = \log 10^2 = 2$

(b) $\log \dfrac{1}{10} = \log 10^{-1} = -1$

(c) $\log \sqrt{1000} = \log (10^3)^{1/2} = \log 10^{3/2} = \dfrac{3}{2}$

(d) It is not obvious how to write 45 as a power of 10. However, we can use a calculator to determine that $\log 45 \approx 1.6532$. Thus $10^{\wedge}(1.6532) \approx 45$. Figure 10.17 supports the answers to parts (a) and (b). Figure 10.18 supports the answers to parts (c) and (d).

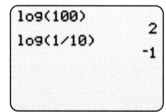

Figure 10.17   Figure 10.18

The points $(10^{-1}, -1)$, $(10^0, 0)$, $(10^{0.5}, 0.5)$, and $(10^1, 1)$ are located on the graph of $y = \log x$, as plotted in Figure 10.19 on the next page. Sketching the graph of $y = \log x$ results in Figure 10.20. Note some important features of this graph.

- The graph of the common logarithm increases very slowly for large values of x. For example, x must be 100 for $\log x$ to reach 2 and x must be 1000 for $\log x$ to reach 3.

- The graph does not exist for negative values of x. The domain of log x includes only positive numbers. The range of log x includes all real numbers.
- When $0 < x < 1$, log x outputs negative values. The y-axis is a vertical asymptote, so as x approaches 0, log x approaches $-\infty$.

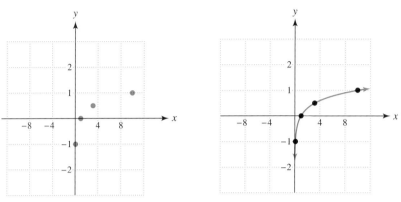

Figure 10.19   **Figure 10.20**

Much like the square root function, the common logarithmic function does not have an easy-to-evaluate formula. For example, we can calculate $\sqrt{4} = 2$ and $\sqrt{100} = 10$ mentally, but for $\sqrt{2}$ we usually rely on a calculator. Similarly, we can find $\log 1000 = \log 10^3 = 3$ mentally, whereas we use a calculator to approximate log 45. Another similarity between the square root function and the common logarithmic function is that their domains do not include negative numbers. If only real numbers are allowed as outputs, both $\sqrt{-3}$ and $\log(-3)$ are undefined expressions.

EXAMPLE 2   Graphing the logarithmic functions

Graph f and compare the graph to $y = \log x$.

(a) $f(x) = \log(x - 2)$ (b) $f(x) = \log(x) + 1$

Solution

(a) The graph of $Y_1 = \log(X)$ is shown in Figure 10.21, and the graph of $Y_1 = \log(X - 2)$ is shown in Figure 10.22. The graph of $y = \log(x - 2)$ is similar to the graph of $y = \log x$, except that it is translated to the right 2 units. Although the calculator graph does not show it, the graph of $y = \log(x - 2)$ continues downward at $x = 2$. That is, there is a vertical asymptote at $x = 2$.

(b) The graph of $Y_1 = \log(X) + 1$ is shown in Figure 10.23. It is similar to the graph of $y = \log x$, except that it is translated upward 1 unit.

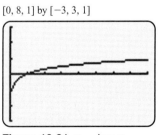

Figure 10.21 $y = \log x$

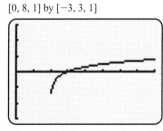

Figure 10.22 $y = \log(x - 2)$

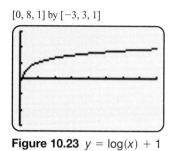

Figure 10.23 $y = \log(x) + 1$

Logarithms are used to model quantities that vary greatly in intensity. For example, the human ear is extremely sensitive and able to detect intensities on the eardrum ranging from 10^{-16} watts per square centimeter (w/cm^2) to 10^{-4} w/cm^2, which usually is painful to the human eardrum.

EXAMPLE 3 *Modeling sound levels*

Sound levels in decibels (db) can be computed by $f(x) = 160 + 10 \log x$, where x is the intensity of the sound in watts per square centimeter. Ordinary conversation has an intensity of 10^{-10} w/cm^2. What decibel level is this? *(Source: Weidner, R., and Sells, R., Elementary Classical Physics, Vol. 2.)*

Solution

To find the decibel level for ordinary conversation evaluate $f(10^{-10})$.

$$\begin{aligned} f(10^{-10}) &= 160 + 10 \log(10^{-10}) && \text{Substitute } x = 10^{-10}. \\ &= 160 + 10(-10) && \text{Evaluate } \log(10^{-10}). \\ &= 60 && \text{Simplify.} \end{aligned}$$

Ordinary conversation corresponds to 60 db, which is supported in Figure 10.24.

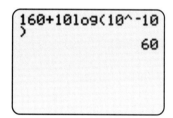

Figure 10.24

Critical Thinking

If the sound level increases by 10 db by what factor does the intensity x increase?

LOGARITHMS WITH OTHER BASES

Common logarithms are base-10 logarithms, but we can define logarithms having other bases. For example, base-2 logarithms are frequently used in computer science. Some values for the base-2 logarithmic function, denoted $f(x) = \log_2 x$, are shown in Table 10.5 on the next page. If x can be expressed as $x = 2^k$ for some real number k, then $\log_2 x = \log_2 2^k = k$.

TABLE 10.5

x	2^{-3}	2^{-2}	2^{-1}	2^0	2^1	2^2	2^3
$\log_2 x$	-3	-2	-1	0	1	2	3

EXAMPLE 4 *Evaluating base-2 logarithms*

Simplify each logarithm.

(a) $\log_2 8$ **(b)** $\log_2 \frac{1}{4}$

Solution

(a) The expression $\log_2 8$ represents the exponent on base 2 that gives 8. Because $8 = 2^3$, $\log_2 8 = \log_2 2^3 = 3$.

(b) Because $\frac{1}{4} = \frac{1}{2^2} = 2^{-2}$, $\log_2 \frac{1}{4} = \log_2 2^{-2} = -2$.

Some values of base-e logarithms are shown in Table 10.6. A base-e logarithm is referred to as a **natural logarithm** and denoted either $\log_e x$ or $\ln x$. Natural logarithms are used in mathematics, science, economics, electronics, and communications.

TABLE 10.6

x	e^{-3}	e^{-2}	e^{-1}	e^0	e^1	e^2	e^3
$\ln x$	-3	-2	-1	0	1	2	3

To evaluate natural logarithms we usually use a calculator.

EXAMPLE 5 *Evaluating natural logarithms*

GCLM Approximate to the nearest hundredth.

(a) $\ln 10$ **(b)** $\ln \frac{1}{2}$

Solution

(a) Figure 10.25 shows that $\ln 10 \approx 2.30$.

(b) Figure 10.25 shows that $\ln \frac{1}{2} \approx -0.69$.

```
ln(10)
         2.302585093
ln(1/2)
         -.6931471806
```

Figure 10.25

10.2 Logarithmic Functions

We now define base-a logarithms.

Base-a Logarithms

The **logarithm with base a of a positive number x,** denoted $\log_a x$, may be calculated as follows. If x can be expressed as $x = a^k$, then

$$\log_a x = k,$$

where $a > 0$, $a \neq 1$, and k is a real number. That is, $\log_a a^k = k$.

The function given by

$$f(x) = \log_a x$$

is called the **logarithmic function with base a.**

Remember that *a logarithm is an exponent*. The expression $\log_a x$ equals the exponent k such that $a^k = x$. In the graph of $y = \log_a x$ shown in Figure 10.26, note that the graph is increasing and passes through the point $(1, 0)$.

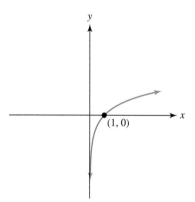

Figure 10.26 $y = \log_a x$

Critical Thinking

Explain why $\log_a 1 = 0$ for any positive base a, $a \neq 1$.

EXAMPLE 6 *Evaluating base-a logarithms*

Simplify each logarithm.

(a) $\log_5 25$ (b) $\log_4 \dfrac{1}{64}$ (c) $\log_7 1$ (d) $\log_3 9^{-1}$

Solution

(a) $25 = 5^2$, so $\log_5 25 = \log_5 5^2 = 2$.

(b) $\dfrac{1}{64} = \dfrac{1}{4^3} = 4^{-3}$, so $\log_4 \dfrac{1}{64} = \log_4 4^{-3} = -3$.

(c) $1 = 7^0$, so $\log_7 1 = \log_7 7^0 = 0$. (Note that the logarithm of 1 is always 0, regardless of the base.)

(d) $9^{-1} = (3^2)^{-1} = 3^{-2}$, so $\log_3 9^{-1} = \log_3 3^{-2} = -2$.

Logarithms occur in many applications. One application is runway length for airplanes, which we discuss in the next example.

EXAMPLE 7 *Calculating runway length*

There is a mathematical relationship between an airplane's weight x and the runway length required at takeoff. For certain types of airplanes, the minimum runway length L in thousands of feet may be modeled by $L(x) = 1.3 \ln x$, where x is in thousands of pounds. (*Source:* **L. Haefner, *Introduction to Transportation Systems*.**

(a) Estimate the runway length needed for an airplane weighing 10,000 pounds.
(b) Does a 20,000-pound airplane need twice the runway length as a 10,000-pound airplane? Explain.

Solution

(a) Because $L(10) = 1.3 \ln (10) \approx 3$, an airplane weighing 10,000 pounds requires a runway 3000 feet long.
(b) Because $L(20) = 1.3 \ln (20) \approx 3.9$, a 20,000-pound airplane does not need twice the runway length of a 10,000-pound airplane. Rather the heavier airplane needs roughly 3900 feet of runway, or an extra 900 feet.

10.2 Putting It All Together

Common logarithms are base-10 logarithms. If a positive number x can be written as $x = 10^k$, then $\log x = k$. The value of $\log x$ represents the exponent on the base 10 that gives x. We can define logarithms having other bases. For example, the natural logarithm is a base-e logarithm that is usually evaluated using a calculator. The following table summarizes some important concepts related to base-a logarithms.

10.2 Logarithmic Functions

Base-a Logarithm	Description	Example
Base-a logarithms	Logarithms can be defined for any base a, where $a > 0$ and $a \neq 1$. Thus $\log_a x = k$ means $x = a^k$. In other words, if we can write x as $x = a^k$, then $$\log_a x = \log_a a^k = k.$$ When $a = 10$, we write $\log x$, which indicates a common logarithm. When $a = e$, we write $\ln x$, which indicates a natural logarithm. The domain of $f(x) = \log_a x$ is all positive numbers, and its range is all real numbers. The graph of $y = \log_a x$ is shown in the following figure. Note that the graph passes through the point $(1, 0)$ and is increasing.	$\log 1000 = \log 10^3 = 3$ $\log_2 16 = \log_2 2^4 = 4$ $\log_3 \dfrac{1}{81} = \log_3 3^{-4} = -4$

10.2 EXERCISES

CONCEPTS

1. What is the base of the common logarithm?
2. What is the base of the natural logarithm?
3. What are the domain and range of $\log x$?
4. What are the domain and range of $\log_a x$?
5. $\log 10^k =$ _____
6. $\ln e^k =$ _____
7. If $\log x = k$, then $10^k =$ _____.
8. $10^{\log x} =$ _____.
9. What does k equal if $10^k = 5$?
10. What does k equal if $2^k = 5$?

EVALUATING AND GRAPHING LOGARITHMIC FUNCTIONS

Exercises 11–28: Evaluate the logarithm by hand.

11. $\log 10^5$
12. $\log 10$
13. $\log 10^{-4}$
14. $\log 10^{-1}$
15. $\log 1$
16. $\log \sqrt[3]{100}$
17. $\log \dfrac{1}{100}$
18. $\log \dfrac{1}{10}$

INVERSE PROPERTIES

The inverse action of opening the door is closing the door, and the inverse action of sitting down is standing up. The concept of an inverse also occurs in mathematics. For example, if we start with a number x, add 7 to it, and then subtract 7 from it, the result is x. That is,

$$x + 7 - 7 = x.$$

We say that addition and subtraction are inverse operations because they undo each other. Similarly, if we multiply x by 5 and then divide by 5 the result is x.

$$\frac{5 \cdot x}{5} = x$$

Multiplication and division are also inverse operations.

Now consider the numerical representations for 10^x and $\log x$ shown in Tables 10.7 and 10.8.

TABLE 10.7

x	-2	-1	0	1	2
10^x	10^{-2}	10^{-1}	10^0	10^1	10^2

TABLE 10.8

x	10^{-2}	10^{-1}	10^0	10^1	10^2
$\log x$	-2	-1	0	1	2

If we start with the number 2, compute 10^2, and then calculate $\log 10^2$, the result is 2. That is,

$$\log(10^2) = 2.$$

In general, $\log 10^x = x$ for any real number x. Now suppose that we perform the calculations in reverse order by taking the common logarithm and then computing a power of 10. For example, suppose that we start with the number $100 = 10^2$, compute $\log 100$ to obtain 2, and then calculate 10^2 to obtain 100, or

$$10^{\log 100} = 10^2 = 100.$$

In general, $10^{\log x} = x$ for all positive numbers x. These results hold for any base-a logarithm and are summarized as follows.

Inverse Properties

The following properties hold for logarithms with base a.

$$\log_a a^x = x, \quad \text{for any real number } x$$

$$a^{\log_a x} = x, \quad \text{for any positive number } x$$

Because of these inverse properties, the exponential and logarithmic functions with base a are called **inverse functions.** Logarithms and exponential expressions compute inverse operations in much the same way that addition and subtraction are inverse operations.

EXAMPLE 9 *Applying inverse properties*

Simplify each expression.

(a) $\ln e^{0.5x}$

(b) $e^{\ln 4}$

(c) $2^{\log_2 7x}$

(d) $10^{\log(9x-3)}$

Solution

(a) $\ln e^{0.5x} = 0.5x$ because $\ln e^k = k$ for all k.
(b) $e^{\ln 4} = 4$ because $e^{\ln k} = k$ for all positive k.
(c) $2^{\log_2 7x} = 7x$ because $a^{\log_a k} = k$ for all positive k.
(d) $10^{\log(9x-3)} = 9x - 3$ because $10^{\log k} = k$ for all positive k.

10.3 Putting It All Together

The following table summarizes some important properties for base-a logarithms. Common and natural logarithms satisfy the same properties.

Type of Property	Description	Example
Change of base	Let x and $a \neq 1$ be positive real numbers. Then $$\log_a x = \frac{\log x}{\log a} \text{ and } \log_a x = \frac{\ln x}{\ln a}.$$	The expression $\log_3 6$ is equivalent to either $\frac{\log 6}{\log 3}$ or $\frac{\ln 6}{\ln 3}$.
Logarithmic	The following properties hold for positive numbers m, n, and $a \neq 1$ and for any real number r. 1. $\log_a mn = \log_a m + \log_a n$ 2. $\log_a \frac{m}{n} = \log_a m - \log_a n$ 3. $\log_a (m^r) = r \log_a m$	1. $\log 20 = \log 10 + \log 2$ 2. $\log \frac{45}{6} = \log 45 - \log 6$ 3. $\ln x^6 = 6 \ln x$
Inverse	The following properties hold for logarithms with base a. $\log_a a^x = x$, for any real number x $a^{\log_a x} = x$, for any positive number x	$\log 10^{7.48} = 7.48$ $2^{\log_2 63} = 63$

10.3 EXERCISES

CONCEPTS

1. $\log 12 = \log 3 + \log (\underline{})$
2. $\ln 5 = \ln 20 - \ln (\underline{})$
3. $\log 8 = (\underline{}) \log 2$
4. $\log mn = \underline{}$
5. $\log \dfrac{m}{n} = \underline{}$
6. $\log (m^r) = \underline{}$
7. Does $\log x + \log y$ equal $\log (x + y)$ for all positive numbers x and y?
8. Does $\log x - \log y$ equal $\log \left(\dfrac{x}{y}\right)$ for all positive numbers x and y?
9. Give the change of base formula.
10. Calculating 10^x and $\log x$ are $\underline{}$ operations, just like adding and subtracting.

BASIC PROPERTIES OF LOGARITHMS

Exercises 11–16: Write the expression as a sum of two or more logarithms.

11. $\ln (3 \cdot 5)$
12. $\log (7 \cdot 11)$
13. $\log_3 xy$
14. $\log_5 y^2$
15. $\ln 10z$
16. $\log x^2 y$

Exercises 17–22: Write the expression as a difference of two logarithms.

17. $\log \dfrac{7}{3}$
18. $\ln \dfrac{11}{13}$
19. $\ln \dfrac{x}{y}$
20. $\log \dfrac{2x}{z}$
21. $\log_2 \dfrac{45}{x}$
22. $\log_7 \dfrac{5x}{4z}$

Exercises 23–30: Write the expression as one logarithm.

23. $\log 45 + \log 5$
24. $\log 30 - \log 10$
25. $\ln x + \ln y$
26. $\ln m + \ln n - \ln n$
27. $\ln 7x^2 + \ln 2x$
28. $\ln x + \ln y - \ln z$
29. $\ln x + \ln y^2 - \ln y$
30. $\ln \sqrt{z} - \ln z^3 + \ln y^3$

Exercises 31–36: Rewrite the expression, using the power rule.

31. $\log 3^6$
32. $\log x^8$
33. $\ln 2^x$
34. $\ln (0.77)^{x+1}$
35. $\log_2 5^{1/4}$
36. $\log_3 \sqrt{x}$

Exercises 37–46: Use properties of logarithms to write the expression as one logarithm.

37. $4 \log z - \log z^3$
38. $2 \log_5 y + \log_5 x$
39. $\log x + 2 \log x + 2 \log y$
40. $\log x^2 + 3 \log z - 5 \log y$
41. $\log x - 2 \log \sqrt{x}$
42. $\ln y^2 - 6 \ln \sqrt[3]{y}$
43. $\ln 2^{x+1} - \ln 2$
44. $\ln 8^{1/2} + \ln 2^{1/2}$
45. $2 \log_3 \sqrt{x} - 3 \log_3 x$
46. $\ln \sqrt[3]{x} + \ln \sqrt{x}$

Exercises 47–56: Use properties of logarithms to write the expression in terms of logarithms of x, y, and z.

47. $\log xy^2$
48. $\log \dfrac{x^2}{y^3}$
49. $\ln \dfrac{x^4 y}{z}$
50. $\ln \dfrac{\sqrt{x}}{y}$
51. $\log_4 \dfrac{\sqrt[3]{z}}{\sqrt{y}}$
52. $\log_2 \sqrt{\dfrac{x}{y}}$

53. $\log(x^4 y^3)$

54. $\log(x^2 y^4 z^3)$

55. $\ln \dfrac{1}{y} - \ln \dfrac{1}{x}$

56. $\ln \dfrac{1}{xy}$

Exercises 57–60: Graph f and g in [−6, 6, 1] by [−4, 4, 1]. If the two graphs appear to be identical, prove that they are using properties of logarithms.

57. $f(x) = \log x^3$, $g(x) = 3 \log x$

58. $f(x) = \ln x + \ln 3$, $g(x) = \ln 3x$

59. $f(x) = \ln(x + 5)$, $g(x) = \ln x + \ln 5$

60. $f(x) = \log(x - 2)$, $g(x) = \log x - \log 2$

INVERSE PROPERTIES OF LOGARITHMS

Exercises 61–68: Simplify the expression.

61. $10^{\log 5}$

62. $\ln e^{3/4}$

63. $\ln e^{-5x}$

64. $e^{\ln 2x}$

65. $\log 10^{(2x-7)}$

66. $\log 10^{(8-4x)}$

67. $5^{\log_5 0.6z}$

68. $7^{\log_7 (x-9)}$

APPLICATIONS

69. *Modeling Sound* (Refer to Example 3, Section 10.2.) The formula $f(x) = 10 \log(10^{16} x)$ can be used to calculate the decibel level of a sound with an intensity x. Use properties of logarithms to simplify this formula to

$$f(x) = 160 + 10 \log x.$$

70. *Cellular Phone Technology* A formula used to calculate the strength of a signal for a cellular phone is

$$L = 110.7 - 19.1 \log h + 55 \log d,$$

where h is the height of the cellular phone tower and d is the distance the phone is from the tower. Use properties of logarithms to write an expression for L that contains only one logarithm. (*Source: C. Smith, Practical Cellular & PCS Design.*)

WRITING ABOUT MATHEMATICS

71. A student simplifies a logarithmic expression *incorrectly* as follows.

$$4 \log(x + 2x) \stackrel{?}{=} 4(\log x + \log 2x)$$
$$\stackrel{?}{=} 4(\log x + \log 2 + \log x)$$
$$\stackrel{?}{=} 4(2 \log x + \log 2)$$
$$\stackrel{?}{=} 8 \log x + 4 \log 2$$

Explain the student's mistake.

72. A student insists that $\log(x - y)$ is equal to $\log x - \log y$. How could you convince the student otherwise? (Use a calculator.)

10.4 EXPONENTIAL AND LOGARITHMIC MODELS

Exponential Equations and Models ~ Logarithmic Equations and Models

INTRODUCTION

Although we have solved many equations throughout this course, one equation that we have not solved *symbolically* is $a^x = k$. This exponential equation occurs frequently in applications and is used to model either exponential growth or decay. Logarithmic equations contain logarithms and are also used in modeling real-world data. To solve exponential equations we use logarithms, and to solve logarithmic equations we use exponential expressions.

EXPONENTIAL EQUATIONS AND MODELS

To solve the equation

$$10 + x = 100$$

we subtract 10 from both sides because addition and subtraction are inverse operations.

$$10 + x - 10 = 100 - 10$$
$$x = 90$$

To solve the equation
$$10x = 100$$
we divide both sides by 10 because multiplication and division are inverse operations.
$$\frac{10x}{10} = \frac{100}{10}$$
$$x = 10$$

Now suppose that we want to solve the exponential equation
$$10^x = 100.$$

What is new about this type of equation is that the variable x is an *exponent*. The inverse operation of 10^x is $\log x$. Rather than subtracting 10 from both sides or dividing both sides by 10, we take the base-10 logarithm of both sides. This results in
$$\log 10^x = \log 100.$$
Because $\log 10^x = x$ for all real numbers x, the equation becomes
$$x = \log 100, \quad \text{or equivalently,} \quad x = 2.$$
These concepts are applied in the next example.

EXAMPLE 1 *Solving exponential equations*

Solve.

(a) $10^x = 150$ **(b)** $e^x = 40$ **(c)** $2^x = 50$ **(d)** $0.9^x = 0.5$

Solution

(a) $\quad 10^x = 150$
$\log 10^x = \log 150$ Take the common logarithm of both sides.
$\quad\quad x = \log 150 \approx 2.18$ Inverse property: $\log 10^x = x$ for all x

(b) The inverse operation of e^x is $\ln x$, so we take the natural logarithm of both sides.
$\quad e^x = 40$
$\ln e^x = \ln 40$ Take the natural logarithm of both sides.
$\quad x = \ln 40 \approx 3.69$ Inverse property: $\ln e^x = x$ for all x

(c) The inverse operation of 2^x is $\log_2 x$. However, because calculators do not usually have a base-2 logarithm key, we take the common logarithm of both sides and then apply the power rule.
$\quad\quad 2^x = 50$
$\log 2^x = \log 50$ Take the common logarithm.
$x \log 2 = \log 50$ Power rule: $\log (m^r) = r \log m$
$\quad\quad x = \dfrac{\log 50}{\log 2} \approx 5.64$ Divide by $\log 2$ and approximate.

(d) We begin by taking the common logarithm of each side.

$$0.9^x = 0.5$$
$$\log 0.9^x = \log 0.5 \qquad \text{Take the common logarithm.}$$
$$x \log 0.9 = \log 0.5 \qquad \text{Power rule: } \log(m^r) = r \log m$$
$$x = \frac{\log 0.5}{\log 0.9} \approx 6.58 \qquad \text{Divide by } \log 0.9 \text{ and approximate.}$$

We check our results in Example 1 with a calculator. Figure 10.31 supports the results for parts (a) and (b). Figure 10.32 supports the results for parts (c) and (d).

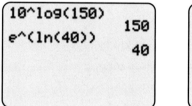

Figure 10.31

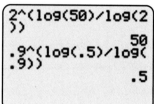

Figure 10.32

■ **MAKING CONNECTIONS**

Logarithms of Quotients and Quotients of Logarithms

The solution in Example 1(c) is $\dfrac{\log 50}{\log 2}$. Note that

$$\frac{\log 50}{\log 2} \neq \log 50 - \log 2.$$

However,

$$\log 50 - \log 2 = \log \frac{50}{2} = \log 25$$

by the quotient rule for logarithms. See Figure 10.33.

```
log(50)/log(2)
           5.64385619
log(50)-log(2)
           1.397940009
log(25)
           1.397940009
```

Figure 10.33

In Section 10.1 you learned that, if $1000 is deposited in a savings account paying 10% annual interest at the end of each year, the amount A in the account after x years is given by

$$A(x) = 1000(1.1)^x.$$

After 10 years there will be

$$A(10) = 1000(1.1)^{10} \approx \$2593.74$$

in the account. To calculate how long it will take for $4000 to accrue in the account, we need to solve the exponential equation

$$1000(1.1)^x = 4000.$$

We do so in the next example.

EXAMPLE 2 *Solving exponential equations*

Solve $1000(1.1)^x = 4000$ symbolically. Give graphical support for your answer.

Solution

Symbolic Solution Begin by dividing each side of the equation by 1000.

$1000(1.1)^x = 4000$	Given equation
$1.1^x = 4$	Divide by 1000.
$\log 1.1^x = \log 4$	Take the common logarithm.
$x \log 1.1 = \log 4$	Power rule for logarithms
$x = \dfrac{\log 4}{\log 1.1} \approx 14.5$	Divide by log 1.1 and approximate.

Interest is paid at the end of the year, so it will take 15 years for $1000 earning 10% annual interest to grow to $4000.

Graphical Solution Graphical support is shown in Figure 10.34, where the graphs of $Y_1 = 1000*1.1\wedge X$ and $Y_2 = 4000$ intersect when $x \approx 14.5$.

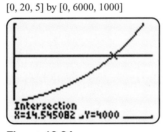

Figure 10.34

As with other types of equations, we can solve exponential equations symbolically, graphically, and numerically. In the next example we model the life span of robins.

EXAMPLE 3 *Modeling the life span of robins*

The life span of 129 robins was monitored over a 4-year period in one study. The formula $f(x) = 10^{-0.42x}$ can be used to calculate the fraction of robins remaining after x years. For example, $f(1) \approx 0.38$ means that after 1 year 38% of the robins were still alive. (*Source:* D. Lack, *The Life Span of a Robin.*)

(a) Evaluate $f(2)$ and interpret the result.
(b) Determine symbolically when 5% of the robins remained.
(c) Support your result from part (b) graphically and numerically.

10.4 Exponential and Logarithmic Models

Solution

(a) $f(2) = 10^{-0.42(2)} \approx 0.145$. After 2 years about 14.5% of the robins were still alive.

(b) *Symbolic Solution*   Use $5\% = 0.05$ and solve the following equation.

$$10^{-0.42x} = 0.05$$
$$\log 10^{-0.42x} = \log 0.05 \qquad \text{Take the common logarithm.}$$
$$-0.42x = \log 0.05 \qquad \text{Inverse property: } \log 10^k = k$$
$$x = \frac{\log 0.05}{-0.42} \approx 3.1 \qquad \text{Divide by } -0.42.$$

After about 3 years only 5% of the robins were still alive.

(c) *Graphical Solution*   Figure 10.35 shows the graphs of $Y_1 = 10\wedge(-0.42X)$ and $Y_2 = 0.05$ intersecting, with the robin population declining—an example of exponential decay. In Figure 10.36 the graphs of Y_1 and Y_2 intersect near $(3.1, 0.05)$.

Numerical Solution   Figure 10.37 reveals that $Y_1 \approx Y_2$ when $x = 3.1$.

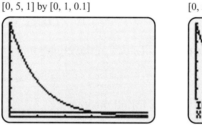
Figure 10.35

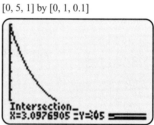
Figure 10.36

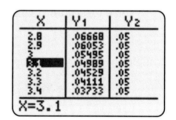
Figure 10.37

EXAMPLE 4   *Solving exponential equations*

Solve each equation symbolically.

(a) $2e^x - 1 = 5$ (b) $3^{x-5} = 15$

Solution

(a) Begin by solving for e^x.

$$2e^x - 1 = 5 \qquad \text{Given equation}$$
$$2e^x = 6 \qquad \text{Add 1 to both sides.}$$
$$e^x = 3 \qquad \text{Divide both sides by 2.}$$
$$\ln e^x = \ln 3 \qquad \text{Take the natural logarithm of both sides.}$$
$$x = \ln 3 \approx 1.10 \qquad \text{Inverse property: } \ln e^k = k$$

(b) Start by taking the common logarithm.

$$3^{x-5} = 15 \qquad \text{Given equation}$$
$$\log 3^{x-5} = \log 15 \qquad \text{Take common logarithms.}$$
$$(x-5)\log 3 = \log 15 \qquad \text{Power rule for logarithms}$$
$$x - 5 = \frac{\log 15}{\log 3} \qquad \text{Divide by log 3.}$$
$$x = \frac{\log 15}{\log 3} + 5 \approx 7.46 \qquad \text{Add 5 to both sides and approximate.}$$

LOGARITHMIC EQUATIONS AND MODELS

To solve an exponential equation we use logarithms. To solve a logarithmic equation we *exponentiate* both sides of the equation. To exponentiate both sides of an equation we use the fact that if $x = y$, then $a^x = a^y$ for any positive base a. For example, to solve

$$\log x = 3$$

we exponentiate both sides of the equation using base 10.

$$10^{\log x} = 10^3$$

Because $10^{\log x} = x$ for all positive x,

$$x = 10^3 = 1000.$$

To solve logarithmic equations we frequently use the inverse property

$$a^{\log_a x} = x.$$

Examples of this inverse property include

$$e^{\ln 2k} = 2k, \qquad 2^{\log_2 x} = x, \quad \text{and} \quad 10^{\log(x+5)} = x + 5.$$

EXAMPLE 5 *Solving logarithmic equations*

Solve and approximate solutions to the nearest hundredth when appropriate. Give graphical support.

(a) $2 \log x = 4$ **(b)** $\ln 3x = 5.5$ **(c)** $\log_2 (x + 4) = 7$

Solution

(a) $2 \log x = 4$ Given equation
      $\log x = 2$ Divide both sides by 2.
      $10^{\log x} = 10^2$ Exponentiate both sides using base 10.
      $x = 100$ Inverse properties: $10^{\log k} = k$

Graphical support is shown in Figure 10.38, where $Y_1 = 2 \log (X)$ and $Y_2 = 4$.

10.4 Exponential and Logarithmic Models

(b) $\ln 3x = 5.5$ Given equation
$e^{\ln 3x} = e^{5.5}$ Exponentiate both sides using base e.
$3x = e^{5.5}$ Inverse property: $e^{\ln k} = k$.
$x = \dfrac{e^{5.5}}{3} \approx 81.56$ Divide both sides by 3 and approximate.

Graphical support is shown in Figure 10.39, where $Y_1 = \ln(3X)$ and $Y_2 = 5.5$.

(c) $\log_2 (x + 4) = 7$ Given equation
$2^{\log_2 (x+4)} = 2^7$ Exponentiate both sides using base 2.
$x + 4 = 2^7$ Inverse property: $2^{\log_2 k} = k$.
$x = 2^7 - 4$ Subtract 4 from both sides.
$= 124$ Simplify.

Graphical support is shown in Figure 10.40. To graph $y = \log_2 (X + 4)$ use the change of base formula. Then $Y_1 = \log(X + 4)/\log (2)$ and $Y_2 = 7$.

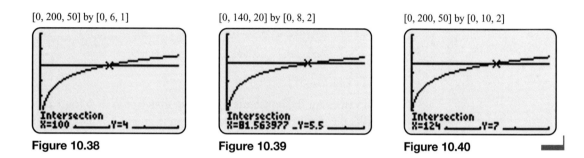

Figure 10.38   Figure 10.39   Figure 10.40

EXAMPLE 6   *Modeling runway length*

For some types of airplanes with weight x, the minimum runway length L required at takeoff is modeled by

$$L(x) = 3 \log x.$$

In this equation L is measured in thousands of feet and x is measured in thousands of pounds. Estimate the weight of the heaviest airplane that can take off from a runway 5100 feet long.   (*Source:* L. Haefner, *Introduction to Transportation Systems.*)

Solution

Runway length is measured in thousands of feet, so we must solve the equation $L(x) = 5.1$.

$3 \log x = 5.1$ Given equation
$\log x = 1.7$ Divide both sides by 3.
$10^{\log x} = 10^{1.7}$ Exponentiate both sides using base 10.
$x = 10^{1.7}$ Inverse property: $10^{\log k} = k$
$x \approx 50.1$ Approximate.

The largest airplane that can take off from this runway weighs 50,000 pounds.

Critical Thinking

In Example 7, Section 10.2, we used the formula $L(x) = 1.3 \ln x$ to model runway length. Are $L(x) = 1.3 \ln x$ and $L(x) = 3 \log x$ equivalent formulas? Explain.

EXAMPLE 7 *Modeling bird populations*

 Near New Guinea there is a relationship between the number of different species of birds and the size of an island. Larger islands tend to have a greater variety of birds. Table 10.9 lists the number of species of birds y found on islands with an area of x square kilometers.

TABLE 10.9

x (km^2)	0.1	1	10	100	1000
y (species)	10	15	20	25	30

Source: B. Freedman, *Environmental Ecology.*

(a) Find values for the constants a and b so that $y = a + b \log x$ models the data.
(b) Predict the number of bird species on an island of 15,000 square kilometers.

Solution

(a) Because $\log 1 = 0$, substitute $x = 1$ and $y = 15$ into the equation to find a.

$$15 = a + b \log 1$$
$$15 = a + b \cdot 0$$
$$15 = a$$

Thus $y = 15 + b \log x$. To find b substitute $x = 10$ and $y = 20$.

$$20 = 15 + b \log 10$$
$$20 = 15 + b \cdot 1$$
$$5 = b$$

The data in Table 10.9 are modeled by $y = 15 + 5 \log x$. This result is supported by Figure 10.41.

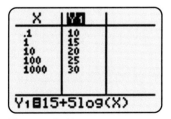

Figure 10.41

(b) To predict the number of species on an island of 15,000 square kilometers, let $x = 15{,}000$ and find y.

$$y = 15 + 5 \log 15{,}000 \approx 36$$

The model estimates about 36 different species of birds on this island.

10.4 Putting It All Together

Exponential and logarithmic equations occur in many real-world applications. When solving exponential equations, we usually take the logarithm of both sides. Similarly, when solving logarithmic equations, we usually *exponentiate* both sides of the equation. That is, if $x = y$ then $a^x = a^y$ for any positive base a. We do so because a^x and $\log_a x$ represent inverse operations, much like adding and subtracting or multiplying and dividing. Basic steps for solving exponential and logarithmic equations are summarized in the following table.

Type of Equation	Procedure	Example	
Exponential	Begin by solving for the exponential expression a^x. Then take a logarithm of both sides.	$4e^x + 1 = 9$ $e^x = 2$ $\ln e^x = \ln 2$ $x = \ln 2$	Given equation Solve for e^x. Take the natural logarithm. Inverse property: $\ln e^k = k$
Logarithmic	Begin by solving for the logarithm in the equation. Then exponentiate both sides of the equation using the same base as the logarithm.	$\frac{1}{3} \log 2x = 1$ $\log 2x = 3$ $10^{\log 2x} = 10^3$ $2x = 1000$ $x = 500$	Given equation Multiply by 3. Exponentiate using base 10. Inverse property: $10^{\log k} = k$ Divide by 2.

10.4 EXERCISES

FOR EXTRA HELP: Student's Solutions Manual, MyMathLab.com, InterAct Math, Math Tutor Center, MathXL, Digital Video Tutor CD 7 Videotape 16

CONCEPTS

1. To solve $x - 5 = 50$, what should be done?

2. To solve $5x = 50$, what should be done?

3. To solve $10^x = 50$, what should be done?

4. To solve $\log x = 5$, what should be done?

5. $\log 10^x = $ _____

6. $10^{\log x} = $ _____

7. $\ln e^{2x} = $ _____

8. $e^{\ln(x+7)} = $ _____

9. Does $\dfrac{\log 5}{\log 4}$ equal $\log \dfrac{5}{4}$? Explain.

10. Does $\dfrac{\log 5}{\log 4}$ equal $\log 5 - \log 4$? Explain.

11. How many solutions are there to the equation $\log x = k$, where k is any real number?

12. How many solutions are there to the equation $10^x = k$, where k is a positive number?

Exponential Equations

Exercises 13–24: Solve the equation symbolically. Approximate answers to the nearest hundredth when appropriate.

13. $10^x = 1000$

14. $10^x = 0.01$

15. $2^x = 64$

16. $3^x = 27$

17. $10^{0.4x} = 124$

18. $0.75^x = 0.25$

19. $e^{-x} = 1$

20. $0.5^{-5x} = 5$

21. $e^x - 1 = 6$

22. $2e^{4x} = 15$

23. $2(10)^{x+2} = 35$

24. $10^{3x} + 10 = 1500$

Exercises 25–28: The symbolic and graphical representations of f and g are given.
(a) Use the graph to estimate solutions to $f(x) = g(x)$.
(b) Solve $f(x) = g(x)$ symbolically.

25. $f(x) = 0.2(10^x)$, $g(x) = 2$

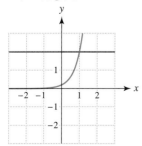

26. $f(x) = e^x$, $g(x) = 7.4$

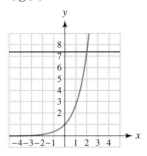

27. $f(x) = 2^{-x}$, $g(x) = 4$

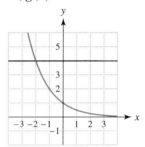

28. $f(x) = 0.1(3^x)$, $g(x) = 0.9$

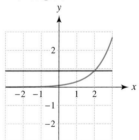

Exercises 29–36: Solve the equation symbolically. Give graphical or numerical support. Approximate answers to the nearest hundredth when appropriate.

29. $10^x = 0.1$

30. $2(10^x) = 2000$

31. $4e^x + 5 = 9$

32. $e^x + 6 = 36$

33. $4^x = 1024$

34. $3^x = 729$

35. $(0.55)^x + 0.55 = 2$

36. $5(0.9)^x = 3$

Exercises 37–40: The given equation cannot be solved symbolically. Find any solutions either graphically or numerically to the nearest hundredth.

37. $e^x - x = 2$

38. $x \log x = 1$

39. $\ln x = e^{-x}$

40. $10^x - 2 = \log(x + 2)$

LOGARITHMIC EQUATIONS

Exercises 41–54: Solve the equation. Appropriate answers to the nearest hundredth when appropriate.

41. $\log x = 2$

42. $\log x = 0.01$

43. $\ln x = 5$

44. $2 \ln x = 4$

45. $\log 2x = 7$

46. $6 \ln 4x = 12$

47. $\log_2 x = 4$

48. $\log_2 x = 32$

49. $\log_2 5x = 2.3$

50. $2 \log_3 4x = 10$

51. $2 \log x + 5 = 7.8$

52. $\ln(x - 1) = 3.3$

53. $5 \ln(2x + 1) = 55$

54. $5 - \log(x + 3) = 2.6$

Exercises 55–58: The symbolic and graphical representations of f and g are given.
    (a) Use the graph to estimate solutions to $f(x) = g(x)$.
    (b) Solve $f(x) = g(x)$ symbolically.

55. $f(x) = \ln x,\ g(x) = 0.7$

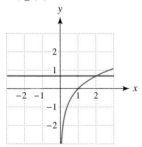

56. $f(x) = \log_2 x,\ g(x) = 1.6$

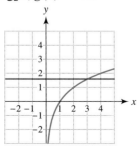

57. $f(x) = 5 \log 2x,\ g(x) = 3$

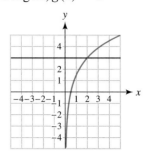

58. $f(x) = 2 \ln(x) - 3,\ g(x) = 0.9$

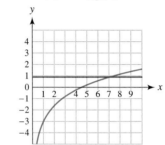

Exercises 59–64: Solve the equation symbolically. Give graphical or numerical support. Approximate answers to the nearest hundredth.

59. $\log x = 1.6$

60. $\ln x = 2$

61. $\ln(x + 1) = 1$

62. $2 \log(2x + 3) = 8$

63. $17 - 6 \log_3 x = 5$

64. $4 \log_2 x + 7 = 12$

APPLICATIONS

65. *Growth of a Mutual Fund* (Refer to Example 2.) An investor deposited $2000 in a mutual fund that returned 15% at the end of 1 year. Determine the length of time required for the investment to triple its value if the annual rate of return remains the same. Make your estimate
    (a) symbolically,
    (b) graphically, and
    (c) numerically.

66. Savings Account (Refer to Example 2.) If a savings account pays 6% annual interest at the end of each year, how many years will it take for the account to double in value?

67. Liver Transplants In the United States the gap between available organs for liver transplants and people who need them has widened. The number of individuals waiting for liver transplants can be modeled by

$$f(x) = 2339(1.24)^{x-1988},$$

where x is the year. (*Source:* United Network for Organ Sharing.)
(a) Evaluate $f(1994)$ and interpret the result.
(b) Determine when the number of individuals waiting for liver transplants was 20,000.

68. Life Span of a Robin (Refer to Example 3.) Determine symbolically when 50% of the robins in the study were still alive. Support your result either graphically or numerically.

69. Runway Length (Refer to Example 6.) Determine the weight of the heaviest airplane that can take off from a runway having a length of $\frac{3}{4}$ mile. (*Hint:* 1 mile = 5280 feet.)

70. Runway Length (Refer to Example 6.)
(a) Suppose that an airplane is 10 times heavier than a second airplane. How much longer should the runway be for the heavier airplane than for the lighter airplane? (*Hint:* Let the heavier airplane have weight 10x.)
(b) If the runway length is increased by 3000 feet, by what factor can the weight of an airplane that uses the runway be increased?

71. The Decline of Bluefin Tuna Bluefin tuna are large fish that can weigh 1500 pounds and swim at a speed of 55 miles per hour. They are used for sushi, and a prime fish can be worth more than $30,000. As a result, the number of western Atlantic bluefin tuna has declined dramatically. Their numbers in thousands between 1974 and 1991 can be modeled by $f(x) = 230(10^{-0.055x})$, where x is the number of years after 1974. See the accompanying graph. (*Source:* B. Freedman.)

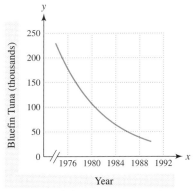

(a) Evaluate $f(1)$ and interpret the result.
(b) Estimate graphically or numerically the year in which they numbered 115 thousand.
(c) Solve part (b) symbolically.

72. Insect Populations (Refer to Example 7.) The table lists species of insects y found on islands having areas of x square miles.

x (square miles)	1	2	4
y (species)	1000	1500	2000

x (square miles)	8	16
y (species)	2500	3000

(a) Find values for constants a and b so that $y = a + b \log_2 x$ models the data. (*Hint:* Start by substituting $x = 1$ and $y = 1000$ into the equation.)
(b) Construct a table of y and verify that your equation models the data. Use the change of base formula to convert $\log_2 x$ to an equivalent common logarithm so that the equation can be entered into a graphing calculator.
(c) Estimate the number of species of insects on an island having an area of 12 square miles.

73. Calories Consumed and Land Ownership In developing countries there is a relationship between the amount of land a person owns and the average number of calories consumed daily. This relationship is modeled by

$$f(x) = 645 \log(x + 1) + 1925,$$

where x is the amount of land owned in acres and $0 \le x \le 4$. (*Source:* D. Grigg, *The World Food Problem.*)

(a) Estimate graphically the acres owned by a typical person consuming 2200 calories per day.
(b) Solve part (a) symbolically.

74. *Population of Industrialized Urban Regions* The number of people living in industrialized urban regions throughout the world has not grown exponentially. Instead it has grown logarithmically and is modeled by
$$f(x) = 0.36 + 0.15 \ln(x - 1949).$$
In this formula the output is billions of people and the input x is the year, where
$$1950 \le x \le 2030.$$
(*Source*: D. Meadows, *Beyond The Limits*.)
(a) Determine either graphically or numerically when this population may reach 1 billion people.
(b) Solve part (a) symbolically.

75. *Greenhouse Gases* If current trends continue, future concentrations of atmospheric carbon dioxide (CO_2) in parts per million (ppm) are expected to increase. This increase in concentration of CO_2 has been accelerated by burning fossil fuels and deforestation. The exponential equation $y = 364(1.005)^x$ may be used to model CO_2 in parts per million, where $x = 0$ corresponds to 2000, $x = 1$ to 2001, and so on. (*Source*: R. Turner, *Environmental Economics*.) Estimate the year when the CO_2 concentration could be double the preindustrial level of 280 parts per million.

76. *Fertilizer Use* Between 1950 and 1980, the use of chemical fertilizers increased worldwide. The table lists worldwide average use y in kilograms per acre of cropland during year x. (*Source*: D. Grigg, *The World Food Problem*.)

x	1950	1963	1972	1979
y	5.0	11.3	22.0	31.2

(a) Are the data linear or nonlinear? Explain.
(b) The equation $y = 5(1.06)^{(x-1950)}$ may be used to model the data. The growth factor is 1.06. What does this growth factor indicate about fertilizer use during this time period?
(c) Estimate the year when fertilizer use was 15 kilograms per acre of cropland.

77. *Modeling Sound* The formula
$$f(x) = 160 + 10 \log x$$
is used to calculate the decibel level of a sound with intensity x measured in watts per square centimeter. The noise level at a basketball game can reach 100 decibels. Find the intensity x of this sound.

78. *Loudness of a Sound* (Refer to Exercise 77.)
(a) Show that, if the intensity of a sound increases by a factor of 10 from x to $10x$, the decibel level increases by 10 decibels. (*Hint:* Show that $160 + 10 \log 10x = 170 + 10 \log x$.)
(b) Find the increase in decibels if the intensity x increases by a factor of 1000.
(c) Find the increase in the intensity x if the decibel level increases by 20.

79. *Hurricanes* (Refer to Exercise 53, Section 10.2.) The barometric air pressure in inches of mercury at a distance of x miles from the eye of a severe hurricane is given by
$$f(x) = 0.48 \ln(x + 1) + 27.$$
(*Source*: A. Miller and R. Anthes, *Meteorology*.)
(a) Determine graphically how far from the eye the pressure is 28 inches of mercury.
(b) Solve part (a) symbolically.

80. *Earthquakes* The Richter scale is used to determine the intensity of earthquakes, which corresponds to the amount of energy released. If an earthquake has an intensity of x, its magnitude, as computed by the Richter scale, is given by $R(x) = \log \frac{x}{I_0}$, where I_0 is the intensity of a small, measurable earthquake.
(a) If x is 1000 times greater than I_0, how large is this increase on the Richter scale?
(b) If the Richter scale increases from 5 to 8, by what factor did the intensity x increase?

WRITING ABOUT MATHEMATICS

81. Explain in words the basics steps for solving the equation $a(10^x) - b = c$ and then write the solution.

82. Explain in words the basics steps for solving the equation $a \log 3x = b$ and then write the solution.

CHECKING BASIC CONCEPTS FOR SECTIONS 10.3 AND 10.4

1. Write the expression in terms of logarithms of x, y, and z.
  (a) $\log xy$
  (b) $\ln \dfrac{x}{yz}$
  (c) $\ln x^2$
  (d) $\log \dfrac{x^2 y^3}{\sqrt{z}}$

2. Write the expression as one logarithm.
  (a) $\log x + \log y$
  (b) $\ln 2x - 3 \ln y$
  (c) $2 \log_2 x + 3 \log_2 y - \log_2 z$

3. Solve the equation symbolically and give graphical support. Approximate answers to the nearest hundredth.
  (a) $2(10^x) = 40$
  (b) $2^{3x} + 3 = 150$
  (c) $\ln x = 4.1$
  (d) $4 \log 2x = 12$

CHAPTER 10 SUMMARY

Section 10.1 Exponential Functions

An exponential function can be represented symbolically by $f(x) = Ca^x$, where $C > 0$, $a > 0$, and $a \neq 1$. When $a > 1$, the graph of $f(x) = Ca^x$ models exponential growth, and when $0 < a < 1$, it models exponential decay. The base a either represents the growth factor or the decay factor. The constant C equals $f(x)$ when $x = 0$.

Example: $f(x) = 1.5(2)^x$ is an exponential function with $a = 2$ and $C = 1.5$. It models exponential growth because $a = 2 > 1$. The growth factor is 2 because for each unit increase in x, the output from $f(x)$ increases by a factor of 2.

Example: $g(x) = e^x$ is the natural exponential function, where $e \approx 2.71828$.

The set of valid inputs (domain) for an exponential function includes all real numbers and the set of corresponding outputs (range) is all positive numbers. Linear growth occurs when a quantity increases by a constant amount for each unit increase in x. Exponential growth (or decay) occurs when a quantity increases (or decreases) by a constant factor equal to a for each unit increase in x.

Section 10.2 Logarithmic Functions

Logarithmic functions output exponents. For example, $\log 10^2 = 2$, $\log 10^{-1} = -1$, and $\log 10^k = k$ for all real numbers k. That is, $\log x$ represents the exponent on base 10 that results in x. Thus $\log 1000 = 3$ because $10^3 = 1000$. Similarly, $\log_2 8$ represents the exponent on base 2 that results in 8. Because $2^3 = 8$, then $\log_2 8 = 3$. The set of valid inputs (domain) for a logarithmic function is all positive numbers and the set of corresponding outputs (range) is all real numbers.

The graph of a logarithmic function passes through the point (1, 0), as illustrated in the following graph. As x becomes large, $\log_a x$ grows very slowly.

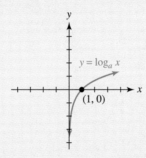

Section 10.3 Properties of Logarithms

Logarithms have several important properties. For positive numbers m, n, and $a \neq 1$ and any real number r,

1. $\log_a mn = \log_a m + \log_a n$
2. $\log_a \dfrac{m}{n} = \log_a m - \log_a n$
3. $\log_a (m^r) = r \log_a m$

Examples:
1. $\log 5 + \log 20 = \log (5 \cdot 20) = \log 100 = 2$
2. $\log 100 - \log 5 = \log \dfrac{100}{5} = \log 20 \approx 1.301$
3. $\ln 2^6 = 6 \ln 2 \approx 4.159$

The following inverse properties are important for solving exponential and logarithmic equations.

1. $\log_a a^x = x$, for any real number x
2. $a^{\log_a x} = x$, for any positive number x

Examples:
1. $\log_2 2^\pi = \pi$
2. $10^{\log 2.5} = 2.5$

Section 10.4 Exponential and Logarithmic Models

Calculating a^x and $\log_a x$ are inverse operations, much like addition and subtraction or multiplication and division. When solving an exponential equation, we usually take a logarithm of both sides. When solving a logarithmic equation, we usually exponentiate both sides.

Examples:

$2(5)^x = 22$	Exponential equation
$5^x = 11$	Divide by 2.
$\log 5^x = \log 11$	Take the common logarithm.
$x \log 5 = \log 11$	Power rule
$x = \dfrac{\log 11}{\log 5}$	Divide by log 5.
$\log 2x = 2$	Logarithmic equation
$10^{\log 2x} = 10^2$	Exponentiate each side.
$2x = 100$	Inverse properties
$x = 50$	Divide by 2.

Chapter 10
Review Exercises

SECTIONS 10.1 AND 10.2

Exercises 1–4: Evaluate the exponential function for the given values of x.

1. $f(x) = 6^x \qquad x = -1, \quad x = 2$
2. $f(x) = 5(2^{-x}) \quad x = 0, \quad x = 3$
3. $f(x) = \left(\dfrac{1}{3}\right)^x \quad x = -1, \quad x = 4$
4. $f(x) = 3\left(\dfrac{1}{6}\right)^x \quad x = 0, \quad x = 1$

Exercises 5–8: Graph f in $[-4, 4, 1]$ by $[0, 8, 1]$. State whether the graph illustrates exponential growth, exponential decay, or logarithmic growth.

5. $f(x) = 1.7^x$
6. $f(x) = \left(\dfrac{1}{5}\right)^x$
7. $f(x) = \ln(x + 1)$
8. $f(x) = 3^{-x}$

Exercises 9 and 10: A table for a function f is given.
  (a) *Determine whether f represents linear or exponential growth.*
  (b) *Find a formula for f.*

9.
x	0	1	2	3	4
$f(x)$	5	10	20	40	80

10.
x	0	1	2	3	4
$f(x)$	5	10	15	20	25

11. Use the graph of $y = Ca^x$ to find C and a.

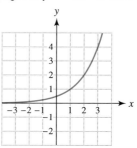

12. Use the graph of $y = k \log_2 x$ to find k.

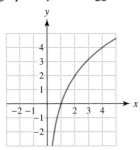

Exercises 13 and 14: If C dollars are deposited in an account that pays r percent annual interest at the end of each year, approximate the amount in the account after x years.

13. $C = \$1200 \quad r = 10\% \quad x = 9$ years

14. $C = \$900 \quad r = 18\% \quad x = 40$ years

Exercises 15–18: Evaluate $f(x)$ for the given value of x. Approximate answers to the nearest hundredth.

15. $f(x) = 2e^x - 1 \quad x = 5.3$

16. $f(x) = 0.85^x \quad x = 2.1$

17. $f(x) = 2 \log x \quad x = 55$

18. $f(x) = \ln(2x + 3) \quad x = 23$

Exercises 19–22: Evaluate the logarithm by hand.

19. $\log 0.001$

20. $\log \sqrt{10{,}000}$

21. $\ln e^{-4}$

22. $\log_4 16$

Exercises 23–26: Approximate the logarithm to the nearest thousandth.

23. $\log 65$ **24.** $\ln 0.85$

25. $\ln 120$ **26.** $\log_2 \frac{2}{5}$

Exercises 27 and 28: Graph f in $[-6, 6, 1]$ by $[-4, 4, 1]$. Compare this graph to the graph of $y = e^x$.

27. $f(x) = e^{(x-2)}$ **28.** $f(x) = e^x - 2$

Exercises 29 and 30: Graph f in $[-6, 6, 1]$ by $[-4, 4, 1]$. Compare this graph to the graph of $y = \log x$.

29. $f(x) = \log(x + 3)$ **30.** $f(x) = \ln x$

Section 10.3

Exercises 31–36: Write the expression by using sums and differences logarithms of x, y, and z.

31. $\ln xy$ **32.** $\log \frac{x}{y}$

33. $\ln x^2 y^3$ **34.** $\log \frac{\sqrt{x}}{z^3}$

35. $\log_2 \frac{x^2 y}{z}$ **36.** $\log_3 \sqrt{\frac{x}{y}}$

Exercises 37–40: Write the expression as one logarithm. Assume x and y are positive.

37. $\log 45 + \log 5 - \log 3$

38. $2 \ln x - 3 \ln y$

39. $\log_4 2x + \log_4 5x$

40. $\log x^4 - \log x^3 + \log y$

Exercises 41–44: Rewrite the expression, using the power rule.

41. $\log 6^3$ **42.** $\ln x^2$

43. $\log_2 5^{2x}$ **44.** $\log_4 (0.6)^{x+1}$

Exercises 45–48: Simplify, using inverse properties of logarithms.

45. $10^{\log 7}$ **46.** $\log_2 2^{5/9}$

47. $\ln e^{6-x}$ **48.** $e^{2 \ln x}$

Section 10.4

Exercises 49–56: Solve the equation symbolically. Give either graphical or numerical support. Approximate answers to the nearest hundredth when appropriate.

49. $10^x = 100$ **50.** $2^{2x} = 256$

51. $3e^x + 1 = 28$ **52.** $0.85^x = 0.2$

53. $5 \ln x = 4$

54. $\ln 2x = 5$

55. $2 \log x = 80$

56. $3 \log x - 5 = 1$

Exercises 57 and 58: The symbolic and graphical representations of f and g are given.
(a) Use the graph to estimate solutions to $f(x) = g(x)$.
(b) Solve $f(x) = g(x)$ symbolically.

57. $f(x) = \frac{1}{2}(2^x), g(x) = 4$

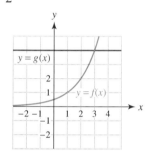

58. $f(x) = \log_2 2x, g(x) = 3$

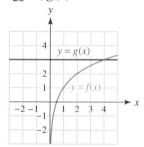

APPLICATIONS

59. *Growth of a Mutual Fund* An investor deposits $1500 in a mutual fund that returns 12% annually. Determine the time required for the investment to double in value. Make your estimate
(a) symbolically,
(b) graphically, and
(c) numerically.

60. *Modeling Data* Find values for the constants a and b so that $y = a + b \log x$ models these data.

x	0.1	1	10	100	1000
y	50	100	150	200	250

61. *Modeling Data* Find values for the constants C and a so that $y = Ca^x$ models these data.

x	0	1	2	3	4
y	3	6	12	24	48

62. *Earthquakes* The Richter scale, used to determine the magnitude of earthquakes, is based on the formula $R(x) = \log \frac{x}{I_0}$, where x is the measured intensity. Let $I_0 = 1$. Find the intensity x for an earthquake with $R = 7$.

63. *Modeling Population* In 1997 the population of Nevada was 1.68 million and growing continuously at an annual rate of 4.8%. The population of Nevada in millions x years after 1997 can be modeled by
$$f(x) = 1.68e^{0.048x}.$$
(a) Graph f in $[0, 10, 2]$ by $[0, 4, 1]$. Does this function represent exponential growth or decay?
(b) Predict the population of Nevada in 2003.

64. *Modeling Bacteria* A growth of bacteria can be modeled by $N(t) = 1000e^{0.0014t}$, where N is measured in bacteria per milliliter and t is in minutes.
(a) Evaluate $N(0)$ and interpret the result.
(b) Estimate how long it takes for N to double.

65. *Modeling Wind Speed* Wind speeds are usually measured at heights between 5 to 10 meters above the ground. For a particular day, $f(x) = 1.2 \ln (x) + 5$ computes the wind speed in meters per second x meters above the ground. (*Source:* A. Miller and R. Anthes, *Meteorology.*)
(a) Find the wind speed at a height of 5 meters.
(b) Estimate the height at which the wind speed is 8 meters per second.

Chapter 10

Test

1. Evaluate $f(x) = 3\left(\frac{1}{4}\right)^x$ at $x = 2$.

2. Graph $f(x) = 1.5^{-x}$ in $[-4, 4, 1]$ by $[-2, 6, 1]$. State whether the graph illustrates exponential growth, exponential decay, or logarithmic growth.

Exercises 3 and 4: A table for a function f is given.
(a) *Determine whether f represents linear or exponential growth.*
(b) *Find a formula for $f(x)$.*

3.
x	-2	-1	0	1	2
$f(x)$	0.75	1.5	3	6	12

4.
x	-2	-1	0	1	2
$f(x)$	-4	-2.5	-1	0.5	2

5. Use the graph of $y = Ca^x$ to find C and a.

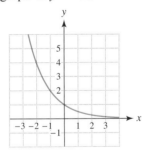

6. If $750 are deposited in an account paying 7% annual interest at the end of each year, approximate the amount in the account after 5 years.

7. Let $f(x) = 1.5 \ln(x - 5)$. Approximate $f(21)$ to the nearest hundredth.

8. Evaluate $\log \sqrt{10}$ by hand.

9. Approximate $\log_2 43$ to the nearest thousandth.

10. Graph $f(x) = \log(x - 2)$ in $[-6, 6, 1]$ by $[-4, 4, 1]$. Compare this graph to the graph of $y = \log x$.

11. Write $\log \dfrac{x^3 y^2}{\sqrt{z}}$, using sums and differences of logarithms of x, y, and z.

12. Write $4 \ln x - 5 \ln y + \ln z$ as one logarithm.

13. Rewrite $\log 7^{2x}$, using the power rule.

14. Simplify $\ln e^{1-3x}$, using inverse properties of logarithms.

Exercises 15–18: Solve the equation symbolically. Give either graphical or numerical support. Approximate answers to the nearest hundredth when appropriate.

15. $2e^x = 50$

16. $3(10)^x - 7 = 143$

17. $5 \log x = 9$

18. $3 \ln 5x = 27$

19. *Modeling Data* Find values for constants a and b so that $y = a + b \log x$ models the data.

x	0.01	0.1	1	10	100
y	-1	2	5	8	11

20. *Modeling Bacteria Growth* A sample of bacteria is growing continuously at a rate of 9% per hour and can be modeled by
$$f(x) = 4e^{0.09x},$$
where the input x represents elapsed time in hours and the output $f(x)$ is in millions of bacteria.
(a) What was the initial number of bacteria in the sample?
(b) Evaluate $f(5)$ and interpret the result.
(c) Graph f in $[0, 10, 2]$ by $[0, 10, 2]$. Does this function represent exponential growth or decay?
(d) Determine graphically the elapsed time when there were 6 million bacteria.
(e) Solve part (d) symbolically. Do your answers agree?

Chapter 10

Extended and Discovery Exercises

Exercises 1–4: Radioactive Carbon Dating While an animal is alive, it breathes both carbon dioxide and oxygen. Because a small portion of normal atmospheric carbon dioxide is made up of radioactive carbon-14, a fixed percentage of the animal's body is composed of carbon-14. When the animal dies, it quits breathing and the carbon-14 disintegrates without being replaced. One method used to determine when an animal died is to estimate the percentage of carbon-14 remaining in its bones. The **half-life** of carbon-14 is 5730 years. That is, half the original amount of carbon-14 in bones of a fossil will remain after 5730 years. The percentage P, in decimal form, of carbon-14 remaining after x years is modeled by $P(x) = a^x$.

1. Find the value of a. (*Hint:* $P(5730) = 0.5$.)

2. Calculate the percentage of carbon-14 remaining after 10,000 years.

3. Estimate the age of a fossil with $P = 0.9$.

4. Estimate the age of a fossil with $P = 0.01$.

Exercises 5–8: Modeling Blood Flow in Animals For medical reasons, dyes may be injected into the bloodstream to determine the health of internal organs. In one study involving animals, the dye BSP was injected to assess blood flow in the liver. The results are listed in the accompanying table, where x represents the elapsed time in minutes and y is the concentration of the dye in the bloodstream in milligrams per milliliter. Scientists modeled the data with $f(x) = 0.133(0.878(0.73^x) + 0.122(0.92^x))$.

x (minutes)	1	2	3	4
y (milligrams/milliliter)	0.102	0.077	0.057	0.045

x (minutes)	5	7	9	13
y (milligrams/milliliter)	0.036	0.023	0.015	0.008

x (minutes)	16	19	22
y (milligrams/milliliter)	0.005	0.004	0.003

Source: F. Harrison, "The measurement of liver blood flow in conscious calves."

5. Graph f together with the data. Comment on the fit.

6. Determine the y-intercept and interpret the result.

7. What happens to the concentration of the dye after a long period of time? Explain.

8. Estimate graphically the time at which the concentration of the dye reached 40% of its initial amount. Would you want to solve this problem symbolically? Explain.

Exercises 9 and 10: Acid Rain Air pollutants frequently cause acid rain. A measure of acidity is pH, which measures the concentration of the hydrogen ions in a solution, and ranges from 1 to 14. Pure water is neutral and has a pH of 7, acid solutions have a pH less than 7, and alkaline solutions have a pH greater than 7. The pH of a substance can be computed by $f(x) = -\log x$, where x represents the hydrogen ion concentration in moles per liter. Pure water exposed to normal carbon dioxide in the atmosphere has a pH of 5.6. If the pH of a lake drops below this level, it is indicative of an acid lake. (**Source:** Howells, G., *Acid Rain and Acid Water.*)

9. In rural areas of Europe, rainwater typically has a hydrogen ion concentration of $x = 10^{-4.7}$. Find its pH. What effect might this rain have on a lake with a pH of 5.6?

10. Seawater has a pH of 8.2. Compared to seawater, how many times greater is the hydrogen ion concentration in rainwater from rural Europe?

Exercises 11 and 12: Investment Account If x dollars are deposited every 2 weeks (26 times per year) in an account paying an annual interest rate r, expressed in decimal form, the amount A in the account after n years can be approximated by the formula

$$A = x\left[\frac{(1 + r/26)^{26n} - 1}{(r/26)}\right].$$

11. If $100 is deposited every 2 weeks in an account paying 9% interest, approximate the amount in the account after 10 years.

12. Suppose that your retirement account pays 12% annual interest. Determine how much you should deposit in this account every 2 weeks, in order to have one million dollars at age 65.

Exercises 13–17: Logistic Functions and Modeling Data Populations of bacteria, insects, and animals do not continue to grow indefinitely. Initially, population growth may be slow. Then, as the numbers of organisms increase, so does the rate of growth. After a region has become heavily populated or saturated, the growth in population usually levels off because of limited resources. This type of growth may be modeled by a **logistic function**. One of the earliest studies of population growth was done with yeast plants in 1913. A small amount of yeast was placed in a container with a fixed amount of nourishment. The units of yeast were recorded every 2 hours, giving the data shown in the table.

Time (hours)	0	2	4	6
Yeast (units)	9.6	29.0	71.1	174.6

Time (hours)	8	10	12
Yeast (units)	350.7	513.3	594.8

Time (hours)	14	16	18
Yeast (units)	640.8	655.9	661.8

Source: D. Brown, *Models in Biology*.

13. Make a scatterplot of the data.

14. Describe the growth of yeast and explain the graph.

15. The data are modeled by the logistic function given by

$$f(x) = \frac{663}{(1 + 71.6(0.579)^x)}.$$

Graph f and the data in the same viewing rectangle.

16. Determine graphically when the amount of yeast equals 400 units.

17. Solve Exercise 16 symbolically.

Chapter 11
Sequences and Series

In this final chapter we present sequences and series, which are essential topics because they are used to model and approximate important quantities. For example, complicated population growth can be modeled with sequences, and accurate approximations for numbers such as π and e are made with series. Series also are essential to the solution of many modern applied mathematics problems.

Although you may not always recognize the impact of mathematics on everyday life, its influence is nonetheless profound. Mathematics is the *language of technology*—it allows experiences to be quantified. In the preceding chapters we showed numerous examples of mathematics being used to model the real world. Computers, CD players, highway design, weather, hurricanes, electricity, government data, cellular phones, medicine, ecology, business, sports, and psychology represent only some of the applications of mathematics. In fact, if a subject is studied in enough detail, mathematics usually appears in one form or another. Although predicting what the future may bring is difficult, one thing *is* certain—mathematics will continue to play an important role in both theoretical research and the real world.

The gains of education are never really lost.
—Franklin Roosevelt

11.1 SEQUENCES

Basic Concepts ~ Representations of Sequences ~ Models and Applications

INTRODUCTION

Sequences are ordered lists. For example, names listed alphabetically represent a sequence. Figure 11.1 shows an insect population in thousands per acre over a 6-year period. Listing populations by year is another example of a sequence. In mathematics a sequence is a function, for which valid inputs must be natural numbers. For example, we can let $f(1)$ represent the insect population after 1 year, $f(2)$ represent the insect population after 2 years, and in general let $f(n)$ represent the population after n years. In this section we discuss sequences and how they can be represented.

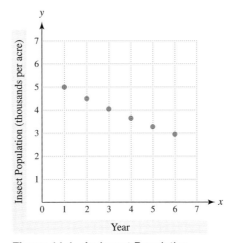

Figure 11.1 An Insect Population

BASIC CONCEPTS

Suppose that an individual's starting salary is $40,000 per year and that the person's salary is increased by 10% each year. This situation is modeled by

$$f(n) = 40,000(1.10)^n.$$

We do not allow the input n to be any real number, but rather limit n to a natural number because the individual's salary is constant throughout a particular year. The first 5 terms of the sequence are

$$f(1), f(2), f(3), f(4), f(5).$$

They can be computed as follows.

$$f(1) = 40,000(1.10)^1 = 44,000$$
$$f(2) = 40,000(1.10)^2 = 48,400$$
$$f(3) = 40,000(1.10)^3 = 53,240$$
$$f(4) = 40,000(1.10)^4 = 58,564$$
$$f(5) = 40,000(1.10)^5 \approx 64,420$$

This sequence is represented numerically in Table 11.1 on the next page.

TABLE 11.1

n	1	2	3	4	5
$f(n)$	44,000	48,400	53,240	58,564	64,420

A graphical representation results when each data point in Table 11.1 is plotted in the *xy*-plane, as illustrated in Figure 11.2. Graphs of sequences are scatterplots.

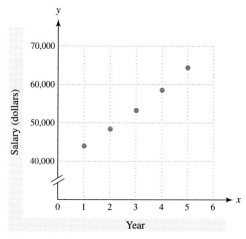

Figure 11.2

The preceding sequence is an example of a *finite sequence* of numbers. The even natural numbers,

$$2, 4, 6, 8, 10, 12, 14, \ldots$$

are an example of an *infinite sequence* represented by $f(n) = 2n$, when n is a natural number. The three dots, or periods (called an *ellipsis*), indicate that the pattern continues indefinitely.

> ### Sequences
>
> A **finite sequence** is a function whose domain is $D = \{1, 2, 3, \ldots, n\}$ for some fixed natural number n.
>
> An **infinite sequence** is a function whose domain is the set of natural numbers.

Because sequences are functions, many of the concepts discussed in previous chapters apply to sequences. Instead of letting y represent the output, however, the convention is to write $a_n = f(n)$, where n is a natural number in the domain of the sequence. The terms of a sequence are

$$a_1, a_2, a_3, \ldots, a_n, \ldots .$$

The first term is $a_1 = f(1)$, the second term is $a_2 = f(2)$, and so on. The ***n*th term**, or **general term**, of a sequence is $a_n = f(n)$.

EXAMPLE 1 Computing terms of a sequence

 Write the first four terms of each sequence.

(a) $f(n) = 2n - 1$ (b) $f(n) = 3(-2)^n$ (c) $f(n) = \dfrac{n}{n+1}$

Solution

(a) For $a_n = f(n) = 2n - 1$, we write
$$a_1 = f(1) = 2(1) - 1 = 1;$$
$$a_2 = f(2) = 2(2) - 1 = 3;$$
$$a_3 = f(3) = 2(3) - 1 = 5;$$
$$a_4 = f(4) = 2(4) - 1 = 7.$$
The first four terms are a_1, a_2, a_3, a_4, or $1, 3, 5, 7$.

(b) For $a_n = f(n) = 3(-2)^n$, we write
$$a_1 = f(1) = 3(-2)^1 = -6;$$
$$a_2 = f(2) = 3(-2)^2 = 12;$$
$$a_3 = f(3) = 3(-2)^3 = -24;$$
$$a_4 = f(4) = 3(-2)^4 = 48.$$
The first four terms are $-6, 12, -24, 48$.

(c) For $a_n = f(n) = \dfrac{n}{n+1}$, we write
$$a_1 = f(1) = \frac{1}{1+1} = \frac{1}{2};$$
$$a_2 = f(2) = \frac{2}{2+1} = \frac{2}{3};$$
$$a_3 = f(3) = \frac{3}{3+1} = \frac{3}{4};$$
$$a_4 = f(4) = \frac{4}{4+1} = \frac{4}{5}.$$
The first four terms are $\dfrac{1}{2}, \dfrac{2}{3}, \dfrac{3}{4}, \dfrac{4}{5}$. Note that, although the input to a sequence is a natural number, the output need not be a natural number.

Technology Note Generating Sequences

Many graphing calculators can generate sequences if you change the MODE from function (Func) to sequence (Seq). In Figures 11.3 and 11.4 on the next page the sequences from Example 1 are generated. On some calculators the sequence utility is found in the LIST OPS menus. The expression
$$\text{seq}(2n - 1, n, 1, 4)$$
represents terms 1 through 4 of the sequence $a_n = 2n - 1$ with the variable n.

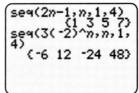
Figure 11.3

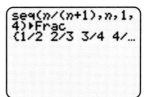
Figure 11.4

Representations of Sequences

Because sequences are functions, they can be represented symbolically, graphically, and numerically. The next two examples illustrate such representations.

EXAMPLE 2 Using a graphical representation

Use Figure 11.5 to write the terms of the sequence.

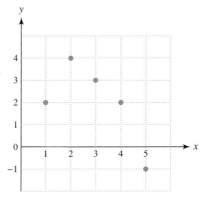

Figure 11.5

Solution

The points $(1, 2)$, $(2, 4)$, $(3, 3)$, $(4, 2)$, $(5, -1)$ are shown in the graph. The terms of the sequence are $a_1 = 2$, $a_2 = 4$, $a_3 = 3$, $a_4 = 2$, and $a_5 = -1$.

EXAMPLE 3 Representing a sequence

The average person in the United States uses 100 gallons of water each day. Give symbolic, numerical, and graphical representations for a sequence that models the total amount of water used over a 7-day period.

Solution

Symbolic Representation Let $a_n = 100n$ for $n = 1, 2, 3, \ldots, 7$.
Numerical Representation Table 11.2 contains the sequence.

TABLE 11.2

n	1	2	3	4	5	6	7
a_n	100	200	300	400	500	600	700

Graphical Representation Plot the points (1, 100), (2, 200), (3, 300), (4, 400), (5, 500), (6, 600), and (7, 700), as shown in Figure 11.6.

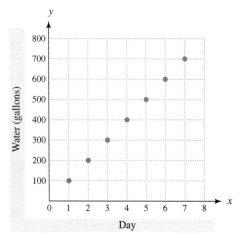

Figure 11.6 Water Use

Technology Note *Graphs and Tables of Sequences*

In the sequence mode, many graphing calculators are capable of representing sequences graphically and numerically, as shown in Figures 11.7 and 11.8, respectively.

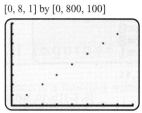

Figure 11.7   **Figure 11.8**

MODELS AND APPLICATIONS

A population model for a species of insect with a life span of 1 year can be described with a sequence. Suppose that each adult female insect produces, on average, r female offspring that survive to reproduce the following year. Let a_n represent the female insect population at the beginning of year n. Then the number of female insects is given by

$$a_n = Cr^{n-1},$$

where C is the initial population of female insects.   (*Source:* D. Brown and P. Rothery, *Models in Biology.*)

EXAMPLE 4 *Modeling numbers of insects*

Suppose that the initial population of adult female insects is 500 per acre and that $r = 1.04$. Then the number of female insects per acre is described by

$$a_n = 500(1.04)^{n-1}.$$

37. *Depreciation* Automobiles usually depreciate in value over time. Often, a newer automobile may be worth only 80% of its previous year's value. Suppose that a car is worth $25,000 new.
    (a) How much is it worth after 1 year? After 2 years?
    (b) Write a formula for a sequence that gives the car's value after n years.
    (c) Make a table that shows how much the car was worth each year during the first 7 years.

38. *Falling Object* The distance d that an object falls during *consecutive* seconds is shown in the table. For example, during the third second an object falls a distance of 80 feet.

n (seconds)	1	2	3	4	5
d (feet)	16	48	80	112	144

    (a) Find values for c and b so that $d = cn + b$ models these data.
    (b) How far does an object fall during the sixth second?

39. *Auditorium Seating* An auditorium has 50 seats in the first row, 55 seats in the second row, 60 seats in the third row, and so on.
    (a) Make a table that shows the number of seats in the first seven rows.
    (b) Write a formula that gives the number of seats in row n.
    (c) How many seats are there in row 23?
    (d) Graph the number of seats in each row for $n = 1, 2, 3, \ldots, 10$.

40. *Modeling Insect Populations* (Refer to Example 4.) Suppose that the initial population of insects is 2048 per acre and that $r = 0.5$. Use a sequence to represent the insect population over a 7-year period
    (a) symbolically,
    (b) numerically, and
    (c) graphically.

WRITING ABOUT MATHEMATICS

41. Compare the graph of $f(x) = 2x + 1$, where x is a real number, with the graph of the sequence $f(n) = 2n + 1$ where n is a natural number.

42. Explain what a sequence is. Describe the difference between a finite and an infinite sequence.

11.2 ARITHMETIC AND GEOMETRIC SEQUENCES

Representations of Arithmetic Sequences ~ Representations of Geometric Sequences ~ Applications and Models

INTRODUCTION

Indoor air pollution has become more hazardous as people spend 80% to 90% of their time in tightly sealed, energy-efficient buildings, which often lack proper ventilation. Many contaminants such as tobacco smoke, formaldehyde, radon, lead, and carbon monoxide are often allowed to increase to unsafe levels. One way to alleviate this problem is to use efficient ventilation systems. Mathematics plays an important role in determining the proper amount of ventilation. In this section we use sequences to model ventilation in classrooms. Before implementing this model, however, we discuss the basic concepts relating to two special types of sequences.

REPRESENTATIONS OF ARITHMETIC SEQUENCES

If a sequence is defined by a linear function, it is an *arithmetic sequence*. For example,

$$f(n) = 2n - 3$$

represents an arithmetic sequence because $f(x) = 2x - 3$ defines a linear function. The first five terms of this sequence are shown in Table 11.3.

TABLE 11.3

n	1	2	3	4	5
$f(n)$	−1	1	3	5	7

Each time n increases by 1, the next term is 2 more than the previous term. We say that the *common difference* of this arithmetic sequence is $d = 2$. That is, the difference between successive terms equals 2. When these terms are graphed, they lie on the line $y = 2x - 3$, as illustrated in Figure 11.15. Arithmetic sequences are represented by linear functions and so their graphical representations consist of collinear points (points that lie on a line).

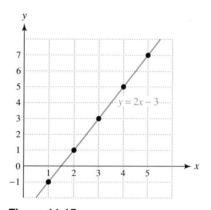

Figure 11.15

Arithmetic Sequence

An **arithmetic sequence** is a linear function given by $a_n = dn + c$ whose domain is the set of natural numbers. The value of d is called the **common difference**.

EXAMPLE 1 *Recognizing arithmetic sequences*

Determine whether f is an arithmetic sequence. If it is, identify the common difference d.
(a) $f(n) = 2 - 3n$.

(b)
n	1	2	3	4	5
$f(n)$	10	5	0	−5	−10

(c) A graph of f is shown in Figure 11.16 on the next page.

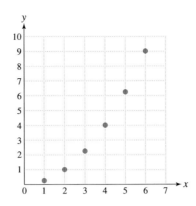

Figure 11.16

Solution

(a) This sequence is linear because $f(n) = -3n + 2$ defines a linear function. The common difference is $d = -3$.
(b) From the table we see that each term is found by adding -5 to the previous term. This represents an arithmetic sequence with common difference $d = -5$.
(c) The sequence shown in Figure 11.16 is not an arithmetic sequence because the points are not collinear. That is, there is no common difference.

■ **MAKING CONNECTIONS**

Common Difference and Slope

The common difference d of an arithmetic sequence equals the slope of the line passing through the collinear points. For example, if $a_n = -2n + 4$, the common difference is -2, and the slope of the line passing through the points on the graph of a_n is also -2 (see Figure 11.17).

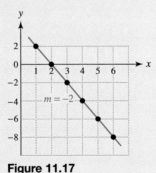

Figure 11.17

EXAMPLE 2 *Finding symbolic representations*

Find the general term a_n for each arithmetic sequence.
(a) $a_1 = 3$ and $d = 4$
(b) $a_1 = 3$ and $a_4 = 12$

Solution

(a) Let $a_n = dn + c$. For $d = 4$, we write $a_n = 4n + c$, and to find c we use $a_1 = 3$.

$$a_1 = 4(1) + c = 3 \quad \text{or} \quad c = -1$$

Thus $a_n = 4n - 1$.

(b) Because $a_1 = 3$ and $a_4 = 12$, the common difference d equals the slope of the line passing through the points $(1, 3)$ and $(4, 12)$, or

$$d = \frac{12 - 3}{4 - 1} = 3.$$

Therefore $a_n = 3n + c$. To find c we use $a_1 = 3$ and obtain

$$a_1 = 3(1) + c = 3 \quad \text{or} \quad c = 0.$$

Thus $a_n = 3n$.

Consider the arithmetic sequence

$$1, 5, 9, 13, 17, 21, 25, 29, \ldots.$$

The common difference is $d = 4$, and the first term is $a_1 = 1$. To find the second term we add $d = 4$ to the first term. To find the third term we add $2d$ to the first term, and to find the fourth term we add $3d$ to the first term a_1. That is,

$$a_1 = 1,$$
$$a_2 = a_1 + 1d = 1 + 1 \cdot 4 = 5,$$
$$a_3 = a_1 + 2d = 1 + 2 \cdot 4 = 9,$$
$$a_4 = a_1 + 3d = 1 + 3 \cdot 4 = 13,$$

and, in general, a_n is determined by

$$a_n = a_1 + (n - 1)d = 1 + (n - 1)4.$$

This result suggests the following formula.

General Term of an Arithmetic Sequence

The nth term a_n of an arithmetic sequence is given by

$$a_n = a_1 + (n - 1)d,$$

where a_1 is the first term and d is the common difference.

EXAMPLE 3   *Finding terms of an arithmetic sequence*

If $a_1 = 5$ and $d = 3$, find a_{54}.

Solution

To find a_{54}, apply the formula $a_n = a_1 + (n - 1)d$.

$$a_{54} = 5 + (54 - 1)3 = 164$$

Representations of Geometric Sequences

If a sequence is defined by an exponential function, it is a *geometric sequence*. For example,

$$f(n) = 3(2)^{n-1}$$

represents a geometric sequence because $f(x) = 3(2)^{x-1}$ defines an exponential function. The first five terms of this sequence are shown in Table 11.4.

TABLE 11.4

n	1	2	3	4	5
$f(n)$	3	6	12	24	48

Successive terms are found by multiplying the previous term by 2. We say that the *common ratio* of this geometric sequence equals 2. Note that the ratios of successive terms are $\frac{6}{3}, \frac{12}{6}, \frac{24}{12},$ and $\frac{48}{24}$ and that they all equal the common ratio 2. When the terms in Table 11.4 are graphed, they do *not* lie on a line. Rather they lie on the exponential curve $y = 3(2)^{x-1}$, as shown in Figure 11.18. Because a geometric sequence is an exponential function, its terms reflect either *exponential growth* or *exponential decay*.

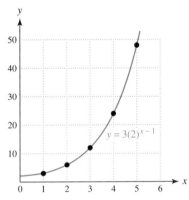

Figure 11.18

Geometric Sequence

A **geometric sequence** is an exponential function given by $a_n = a_1(r)^{n-1}$ whose domain is the set of natural numbers. The value of r is called the **common ratio**, and a_1 is the first term of the sequence.

EXAMPLE 4 *Recognizing geometric sequences*

Determine whether f is a geometric sequence. If it is, identify the common ratio.

(a) $f(n) = 2(0.9)^{n-1}$

(b)

n	1	2	3	4	5
$f(n)$	8	4	2	1	$\frac{1}{2}$

(c) A graph of f is shown in Figure 11.19.

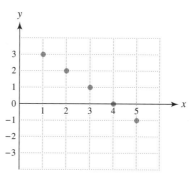

Figure 11.19

Solution

(a) This sequence is geometric because $f(n) = 2(0.9)^{n-1}$ is an exponential function. The common ratio is $r = 0.9$.
(b) The table shows that each successive term is half the previous term. This progression represents a geometric sequence with a common ratio of $r = \frac{1}{2}$.
(c) The sequence shown in Figure 11.19 is not a geometric sequence because the points are collinear. There is no common ratio.

■ **MAKING CONNECTIONS**

Common Ratios and Growth or Decay Factors

If the common ratio r of a geometric sequence is positive, then r equals either the growth factor or the decay factor for an exponential function.

EXAMPLE 5 *Finding symbolic representations*

Find a general term a_n for each geometric sequence.

(a) $a_1 = \frac{1}{2}$ and $r = 5$ (b) $a_1 = 2$, $a_3 = 18$, and $r < 0$.

Solution

(a) Let $a_n = a_1(r)^{n-1}$. Since $a_1 = \frac{1}{2}$ and $r = 5$, we write $a_n = \frac{1}{2}(5)^{n-1}$.
(b) Since $a_1 = 2$ and $a_3 = 18$,
$$\frac{a_3}{a_1} = \frac{18}{2} = 9.$$
Now $a_3 = a_1 \cdot r^2$ because a_3 is obtained from multiplying a_1 by r twice. Thus
$$r^2 = 9 \quad \text{or} \quad r = \pm 3.$$
It is specified that $r < 0$, so $r = -3$ and $a_n = 2(-3)^{n-1}$.

Critical Thinking

If we are given a_1 and a_5, can we determine the common ratio of a geometric series? Explain.

EXAMPLE 6 *Finding a term of a geometric sequence*

If $a_1 = 5$ and $r = 3$, find a_{10}.

Solution

To find a_{10}, apply the formula $a_n = a_1(r)^{n-1}$ with $a_1 = 5$, $r = 3$, and $n = 10$.
$$a_{10} = 5(3)^{10-1} = 5(3)^9 = 98{,}415$$

APPLICATIONS AND MODELS

Sequences are frequently used to describe a variety of situations. In the next example, we use a sequence to model classroom ventilation.

EXAMPLE 7 *Modeling classroom ventilation*

Ventilation is an effective means for removing indoor air pollutants. According to the American Society of Heating, Refrigerating, and Air-Conditioning Engineers (ASHRAE), a classroom should have a ventilation rate of 900 cubic feet per hour per person.

(a) Write a sequence that gives the hourly ventilation necessary for 1, 2, 3, 4, and 5 people in a classroom. Is this sequence arithmetic, geometric, or neither?
(b) Write a symbolic representation for this sequence. Why is it reasonable to limit the domain to natural numbers?
(c) Find a_{30} and interpret the result.

Solution

(a) One person requires 900 cubic feet of air circulated per hour, two people require 1800, three people 2700, and so on. The first five terms of this sequence are
$$900, 1800, 2700, 3600, 4500.$$
This sequence is arithmetic, with a common difference of 900.
(b) The nth term equals $900n$, so we let $a_n = 900n$. Because we cannot have a fraction of a person, limiting the domain to the natural numbers is reasonable.
(c) The result $a_{30} = 900(30) = 27{,}000$ indicates that a classroom with 30 people should have a ventilation rate of 27,000 cubic feet per hour.

Chlorine is frequently added to the water to disinfect swimming pools. The chlorine concentration should remain between 1.5 and 2.5 parts per million (ppm). On a warm, sunny day 30% of the chlorine may dissipate from the water. In the next example we use a sequence to model the amount of chlorine in a pool at the beginning of each day. (*Source:* D. Thomas, *Swimming Pool Operators Handbook.*)

11.2 Arithmetic and Geometric Sequences

EXAMPLE 8 *Modeling chlorine in a swimming pool*

A swimming pool on a warm, sunny day begins with a high chlorine content of 4 parts per million.
(a) Write a sequence that models the amount of chlorine in the pool at the beginning of the first 3 days, assuming that no additional chlorine is added and that the days are warm and sunny. Is this sequence arithmetic, geometric, or neither?
(b) Write a symbolic representation for this sequence.
(c) At the beginning of what day does the chlorine first drop below 1.5 parts per million?

Solution

(a) Because 30% of the chlorine dissipates, 70% remains in the water at the beginning of the next day. If the concentration at the beginning of the first day is 4 parts per million, then at the beginning of the second day it is

$$4 \cdot 0.70 = 2.8 \text{ parts per million,}$$

and at the start of the third day it is

$$2.8 \cdot 0.70 = 1.96 \text{ parts per million.}$$

The first three terms are

$$4, 2.8, 1.96.$$

Successive terms are found by multiplying the previous term by 0.7. Thus the sequence is geometric, with a common ratio of 0.7.

(b) The initial amount is $a_1 = 4$ and the common ratio is $r = 0.7$, so the sequence can be represented by $a_n = 4(0.7)^{n-1}$.

(c) The table shown in Figure 11.20 reveals that $a_4 = 4(0.7)^{4-1} \approx 1.372 < 1.5$. Thus, at the beginning of the fourth day, the chlorine level in the swimming pool will have dropped below the recommended minimum of 1.5 parts per million.

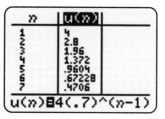

Figure 11.20

11.2 Putting It All Together

In this section we discussed two types of sequences: arithmetic and geometric. Arithmetic sequences are linear functions, and geometric sequences are exponential functions. The inputs for both are limited to the natural numbers. Each successive term in an arithmetic sequence is found by adding the common difference d to the previous term. For a geometric sequence each successive term is found by multiplying the previous term by the common ratio r. The graph of an arithmetic sequence consists of points that lie on a line, whereas the graph of a geometric sequence (with a positive common ratio) consists of points that lie on an exponential curve. Examples are shown in the following table.

Sequence	Formula	Example
Arithmetic	$a_n = dn + c$ or $a_n = a_1 + (n-1)d$, where d is the common difference and a_1 is the first term.	If $a_n = 5n + 2$, then the common difference is $d = 5$ and the terms of the sequence are $7, 12, 17, 22, 27, 32, 37, \ldots$ Each term is found by adding 5 to the previous term. This sequence can also be written as $a_n = 7 + 5(n-1)$.
Geometric	$a_n = a_1(r)^{n-1}$, where r is the common ratio and a_1 is the first term.	If $a_n = 4(-2)^{n-1}$, then the common ratio is $r = -2$ and the first term is $a_1 = 4$. The terms of the sequence are $4, -8, 16, -32, 64, -128, 256, \ldots$ Each term is found by multiplying the previous term by -2.

11.2 EXERCISES

CONCEPTS

1. Give an example of an arithmetic sequence. State the common difference.

2. Give an example of a geometric sequence. State the common ratio.

3. To find successive terms in an arithmetic sequence, _____ the common difference to the _____ term.

4. To find successive terms in a geometric sequence, _____ the previous term by the _____.

5. Find the next term in the arithmetic sequence 3, 7, 11, 15. What is the common difference?

6. Find the next term in the geometric sequence 2, −4, 8, −16. What is the common ratio?

7. Write the general term a_n for a geometric sequence, using a_1 and r.

8. Write the general term a_n for an arithmetic sequence using a_1 and d.

ARITHMETIC SEQUENCES

Exercises 9–26: (Refer to Example 1.) Determine whether f is an arithmetic sequence. Identify the common difference when possible.

9. $f(n) = 10n - 5$ 10. $f(n) = -3n - 5$

11. $f(n) = 6 - n$ 12. $f(n) = 6 + \dfrac{1}{2}n$

11.2 Arithmetic and Geometric Sequences

13. $f(n) = n^3 + 1$

14. $f(n) = 5\left(\dfrac{1}{3}\right)^{n-1}$

15.
n	1	2	3	4
$f(n)$	3	6	9	12

16.
n	1	2	3	4
$f(n)$	-7	-5	-3	-1

17.
n	1	2	3	4
$f(n)$	10	7	4	1

18.
n	1	2	3	4
$f(n)$	1	2	4	8

19.
n	1	2	3	4
$f(n)$	-4	0	8	12

20.
n	1	2	3	4
$f(n)$	1	2.5	4	5.5

21.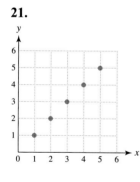

22.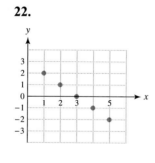

23.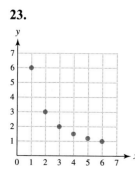

24.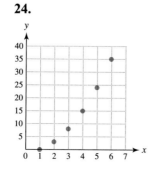

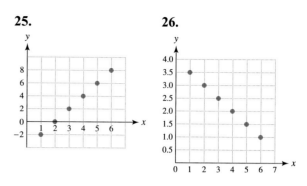

25.

26.

Exercises 27–32: (Refer to Example 2.) Find the general term a_n for the arithmetic sequence.

27. $a_1 = 7$ and $d = -2$

28. $a_1 = 5$ and $a_2 = 9$

29. $a_1 = -2$ and $a_3 = 6$

30. $a_2 = 7$ and $a_3 = 10$

31. $a_8 = 16$ and $a_{12} = 8$

32. $a_3 = 7$ and $d = -5$

Exercises 33–36: (Refer to Example 3.)

33. If $a_1 = -3$ and $d = 2$, find a_{32}.

34. If $a_1 = 2$ and $d = -3$, find a_{19}.

35. If $a_1 = -3$ and $a_2 = 0$, find a_9.

36. If $a_3 = -3$ and $d = 4$, find a_{62}.

GEOMETRIC SEQUENCES

Exercises 37–52: (Refer to Example 4.) Determine whether f is a geometric sequence. Identify the common ratio when possible.

37. $f(n) = 3^n$

38. $f(n) = 2(4)^n$

39. $f(n) = \dfrac{2}{3}(0.8)^{n-1}$

40. $f(n) = 7 - 3n$

41. $f(n) = 2(n-1)^2$

42. $f(n) = 2\left(-\dfrac{3}{4}\right)^{n-1}$

43.
n	1	2	3	4
$f(n)$	2	4	8	16

44.
n	1	2	3	4
$f(n)$	-6	3	-1.5	0.75

45.

n	1	2	3	4
f(n)	1	4	9	16

46.

n	1	2	3	4
f(n)	7	4	-1	-8

47.

n	1	2	3	4
f(n)	2	8	32	128

48.

n	1	2	3	4
f(n)	1	$\frac{1}{2}$	$\frac{1}{4}$	$\frac{1}{8}$

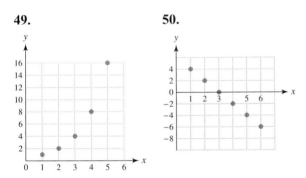

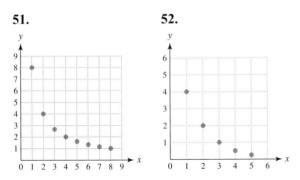

Exercises 53–58: (Refer to Example 5.) Find the general term a_n for the geometric sequence.

53. $a_1 = 1.5$ and $r = 4$ **54.** $a_1 = 3$ and $r = \frac{1}{4}$

55. $a_1 = -3$ and $a_2 = 6$

56. $a_1 = 2$ and $a_4 = 54$

57. $a_1 = 1$, $a_3 = 16$, and $r > 0$

58. $a_2 = 3$, $a_4 = 12$, and $r < 0$

Exercises 59–62: (Refer to Example 6.)

59. If $a_1 = 2$ and $r = 3$, find a_8.

60. If $a_1 = 4$ and $a_2 = 2$, find a_9.

61. If $a_1 = -1$ and $a_2 = 3$, find a_6.

62. If $a_3 = 5$ and $r = -3$, find a_7.

APPLICATIONS

63. *Room Ventilation* (Refer to Example 7.) In areas such as bars and lounges that allow smoking, the ventilation rate should be 3000 cubic feet per hour per person. (*Source:* **ASHRAE.**)
 (a) Write a sequence that gives the hourly ventilation necessary for 1, 2, 3, 4, and 5 people in a barroom. Is this sequence arithmetic, geometric, or neither?
 (b) Write a symbolic representation for this sequence.
 (c) Find a_{20} and interpret the result.
 (d) Give a graphical representation for this sequence, using $n = 1, 2, 3, \ldots, 8$. Are the points collinear?

64. *Chlorine in Swimming Pools* (Refer to Example 8.) Suppose that the water in a swimming pool initially has a chlorine content of 3 parts per million and that 20% of the chlorine dissipates each day.
 (a) If no additional chlorine is added, write a symbolic representation for a sequence that gives the chlorine concentration at the beginning of each day.
 (b) Give a graphical representation for this sequence, using $n = 1, 2, 3, \ldots, 8$. Are the points collinear? Is this sequence arithmetic, geometric, or neither?

65. *Salary* Suppose that an employee receives a $2000 raise each year and that the sequence a_n models the employee's salary after n years. Is this sequence arithmetic, geometric, or neither? Explain.

66. *Salary* Suppose that an employee receives a 7% increase in salary each year and that the sequence a_n models the employee's salary after n years. Is this sequence arithmetic, geometric, or neither? Explain.

11.2 Arithmetic and Geometric Sequences

67. *Appreciation of Lake Property* A certain type of lake home in northern Minnesota is increasing in value by 15% per year. Let the sequence a_n give the value of this type of lake home at the beginning of year n.
(a) Is a_n arithmetic, geometric, or neither? Explain your reasoning.
(b) Write a symbolic representation for this sequence if $a_1 = \$100{,}000$.
(c) Find a_7 and interpret the result.
(d) Give a graphical representation for a_n, where $n = 1, 2, 3, \ldots, 10$.

68. *Falling Object* The total distance D_n that an object falls in n seconds is shown in the table. Is the sequence arithmetic, geometric, or neither? Explain your reasoning.

n (seconds)	1	2	3	4	5
D_n (feet)	16	64	144	256	400

69. *Theater Seating* A theater has 40 seats in the first row, 42 seats in the second row, 44 seats in the third row, and so on.
(a) Can the number of seats in each row be modeled by an arithmetic or geometric sequence? Explain.
(b) Find a symbolic representation for a sequence a_n that gives the number of seats in row n.
(c) How many seats are there in row 20?

70. *Bouncing Ball* A tennis ball bounces back to 85% of the height from which it was dropped and then to 85% of the height of each successive bounce.
(a) Write a symbolic representation of a sequence a_n that gives the maximum height of the ball on the nth bounce. Let $a_1 = 5$ feet.
(b) Is the sequence arithmetic or geometric? Explain.
(c) Find a_8 and interpret the result.

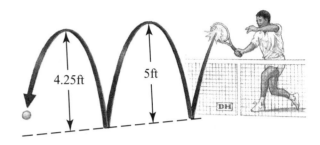

WRITING ABOUT MATHEMATICS

71. If you have a numerical representation for a sequence, how can you determine whether it is geometric? Give an example.

72. If you have a graphical representation for a sequence, how can you determine whether it is arithmetic? Give an example.

CHECKING BASIC CONCEPTS FOR SECTIONS 11.1 AND 11.2

1. Write the first four terms of the sequence defined by $a_n = \dfrac{n}{n+4}$.

2. Represent the sequence $a_n = n + 1$ graphically and numerically for $n = 1, 2, 3, 4, 5$.

3. Use the table to determine whether the sequence is arithmetic or geometric. Write a symbolic representation for the sequence.

(a)

n	1	2	3	4	5
a_n	−2	1	4	7	10

(b)

n	1	2	3	4	5
a_n	3	−6	12	−24	48

4. Find the general term a_n for an arithmetic sequence with $a_1 = 5$ and $d = 2$.

5. Find the general term a_n for a geometric sequence with $a_1 = 5$ and $r = 2$.

11.3 SERIES

Basic Concepts ~ Arithmetic Series ~ Geometric Series ~ Summation Notation

INTRODUCTION

Although the terms *sequence* and *series* are sometimes used interchangeably in everyday life, they represent different mathematical concepts. In mathematics a sequence is a function whose domain is the set of natural numbers, whereas a series is a summation of the terms in a sequence. Series have played a central role in the development of modern mathematics. Today series are often used to approximate functions that are too complicated to have formulas. Series are also used to calculate accurate approximations for numbers such as π and e.

BASIC CONCEPTS

Suppose that a person has a starting salary of $30,000 per year and receives a $2,000 raise each year. Then the sequence

$$30{,}000,\ 32{,}000,\ 34{,}000,\ 36{,}000,\ 38{,}000$$

models these salaries over a five-year period. The total amount earned is given by the *series*

$$30{,}000 + 32{,}000 + 34{,}000 + 36{,}000 + 38{,}000,$$

whose sum is $170,000. We now define the concept of a series.

Finite Series

A **finite series** is an expression of the form

$$a_1 + a_2 + a_3 + \cdots + a_n.$$

EXAMPLE 1 *Computing total reported AIDS cases*

Table 11.5 presents a numerical representation of a sequence a_n that computes the number of AIDS cases diagnosed each year from 1991 through 1997, where $n = 1$ corresponds to 1991.

TABLE 11.5

n	1	2	3	4	5	6	7
a_n	60,124	79,054	79,049	71,209	66,233	54,656	31,153

Source: Department of Health and Human Services.

(a) Write a series whose sum represents the total number of AIDS cases diagnosed from 1991 to 1997. Find its sum.

(b) Interpret the sum $a_1 + a_2 + a_3 + \cdots + a_{10}$.

Solution

(a) A series that represents the total number of AIDS cases diagnosed from 1991 through 1997 is

$$60{,}124 + 79{,}054 + 79{,}049 + 71{,}209 + 66{,}233 + 54{,}656 + 31{,}153 = 441{,}478.$$

Thus 441,478 AIDS cases were diagnosed from 1991 through 1997.

(b) The series $a_1 + a_2 + a_3 + \cdots + a_{10}$ represents the total number of AIDS cases diagnosed from 1991 through 2000.

ARITHMETIC SERIES

Summing the terms of an arithmetic sequence results in an **arithmetic series**. For example, $a_n = 2n - 1$ for $n = 1, 2, 3, \ldots, 7$ defines the arithmetic sequence

$$1, 3, 5, 7, 9, 11, 13.$$

The corresponding arithmetic *series* is

$$1 + 3 + 5 + 7 + 9 + 11 + 13,$$

whose sum is 49. The following formula gives the sum of the first n terms of an arithmetic sequence.

Sum of the First n Terms of an Arithmetic Sequence

The **sum of the first n terms of an arithmetic sequence**, denoted S_n, is found by averaging the first and nth terms and then multiplying by n. That is,

$$S_n = a_1 + a_2 + a_3 + \cdots + a_n = n\left(\frac{a_1 + a_n}{2}\right).$$

The series $1 + 3 + 5 + 7 + 9 + 11 + 13$ consists of 7 terms, where the first term is 1 and the last term is 13. Substituting into the formula gives

$$S_7 = 7\left(\frac{1 + 13}{2}\right) = 49,$$

which agrees with the sum obtained by adding the 7 terms.

Because $a_n = a_1 + (n - 1)d$ for an arithmetic sequence, S_n can also be written

$$S_n = n\left(\frac{a_1 + a_n}{2}\right)$$

$$= \frac{n}{2}(a_1 + a_n)$$

$$= \frac{n}{2}(a_1 + a_1 + (n - 1)d)$$

$$= \frac{n}{2}(2a_1 + (n - 1)d).$$

EXAMPLE 2 *Finding the sum of a finite arithmetic series*

GCLM Suppose that a person has a starting annual salary of $30,000 and receives a $1500 raise each year.
(a) Calculate the total amount earned after 10 years.
(b) Verify this value with a calculator.

Solution

(a) The sequence describing the salaries during year n is given by
$$a_n = 30{,}000 + 1500(n - 1).$$
One way to calculate the sum of the first 10 terms, denoted S_{10}, is to find a_1 and a_{10}, or
$$a_1 = 30{,}000 + 1500(1 - 1) = 30{,}000$$
$$a_{10} = 30{,}000 + 1500(10 - 1) = 43{,}500.$$
Thus the total amount earned during this 10-year period is
$$S_{10} = 10\left(\frac{a_1 + a_{10}}{2}\right)$$
$$= 10\left(\frac{30{,}000 + 43{,}500}{2}\right)$$
$$= \$367{,}500.$$

This sum can also be found using the second formula, or
$$S_n = \frac{n}{2}(2a_1 + (n - 1)d)$$
$$= \frac{10}{2}(2 \cdot 30{,}000 + (10 - 1)1500)$$
$$= 5(60{,}000 + 9 \cdot 1500)$$
$$= \$367{,}500.$$

(b) To verify this result with a calculator, let $a_n = 30{,}000 + 1500(n - 1)$. The value 367,500 for S_{10} is shown in Figure 11.21. The "seq(" utility, found on some calculators under the LIST OPS menus, generates a sequence of values. The "sum(" utility found on some calculators under the LIST MATH menus, calculates the sum of the sequence inside the parentheses.

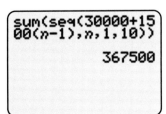

Figure 11.21

36. $1 + \dfrac{1}{10} + \dfrac{1}{100} + \dfrac{1}{1000} + \dfrac{1}{10{,}000}$

37. Verify that $\displaystyle\sum_{k=1}^{n} k = \dfrac{n(n+1)}{2}$ with a formula for the sum of the first n terms of a finite arithmetic series.

38. Use Exercise 37 to find the sum of the series $\displaystyle\sum_{k=1}^{200} k$.

APPLICATIONS

39. *Prison Escapees* The table lists the number of escapees from state prisons each year.

Year	1990	1991	1992
Escapees	8518	9921	10,706

Year	1993	1994	1995
Escapees	14,035	14,307	12,249

Source: Bureau of Justice Statistics.

(a) Write a series whose sum is the total number of escapees from 1990 to 1995.
(b) Find its sum.

40. *Captured Prison Escapees* (Refer to Exercise 39.) The table lists the number of escapees from state prisons who were captured, including inmates who may have escaped during a previous year.

Year	1990	1991	1992
Captured	9324	9586	10,031

Year	1993	1994	1995
Captured	12,872	13,346	12,166

Source: Bureau of Justice Statistics.

(a) Write a series whose sum is the total number of escapees captured from 1990 to 1995.
(b) Find its sum.
(c) Compare the number of escapees to the number captured during this time period.

41. *Area* A sequence of smaller squares is formed by connecting the midpoint of the sides of a larger square as shown in the figure.

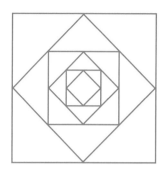

(a) If the area of the largest square is 1 square unit, give the first five terms of a sequence that describes the area of each successive square.
(b) Use a geometric series to sum the areas of the first 10 squares.

42. *Perimeter* (Refer to Exercise 41.) Use a geometric series to find the sum of the perimeters of the first 10 squares.

43. *Stacking Logs* A stack of logs is made in layers, with one log less in each layer, as shown in the accompanying figure. If the top layer has 6 logs and the bottom row has 14 logs, what is the total number of logs in the pile? Use a formula to find this sum.

44. *Stacking Logs* (Refer to Exercise 43.) Suppose that a stack of logs has 15 logs in the top layer and a total of 10 layers. How many logs are in the stack?

45. *Salaries* Suppose that an individual's starting salary is $35,000 per year and that the individual receives a $2000 raise each year. Find the total amount earned over 20 years.

46. *Salaries* Suppose that an individual's starting salary is $35,000 per year and that the individual receives a 10% raise each year. Find the total amount earned over 20 years.

47. *Bouncing Ball* A tennis ball first bounces to 75% of the height from which it was dropped and then to 75% of the height of each successive bounce. If it is dropped from a height of 10 feet, find the distance it *falls* between the fourth and fifth bounce.

48. *Bouncing Ball* A tennis ball first bounces to 75% of the height from which it was dropped and then to 75% of the height of each successive bounce. If it is dropped from a height of 10 feet, find the *total* distance it travels before it reaches its fifth bounce. (*Hint:* Make a sketch.)

WRITING ABOUT MATHEMATICS

49. Discuss the difference between a sequence and a series. Give an example of each.

50. Suppose that an arithmetic series has $a_1 = 1$ and a common difference of $d = 2$, whereas a geometric series has $a_1 = 1$ and a common ratio of $r = 2$. Discuss how their sums compare as the number of terms n becomes large. (*Hint:* Calculate each sum for $n = 10, 20, 30$.)

GROUP ACTIVITY
Working with Real Data

Directions: Form a group of 2 to 4 people. Select someone to record the group's responses for this activity. All members of the group should work cooperatively to answer the questions. If your instructor asks for your results, each member of the group should be prepared to respond. If the group is asked to turn in its work, be sure to include each person's name on the paper.

Depreciation For tax purposes, businesses frequently depreciate equipment. Two different methods of depreciation are called *straight-line depreciation* and *sum-of-the-years'-digits*. Suppose that a college student buys a $3000 computer to start a business that provides Internet services. This student estimates the life of the computer at 4 years, after which its value will be $200. The difference between $3000 and $200, or $2800, may be deducted from the student's taxable income over a 4-year period.

In straight-line depreciation, equal portions of $2800 are deducted each year over the 4 years. The sum-of-the-years'-digits method calculates depreciation differently. For a computer having a useful life of 4 years, the sum of the years is computed by

$$1 + 2 + 3 + 4 = 10.$$

With this method, $\frac{4}{10}$ of $2800 is deducted the first year, $\frac{3}{10}$ the second year, and so on, until $\frac{1}{10}$ is deducted the fourth year. Both depreciation methods yield a total deduction of $2800 over the 4 years. (*Source:* Sharp Electronics Corporation, *Conquering the Sciences.*)

(a) Find an arithmetic sequence that gives the amount depreciated each year by each method.

(b) Write a series whose sum is the amount depreciated over 4 years by each method.

11.4 THE BINOMIAL THEOREM

Pascal's Triangle ~ Factorial Notation and Binomial Coefficients ~ Using the Binomial Theorem

INTRODUCTION

In this section we demonstrate how to expand expressions of the form $(a + b)^n$, where n is a natural number. These expressions occur in statistics, finite mathematics, computer science, and calculus. The two methods that we discuss are Pascal's triangle and the binomial theorem.

PASCAL'S TRIANGLE

Expanding $(a + b)^n$ for increasing values of n gives the following results.

$$
\begin{aligned}
(a + b)^0 &= 1 \\
(a + b)^1 &= 1a + 1b \\
(a + b)^2 &= 1a^2 + 2ab + 1b^2 \\
(a + b)^3 &= 1a^3 + 3a^2b + 3ab^2 + 1b^3 \\
(a + b)^4 &= 1a^4 + 4a^3b + 6a^2b^2 + 4ab^3 + 1b^4 \\
(a + b)^5 &= 1a^5 + 5a^4b + 10a^3b^2 + 10a^2b^3 + 5ab^4 + 1b^5
\end{aligned}
$$

Note that $(a + b)^1$ has two terms, starting with a and ending with b; $(a + b)^2$ has three terms, starting with a^2 and ending with b^2; and in general, $(a + b)^n$ has $n + 1$ terms, starting with a^n and ending with b^n. The exponent on a decreases by 1 each successive term, and the exponent on b increases by 1 each successive term.

The triangle formed by the highlighted numbers is called **Pascal's triangle**. This triangle consists of 1s along the sides, and each element inside the triangle is the sum of the two numbers above it, as shown in Figure 11.25. Pascal's triangle is usually written without variables and can be extended to include as many rows as needed.

$$
\begin{array}{c}
1 \\
1 \quad 1 \\
1 \quad 2 \quad 1 \\
1 \quad 3 \quad 3 \quad 1 \\
1 \quad 4 \quad 6 \quad 4 \quad 1 \\
1 \quad 5 \quad 10 \quad 10 \quad 5 \quad 1
\end{array}
$$

Figure 11.25 Pascal's Triangle

We can use this triangle to expand $(a + b)^n$, where n is a natural number. The expression $(m + n)^4$ consists of five terms written as

$$(m + n)^4 = \underline{}m^4 + \underline{}m^3n^1 + \underline{}m^2n^2 + \underline{}m^1n^3 + \underline{}n^4.$$

Because there are five terms, the coefficients can be found in the fifth row of Pascal's triangle, which is

$$1 \quad 4 \quad 6 \quad 4 \quad 1.$$

Thus

$$
\begin{aligned}
(m + n)^4 &= \underline{1}\,m^4 + \underline{4}\,m^3n^1 + \underline{6}\,m^2n^2 + \underline{4}\,m^1n^3 + \underline{1}\,n^4 \\
&= m^4 + 4m^3n + 6m^2n^2 + 4mn^3 + n^4.
\end{aligned}
$$

EXAMPLE 1 *Expanding a binomial*

Expand each binomial, using Pascal's triangle.
(a) $(x + 2)^5$ (b) $(2m - n)^3$

Solution

(a) To find the coefficients, use the sixth row in Pascal's triangle.
$$(x + 2)^5 = \underline{1}\,x^5 + \underline{5}\,x^4 \cdot 2^1 + \underline{10}\,x^3 \cdot 2^2 + \underline{10}\,x^2 \cdot 2^3 + \underline{5}\,x^1 \cdot 2^4 + \underline{1}\,2^5$$
$$= x^5 + 10x^4 + 40x^3 + 80x^2 + 80x + 32$$

(b) To find the coefficients, use the fourth row in Pascal's triangle.
$$(2m - n)^3 = \underline{1}\,(2m)^3 + \underline{3}\,(2m)^2(-n)^1 + \underline{3}\,(2m)^1(-n)^2 + \underline{1}\,(-n)^3$$
$$= 8m^3 - 12m^2n + 6mn^2 - n^3$$

FACTORIAL NOTATION AND BINOMIAL COEFFICIENTS

An alternative to Pascal's triangle is the binomial theorem, which makes use of **factorial notation**.

---

n Factorial ($n!$)

For any positive integer,
$$n! = 1 \cdot 2 \cdot 3 \cdot \cdots \cdot n.$$
We also define $0! = 1$.

Examples include the following.
$$0! = 1$$
$$1! = 1$$
$$2! = 1 \cdot 2 = 2$$
$$3! = 1 \cdot 2 \cdot 3 = 6$$
$$4! = 1 \cdot 2 \cdot 3 \cdot 4 = 24$$
$$5! = 1 \cdot 2 \cdot 3 \cdot 4 \cdot 5 = 120$$

Figures 11.26 and 11.27 support these results. On some calculators, factorial (!) can be accessed in the MATH PRB menus.

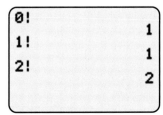

Figure 11.26

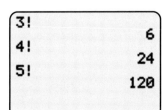

Figure 11.27

11.4 The Binominal Theorem

EXAMPLE 2 *Evaluating factorial expressions*

Simplify the expression.

(a) $\dfrac{5!}{3!\,2!}$ (b) $\dfrac{4!}{4!\,0!}$

Solution

(a) $\dfrac{5!}{3!\,2!} = \dfrac{1 \cdot 2 \cdot 3 \cdot 4 \cdot 5}{(1 \cdot 2 \cdot 3)(1 \cdot 2)} = \dfrac{120}{6 \cdot 2} = 10$

(b) $\dfrac{4!}{4!\,0!} = \dfrac{24}{24 \cdot 1} = 1$ because $0! = 1$

The expression $_nC_r$ represents the **binomial coefficient** that can be used to calculate the numbers in Pascal's triangle.

The Binomial Coefficient $_nC_r$

For n and r nonnegative integers, $n \geq r$,

$$_nC_r = \dfrac{n!}{(n-r)!\,r!}.$$

Values of $_nC_r$ for $r = 0, 1, 2, \ldots, n$ correspond to the $n+1$ numbers in row $n+1$ of Pascal's triangle.

EXAMPLE 3 *Calculating $_nC_r$*

Calculate $_3C_r$ for $r = 0, 1, 2, 3$ by hand. Check your results on a calculator. Compare these numbers with the fourth row in Pascal's triangle.

Solution

$$_3C_0 = \dfrac{3!}{(3-0)!\,0!} = \dfrac{6}{6 \cdot 1} = 1$$

$$_3C_1 = \dfrac{3!}{(3-1)!\,1!} = \dfrac{6}{2 \cdot 1} = 3$$

$$_3C_2 = \dfrac{3!}{(3-2)!\,2!} = \dfrac{6}{1 \cdot 2} = 3$$

$$_3C_3 = \dfrac{3!}{(3-3)!\,3!} = \dfrac{6}{1 \cdot 6} = 1$$

These results are supported in Figures 11.28 and 11.29 on the next page. The fourth row of Pascal's triangle is

$$1 \quad 3 \quad 3 \quad 1,$$

which agrees with the calculated values for $_3C_r$. On some calculators, the MATH PRB menus are used to calculate $_nC_r$.

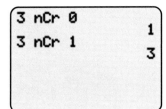

Figure 11.28

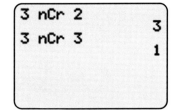

Figure 11.29

USING THE BINOMIAL THEOREM

The binomial coefficients can be used to expand expressions of the form $(a + b)^n$. To do so, we use the **binomial theorem**.

> **Binomial Theorem**
>
> For any positive integer n and any numbers a and b,
> $$(a + b)^n = {_nC_0}a^n + {_nC_1}a^{n-1}b^1 + \cdots + {_nC_{n-1}}a^1b^{n-1} + {_nC_n}b^n.$$

Using the results of Example 3 we write
$$(a + b)^3 = {_3C_0}a^3 + {_3C_1}a^2b^1 + {_3C_2}a^1b^2 + {_3C_3}b^3$$
$$= 1a^3 + 3a^2b + 3ab^2 + 1b^3$$
$$= a^3 + 3a^2b + 3ab^2 + b^3.$$

EXAMPLE 4 *Expanding a binomial*

Use the binomial theorem to expand each expression.
(a) $(x + y)^5$ (b) $(3 - 2x)^4$

Solution

(a) The coefficients are calculated as follows.

$${_5C_0} = \frac{5!}{(5-0)!0!} = 1, \quad {_5C_1} = \frac{5!}{(5-1)!1!} = 5, \quad {_5C_2} = \frac{5!}{(5-2)!2!} = 10$$

$${_5C_3} = \frac{5!}{(5-3)!3!} = 10, \quad {_5C_4} = \frac{5!}{(5-4)!4!} = 5, \quad {_5C_5} = \frac{5!}{(5-5)!5!} = 1$$

Using the binomial theorem, we arrive at the following result.
$$(x + y)^5 = {_5C_0}x^5 + {_5C_1}x^4y^1 + {_5C_2}x^3y^2 + {_5C_3}x^2y^3 + {_5C_4}x^1y^4 + {_5C_5}y^5$$
$$= 1x^5 + 5x^4y + 10x^3y^2 + 10x^2y^3 + 5xy^4 + 1y^5$$
$$= x^5 + 5x^4y + 10x^3y^2 + 10x^2y^3 + 5xy^4 + y^5$$

(b) The coefficients are calculated as follows.

$$_4C_0 = \frac{4!}{(4-0)!0!} = 1, \quad _4C_1 = \frac{4!}{(4-1)!1!} = 4, \quad _4C_2 = \frac{4!}{(4-2)!2!} = 6,$$

$$_4C_3 = \frac{4!}{(4-3)!3!} = 4, \quad _4C_4 = \frac{4!}{(4-4)!4!} = 1$$

Using the binomial theorem with $a = 3$ and $b = (-2x)$, we arrive at the following result.

$$(3 - 2x)^4 = {_4C_0}(3)^4 + {_4C_1}(3)^3(-2x) + {_4C_2}(3)^2(-2x)^2 + {_4C_3}(3)(-2x)^3$$
$$+ {_4C_4}(-2x)^4$$
$$= 1(81) + 4(27)(-2x) + 6(9)(4x^2) + 4(3)(-8x^3) + 1(16x^4)$$
$$= 81 - 216x + 216x^2 - 96x^3 + 16x^4$$

11.4 Putting It All Together

In this section we learned how to expand the expression $(a + b)^n$ by using Pascal's triangle and the binomial theorem. The following table outlines important topics from this section.

Topic	Explanation	Example
Pascal's triangle	$$\begin{array}{c}1\\1\quad1\\1\quad2\quad1\\1\quad3\quad3\quad1\\1\quad4\quad6\quad4\quad1\\1\quad5\quad10\quad10\quad5\quad1\end{array}$$ To expand $(a + b)^n$, use row $n + 1$ in the triangle.	$(a + b)^3 = 1a^3 + 3a^2b + 3ab^2 + 1b^3$ (Row 4)
Factorial notation	The expression $n!$ equals $1 \cdot 2 \cdot 3 \cdot \cdots \cdot n$.	$5! = 1 \cdot 2 \cdot 3 \cdot 4 \cdot 5 = 120.$
Binomial coefficient $_nC_r$	$_nC_r = \dfrac{n!}{(n-r)!r!}$	$_6C_4 = \dfrac{6!}{(6-4)!4!} = \dfrac{6!}{2!4!} = \dfrac{720}{2 \cdot 24} = 15$
Binomial theorem	$(a + b)^n = {_nC_0}a^n + {_nC_1}a^{n-1}b^1$ $+ \cdots + {_nC_{n-1}}a^1b^{n-1} + {_nC_n}b^n$	$(a + b)^4 = {_4C_0}a^4 + {_4C_1}a^3b + {_4C_2}a^2b^2$ $+ {_4C_3}ab^3 + {_4C_4}b^4$ $= 1a^4 + 4a^3b + 6a^2b^2 + 4ab^3 + 1b^4$ $= a^4 + 4a^3b + 6a^2b^2 + 4ab^3 + b^4$

11.4 EXERCISES

CONCEPTS

1. How many terms result from expanding $(a + b)^4$?

2. How many terms result from expanding $(a + b)^n$?

3. To find the coefficients for the expansion of $(a + b)^3$, what row of Pascal's triangle do you use?

4. Write down the first 5 rows of Pascal's triangle.

5. $4! = $ _____

6. $1 \cdot 2 \cdot 3 \cdot 4 \cdot 5 \cdot 6 = $ _____

7. $_nC_r = $ _____

8. $(a + b)^2 = $ _____

USING PASCAL'S TRIANGLE

Exercises 9–16: Use Pascal's triangle to expand the expression.

9. $(x + y)^3$
10. $(x + y)^4$
11. $(2x + 1)^4$
12. $(2x - 1)^4$
13. $(a - b)^5$
14. $(3x + 2y)^3$
15. $(x^2 + 1)^3$
16. $\left(\dfrac{1}{2} - x^2\right)^5$

FACTORIALS AND BINOMIAL COEFFICIENTS

Exercises 17–32: Evaluate the expression.

17. $3!$
18. $6!$
19. $\dfrac{4!}{3!}$
20. $\dfrac{6!}{3!}$
21. $\dfrac{2!}{0!}$
22. $\dfrac{5!}{1!}$
23. $\dfrac{6!}{5!\,1!}$
24. $\dfrac{3!}{2!\,1!}$
25. $\dfrac{5!}{2!\,3!}$
26. $\dfrac{6!}{4!\,2!}$
27. $_5C_4$
28. $_3C_1$
29. $_6C_5$
30. $_2C_2$
31. $_4C_0$
32. $_4C_3$

Exercises 33–38: Evaluate the binomial coefficient with a calculator.

33. $_{12}C_7$
34. $_{13}C_8$
35. $_9C_5$
36. $_{25}C_{14}$
37. $_{19}C_{11}$
38. $_{10}C_6$

THE BINOMIAL THEOREM

Exercises 39–48: Use the binomial theorem to expand the expression.

39. $(m + n)^3$
40. $(m + n)^5$
41. $(x - y)^4$
42. $(1 - 3x)^4$
43. $(2a + 1)^3$
44. $(x^2 - 1)^3$
45. $(x + 2)^5$
46. $(a - 3)^5$
47. $(2x - y)^3$
48. $(2a + 3b)^4$

Exercises 49–54: The $(r + 1)$st term of $(a + b)^n$, $0 \leq r \leq n$, is given by $_nC_r a^{n-r} b^r$. Find the specified term.

49. The first term of $(a + b)^8$
50. The second term of $(a - b)^{10}$
51. The fourth term of $(x + y)^7$
52. The sixth term of $(a + b)^9$
53. The first term of $(2m + n)^9$
54. The eighth term of $(2a - b)^8$

WRITING ABOUT MATHEMATICS

55. Explain how to find the numbers in Pascal's triangle.

56. Compare the expansion of $(a + b)^n$ to the expansion of $(a - b)^n$. Give an example.

CHECKING BASIC CONCEPTS FOR SECTIONS 11.3 AND 11.4

1. Determine whether the series is arithmetic or geometric.
   (a) $\dfrac{1}{2} + \dfrac{1}{4} + \dfrac{1}{8} + \cdots + \dfrac{1}{256}$
   (b) $\dfrac{1}{2} + \dfrac{5}{2} + \dfrac{9}{2} + \dfrac{13}{2} + \dfrac{17}{2}$

2. Use a formula to find the sum of the arithmetic series
   $$4 + 8 + 12 + \cdots + 48.$$

3. Use a formula to find the sum of the geometric series
   $$1 - 2 + 4 - 8 + 16 - 32 + 64 - 128 + 256 - 512.$$

4. Use Pascal's triangle to expand $(x - y)^4$.

5. Use the binomial theorem to expand $(x + 2)^3$.

Chapter 11 Summary

Section 11.1 *Sequences*

An infinite sequence is a function whose domain is the natural numbers. A finite sequence is a function whose domain is $D = \{1, 2, 3, \ldots, n\}$ for some natural number n. Because sequences are functions, they have symbolic, graphical, and numerical representations. Graphing calculators can be used to create graphical and numerical representations for sequences.

Example: $a_n = 2n$ is a symbolic representation of the even natural numbers. The first six terms of this sequence are represented numerically and graphically in the table and figure.

n	1	2	3	4	5	6
a_n	2	4	6	8	10	12

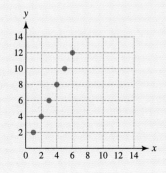

Section 11.2 *Arithmetic and Geometric Sequences*

Two common types of sequences are arithmetic and geometric.

Arithmetic Sequence An arithmetic sequence is determined by a linear function of the form $f(n) = dn + c$ or $f(n) = a_1 + (n - 1)d$. Successive terms in an arithmetic sequence are found by adding the common difference d to the previous term. The sequence $1, 3, 5, 7, 9, 11, \ldots$ is an arithmetic sequence with its first term $a_1 = 1$, common difference $d = 2$, and general term $a_n = 2n - 1$.

Geometric Sequence A geometric sequence is determined by an exponential function of the form $f(n) = a_1 r^{n-1}$. Successive terms in a geometric sequence are found by multiplying the previous term by the common ratio r. The sequence $3, 6, 12, 24, 48, \ldots$ is a geometric sequence with its first term $a_1 = 3$, common ratio $r = 2$, and general term $a_n = 3(2)^{n-1}$.

Section 11.3 *Series*

Series A series results when the terms of a sequence are summed. The series associated with the sequence $2, 4, 6, 8, 10$ is

$$2 + 4 + 6 + 8 + 10,$$

and its sum is 30. An arithmetic series results when the terms of an arithmetic sequence are summed, and a geometric series results when the terms of a geometric sequence are summed. In this chapter, we discussed formulas for finding sums of arithmetic and geometric series. See Putting It All Together for Section 11.3.

Summation Notation Summation notation can be used to write series efficiently. For example,

$$1^2 + 2^2 + 3^2 + 4^2 + 5^2 = \sum_{k=1}^{5} k^2.$$

Section 11.4 *The Binomial Theorem*

Pascal's triangle may be used to find the coefficients for the expansion of $(a + b)^n$, where n is a natural number.

$$
\begin{array}{c}
1 \\
1 \quad 1 \\
1 \quad 2 \quad 1 \\
1 \quad 3 \quad 3 \quad 1 \\
1 \quad 4 \quad 6 \quad 4 \quad 1 \\
1 \quad 5 \quad 10 \quad 10 \quad 5 \quad 1
\end{array}
$$

The binomial theorem can also be used to expand powers of binomials.

Example: To expand $(x + y)^4$, use the fifth row of Pascal's triangle.

$$(x + y)^4 = 1x^4 + 4x^3y + 6x^2y^2 + 4xy^3 + 1y^4$$
$$= x^4 + 4x^3y + 6x^2y^2 + 4xy^3 + y^4$$

CHAPTER 11
REVIEW EXERCISES

SECTION 11.1

Exercises 1–4: Write the first four terms of the sequence.

1. $f(n) = n^3$

2. $f(n) = 5 - 2n$

3. $f(n) = \dfrac{2n}{n^2 + 1}$

4. $f(n) = (-2)^n$

Exercises 5–6: Use the graph to write the terms of the sequence.

5.

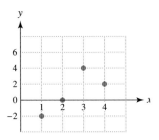

6.

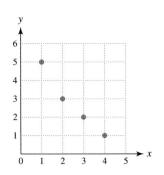

Exercises 7–10: Represent the first seven terms of the sequence numerically and graphically.

7. $a_n = 2n$

8. $a_n = n^2 - 4$

9. $a_n = 4\left(\dfrac{1}{2}\right)^n$

10. $a_n = \sqrt{n}$

SECTION 11.2

Exercises 11–18: Determine whether f is an arithmetic sequence. Identify the common difference when possible.

11. $f(n) = 5n - 1$

12. $f(n) = 4 - n^2$

13. $f(n) = 2^n$

14. $f(n) = 4 - \dfrac{1}{3}n$

15.

n	1	2	3	4
f(n)	20	17	14	11

16.

n	1	2	3	4
f(n)	−3	0	6	12

17.

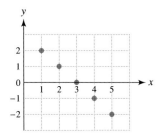

18.

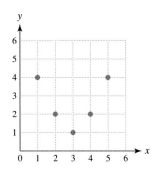

Exercises 19 and 20: Find the general term a_n for the arithmetic sequence.

19. $a_1 = -3$ and $d = 4$

20. $a_1 = 2$ and $a_2 = -3$

Exercises 21–28: Determine whether f is a geometric sequence. Identify the common ratio when possible.

21. $f(n) = 2(4)^n$

22. $f(n) = 2n^4$

23. $f(n) = 1 - 2n$

24. $f(n) = 5(0.7)^n$

25.

n	1	2	3	4
f(n)	5	4	3	1

26.

n	1	2	3	4
$f(n)$	27	-9	3	-1

27.

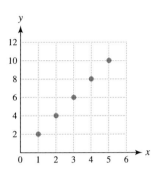

28.

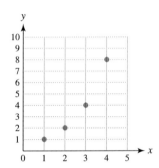

Exercises 29 and 30: Find the general term a_n for the geometric sequence.

29. $a_1 = 5$ and $r = 0.9$

30. $a_1 = 2$ and $a_2 = 8$

Section 11.3

Exercises 31–34: Find the sum, using a formula.

31. $4 + 9 + 14 + 19 + 24 + 29 + 34 + 39 + 44$

32. $4.5 + 3.0 + 1.5 + 0 - 1.5$

33. $1 - 4 + 16 - 64 + \cdots + 4096$

34. $1 + \dfrac{1}{2} + \dfrac{1}{4} + \dfrac{1}{8} + \dfrac{1}{16} + \cdots + \dfrac{1}{256}$

Exercises 35–38: Write the terms of the series.

35. $\displaystyle\sum_{k=1}^{5} 2k + 1$

36. $\displaystyle\sum_{k=1}^{4} \dfrac{1}{k+1}$

37. $\displaystyle\sum_{k=1}^{4} k^3$

38. $\displaystyle\sum_{k=2}^{7} (1-k)$

Exercises 39–42: Write the series in summation notation.

39. $1 + 2 + 3 + \cdots + 20$

40. $1 + \dfrac{1}{2} + \dfrac{1}{3} + \cdots + \dfrac{1}{20}$

41. $\dfrac{1}{2} + \dfrac{2}{3} + \dfrac{3}{4} + \cdots + \dfrac{9}{10}$

42. $1^2 + 2^2 + 3^2 + 4^2 + 5^2 + 6^2 + 7^2$

Section 11.4

Exercises 43–46: Use Pascal's triangle to expand the expression.

43. $(x + 4)^3$

44. $(2x + 1)^4$

45. $(x - y)^5$

46. $(a - 1)^6$

Exercises 47–50: Evaluate the expression.

47. $3!$

48. $\dfrac{5!}{3!\,2!}$

49. $_6C_3$

50. $_4C_3$

Exercises 51–54: Use the binomial theorem to expand the expression.

51. $(m + 2)^4$

52. $(a + b)^5$

53. $(x - 3y)^4$

54. $(3x - 2)^3$

Applications

55. *Salaries* An individual's starting salary is $45,000, and the individual receives a 10% raise each year. Give symbolic, numerical, and graphical representations for this person's salary over 7 years. What type of sequence is it?

56. *Salaries* An individual's starting salary is $45,000, and the individual receives an increase of $5000 each year. Give symbolic, numerical, and graphical representations for this person's salary over 7 years. What type of sequence is it?

57. *Rain Forests* Rain forests are defined as forests that grow in regions that receive more than 70 inches of rain each year. The world is losing an estimated 49 million acres of rain forests annually. Give symbolic, graphical, and numerical representations for a sequence that models the total number of acres (in millions) lost over a 7-year period. *(Source: New York Times Almanac, 1999.)*

58. *Home Mortgage Payments* The average home mortgage payment in 1996 was $1087 per month. Since then, mortgage payments have risen, on average, 2.5% per year.

(a) Write a sequence a_n that models the average mortgage payment in year n, where $n = 1$ corresponds to 1996, $n = 2$ to 1997, and so on.
(b) Is a_n arithmetic, geometric, or neither? Explain your reasoning.
(c) Find a_5 and interpret the result.
(d) Give a graphical representation for a_n, where $n = 1, 2, 3, \ldots, 10$.

Chapter 11

Test

1. Write the first four terms of the sequence $f(n) = \dfrac{n^2}{n+1}$.

2. Use the graph to write the terms of the sequence.

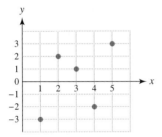

3. Represent the first seven terms of $a_n = n^2 - n$ in a table.

4. Expand the expression $(2x - 1)^4$.

Exercises 5 and 6: Determine whether the sequence is arithmetic or geometric. Identify either the common difference or the common ratio.

5. $f(n) = 7 - 3n$

6.

7. Find the general term a_n for the arithmetic sequence if $a_1 = 2$ and $d = -3$.

8. Find the general term a_n for the geometric sequence if $a_1 = 2$ and $a_3 = 4.5$.

Exercises 9 and 10: Determine whether f is a geometric sequence. Identify the common ratio when possible.

9. $f(n) = -3(2.5)^n$

10.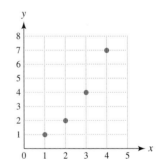

Exercises 11–12: Find the sum, using a formula.

11. $-1 + 2 + 5 + 8 + 11 + 14 + 17 + 20 + 23$

12. $1 - \dfrac{2}{3} + \dfrac{4}{9} - \dfrac{8}{27} + \dfrac{16}{81} - \dfrac{32}{243} + \dfrac{64}{729}$

13. Write the terms of the series $\sum_{k=2}^{7} 3k$

14. Write the series $1^3 + 2^3 + 3^3 + \cdots + 60^3$ in summation notation.

15. Evaluate $\dfrac{7!}{4!\,3!}$

16. Evaluate $_5C_3$

17. *Auditorium Seating* An auditorium has 50 seats in the first row, 57 seats in the second row, 64 seats in the third row, and so on. Use a formula to find the total number of seats in the first 45 rows.

18. *Median Home Price* In 1997 the median price of a single family home was $159,700 and was increasing at a rate of 4% per year. Give symbolic, numerical, and graphical representations for the median home price over a 7-year period, starting in 1997. What type of sequence is it?

19. *Tent Worms* Large numbers of tent worms can defoliate trees and ruin crops. After they mature, they spin a cocoon and develop into moths that lay eggs. Suppose that an initial population of 2000 tent worms doubles every 5 days.
   (a) Write a formula for a_n that models the number of tent worms after $n - 1$ five-day time periods. (*Hint*: $a_1 = 2000$, $a_2 = 4000$, and $a_3 = 8000$.
   (b) Is a_n arithmetic, geometric, or neither? Explain your reasoning.
   (c) Find a_6 and interpret the result.
   (d) Give a graphical representation for a_n, where $n = 1, 2, 3, 4, 5, 6$.

CHAPTER 11
EXTENDED AND DISCOVERY EXERCISES

SEQUENCES AND SERIES

Exercises 1–2: Recursive Sequences Some sequences are not defined by a formula for a_n. Instead they are defined recursively. With a **recursive formula** we must find terms a_1 through a_{n-1} before we can find a_n. For example, let

$$a_1 = 2$$
$$a_n = a_{n-1} + 3, \quad \text{for } n \geq 2.$$

To find $a_2, a_3,$ and a_4, we let $n = 2, 3, 4$.

$$a_2 = a_1 + 3 = 2 + 3 = 5$$
$$a_3 = a_2 + 3 = 5 + 3 = 8$$
$$a_4 = a_3 + 3 = 8 + 3 = 11$$

The first four terms of the sequence are 2, 5, 8, 11.

1. *Fibonacci Sequence* The Fibonacci sequence dates back to 1202 and is one of the most famous sequences in mathematics. It can be defined recursively as follows.

   $$a_1 = 1, \quad a_2 = 1$$
   $$a_n = a_{n-1} + a_{n-2}, \quad \text{for } n \geq 3$$

   Find the first 12 terms of this sequence.

2. *Insect Populations* Frequently the population of a particular insect does not continue to grow indefinitely. Instead, its population grows rapidly at first and then levels off because of competition for limited resources. In one study, the behavior of the winter moth was modeled with a sequence similar to the following, where a_n gives the population density in thousands per acre during year n.

   $$a_1 = 1$$
   $$a_n = 2.85a_{n-1} - 0.19a_{n-1}^2, \quad n \geq 2$$

   (*Source*: G. Varley and G. Gradwell, "Population models for the winter moth.")
   (a) Give a numerical representation for $n = 1, 2, 3, \ldots, 7$. Describe what happens to the population density of the winter moth.

   Note: Many graphing calculators are capable of generating numerical and graphical representations of a recursive sequence.

   (b) Graph the sequence for $n = 1, 2, 3, \ldots, 20$. Discuss the graph.

3. *Bode's Law* The average distances of the planets from the sun display a pattern first described by Johann Bode in 1772. This relationship is called Bode's law and was proposed before Uranus, Neptune, and Pluto were discovered. It is a sequence defined by

   $$a_1 = 0.4$$
   $$a_n = 0.3(2)^{n-2} + 0.4$$
   $$\text{for } n = 2, 3, 4, \ldots, 10.$$

In this sequence, a distance of 1 unit corresponds to the average Earth–sun distance of 93 million miles. The number n represents the nth planet. The actual distances of the planets, including an average distance for the asteroids, are listed in the accompanying table. (*Source:* **M. Zeilik,** *Introductory Astronomy and Astrophysics.*)

(a) Find a_4 and interpret the result.
(b) Calculate the terms of Bode's sequence. Compare them with the values in the table.
(c) If there is another planet beyond Pluto, use Bode's Law to predict its distance from the sun.

Planet	Distance
Mercury	0.39
Venus	0.72
Earth	1.00
Mars	1.52
Asteroids	2.8
Jupiter	5.20
Saturn	9.54
Uranus	19.2
Neptune	30.1
Pluto	39.5

4. *Calculating* π The quest for an accurate estimation for π is a fascinating story covering thousands of years. Because π is an irrational number, it cannot be represented exactly by a fraction. Its decimal expansion neither repeats nor has a pattern. The ability to compute π was essential to the development of societies because π appears in formulas used in construction, surveying, and geometry. In early historical records, π was given the value of 3. Later the Egyptians used a value of

$$\frac{256}{81} \approx 3.1605$$

Not until the discovery of series was an exceedingly accurate decimal approximation of π possible. In 1989, π was computed to 1,073,740,000 digits, which required 100 hours of supercomputer time. Why would anyone want to compute π to so many decimal places? One practical reason is to test electrical circuits in new computers. If a computer has a small defect in its hardware, there is a good chance that an error will appear after it has performed trillions of arithmetic calculations during the computation of π. (*Source:* **P. Beckmann,** *A History of PI.*) The series given by

$$\frac{\pi^4}{90} \approx \frac{1}{1^4} + \frac{1}{2^4} + \frac{1}{3^4} + \frac{1}{4^4} + \frac{1}{5^4} + \cdots + \frac{1}{n^4}$$

can be used to estimate π, where larger values of n give better approximations.

(a) Approximate π by finding the sum of the first four terms.
(b) Use a calculator to approximate π by summing the first 50 terms. Compare the result to the actual value of π.

5. *Infinite Series* The sum S of an infinite geometric series can be found if its common ratio r satisfies $|r| < 1$. It is given by

$$S = \frac{a_1}{1 - r},$$

If $|r| \geq 1$, this sum does not exist. For example, the infinite geometric series

$$1 + \frac{1}{2} + \frac{1}{4} + \frac{1}{8} + \frac{1}{16} + \cdots$$

has $a_1 = 1$ and $r = \frac{1}{2}$. Therefore its sum S equals

$$S = \frac{1}{1 - \frac{1}{2}} = 2.$$

You might want to add terms of this series to see how increasing the number of terms results in a number closer to 2. Find the sum of each infinite geometric series.

(a) $2 - 1 + \frac{1}{2} - \frac{1}{4} + \frac{1}{8} - \frac{1}{16} + \cdots$
(b) $1 + \frac{1}{3} + \frac{1}{9} + \frac{1}{27} + \frac{1}{81} + \cdots$
(c) $0.1 + 0.01 + 0.001 + 0.0001 + \cdots$
(d) $0.12 + 0.0012 + 0.000012 + 0.00000012 + \cdots$

Answers to Selected Exercises

CHAPTER 1: REAL NUMBERS AND ALGEBRA

Section 1.1

1. $N = \{1, 2, 3, \ldots\}$; $W = \{0, 1, 2, 3, \ldots\}$
3. $\frac{p}{q}$, where p and q are integers and $q \neq 0$; $\frac{3}{4}$ (*answers may vary*)
5. $3 + 2 = 2 + 3$ (*answers may vary*)
7. $a \cdot 1 = a$ for any a
9. $2 \cdot (3 \cdot 4) = (2 \cdot 3) \cdot 4$ (*answers may vary*)
11. Natural, integer, rational, and real
13. Rational and real   15. Rational and real
17. Real
19. Natural numbers: 6; Whole numbers: 6; Integers: $-5, 6$; Rational numbers: $-5, 6, \frac{1}{7}, 0.2$; Irrational numbers: $\sqrt{7}$
21. Natural numbers: $\frac{3}{1} = 3$; Whole numbers: $\frac{3}{1} = 3$; Integers: $\frac{3}{1} = 3$; Rational numbers: $\frac{3}{1} = 3, -\frac{5}{8}, 0.\overline{45}$; Irrational numbers: $\sqrt{5}, \pi$
23. Rational   25. Rational   27. Integers
29. Identity   31. Commutative   33. Distributive
35. Associative   37. Commutative   39. $a + 4$
41. $\frac{1}{3}a$ 43. $9 + b$ 45. $50x$ 47. $4x + 4y$
49. $5x - 35$ 51. $-x - 1$ 53. $a(x - y)$
55. $3(4 + x)$ 57. 507   59. -60
61. 5, natural, integer, and rational   63. $43.\overline{3}$, rational
65. 78.1, rational   67. Both 86,400; commutative
69. (a) 13.9 million   (b) About 14.1 million (*answers may vary*)   (c) 14.125 million

Section 1.2

1. Right   3. Positive   5. Negative   7. $-a$
9.

11.

13. 6.1   15. 3   17. x 19. $x - y$ 21. Positive
23. Both   25. Positive   27. -56 29. 6.9
31. $\pi - 2$ 33. $-a + b$ 35. $\frac{1}{3}$ 37. $-\frac{3}{2}$
39. $\frac{1}{\pi}$ 41. $\frac{1}{a+3}$ 43. -2 45. -16.6
47. $-\frac{2}{4} = -\frac{1}{2}$ 49. -20 51. -3 53. -63
55. 32   57. $-\frac{5}{42}$ 59. $\frac{4}{6} = \frac{2}{3}$ 61. $-\frac{3}{8}$
63. -4 65. 5   67. 140   69. -2 71. 57.2
73. $\frac{19}{42}$ 75. $-\frac{9}{25}$ 77. $\frac{21}{55}$ 79. 5   81. 59.75
83. 1   85. $\frac{2}{6} = \frac{1}{3}$ 87. $2600; $2577.76 (*answers may vary*)
89. (a) $5.3 billion   (b) $4.6 billion; Average 1992 and 1994 (*answers may vary*)   (c) $6.25 billion (*answers may vary*)
91. (a) They are increasing.   (b) About 5.5 million (*answers may vary*)

Checking Basic Concepts 1.1 & 1.2

1. (a) Integer, rational, and real   (b) Natural, integer, rational, and real   (c) Real   (d) Rational and real
2. (a) Commutative; $a + b = b + a$ (b) Associative; $a \cdot (b \cdot c) = (a \cdot b) \cdot c$ (c) Distributive; $a(b + c) = ab + ac$
3. (a) -4 (b) -40.8 (c) $-\frac{5}{12}$
4. About 12 billion gallons

Section 1.3

1. Base: 8; exponent: 3   3. 7^3
5. No. $2^3 = 8$ and $3^2 = 9$ 7. $\frac{1}{7^n}$ 9. 5^{m-n}
11. 2^{mk} 13. 5000   15. 2^3 17. 4^4 19. 6^0

A-1

21. 16   **23.** −81   **25.** 1   **27.** $\frac{8}{27}$ **29.** $\frac{1}{16}$
31. $3^2 = 9$ **33.** $10^{-3} = 0.001$ **35.** y^{-1}
37. $10^8 = 100{,}000{,}000$ **39.** $5 \cdot 2^{-1} = \frac{5}{2}$
41. $8a^{-1}b^{-3}$ **43.** $4^1 = 4$ **45.** $10^2 = 100$
47. $\frac{1}{b^5}$ **49.** $4x^2$ **51.** $\frac{2b}{3a^2}$ **53.** $\frac{3y^6}{x^7}$
55. $3^8 = 6561$ **57.** $\frac{1}{x^6}$ **59.** $4^3y^6 = 64y^6$
61. $\frac{4^3}{x^3} = \frac{64}{x^3}$ **63.** $\frac{z^{20}}{2^5x^5} = \frac{z^{20}}{32x^5}$ **65.** 34   **67.** −8
69. 40   **71.** $-\frac{11}{2}$ **73.** −155   **75.** 5
77. 2.322×10^6 **79.** 2.69×10^{10}
81. 5.1×10^{-2} **83.** 1.0×10^{-6} **85.** 500,000
87. 9,300,000   **89.** −0.006   **91.** 0.00005876
93. 6×10^6; 6,000,000   **95.** 8×10^{-6}; 0.000008
97. 2×10^5; 200,000
99. (a) $k = 10$ (b) $\frac{1024}{52} \approx 20$ years
101. (a) $551.25   (b) $1215.51
103. (a) About 5.866×10^{12} miles
(b) 5,866,000,000,000 miles   (c) About 2.5×10^{13} miles   (d) 57,000 years
105. About $8795 per person
107. $256 \times 2^{20} = 268{,}435{,}456$ bytes
109.

Country	1996	2025
China	1.2551×10⁹	1.48×10⁹
Germany	8.24×10⁷	8.09×10⁷
India	9.758×10⁸	1.3302×10⁹
Mexico	9.58×10⁷	1.302×10⁸
U.S.	2.65×10⁸	3.325×10⁸

Section 1.4

1. x and y **3.** $3x = 15$ (answers may vary)
5. $y = 5280x$ **7.** $A = s^2$ **9.** $y = 3600x$
11. $y = 30$ **13.** $y = 1.9$ **15.** $d = 10$
17. $z = 6$ **19.** $y = -\frac{1}{4}$ **21.** $N = -\frac{8}{9}$
23. $P = 0.3$ **25.** (ii)   **27.** (iii)   **29.** (iii)
31. −3   **33.** 2   **35.** 2
37.

x	0	2	4	6	8
y	−0.5	4.5	9.5	14.5	19.5

39.

x	−3	−1	1	3
y	15	5	5	15

41.

x	−1	0	1	8
y	−3	−2	−1	0

43. $y = 60t$
45. (a) 178 pounds per square inch   (b) 3115 pounds per square inch
47. (a) About 0.464 kilograms   (b) About 3.7125 kilograms   (c) It increases 8 times.
49. Venus: $E \approx 22{,}995$ miles per hour
Earth: $C \approx 17{,}706$ miles per hour
Moon: $C \approx 3790$ miles per hour
Mars: $E \approx 11{,}384$ miles per hour
51. About 80,610 miles per hour
53. (a) About 162 beats per minute   (b) About 20 beats per minute
55. (a) About 88.6 inches   (b) About 8.6 miles
57. (a) $C = 336x$ (b) 10,080 calories

Checking Basic Concepts 1.3 & 1.4

1. (a) 16   (b) $\frac{1}{9}$ (c) 32   (d) x (e) $4x^6y^8$
2. (a) −6   (b) −4   (c) 58
3. (a) 1.03×10^5 (b) 5.23×10^{-4} (c) 6.7×10^0
4. (a) 5,430,000   (b) 0.0098
5. 900 cubic feet per person

People	10	20	30	40
Ventilation (ft³/hr)	9000	18,000	27,000	36,000

Section 1.5

1. A set of ordered pairs (x, y)
3.

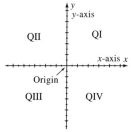

5. $D = \{1, 3, 5\}; R = \{−4, 2, 6\}$
7. $D = \{−2, −1, 0, 1, 2\}; R = \{0, 1, 2, 3\}$
9. $D = \{41, 87, 96\}; R = \{24, 53, 67, 88\}$
11. $S = \{(1, 3), (3, 7), (5, 11), (7, 15), (9, 19)\}$
$D = \{1, 3, 5, 7, 9\}; R = \{3, 7, 11, 15, 19\}$

13. $S = \{(1994, 6.1), (1995, 5.6), (1996, 5.5), (1997, 4.9), (1998, 4.5), (1999, 4.1)\}$
$D = \{1994, 1995, 1996, 1997, 1998, 1999\}$;
$R = \{4.1, 4.5, 4.9, 5.5, 5.6, 6.1\}$
15. $S = \{(-3, 2), (-2, 1), (2, -3), (3, 3)\}$
17. $S = \{(-4, 4), (-3, 2), (-2, 0), (0, -3), (2, 4), (4, 4)\}$
19. $S \approx \{(1970, 29), (1980, 41), (1990, 79), (1998, 70)\}$
21. $(1, 2)$ is in QI; $(-1, 3)$ is in QII

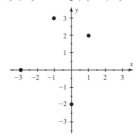

23. $(10, 50)$ is in QI; $(-30, 20)$ is in QII; $(-20, -25)$ is in QIII; $(50, -25)$ is in QIV.

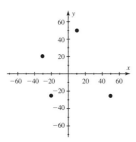

25. (a) $D = \{-3, -2, 0, 1\}$; $R = \{-4, -3, 0, 2, 4\}$
(b) Xmin $= -3$, Xmax $= 1$, Ymin $= -4$, Ymax $= 4$ (c) and (d) See the graph.

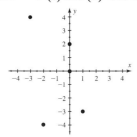

27. (a) $D = \{-30, 10, 20, 30\}$; $R = \{-50, 20, 40, 50\}$ (b) Xmin $= -30$, Xmax $= 30$, Ymin $= -50$, Ymax $= 50$ (c) and (d) See the graph.

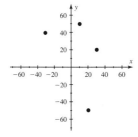

29.

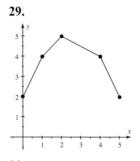

31.

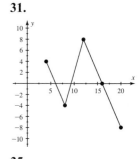

33.

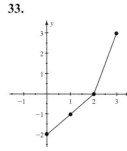

35.

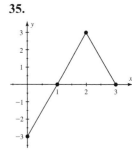

37. 10; 10
$[-10, 10, 1]$ by $[-10, 10, 1]$

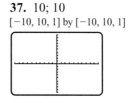

39. 10; 5
$[0, 100, 10]$ by $[-50, 50, 10]$

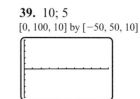

41. 15; 4
$[1980, 1995, 1]$ by $[12{,}000, 16{,}000, 1000]$

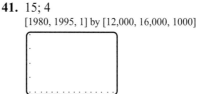

43. d. **45.** c.
47. $S = \{(-2, 1), (-1, 0), (1, -1), (2, 1)\}$
49. $S = \{(-4, -2), (-2, -2), (0, 2), (2, 1), (4, -1)\}$
51. $S = \{(1985, 20), (1990, 30), (1995, 50)\}$

53.
$[-6, 6, 1]$ by $[-6, 6, 1]$

55.
$[-30, 30, 5]$ by $[-50, 50, 5]$

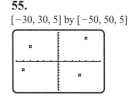

57.
$[-8, 8, 1]$ by $[-5, 15, 1]$

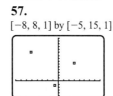

59. (a) and (b) See the graph.
[1965, 2000, 5] by [0, 900, 100]

(c) Participation decreased during the first 10 years, and then participation increased.

61. (a) and (b) See the graph.
[1989, 1999, 2] by [0, 16, 2]

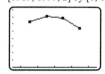

(c) The number of welfare beneficiaries first increased and then decreased.

63. (a) and (b) See the graph.
[1996, 2006, 2] by [0, 14, 2]

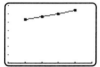

(c) The Asian-American population has increased.

CHECKING BASIC CONCEPTS 1.5

1. $D = \{-5, 1, 2\}$; $R = \{-1, 3, 4\}$
2. $(1, 4)$ is in QI; $(-2, 3)$ is in QII; $(2, -2)$ is in QIV.

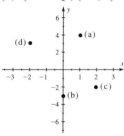

3. [1950, 2000, 10] by [0, 50, 5]

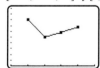

The number of people below the poverty level decreased for about 10 years and then began to increase.

CHAPTER 1 REVIEW EXERCISES

1. Natural numbers: 9; Whole numbers: 9; Integers: $-2, 9$; Rational numbers: $-2, 9, \frac{2}{5}, 2.68$; Irrational numbers: $\sqrt{11}, \pi$

2. Natural numbers: $\frac{6}{2} = 3$;
Whole numbers: $\frac{6}{2} = 3, \frac{0}{4} = 0$
Integers: $\frac{6}{2} = 3, \frac{0}{4} = 0$
Rational numbers: $\frac{6}{2} = 3, -\frac{2}{7}, 0.\overline{3} = \frac{1}{3}, \frac{0}{4} = 0$
Irrational numbers: $\sqrt{6}$

3. Identity **4.** Commutative **5.** Associative
6. Distributive **7.** a **8.** $\frac{1}{4}x$ **9.** $(8 \cdot 10)x = 80x$
10. $a \cdot 10 + a \cdot b = 10a + ab$ **11.** 95 **12.** 12
13. 9 **14.** 5.72
15.

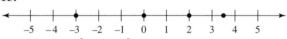

16. 3.2 **17.** $\frac{2}{3}$ **18.** $\frac{5}{4}$ **19.** -4 **20.** 2 **21.** -4
22. $-\frac{3}{4}$ **23.** 36 **24.** $-\frac{53}{30}$ **25.** $-\frac{1}{2}$ **26.** $\frac{29}{110}$
27. Base: 4; exponent: -2
28. $3^\pi \approx 31.54$; $\pi^3 \approx 31.01$; not equal
29. 25 **30.** $\frac{1}{9}$ **31.** -16 **32.** 16 **33.** 1
34. $\frac{27}{8}$ **35.** $\frac{1}{4^2} = \frac{1}{16}$ **36.** $10^2 = 100$ **37.** x^5
38. 3^{11} **39.** $\frac{1}{2a^6}$ **40.** $\frac{5a^2}{b^3}$ **41.** 2^8 **42.** $\frac{1}{x^{15}}$
43. $\frac{16y^6}{x^4}$ **44.** $4^5 a^5 = 1024a^5$ **45.** $\frac{125x^9}{27z^{12}}$
46. $\frac{z^2}{9x^8 y^6}$ **47.** 29 **48.** -3 **49.** 40 **50.** $\frac{11}{3}$
51. $\frac{5}{2}$ **52.** 11 **53.** 1.86×10^5 **54.** 3.4×10^{-4}
55. 45,000 **56.** 0.00923 **57.** $y = 36$
58. $d = 8$ **59.** $N = \frac{3}{2}$ **60.** $P = -10$
61. (iii) **62.** $a = \frac{3}{2}$
63. $D = \{-1, 2, 3\}$; $R = \{-6, 1, 3, 7\}$
64. $S = \{(-8, 4), (-4, -4), (4, 0), (8, 4)\}$

65. $(-2, 2)$ is in QII; $(-1, -3)$ is in QIII; $(2, -1)$ is in QIV.

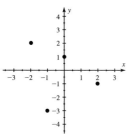

66. $(10, 20)$ is in QI; $(-15, -5)$ is in QIII; $(20, -10)$ is in QIV.

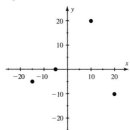

67. 9; 2
$[-9, 9, 1]$ by $[-6, 6, 3]$

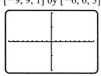

68. 4; 3
$[-20, 20, 5]$ by $[-12, 12, 4]$

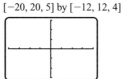

69.
$[-5, 5, 1]$ by $[-5, 5, 1]$

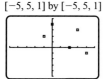

70.
$[-60, 60, 10]$ by $[-40, 40, 10]$

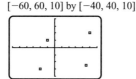

71. (a) $156,250 (b) $156,250
72. (a) $5397.31 (b) $1371.06
73. (a) 5.8×10^8 miles (b) 66,700 miles per hour
74. $d = 40t$
75. Cat: 280 beats per minute; Person: 70 beats per minute
76. $[1950, 2000, 10]$ by $[0, 10,000, 1000]$

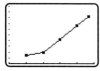

The poverty threshold has been increasing.

Chapter 1 Test

1. Natural numbers: $\sqrt{9}$; Whole numbers: $\sqrt{9}$; Integers: $-5, \sqrt{9}$; Rational numbers: $-5, \frac{2}{3}, \sqrt{9}, -1.83$; Irrational numbers: $-\frac{1}{\sqrt{5}}, \pi$

2. (a) Identity (b) Associative (c) Commutative (d) Distributive **3.** 31.2

4.

5. $\frac{3}{2}$ **6.** $-\frac{4}{5}$ **7.** (a) $-\frac{5}{18}$ (b) $-\frac{20}{3}$ (c) -5

8. (a) $\frac{1}{25}$ (b) 1 (c) $\frac{16}{625}$

9. (a) x^2y^3 (b) $\frac{8y^{15}}{3x^3}$ (c) $\frac{8y^3}{z^6}$ (d) $\frac{4}{9x^6y^4}$

10. 0.00052
11. $S = \{(-200, 100), (-100, 200), (0, -200), (200, -100), (400, 0)\}$
$D = \{-200, -100, 0, 200, 400\}$
$R = \{-200, -100, 0, 100, 200\}$

12. $y = 1.25x$
$[0, 5, 1]$ by $[0, 6, 1]$

13. *Window sizes may vary.*
$[-20, 20, 5]$ by $[-5, 40, 5]$

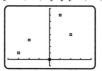

14. $-15°C$ **15.** 4.6631×10^8 gallons
16. $\sqrt{\frac{25}{\pi}} \approx 2.82$ feet

CHAPTER 2: LINEAR FUNCTIONS AND INEQUALITIES

Section 2.1

1. Domain **3.** One
5. Verbal, numerical, symbolic, graphical **7.** Yes
9. No **11.** $-6; -2$ **13.** $0; \frac{3}{2}$ **15.** $25; \frac{9}{4}$
17. $3; 3$ **19.** $-\frac{1}{2}; \frac{2}{5}$

21. 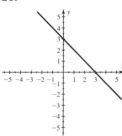   **23.**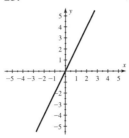

25. 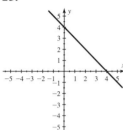   **27.**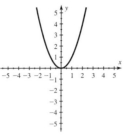

29. 3; −1   **31.** 1; 0   **33.** −4; −3   **35.** 5.5; 3.7
37. 26.9
39. Numerical: Graphical: [−3, 3, 1] by [−10, 10, 1]

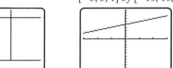

Symbolic: $y = x + 5$

41. Numerical: Graphical: [−3, 3, 1] by [−20, 20, 2]

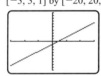

Symbolic: $y = 5x - 2$

43. Symbolic: $y = 3.785x$ **45.** Symbolic: $y = \dfrac{x}{1.609}$

Graphical: [0, 10, 1] by [0, 40, 4]   Graphical: [0, 10, 1] by [0, 10, 1]

 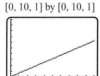

47. Subtract $\dfrac{1}{2}$ from the input x to obtain the output y.

49. Divide the input x by 3 to obtain the output y.

51. Symbolic: $f(x) = 0.41x$
Graphical: Numerical:
[0, 70, 10] by [0, 40, 5]

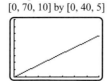

53. −2   **55.** 0.5
57. $D: -2 \le x \le 2; R: 0 \le y \le 2$
59. $D: -2 \le x \le 4; R: -2 \le y \le 2$
61. D: All real numbers; $R: y \ge -1$
63. $D: -3 \le x \le 3; R: -3 \le y \le 2$
65. $D: \{1, 2, 3, 4\}; R: \{5, 6, 7\}$
67. (a) 60.3 **(b)** $D: \{1910, 1930, 1950, 1970, 1990\}$;
$R: \{36.9, 56.2, 60.3, 80.5, 84.4\}$ **(c)** Decreased
69. (a) 0.2 **(b)** Yes. Each month has one average amount of precipitation. **(c)** 2, 3, 7, 11   **71.** No
73. Yes   **75.** Yes   **77.** No   **79.** Yes   **81.** Yes
83. No   **85.** Yes   **87.** No

Section 2.2

1. $ax + b$ **3.** line   **5.** 7
7. Linear: $a = \dfrac{1}{2}, b = -6$ **9.** Nonlinear
11. Linear: $a = 0, b = -9$
13. Linear: $a = -9, b = 0$ **15.** Yes   **17.** No
19. Yes   **21.** No   **23.** −16; 20   **25.** $\dfrac{17}{3}$; 2
27. −22; −22   **29.** −2; 0   **31.** −1; −4
33. −2; 2   **35.** 18   **37.** 8.7   **39.** d.   **41.** b.
43. (a)

x	−2	0	2
$f(x)$	−4	0	4

(b)   **(c)**

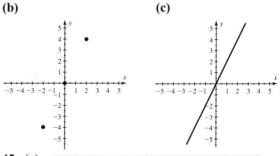

45. (a)

x	−2	0	2
$f(x)$	−1	1	3

Answers to Selected Exercises   A-7

(b) 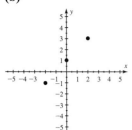 (c)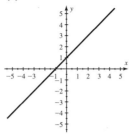

47. (a)

x	-2	0	2
$f(x)$	-9	-3	3

(b) 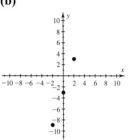 (c)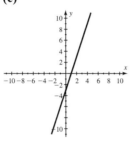

49. (a)

x	-2	0	2
$f(x)$	5	3	1

(b) 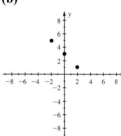 (c)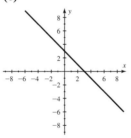

51. $f(x) = \dfrac{1}{16}x$ **53.** $f(t) = 65t$ **55.** $f(x) = 24$
57. b.   **59.** c.
61. (a) Symbolic: $f(x) = 70$
    Graphical:
    $[0, 24, 4]$ by $[0, 100, 10]$
       (b)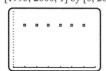

(c) Constant
63. (a) $f(x) = 15$
    (b) $[1993, 2000, 1]$ by $[0, 20, 1]$

65. (a) $[1990, 2000, 1]$ by $[0, 70, 10]$
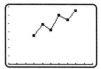

(b) No. The graph does not form a straight line.
67. (a) $f(x) = 4.3x$
    (b) $f(60) = 258$
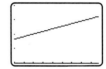

69. (a) Median age is increasing.
$[1820, 1995, 20]$ by $[0, 40, 10]$ **(b)** $f(1900) = 23.9$

(c) Each year the median age increased by 0.09 year.
71. $f(x) = 40x$; About 2.92 pounds

CHECKING BASIC CONCEPTS 2.1 & 2.2

1. Symbolic: $f(x) = x^2 - 1$
Graphical:
$[-3, 3, 1]$ by $[-10, 10, 1]$ Numerical:
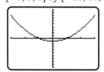

2. (a) D: $-3 \leq x \leq 3$; R: $-4 \leq y \leq 4$ **(b)** 0; 4
(c) No. The graph is not a line.
3. (a) Yes **(b)** No **(c)** Yes **(d)** Yes
4. $f(-2) = 10$;
$[-10, 10, 1]$ by $[-10, 10, 1]$

SECTION 2.3

1. 2   **3.** $-\dfrac{2}{3}$ **5.** 0   **7.** 2   **9.** $-\dfrac{2}{3}$ **11.** 0
13. Undefined   **15.** $\dfrac{3}{5}$ **17.** $-0.\overline{21}$

19. (a) $-3; 2$
(b)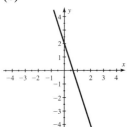

21. (a) $\frac{1}{3}; 0$
(b)

23. (a) $0; 2$
(b)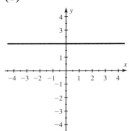

25. (a) $-1; 3$
(b)

27. $2; -2$ **29.** $-7; 4$ **31.** $y = -x + 4$
33. $y = 2$ **35.** $y = x - 2$ **37.** $y = 3x - 5$
39. $y = 2x + 4$
41. (a) 3 (b) $f(x) = 2x - 1$
43. (a) 11 (b) $f(x) = -\frac{3}{2}x + 5$
45. c. **47.** b.
49. (a) $m_1 = 50, m_2 = 0, m_3 = 150$ (b) m_1: Water is being added at the rate of 50 gallons per hour. m_2: The pump is neither adding nor removing water. m_3: Water is being added at the rate of 150 gallons per hour. (c) The pool initially contained 100 gallons of water. Water was added at the rate of 50 gallons per hour for the first 4 hours. Then the pump was turned off for 2 hours. Finally, water was added at the rate of 150 gallons per hour for 2 hours. The pool contains 600 gallons after 8 hours.
51. (a) $m_1 = 100, m_2 = 25, m_3 = -100$ (b) m_1: Water is being added at the rate of 100 gallons per hour. m_2: Water is being added at the rate of 25 gallons per hour. m_3: Water is being removed at the rate of 100 gallons per hour. (c) The pool initially contained 100 gallons of water. Water was added at the rate of 100 gallons per hour for the first 2 hours. Then water was added at the rate of 25 gallons per hour for 4 hours. Finally, water was removed at the rate of 100 gallons per hour for 2 hours. The pool contains 200 gallons after 8 hours.

53. (a) $m_1 = 50, m_2 = 0, m_3 = -20, m_4 = 0$
(b) m_1: The car is moving away from home at the rate of 50 miles per hour. m_2: The car is not moving. m_3: The car is moving toward home at the rate of 20 miles per hour. m_4: The car is not moving. (c) The car starts at home and moves away from home at 50 miles per hour for 1 hour. Then the car does not move for 1 hour. The car then moves toward home at the rate of 20 miles per hour for 2 hours. Finally, the car stops and does not move for 1 hour. The car is 10 miles from home at the end of the trip.
55. (a) $m_1 = -50, m_2 = 0, m_3 = -50$ (b) m_1: The car is moving toward home at the rate of 50 miles per hour. m_2: The car is not moving. m_3: The car is moving toward home at the rate of 50 miles per hour. (c) The car starts at 300 miles from home and moves toward home at 50 miles per hour for 2 hours. Then the car does not move for 1 hour. The car then moves toward home at 50 miles per hour for 4 hours. The car is at home at the end of the trip.

57. 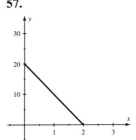   **59.**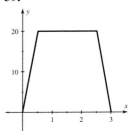

61. (a) 62 thousand
(b) Increasing
[1990, 2000, 1] by [0, 100, 10]

(c) 4 (d) The number of children born to older mothers is increasing by about 4000 each year.
63. (a) 8748 (b) 214.2 (c) The number of radio stations increased by about 214 stations per year.
65. (a) $\frac{24}{25} = 0.96$ pint
(b) [0, 25, 1] by [0, 5, 1]

(c) $\frac{4}{25} = 0.16$ (d) 0.16 pint of oil should be added per gallon of gasoline.

67. (a) $a = \dfrac{14{,}000}{9}$, $b = 41{,}000$ (b) About $48,778

SECTION 2.4

1. 1 **3.** $y = mx + b$ **5.** $y = b$ **7.** $m_1 = m_2$
9. d. **11.** a. **13.** Yes **15.** Yes **17.** No
19. $y = -\dfrac{3}{4}(x + 3) + 2$ **21.** $y = \dfrac{1}{3}(x - 1) + 3$
23. $y = -2(x - 2) - 3$
25. $y = 1.3(x - 1990) + 25$
27. $y = \dfrac{2}{3}(x - 1) + 3$ or $y = \dfrac{2}{3}(x + 5) - 1$
29. $y = -\dfrac{2}{3}(x - 6) + 0$ or $y = -\dfrac{2}{3}(x - 0) + 4$
31. $y = 2x - 4$ **33.** $y = \dfrac{1}{2}x + 3$
35. $y = 22x - 43$ **37.** $y = -\dfrac{1}{3}x - 5$
39. $y = -x + 1$ **41.** $y = \dfrac{1}{3}x - \dfrac{2}{3}$
43. $y = 4x - 1$ **45.** $y = 3x + 14$ **47.** $x = -1$
49. $y = -\dfrac{5}{6}$ **51.** $x = 4$ **53.** $x = -\dfrac{2}{3}$
55. (a) Away (b) After 1 hour the person is 35 miles from home. After 3 hours the person is 95 miles from home. (c) $y = 30x + 5$; the person is traveling at 30 mph. (d) 125
57. b. **59.** a. **61.** f. **63.** Yes; $y = 4x - 8$
65. (a) 628,000
(b) [1988, 1995, 1] by [600, 1200, 100]

(c) 70; the number of inmates is increasing by 70,000 per year.
67. (a) In 1984, average tuition was $1225; in 1987, average tuition was $1621.
(b) $y = 132(x - 1984) + 1225$ or $y = 132(x - 1987) + 1621$; tuition is increasing by $132 per year.
69. (a) $a = 0.30$, $b = 189.20$ (b) The fixed cost of owning the car
71. (a) $m = 3.1$, $h = 1996$, $k = 22$ (b) 43.7 million
73. (a) [1950, 2000, 10] by [0, 300, 10]

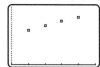

(b) $m = 2.4$, $h = 1960$, $k = 179$ (c) 275 million

CHECKING BASIC CONCEPTS 2.3 & 2.4

1. (a) $m = -3$; (5, 7) (answers may vary) (b) $m = 0$; (1, 10) (answers may vary) (c) m is undefined; $(-5, 0)$ (answers may vary) (d) $m = 5$; (0, 3) (answers may vary)
2. (a) 2 (b) $y = 2(x - 2) - 4$ or $y = 2(x - 5) + 2$; $y = 2x - 8$ (c) x-int: 4; y-int: -8
3. (a) Toward home (b) -50; the car is moving toward home at 50 mph. (c) x-int: 5; after 5 hours the driver is home. y-int: 250; the driver is initially 250 miles from home. (d) $a = -50$, $b = 250$
4. $x = -2$; $y = 5$
5. $y = 2x - 8$; $y = -\dfrac{1}{2}x - 3$

CHAPTER 2 REVIEW EXERCISES

1. -7; 0 **2.** -22; 2 **3.** -2; 1 **4.** 5; 5
5.

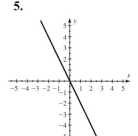

6.

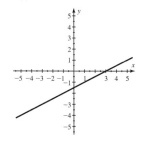

7.

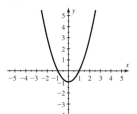

8.

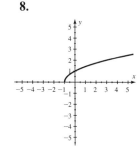

9. 1; 4 **10.** 1; -2 **11.** 7; -1
12. Numerical: Graphical: [$-3, 3, 1$] by [$-12, 12, 1$]

Symbolic: $f(x) = 3x - 2$
13. D: All real numbers, R: $y \leq 4$
14. D: $-4 \leq x \leq 4$, R: $-4 \leq y \leq 0$ **15.** Yes
16. No **17.** $D = \{-3, -1, 2, 4\}$, $R = \{-1, 3, 4\}$; yes **18.** $D = \{-1, 0, 1, 2\}$, $R = \{-2, 2, 3, 4, 5\}$; no
19. No **20.** Yes **21.** -2 **22.** 0 **23.** $\dfrac{1}{3}$

24. Undefined **25.** $\frac{3}{2}$ **26.** $-\frac{3}{4}$ **27.** 0
28. Undefined
29. (a) $\frac{1}{2}$; -2 (b)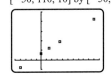

30. $-2; 2; 4$ **31.** $y = \frac{3}{2}(x - 2) + 0$ or $y = \frac{3}{2}(x - 0) - 3$; $y = \frac{3}{2}x - 3$
32. $y = -2(x + 1) + 4$ or $y = -2(x - 2) - 2$; $y = -2x + 2$
33. $y = 4\left(x + \frac{3}{5}\right) + \frac{1}{5}$; $y = 4x + \frac{13}{5}$
34. $y = -2(x + 1) + 1$; $y = -2x - 1$
35. $y = -x + 2$ **36.** $y = 2x - 3$ **37.** No
38. Yes   **39.** $x = -4$ **40.** $y = -\frac{7}{13}$ **41.** No
42. Yes, $y = 2x + 5$
43. (a) $m_1 = 1.3$, $m_2 = 2.35$, $m_3 = 2.4$, $m_4 = 2.7$
(b) m_1: From 1920 to 1940 the population increased on average by 1.3 million per year. The other slopes may be interpreted similarly.
44. For the first 2 minutes the inlet pipe is open. For the next 3 minutes both pipes are open. For the next 2 minutes only the outlet pipe is open. Finally, for the last 3 minutes both pipes are closed.
45. (a) About 25.1   (b) Decreased
[1885, 1965, 10] by [22, 26, 1]

(c) -0.0492; the median age decreased by about 0.0492 year per year.
46. (a) 2.4 million (b) The number of marriages each year did not change.
47. (a) $f(x) = 8x$ (b) 8 (c) The total fat increases at the rate of 8 grams per cup.
48. (a) [1989, 1998, 1] by [0, 20, 2]
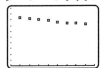

(b) $m = -0.31$, $h = 1990$, $k = 16.7$ (answers may vary) (c) About 13.6 per 1000 people (answers may vary)
49. (a) 103; In 1995, there were 103 unhealthy days.
(b) $D = \{1992, 1993, 1994, 1995, 1996\}$, $R = \{88, 103, 136, 146, 185\}$ (c) The number of unhealthy days is decreasing.
50. (a) Linear
[−50, 110, 10] by [−50, 250, 50]
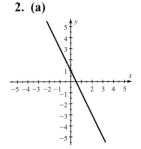

(b) $f(x) = \frac{9}{5}x + 32$ (c) 68°F
51.
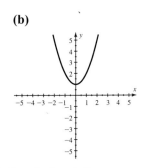

52. (a) $f(x) = 40{,}000(x - 1996) + 775{,}000$
(b) 935,000; in 2000 there were 935,000 HIV infections.
53. (a) $m = 0.4$; $h = 1998$, $k = 10.5$ or $h = 2003$, $k = 12.5$ (b) 12.1 million
54. $y = -1.2x + 3$

Chapter 2 Test

1. 46
2. (a)   (b)
 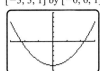

3. $0, -3$; D: $-3 \leq x \leq 3$, R: $-3 \leq y \leq 0$
4. Symbolic: $f(x) = x^2 - 5$
   Numerical: Graphical:
   [−3, 3, 1] by [−6, 6, 1]

5. No, it fails the vertical line test.   **6.** $-\frac{1}{2}$
7. $y = -\frac{7}{12}x - \frac{17}{12}$; x-int: $-\frac{17}{7}$; y-int: $-\frac{17}{12}$
8. $m = -2$; x-int: $\frac{5}{2}$; y-int: 5
9. $y = -3\left(x - \frac{1}{3}\right) + 2$; $y = -3x + 3$
10. $y = \frac{2}{3}x - 2$ **11.** $x = \frac{2}{3}$
12. (a) $m_1 = 0.4$; $m_2 = 0$, $m_3 = 0.6$ (b) From 1970 to 1980, beneficiaries increased by 0.4 million per year. From 1980 to 1990, there was no change. From 1990 to 1995, beneficiaries increased by 0.6 million per year.
13.

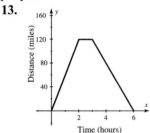

14. (a) [1988, 2002, 1] by [20, 40, 5]

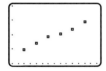

(b) $m = 0.96$, $h = 1990$, $k = 24.8$ thus $y = 0.96(x - 1990) + 24.8$ (answers may vary)
(c) About 36.3 million

CHAPTER 3: LINEAR EQUATIONS AND INEQUALITIES

Section 3.1

1. $ax + b = 0$ **3.** Yes   **5.** An equals sign (=)
7. No   **9.** Yes   **11.** $x = -3$ **13.** $x = -1.5$
15. $x = 1$
17. $x = 2$

x	1	2	3	4	5
$-4x + 8$	4	0	-4	-8	-12

19. $x = -1$

x	-2	-1	0	1	2
$4 - 2x$	8	6	4	2	0
$x + 7$	5	6	7	8	9

21. $x = 10$ **23.** $x = 1.75$ **25.** $x = -2$
27. $x = -2$ **29.** $x = -1$ **31.** $x = -1$
33. $x = 3.2$ **35.** $x = -1.375$ **37.** $x = 2007$
39. $x = 5$ **41.** $x = \frac{16}{5}$ **43.** $x = \frac{6}{7}$
45. $x = -\frac{3}{2}$ **47.** $x = \frac{7}{4}$ **49.** $x = \frac{5}{2}$
51. $x = \frac{72}{53} \approx 1.36$ **53.** $x = 11$ **55.** $x = 1983$
57. $x = 7$ **59.** $x = 3$
61. (a) $x + 2 = 12$ (b) $x = 10$
63. (a) $\frac{x}{5} = x + 1$ (b) $x = -\frac{5}{4}$
65. (a) $\frac{x + 5}{2} = 7$ (b) $x = 9$
67. (a) $\frac{x}{2} = 17$ (b) $x = 34$
69. (a) $x + (x + 1) + (x + 2) = 30$ (b) $x = 9$
71. $x = 8$ feet   **73.** 8.5 hours   **75.** 5 years
77. (a) 36,000 cubic feet per hour   (b) 66 people
79. 51.6 feet   **81.** About 49,412 deaths
83. 1500 people
85. (a) $x = 1992$ (b) $x = 1992$ (c) $x = 1992$
87. (a) $456.8 million; $588 million   (b) $x = 1994$
(c) $x = 1994$

Section 3.2

1. $3x + 2 < 5$ (answers may vary)
3. Yes, they have the same solution set.
5. An equation has an equals sign, whereas an inequality has an inequality sign.
7. Yes   **9.** No   **11.** $\{x | x > -2\}$
13. $\{x | x < 1\}$ **15.** $\{x | x \leq 1\}$
17. $\{x | x \geq 3\}$

x	1	2	3	4	5
$-2x + 6$	4	2	0	-2	-4

19. $\{x | x < -1\}$

x	-3	-2	-1	0	1
$5 - x$	8	7	6	5	4
$x + 7$	4	5	6	7	8

21. $\{x | x > 3\}$ **23.** $\{x | x \geq 2\}$ **25.** $\{x | x > 1\}$
27. $\{x | x \geq 1\}$ **29.** $\{x | x > 2\}$
31. (a) Car 1; In the first 5 hours Car 1 travels 300 miles while Car 2 travels 200 miles.   (b) 5 hours; 400 miles   (c) $0 \leq x < 5$
33. $\{x | x \geq -2\}$ **35.** $\{x | x < -4\}$

37. $\{x|x < -3\}$ **39.** $\{x|x \le 1991\}$
41. $\{x|x \le 2\}$ **43.** $\{x|x > 36\}$ **45.** $\{x|x \ge \frac{9}{8}\}$
47. $\{x|x < \frac{17}{3}\}$ **49.** $\{x|x < \frac{27}{14}\}$
51. $\{x|x \le \frac{2}{13}\}$ **53.** $\{x|x < 2010\}$
55. $\{x|x < 2\}$ **57.** $\{x|x \ge -13\}$
59. 1980 or before
61. (a) $1000   (b) More than $1000   (c) $1000 or less
63. $x < 10$ feet
65. (a) First car: 70 miles per hour; second car: 60 miles per hour   (b) When $x = 3.5$ hours   (c) Elapsed times after 3.5 hours
67. Below about 2.07 miles
69. (a) Increased by $0.083 billion per year   (b) 1997   (c) 1997 or after
71. (a) About 22 inches   (b) Less than 22 inches

Checking Basic Concepts 3.1 & 3.2

1. $x = \frac{1}{4}$
2. (a) $x = -2$ (b) $x = -2$ (c) $x = -2$ Yes
3. $\{x|x > \frac{8}{7}\}$
4. (a) $\{x|x \ge -1\}$ (b) $\{x|x \ge -1\}$
(c) $\{x|x \ge -1\}$ Yes

Section 3.3

1. $x > 1$ and $x \le 7$ (answers may vary)
3. No   **5.** Yes   **7.** Yes, no   **9.** No, yes
11. No, yes
13. $\{x|-1 \le x \le 3\}$
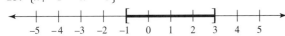
15. $\{x|-2 < x < 2.5\}$

17. $\{x|x \le -1$ or $x \ge 2\}$

19. All real numbers

21. $\{x|-6 \le x \le 7\}$ **23.** $\{x|x > \frac{13}{4}\}$

25. All real numbers   **27.** $\{x|x < -\frac{11}{5}$ or $x > -1\}$
29. No solutions   **31.** $\{x|-9 \le x \le 3\}$
33. $\{x|0 < x < 2\}$ **35.** $\{x|9 < x \le 21\}$
37. $\{x|\frac{12}{5} \le x \le \frac{22}{5}\}$ **39.** $\{x|-1 \le x \le 2\}$
41. $\{x|-1 < x < 2\}$ **43.** $\{x|-3 \le x \le 1\}$
45. $\{x|x < -2$ or $x > 0\}$
47. (a) Toward, because distance is decreasing
(b) 4 hours, 2 hours   (c) From 2 to 4 hours   (d) During the first 2 hours
49. $\{x|1 \le x \le 4\}$ **51.** $\{x|x < -2$ or $x > 0\}$
53. $[2, 10]$ **55.** $(5, 8]$ **57.** $(-\infty, 4)$ **59.** $(-2, \infty)$
61. $(-\infty, -2] \cup [4, \infty)$ **63.** $(-\infty, 1) \cup [5, \infty)$
65. $(-3, 5]$ **67.** $(-\infty, -2)$ **69.** $[1, 3]$
71. $(-\infty, \infty)$ **73.** From 1950 to 1980
75. (a) From 2000 to 2005   (b) From 2000 to 2005
77. From $5.\overline{6}$ to 9 feet
79. (a) $[6, 9]$; the car is from 470 to 680 miles from the rest stop for times between 6 and 9 hours.
(b) $[6, 9]$
81. $-67.\overline{7}°C$ to $36.\overline{6}°C$

Section 3.4

1. $|3x + 2| = 6$ (answers may vary)
3. Yes   **5.** Yes   **7.** No, yes   **9.** No, yes
11. Yes, yes
13. (a) $x = 0$ or $x = 4$ (b) $\{x|0 < x < 4\}$
(c) $\{x|x < 0$ or $x > 4\}$
15. $x = -7$ or $x = 7$ **17.** $x = 0$
19. $x = -\frac{9}{4}$ or $x = \frac{9}{4}$ **21.** $x = -4$ or $x = 4$
23. $x = -6$ or $x = 5$ **25.** $x = 1$ or $x = 2$
27. $x = -8$ or $x = 12$ **29.** No solutions
31. (a) $x = -4$ or $x = 4$ (b) $\{x|-4 < x < 4\}$
(c) $\{x|x < -4$ or $x > 4\}$
33. (a) $x = \frac{1}{2}$ or $x = 2$ (b) $\{x|\frac{1}{2} \le x \le 2\}$
(c) $\{x|x \le \frac{1}{2}$ or $x \ge 2\}$
35. $\{x|x < -\frac{7}{2}$ or $x > \frac{7}{2}\}$ **37.** $\{x|-3 < x < 5\}$
39. $\{x|x \le -9$ or $x \ge -1\}$ **41.** $\{x|\frac{5}{6} \le x \le \frac{11}{6}\}$
43. $\{x|-10 \le x \le 14\}$ **45.** No solutions
47. All real numbers
49. (a) $x = -1$ or $x = 3$ (b) $\{x|-1 < x < 3\}$
51. (a) $x = -1$ or $x = 1$ (b) $\{x|x \le -1$ or $x \ge 1\}$
53. (a) $x = 0$ or $x = 1$ (b) $\{x|0 \le x \le 1\}$
(c) $\{x|x \le 0$ or $x \ge 1\}$

55. $\{x | x \leq -1 \text{ or } x \geq 1\}$ **57.** $\{x | -2 \leq x \leq 4\}$
59. $\{x | x < 1 \text{ or } x > 3\}$ **61.** $\{x | x \leq -2 \text{ or } x \geq 6\}$
63. All real numbers **65.** $\{x | -3 \leq x \leq 3\}$
67. $\{x | x < 2 \text{ or } x > 3\}$
69. (a) $\{T | 19 \leq T \leq 67\}$ (b) Monthly average temperatures vary from 19°F to 67°F.
71. (a) $\{T | -26 \leq T \leq 46\}$ (b) Monthly average temperatures vary from –26°F to 46°F.
73. (a) About 19,058 feet (b) Africa and Europe (c) South America, North America, Africa, Europe, and Antarctica
75. $\{d | 2.498 \leq d \leq 2.502\}$; the diameter can vary from 2.498 to 2.502 inches.

CHECKING BASIC CONCEPTS 3.3 & 3.4

1. (a) Yes (b) No
2. (a) $\{x | -3 \leq x \leq 1\}$ (b) $\{x | x < -1 \text{ or } x \geq 3\}$
3. (a) $x = -2$ or $x = 8$ (b) $x = -2$ or $x = 8$
(c) $x = -2$ or $x = 8$
4. (a) $x = -\dfrac{2}{3}$ or $x = \dfrac{14}{3}$ (b) $\left\{x \mid -\dfrac{2}{3} < x < \dfrac{14}{3}\right\}$
(c) $\left\{x \mid x < -\dfrac{2}{3} \text{ or } x > \dfrac{14}{3}\right\}$

CHAPTER 3 REVIEW

1. $x = 2$

x	0	1	2	3	4
3x − 6	−6	−3	0	3	6

2. $x = 1$

x	−1	0	1	2	3
5 − 2x	7	5	3	1	−1

3. $x = -3$ **4.** $x = 1$ **5.** $x = 1.5$ **6.** $x = 4$
7. $x = 14$ **8.** $x = \dfrac{1}{2}$ **9.** $x = \dfrac{1}{7}$ **10.** $x = -\dfrac{1}{3}$
11. $x = 10$ **12.** $x = \dfrac{21}{8}$
13. $2x + 25 = 19$; $x = -3$
14. $2x - 5 = x + 1$; $x = 6$
15. $\{x | x \leq 0\}$ **16.** $\{x | x > 5\}$ **17.** $\{x | x \geq 2\}$
18. $\{x | x \geq 1.75\}$ **19.** $\{x | x \geq -1\}$
20. $\{x | x \leq -8\}$ **21.** $\{x | x > 1\}$ **22.** $\left\{x \mid x > \dfrac{4}{3}\right\}$
23. $\{x | -2 \leq x \leq 2\}$

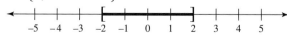

24. $\{x | x \leq -3\}$

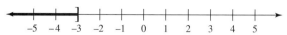

25. $\left\{x \mid x \leq \dfrac{4}{5} \text{ or } x > 2\right\}$

26. All real numbers

27. $\{x | -2 \leq x \leq 1\}$
28. (a) $x = -4$ (b) $x = 2$ (c) $\{x | -4 \leq x \leq 2\}$
(d) $\{x | x < 2\}$
29. (a) $x = 2$ (b) $\{x | x > 2\}$ (c) $\{x | x < 2\}$
30. (a) $x = 4$ (b) $x = 2$ (c) $\{x | 2 < x < 4\}$
31. $\left[-3, \dfrac{2}{3}\right]$ **32.** $(-6, 45]$ **33.** $\left(-\infty, \dfrac{7}{2}\right)$
34. $[1.8, \infty)$ **35.** $(-3, 4)$ **36.** $(-\infty, 4) \cup (10, \infty)$
37. $(-5, 5)$ **38.** $[8, 28]$ **39.** $(-9, 21)$ **40.** $[8, 28]$
41. No, no **42.** No, yes **43.** No, yes
44. No, yes
45. (a) $x = 0$ or $x = 4$ (b) $\{x | 0 < x < 4\}$
(c) $\{x | x < 0 \text{ or } x > 4\}$
46. (a) $x = -1$ or $x = 2$ (b) $\{x | -1 \leq x \leq 2\}$
(c) $\{x | x \leq -1 \text{ or } x \geq 2\}$
47. $x = -22$ or $x = 22$ **48.** $x = 1$ or $x = 8$
49. $x = -26$ or $x = 42$ **50.** $x = -\dfrac{23}{3}$ or $x = \dfrac{25}{3}$
51. (a) $x = -8$ or $x = 6$ (b) $\{x | -8 \leq x \leq 6\}$
(c) $\{x | x \leq -8 \text{ or } x \geq 6\}$
52. (a) $x = -\dfrac{5}{2}$ or $x = \dfrac{7}{2}$ (b) $\left\{x \mid -\dfrac{5}{2} \leq x \leq \dfrac{7}{2}\right\}$
(c) $\left\{x \mid x \leq -\dfrac{5}{2} \text{ or } x \geq \dfrac{7}{2}\right\}$
53. $\{x | x < -3 \text{ or } x > 3\}$ **54.** $\{x | -4 < x < 4\}$
55. $\{x | -3 \leq x \leq 4\}$ **56.** $\left\{x \mid x \leq -\dfrac{5}{2} \text{ or } x \geq 5\right\}$
57. $\{x | x \leq -1.5 \text{ or } x \geq 1.5\}$ **58.** $\{x | -2 \leq x \leq 6\}$
59. (a) 3 hours (b) 2 hours and 4 hours
(c) Between 2 and 4 hours, exclusively (d) 20 miles per hour
60. (a) $2000 (b) More than $2000
(c) Less than $2000
61. (a) Car 1; its graph has the steeper slope.
(b) 3 hours; 200 miles (c) Before 3 hours
62. 1989, 1990, 1991, 1993, and 1995

63. (a) Median age is increasing at 0.09 year per year.
[1820, 1995, 20] by [0, 40, 10]

(b) 1890 **(c)** $x = 1890$
64. (a) Approximately from 1985 to 1995
(b) Approximately from 1985 to 1995
65. (a) About 31.8 inches
(b) Lengths less than 31.8 inches
66. Approximately from 0.69 mile to 1.38 miles
67. 8.5 hours **68.** 13 feet by 31 feet
69. $-44.\overline{4}°C$ to $41.\overline{6}°C$
70. $|L - 160| < 1; 159 < L < 161$
71. (a) $|A - 3.9| < 1.7$ **(b)** $2.2 < A < 5.6$

Chapter 3 Test

1. -3
2. (a) 2 **(b)** $\{x | x \leq 2\}$ **(c)** $\{x | x \geq 2\}$
3. $\frac{4}{9} = 0.\overline{4}$ **4.** $\frac{7}{9}$ **5.** $2 + 5x = x - 4$; $-\frac{3}{2}$
6. $x \geq 1.25$ **7. (a)** $x \geq -\frac{3}{5}$ **(b)** $x > \frac{31}{17}$
8.

9. $x < -2$ or $x > 1$
10. (a) -5 **(b)** 5 **(c)** $\{x | -5 \leq x \leq 5\}$
(d) $\{x | x < 5\}$
11. $(-8, 0)$ **12.** $-12, 24$
13. (a) $-\frac{2}{5}, \frac{4}{5}$ **(b)** $-\frac{2}{5} \leq x \leq \frac{4}{5}$
(c) $x \leq -\frac{2}{5}$ or $x \geq \frac{4}{5}$
14. (a) 65 calories **(b)** More than 65 calories
(c) Less than 65 calories
15. (a) $f(x) = 0.4x + 75$ **(b)** 35 minutes
(c) 35 minutes

CHAPTER 4: SYSTEMS OF LINEAR EQUATIONS

Section 4.1

1. No. Two straight lines cannot have exactly two points of intersection.
3. Numerically and graphically
5. There are no solutions. **7.** $(1, -2)$
9. $\left(-1, \frac{13}{3}\right)$ **11.** $(3, 1)$ **13.** No solutions
15. (a) $(1, 2)$ **(b)** $x + y = 3$; $x - y = -1$
17. (a) $(-4, 16)$ **(b)** $2x + 3y = 40$; $x - y = -20$
19. (a) No solutions **(b)** $x - y = 2$; $x - y = -2$
21. Unique solution: $(4, 2)$
23. Inconsistent: no solutions
25. Dependent: $\{(x, y) | -3x + 2y = 1\}$
27. Unique solution: $(2, -2)$
29. Unique solution: $(4, 2)$
31. Unique solution: $(1, 2)$ **33.** $(2, 3)$ **35.** $(1, 1)$
37. $(5, -2)$ **39.** $(-5, 10)$ **41.** $(0.5, 1.5)$
43. $(1, 1)$ **45.** $(1, -4)$ **47.** $(100, 200)$
49. No solutions
51. (a) $x + y = 18$; $x - y = 6$ **(b)** $x = 12, y = 6$
53. (a) $x + y = 1$; $6x + 9y = 8$ **(b)** 6 miles per hour for 20 minutes and 9 miles per hour for 40 minutes
55. (a) $2x + 2y = 76$; $x - y = 4$ **(b)** 21 by 17 inches
57. (a) $x + 2y = 180$; $x - y = 60$ **(b)** $100°, 40°, 40°$
59. 6,295,003 in 1998; 6,539,864 in 1999
61. 1,484,477 in 1996; 1,406,505 in 1997
63. (a) and **(b)** Approximately 115.5 pounds on each rafter

Section 4.2

1. Substitution and elimination
3. Elimination yields equations that are always false.
5. Solve the first equation for y. **7.** $(1, 2)$
9. $(5, 3)$ **11.** $(2, 1)$ **13.** $\left(\frac{1}{4}, \frac{3}{4}\right)$ **15.** $(0, 1)$
17. $(-3, -1)$ **19.** $\left(\frac{1}{2}, -\frac{1}{2}\right)$ **21.** $(4, -6)$
23. $(5, 2)$
25. For $x = y + 5$, the result is $10 = 10$, which is always true.
27. $(7, 2)$ **29.** $\left(\frac{1}{2}, 3\right)$ **31.** $\left(\frac{6}{7}, -\frac{12}{7}\right)$
33. Inconsistent: no solutions
35. Dependent: $\{(x, y) | 2x + y = 2\}$
37. $\left(-\frac{5}{9}, -\frac{47}{9}\right)$ **39.** $(2, 1)$ **41.** $(5, 1)$
43. $(1, 2)$ **45.** $(2, -3)$
47. The heavier athlete burned 174 calories or about 19.3 fat grams; the lighter athlete burned 116 calories or about 12.9 fat grams.
49. $\frac{16}{7}$ gallons
51. (a) $a = -5.125, b = 10,278.75$ **(b)** Approximately $54.38

53. Plane: 608 miles per hour; jet stream: 32 miles per hour
55. $2100 at 8% and $1400 at 9%
57. $2x + 2y = 296$; $x - y = 44$ Length = 96 feet, width = 52 feet

CHECKING BASIC CONCEPTS 4.1 & 4.2

1. (3, 1); Yes   **2.** (2, 3)   **3.** $\left(-\dfrac{7}{2}, 1\right)$; no

SECTION 4.3

1. Two   **3.** Yes
5.   **7.**

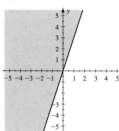 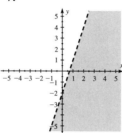

9.   **11.**

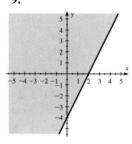 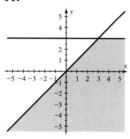

13.   **15.**

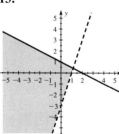 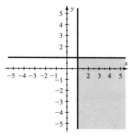

17.   **19.**

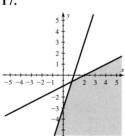 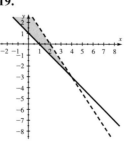

21. b.   **23.** a.   **25.** $y \geq 2$
27. $y \geq x$; $x \geq -2$
29.   **31.**
[−10, 10, 1] by [−10, 10, 1]   [−10, 10, 1] by [−10, 10, 1]

33.
[−10, 10, 1] by [−10, 10, 1]

35. (a) About 105 to 134 beats per minute
(b) $y \leq -0.6(x - 20) + 140$; $y \geq -0.5(x - 20) + 110$
37. The person weighs less than recommended.
39. $25h - 7w \leq 800$; $5h - w \geq 170$

SECTION 4.4

1. No; three planes cannot intersect at exactly 2 points.
3. Yes   **5.** Two   **7.** (1, 2, 3)   **9.** (−1, 1, 2)
11. (3, −1, 1)   **13.** $\left(\dfrac{17}{2}, -\dfrac{3}{2}, 2\right)$
15. (5, −2, −2)   **17.** (11, 2, 2)
19. (1, −1, 2)   **21.** (0, −3, 2)   **23.** (−1, 3, −2)
25. $\left(-\dfrac{3}{2}, 5, -10\right)$ **27.** $\left(\dfrac{3}{2}, 1, -\dfrac{1}{2}\right)$
29. (a) $x + 2y + 4z = 10$; $x + 4y + 6z = 15$; $3y + 2z = 6$ **(b)** (2, 1, 1.5); a hamburger costs $2, fries $1, and a soft drink $1.50.
31. (a) $x + y + z = 180$; $x - z = 55$; $x - y - z = -10$ **(b)** $x = 85°, y = 65°$, and $z = 30°$ **(c)** These values check.
33. (a) $a + 600b + 4c = 525$; $a + 400b + 2c = 365$; $a + 900b + 5c = 805$
(b) $a = 5, b = 1, c = -20$; $F = 5 + A - 20W$
(c) 445 fawns
35. $7500 at 8%, $8500 at 10%, and $14,000 at 15%

Checking Basic Concepts 4.3 & 4.4

1. $y \leq -2x + 3$
2.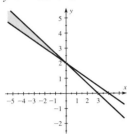

3. $(1, 3, -1)$ 4. $(1, 2, 3)$

Section 4.5

1. A rectangular array of numbers
3. $\begin{bmatrix} 1 & 3 & | & 10 \\ 2 & -6 & | & 4 \end{bmatrix}$; 2×3 (answers may vary)
5. $\begin{bmatrix} 1 & 0 & | & -3 \\ 0 & 1 & | & 5 \end{bmatrix}$ (answers may vary)
7. 3×3 9. 3×2 11. $\begin{bmatrix} 1 & -3 & | & 1 \\ -1 & 3 & | & -1 \end{bmatrix}$
13. $\begin{bmatrix} 2 & -1 & 2 & | & -4 \\ 1 & -2 & 0 & | & 2 \\ -1 & 1 & -2 & | & -6 \end{bmatrix}$
15. $x + 2y = -6$; $5x - y = 4$
17. $x - y + 2z = 6$; $2x + y - 2z = 1$; $-x + 2y - z = 3$
19. $x = 4$; $y = -2$; $z = 7$
21. $\begin{bmatrix} 1 & -1 & | & 5 \\ 1 & 3 & | & -1 \end{bmatrix}$; $\left(\frac{7}{2}, -\frac{3}{2}\right)$
23. $\begin{bmatrix} 4 & -8 & | & -10 \\ 1 & 1 & | & 2 \end{bmatrix}$; $\left(\frac{1}{2}, \frac{3}{2}\right)$
25. $\begin{bmatrix} 1 & 1 & 1 & | & 0 \\ 2 & 1 & 2 & | & -1 \\ 1 & 1 & 0 & | & 0 \end{bmatrix}$; $(-1, 1, 0)$
27. $\begin{bmatrix} 1 & 1 & 1 & | & 3 \\ -1 & 0 & -1 & | & -2 \\ 1 & 1 & 2 & | & 4 \end{bmatrix}$; $(1, 1, 1)$
29. $\begin{bmatrix} 1 & 2 & 1 & | & 3 \\ 2 & 1 & -1 & | & -6 \\ -1 & -1 & 2 & | & 5 \end{bmatrix}$; $(-3, 2, 2)$
31. $(-7, 5)$ 33. $(1, 2, 3)$ 35. $(1, 0.5, -3)$
37. $(0.5, 0.25, -1)$ 39. $(0.5, -0.2, 1.7)$
41. 214 pounds
43. (a) $a + 2b + 1.4c = 3$; $a + 1.5b + 0.65c = 2$; $a + 4b + 3.4c = 6$
(b) $a = 0.6$, $b = 0.5$, and $c = 1$ (c) $4.1 \approx 4$ people
45. $\frac{1}{2}$ hour at 5 miles per hour, 1 hour at 6 miles per hour, and $\frac{1}{2}$ hour at 8 miles per hour
47. $500 at 5%, $1000 at 8%, and $1500 at 12%

Section 4.6

1. -2 3. -53 5. -323 7. -5 9. -36
11. -42 13. -50 15. 0 17. -3555
19. -7466.5 21. 15 feet^2 23. 52 feet^2
25. 25.5 feet^2 27. $x = 2, y = -2$
29. $x = -\frac{29}{11}, y = -\frac{34}{11}$ 31. $x = -5, y = 7$

Checking Basic Concepts 4.5 & 4.6

1. $(-2, 2, -1)$ 2. (a) -1 (b) 17
3. $x = -4, y = 6$ 4. 21 square units

Chapter 4 Review Exercises

1. $(3, 2)$ 2. $(4, -3)$ 3. $(-1, -3)$ 4. $(3.5, 8.5)$
5. Unique solution: $(1, 5)$
6. Dependent: $\{(x, y) | x - y = -2\}$
7. Inconsistent: no solutions
8. Unique solution: $(2, -1)$
9. Unique solution: $(-0.5, 1)$
10. Unique solution: $(1.5, 3)$ 11. $(3, -2)$
12. $(-2, 1)$
13. (a) $x + y = 25$; $x - y = 10$ (b) $(17.5, 7.5)$
14. (a) $3x - 2y = 19$; $x + y = 18$ (b) $(11, 7)$
15. $(-3, 1)$ 16. $(2, 0)$ 17. $\left(\frac{2}{5}, \frac{14}{5}\right)$
18. $(-2, -3)$ 19. $\{(x, y) | 3x - y = 5\}$
20. No solution
21.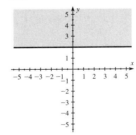
22.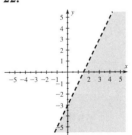

23. 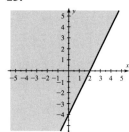   **24.**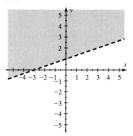

25. 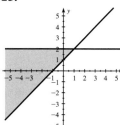   **26.**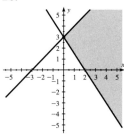

27. 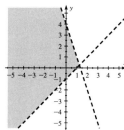   **28.**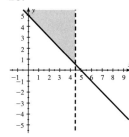

29. $x > 1$; $y < -1$ **30.** $y \leq -x + 4$; $y \geq 2x + 1$
31. Yes   **32.** $(1, -1, 2)$ **33.** $(-4, 3, 2)$
34. $(-1, 5, -3)$ **35.** $(-1, 3, 2)$ **36.** $(1, 1, 1)$

37. $\begin{bmatrix} 1 & 1 & 1 & | & -6 \\ 1 & 2 & 1 & | & -8 \\ 0 & 1 & 1 & | & -5 \end{bmatrix}$; $(-1, -2, -3)$

38. $\begin{bmatrix} 1 & 1 & 1 & | & -3 \\ -1 & 1 & 0 & | & 5 \\ 0 & 1 & 1 & | & -1 \end{bmatrix}$; $(-2, 3, -4)$

39. $\begin{bmatrix} 1 & 2 & -1 & | & 1 \\ -1 & 1 & -2 & | & 5 \\ 0 & 2 & 1 & | & 10 \end{bmatrix}$; $(-5, 4, 2)$

40. $\begin{bmatrix} 2 & 2 & -2 & | & -14 \\ -2 & -3 & 2 & | & 12 \\ 1 & 1 & -4 & | & -22 \end{bmatrix}$; $(-4, 2, 5)$

41. $(-7, 4, 2)$ **42.** $(-3, 1.5, 3)$
43. $(5.5, 7, -2.5)$ **44.** $(5.4, 2.1, 9.7)$ **45.** -8
46. 30   **47.** 89   **48.** 130   **49.** 181,845
50. 67.688   **51.** 46 feet^2 **52.** 128 feet^2

53. $x = 2, y = -1$ **54.** $x = -5, y = 3$
55. $x = \dfrac{3}{2}, y = \dfrac{1}{2}$ **56.** $x = 7, y = -3$
57. 6870 in 1988; 5220 in 1998
58. Stair climber: 8 minutes; bicycle: 22 minutes
59. (a) $1996a + b = 42$; $1999a + b = 58$; $a = 5.\overline{3}$; $b = -10,603.\overline{3}$ (b) $52.\overline{6}$ million
60. 30% solution: 2.4 gallons; 55% solution: 1.6 gallons
61. Boat: 15 miles per hour; current: 3 miles per hour
62. $8 tickets: 285; $12 tickets: 195
63. (a) $x + 3y + 5z = 14$; $x + 2y + 4z = 11$; $y + 3z = 5$ (b) Malts: $3; cones: $2; bars: $1
64. 100°, 65°, and 15°
65. 2 pounds of $1.50 candy; 4 pounds of $2 candy; 6 pounds of $2.50 candy
66. (a) $a + 202b + 63c = 40$; $a + 365b + 70c = 50$; $a + 446b + 77c = 55$ (b) $a \approx 27.134$; $b \approx 0.061$; $c \approx 0.009$ (c) About 46 inches

Chapter 4 Test

1. Unique solution: $\left(\dfrac{3}{2}, -\dfrac{3}{2}\right)$
2. Inconsistent: no solutions   **3.** $(-3, 1)$
4. (a) $x - y = 34$; $x - 2y = 0$ (b) $(68, 34)$
(c) $(68, 34)$
5.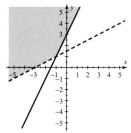

6. $(-4, 2, -5)$ **7.** $y \leq x + 2$; $y \geq 3x - 1$

8. (a) $\begin{bmatrix} 1 & 1 & 1 & | & 2 \\ 1 & -1 & -1 & | & 3 \\ 2 & 2 & 1 & | & 6 \end{bmatrix}$ (b) $\left(\dfrac{5}{2}, \dfrac{3}{2}, -2\right)$

(c) The solution checks.
9. 114   **10.** $\left(-\dfrac{47}{2}, -\dfrac{83}{2}\right)$
11. (a) $x - y = 11,500$; $x - 2.5y = 0$ (b) Approximately $(19,167, 7667)$; private tuition was $19,167, and public tuition was $7667.
12. Running: 43 minutes; rowing: 17 minutes
13. Airplane: 270 miles per hour; wind: 30 miles per hour
14. 85°, 60°, and 35°

CHAPTER 5: POLYNOMIAL EXPRESSIONS AND FUNCTIONS

Section 5.1

1. $3x^2$ (answers may vary) 3. $3; -1$
5. $x^4 - 3x + 5$ (answers may vary)
7. No, the opposite is $-x^2 - 1$. 9. Yes
11. No 13. Yes 15. No 17. x^2 19. πr^2
21. xy 23. $7; 3$ 25. $7; -3$ 27. $6; -1$
29. $5x^2$ 31. $3y^4$ 33. Not possible 35. $3x^2 + 3x$
37. $5x^2 + 9xy$ 39. $5x^3y^7 + 13x$ 41. $2; 5$
43. $3; -\dfrac{2}{5}$ 45. $5; 1$ 47. $2x + 2$
49. $-2x^2 + 3x + 8$ 51. $-0.5x + 1$
53. $-7x^4 - 2x^2 - \dfrac{9}{2}$ 55. $-6x^5$
57. $-19x^5 + 5x^3 - 3x$ 59. $7z^4 - z^2 + 8$
61. $3x - 7$ 63. $6x^2 - 5x + 5$
65. $3x^4 + 4x^2 - 4$ 67. $-3x^4 - 3x - 8$
69. $7x - 10$ 71. $-10x - 5$
73. (a) 1250; 7250 (b) 1268; 7276; they are similar
(c) 8188
75. About 180 million
77. $0.0763x - 146.6$

79. $x^2 + \pi x^2$; $100 + 100\pi \approx 414.2$ square inches

Section 5.2

1. Linear data lie on a line.
3. No. A function has only one output for each input.
5. The highest power of the variable 7. Yes
9. No 11. No 13. Yes 15. 12 17. 7
19. 5.8 21. 2 23. $-1; 2$ 25. $-4; 2$
27. $-4; 2$ 29. 1 31. 0 33. 1.96 35. -6
37. $f(10) = 206$ thousand, which is close to the true value.
39. From 4 to 8 minutes

41. The deaths oscillate with lower values in 1988, 1992, and 1998 and peak values in 1989 and 1996.

43. (a) About $53,770; about $1,318,100

(b) 1985
[6, 29, 1] by [0, 2, 0.5]

45. $f(x)$ is the best, which can be determined from a table of values.

Checking Basic Concepts 5.1 & 5.2

1. (a) $3x^2 + 7x$ (b) $5x^3 + 2x^2 - 3x + 1$
2. $x(x + 120)$; 1300, when tickets cost $10 each, the revenue will be $1300.
3. -27
4. (a) 160 beats per minute (b) 92 beats per minute
(c) As time passes, the heart rate returns to a normal, slower rate.
[0, 6, 1] by [0, 180, 20]

Section 5.3

1. Distributive 3. x^8 5. $a^2 - b^2$ 7. $5y + 10$
9. $-10x - 18$ 11. $-6y^2 + 18y$ 13. $27x - 12x^2$
15. x^{12} 17. $-20y^8$ 19. $-4x^4y^6$ 21. $20x^2y^3z^6$
23. $x^2 + 3x + 2$; 42 square inches
25. $2x^2 + 3x + 1$; 66 square inches
27. $x^2 + 11x + 30$
29. $4x^2 + 4x + 1$ 31. No 33. Yes 35. No
37. $x^2 + 15x + 50$ 39. $x^2 - 7x + 12$
41. $2x^2 + 3x - 2$ 43. $y^2 - y - 12$
45. $-36x^2 + 43x - 12$ 47. $-2x^2 + 7x - 6$
49. $x^2 - \dfrac{1}{4}x - \dfrac{1}{8}$ 51. $2x^4 + x^2 - 1$
53. $x^2 - xy - 2y^2$ 55. $4x^3 - 8x^2 - 12x$
57. $-x^5 + 3x^3 - x$ 59. $6x^4 - 12x^3 + 3x^2$
61. $x^3 + 3x^2 - x - 3$ 63. $2^6 = 64$ 65. $2z^{18}$
67. $25x^2$ 69. $-8x^3y^6$ 71. $x^2 - 9$ 73. $4x^2 - 9$
75. $x^2 - y^2$ 77. $x^2 + 4x + 4$ 79. $4x^2 + 4x + 1$
81. $x^2 - 2x + 1$ 83. $9x^2 - 12x + 4$
85. $3x^3 - 3x$ 87. $x(x + 4)$; 480 square feet
89. $(x + 3)^2$; 529 square feet
91. (a) $Nr^2 + 2Nr + N$ (b) $242; the answers agree.

93. $x(x+1) = x^2 + x$
95. $x(50-x) = 50x - x^2$

Section 5.4

1. To solve equations
3. Yes; $2x$ is a factor of each term.
5. No; because $\frac{1}{2}(4) = 2$, and $\frac{1}{2} \neq 2$ and $4 \neq 2$
7. $5(2x-3)$ 9. $x(2x^2-5)$
11. $4x(2x^2 - x + 4)$ 13. $5x^2y^2(1-3y)$
15. $5xy(3x + 2 - 5xy)$ 17. $0, 2$ 19. $0, \frac{1}{4}$
21. $0, 1$ 23. $0, \frac{1}{5}$ 25. $0, \frac{25}{8}$ 27. $0, \frac{1}{2}$ 29. -2
31. $0, 1, 5$ 33. $(x+3)(x^2+2)$
35. $(3x-2)(2x^2+3)$ 37. $(x-5)(3x^2+5)$
39. $(y+1)(x+3)$
41. **(a)** After 8 seconds
   (b) It does.
   [0, 9.4, 1] by [−10, 300, 100]

   (c) Yes

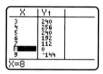

   (d) 256 ft; after 4 seconds
43. **(a)** The function models the data quite well.

   (b) About 54,000 feet
   (c) $t = 0$ or $t = -\frac{6}{11}$; when $t = 0$, the shuttle has not left the ground; $t = -\frac{6}{11}$ has no physical meaning.
45. No; there are many possibilities such as 9×24 and 8×27

Checking Basic Concepts 5.3 & 5.4

1. **(a)** $-5x + 30$ **(b)** $12x^5 - 20x^4$
   (c) $2x^2 + 5x - 3$
2. **(a)** $25x^2 - 36$ **(b)** $9x^2 - 24x + 16$

3.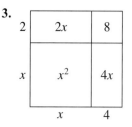

4. **(a)** $3x(x-2)$ **(b)** $4x(4x^2 - 2x + 1)$
5. **(a)** $0, 2$ **(b)** $0, 9$

Section 5.5

1. A polynomial with three terms: $x^2 - x + 1$ (answers may vary)
3. First, Outside, Inside, Last
5. No, $(x-2)(x-3) = x^2 - 5x + 6$ 7. Yes
9. No   11. Yes   13. No   15. $(x+2)(x+5)$
17. $(x+2)(x+6)$ 19. $(x-9)(x-4)$
21. $(x-8)(x+1)$ 23. $(x+3)(2x+1)$
25. $(2x+1)(3x-2)$ 27. $(1-x)(1+2x)$
29. $(5-2x)(4+3x)$ 31. $x(x-1)(5x+6)$
33. $3x(x+3)(2x+1)$ 35. $(x+1)(x-4)$
37. $2(x+1)(x-2)$ 39. $(x-2)(x-4)$
41. $2(x+1)(x-2)$ 43. $(x-2)(x+5)$
45. $(x-7)(x+4)$ 47. $2(x-5)(x-2)$
49. $5(x-10)(x+4)$ 51. $4(x-5)(2x-1)$
53. **(a)** $(35-x)(1200+100x)$ **(b)** $20 or $27
   (c) $20 or $27   **(d)** $23.50

Section 5.6

1. $x^2 - 9$ (answers may vary)
3. $x^3 + 8$ (answers may vary)   5. $(a+b)^2$
7. Yes; $(x-5)(x+5)$
9. No; $(x+y)(x^2 - xy + y^2)$
11. $(x-6)(x+6)$ 13. $(2x-5)(2x+5)$
15. $4(3x-5)(3x+5)$ 17. $z^2(8-5z)(8+5z)$
19. $(2x-y)(2x+y)(4x^2+y^2)$ 21. No
23. Yes; $(x+4)^2$ 25. Yes; $(2z-1)^2$ 27. No
29. $(x+1)^2$ 31. $(2x+5)^2$ 33. $(x-6)^2$
35. $z(3z-1)^2$ 37. $(x-2)(x^2+2x+4)$
39. $(y+z)(y^2 - yz + z^2)$
41. $(3x-2)(9x^2+6x+4)$
43. $x(2x+5)(4x^2 - 10x + 25)$
45. $y(3-2x)(9+6x+4x^2)$
47. $(5x-8)(5x+8)$ 49. $(x+3)(x^2-3x+9)$
51. $(8x+1)^2$ 53. $(x+4)(3x+2)$
55. $x(x+2)(x^2-2x+4)$
57. $8(2x+y)(4x^2-2xy+y^2)$ 59. $-3, 2$
61. $-2, 4$ 63. $-3, 5$ 65. $-\frac{1}{2}, 2$ 67. $-5, -1$

69. 0, 2
71. **(a)** The elevation begins at 500 feet, decreases, and then increases back to 500 feet.

(b) 500 feet and 1000 feet

CHECKING BASIC CONCEPTS 5.5 & 5.6

1. **(a)** $(x - 2)(x + 5)$ **(b)** $(x - 5)(x + 2)$
(c) $(2x + 3)(4x + 1)$
2. **(a)** $(5x - 4)(5x + 4)$ **(b)** $(x + 6)^2$
(c) $(x - 3)(x^2 + 3x + 9)$ **(d)** $(x - 3)(x^2 + 1)$
3. $-1, -2$ **4.** After 1 and 3 seconds

CHAPTER 5 REVIEW EXERCISES

1. $x + 5$; $x^2 - 3x + 1$ (answers may vary)
2. $10xy^2$; 3200 square inches **3.** 5; -4
4. 3; 1 **5.** 7; 5 **6.** 10; -9 **7.** $11x$ **8.** $3y$
9. $4x^3 + x^2$ **10.** $14x^3y + x^3$ **11.** 2; 5
12. 3; -9 **13.** 2; 8 **14.** 4; 2 **15.** $8x^2 + 3x - 1$
16. $23z^3 - 4z^2 + z$ **17.** $-x^2 + x$
18. $-5x^3 - x^2 - 3x + 6$ **19.** -32 **20.** -21
21. 7 **22.** -259 **23.** -4 **24.** 0.75
25. $15x - 20$ **26.** $-2x - 2x^2 + 8x^3$ **27.** x^8
28. $-6x^4$ **29.** $-42x^2y^8$ **30.** $60x^3y^5$
31. $x^2 + 9x + 20$ **32.** $x^2 - 15x + 56$
33. $12x^2 - 48x - 27$ **34.** $2x^2 - 5x - 12$
35. $x^2 - \frac{1}{9}$ **36.** $y^2 - y + \frac{6}{25}$
37. $8x^4 - 12x^3 - 4x^2$ **38.** $-4x - 5x^2 + 7x^3$
39. $16x^2 - y^2$ **40.** $x^2 + 6x + 9$
41. $4y^2 - 20y + 25$ **42.** $a^3 - b^3$
43. $5x(5x - 6)$ **44.** $6x(3x^2 + 1)$
45. $5y(1 + 3y)$ **46.** $4x(3x^2 + 2x - 4)$
47. $-3, 0$ **48.** $-2, 0$ **49.** $-2, 0, 2$ **50.** 0, 3
51. $\frac{1}{2}, 1$ **52.** $-1, 3$ **53.** $(x + 1)(2x^2 - 3)$
54. $(2x + 3)(x^2 + 3)$ **55.** $(z + 1)(z^2 + 1)$
56. $(a - b)(x + y)$ **57.** $(x + 2)(x + 6)$
58. $(x - 10)(x + 5)$ **59.** $(x + 3)(9x - 2)$
60. $2(x - 5)(2x - 1)$ **61.** $(t - 7)(t + 7)$
62. $(2y - 3x)(2y + 3x)$ **63.** $(x + 2)^2$
64. $(4x - 1)^2$ **65.** $(x - 3)(x^2 + 3x + 9)$
66. $(2y + 1)(4y^2 - 2y + 1)$
67. $(5a - 4)(25a^2 + 20a + 16)$
68. $(4x + 3y)(16x^2 - 12xy + 9y^2)$
69. $(x + 3)(x - 5)$ **70.** $(x - 11)(x - 13)$
71. $(x + 7)(x - 4)$ **72.** 8, 13

73. **(a)** 30.304 million **(b)** 1996
(c) About 1992

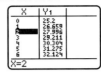

74. **(a)** 168.5 million in 1995; 419.5 million in 1997; close to actual sales shown in the graph
(b) The table shows 850.5 million, which is too high, because f continues to increase after 1997, whereas the actual data decrease.

75. **(a)** 73.6°F
(b) July

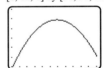

(c) Temperatures increase from January through July and then decrease from July through December.

[1, 12, 1] by [30, 90, 10]

76. **(a)** $1 - \frac{x}{50} + \frac{x^2}{10,000}$ **(b)** 0.09
77. $(x - 2)(x - 1)(x) = x^3 - 3x^2 + 2x$
78.

7	$7x$	14
x	x^2	$2x$
	x	2

79. 9 feet by 16 feet

CHAPTER 5 TEST

1. $5x - 4x^2y^2$ **2.** $-7x^3 + x^2 - 7x + 11$ **3.** -8
4. -2 **5.** $-4x^3 + 2x^2$ **6.** $14x^2y^8$
7. $10x^2 - 9x - 7$ **8.** $9x^2 - 30x + 25$
9. $(x + 4)(3x - 5)$ **10.** $5x^2(x - 1)(x + 1)$
11. $(2x + 1)(x^2 - 5)$ **12.** $(7x - 1)^2$
13. $(x + 2)(x^2 - 2x + 4)$ **14.** 3; 1
15. $2(x + 1)\left(x - \frac{1}{2}\right) = (x + 1)(2x - 1)$

16. $(x + 8)(x - 6)$ **17.** $x(x + 2)$ **18.** $0, 3$
19. $-5, \frac{1}{4}$ **20.** $-2, 0, 2$
21. (a) $x(x + 4) = 221$ **(b)** $13, -17$; width is 13 inches. **(c)** 60 inches
22. (a) About 57.5°F
(b) August

(c) From January to August the monthly average dew point increases, and then it decreases from August to December.
$[1, 12, 1]$ by $[25, 75, 10]$

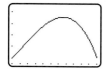

CHAPTER 6: QUADRATIC FUNCTIONS AND EQUATIONS

Section 6.1

1. Parabola **3.** Axis of symmetry **5.** $(0, 0)$
7. $-\frac{b}{2a}$ **9.** Linear
11. Quadratic; $(1, -2)$; $x = 1$; opens upward
13. Neither
15. Quadratic; $(-2, 3)$; $x = -2$; opens downward
17. (a) $[-5, 5, 1]$ by $[-5, 5, 1]$

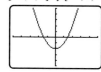

(b) $(0, -2)$; $x = 0$
(c)

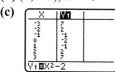

(d) $2; 7$
19. (a) $[-5, 5, 1]$ by $[-5, 5, 1]$

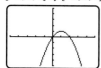

(b) $(1, 1)$; $x = 1$
(c)

(d) $-8; -3$
21. (a) $[-5, 5, 1]$ by $[-5, 5, 1]$

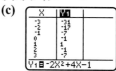

(b) $(1, 1)$; $x = 1$
(c)

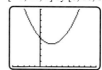

(d) $-17; -7$
23. (a) $[-5, 10, 1]$ by $[0, 10, 1]$

(b) $(2, 4)$; $x = 2$
(c)

(d) $8; 4.25$
25. $0, -4$ **27.** $-2, -2$ **29.** $(2, -6)$ **31.** $(-3, 4)$
33. $(0, 3)$ **35.** $(1, 1.4)$ **37.** $a = 2$ **39.** $a = 0.3$
41. (a) $[-10, 10, 1]$ by $[-10, 10, 1]$

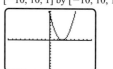

(b) The graph of f is shifted 4 units downward compared to $y = x^2$.
43. (a) $[-10, 10, 1]$ by $[-10, 10, 1]$

(b) The graph of f is shifted 3 units to the right compared to $y = x^2$.

45. (a) $[-10, 10, 1]$ by $[-10, 10, 1]$

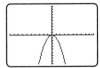

(b) The graph of f is reflected across the x-axis compared to $y = x^2$.

47. (a) $[-10, 10, 1]$ by $[-10, 10, 1]$

(b) The graph of f is shifted 1 unit to the left and 2 units downward compared to $y = x^2$.

49. (a) $[-10, 10, 1]$ by $[-10, 10, 1]$

(b) The graph of f is shifted 1 unit to the right and 2 units upward compared to $y = x^2$.

51. $(0, 0)$ **53.** $(0, 2)$

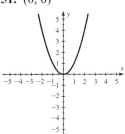 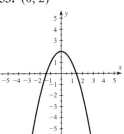

55. $(-2, 0)$

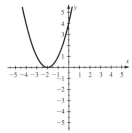

57. $y = 3(x - 3)^2 + 4$; $y = 3x^2 - 18x + 31$
59. $y = -\frac{1}{2}(x - 5)^2 - 2$; $y = -\frac{1}{2}x^2 + 5x - \frac{29}{2}$
61. $y = (x - 1)^2 + 2$ **63.** $y = -(x - 0)^2 - 3$
65. $y = (x - 0)^2 - 3$ **67.** $y = -(x + 1)^2 + 2$
69. d.   **71.** a.
73. (a) $f(x) = x(30 - x)$
(b) 15 feet by 15 feet; 225 square feet

75. $\frac{66}{32} \approx 2$ seconds; about 74 feet
77. (a) $f(x) = x(300 - 1.5x)$
(b) 100 tickets; $15,000
79. $y = 2(x - 1)^2 - 3$
81. $y = 0.5(x - 1980)^2 + 6$

Section 6.2

1. $x^2 + 3x - 2 = 0$ (*answers may vary*); 0, 1, or 2 solutions
3. Factoring, square root property, completing the square
5. (*answers may vary*)

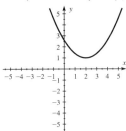

7. $x = \pm 8$; the square root property   **9.** Yes
11. No   **13.** Yes   **15.** No   **17.** $-2, 1$
19. No real solutions   **21.** $-2, 3$ **23.** -0.5
25. $-1, 5$ **27.** $-\frac{1}{2}, \frac{3}{2}$ **29.** $-3, 3$
31. No real solutions   **33.** $-7, 5$ **35.** $-\frac{1}{3}, \frac{1}{2}$
37. $1, \frac{3}{2}$ **39.** $\frac{6}{5}, \frac{3}{2}$ **41.** ± 12 **43.** $\pm\frac{8}{\sqrt{5}}$
45. $-6, 4$ **47.** No real solutions   **49.** $\frac{1 \pm \sqrt{5}}{2}$
51. 4   **53.** $\frac{25}{4}$ **55.** 16; $(x - 4)^2$
57. $\frac{81}{4}$; $\left(x + \frac{9}{2}\right)^2$ **59.** $-4, 6$ **61.** $-3 \pm \sqrt{11}$
63. $\frac{3 \pm \sqrt{41}}{4}$ **65.** $\frac{2 \pm \sqrt{11}}{2}$ **67.** $-3, 6$
69. $3, 5$ **71.** $5, 7$
73. (a) 30 miles per hour   **(b)** 40 miles per hour
75. About 23°C and 34°C
77. (a) The rate of increase each 20-year period is not constant.
(b) $(1800, 5)$; in 1800 the population was 5 million.
(c) 1910   **(d)** 1910
79. (a) 15 miles per hour   **(b)** 57 miles per hour
(c) 84 miles per hour
81. About 1.9 seconds; no

Checking Basic Concepts 6.1 & 6.2

1. $f(x) = x^2 - 1$
[−10, 10, 1] by [−10, 10, 1]

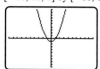 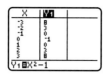

2. $(1, -2)$ 3. $y = (x + 2)^2 + 1$ 4. $\frac{1}{2}, 3$
5. (a) $\pm\sqrt{5}$ (b) $2 \pm \sqrt{3}$

Section 6.3

1. Solve $ax^2 + bx + c = 0$ 3. $b^2 - 4ac$
5. Factoring, square root property, completing the square, and the quadratic formula
7. $-6, \frac{1}{2}$ 9. 1 11. $-2, 8$ 13. $\frac{1 \pm \sqrt{17}}{8}$
15. No real solutions 17. $\frac{1}{2}$ 19. $\frac{3 \pm \sqrt{13}}{2}$
21. $2 \pm \sqrt{3}$ 23. $\frac{1 \pm \sqrt{37}}{6}$
25. (a) $a > 0$ (b) $-1, 2$ (c) Positive
27. (a) $a > 0$ (b) No real solutions (c) Negative
29. (a) $a < 0$ (b) $0, 3$ (c) Positive
31. (a) 25 (b) 2
33. (a) 0 (b) 1
35. (a) $-\frac{7}{4}$ (b) 0
37. (a) 21 (b) 2
39. 1, 2; (answers may vary)
41. $-\frac{1}{2}, 4$; (answers may vary)
43. $\frac{5 \pm \sqrt{17}}{2}$; quadratic formula or completing the square 45. 9 miles per hour 47. 45 miles per hour
49. $x \approx 8.04$, or about 1992

51. (a) The rate of change is not constant
(b) 75 seconds; (answers may vary)
(c) 75 seconds

Section 6.4

1. It has an inequality symbol rather than an equals sign.
3. No 5. $-2 < x < 4$ 7. Yes 9. Yes
11. No 13. Yes 15. No 17. No
19. (a) $-3, 2$ (b) $-3 < x < 2$ (c) $x < -3$ or $x > 2$
21. (a) $-2, 2$ (b) $-2 < x < 2$ (c) $x < -2$ or $x > 2$
23. (a) $-10, 5$ (b) $x < -10$ or $x > 5$
(c) $-10 < x < 5$
25. (a) $-2, 2$ (b) $-2 < x < 2$ (c) $x < -2$ or $x > 2$
27. (a) $-4, 0$ (b) $-4 < x < 0$
(c) $x < -4$ or $x > 0$
29. $-3 < x < -1$ 31. $x \leq -2.5$ or $x \geq 3$
33. $-2 \leq x \leq 2$ 35. All real numbers
37. $0 < x < 3$
39. (a) $-2, 2$ (b) $-2 < x < 2$ (c) $x < -2$ or $x > 2$
41. (a) $-5, 1$ (b) $-5 \leq x \leq 1$ (c) $x \leq -5$ or $x \geq 1$
43. $-7 \leq x \leq -3$ 45. $x < 1$ or $x > 2$
47. $-\sqrt{10} < x < \sqrt{10}$ 49. $x \leq 0$ or $x \geq 6$
51. $x \leq 2 - \sqrt{2}$ or $x \geq 2 + \sqrt{2}$
53. (a) From 1131 feet to 3535 feet (approximately)
(b) Before 1131 feet or after 3535 feet (approximately)
55. (a) About 183; they agree.
(b) About 1970 or after
(c) About 1969 or after
[1910, 1996, 10] by [0, 100, 10]

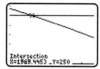

57. From 11 feet to 20 feet

Checking Basic Concepts 6.3 & 6.4

1. (a) $\frac{3 \pm \sqrt{17}}{4}$ (b) $\frac{4}{3}$
2. (a) 5; two real solutions
(b) -7; no real solutions (c) 0; one real solution
3. $x < -2$ or $x > 3$ 4. $-1 \leq x \leq -\frac{2}{3}$

Chapter 6 Review Exercises

1. Linear
2. Quadratic; $(1, 0)$; $x = 1$; opens upward
3. Quadratic; $(-3, 4)$; $x = -3$; opens downward
4. Neither
5. (a) $[-4.7, 4.7, 1]$ by $[-3.1, 3.1, 1]$

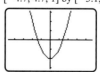

(b) $(0, -2)$; $x = 0$ (c) -1
6. (a) $[-4.7, 4.7, 1]$ by $[-3.1, 3.1, 1]$

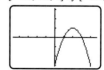

(b) $(2, 1)$; $x = 2$ (c) 0

7. (a) $[-4.7, 4.7, 1]$ by $[-3.1, 3.1, 1]$

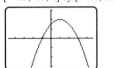

(b) $(1, 2)$; $x = 1$ **(c)** -2.5

8. (a) $[-4.7, 4.7, 1]$ by $[-3.1, 3.1, 1]$

(b) $(-2, -3)$; $x = -2$ **(c)** -1
9. $(2, -6)$ **10.** $(0, 5)$ **11.** $(2, 2)$ **12.** $(-1, 1)$
13. $a = 3$ **14.** $a = \dfrac{1}{4}$

15. (a) $[-10, 10, 1]$ by $[-10, 10, 1]$

(b) The graph of f is shifted 2 units upward compared to $y = x^2$.

16. (a) $[-10, 10, 1]$ by $[-10, 10, 1]$

(b) The graph of f is narrower, compared to $y = x^2$.

17. (a) $[-10, 10, 1]$ by $[-10, 10, 1]$

(b) The graph of f is shifted 2 units to the right compared to $y = x^2$.

18. (a) $[-10, 10, 1]$ by $[-10, 10, 1]$

(b) The graph of f is shifted 1 unit to the left and 3 units downward compared to $y = x^2$.

19. (a) $[-10, 10, 1]$ by $[-10, 10, 1]$

(b) The graph of f is wider, shifted 1 unit to the left and 2 units upward compared to $y = x^2$.

20. (a) $[-10, 10, 1]$ by $[-10, 10, 1]$

(b) The graph of f is narrower, shifted 1 unit to the right and 3 units downward compared to $y = x^2$.
21. $-2, 3$ **22.** -1 **23.** No real solutions
24. $-4, 6$ **25.** $-10, 5$ **26.** $-0.5, 0.25$
27. $-5, 10$ **28.** $-3, 1$ **29.** $-4, 2$ **30.** $-1, 3$
31. $-5, 4$ **32.** $-8, -3$ **33.** $-\dfrac{2}{5}, \dfrac{2}{3}$ **34.** $\dfrac{4}{7}, 3$
35. ± 10 **36.** $\pm \dfrac{1}{3}$ **37.** $\pm \dfrac{\sqrt{6}}{2}$
38. No real solutions **39.** $-3 \pm \sqrt{7}$ **40.** $2 \pm \sqrt{10}$
41. $1 \pm \sqrt{6}$ **42.** $\dfrac{-3 \pm \sqrt{11}}{2}$ **43.** $3, 6$
44. $11, 13$ **45.** $-\dfrac{1}{2}, \dfrac{1}{3}$ **46.** $\dfrac{5 \pm \sqrt{5}}{10}$
47. $4 \pm \sqrt{21}$ **48.** $\dfrac{3 \pm \sqrt{3}}{2}$
49. (a) $a > 0$ **(b)** $-2, 3$ **(c)** Positive
50. (a) $a > 0$ **(b)** 2 **(c)** 0
51. (a) $a < 0$ **(b)** No real solutions **(c)** Negative
52. (a) $a < 0$ **(b)** $-4, 2$ **(c)** Positive
53. (a) 1 **(b)** 2 **54. (a)** 144 **(b)** 2
55. (a) -23 **(b)** 0 **56. (a)** 0 **(b)** 1
57. (a) $-2, 6$ **(b)** $-2 < x < 6$ **(c)** $x < -2$ or $x > 6$
58. (a) $-2, 0$ **(b)** $x < -2$ or $x > 0$
(c) $-2 < x < 0$
59. (a) $-4, 4$ **(b)** $-4 < x < 4$ **(c)** $x < -4$ or $x > 4$
60. (a) $-1, 2$ **(b)** $-1 < x < 2$ **(c)** $x < -1$ or $x > 2$
61. (a) $-1, 3$ **(b)** $-1 < x < 3$ **(c)** $x < -1$ or $x > 3$
62. (a) $-\dfrac{3}{2}, 5$ **(b)** $-\dfrac{3}{2} \le x \le 5$
(c) $x \le -\dfrac{3}{2}$ or $x \ge 5$
63. $-3 \le x \le -1$ **64.** $\dfrac{1}{5} < x < 3$
65. $x < \dfrac{1}{6}$ or $x > 2$ **66.** $x \le -\sqrt{5}$ or $x \ge \sqrt{5}$
67. (a) $f(x) = x(12 - 2x)$
(b) $[0, 7, 1]$ by $[0, 20, 5]$

(c) and (d) 6 inches by 3 inches

68. (a) [0, 3, 1] by [0, 40, 10]

(b) 1 second and 1.75 seconds
(c) and **(d)** 1.375 seconds; 34.25 feet
69. (a) $f(x) = x(90 - 3x)$ **(b)** [0, 30, 5] by [0, 800, 100]

(c) 10 or 20 rooms **(d)** and **(e)** 15 rooms
70. (a) 2.4; in 1999, there were 2.4 complaints per 100,000 passengers
(b) Complaints have increased
[1997, 1999, 1] by [0.5, 3, 0.5]

71. (a) $\sqrt{1728} \approx 41.6$ miles per hour **(b)** 60 miles per hour
72. (a) $x(x + 2) = 143$
(b), (c), and **(d)** $x = -13$ or $x = 11$; the numbers are -13 and -11 or 11 and 13.
73. (a) [1935, 1995, 10] by [0, 100, 10]

(b) Yes, the data are nearly linear.
(c) $f(x) = 1.04(x - 1940) + 25$ (*answers may vary*)
74. (a) (1950, 220); in 1950, the per capita consumption was at a low of 220 million Btu.
(b) It increased.
[1950, 1970, 5] by [200, 350, 25]

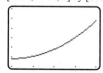

(c) $f(1996) = 749$; the trend represented by this model did not continue after 1970.
75. About 11.1 inches by 11.1 inches **76.** 50 feet
77. About 1.5 feet **78.** About 6.0 to 9.0 inches

Chapter 6 Test

1. $\left(1, \dfrac{3}{2}\right); x = 1$ **2.** $a = -\dfrac{1}{2}$
3. It is wider, translated right 3 units and translated upward 2 units.
[−10, 10, 1] by [−10, 10, 1]

4. $-1, 2$ **5.** $-4, \dfrac{1}{3}$ **6.** $-\dfrac{1}{2}, \dfrac{1}{2}$ **7.** $\dfrac{5 \pm \sqrt{29}}{2}$
8. $\dfrac{3 \pm \sqrt{17}}{4}$
9. (a) $a < 0$ **(b)** $-3, 1$ **(c)** Positive
10. (a) -44 **(b)** No real solutions
(c) The graph of $y = -3x^2 + 4x - 5$ does not intersect the x-axis.
11. (a) $-1, 1$ **(b)** $-1 < x < 1$ **(c)** $x < -1$ or $x > 1$
12. (a) $-10, 20$ **(b)** $x < -10$ or $x > 20$
(c) $-10 < x < 20$
13. (a) $-\dfrac{1}{2}, \dfrac{3}{4}$ **(b)** $-\dfrac{1}{2} \leq x \leq \dfrac{3}{4}$
(c) $x \leq -\dfrac{1}{2}$ or $x \geq \dfrac{3}{4}$
14. $\sqrt{2250} \approx 47.4$ miles per hour
15. (a) $f(x) = (x + 20)(90 - x)$
(b) [−100, 150, 50] by [−5000, 5000, 1000]

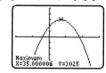

(c) 35
16. (a) [0, 6, 1] by [0, 150, 50]

(b) After about 5.6 seconds
(c) 2.75 seconds; 129 feet

Chapter 7: Rational Expressions and Functions

Section 7.1

1. A polynomial divided by a nonzero polynomial; $\dfrac{x^2 + 1}{3x}$ (*answers may vary*).

3. Multiply both sides by $x + 7$.
5. No, it simplifies to $5 + x$.   **7.** Yes   **9.** Yes
11. No
13. Symbolic: $f(x) = \dfrac{x}{x + 1}$

Graphical: Numerical:
$[-9.4, 9.4, 1]$ by $[-6.2, 6.2, 1]$

15. Symbolic: $f(x) = \dfrac{x^2}{x - 2}$

Graphical: Numerical:
$[-9.4, 9.4, 1]$ by $[-18.8, 18.8, 2]$

17. $x = 1$ **19.** $x = 1$

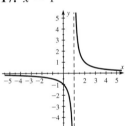 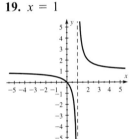

21. None

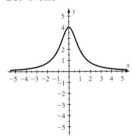

23. $-\dfrac{1}{3}$ **25.** $\dfrac{1}{2}$ **27.** 3   **29.** $2, 0; x = -1$
31. $0,$ undefined; $x = -2, x = 2$
33. $-0.4, 2$ **35.** Undefined, 0

37. $\dfrac{3}{5}$ **39.** 1   **41.** -2 **43.** 3   **45.** 0
47. No solutions   **49.** -4 **51.** -2 **53.** $-4, 1$
55. $0, 2$ **57.** -1 **59.** 2   **61.** c   **63.** d
65. (a) 6.35; a curve with radius of 400 feet will have an outer rail elevation of 6.35 inches.
(b)

(c) It is halved.
67. (a) 75; the braking distance is 75 feet when the uphill grade is 0.05.   **(b)** 0.15
69. (a) 1; when cars are leaving the ramp at a rate of 4 vehicles per minute, the wait is 1 minute.
(b) $[0, 5, 1]$ by $[0, 5, 1]$

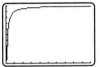

(c) As more cars try to exit, the waiting time increases.
(d) $4.\overline{6}$ vehicles per minute
71. (a) $P(1) = 0, P(50) = 0.98$; when there is only 1 ball there is no chance of losing. With 50 balls there is a 98% chance of losing.

(b) $[0, 100, 10]$ by $[0, 1, 0.1]$

(c) It increases. Yes, there are more balls without winning numbers.   **(d)** 40 balls

Section 7.2

1. 1
3. No; no; it is equal to $1 + \dfrac{1}{x}$.   **5.** $\dfrac{2}{3}; \dfrac{7}{5}$
7. $\dfrac{ad}{bc}$ **9.** $\dfrac{2}{5}$ **11.** $-\dfrac{7}{2}$ **13.** $\dfrac{2}{3}$ **15.** -18
17. $\dfrac{5}{x}$ **19.** $x + 1$ **21.** $x - 2$ **23.** $\dfrac{1}{x - 3}$
25. 1   **27.** $\dfrac{15}{4}$ **29.** $\dfrac{x + 1}{x}$ **31.** $x(x + 3)$
33. $\dfrac{(x - 1)(x + 5)}{2(2x - 3)}$ **35.** $\dfrac{4}{b^3}$
37. $\dfrac{3a(3a + 1)}{a + 1}$ **39.** $\dfrac{x^2}{(x - 5)(x^2 - 1)}$
41. $5x$ **43.** $x^2 - 1$ **45.** $2(2x + 1)(x - 2)$

47. 1 **49.** 5 **51.** 6 **53.** $1, -\frac{1}{2}$ **55.** $-1, 8$
57. 2.5 **59.** $-0.4, 0.67$ **61.** ± 1 **63.** 3
65. $-\frac{3}{2}, 1$
67. (a) $\frac{x}{5} + \frac{x}{2} = 1$ (b) $\frac{10}{7}$ hours (c) $\frac{10}{7}$ hours
69. Winner: 10 miles per hour; second place: 8 miles per hour **71.** 6250 feet **73.** 3 miles per hour
75. 250 miles per hour

CHECKING BASIC CONCEPTS 7.1 & 7.2

1. (a) 2 (b) 1 (c) $x = 1$
 $[-4.7, 4.7, 1]$ by $[-3.1, 3.1, 1]$

2. 4   **3.** (a) $\frac{x}{2(x-1)}$ (b) $\frac{x+3}{3}$ **4.** $-\frac{7}{2}$

SECTION 7.3

1. A common denominator **3.** $\frac{a+b}{c}$ **5.** 18
7. Multiply numerator and denominator by $x - 1$.
9. $\frac{5}{7}$ **11.** $\frac{7}{4}$ **13.** $-\frac{1}{5}$ **15.** $\frac{11}{8}$ **17.** $\frac{4}{x}$
19. $\frac{1-x}{x^2-4}$ **21.** $\frac{-5x-4}{x(x+4)}$ **23.** $\frac{-4x^2+x+2}{x^2}$
25. $\frac{x^2+5x-34}{(x-5)(x-3)}$ **27.** $\frac{-2(x^2-4x+2)}{(x-5)(x-3)}$
29. $\frac{x(5x+16)}{(x-3)(x+3)}$ **31.** $-\frac{1}{2}$
33. $\frac{6x^2-1}{(x-5)(3x-1)}$
35. $\frac{4x^2+4xy-9x+9y}{(x-y)(x+y)}$
37. $\frac{4x-7}{(x-2)(x-1)^2}$ **39.** $\frac{32}{21}$ **41.** $\frac{1}{x}$
43. $\frac{-x}{2x+3}$ **45.** $\frac{x+2}{3x-1}$
47. $\frac{3(x+1)}{(x+3)(2x-7)}$ **49.** $\frac{4x(x+5)}{(x-5)(2x+5)}$
51. $\frac{1200}{19} \approx 63.2$ ohms
53. $S \approx 0.256$ foot or about 3 inches

SECTION 7.4

1. A statement that two ratios are equal
3. It doubles. **5.** Constant **7.** kxy
9. Directly; if the number of people doubled, the food bill would double.
11. 10 **13.** 12 **15.** 72
17. (a) $\frac{7}{9} = \frac{10}{x}$ (b) $x = \frac{90}{7}$
19. (a) $\frac{5}{3} = \frac{x}{6}$ (b) $x = 10$
21. (a) $\frac{78}{6} = \frac{x}{8}$ (b) $x = \$104$
23. (a) $\frac{2}{90} = \frac{5}{x}$ (b) $x = 225$ minutes
25. (a) $k = 2$ (b) $y = 14$
27. (a) $k = 2.5$ (b) $y = 17.5$
29. (a) $k = 20$ (b) $y = 2$
31. (a) $k = 50$ (b) $y = 5$
33. (a) $k = 0.25$ (b) $z = 8.75$
35. (a) $k = 11$ (b) $z = 385$
37. (a) Direct (b) $y = 1.5x$
(c) $[0, 10, 1]$ by $[0, 10, 1]$

39. (a) Neither (b) N/A (c) N/A
41. (a) Neither (b) N/A (c) N/A
43. Direct; $k = 1$ **45.** Neither **47.** Direct; $k = 2$
49. 40.8 minutes **51.** 1.375 inches
53. (a) Direct; the ratios $\frac{R}{W}$ always equal 0.012.
(b) $R = 0.012W$
 $[0, 5000, 1000]$ by $[0, 50, 10]$

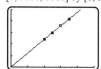

(c) 38.4 pounds
55. (a) Direct; the ratios $\frac{G}{A}$ always equal 27.
(b) $G = 27A$
 $[0, 0.6, 0.1]$ by $[0, 15, 1]$

(c) For each square-inch increase in the cross-sectional

CHAPTER 8 REVIEW EXERCISES

1. 2   2. 6   3. $3|x|$ 4. $|x-1|$ 5. 4
6. -5 7. x^2 8. $3x$ 9. 2   10. -1 11. x^2
12. $x+1$ 13. $\sqrt{14}$ 14. $\sqrt[3]{-5}$
15. $\left(\sqrt{\dfrac{x}{y}}\right)^3$ or $\sqrt{\left(\dfrac{x}{y}\right)^3}$ 16. $\dfrac{1}{\sqrt[3]{(xy)^2}}$ or $\dfrac{1}{(\sqrt[3]{xy})^2}$
17. 9   18. 2   19. 64   20. 27   21. z^2
22. xy^2 23. $\dfrac{x^3}{y^9}$ 24. $\dfrac{y^2}{x}$
25. $[-3, 7, 1]$ by $[-5, 5, 1]$ 26. $[-3, 7, 1]$ by $[-5, 5, 1]$

27. Shifted 2 units downward
$[-3, 7, 1]$ by $[-5, 5, 1]$

28. Shifted 3 units left   29. Shifted 1 unit right
$[-4, 6, 1]$ by $[-5, 5, 1]$ $[-4, 6, 1]$ by $[-5, 5, 1]$

30. This is a vertical stretch; the graph increases faster.
$[-5, 5, 1]$ by $[-5, 5, 1]$

31. $\dfrac{1}{4}$
32. $[-6, 6, 1]$ by $[-6, 6, 1]$

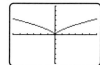

33. 8   34. -2 35. x^2 36. 2   37. $-\dfrac{\sqrt[3]{x}}{2}$
38. $\dfrac{1}{3}$ 39. $4\sqrt{3}$ 40. $3\sqrt[3]{x}$ 41. $-3\sqrt[3]{5}$
42. $-\sqrt[4]{y}$ 43. $11\sqrt{3}$ 44. $7\sqrt{2}$ 45. $13\sqrt[3]{2}$
46. $3\sqrt{x+1}$ 47. 2   48. 4   49. 4.5   50. 1.62
51. 5   52. 9   53. 8   54. $\dfrac{1}{4}$ 55. $c = \sqrt{65}$
56. $b = \sqrt{39}$ 57. $\sqrt{41}$ 58. $\sqrt{52} = 2\sqrt{13}$
59. -2 60. $-2 + 4i$ 61. $5 + i$ 62. $1 + 2i$
63. $\pm 3i$ 64. $-\dfrac{1}{2} \pm \dfrac{\sqrt{3}}{2}i$ 65. $\dfrac{1}{6} \pm \dfrac{\sqrt{11}}{6}i$
66. $\dfrac{1}{2} \pm \dfrac{\sqrt{7}}{2}i$ 67. About 85 feet
68. $\sqrt{16{,}200} \approx 127.3$ feet   69. About 0.79 second
70. (a) About 43 miles per hour   (b) About 34 miles per hour; a steeper bank allows for a higher speed limit; yes
71. $\sqrt{7} \approx 2.65$ feet
72. (a) 5 square units   (b) $5\sqrt{5}$ cubic units
(c) $\sqrt{10}$ units   (d) $\sqrt{15}$ units
73. About 0.82 foot
74. About 0.13 foot; it is shorter.
75. $r = \sqrt[200]{\dfrac{249}{4}} - 1 \approx 0.021$; from 1790 through 1990 the annual percentage growth rate was about 2.1%.
76. About 2108 square inches
77. (a) $0.5 = \dfrac{1}{2}$ (b) About 0.09, or $\dfrac{9}{100}$

CHAPTER 8 TEST

1. $5x^2$ 2. $2z^2$ 3. $\sqrt[5]{7^2}$ or $\left(\sqrt[5]{7}\right)^2$
4. $\sqrt[3]{\left(\dfrac{y}{x}\right)^2}$ or $\left(\sqrt[3]{\dfrac{y}{x}}\right)^2$ 5. 16   6. $\dfrac{1}{216}$
7. $[-5, 5, 1]$ by $[-5, 5, 1]$

8. The graph of f is shifted to the right 3 units and upward 1 unit.
9. $8z^{3/2}$ 10. $\dfrac{z}{y^{2/3}}$ 11. 9   12. $\dfrac{y}{2}$
13. $4\sqrt{7} + \sqrt{5}$ 14. $6\sqrt[3]{x}$ 15. 17   16. 5
17. $\sqrt{120} \approx 10.95$ 18. $\sqrt{8} = 2\sqrt{2}$
19. $2 - 19i$ 20. $-6 + 8i$ 21. $\dfrac{5}{4}$
22. $\dfrac{4}{29} + \dfrac{10}{29}i$ 23. $\dfrac{1}{4} \pm i\dfrac{\sqrt{23}}{4}$
24. (a) $r = \sqrt[3]{\dfrac{3V}{4\pi}}$ (b) About 2.29 inches

23. 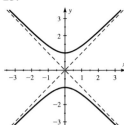 **24.**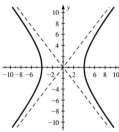

25. Parabola
[−4.7, 4.7, 1] by [−3.1, 3.1, 1]
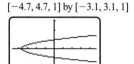

26. Parabola
[−9.4, 9.4, 1] by [−6.2, 6.2, 1]
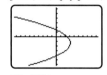

27. Ellipse
[−4.7, 4.7, 1] by [−3.1, 3.1, 1]

28. Ellipse
[−9.4, 9.4, 1] by [−6.2, 6.2, 1]
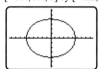

29. Hyperbola
[−9.4, 9.4, 1] by [−6.2, 6.2, 1]

30. Hyperbola
[−9.25, 9.25, 1] by [−6.1, 6.1, 1]

31. (0, 3), (3, 0) **32.** (−1, −2), (1, 2)
33. (0, 0), (2, 2)
34. (−2, −1), (−2, 1), (2, −1), (2, 1)
35. (−4, −4), (4, 4) **36.** (0, −4), (4, 0)
37. (−1, 1), (1, 1) **38.** (1, 2), (2, 5)
39. (−4, −12), (2, 0)
40. (−2, 0), (−1.73, 1), (1.73, 1), (2, 0)
41. (2, 2), (−4, −4) **42.** (1, 1), (0, 0)
43. 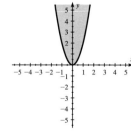 **44.**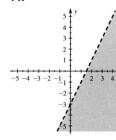

47.
[−10, 10, 1] by [−

51. (a) $A \approx 6$
(b) $A \approx 11.75$
53. (a) [−60, 6

(b) $P \approx 243.9$
$A \approx 4733$ squ
miles
55. Maximum
57. Height: 2(

CHECKING BA

1. (1, 2); $y =$
[−5, 5, 1] by

2. $(x - 1)^2$
[−4.7, 4.7, 1

3. x-intercep
4. (a) Parab

25. (a) −67°F (b) As the wind velocity increases, the windchill decreases—that is, the air feels colder.
[0, 50, 10] by [−40, 40, 10]
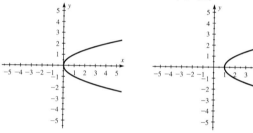

26. 1.31 pounds

CHAPTER 9: CONIC SECTIONS

SECTION 9.1

1. Parabola, ellipse, hyperbola **3.** No
5. No; it does not pass the vertical line test.
7. left **9.** circle; (h, k)
11. (0, 0); $y = 0$ **13.** (1, 0); $y = 0$
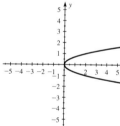

15. (0, 0); $y = 0$ **17.** (2, 1); $y = 1$
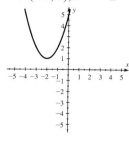 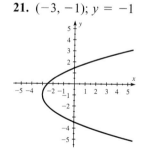

19. (−2, 1); $x = -2$ **21.** (−3, −1); $y = -1$

23. (0, 1); $y = 1$
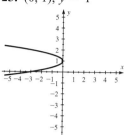

25. (0, 0); $y = 0$
[−5, 5, 1] by [−5, 5, 1]

27. (−3, 0); $y = 0$
[−5, 5, 1] by [−5, 5, 1]

29. $\left(-\dfrac{1}{4}, -\dfrac{3}{2}\right); y = -\dfrac{3}{2}$
[−5, 5, 1] by [−5, 5, 1]

31. $\left(\dfrac{1}{4}, \dfrac{7}{8}\right); x = \dfrac{1}{4}$
[−5, 5, 1] by [−5, 5, 1]

33. $\left(-\dfrac{3}{2}, -1\right); y = -1$
[−5, 5, 1] by [−5, 5, 1]

35. $\left(-\dfrac{1}{12}, -\dfrac{1}{6}\right); y = -\dfrac{1}{6}$
[−1, 3, 1] by [−2, 2, 1]

37. $y = x^2$ **39.** $x = (y + 1)^2 - 2$
41. Upward **43.** Downward **45.** $x \ge 0$
47. Two **49.** 1 **51.** $x^2 + y^2 = 1$
53. $(x + 1)^2 + (y - 5)^2 = 9$
55. $(x + 4)^2 + (y + 6)^2 = 2$ **57.** $x^2 + y^2 = 16$
59. $(x + 3)^2 + (y - 2)^2 = 1$
61. 3; (0, 0)
[−4.7, 4.7, 1] by [−3.1, 3.1, 1]

63. 3; (1, 3)
[−9.4, 9.4, 1] by [−6.2, 6.2, 1]

65. 5; (−5, 5)
[−18.8, 18.8, 2] by [−12.4, 12.4, 2]

67. 3; $(-3, 1)$
$[-9.4, 9.4, 1]$ by $[-6.2,$

71. (a) $[-40, 40, 10$

(b) 32 ft
73. (a) $[-1.5, 1.5, 0$
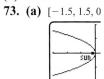

(b) 2.6 A.U., or 24

Section 9.2

1. (Answers may
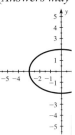

3. Horizontal
9. They are the d
11.
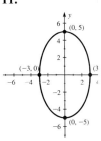

15.
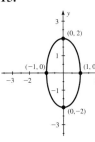

37. (1, 0)

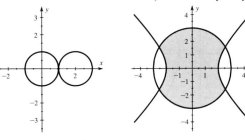

39. (0, 0) (answers may vary)

41. a. **43.** $y \geq x^2$; $y < 4 - x$
45. $r = 1.6$ inches; $h \approx 4.97$ inches
47. (a) $h = \dfrac{3V}{\pi r^2}$; $h = \sqrt{\left(\dfrac{S}{\pi r}\right)^2 - r^2}$
(b) $r \approx 2.02$ feet, $h \approx 7.92$ ft.; $r \approx 3.76$ feet, $h \approx 2.30$ feet

Checking Basic Concepts 9.3

1. $(1, -1), (3, 3)$ **2.** 2
3. (a) $(0, 3), (4, 4)$ (answers may vary)
(b) $y \geq 2 - x$ and $y \leq 4 - x^2$
4.
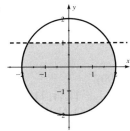

Chapter 9 Review Exercises

1. $(0, 0); y = 0$
$[-5, 5, 1]$ by $[-5, 5, 1]$
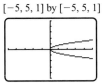

2. $(0, -1); y = -1$
$[-5, 5, 1]$ by $[-5, 5, 1]$
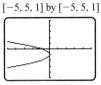

3. $(0, 2); y = 2$
$[-5, 5, 1]$ by $[-5, 5, 1]$

4. $(-1, -2); y = -2$
$[-5, 5, 1]$ by $[-5, 5, 1]$

Chapter 9 Test

1. $(-2, 4); y = 4$
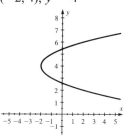

2. $x = -y^2 + 1$ **3.** $(x - 2)^2 + (y + 4)^2 = 4$
4. $(x + 5)^2 + (y - 2)^2 = 100$
5. $r = 4$, center $= (-2, 3)$ **6.**
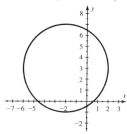
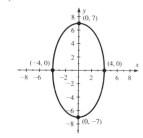

7. $\dfrac{x^2}{100} + \dfrac{y^2}{64} = 1$

8.
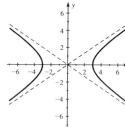

9. Ellipse
$[-4.7, 4.7, 1]$ by $[-3.1, 3.1, 1]$

10. $(0, -4), (4, 0)$ **11.** $(-1, -4), (4, 1)$
12. $(-0.57, 1.22), (1, 1)$
13.
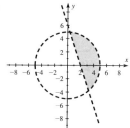

14. $y \leq 4 - x^2$; $y \geq x^2 - 4$
15. (a) $xy = 5000$; $2x + 2y = 300$
(b) 50 feet by 100 feet (c) 50 feet by 100 feet
16. Either $x \approx 22.08$, $y \approx 2.43$ or $x \approx 7.29$, $y \approx 22.26$; no
17. (a) $[-30, 30, 10]$ by $[-20, 20, 10]$

(b) 18.28 A.U., or about 1,700,040,000 miles

CHAPTER 10: EXPONENTIAL AND LOGARITHMIC FUNCTIONS

Section 10.1

1. $f(x) = Ca^x$ **3.** Growth **5.** 2.718
7. Factor **9.** $\dfrac{1}{9}$; 9 **11.** 5; 160 **13.** 4; $\dfrac{1}{8}$
15. 15; $\dfrac{5}{9}$ **17.** 0.17; 2.41 **19.** 5; 1.08
21. Growth
$[-4, 4, 1]$ by $[0, 8, 1]$

23. Decay
$[-4, 4, 1]$ by $[0, 8, 1]$

25. Decay
$[-4, 4, 1]$ by $[0, 8, 1]$

27. Growth
$[-4, 4, 1]$ by $[0, 8, 1]$
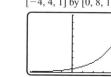

29. (a) Exponential decay (b) $f(x) = 64\left(\dfrac{1}{4}\right)^x$
31. (a) Linear growth (b) $f(x) = 3x + 8$
33. (a) Exponential growth (b) $f(x) = 4(1.25)^x$
35. $C = 1$, $a = 2$ **37.** $C = 4$, $a = \dfrac{1}{4}$ **39.** c
41. d **43.** $3551.05 **45.** $1,820,087.63
47. $792.75
49. Yes; this is equivalent to having two accounts, each containing $1000 initially.
51. About $778.7 billion **53.** 3.32 **55.** 0.86
57. Growth
$[-4, 4, 1]$ by $[0, 8, 1]$

59. Decay
$[-4, 4, 1]$ by $[0, 8, 1]$
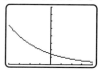

61. (a) $f(x) = 4.56e^{0.031x}$
(b) [0, 10, 1] by [4, 7, 1]

(c) About 5.49 million
63. (a) $C = 500$, $a = 2$ **(b)** About 5278 thousand per milliliter
(c) The growth is exponential.
[0, 300, 50] by [0, 35,000, 5000]

65. (a) About 11.3; in 1995 there were about 11.3 million cellular phone subscribers.
(b) 1.495; each year from 1985 to 2000 the number of subscribers increased by a factor of 1.495, or by 49.5%.
67.

(a) 1; the probability that no vehicle will enter the intersection during a period of 0 seconds is 1.
(b) About 30 seconds

Section 10.2

1. 10 **3.** D: $\{x \mid x > 0\}$; R: all real numbers
5. k **7.** x **9.** log 5 **11.** 5 **13.** -4
15. 0 **17.** -2 **19.** $-\dfrac{3}{2}$ **21.** 2 **23.** -4
25. -2 **27.** 0 **29.** 1.398 **31.** 0.161
33. 1.946 **35.** -0.560
37. Shifted downward 1 unit
[-4, 4, 1] by [-4, 4, 1]

39. Shifted 3 units to the left
[-4, 4, 1] by [-4, 4, 1]

41. This is a reflection across the y-axis together with the graph of $y = \ln x$.
[-4, 4, 1] by [-4, 4, 1]

43. Shifted 2 units to the left
[-4, 4, 1] by [-4, 4, 1]

45.

x	$\frac{1}{4}$	$\frac{1}{2}$	1	$\sqrt{2}$	64
$\log_2 x$	-2	-1	0	$\frac{1}{2}$	6

47. d **49.** a **51.** 120 db; yes
53. (a) $f(0) = 27$; the air pressure at the eye of the hurricane is 27 inches of mercury. $f(50) \approx 28.9$; the air pressure 50 miles from the eye is about 28.9 inches of mercury. **(b)** It increases.
[0, 150, 50] by [26, 30, 1]

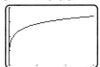

(c) Low
55. (a) 10^6; 10^8 **(b)** 100 times
57. (a) About 2119 calories **(b)** Caloric intake increases as the amount of land increases.
[0, 4, 1] by [1800, 2500, 100]

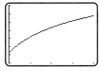

(c) No; the growth levels off and is not linear.
59. (a) $f(50) \approx 1.89$; in 2000 the population of less industrialized regions increased by about 1.89 billion. $g(50) \approx 0.95$; in 2000 the population of industrialized regions increased by about 0.95 billion.
(b) The population of less industrialized regions grew faster than industrialized regions.
[0, 80, 10] by [0, 5, 1]

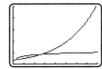

CHECKING BASIC CONCEPTS 10.1 & 10.2

1. (a) 16 (b) 17.93 (c) 2.18 (d) 4.61
2. $y = a^x$ represents exponential growth when $a > 1$ and decay when $0 < a < 1$.

$[-4, 4, 1]$ by $[-2, 6, 1]$

3. $1105.34
4. (a) 4 (b) x (c) -3 (d) $\frac{1}{2}$
5. $[-6, 6, 1]$ by $[-4, 4, 1]$

(a) $D: \{x \mid x > 0\}$; R: all real numbers   (b) 0
(c) Yes; for example, $\log \frac{1}{10} = -1$.
(d) No; negative numbers are not in the domain of $\log x$.

SECTION 10.3

1. 4   3. 3   5. $\log m - \log n$ 7. No
9. $\log_a x = \frac{\log x}{\log a}$ or $\log_a x = \frac{\ln x}{\ln a}$
11. $\ln 3 + \ln 5$ 13. $\log_3 x + \log_3 y$
15. $\ln 2 + \ln 5 + \ln z$ 17. $\log 7 - \log 3$
19. $\ln x - \ln y$ 21. $\log_2 45 - \log_2 x$ 23. $\log 225$
25. $\ln xy$ 27. $\ln 14x^3$ 29. $\ln xy$ 31. $6 \log 3$
33. $x \ln 2$ 35. $\frac{1}{4} \log_2 5$ 37. $\log z$ 39. $\log x^3 y^2$
41. 0   43. $\ln 2^x$ 45. $\log_3 \frac{1}{x^2}$ 47. $\log x + 2 \log y$
49. $4 \ln x + \ln y - \ln z$ 51. $\frac{1}{3} \log_4 z - \frac{1}{2} \log_4 y$
53. $4 \log x + 3 \log y$ 55. $\ln x - \ln y$
57. By the power rule, $\log x^3 = 3 \log x$
$[-6, 6, 1]$ by $[-4, 4, 1]$ $[-6, 6, 1]$ by $[-4, 4, 1]$

59. Not the same
$[-6, 6, 1]$ by $[-4, 4, 1]$ $[-6, 6, 1]$ by $[-4, 4, 1]$

 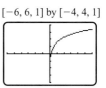

61. 5   63. $-5x$ 65. $2x - 7$ 67. $0.6z$
69. $10 \log (10^{16} x) = 10(\log 10^{16} + \log x) = 10(16 + \log x) = 160 + 10 \log x$

SECTION 10.4

1. Add 5 to both sides.
3. Take the common logarithm of both sides.
5. x 7. $2x$ 9. No; $\log \frac{5}{4} = \log 5 - \log 4$
11. 1   13. 3   15. 6   17. 5.23   19. 0   21. 1.95
23. -0.76 25. (a) 1 (b) 1   27. (a) -2 (b) -2
29. -1 31. 0   33. 5   35. $\frac{\log 1.45}{\log 0.55} \approx -0.62$
37. $-1.84, 1.15$ 39. 1.31
41. 100
43. $e^5 \approx 148.41$ 45. $5{,}000{,}000$ 47. 16
49. $\frac{2^{2.3}}{5} \approx 0.98$ 51. $10^{1.4} \approx 25.12$
53. $\frac{(e^{11} - 1)}{2} \approx 29{,}936.57$
55. (a) 2   (b) $e^{0.7} \approx 2.01$
57. (a) 2   (b) $\frac{1}{2}(10^{0.6}) \approx 1.99$
59. $10^{1.6} \approx 39.81$ 61. $e - 1 \approx 1.72$ 63. 9
65. 8 years
67. (a) About 8503; in 1994 about 8503 people were waiting for liver transplants. (b) About 1998
69. About 20,893 pounds
71. (a) About 203; in 1975 there were about 203 thousand bluefin tuna. (b) and (c) In 1979
73. About 1.67 acres   75. In 2086
77. 10^{-6} watts/square centimeter
79. About 7 miles

CHECKING BASIC CONCEPTS 10.3 & 10.4

1. (a) $\log x + \log y$ (b) $\ln x - \ln y - \ln z$
(c) $2 \ln x$ (d) $2 \log x + 3 \log y - \frac{1}{2} \log z$
2. (a) $\log xy$ (b) $\ln \frac{2x}{y^3}$ (c) $\log_2 \frac{x^2 y^3}{z}$
3. (a) $\log 20 \approx 1.30$ (b) $\frac{\log 147}{3 \log 2} \approx 2.40$
(c) $e^{4.1} \approx 60.34$ (d) 500

Chapter 10 Review

1. $\frac{1}{6}$; 36 2. 5; $\frac{5}{8}$ 3. 3; $\frac{1}{81}$ 4. 3; $\frac{1}{2}$
5. Exponential growth
 [−4, 4, 1] by [0, 8, 1]
6. Exponential decay
 [−4, 4, 1] by [0, 8, 1]
7. Logarithmic growth
 [−4, 4, 1] by [0, 8, 1]
8. Exponential decay
 [−4, 4, 1] by [0, 8, 1]
9. (a) Exponential growth (b) $f(x) = 5(2)^x$
10. (a) Linear growth (b) $f(x) = 5x + 5$
11. $C = \frac{1}{2}$, $a = 2$ 12. $k = 2$ 13. $2829.54
14. $675,340.51 15. 399.67 16. 0.71 17. 3.48
18. 3.89 19. −3 20. 2 21. −4 22. 2
23. 1.813 24. −0.163 25. 4.787 26. −1.322
27. Shifted to the right 2 units
 [−6, 6, 1] by [−4, 4, 1]

28. Shifted downward 2 units
 [−6, 6, 1] by [−4, 4, 1]

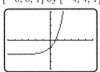

29. Shifted to the left 3 units
 [−6, 6, 1] by [−4, 4, 1]

30. Increases more rapidly
 [−6, 6, 1] by [−4, 4, 1]

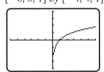

31. $\ln x + \ln y$ 32. $\log x - \log y$
33. $2 \ln x + 3 \ln y$ 34. $\frac{1}{2} \log x - 3 \log z$
35. $2 \log_2 x + \log_2 y - \log_2 z$
36. $\frac{1}{2} \log_3 x - \frac{1}{2} \log_3 y$ 37. $\log 75$ 38. $\ln \frac{x^2}{y^3}$
39. $\log_4 (10x^2)$ 40. $\log xy$ 41. $3 \log 6$
42. $2 \ln x$ 43. $(2x) \log_2 5$ 44. $(x + 1) \log_4 0.6$
45. 7 46. $\frac{5}{9}$ 47. $6 - x$ 48. x^2 49. 2
50. 4 51. $\ln 9 \approx 2.20$ 52. $\frac{\log (0.2)}{\log (0.85)} \approx 9.90$
53. $e^{0.8} \approx 2.23$ 54. $\frac{1}{2} e^5 \approx 74.21$ 55. 10^{40}
56. 100 57. (a) 3 (b) 3
58. (a) 4 (b) 4
59. 7 years 60. $a = 100$, $b = 50$
61. $C = 3$, $a = 2$ 62. 10^7
63. (a) Growth
 [0, 10, 2] by [0, 4, 1]

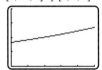

(b) About 2.24 million
64. (a) 1000; there were 1000 bacteria initially.
(b) About 495.11 minutes
65. (a) About 6.93 meters/second (b) 12.18 meters

Chapter 10 Test

1. $\frac{3}{16}$
2. Exponential decay
 [−4, 4, 1] by [−2, 6, 1]

3. (a) Exponential growth (b) $f(x) = 3(2)^x$
4. (a) Linear growth (b) $f(x) = 1.5x - 1$
5. $C = 1$, $a = \frac{1}{2}$ 6. $1051.91 7. 4.16 8. $\frac{1}{2}$
9. 5.426
10. Shifted to the right 2 units
 [−6, 6, 1] by [−4, 4, 1]

11. $3 \log x + 2 \log y - \frac{1}{2} \log z$ 12. $\ln \frac{x^4 z}{y^5}$
13. $2x \log 7$ 14. $1 - 3x$ 15. $\ln 25 \approx 3.22$
16. $\log 50 \approx 1.70$ 17. $10^{1.8} \approx 63.10$

18. $\frac{1}{5}e^9 \approx 1620.62$ **19.** $a = 5$, $b = 3$

20. (a) 4 million **(b)** $4e^{0.45} \approx 6.27$; after 5 hours there were about 6.27 million bacteria.
(c) Growth
[0, 10, 2] by [0, 10, 2]

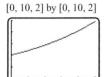

(d) After 4.51 hours **(e)** After 4.51 hours; yes

35. (a) 1, 4, 9, 16 **(b)** 4, 8, 12, 16
37. (a) $20,000; $16,000 **(b)** $a_n = 25{,}000(0.8)^n$
(c)

39. (a)

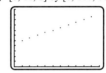

(b) $a_n = 50 + 5(n - 1)$ or $a_n = 45 + 5n$
(c) 160 seats
(d) [0, 11, 1] by [0, 110, 10]

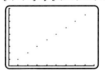

CHAPTER 11: SEQUENCES AND SERIES

Section 11.1

1. 1, 2, 3, 4; answers may vary.
3. function; the set of natural numbers **5.** 6
7. $f(2)$ **9.** 1, 4, 9, 16 **11.** $\frac{1}{6}, \frac{1}{7}, \frac{1}{8}, \frac{1}{9}$
13. $\frac{5}{2}, \frac{5}{4}, \frac{5}{8}, \frac{5}{16}$ **15.** 1, 8, 27 **17.** 1, $\frac{8}{5}$, 2
19. 2, 9, 20 **21.** 7 **23.** 3, 4, 5, 3, 1
25. 6, 5, 4, 3, 2, 1
27. Numerical   Graphical
                               [0, 8, 1] by [0, 9, 1]

 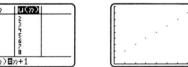

29. Numerical   Graphical
                               [0, 8, 1] by [0, 45, 5]

31. Numerical   Graphical
                               [0, 8, 1] by [0, 140, 10]

33. $a_n = 30n$ for $n = 1, 2, 3, \ldots, 7$
Graphical   Numerical
[0, 8, 1] by [0, 220, 20]

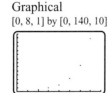

Section 11.2

1. $a_n = 3n + 1$; 3 **3.** add; previous **5.** 19; 4
7. $a_n = a_1(r)^{n-1}$ **9.** Yes; 10 **11.** Yes; -1
13. No **15.** Yes; 3 **17.** Yes; -3 **19.** Yes; 4
21. Yes; 1 **23.** No **25.** Yes; 2
27. $a_n = -2n + 9$ **29.** $a_n = 4n - 6$
31. $a_n = -2n + 32$ **33.** 59 **35.** 21 **37.** Yes; 3
39. Yes; 0.8 **41.** No **43.** Yes; 2 **45.** No
47. Yes; 4 **49.** Yes; 2 **51.** No
53. $a_n = 1.5(4)^{n-1}$ **55.** $a_n = -3(-2)^{n-1}$
57. $a_n = 1(4)^{n-1}$ **59.** 4374 **61.** 243
63. (a) 3000, 6000, 9000, 12,000, 15,000; arithmetic
(b) $a_n = 3000n$ **(c)** 60,000; when there are 20 people, the ventilation should be 60,000 cubic feet per hour.
(d) Yes
   [0, 9, 1] by [0, 27,000, 3000]

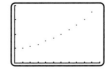

65. Arithmetic; the common difference is 2000.
67. (a) Geometric; the common ratio is 1.15.
(b) $a_n = 100{,}000(1.15)^{n-1}$
(c) $a_7 \approx 231{,}306$; at the beginning of the seventh year, it will be worth about $231,306.

(d) [0, 11, 1] by [0, 400,000, 100,000]

69. (a) Arithmetic; the common difference is 2.
(b) $a_n = 40 + 2(n - 1)$ or $a_n = 38 + 2n$ **(c)** 78

CHECKING BASIC CONCEPTS 11.1 & 11.2

1. $\dfrac{1}{5}, \dfrac{1}{3}, \dfrac{3}{7}, \dfrac{1}{2}$

2. Graphical   Numerical
[0, 6, 1] by [0, 8, 1]

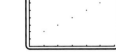

3. (a) Arithmetic; $a_n = 3n - 5$
(b) Geometric; $a_n = 3(-2)^{n-1}$
4. $a_n = 2n + 3$ **5.** $a_n = 5(2)^{n-1}$

SECTION 11.3

1. series   **3.** arithmetic
5. $S_n = n\left(\dfrac{a_1 + a_n}{2}\right)$ or $\dfrac{n}{2}(2a_1 + (n-1)d)$
7. $a_1 + a_2 + a_3 + a_4$ **9.** 48   **11.** 820   **13.** -5
15. 3279   **17.** -85 **19.** 182   **21.** $91,523.93
23. $62,278.01   **25.** $2 + 4 + 6 + 8$; 20
27. $4 + 4 + 4 + 4 + 4 + 4 + 4 + 4$; 32
29. $1 + 4 + 9 + 16 + 25 + 36 + 49$; 140
31. $12 + 20$; 32   **33.** $\displaystyle\sum_{k=1}^{6} k^4$ **35.** $\displaystyle\sum_{k=1}^{5} \dfrac{1}{k^2}$
37. $\displaystyle\sum_{k=1}^{n} k = n\left(\dfrac{a_1 + a_n}{2}\right) = n\left(\dfrac{1 + n}{2}\right) = \dfrac{n(n+1)}{2}$
39. (a) $8518 + 9921 + 10{,}706 + 14{,}035 + 14{,}307 + 12{,}249$ **(b)** 69,736
41. (a) $1, \dfrac{1}{2}, \dfrac{1}{4}, \dfrac{1}{8}, \dfrac{1}{16}$ **(b)** $\dfrac{1023}{512}$
43. 90 logs   **45.** $1,080,000   **47.** About 3.16 feet

SECTION 11.4

1. 5   **3.** 4   **5.** 24   **7.** $\dfrac{n!}{(n-r)!\,r!}$
9. $x^3 + 3x^2y + 3xy^2 + y^3$
11. $16x^4 + 32x^3 + 24x^2 + 8x + 1$
13. $a^5 - 5a^4b + 10a^3b^2 - 10a^2b^3 + 5ab^4 - b^5$
15. $x^6 + 3x^4 + 3x^2 + 1$ **17.** 6   **19.** 4   **21.** 2
23. 6   **25.** 10   **27.** 5   **29.** 6   **31.** 1   **33.** 792
35. 126   **37.** 75,582
39. $m^3 + 3m^2n + 3mn^2 + n^3$
41. $x^4 - 4x^3y + 6x^2y^2 - 4xy^3 + y^4$
43. $8a^3 + 12a^2 + 6a + 1$
45. $x^5 + 10x^4 + 40x^3 + 80x^2 + 80x + 32$

47. $8x^3 - 12x^2y + 6xy^2 - y^3$ **49.** a^8
51. $35x^4y^3$ **53.** $512m^9$

CHECKING BASIC CONCEPTS 11.3 & 11.4

1. (a) Geometric **(b)** Arithmetic
2. 312   **3.** -341
4. $x^4 - 4x^3y + 6x^2y^2 - 4xy^3 + y^4$
5. $x^3 + 6x^2 + 12x + 8$

CHAPTER 11 REVIEW

1. 1, 8, 27, 64   **2.** 3, 1, -1, -3 **3.** $1, \dfrac{4}{5}, \dfrac{3}{5}, \dfrac{8}{17}$
4. $-2, 4, -8, 16$ **5.** $-2, 0, 4, 2$ **6.** 5, 3, 2, 1
7. Numerical   Graphical
                               [0, 8, 1] by [0, 16, 2]

8. Numerical   Graphical
                               [0, 8, 1] by [-5, 50, 5]

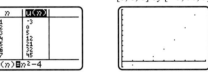

9. Numerical   Graphical
                               [0, 8, 1] by [0, 2.1, 0.1]

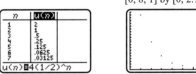

10. Numerical   Graphical
                               [0, 8, 1] by [0, 3, 1]

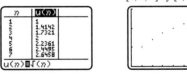

11. Yes; 5   **12.** No   **13.** No   **14.** Yes; $-\dfrac{1}{3}$
15. Yes; -3 **16.** No   **17.** Yes; -1 **18.** No
19. $a_n = 4n - 7$ **20.** $a_n = -5n + 7$ **21.** Yes; 4
22. No   **23.** No   **24.** Yes; 0.7   **25.** No
26. Yes; $-\dfrac{1}{3}$ **27.** No   **28.** Yes; 2
29. $a_n = 5(0.9)^{n-1}$ **30.** $a_n = 2(4)^{n-1}$ **31.** 216
32. 7.5   **33.** 3277   **34.** $\dfrac{511}{256}$
35. $3 + 5 + 7 + 9 + 11$ **36.** $\dfrac{1}{2} + \dfrac{1}{3} + \dfrac{1}{4} + \dfrac{1}{5}$

37. $1 + 8 + 27 + 64$
38. $-1 + (-2) + (-3) + (-4) + (-5) + (-6)$
39. $\sum_{k=1}^{20} k$ **40.** $\sum_{k=1}^{20} \frac{1}{k}$ **41.** $\sum_{k=1}^{9} \frac{k}{k+1}$ **42.** $\sum_{k=1}^{7} k^2$
43. $x^3 + 12x^2 + 48x + 64$
44. $16x^4 + 32x^3 + 24x^2 + 8x + 1$
45. $x^5 - 5x^4y + 10x^3y^2 - 10x^2y^3 + 5xy^4 - y^5$
46. $a^6 - 6a^5 + 15a^4 - 20a^3 + 15a^2 - 6a + 1$
47. 6   **48.** 10   **49.** 20   **50.** 4
51. $m^4 + 8m^3 + 24m^2 + 32m + 16$
52. $a^5 + 5a^4b + 10a^3b^2 + 10a^2b^3 + 5ab^4 + b^5$
53. $x^4 - 12x^3y + 54x^2y^2 - 108xy^3 + 81y^4$
54. $27x^3 - 54x^2 + 36x - 8$
55. $a_n = 45{,}000(1.1)^{n-1}$ for $n = 1, 2, 3, \ldots, 7$; geometric

Numerical

Graphical
[0, 8, 1] by [40,000, 85,000, 5000]

56. $a_n = 45{,}000 + 5000(n - 1)$ for $n = 1, 2, 3, \ldots, 7$; arithmetic

Numerical

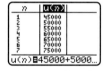

Graphical
[0, 8, 1] by [40,000, 80,000, 5000]

57. $a_n = 49n$ for $n = 1, 2, 3, \ldots, 7$

Graphical
[0, 8, 1] by [0, 400, 50]

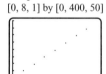

Numerical

58. (a) $a_n = 1087(1.025)^{n-1}$
(b) Geometric; the common ratio is 1.025.
(c) About 1200; the average mortgage payment in 2000 was about $1200.
(d) [0, 11, 1] by [1000, 1400, 50]

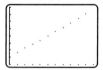

CHAPTER 11 TEST

1. $\frac{1}{2}, \frac{4}{3}, \frac{9}{4}, \frac{16}{5}$
2. $-3, 2, 1, -2, 3$
3.

n	$u(n)$
1	0
2	2
3	6
4	12
5	20
6	30
7	42

$u(n) \blacksquare n^2 - n$

4. $16x^4 - 32x^3 + 24x^2 - 8x + 1$
5. Arithmetic; -3 **6.** Geometric; -2
7. $a_n = 2 - 3(n - 1)$ or $a_n = 5 - 3n$
8. $a_n = 2(1.5)^{n-1}$ **9.** Yes; 2.5   **10.** No
11. 99   **12.** $\frac{463}{729}$
13. $6 + 9 + 12 + 15 + 18 + 21$
14. $\sum_{k=1}^{60} k^3$ **15.** 35   **16.** 10   **17.** 9180
18. $a_n = 159{,}700(1.04)^{n-1}$ for $n = 1, 2, 3, \ldots, 7$; geometric

Numerical

Graphical
[0, 8, 1] by [150,000, 210,000, 10,000]

19. (a) $a_n = 2000(2)^{n-1}$
(b) Geometric; the common ratio is 2.   (c) 64,000; after 30 days there are 64,000 worms.
(d) [0, 7, 1] by [0, 70,000, 10,000]

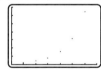

GLOSSARY

absolute value of a real number a, written $|a|$, is equal to its distance from the origin on the number line.

absolute value equation An equation that contains an absolute value.

absolute value function The function defined by $f(x) = |x|$.

absolute value inequality An inequality that contains an absolute value.

addends In an addition problem, the two numbers that are added.

addition property of equality If a, b, and c are real numbers, then $a = b$ is equivalent to $a + c = b + c$.

additive identity The number 0.

additive inverse (opposite) The additive inverse or opposite of a real number a is $-a$.

annuity A sequence of deposits made at regular time intervals.

approximately equal The symbol $\approx$ indicates that two quantities are nearly equal.

arithmetic sequence A linear function given by $a_n = dn + c$ whose domain is the set of natural numbers.

arithmetic series The sum of the terms of an arithmetic sequence.

associative property for addition For any real numbers a, b, and c, $(a + b) + c = a + (b + c)$.

associative property for multiplication For any real numbers a, b, and c, $(a \cdot b) \cdot c = a \cdot (b \cdot c)$.

asymptotes of a hyperbola The two lines determined by the diagonals of the hyperbola's fundamental rectangle.

augmented matrix A matrix used to represent a system of linear equations; a vertical line is positioned in the matrix where the equal signs occur in the system of equations.

average The result of adding up the numbers of a set and then dividing the sum by the number of elements in the set.

axis of symmetry of a parabola The line passing through the vertex of the parabola that divides the parabola into two symmetric parts.

base The value of a in the expression a^x.

binomial A polynomial with two terms.

binomial theorem A theorem that provides a formula to expand expressions of the form $(a + b)^n$.

branches A hyperbola has two branches, a left branch and a right branch, or an upper branch and a lower branch.

byte A unit of computer memory, capable of storing one letter of the alphabet.

Cartesian coordinate plane The xy-plane used to plot points and visualize a relation; points are identified by ordered pairs.

center of a circle The point that is a fixed distance from all the points on a circle.

circle The set of points in a plane that are a constant distance from a fixed point called the center.

clearing fractions The process of multiplying both sides of an equation by a common denominator to eliminate the fractions.

coefficient The numeric constant in a monomial.

common denominator A number or expression that each denominator can divide into *evenly* (without a remainder).

common difference The value of d in an arithmetic sequence, $a_n = dn + c$.

common logarithmic function The function given by $f(x) = \log x$.

common logarithm of a positive number x Denoted $\log x$, it may be calculated as follows: if x is expressed as $x = 10^k$, then $\log x = k$, where k is a real number. That is, $\log 10^k = k$.

common ratio The value of r in a geometric sequence, $a_n = a_1(r)^{n-1}$.

commutative property for addition For any real numbers a and b, $a + b = b + a$.

commutative property for multiplication For any real numbers a and b, $a \cdot b = b \cdot a$.

completing the square An important technique in mathematics that involves adding a constant to a binomial so that a perfect square trinomial results.

complex conjugate The complex conjugate of $a + bi$ is $a - bi$.

complex fraction A rational expression that contains fractions in its numerator, denominator, or both.

complex number A complex number can be written in standard form as $a + bi$, where a and b are real numbers and i is the imaginary unit.

compound inequality Two inequalities joined by the words *and* or *or*.

conic section The curve formed by the intersection of a plane and a cone.

consistent A system of linear equations with at least one solution.

constant function A linear function of the form $f(x) = b$, where b is a constant.

constant of proportionality (constant of variation) In the equation $y = kx$, the number k.

G-1

continuous growth Growth in a quantity that is directly proportional to the amount present.

Cramer's rule A method that uses determinants to solve linear systems of equations.

cubed Process of raising a number or variable to the third power.

cube root The number b is a cube root of a number a if $b^3 = a$.

cube root function The function defined by $f(x) = \sqrt[3]{x}$.

cubic polynomial A polynomial of degree 3 that can be written as $ax^3 + bx^2 + cx + d$, where $a \neq 0$.

decay factor The value of a in an exponential function, $f(x) = Ca^x$, where $0 < a < 1$.

degree of a monomial The sum of the exponents of the variables.

degree of a polynomial The degree of the monomial with highest degree.

dependent equations Equations in a linear system that have the same solution set.

dependent variable The variable that represents the output of a function.

determinant A real number associated with a square matrix.

diagrammatic representation A function represented by a diagram.

difference The answer to a subtraction problem.

difference of two cubes Expression in the form $a^3 - b^3$, which can be factored as $(a - b)(a^2 + ab + b^2)$.

difference of two squares Expression in the form $a^2 - b^2$, which can be factored as $(a - b)(a + b)$.

dimension of a matrix The size expressed in number of rows and columns. For example, if a matrix has n rows and m columns, its dimension is $n \times m$ (n by m).

directly proportional A quantity y is directly proportional to x if there is a nonzero number k such that $y = kx$.

discriminant The expression $b^2 - 4ac$ in the quadratic formula.

distance formula The distance between the points (x_1, y_1) and (x_2, y_2) in the xy-plane is $d = \sqrt{(x_2 - x_1)^2 + (y_2 - y_1)^2}$.

distributive properties For any real numbers a, b, and c, $a(b + c) = ab + ac$ and $a(b - c) = ab - ac$.

dividend In a division problem, the number being divided.

divisor In a division problem, the number being divided *into* another.

domain The set of all x-values of the ordered pairs in a relation.

element Each number in a matrix.

elimination (or addition) method A symbolic method used to solve a system of equations that is based on the property that "equals added to equals are equal."

ellipse The set of points in a plane, the sum of whose distances from two fixed points is constant.

equation A statement that says two mathematical expressions are equal.

equivalent equations Two equations that have the same solution set.

even root The nth root, $\sqrt[n]{a}$, where n is even.

expansion of a determinant by minors A method of finding a 3×3 determinant by using determinants of 2×2 matrices.

exponent The value of a in the expression b^a.

exponential equation An equation that has a variable as an exponent.

exponential expression An expression that has a variable occurring as an exponent.

exponential function A function represented by $f(x) = Ca^x$, where $a > 0$, $C > 0$, and $a \neq 1$.

extraneous solution A solution that does not satisfy the given equation.

factors In a multiplication problem, the two numbers multiplied.

factorial notation $n! = 1 \cdot 2 \cdot 3 \cdot \cdots \cdot n$ for any positive integer.

factoring by grouping A technique that uses the distributive property by grouping the terms of a polynomial in such a way that the polynomial can be factored even though its greatest common factor is 1.

finite sequence A function with domain $D = \{1, 2, 3, \ldots, n\}$ for some fixed natural number n.

finite series A series that contains a finite number of terms, and can be expressed in the form $a_1 + a_2 + a_3 + \cdots + a_n$ for some n.

focus (plural foci) A fixed point used to determine the points that form a parabola, an ellipse, or a hyperbola.

FOIL A method for multiplying two binomials $(A + B)$ and $(C + D)$. Multiply First terms AC, Outer terms AD, Inner terms BC, and Last terms BD; then combine like terms.

formula An equation that can be used to calculate one quantity by using a known value of another quantity.

function A relation where each element in the domain corresponds to exactly one element in the range.

function notation $y = f(x)$ is read "y equals f of x." This means that function f with input x produces output y.

fundamental rectangle of a hyperbola The rectangle whose four vertices are determined by either $(\pm a, \pm b)$ or $(\pm b, \pm a)$, where $\frac{x^2}{a^2} - \frac{y^2}{b^2} = 1$ or $\frac{y^2}{a^2} - \frac{x^2}{b^2} = 1$.

Gaussian elimination A numerical method used to solve a linear system in which matrix row transformations are used.

general term (nth term) of a sequence a_n where n is a natural number in the domain of a sequence $a_n = f(n)$.

geometric sequence An exponential function given by $a_n = a_1(r)^{n-1}$ whose domain is the set of natural numbers.

graphical representation A graph of a function.

graphical solution A solution to an equation obtained by graphing.

greater than If a real number a is located to the right of a real number b on the number line, a is greater than b, or $a > b$.

greatest common factor (GCF) The largest term that is a factor of all terms in the polynomial.

growth factor The value of a in the exponential function, $f(x) = Ca^x$, where $a > 1$.

half-life The time it takes for a radioactive sample to decay to half its original amount.

hyperbola The set of points in a plane, the difference of whose distances from two fixed points is constant.

identity property of 1 If any number a is multiplied by 1, the result is a, that is, $a \cdot 1 = 1 \cdot a = a$.

identity property of 0 If 0 is added to any real number, a, the result is a, that is, $a + 0 = 0 + a = a$.

imaginary number A complex number $a + bi$ with $b \neq 0$.

imaginary part The value of b in the complex number $a + bi$.

imaginary unit A number denoted i whose properties are $i = \sqrt{-1}$ and $i^2 = -1$.

inconsistent system of linear equations A system that has no solution.

independent variable The variable that represents the input of a function.

index The value of n in the expression $\sqrt[n]{a}$.

index of summation The variable k in the expression $\sum_{k=1}^{n}$.

infinite sequence A function whose domain is the set of natural numbers.

input An element of the domain of a function.

integers A set I of numbers given by $I = \{\ldots, -3, -2, -1, 0, 1, 2, 3, \ldots\}$.

intersection The set of elements found in *both* sets; for example, if $A = \{1, 2, 3\}$ and $B = \{2, 3, 4\}$, then the intersection is $\{2, 3\}$.

interval notation A notation for number line graphs that eliminates the need to draw the entire line. A **closed interval** is expressed by the endpoints in brackets, such as $[-2, 7]$; an **open interval** is expressed in parentheses, such as $(-2, 7)$, and indicates that the endpoints are not included in the solution. A **half-open interval** has a bracket and a parenthesis, such as $[-2, 7)$.

inversely proportional A quantity y is inversely proportional to x if there is a nonzero number k such that $y = \dfrac{k}{x}$.

irrational numbers Real numbers that cannot be expressed as fractions, such as π or $\sqrt{2}$.

joint variation A quantity z varies jointly with x and y if there is a nonzero k such that $z = kxy$.

leading coefficient of a polynomial of one variable The coefficient of the monomial with highest degree.

least common denominator (LCD) The common denominator with the fewest factors.

less than If a real number a is located to the left of a real number b on the number line, we say that a is less than b, or $a < b$.

like radicals Radicals that have the same index and the same radicand.

like terms Two terms that contain the same variables raised to the same powers.

linear equation in one variable An equation that can be written in the form $ax + b = 0$, where $a \neq 0$.

linear function A function f represented by $f(x) = ax + b$, where a and b are constants.

linear inequality in one variable An inequality that can be written in the form $ax + b > 0$, where $a \neq 0$. (The symbol $>$ may be replaced with $\geq$, $<$, or $\leq$.)

linear inequality in two variables When the equals sign in a linear equation of two variables is replaced with $<$, $\leq$, $>$, or $\geq$, a linear inequality in two variables results.

linear polynomial A polynomial of degree 1 that can be written as $ax + b$, where $a \neq 0$.

line graph The resulting graph when the data points in a scatterplot are connected with straight-line segments.

logarithm with base a of a positive number x Denoted $\log_a x$, it may be calculated as follows: if x can be expressed as $x = a^k$, then $\log_a x = k$, where $a > 0$, $a \neq 1$, and k is a real number. That is, $\log_a a^k = k$.

logarithmic equation An equation that contains logarithms.

logarithmic function with base a The function denoted $f(x) = \log_a x$.

lower limit In summation notation, the number representing the subscript of the first term of the series.

main diagonal The elements $a_{11}, a_{22}, a_{33}, \ldots, a_{nn}$ in a matrix with n rows.

major axis The longest axis of an ellipse, which connects the vertices.

matrix A rectangular array of numbers.

minor axis The shortest axis of an ellipse.

monomial A term whose variables have only nonnegative integer exponents.

multiplication property of equality If a, b, and c are real numbers with $c \neq 0$, then $a = b$ is equivalent to $ac = bc$.

multiplicative identity The number 1.

multiplicative inverse (reciprocal) The multiplicative inverse of a real number a is $\dfrac{1}{a}$.

natural exponential function The function represented by $f(x) = e^x$, where $e \approx 2.71828$.

natural logarithm The base-e logarithm, denoted either $\log_e x$ or $\ln x$.

natural numbers The set of numbers given by $N = \{1, 2, 3, 4, 5, 6, \ldots\}$.

negative slope On a graph, the slope of a line that falls from left to right.

negative square root The negative square root is denoted $-\sqrt{a}$.

nonlinear data If data points do not lie on a (straight) line, the data are nonlinear.

nth root The number b is an nth root of a if $b^n = a$, where n is a positive integer, and is denoted $\sqrt[n]{a} = b$.

nth term (general term) of a sequence Denoted $a_n = f(n)$.

numerical representation A table of values for a function.

numerical solution A solution often obtained by using a table of values.

odd root The nth root, $\sqrt[n]{a}$, where n is odd.

opposite (additive inverse) The opposite of a real number a is $-a$.

opposite of a polynomial The polynomial obtained by negating each term in a given polynomial.

ordered pair A pair of numbers written in parentheses (x, y), in which the order of the numbers is important.

ordered triple Can be expressed as (x, y, z) where $x, y,$ and z are numbers.

origin On the number line, the point associated with the real number 0; in the xy-plane, the point $(0, 0)$.

output An element of the range of a function.

parabola The U-shaped graph of a quadratic function.

parallel lines Two or more lines in the same plane that never intersect.

Pascal's triangle A triangle made up of numbers in which there are 1s along the sides and each element inside the triangle is the sum of the two numbers above it.

perfect square trinomial A trinomial that can be factored as the square of a binomial, for example, $a^2 + 2ab + b^2 = (a + b)^2$.

perpendicular lines Two lines in a plane that intersect to form a right (90°) angle.

pixels The tiny units that comprise screens for computer terminals or graphing calculators.

point–slope form The equation of a line with slope m passing through the point (h, k) given by $y = m(x - h) + k$, or equivalently, $y - k = m(x - h)$.

polynomial A monomial or a sum of monomials.

polynomials in one variable Polynomials that contain one variable.

positive slope On a graph, the slope of a line that rises from left to right.

power function A function that can be represented by $f(x) = x^p$, where p is a rational number.

principle square root The positive square root, denoted $\sqrt{a}$.

product The answer to a multiplication problem.

proportion A statement that two ratios are equal.

Pythagorean theorem If a right triangle has legs a and b with hypotenuse c, then $a^2 + b^2 = c^2$.

quadrants The four regions determined by a Cartesian coordinate plane.

quadratic equation An equation that can be written as $ax^2 + bx + c = 0$, where $a, b,$ and c are real numbers, with $a \neq 0$.

quadratic formula The solutions of the quadratic equation, $ax^2 + bx + c = 0, a \neq 0$, are given by the formula $x = \dfrac{-b \pm \sqrt{b^2 - 4ac}}{2a}$.

quadratic function A function f represented by $f(x) = ax^2 + bx + c$, where $a, b,$ and c are real numbers with $a \neq 0$.

quadratic inequality If the equals sign in a quadratic equation is replaced with $>, \geq, <,$ or $\leq$, a quadratic inequality results.

quadratic polynomial A polynomial of degree 2 that can be written as $ax^2 + bx + c$ with $a \neq 0$.

quotient The answer to a division problem.

radical expression An expression that contains a radical sign.

radical sign The symbol, $\sqrt{}$ or $\sqrt[n]{}$ for some n.

radicand The expression under the radical sign.

radius The fixed distance between the center and any point on the circle.

range The set of all y-values of the ordered pairs in a relation.

rate of change The value of a for the linear function, $f(x) = ax + b$; slope can be interpreted as a rate of change.

rational equation An equation that involves a rational expression.

rational expression A polynomial divided by a nonzero polynomial.

rational function A function defined by $f(x) = \dfrac{p(x)}{q(x)}$, where $p(x)$ and $q(x) \neq 0$ are polynomials.

rational number Any number that can be expressed as the ratio of two integers $\dfrac{p}{q}$, where $q \neq 0$; a fraction.

real numbers All rational and irrational numbers; any number that can be written using decimals.

real part The value of a in a complex number $a + bi$.

reciprocal (multiplicative inverse) The reciprocal of a real number a is $\dfrac{1}{a}$.

reduced row–echelon form A matrix form for representing a system of linear equations in which there are 1s on the main diagonal with 0s above and below each 1.

relation A set of ordered pairs.

rise The vertical change between two points on a line, that is, the change in the y-values.

root function In the power function, $f(x) = x^p$, if $p = \frac{1}{n}$, where $n \geq 2$ is an integer, then f is also a root function, which is given by $f(x) = \sqrt[n]{x}$.

run The horizontal change between two points on a line, that is, the change in the x-values.

scatterplot A graph of distinct points plotted in the xy-plane.

scientific notation A real number a written as $b \times 10^n$, where $1 < |b| < 10$ and n is an integer.

set braces { }, used to enclose the elements of a set.

set-builder notation Notation to describe a set of numbers without having to list all of the elements. For example, $\{x | x > 5\}$ is read as "the set of all real numbers x such that x is greater than 5."

slope The ratio of the change in y (rise) to the change in x (run) along a line. Slope m of a line equals $\frac{y_2 - y_1}{x_2 - x_1}$, where (x_1, y_1) and (x_2, y_2) are points on the line.

slope–intercept form The equation of a line with slope m and y-intercept b is given by $y = mx + b$.

solutions All of the values for a variable that make an equation a true statement.

solution set The set of all solutions to an equation.

squared The process of raising a number or variable to the second power.

square matrix A matrix in which the number of rows and columns are equal.

square root The number b is a square root of a number a if $b^2 = a$.

square root function The function given by $f(x) = \sqrt{x}$, where $x \geq 0$.

square root property If k is a nonnegative number, then the solutions to the equation $x^2 = k$ are $x = \pm\sqrt{k}$. If $k < 0$, then this equation has no real solution.

standard equation of a circle The standard equation of a circle with center (h, k) and radius r is $(x - h)^2 + (y - k)^2 = r^2$.

standard viewing rectangle of a graphing calculator Xmin = –10, Xmax = 10, Xscl = 1, Ymin = –10, Ymax = 10, and Yscl = 1, denoted [–10, 10, 1] by [–10, 10, 1].

substitution method A symbolic method for solving a system of equations in which one equation is solved for one of the variables and then the result is substituted into the other equation.

sum The answer to an addition problem.

summation notation Notation in which the uppercase Greek letter sigma represents the sum, for example $\sum_{k=1}^{n} k^2$.

sum of two cubes Expression in the form $a^3 + b^3$, and which can be factored as $(a + b)(a^2 - ab + b^2)$.

symbolic representation Representing a function with a formula; for example, $f(x) = x^2 - 2x$.

symbolic solution A solution to an equation obtained by using properties of equations, and the resulting solution set is exact.

synthetic division A shortcut that can be used to divide $x - k$, where k is a number, into a polynomial.

system of two linear equations in two variables A system of equations in which two equations can be written in the form $ax + by = c$; an ordered pair (x, y) is a solution to the system of equations if the values for x and y make *both* equations true.

system of linear inequalities Two or more inequalities to be solved at the same time, the solution to which must satisfy both inequalities.

system of nonlinear equations Two or more equations, at least one of which is nonlinear.

system of nonlinear inequalities Two or more inequalities, at least one of which is nonlinear.

table of values An organized way to display the inputs and outputs of a function; a numerical representation.

term A number, a variable, or a product of numbers and variables raised to powers.

terms of a sequence $a_1, a_2, a_3, \ldots, a_n, \ldots$, where the first term is $a_1 = f(1)$, the second term is $a_2 = f(2)$, and so on.

test point When graphing the solution set of a linear inequality, a point chosen to determine which region of the xy-plane to shade.

three-part inequality A compound inequality written in the form $a < x < b$.

translation The shifting of a graph upward, downward, to the right, or to the left in such a way that the shape of the graph stays the same.

transverse axis In a hyperbola, the line segment that connects the vertices.

trinomial A polynomial with three terms.

upper limit In summation notation, the number that represents the subscript of the last term of the series.

variable A symbol, such as x, y, or z, used to represent an unknown number or quantity.

varies directly A quantity y varies directly with x if there is a nonzero number k such that $y = kx$.

varies inversely A quantity y varies inversely with x if there is a nonzero number k such that $y = \frac{k}{x}$.

varies jointly A quantity z varies jointly as x and y if there is a nonzero number k such that $z = kxy$.

verbal representation A description of what a function computes in words.

vertex The lowest point on the graph of a parabola that opens upward or the highest point on the graph of a parabola that opens downward.

vertex form of a parabola The vertex form of a parabola with vertex (h, k) is $y = a(x - h)^2 + k$, where $a \neq 0$ is a constant.

vertical asymptote A vertical asymptote occurs in the graph of a rational function when the denominator of the rational expression is 0, but the numerator is not 0; it can be represented by a vertical line in the graph of a rational function.

vertical line test If a vertical line is drawn through a graph and it intersects the graph in at most one point, then it is a graph of a function.

vertices of an ellipse The endpoints of the major axis.

vertices of a hyperbola The endpoints of the transverse axis.

viewing rectangle On a graphing calculator, the window that determines the x- and y-values shown in the graph.

whole numbers The set of numbers given by $W = \{0, 1, 2, 3, 4, 5, \ldots\}$.

x-axis The horizontal axis in a Cartesian coordinate plane.

x-intercept The x-coordinate of a point where a graph intersects the x-axis.

Xmax Regarding the viewing rectangle of a graphing calculator, Xmax is the maximum x-value along the x-axis.

Xmin Regarding the viewing rectangle of a graphing calculator, Xmin is the minimum x-value along the x-axis.

Xscl The distance represented by consecutive tick marks on the x-axis.

y-axis The vertical axis in a Cartesian coordinate plane.

y-intercept The y-coordinate of a point where a graph intersects the y-axis.

Ymax Regarding the viewing rectangle of a graphing calculator, Ymax is the maximum y-value along the y-axis.

Ymin Regarding the viewing rectangle of a graphing calculator, Ymin is the minimum y-value along the y-axis.

Yscl The distance represented by consecutive tick marks on the y-axis.

zero-product property If the product of two numbers is 0, then at least one of the numbers must equal 0, that is, $ab = 0$ implies $a = 0$ or $b = 0$.

zero of a polynomial An x-value that results in an output of 0 when it is substituted into a polynomial; for example, the zeros of $f(x) = x^2 - 4$ are 2 and -2.

INDEX

Absolute value, 13, 20, 471
   equation, 180, 184
   function, 178
   inequality, 182, 184
Addends, 14
Addition
   associative property of, 7, 9
   commutative property of, 6–7, 9
   identity for, 6, 9
   of complex numbers, 513, 517
   of fractions, 424–425
   of polynomials, 279, 282
   of radical expressions, 493, 495
   of rational expressions, 426, 431
   of real numbers, 15, 21
Addition property
   of equality, 146
   of inequality, 159
Additive
   identity, 6, 9
   inverse, 14, 279
Alpha Centauri, 24, 39
Annuity, 654
Appolonius, 527
Approximately equal symbol, 4
Area of a triangle, 259–260, 263
Arithmetic sequence, 639, 646
   common difference of, 639
   general term of, 641
   sum of the terms of, 651, 657
Arithmetic series, 651, 657
Arrhenius, Svante, 573
Associative property
   for addition, 7, 9
   for multiplication, 7, 9
Astronomical unit (A.U.), 539, 544
Asymptote, vertical, 402
Asymptote of a hyperbola, 546
Atanasoff, John, 198
Augmented matrix, 245
Average, 5, 656
Axis
   of a coordinate system, 53
   of a parabola, 341, 529
Axis of symmetry, 341, 529

Base, 574
   of an exponential expression, 25, 37
Base-a logarithm, 593, 595
Base-a logarithmic function, 593, 595
Binomial, 298
   multiplying, 298–299
   product of sum and difference, 301, 303
   squaring, 301–302, 303
Binomial coefficient, 663, 665
Binomial theorem, 664, 665
Bode's law, 672–673
Boole, George, 467
Boolean algebra, 467
Brahe, Tyco, 527
Branches of a hyperbola, 545

Calculating π, 673
Cardano, Girolamo, 467
Cartesian coordinate plane, 55
   origin, 53
   quadrants, 53
   x-axis, 55
   xy-plane, 55
   y-axis, 55
Center
   of a circle, 532
   of a hyperbola, 545
   of an ellipse, 542
Change
   in x, 103
   in y, 103
Change of base formula, 603, 605
Circle, 528, 532
   center, 532
   equation, 532
   graphing with technology, 534–535
   radius, 532
Coefficient of a monomial, 278
Common denominator, 416
Common difference, 639
   and slope, 640
Common factors, 305–307
   greatest, 306, 311
Common logarithm, 589
Common logarithmic function, 589–590

Common ratio, 642
Commutative property
   for addition, 6–7, 9
   for multiplication, 7, 9
Completing the square, 360, 364, 534
Complex conjugates, 514, 517
Complex fraction, 430, 432
Complex numbers, 467, 512, 517
   addition of, 513, 517
   conjugate, 514, 517
   division of, 515, 517
   imaginary, 467, 512
   imaginary part of, 512
   multiplication of, 514, 517
   real part of, 512
   standard form of, 512
   subtraction of, 513, 517
   and technology, 513
Compound inequality, 167
   solving, 174
Compound interest, 577, 583
Conic sections, 527
   types of, 528
Consistent system of linear equations, 202
Constant
   of proportionality, 437, 439
   of variation, 437, 439
Constant (linear) function, 95, 292
Continuous growth, 582
Coordinate system, 55
Counting numbers, 2, 9
Cramer, Gabriel, 260
Cramer's rule
   in the real world, 261–262
   in two variables, 260–261, 263
Cube root, 43
   function, 482, 486
Cube, 25
Cubic function, 287, 292
Cubic polynomial, 279

Data
   linear, 286
   nonlinear, 275, 286
   visualization of, 50–51

I-1

Matrix *(continued)*
  solution of systems, 247–249
  square, 245
Method of substitution, 553, 559
Minor axis, 541
Minor of a determinant, 258
Modeling data with
  direct variation, 438
  exponential functions, 587–588
  formulas, 45
  inverse variation, 440
  linear functions, 93–96
  linear systems, 238–240
  logarithmic functions, 614
  power functions, 485
  quadratic functions, 344–345, 349–350
  a sequence, 634–635
Monomial, 275, 298
  coefficient of, 278
  degree of, 278
  division by, 449, 455
Multiplication
  of binomials, 298–299
  of complex numbers, 514, 517
  of fractions, 412–413
  of monomials, 297
  of polynomials, 299, 303
  property of equality, 146
  property of inequalities, 159
  of radical expressions, 489–491, 494
  of rational expressions, 413, 420
  of real numbers, 17, 21
Multiplicative
  identity, 6, 9
  inverse, 16

Napier, John, 599
Natural logarithm, 592
Natural numbers, 2, 9
Negative exponent, 27
Negative slope, 104
Newton, Sir Isaac, 527, 599
n factorial, 662, 665
Nonlinear systems
  of equations, 552, 559
  of inequalities, 557
  solving with technology, 558–559
nth root, 471, 472
nth term of a sequence, 630
Number line, 13
Numbers
  complex, 467, 512, 517
  counting, 2, 9

  imaginary, 467
  integers, 3, 9
  irrational, 4
  natural, 2, 9
  rational, 3, 9
  real, 4, 9
  whole, 2, 9
Numerical representation, 74
Numerical solution, 142

Odd root, 471
Ohms, 431
Old Faithful Geyser, 255
Opposite, 14, 279
Opposite of a polynomial, 280, 282
Order of operations, 32
Ordered
  pair, 51
  triple, 233
Origin
  of a number line, 13
  of the xy-plane, 53
Output, 73

Pairs, ordered, 51
Parabola, 341, 528
  axis of symmetry, 341
  focus of, 571
  with horizontal axis, 529, 536
  translations of, 346
  vertex, 341, 351
  vertex form of, 347, 351, 529
Paraboloid, 571
Parallel lines, 122, 125
Partial numerical representation, 74
Pascal's triangle, 661, 665
Percent problems, 149–150
Perfect square trinomials, 323–324
Perpendicular lines, 123, 125
pH, 626
Pixels, 3
Point–slope form, 116–117, 125
Point of curvature, 504
Polynomial, 279
  addition of, 279, 282
  binomial, 298
  cubic, 279
  degree of, 279
  division, 451, 455
  equations, 307–311
  evaluation, 280–282, 287–288
  function, 287, 292
  leading coefficient of, 279
  linear, 279

  monomial, 275, 298
  multiplication of, 299, 303
  of one variable, 279
  quadratic, 279
  subtraction, 280, 282
  trinomial, 298
  zeros, 307
Polynomial functions, 287, 292
  cubic, 287, 292
  linear, 287, 292
  quadratic, 287, 292
Positive slope, 104
Power function, 483–484, 486
  modeling with, 485
Power rule
  for exponents, 30–31, 37
  for logarithms, 601, 605
  for solving equations, 501, 505
Powers of i, 526
Principal square root, 43, 46, 468
Product, 16
Product rule
  for exponents, 29, 37
  for logarithms, 599, 605
  for radicals, 490
Properties of
  equality, 146
  exponents, 28–31, 37, 300, 303, 475–477
  inequalities, 159
Properties of logarithms
  inverse, 604, 605
  power rule, 601, 605
  product rule, 599, 605
  quotient rule, 600, 605
Proportion, 434
Pythagorean theorem, 502, 505

Quadrants, 53
Quadratic equation, 356
  with complex solutions, 515–517
Quadratic formula, 369, 375
  derivation of, 374
Quadratic function, 287, 292
  graphs of, 341
Quadratic inequality, 378
Quadratic polynomial, 279
Quotient, 17, 451, 452
Quotient rule for exponents, 29, 37

Radical equations, 497–502
  power rule for solving, 501
Radical expressions, 468
  addition of, 493

division of, 491
  graphing, 500
  index, 471
  like, 492, 494
  multiplication of, 489–491
  product rule for, 490, 494
  quotient rule for, 491, 494
  simplifying, 492–493
  subtraction, 493–494
Radical notation, 472–473
Radical sign, 468
Radicand, 468
Radius of a circle, 532
Raising powers to powers, 30, 37
Raising products to powers, 31, 37
Raising quotients to powers, 31, 37
Range
  of a function, 77
  of a relation, 52, 60
Rate of change, 108
Ratio, 434
Rational approximation, 464
Rational equation, 404, 407, 415
Rational exponents, 472–474
  converting to radical notation, 473
  evaluating with technology, 474
  properties, 475
Rational expressions, 400
  addition of, 426
  division of, 413
  multiplication of, 413
  reducing, 414
  subtraction of, 426
Rational function, 400, 407
  domain, 400
Rational numbers, 3, 9
Real number line, 13
Real numbers, 4–5, 9
  addition of, 15, 21
  division of, 17, 21
  multiplication of, 17, 21
  properties of, 6–9
  reciprocal, 16
  subtraction of, 16, 21
Real part, 512
Reciprocal, 16, 413
Recursive
  formula, 672
  sequence, 672
Reduced row–echelon form, 246
Reducing
  fractions, 414
  rational expressions, 414

Relation, 52, 60
  domain, 52, 60
  range, 52, 60
Remainder, 451, 452
Repeating decimal, 3
Representations of a function, 73–76, 83
  diagrammatic, 74, 83
  graphical, 74, 83
  numerical, 74, 83
  symbolic, 74, 83
  verbal, 73, 83
Representations of linear functions, 90–93, 96
Representations of sequences, 632, 635–636
Rise, 103
Root function, 483
Roots
  calculator approximation of, 469, 470
  cube, 43, 469
  even, 471
  nth, 471, 472
  odd, 471
  principal square, 43, 46, 468
  square, 43, 46, 468
Row of a matrix, 245
Row transformations, 246
Rules for exponents
  product rule, 29, 37
  quotient rule, 29, 37
  raising powers to powers, 30, 37
  raising products to powers, 31, 37
  raising quotients to powers, 31, 37
Run, 103

Saint Augustine, 1
Scatterplot, 55, 60
Scientific notation, 33–34
Sequence, 630
  arithmetic, 639
  common difference of, 639
  Fibonoacci, 672
  finite, 630
  geometric, 642, 646
  infinite, 630
  recursive, 672
  terms of, 630
Series
  finite, 650
  infinite, 673
Set braces, 2
Set builder notation, 156

Sigma (Σ), 655
Similar figures, 436
Similar triangles, 436
Skinner, B.F., 198
Slope, 103, 110
  interpreting, 108
  negative, 104
  of parallel lines, 122
  of perpendicular lines, 123
  positive, 104
  as a rate of change, 108–110
  undefined, 104
  zero, 104
Slope–intercept form, 106, 110, 125
Solution set, 141, 156
Solutions
  to equations, 141
  extraneous, 405
  to inequalities, 156
Solving polynomial equations
  by factoring, 307–308, 309–311, 325–327
  by grouping, 310–311
  graphically, 308–310, 325–326
  numerically, 308–310, 325–326
Solving quadratic
  equations, 357–361, 369–371
  inequalities, 379–385
Square matrix, 245
Square root, 43, 46, 468
  function, 480–481
  negative, 468
  principal, 43, 46, 468
Square root property, 359, 364
Square viewing rectangle, 124
Square, 25
Squaring a binomial, 301–303
Standard equation
  of a circle, 533, 536
  of an ellipse, 542, 548
  of a hyperbola, 545, 548
Standard form
  of a complex number, 512
  of a real number, 35
Standard viewing rectangle, 58
Substitution and elimination, 236–238
Substitution method, 211–213, 218, 553, 559
Subtraction
  of complex numbers, 513, 517
  of fractions, 425–426
  of polynomials, 280, 282
  of radical expressions, 493–494, 495

Subtraction *(continued)*
  of rational expressions, 426, 431
  of real numbers, 16, 21
Sum, 14
  of the terms of an arithmetic sequence, 651, 657
  of the terms of a geometric sequence, 653, 657
  of two cubes, 324–325, 327
Summation notation, 655
  index, 655
  lower limit, 655
  upper limit, 655
Symbolic representation, 74
Synthetic division, 454–455
System of equations
  dependent, 202
  inconsistent, 202
  linear in three variables, 233
  linear in two variables, 199
  nonlinear, 552, 559
  solving with technology, 249–252
  unique solution to, 202
System of inequalities, 225
  nonlinear, 557, 560
  solving with technology, 558–559

Table of values, 74
Technology, 339
Term, 275
  general, 630
  like, 278
  nth, 630
Terminating decimal, 3
Test point, 224

Three-part inequality, 169
  graphical solutions, 171
  number line solutions, 169
  numerical solutions, 171
  symbolic solutions, 169–170
Translations
  of an ellipse, 572
  of a hyperbola, 572
  of a parabola, 346
Transverse axis, 545, 548
  horizontal, 545, 548
  vertical, 545, 548
Trinomial, 298
  factoring, 314–320
  perfect square, 323–324
Types of linear systems, 202

Undefined slope, 104
Union, 170
Upper limit, 655

Variable, 41
  dependent, 73
  independent, 73
Variation
  constant of, 437
  direct, 437, 444, 464–465
  inverse, 439, 444, 465–466
  joint, 442, 444
Verbal representation, 73–74
Vertex
  formula, 342
  of a parabola, 341, 351
  locating with technology, 344
Vertex form of a parabola, 347, 351, 529, 536

Vertical asymptote, 403
Vertical line, 121, 125
Vertical line test, 79
Vertical major axis, 541, 548
Vertical transverse axis, 545, 548
Vertices
  of an ellipse, 541
  of a hyperbola, 545
Viewing rectangle, 57
  decimal, 406
  friendly, 406
  square, 124, 535
  standard, 58

Whole numbers, 2, 9
Window, 57
  decimal, 406
  friendly, 406

x-axis, 53
x-intercept, 119
Xmax, 57
Xmin, 57
Xscl, 57

y-axis, 53
Yeats, William Butler, 72
y-intercept, 106
Ymax, 57
Ymin, 57
Yscl, 58

Zero exponent, 27
Zero-product property, 307, 311
Zero slope, 104
Zeros of a polynomial, 307

BIBLIOGRAPHY

Acker, A., and C. Jaschek. *Astronomical Methods and Calculations.* New York: John Wiley and Sons, 1986.

Baase, S. *Computer Algorithms: Introduction to Design and Analysis.* 2nd ed. Reading, Mass.: Addison-Wesley Publishing Company, 1988.

Battan, L. *Weather in Your Life.* San Francisco: W. H. Freeman, 1983.

Beckmann, P. *A History of PI.* New York: Barnes and Noble, Inc., 1993.

Brown, D., and P. Rothery. *Models in Biology: Mathematics, Statistics and Computing.* West Sussex, England: John Wiley and Sons Ltd, 1993.

Callas, D. *Snapshots of Applications in Mathematics.* Deli, New York: State University College of Technology, 1994.

Carr, G. *Mechanics of Sport.* Champaign, Ill.: Human Kinetics, 1997.

Cheney, W., and D. Kincaid. *Numerical Mathematics and Computing.* 3rd ed. Pacific Grove, Calif.: Brooks/Cole Publishing Company, 1994.

Conquering the Sciences. Sharp Electronics Corporation, 1986.

Elton, C. S., and M. Nicholson. "The ten year cycle in numbers of lynx in Canada." *J. Anim. Ecol.* 11, (1942): 215–244.

Eves, H. *An Introduction to the History of Mathematics,* 5th ed. Philadelphia: Saunders College Publishing, 1983.

Freedman, B. *Environmental Ecology: The Ecological Effects of Pollution, Disturbance, and Other Stresses.* 2nd ed. San Diego: Academic Press, 1995.

Friedhoff, M., and W. Benzon. *The Second Computer Revolution: Visualization.* New York: W. H. Freeman, 1991.

Garber, N., and L. Hoel. *Traffic and Highway Engineering.* Boston, Mass.: PWS Publishing Co., 1997.

Goldstein, M., and J. Larson. *Jackie Joyner-Kersee: Superwoman.* Minneapolis: Lerner Publications Company, 1994.

Grigg, D. *The World Food Problem.* Oxford: Blackwell Publishers, 1993.

Haefner, L. *Introduction to Transportation Systems.* New York: Holt, Rinehart and Winston, 1986.

Harrison, F., F. Hills, J. Paterson, and R. Saunders. "The measurement of liver blood flow in conscious calves." *Quarterly Journal of Experimental Physiology* 71: 235–247.

Historical Topics for the Mathematics Classroom, Thirty-first Yearbook. National Council of Teachers of Mathematics, 1969.

Horn, D. *Basic Electronics Theory.* Blue Ridge Summit, Penn.: TAB Books, 1989.

Howells, G. *Acid Rain and Acid Waters.* 2nd ed. New York: Ellis Horwood, 1995.

Huffman, R. *Atmospheric Ultraviolet Remote Sensing.* San Diego: Academic Press, 1992.

Karttunen, H., P. Kroger, H. Oja, M. Poutanen, K. Donner, eds. *Fundamental Astronomy.* 2nd ed. New York: Springer-Verlag, 1994.

Kincaid, D., and W. Cheney. *Numerical Analysis.* Pacific Grove, Calif.: Brooks/Cole Publishing Company, 1991.

Kraljic, M. *The Greenhouse Effect.* New York: The H. W. Wilson Company, 1992.

Lack, D. *The Life of a Robin.* London: Collins, 1965.

Lancaster, H. *Quantitative Methods in Biological and Medical Sciences: A Historical Essay.* New York: Springer-Verlag, 1994.

Mannering, F., and W. Kilareski. *Principles of Highway Engineering and Traffic Analysis.* New York: John Wiley and Sons, 1990.

Mar, J., and H. Liebowitz. *Structure Technology for Large Radio and Radar Telescope Systems.* Cambridge, Mass.: The MIT Press, 1969.

Mason, C. *Biology of Freshwater Pollution.* New York: Longman and Scientific and Technical, John Wiley and Sons, 1991.

Meadows, D. *Beyond the Limits.* Post Mills, Vermont: Chelsea Green Publishing Co., 1992.

Miller, A., and J. Thompson. *Elements of Meteorology.* 2nd ed. Columbus, Ohio: Charles E. Merrill Publishing Company, 1975.

Miller, A., and R. Anthes. *Meteorology.* 5th ed. Columbus, Ohio: Charles E. Merrill Publishing Company, 1985.

Monroe, J. *Steffi Graf.* Mankato, Minn.: Crestwood House, 1988.

Motz, L., and J. Weaver. *The Story of Mathematics.* New York: Plenum Press, 1993.

Nemerow, N., and A. Dasgupta. *Industrial and Hazardous Waste Treatment.* New York: Van Nostrand Reinhold, 1991.

Nicholson, A. J. "An Outline of the dynamics of animal populations." *Austr. J. Zool.* 2 (1935): 9–65.

Nielson, G., and B. Shriver, eds. *Visualization in Scientific Computing.* Los Alamitos, Calif.: IEEE Computer Society Press, 1990.

Nilsson, A. *Greenhouse Earth.* New York: John Wiley and Sons, 1992.

Paetsch, M. *Mobile Communications in the U.S. and Europe: Regulation, Technology, and Markets.* Norwood, Mass.: Artech House, Inc., 1993.

Pearl, R., T. Edwards, and J. Miner. "The growth of *Cucumis melo* seedlings at different temperatures." *J. Gen. Physiol.* 17: 687–700.

Pennycuick, C. *Newton Rules Biology.* New York: Oxford University Press, 1992.

Pielou, E. *Population and Community Ecology: Principles and Methods.* New York: Gordon and Breach Science Publishers, 1974.

Pokorny, C., and C. Gerald. *Computer Graphics: The Principles behind the Art and Science.* Irvine, Calif.: Franklin, Beedle, and Associates, 1989.

Ronan, C. *The Natural History of the Universe.* New York: MacMillan Publishing Company, 1991.

Sharov, A., and I. Novikov. *Edwin Hubble, The Discoverer of the Big Bang Universe.* New York: Cambridge University Press, 1993.

Smith, C. *Practical Cellular and PCS Design.* New York: McGraw-Hill, 1998.

Stent, G. S. *Molecular Biology of Bacterial Viruses.* San Francisco: W. H. Freeman, 1963.

Taylor, J. *DVD Demystified.* New York: McGraw-Hill, 1998.

Taylor, W. *The Geometry of Computer Graphics.* Pacific Grove, Calif.: Wadsworth and Brooks/Cole, 1992.

Thomas, D. *Swimming Pool Operators Handbook.* National Swimming Pool Foundation of Washington, D.C., 1972.

Thomas, V. *Science and Sport.* London: Faber and Faber, 1970.

Thomson, W. *Introduction to Space Dynamics.* New York: John Wiley and Sons, 1961.

Toffler, A., and H. Toffler. *Creating a New Civilization: The Politics of the Third Wave.* Kansas City, Mo.: Turner Publications, 1995.

Triola, M. *Elementary Statistics.* 7th ed. Reading, Mass.: Addison-Wesley Publishing Company, 1998.

Tucker, A., A. Bernat, W. Bradley, R. Cupper, and G. Scragg. *Fundamentals of Computing I Logic: Problem Solving, Programs, and Computers.* New York: McGraw-Hill, 1995.

Turner, R. K., D. Pierce, and I. Bateman. *Environmental Economics, An Elementary Approach.* Baltimore: The Johns Hopkins University Press, 1993.

Varley, G., and G. Gradwell. "Population models for the winter moth." *Symposium of the Royal Entomological Society of London* 4: 132–142.

Wang, T. *ASHRAE Trans.* 81, Part 1 (1975): 32.

Weidner, R., and R. Sells. *Elementary Classical Physics,* Vol. 2. Boston: Allyn and Bacon, Inc., 1965.

Williams, J. *The Weather Almanac 1995.* New York: Vintage Books, 1994.

Wright, J. *The New York Times Almanac 1999.* New York: Penguin Group, 1998.

Zeilik, M., S. Gregory, and D. Smith. *Introductory Astronomy and Astrophysics.* 3rd ed. Philadelphia: Saunders College Publishers, 1992.

List of Graphing Calculator Lab Manual Topics

Section	Example	Topic name
1.2	Ex. 4	Balancing a checking account 15
	Ex. 8	Performing arithmetic operations with technology 18
1.3	Ex. 1	Writing numbers in exponential notation 25
	Ex. 9	Evaluating arithmetic expressions 32
	Ex. 10	Writing a number in scientific notation 34
	Ex. 11	Writing a number in standard form 35
1.4	Ex. 5	Using the table feature 45
1.5	Ex. 6	Setting the viewing rectangle 58
	Ex. 7	Making a scatterplot with a graphing calculator 59
	Ex. 8	Making a line graph with a graphing calculator 59
2.1	Ex. 8	Using a graphing calculator 82
2.4	Ex. 5	Finding equations of horizontal and vertical lines 122
	Ex. 7	Finding perpendicular lines 124
3.1	Ex. 1	Solving a linear equation numerically 142
	Ex. 3	Solving a linear equation graphically 144
	Ex. 6	Solving a linear equation symbolically 146
3.2	Ex. 1	Solving an inequality numerically 156
3.4	Ex. 2	Solving an absolute value equation 180
	Ex. 5	Modeling temperature in Boston 183
4.1	Ex. 2	Solving a system of equations graphically and numerically 200
4.3	Ex. 4	Solving a system of linear inequalities with technology 228
4.5	Ex. 5	Using technology 249
4.6	Ex. 3	Using technology to find a determinant 259
5.2	Ex. 2	Evaluating a polynomial function 287

Section	Example	Topic name
	Ex. 3	Evaluating a polynomial function graphically 288
5.3	Ex. 3	Multiplying binomials 298
5.4	Ex. 4	Modeling the flight of a baseball 309
5.5	Ex. 5	Factoring with technology 318
6.1	Ex. 3	Maximizing revenue 343
	Ex. 6	Graphing a parabola 347
	Ex. 8	Modeling AIDS cases 350
6.3	Ex. 5	Using the discriminant 372
7.2	Ex. 6	Solving rational equations 416
7.4	Ex. 3	Modeling college tuition 438
	Ex. 4	Loosening a nut on a bolt 440
	Ex. 5	Analyzing data 441
8.1	Ex. 3	Finding cube roots 469
	Ex. 6	Interpreting rational exponents 472
8.2	Ex. 3	Graphing $x = y^2$ 481
8.5	Ex. 3	Multiplying complex numbers 514
	Ex. 4	Dividing complex numbers 515
9.1	Ex. 3	Graphing a parabola with a graphing calculator 531
	Ex. 7	Graphing a circle with a graphing calculator 534
9.2	Ex. 2	Modeling the orbit of Mercury 544
9.3	Ex. 6	Solving a system of inequalities with a graphing calculator 558
	Ex. 7	Solving a system of inequalities with a graphing calculator 559
10.1	Ex. 7	Modeling population 582
10.2	Ex. 1	Evaluating common logarithms 589
	Ex. 5	Evaluating natural logarithms 592
10.3	Ex. 8	Change of base formula 603
10.4	Ex. 7	Modeling bird populations 614
11.1	Ex. 1	Computing terms of a sequence 631
	Ex. 4	Modeling numbers of insects 633
11.3	Ex. 2	Finding the sum of a finite arithmetic series 652
11.4	Ex. 3	Calculating $_nC_r$ 663

GC–1

Videotape Index

Video #	Section #	Section Title	Exercise #
1	1.1	Describing Data with Sets of Numbers	21, 39, 41, 43, 45, 47, 49, 51, 55, 63, 69
1	1.2	Operations on Real Numbers	61, 87
1	1.3	Integer Exponents	25, 31, 59, 61, 75
2	1.4	Modeling Data with Formulas	1, 2, 3, 5, 9, 13, 15, 17, 19, 31, 35, 49
2	1.5	Visualization of Data	59
3	2.1	Functions and Their Representations	15
3	2.2	Linear Functions	7, 9, 11, 19, 29, 35, 37, 47, 49, 65, 69
3	2.3	The Slope of a Line	19
3	2.4	Equations of Lines and Linear Models	27, 43
4	3.1	Linear Equations	9, 19, 23, 25, 41, 49, 83
4	3.2	Linear Inequalities	47
4	3.3	Compound Inequalities	17, 19
4	3.4	Absolute Value Equations and Inequalities	7, 15, 21, 29, 31, 37, 67
5	4.1	Systems of Linear Equations in Two Variables	45
5	4.2	The Substitution and Elimination Methods	11, 31
5	4.3	Systems of Linear Inequalities	7, 9, 13, 17, 29, 31
6	4.4	Systems of Equations in Three Variables	23
6	4.5	Matrix Solutions of Linear Systems	13
6	4.6	Determinants	3, 9, 11, 19, 23, 31
7	5.1	Polynomial Expressions	63
7	5.2	Polynomial Functions and Models	7, 11, 15, 17, 19
7	5.3	Multiplication of Polynomials	11, 19, 27, 31, 43, 47, 57, 67, 75
8	5.4	Factoring Polynomials	35